AF702970

functional analysis

PURE AND APPLIED MATHEMATICS

A Program of Monographs, Textbooks, and Lecture Notes

LECTURE NOTES IN PURE AND APPLIED MATHEMATICS

1. *N. Jacobson,* Exceptional Lie Algebras
2. *L.-Å. Lindahl and F. Poulsen,* Thin Sets in Harmonic Analysis
3. *I. Satake,* Classification Theory of Semi-Simple Algebraic Groups
4. *F. Hirzebruch, W. D. Newmann, and S. S. Koh, Differentiable Manifolds and Quadratic Forms*
5. *I. Chavel,* Riemannian Symmetric Spaces of Rank One
6. *R. B. Burckel,* Characterization of C(X) Among Its Subalgebras
7. *B. R. McDonald, A. R. Magid, and K. C. Smith,* Ring Theory: Proceedings of the Oklahoma Conference
8. *Y.-T. Siu,* Techniques of Extension on Analytic Objects
9. *S. R. Caradus, W. E. Pfaffenberger, and B. Yood,* Calkin Algebras and Algebras of Operators on Banach Spaces
10. *E. O. Roxin, P.-T. Liu, and R. L. Sternberg,* Differential Games and Control Theory
11. *M. Orzech and C. Small,* The Brauer Group of Commutative Rings
12. *S. Thomier,* Topology and Its Applications
13. *J. M. Lopez and K. A. Ross,* Sidon Sets
14. *W. W. Comfort and S. Negrepontis,* Continuous Pseudometrics
15. *K. McKennon and J. M. Robertson,* Locally Convex Spaces
16. *M. Carmeli and S. Malin,* Representations of the Rotation and Lorentz Groups: An Introduction
17. *G. B. Seligman,* Rational Methods in Lie Algebras
18. *D. G. de Figueiredo,* Functional Analysis: Proceedings of the Brazilian Mathematical Society Symposium
19. *L. Cesari, R. Kannan, and J. D. Schuur,* Nonlinear Functional Analysis and Differential Equations: Proceedings of the Michigan State University Conference
20. *J. J. Schäffer,* Geometry of Spheres in Normed Spaces
21. *K. Yano and M. Kon,* Anti-Invariant Submanifolds
22. *W. V. Vasconcelos,* The Rings of Dimension Two
23. *R. E. Chandler,* Hausdorff Compactifications
24. *S. P. Franklin and B. V. S. Thomas,* Topology: Proceedings of the Memphis State University Conference
25. *S. K. Jain,* Ring Theory: Proceedings of the Ohio University Conference
26. *B. R. McDonald and R. A. Morris,* Ring Theory II: Proceedings of the Second Oklahoma Conference
27. *R. B. Mura and A. Rhemtulla,* Orderable Groups
28. *J. R. Graef,* Stability of Dynamical Systems: Theory and Applications
29. *H.-C. Wang,* Homogeneous Branch Algebras
30. *E. O. Roxin, P.-T. Liu, and R. L. Sternberg,* Differential Games and Control Theory II
31. *R. D. Porter,* Introduction to Fibre Bundles
32. *M. Altman,* Contractors and Contractor Directions Theory and Applications
33. *J. S. Golan,* Decomposition and Dimension in Module Categories
34. *G. Fairweather,* Finite Element Galerkin Methods for Differential Equations
35. *J. D. Sally,* Numbers of Generators of Ideals in Local Rings
36. *S. S. Miller,* Complex Analysis: Proceedings of the S.U.N.Y. Brockport Conference
37. *R. Gordon,* Representation Theory of Algebras: Proceedings of the Philadelphia Conference
38. *M. Goto and F. D. Grosshans,* Semisimple Lie Algebras
39. *A. I. Arruda, N. C. A. da Costa, and R. Chuaqui,* Mathematical Logic: Proceedings of the First Brazilian Conference
40. *F. Van Oystaeyen,* Ring Theory: Proceedings of the 1977 Antwerp Conference
41. *F. Van Oystaeyen and A. Verschoren,* Reflectors and Localization: Application to Sheaf Theory
42. *M. Satyanarayana,* Positively Ordered Semigroups
43. *D. L Russell,* Mathematics of Finite-Dimensional Control Systems
44. *P.-T. Liu and E. Roxin,* Differential Games and Control Theory III: Proceedings of the Third Kingston Conference, Part A
45. *A. Geramita and J. Seberry,* Orthogonal Designs: Quadratic Forms and Hadamard Matrices
46. *J. Cigler, V. Losert, and P. Michor,* Banach Modules and Functors on Categories of Banach Spaces

47. *P.-T. Liu and J. G. Sutinen,* Control Theory in Mathematical Economics: Proceedings of the Third Kingston Conference, Part B
48. *C. Byrnes,* Partial Differential Equations and Geometry
49. *G. Klambauer,* Problems and Propositions in Analysis
50. *J. Knopfmacher,* Analytic Arithmetic of Algebraic Function Fields
51. *F. Van Oystaeyen,* Ring Theory: Proceedings of the 1978 Antwerp Conference
52. *B. Kadem,* Binary Time Series
53. *J. Barros-Neto and R. A. Artino,* Hypoelliptic Boundary-Value Problems
54. *R. L. Sternberg, A. J. Kalinowski, and J. S. Papadakis,* Nonlinear Partial Differential Equations in Engineering and Applied Science
55. *B. R. McDonald,* Ring Theory and Algebra III: Proceedngs of the Third Oklahoma Conference
56. *J. S. Golan,* Structure Sheaves Over a Noncommutative Ring
57. *T. V. Narayana, J. G. Williams, and R. M. Mathsen,* Combinatorics, Representation Theory and Statistical Methods in Groups: YOUNG DAY Proceedings
58. *T. A. Burton,* Modeling and Differential Equations in Biology
59. *K. H. Kim and F. W. Roush,* Introduction to Mathematical Consensus Theory
60. *J. Banas and K. Goebel,* Measures of Noncompactness in Banach Spaces
61. *O. A. Nielson,* Direct Integral Theory
62. *J. E. Smith, G. O. Kenny, and R. N. Ball,* Ordered Groups: Proceedings of the Boise State Conference
63. *J. Cronin,* Mathematics of Cell Electrophysiology
64. *J. W. Brewer,* Power Series Over Commutative Rings
65. *P. K. Kamthan and M. Gupta,* Sequence Spaces and Series
66. *T. G. McLaughlin,* Regressive Sets and the Theory of Isols
67. *T. L. Herdman, S. M. Rankin III, and H. W. Stech,* Integral and Functional Differential Equations
68. *R. Draper,* Commutative Algebra: Analytic Methods
69. *W. G. McKay and J. Patera,* Tables of Dimensions, Indices, and Branching Rules for Representations of Simple Lie Algebras
70. *R. L. Devaney and Z. H. Nitecki,* Classical Mechanics and Dynamical Systems
71. *J. Van Geel,* Places and Valuations in Noncommutative Ring Theory
72. *C. Faith,* Injective Modules and Injective Quotient Rings
73. *A. Fiacco,* Mathematical Programming with Data Perturbations I
74. *P. Schultz, C. Praeger, and R. Sullivan,* Algebraic Structures and Applications: Proceedings of the First Western Australian Conference on Algebra
75. *L Bican, T. Kepka, and P. Nemec,* Rings, Modules, and Preradicals
76. *D. C. Kay and M. Breen,* Convexity and Related Combinatorial Geometry: Proceedings of the Second University of Oklahoma Conference
77. *P. Fletcher and W. F. Lindgren,* Quasi-Uniform Spaces
78. *C.-C. Yang,* Factorization Theory of Meromorphic Functions
79. *O. Taussky,* Ternary Quadratic Forms and Norms
80. *S. P. Singh and J. H. Burry,* Nonlinear Analysis and Applications
81. *K. B. Hannsgen, T. L. Herdman, H. W. Stech, and R. L. Wheeler,* Volterra and Functional Differential Equations
82. *N. L. Johnson, M. J. Kallaher, and C. T. Long,* Finite Geometries: Proceedings of a Conference in Honor of T. G. Ostrom
83. *G. I. Zapata,* Functional Analysis, Holomorphy, and Approximation Theory
84. *S. Greco and G. Valla,* Commutative Algebra: Proceedings of the Trento Conference
85. *A. V. Fiacco,* Mathematical Programming with Data Perturbations II
86. *J.-B. Hiriart-Urruty, W. Oettli, and J. Stoer,* Optimization: Theory and Algorithms
87. *A. Figa Talamanca and M. A. Picardello,* Harmonic Analysis on Free Groups
88. *M. Harada,* Factor Categories with Applications to Direct Decomposition of Modules
89. *V. I. Istrăţescu,* Strict Convexity and Complex Strict Convexity
90. *V. Lakshmikantham,* Trends in Theory and Practice of Nonlinear Differential Equations
91. *H. L. Manocha and J. B. Srivastava,* Algebra and Its Applications
92. *D. V. Chudnovsky and G. V. Chudnovsky,* Classical and Quantum Models and Arithmetic Problems
93. *J. W. Longley,* Least Squares Computations Using Orthogonalization Methods
94. *L. P. de Alcantara,* Mathematical Logic and Formal Systems
95. *C. E. Aull,* Rings of Continuous Functions
96. *R. Chuaqui,* Analysis, Geometry, and Probability
97. *L. Fuchs and L. Salce,* Modules Over Valuation Domains

98. *P. Fischer and W. R. Smith,* Chaos, Fractals, and Dynamics
99. *W. B. Powell and C. Tsinakis,* Ordered Algebraic Structures
100. *G. M. Rassias and T. M. Rassias,* Differential Geometry, Calculus of Variations, and Their Applications
101. *R.-E. Hoffmann and K. H. Hofmann,* Continuous Lattices and Their Applications
102. *J. H. Lightbourne III and S. M. Rankin III,* Physical Mathematics and Nonlinear Partial Differential Equations
103. *C. A. Baker and L, M. Batten,* Finite Geometrics
104. *J. W. Brewer, J. W. Bunce, and F. S. Van Vleck,* Linear Systems Over Commutative Rings
105. *C. McCrory and T. Shifrin,* Geometry and Topology: Manifolds, Varieties, and Knots
106. *D. W. Kueker, E. G. K. Lopez-Escobar, and C. H. Smith,* Mathematical Logic and Theoretical Computer Science
107. *B.-L. Lin and S. Simons,* Nonlinear and Convex Analysis: Proceedings in Honor of Ky Fan
108. *S. J. Lee,* Operator Methods for Optimal Control Problems
109. *V. Lakshmikantham,* Nonlinear Analysis and Applications
110. *S. F. McCormick,* Multigrid Methods: Theory, Applications, and Supercomputing
111. *M. C. Tangora,* Computers in Algebra
112. *D. V. Chudnovsky and G. V. Chudnovsky,* Search Theory: Some Recent Developments
113. *D. V. Chudnovsky and R. D. Jenks,* Computer Algebra
114. *M. C. Tangora,* Computers in Geometry and Topology
115. *P. Nelson, V. Faber, T. A. Manteuffel, D. L. Seth, and A. B. White, Jr.,* Transport Theory, Invariant Imbedding, and Integral Equations: Proceedings in Honor of G. M. Wing's 65th Birthday
116. *P. Clément, S. Invernizzi, E. Mitidieri, and I. I. Vrabie,* Semigroup Theory and Applications
117. *J. Vinuesa,* Orthogonal Polynomials and Their Applications: Proceedings of the International Congress
118. *C. M. Dafermos, G. Ladas, and G. Papanicolaou,* Differential Equations: Proceedings of the EQUADIFF Conference
119. *E. O. Roxin,* Modern Optimal Control: A Conference in Honor of Solomon Lefschetz and Joseph P. Lasalle
120. *J. C. Díaz,* Mathematics for Large Scale Computing
121. *P. S. Milojević,* Nonlinear Functional Analysis
122. *C. Sadosky,* Analysis and Partial Differential Equations: A Collection of Papers Dedicated to Mischa Cotlar
123. *R. M. Shortt,* General Topology and Applications: Proceedings of the 1988 Northeast Conference
124. *R. Wong,* Asymptotic and Computational Analysis: Conference in Honor of Frank W. J. Olver's 65th Birthday
125. *D. V. Chudnovsky and R. D. Jenks,* Computers in Mathematics
126. *W. D. Wallis, H. Shen, W. Wei, and L. Zhu,* Combinatorial Designs and Applications
127. *S. Elaydi,* Differential Equations: Stability and Control
128. *G. Chen, E. B. Lee, W. Littman, and L. Markus,* Distributed Parameter Control Systems: New Trends and Applications
129. *W. N. Everitt,* Inequalities: Fifty Years On from Hardy, Littlewood and Pólya
130. *H. G. Kaper and M. Garbey,* Asymptotic Analysis and the Numerical Solution of Partial Differential Equations
131. *O. Arino, D. E. Axelrod, and M. Kimmel,* Mathematical Population Dynamics: Proceedings of the Second International Conference
132. *S. Coen,* Geometry and Complex Variables
133. *J. A. Goldstein, F. Kappel, and W. Schappacher,* Differential Equations with Applications in Biology, Physics, and Engineering
134. *S. J. Andima, R. Kopperman, P. R. Misra, J. Z. Reichman, and A. R. Todd,* General Topology and Applications
135. *P Clément, E. Mitidieri, B. de Pagter,* Semigroup Theory and Evolution Equations: The Second International Conference
136. *K. Jarosz,* Function Spaces
137. *J. M. Bayod, N. De Grande-De Kimpe, and J. Martínez-Maurica,* *p*-adic Functional Analysis
138. *G. A. Anastassiou,* Approximation Theory: Proceedings of the Sixth Southeastern Approximation Theorists Annual Conference
139. *R. S. Rees,* Graphs, Matrices, and Designs: Festschrift in Honor of Norman J. Pullman
140. *G. Abrams, J. Haefner, and K. M. Rangaswamy,* Methods in Module Theory

141. *G. L. Mullen and P. J.-S. Shiue*, Finite Fields, Coding Theory, and Advances in Communications and Computing
142. *M. C. Joshi and A. V. Balakrishnan*, Mathematical Theory of Control: Proceedings of the International Conference
143. *G. Komatsu and Y. Sakane*, Complex Geometry: Proceedings of the Osaka International Conference
144. *I. J. Bakelman*, Geometric Analysis and Nonlinear Partial Differential Equations
145. *T. Mabuchi and S. Mukai*, Einstein Metrics and Yang–Mills Connections: Proceedings of the 27th Taniguchi International Symposium
146. *L. Fuchs and R. Göbel*, Abelian Groups: Proceedings of the 1991 Curaçao Conference
147. *A. D. Pollington and W. Moran*, Number Theory with an Emphasis on the Markoff Spectrum
148. *G. Dore, A. Favini, E. Obrecht, and A. Venni*, Differential Equations in Banach Spaces
149. *T. West*, Continuum Theory and Dynamical Systems
150. *K. D. Bierstedt, A. Pietsch, W. Ruess, and D. Vogt*, Functional Analysis
151. *K. G. Fischer, P. Loustaunau, and J. Shapiro*, Computational Algebra
152. *K. D. Elworthy, W. N. Everitt, and E. B. Lee*, Differential Equations, Dynamical Systems, and Control Science

Additional Volumes in Preparation

functional analysis

proceedings of the Essen conference

edited by
Klaus D. Bierstedt
University of Paderborn
Paderborn, Germany

Albrecht Pietsch
University of Jena
Jena, Germany

Wolfgang M. Ruess
University of Essen
Essen, Germany

Dietmar Vogt
University of Wuppertal
Wuppertal, Germany

Marcel Dekker, Inc. **New York • Basel • Hong Kong**

Library of Congress Cataloging-in-Publication Data

Functional analysis: proceedings of the Essen conference / edited by Klaus D. Bierstedt ... [et al.].

p. cm. -- (Lecture notes in pure and applied mathematics; v. 150)

Proceedings of the Symposium on Functional Analysis, held in Essen, Germany, Nov. 24–30, 1991.

Includes bibliographical references and index.

ISBN 0-8247-9066-9 (acid-free)

1. Functional analysis--Congresses. I. Bierstedt, Klaus Dieter. II. Symposium on Functional Analysis (1991: Essen, Germany). III. Series.

QA319.F86 1993

515--dc20 93-28421

CIP

The publisher offers discounts on this book when ordered in bulk quantities. For more information, write to Special Sales/Professional Marketing at the address below.

This book is printed on acid-free paper.

MARCEL DEKKER, INC.
270 Madison Avenue, New York, New York 10016

Current printing (last digit):
10 9 8 7 6 5 4 3 2 1

PRINTED IN THE UNITED STATES OF AMERICA

Preface

The international Symposium on Functional Analysis, organized by the University of Essen, was held in Essen, Germany, November 24 – 30, 1991. The following were the two main objectives of the symposium:

(a) to emphasize and deepen the interaction between three currently highly active neighboring branches of functional analysis, viz.,

i. Geometry of Banach spaces and operator ideals,
ii. Fréchet spaces with applications to analysis and partial differential equations,
iii. Semigroups of operators and evolution equations;

(b) to take advantage of the recent political opening of Eastern Europe to vitalize the longstanding scientific contacts between functional analysts in the East and in the West. (The members of the Organizing Committee previously had been cooperating in the mathematical project within the scope of the Cultural Agreement between the Federal Republic of Germany and the German Democratic Republic. The symposium in Essen followed up a joint meeting in Georgenthal, GDR, organized by Jena University in 1989.)

The Essen conference was structured into three (main) lecture series by Stefan Heinrich (from area i), Reinhold Meise (ii), and Philippe Bénilan (iii), providing state-of-the-art surveys on aspects of current interest in the respective fields, and about 30 further contributions to the main (and related) subjects. For the spectrum of the topics discussed by the speakers, we refer to the conference program that follows; a list of participants is also included. The main lecture series and nearly all the topics covered at the meeting, plus some additional ones, are reported in this Proceedings volume, which contains a total of 29 articles by 39 authors. The original deadline for contributions was October 1, 1992; the first articles arrived long before this date, the last one in early January 1993. The articles were refereed, and we thank the referees for their cooperation and valuable comments.

The conference was fully sponsored by the Volkswagen Foundation, Hannover, Germany. The organizers – and the editors of this volume – express their deep gratitude for this generous support. Further institutions and individuals have contributed to the success of the conference. We gratefully acknowledge financial support by the University of Essen and the association of Freunde und Förderer der Universität Essen. We extend our sincere gratitude to Professor C. Streffer, then Rector of the University of Essen, whose constant support was invaluable to the local organization. We express our thanks to the staff of the conference site, the ZAWA-Educational Centre in Essen, who provided the participants of the conference with a most pleasant atmosphere throughout their stay. Finally, we thank the contributors and Ms. Maria Allegra of Marcel Dekker, Inc., for their cooperation during the preparation of this volume.

Klaus D. Bierstedt
Albrecht Pietsch
Wolfgang M. Ruess
Dietmar Vogt

Contents

Contributors

ERNST ALBRECHT, Fachbereich Mathematik, Universität des Saarlandes
D-W-6600 Saarbrücken, Germany

W. ARENDT, Equipe de Mathématiques, C.N.R.S.-URA 741
Université de Franche-Comté, F-25030 Besançon Cedex, France

PHILIPPE BÉNILAN, Equipe de Mathématiques, C.N.R.S.-URA 741
Université de Franche-Comté, F-25030 Besançon Cedex, France

JOSÉ BONET, Departamento de Matemática Aplicada, E.T.S. Arquitectura
Universidad Politécnica de Valencia, E-46071 Valencia, Spain

BERNARD CHEVREAU, U.E.R. de Mathématiques et Informatique
Université de Bordeaux-I, 351, cours de la Libération, F-33405 Talence, France

ANDREAS DEFANT, Fachbereich Mathematik, Carl v. Ossietzky-Universität
D-W-2900 Oldenburg, Germany

LECH DREWNOWSKI, Institute of Mathematics, A. Mickiewicz University
ul. Matejki 48/49, PL-60-769 Poznań, Poland

PAWEŁ DOMAŃSKI, Institute of Mathematics, A. Mickiewicz University
ul. Matejki 48/49, PL-60-769 Poznań, Poland

STEFAN GEISS, Mathematisches Institut, Friedrich-Schiller-Universität Jena
D-O-6900 Jena, Germany

STEFAN HEINRICH, Fachbereich Informatik, Universität Kaiserslautern
Erwin-Schrödinger-Str., D-W-6750 Kaiserslautern, Germany

HANS JARCHOW, Mathematisches Institut, Universität Zürich, Rämistr. 74
CH-8001 Zürich, Switzerland

H. JUNEK, Fachbereich Mathematik, Universität Potsdam, Am Neuen Palais 10
D-O-1571 Potsdam, Germany

S.V. KISLIAKOV, Steklov Math. Institute, St. Petersburg Branch, Fontanka 27
191011, St. Petersburg, Russia

HERMANN KÖNIG, Mathematisches Seminar, Universität Kiel, Ludewig-Meyn-Str. 4
D-W-2300 Kiel 1, Germany

KLAUS-DETLEF KÜRSTEN, Fachbereich Mathematik-Informatik, Universität Leipzig
Augustusplatz, D-O-7010 Leipzig, Germany

MICHAEL LANGENBRUCH, Fachbereich Mathematik
Bergische Universität-GH Wuppertal, Gaußstr. 20, D-W-Wuppertal 1, Germany

MIKAEL LINDSTRÖM, Department of Mathematics, Åbo Akademi
SF-20500 Åbo, Finland

REINHOLD MEISE, Mathematisches Institut, Heinrich-Heine-Universität Düsseldorf
Universitätsstr. 1, D-W-4000 Düsseldorf, Germany

SIEGFRIED MOMM, Mathematisches Institut, Heinrich-Heine-Universität Düsseldorf
Universitätsstr. 1, D-W-4000 Düsseldorf, Germany

RAINER NAGEL, Mathematisches Institut, Universität Tübingen
Auf der Morgenstelle 10, D-W-7400 Tübingen, Germany

V.P. PALAMODOV, Mech.-Math. Faculty, University of Moscow
Moscow, Russia

ALEKSANDER PEŁCZYŃSKI, Institute of Mathematics, Polish Academy of Sciences
Śniadeckich 8, PL-00-950 Warsaw, Poland

VŨ QUÔC PHÓNG, Department of Mathematics, University of Iowa
Iowa City, IA 52242, U.S.A. and
Institute of Mathematics, P.O. Box 631, Bo Ho, Hanoi, Vietnam

ALBRECHT PIETSCH, Mathematisches Institut, Friedrich-Schiller-Universität Jena
D-O-6900 Jena, Germany

MARKUS POPPENBERG, Fachbereich Mathematik, Universität Dortmund
Postfach 500500, D-W-4600 Dortmund, Germany

M.S. RAMANUJAN, Department of Mathematics, University of Michigan
Ann Arbor, MI 48109, U.S.A.

T. RUNST, Mathematisches Institut, Friedrich-Schiller-Universität Jena
D-O-6900 Jena, Germany

GABRIELE SCHREIECK, Fachbereich Mathematik, Carl v. Ossietzky-Universität
D-W-2900 Oldenburg, Germany

EUGENIO SINESTRARI, Dipartimento di Matematica, Universita di Roma "La Sapienza"
P. Aldo Moro 2, I-00185 Roma, Italy

CHARLES STEGALL, Institut für Mathematik, Johannes Kepler Universität
A-4040 Linz, Austria

B.A. TAYLOR, Department of Mathematics, University of Michigan, West Engineering Ann Arbor, MI 48109, U.S.A.

KEITH F. TAYLOR, Department of Mathematics, University of Saskatchewan Saskatoon, Saskatchewan, S7N 0W0, Canada

DIETMAR VOGT, Fachbereich Mathematik, Bergische Universität-GH Wuppertal Gaußstr. 20, D-W-5600 Wuppertal 1, Germany

JÜRGEN VOIGT, Fachbereich Mathematik, Carl v. Ossietzky-Universität D-W-2900 Oldenburg, Germany

JÖRG WENZEL, Mathematisches Institut, Friedrich-Schiller-Universität Jena D-O-6900 Jena, Germany

DIRK WERNER, I. Mathematisches Institut, Freie Universität Berlin, Arnimallee 3 D-W-1000 Berlin 33, Germany

WEND WERNER, Fachbereich 17, Universität-GH-Paderborn, Postfach 1621 D-W-4790 Paderborn, Germany

PETRA WITTBOLD, Equipe de Mathématiques, C.N.R.S.-URA 741 Université de Franche-Comté, F-25030 Besançon Cedex, France and Fachbereich Mathematik, Universität Essen, Universitätsstr. 2 D-W-4300 Essen, Germany

V. ZAHARIUTA, Mechanics and Mathematics Department, Rostov State University Rostov-on-Don, Russia

Conference Program

Monday

Morning Session
Chair: D. Vogt

8.30 A. PEŁCZYŃSKI, Non-isomorphism of the disc algebra with spaces of differentiable functions

9.30 R. MEISE, Continuous linear right inverses for linear partial differential operators, Part I

10.45 M.S. RAMANUJAN, Fréchet spaces between which all continuous linear maps are compact

11.20 J. BONET, The theorem of Josefson-Nissenzweig for Fréchet spaces

Afternoon Session
Chair: M.S. Ramanujan

14.00 T. TERZIOĞLU, Existence of Fréchet subspaces and quotient spaces of locally convex spaces

14.55 E. ALBRECHT, Invariant subspaces for certain representations of $H^\infty(G)$

16.00 H. JARCHOW, Order bounded composition operators

16.30 D. WERNER, New results on M-ideals of operators

17.00 A. DEFANT, On Gelfand numbers of tensor product operators

Tuesday

Morning Session
Chair: K.D. Bierstedt

8.30 R. MEISE, Continuous linear right inverses for linear partial differential operators, Part II

9.30 S. HEINRICH, Random approximation in numerical analysis, Part I

10.45 M. LANGENBRUCH, Splitting of the $\overline{\partial}$-complex in weighted spaces of square integrable functions

11.20 M. POPPENBERG, A tame splitting theorem for short exact sequences of Fréchet spaces

Afternoon/Evening Session
Chair: A. Pietsch

15.00 H. KÖNIG, Vector-valued L_p-convergence of orthogonal sequences and Lagrange interpolation

15.55 S. GEISS, Random unconditional constants of Walsh-Paley martingales

17.00 C. SCHÜTT, Floating bodies of a polytope

18.45 K.-D. KÜRSTEN, On unbounded Calkin representations

19.20 S. MOMM, Convex univalent functions and continuous linear right inverses

Wednesday

Morning Session
Chair: H. König

8.30 V.P. PALAMODOV, F-space methods for some problems of analysis in presence of a group action

9.30 S. HEINRICH, Random approximation in numerical analysis, Part II

10.30 R. NAGEL, Superstable operators

11.30 C. STEGALL, Weak compactness today

14.00 **Visit to Villa Hügel**

17.00 Reception by the Rector of the University of Essen

18.00 A. PEŁCZYŃSKI, Number of variables and linear topological invariants
(Special Colloquium Lecture at the University of Essen)

Thursday

Morning Session
Chair: E. Albrecht

8.30 R. Meise, Continuous linear right inverses for linear partial differential operators, Part III

Chair: W.M. Ruess

9.30 A. Haraux, Stable positive bounded solutions of some parabolic problems

10.45 Ph. Bénilan, Main results and open problems in the theory of nonlinear evolution equations, Part I

Afternoon/Evening Session
Chair: R. Nagel

15.00 B. de Pagter, Positive operators and renewal sequences

15.55 W. Arendt, Integrated semigroups: evolution with jumps of regularity

17.00 V.Q. Phóng, Spectrum and asymptotic behavior of commuting semigroups

18.45 T. Runst, Singularity theory and nonlinear partial differential equations

19.20 W. Werner, Smooth points in spaces of bounded operators

Friday

Morning Session
Chair: V.P. Palamodov

8.30 S. Heinrich, Random approximation in numerical analysis, Part III

9.30 S.V. Kisliakov, Interpolation of H^p-spaces

10.45 Ph. Bénilan, Main results and open problems in the theory of nonlinear evolution equations, Part II

Afternoon Session
Chair: R. Meise

14.00 V. Zahariuta, Spaces of harmonic functions

15.00 H. Junek, Bounded sets in F- and DF-spaces

16.00 P. Domański, Banach and Fréchet spaces of weakly continuous vector-valued functions

16.40 J. Voigt, Perturbations of Dirichlet forms by measures

Conference Participants

E. Albrecht (Saarbrücken, Germany)
W. Arendt (Besançon, France)
Ph. Bénilan (Besançon, France)
K.D. Bierstedt (Paderborn, Germany)
J. Bonet (Universidad Politécnica de Valencia, Spain)
A. Defant (Oldenburg, Germany)
P. Domański (Poznań, Poland / Wuppertal, Germany)
S. Geiß (Jena, Germany)
A. Haraux (Université Pierre et Marie Curie, Paris, France)
S. Heinrich (Kaiserslautern, Germany)
H. Jarchow (Universität Zürich, Switzerland)
H. Junek (Potsdam, Germany)
S.V. Kisliakov (St. Petersburg, USSR)
H. König (Kiel, Germany)
K.-D. Kürsten (Leipzig, Germany)
M. Langenbruch (Wuppertal, Germany)
R. Meise (Düsseldorf, Germany)
S. Momm (Düsseldorf, Germany)
R. Nagel (Tübingen, Germany)
B. de Pagter (Delft, Netherlands)
V.P. Palamodov (Moscow, USSR)
A. Pełczyński (Warszawa, Poland)
V.Q. Phóng (Hanoi, Vietnam)
A. Pietsch (Jena, Germany)
M. Poppenberg (Dortmund, Germany)
M.S. Ramanujan (Ann Arbor, USA)
W.M. Ruess (Essen, Germany)

T. Runst (Jena, Germany)
C. Schütt (Kiel, Germany)
C. Stegall (Linz, Austria)
T. Terzioğlu (Ankara, Turkey)
D. Vogt (Wuppertal, Germany)
J. Voigt (Oldenburg, Germany)
D. Werner (Freie Universität Berlin, Germany)
W. Werner (Paderborn, Germany)
V. Zahariuta (Rostov on Don, USSR)

Nonlinear Evolution Equations in Banach Spaces: Basic Results and Open Problems

Philippe BENILAN and Petra WITTBOLD Equipe de Mathématiques U.A. C.N.R.S 741 , Université de Franche-Comté , 25030 Besançon Cedex, France

1 BASIC RESULTS OF THE THEORY

The main object of the theory is the study of evolution problems of the form

$$(CP) \qquad \begin{cases} \dfrac{du}{dt} + Au & \ni \; f \qquad \text{on} \quad (0,T) \\ \quad u(0) & = \; u_0 \end{cases}$$

where $f : (0,T) \to X, u_0 \in X, A : X \to \mathcal{P}(X)$ is a multivalued operator in X , X a Banach space, $0 < T \leq \infty$.

To emphasize the dependence on the data u_0, f, we will eventually write $(CP)_{u_0,f}$ or, though not consistent, but for the sake of simplicity, $(CP)_{u_0}$ to denote $(CP)_{u_0,0}$, the homogeneous evolution equation.

1.1 Multivalued operators

The use of multivalued nonlinear operators does not only permit to obtain a coherent theory but is often also necessary with regard to applications. So let us recall

---

All the material of this part comes from the book in preparation (see [9]) by the first author and M.G.CRANDALL and A.PAZY to whom we are deeply indebted.

some notations and basic facts concerning multivalued operators (and we refer to [9] for more details).

An operator A in X in our context is an application $A : X \to \mathcal{P}(X)$; for $x \in X, Ax$ denotes the value of A at x , $D(A) = \{x \in X; Ax \neq \emptyset\}$ will be called the (effective) domain of A , $R(A) = \bigcup\{Ax; x \in D(A)\}$ its range.

The operator A uniquely determines the subset $\{(x, y); y \in Ax\}$ of $X \times X$, called the graph of A ; conversely, each set $G \subset X \times X$ determines a unique operator in X whose graph is G and therefore A may be identified with its graph.

Let us emphasize that some well-known difficulties of classical linear operator theory do not arise in " multivalued " theory:

Each operator A in X admits a natural inverse A^{-1} , the operator with graph $\{(y, x); y \in Ax\}$; its closure , denoted by $\bar{A}$, is defined to be the closure of the graph of A in $X \times X$.

Just as well it is worth noting that within the multivalued concept one may associate to each linear , but possibly non-densely defined operator A a transpose A^* in the dual space X^* of X. Before making this precise, however, we still have to introduce some notions:

An operator A will be called singlevalued if Ax is a singleton for each $x \in D(A)$, and we say that A is a linear operator whenever its graph is a linear subspace of $X \times X$; in this event, also A^{-1} and $\bar{A}$ are linear operators , $D(A)$ and $A0$ are linear subspaces of X , $Ax = Ax + A0$ for each $x \in D(A)$ and hence A is singlevalued iff $A0 = \{0\}$. So, in our context, the classical linear operator $A : D(A) \subset X \to X$ is what we call a linear, singlevalued operator, and A is closable in the classical sense iff $\bar{A}$ is singlevalued.

Now, given any linear operator A, we define its (possibly multivalued) transpose A^* in X^* in the following natural way :

$$y^* \in A^*x^* \quad \Leftrightarrow \quad < y^*, x > = < x^*, y >$$

for all $(x, y) \in A$ where $< ., . >$ denotes the natural pairing between X and X^* . These and other characteristics of the nonlinear concept turn out to be extremely profitable as we shall see in the sequel.

1.2 Solutions of (CP)

1.2.1 The classical solution

One standard notion of a solution for (CP) is that of a classical solution, i.e. a function u satisfying

$$\begin{cases} u \in C([0, T[; X) \cap C^1(]0, T[; X) \\ \dot{u}(t) + Au(t) \ni f(t) \quad \text{for all} \quad t \in]0, T[\\ u(0) = u_0 \end{cases}$$

In fact, this is a common and reasonable notion of a solution in the classical theory of ordinary differential equations (which, in our context, corresponds to the particular case where both the operator A and the function f are continuous) and it is also adequate in basic linear semigroup theory (that is, in the case where $-A$ is the infinitesimal generator of a C_0-semigroup of bounded linear operators and f in

$C^1([0,T[;X)$). But as soon as discontinuities arise, the notion of a classical solution turns out to be too restrictive as may be illustrated by the following examples:

Example 1: Non-continuous operators
Let $X = \mathbb{R}, f \equiv 0, u_0 = 1$ and consider

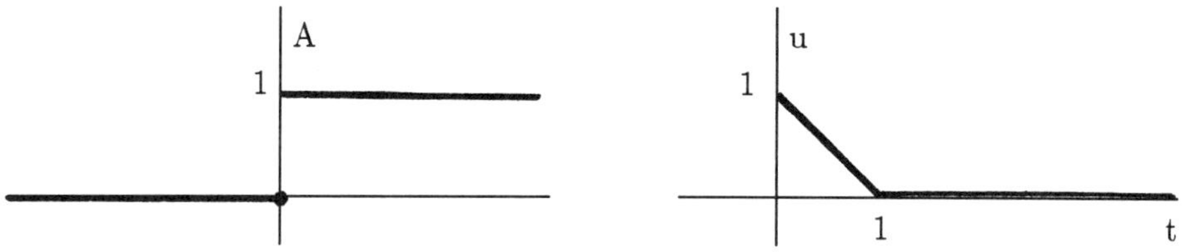

In other words, we consider the ordinary differential equation

$$(CP) \qquad \begin{cases} \dot{u} + Au &= 0 \\ u(0) &= 1 \end{cases}$$

in $\mathbb{R}$ where A is the Heaviside function.

Example 2: Non-continuous right-hand side
Consider the evolution problem in $X = \mathbb{R}$ corresponding to the data
$A \equiv 0, u_0 = 0, f$ as outlined

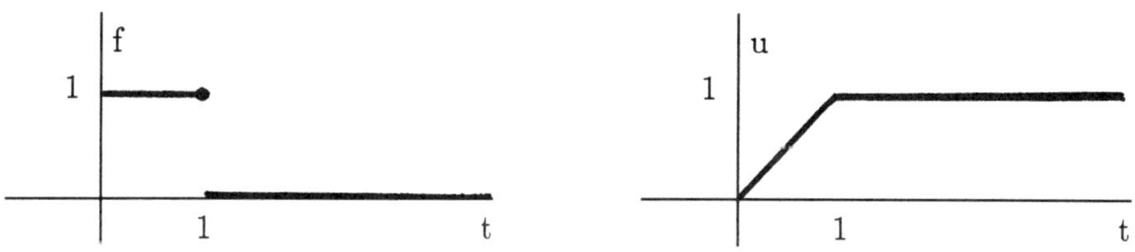

Evidently, in neither of the two examples (CP) admits a classical solution. Nevertheless, the functions u outlined above are, in a sense, the "natural" solutions of the evolution problem.The same remark clearly also applies to the following setting which is merely a generalization of Example 2.If $A \equiv 0$, then $(CP)_{u_0,f}$ has in general no classical solution unless $f \in C([0,T[;X)$ which is then given by $u(t) = u_0 + \int_0^t f(s)ds$.However, even if f only belongs to $L^1_{loc}([0,T[;X)$, this formula still makes sense and defines a function $u \in W^{1,1}_{loc}([0,T[;X)$ which clearly should be considered as a solution of $(CP)_{u_0,f}$.

These observations motivate the following weaker notion of solution for (CP)

1.2.2. The Strong Solution

A function u is called a strong solution of $(CP)_{u_0,f}$ if

$$\begin{cases} u \in C([0,T[;X) \cap W^{1,1}_{loc}(]0,T[;X) \\ \dot{u}(t) + Au(t) \ni f(t) \quad \text{a.e. on }]0,T[\\ u(0) = u_0 \end{cases}$$

Clearly, the examples we have just considered are covered by this notion of solution. In fact, it rather turns out to be an appropriate notion for multivalued ODE's (with the operator A being upper semicontinuous) as well as for nonlinear semigroup theory in reflexive Banach spaces (or, more generally, Banach spaces with the Radon-Nikodym property).

However, it is still not sufficient in general. Indeed, this is already well-known indeed in classical linear semigroup theory: if $-A$ is the generator of a C_o-semigroup of bounded linear operators $(S(t))_{t\geq 0}, f \in L^1([0,T[;X), u_0 \in X$ and $(CP)_{u_0,f}$ has a strong solution u, then it is given by the classical Duhamel-formula:

$$u(t) = S(t)u_0 + \int_0^t S(t-s)f(s)ds$$

But, if only $f \in L^1_{loc}([0,T[;X)$, this formula always defines a continuous function and hence it is natural and has become standard in linear theory to interpret it as a generalized (or so-called "mild") solution of (CP) (see e.g. [33] and references therein). However, even if $u_0 \in D(A), f \in C([0,T[;X)$, this "mild" solution need not be differentiable at any point - indeed, let $x \in X\backslash D(A)$ such that $t \mapsto S(t)x$ is nowhere differentiable and $f(t) = S(t)x, u_0 \equiv 0$, then u, given by the Duhamel-formula, is the function $t \mapsto u(t) = t\ S(t)x$ which, obviously, is not a strong solution.

Actually, the inadequacy of strong solutions already turns out if one reflects upon the following property of this class of solutions:
If $T < \infty$ and $u \in C([0,T];X)$ is a (strong) solution of $(CP)_{u_0,f}$, then the function $v \in C([0,T];X)$ defined by $v(t) = u(T-t)$ on $[0,T]$ is a strong solution of the (inverse) problem:

$$\overleftarrow{(CP)} \qquad \begin{cases} \dfrac{dv}{dt} + (-A)v \ni g & \text{on } (0,T) \\ v(0) = u(T) & \end{cases}$$

where $g(t) := f(T-t)$ on $(0,T)$. In other words, strong solutions are, what we call, "reversible in time".

As one learns already in basic thermodynamics, there are physical processes, characterized by an increase of entropy, which are irreversible in principle. An adequate notion of solution should reflect this property, i.e. solutions of equations modeling physically irreversible processes likewise should not be time-reversible.

The notion of solution we introduce now and which will play the fundamental role in all what follows, meets, among other things, this basic requirement.

1.2.3. The Mild Solution

A **mild solution of** $(CP)_{u_0,f}$ is defined to be a function $u \in C([0,T[;X)$ with $u(0) = u_0$ satisfying:
for all $0 < T_0 < T$ and $\epsilon > 0$, there exist

i) $0 = t_0 < t_1 < \cdots < t_{n-1} < T_0 \leq t_n < T, \quad t_i - t_{i-1} \leq \epsilon \quad \text{for} \quad i = 1, \ldots, n,$

ii) $f_1, \ldots, f_n \in X$ such that $\sum_{i=1}^{n} \int_{t_{i-1}}^{t_i} \|f(t) - f_i\| dt \leq \epsilon,$

iii) $x_0, \ldots, x_n \epsilon X$ verifying $\|u_0 - x_0\| \leq \epsilon$, and

$$(IS) \qquad \frac{x_i - x_{i-1}}{t_i - t_{i-1}} + Ax_i \ni f_i \qquad \text{for } i = 1, \ldots, n$$

such that

$$\|u(t) - x_i\| \leq \epsilon \qquad \text{for} \quad t \in]t_{i-1}, t_i], \quad i = 1, \ldots, n.$$

Thus, roughly speaking, a continuous function satisfying the initial condition is a mild solution iff it is the uniform limit of solutions of time-discretized "near-by problems" (where by a solution of the discretized problem, given by the implicit scheme (IS) of difference relations corresponding to a discretization $D_A(t_0, \ldots, t_n; f_1, \ldots, f_n)$ as above, we mean a piecewise constant function $v : [0, t_n] \longrightarrow X$ with $v(0) = x_0, v(t) = v(t_i) = x_i$ on $]t_{i-1}, t_i], i = 1, \ldots, n$, such that the x_i satisfy (IS)). The underlying idea of the definition of this new notion of solution thus is simple and, from the viewpoint of numerical analysis, even classical.

The first fundamental results of the theory of mild solutions for the evolution problem (CP) may be found in [22] and [5]; their results have been generalized and specified further on by several mathematicians (cf. [34] , [28] , ...) and themselves (for a systematic development of the theory we refer to [9]). Extensions to time-dependent evolution problems of the form $\frac{du}{dt} + A(t)u(t) \ni f(t)$ have also been carried out (cf. [23] , [27] ...).

Once we are given the definition, the natural questions for existence of mild solutions (which, by definition, requires first of all existence and convergence of solutions for discretized problems), for uniqueness and the properties of this class of solutions in general arise.

It is the intention of this introduction to the theory of nonlinear evolution equations to show that the mild solution is indeed the appropriate extension of a notion of a solution of (CP), still meeting all the basic requirements a solution of (CP) is expected to verify, but overcoming deficiencies, such as the principle of time-reversibility, of standard notions at the same time.

First of all, we observe that if $(CP)_{u_0,f}$ has a mild solution, then by the very definition, we have $u_0 \in \overline{D(A)}$, $f \in L^1_{loc}([0, T[; X)$ which we will therefore assume from now on. Next it is worth to make a remark upon the way we chose to approximate (CP) which may seem to be arbitrary at first glance as, clearly, instead of the implicit scheme (IS) chosen in the definition of mild solutions, one could also think of an explicit one such as

$$(ES) \qquad \frac{x_i - x_{i-1}}{t_i - t_{i-1}} + Ax_{i-1} \ni f_i \quad i = 1, \ldots, n.$$

Solving (ES) instead of (IS) however would oblige us to deal with powers of the "unkind" multivalued operator A and we will see later on that in our framework the scheme (IS) is more profitable. We emphasize that, in fact, the use of (ES) instead of (IS) would lead to a different definition of mild solutions - nevertheless, in "smooth" cases (for instance, if $A : D(A) \subset X \to X$ is continuous with $D(A)$ closed) both definitions coincide.

1.2.4. Properties of mild solutions

Let us now state the basic properties of mild solutions.Recalling that our intention was to provide an appropriate extension of the standard notions of solutions, let us mention first of all that :

1) Strong solutions of (CP) are mild solutions.

Moreover, we have:

2) The notion of a mild solution in the classical linear setting coincides with our notion of a mild solution, i. e. if $-A$ is the infinitesimal generator of a C_0 - semigroup of bounded linear operators $(S(t))_{t\geq 0}$, then u is a mild solution of $(CP)_{u_0,f}$ iff

$$u(t) = S(t)u_0 + \int_0^t S(t-s)f(s)ds.$$

Note that this implies that the notion of a mild solution in fact properly extends the one of a strong solution.In particular, a mild solution, though continuous by definition, may happen to be nowhere differentiable (see example in 1.2.2.).

However, it does not weaken the notion of a solution unnecessarily in "smooth" cases, more precisely :

3) If $A : D(A) \subset X \to X$ is continuous with $D(A)$ closed, then u is a mild solution of $(CP)_{u_0,f}$ iff it is a strong solution of $(CP)_{u_0,f}$ and in this case u satisfies:

$$u(t) = u_0 + \int_0^t (f(s) - Au(s))ds.$$

In particular, if $f \in C([0,T[;X)$, then u is even a classical solution.

4) If $A : X \to \mathcal{P}(X)$ is uppersemicontinuous from the strong into the weak topology of X with $D(A)$ closed and with convex, weakly compact image sets, then u is a mild solution of (CP) iff u is a strong solution of (CP).

Nevertheless, in general, due to the lack of regularity and the fact that a mild solution need not take its values in $D(A)$, one cannot interpret this solution in terms of the differential inclusion (in a pointwise sense). But, on the positive side, we note the following properties :

5) Restrictability :

If u is a mild solution of $(CP)_{u_0,f}$, then $u\backslash_{[a,b[}$ is a mild solution of $(CP)_{u(a),f\backslash[a,b[}$ for all $0 \leq a < b \leq T$.

6) Continuation property :

If $u \in C([0,T[;X)$, $a \in]0,T[$, $u\backslash_{[0,a[}$ is a mild solution of $(CP)_{u_0,f\backslash[0,a[}$ and $u\backslash_{[a,T[}$ is a mild solution of $(CP)_{u(a),f\backslash[a,T[}$, then u is a mild solution of $(CP)_{u_0,f}$ on $[0,T[$.

7) Translation invariance :

If u is a mild solution of $(CP)_{u_0,f}$, $h \in \mathbb{R}$, then $u_h(t) := u(t-h)$ is a mild solution of $(CP)_{u_0,f_h}$ on $[h, T+h[$, where $f_h(t) := f(t-h)$.

8) "Closedness" property :
If $(u_n)_n$ are mild solutions of $(CP)_{u_{0n},f_n}, n \in \mathbb{N}, f_n \to f$ in $L^1_{loc}([0,T[;X)$, $u_{0n} \to u_0$ in X, and $u_n \to u$ in $C([0,T[;X)$, then u is a mild solution of $(CP)_{u_0,f}$.

And finally we note

9) u is a mild solution of $(CP)_{u_0,f}$ iff u is a mild solution of $\frac{du}{dt} + \bar{A}u \ni f$ on $(0,T), u(0) = u_0$.

In many cases it is also useful to interpret a given evolution problem (CP) in X in some larger Banach space Y such that $X \hookrightarrow Y$ with continuous injection (we will return to this in one of the following sections). It is therefore important, that , by its definition, the notion of a mild solution only depends on the topology of the space ; hence a mild solution of $(CP)_{u_0,f}$ in X is also one in Y (and it even may turn out to be a strong solution of the evolution problem corresponding to some extension $\mathcal{A}$ in Y of the operator A in X). In particular, one may often prove that the mild solution is a weak solution in the sense of distributions for the partial differential equation underlying the abstract evolution problem, but without sharing the main drawback of these, the lack of non-uniqueness in general, under merely mild assumptions on the operator A (cf. [5] , Chapter 2, [9] for concrete examples).

However, we emphasize that mild solutions, in contrast to the standard notions of solutions (including, solutions in the sense of distributions) fail to be time-reversible in general which is a crucial point as already mentioned. The complete development of a concrete example of an evolution problem to demonstrate this phenomenon would be beyond the scope of our lecture and we have to refer to, among others, [10] (for the "Burger equation") and [9] for more general examples.

We remark, however, at this place that from the theoretical point of view this property is reflected by the fact that a discretization of the retrograde problem by an implicit scheme corresponds to a discretization of (CP) by an explicit scheme. But, as already underlined, in general the notions of a mild solution based upon a discretization by explicit and implicit scheme, respectively, are different.

1.3 The nonlinear semigroup

One may now naturally associate to every nonlinear operator $A : X \to \mathcal{P}(X)$ a nonlinear semigroup $(S^A(t))_{t\geq 0}$ by the following definition:

$D(S^A) = \{x \in X; \exists!$ mild solution u_x of $(CP)_x$ on $(0,\infty)$ with $u_x(0) = x\}$ and, for $t \geq 0, x \in D(S^A)$, we set $S^A(t)x = u_x(t)$.

It is an immediate consequence of the properties of mild solutions that, in fact, $(S^A(t))_{t\geq 0}$ is a strongly continuous semigroup on $D(S^A)$, i. e.

1) $S^A(t) : D(S^A) \to D(S^A) \quad \forall t \geq 0$

2) $S^A(t+s) = S^A(t)S^A(s) \quad \forall t,s \geq 0,\ S^A(0)x = x\ \forall x \in D(S^A)$

3) $t \in [0,\infty[\longmapsto S^A(t)x \in X$ is continuous for all $x \in D(S^A)$.

Observe that this definition is consistent with classical linear semigroup theory ; more precisely, we have :

$-A$ is the infinitesimal generator of a C_0-semigroup of bounded linear operators $(S(t))_{t\geq 0}$ on X , iff A is linear, closed and $D(S^A) = X$, and then $S^A \equiv S$.

This motivates the development of a "nonlinear semigroup theory" analogous to the classical linear one. We will enter into this discussion later on, especially in part 2 of the lecture, demonstrating difficulties that arise in doing so. First of all, however, we have to introduce a class of operators to which we restrict our considerations in the sequel to obtain existence, uniqueness and further more precise results concerning mild solutions of (CP).

1.4 Accretivity

DEFINITION 1 *Let A be an operator in X.*

1) A is called **accretive** *if and only if*

$$\|x-\hat{x}\| \leq \|x-\hat{x}+\lambda(y-\hat{y})\| \text{ for all } (x,y),(\hat{x},\hat{y}) \in A, \lambda > 0$$

or, equivalently, iff

for all $\lambda > 0$ the operator $\mathcal{J}_\lambda^A = (I+\lambda A)^{-1}$ is a (singlevalued) contraction on $D(\mathcal{J}_\lambda^A) = R(I+\lambda A)$.

2) A is called **m-accretive** *in X iff A is accretive and $R(I+\lambda A) = X$ for $\lambda > 0$.*

We will use the following notations for $\omega \in \mathbb{R}$:

$$\mathcal{A}(\omega) := \{A \subset X \times X; A + \omega I \text{ is accretive }\}$$

$$\mathcal{M}(\omega) := \{A \in \mathcal{A}(\omega); R(I+\lambda A) = X \ \forall \lambda > 0, \lambda\omega < 1\}$$

In order to verify accretivity of a given operator A, it is useful to take into account alternative characterizations of this property:

1) Characterization by means of the "bracket" .

The bracket $[.,.]$ is defined as follows:

$$[.,.] : X \times X \to \mathbb{R}$$

$$(x,y) \to \inf_{\lambda>0} \frac{\|x+\lambda y\| - \|x\|}{\lambda} = \lim_{\lambda \downarrow 0} \frac{\|x+\lambda y\| - \|x\|}{\lambda} \ ;$$

in other words, for $x, y \in X, [x,y]$ is the righthand derivative of $\|.\|$ at x in the direction y.

It follows that A is accretive iff

$$[x-\hat{x}, y-\hat{y}] \geq 0 \quad \text{for all} \quad (x,y),(\hat{x},\hat{y}) \in A.$$

2) Characterization using the duality map.

Let $\mathcal{J} : X \to \mathcal{P}(X^*)$ be the duality map defined by

$$\mathcal{J}(x) = \{x^* \in X^*; \|x^*\| \leq 1, \langle x^*, x \rangle = \|x\|\};$$

we note that A is accretive iff

for all $(x,y),(\hat{x},\hat{y}) \in A$ there exists $w \in \mathcal{J}(x-\hat{x})$ such that $\langle w, y-\hat{y} \rangle \geq 0$.

An accretiveness condition of some kind is natural in many ways:

1) First of all, we recall that by definition, the existence of mild solutions required the existence of solutions of discretized equations of the form

$$(*)\qquad \frac{x_i - x_{i-1}}{t_i - t_{i-1}} + Ax_i \ni f_i \quad i = 1, \cdots, n,$$

or, equivalently,

$$(**)\qquad x_i + (t_i - t_{i-1})Ax_i \ni (t_i - t_{i-1})f_i + x_{i-1} \quad i = 1, \cdots, n.$$

From (**) , we immediatly observe that accretiveness ensures uniqueness of the solutions x_i , which, if they exist, are then given by

$$x_i = \mathcal{J}^A_{(t_i - t_{i-1})}((t_i - t_{i-1})f_i + x_{i-1}) \quad i = 1, \cdots, n.$$

This formula already indicates that apart from accretivity one should expect a range condition (i.e. a condition on $R(I + \lambda A), \lambda > 0$) to hold in order to get existence of solutions as well.

2) In linear theory the main basic results contain (at least implicitly) accretiveness conditions - such as the classical **Lumer-Phillips-theorem** which we recall, using our terminology:

THEOREM *$-A$ is the infinitesimal generator of a C_0-semigroup of linear contractions on X iff A is linear, m-accretive and $\overline{D(A)} = X$.*

Respectively, more generally, we should also cite the following extension of the classical **Hille-Yosida-theorem**

THEOREM *$-A$ is the infinitesimal generator of a C_0-semigroup of bounded linear operators on X iff A is linear, $\overline{D(A)} = X$ and there exists $\omega \in \mathbb{R}$ and an equivalent norm on X such that $A \in \mathcal{M}(\omega)$.*

We want to remark at this place that, in the nonlinear setting, it is not clear in how far one can overcome an accretiveness condition without loosing existence and uniqueness results for mild solutions. We will eventually return to this point in part 2 of the lecture.

1.5. Existence and uniqueness theorems

We assume $\omega \in \mathbb{R}, A \in \mathcal{A}(\omega)$ in the sequel.

1.5.1. Uniqueness and a-priori-estimates

Though all results on mild solutions mentioned so far can be deduced directly by using the, however unwieldy, definition, it is of particular importance that, in the accretive case, the mild solution actually satisfies certain characteristic differential inequalities in the weak sense. These do not only permit to prove uniqueness of mild solutions - they represent a mean to verify, given some $u \in C([0,T[;X)$ such that $u(0) = u_0$, that this function is a mild solution of $(CP)_{u_0,f}$ at all and, moreover, to obtain further properties, in particular, regularity results.

The following theorem as well as its immediate consequences listed below have been essentially proved in [5]:

THEOREM 1

i) Let u be a mild solution of $(CP)_{u_0,f}$. *Then*

$$(1)\qquad \frac{d}{dt}e^{-\omega t}\|u(t)-x\| \leq e^{-\omega t}[u(t)-x, f(t)-y] \text{ in } \mathcal{D}'(]0,T[) \text{ for all } (x,y)\in A.$$

ii) Let $u \in C([0,T[;X)$ *satisfy (1) and* $\hat{u}$ *be a mild solution of* $(CP)_{\hat{u}_0,\hat{f}}$. *Then*

$$(2)\qquad \frac{d}{dt}e^{-\omega t}\|u(t)-\hat{u}(t)\| \leq e^{-\omega t}[u(t)-\hat{u}(t), f(t)-\hat{f}(t)] \text{ in } \mathcal{D}'(]0,T[).$$

It should be mentioned that one may state (1) (resp. (2)) in an equivalent integrated form; (1) then reads as

$$(1)^*\qquad \|u(t)-x\| \leq e^{\omega(t-s)}\|u(s)-x\| + \int_s^t e^{\omega(t-\tau)}[u(\tau)-x, f(\tau)-y]d\tau$$

for all $(x,y)\in A, 0\leq s\leq t<T$.

COROLLARY 1 *If there exists a mild solution u of* $(CP)_{u_0,f}$, *then it is unique and characterized by :*

$u \in C([0,T[;X), u(0)=u_0$ *and u satisfies (1).*

COROLLARY 2 *If u is a mild solution of* $(CP)_{u_0,f}$, *and* $\hat{u}$ *is a mild solution of* $(CP)_{\hat{u}_0,\hat{f}}$, *then*

$$\|u(t)-\hat{u}(t)\| \leq e^{\omega t}\|u_0-\hat{u}_0\| + \int_0^t e^{\omega(t-s)}\|f(s)-\hat{f}(s)\|ds \text{ for all } 0\leq s\leq t<T.$$

1.5.2. Existence theorems

The first basic result proving existence of a mild solution for the homogeneous equation is due to [22]. Though too restrictive in its assumptions if one is merely interested in proving existence, it is worth noting as it was the starting point for all further generalizations (cf.[28] in the homogeneous case $f = 0$ under the weaker assumption of quasi-accretivity on A , cf.[24] ,[5] , [21] for existence results for the inhomogeneous equation, see also [9])

CRANDALL-LIGGETT-THEOREM *Let $A \in \mathcal{A}(\omega)$ and suppose*

$$(RC) \qquad R(I + \lambda A) \supseteq \overline{D(A)} \quad \text{for } \lambda > 0 \text{ small.}$$

Then for all $u_0 \in \overline{D(A)}$

$$(3) \qquad S^A(t)u_0 = \lim_n (I + \frac{t}{n}A)^{-n} u_0$$

exists for all $t \geq 0$ uniformly on compact subsets of $\mathbb{R}^+$.
Moreover,

$$\|(\mathcal{J}_\lambda^A)^n u_0 - S^A(t)u_0\| \leq \|u_0 - x\| + \|y\|\sqrt{(n\lambda - t)^2 + n\lambda^2}$$

for all $(x, y) \in A$, $\lambda > 0$, $t > 0$.

Recalling our remarks on solvability of discretized equations (cf. 1.4), it is obvious that assumption (RC) , the "range condition " , enables us to solve (uniquely) for all $u_0 \in \overline{D(A)}$, $\lambda > 0$, the following particular discretization schemes

$$\begin{cases} \dfrac{x_i - x_{i-1}}{\lambda} + Ax_i \ni 0 & i = 1, \cdots, n \\ x_0 = u_0 \end{cases}$$

where $0 < T - n\lambda \leq \lambda$; therefore in this situation the mild solution of (CP) can be obtained via the nice representation formula (3) , classically referred to as the "exponential formula". However, in view of the definitions, it should be clear that, without restricting to such particular discretizations, one may get existence results under somewhat weaker assumptions than (RC).

In order to specify notations, let us call for $\epsilon > 0$ a function u an ϵ **- approximate solution of** (CP) **on** $[0, T_0]$ if it is the solution of an ϵ-discretization of (CP) (in the sense of the definition in 1.2.3).

The fundamental result we state below will show that merely the existence of some sequence of ϵ_n- approximate solutions such that $\epsilon_n \downarrow 0$ of (CP) is enough to obtain existence of a mild solution . Moreover , this approximating sequence then converges uniformly to the (unique) mild solution on compact subsets of [0,T] as does any other sequence of $\hat{\epsilon}_n-$ approximate solutions of (CP) such that $\hat{\epsilon}_n \downarrow 0$.

THEOREM 2 *Let $A \in \mathcal{A}(\omega)$, $u_0 \in \overline{D(A)}$, $f \in L^1_{loc}([0, T[; X)$, $0 < T_0 < T$, $\omega \in \mathbb{R}$*

Then there exists a continuous function $\delta : \mathbb{R}^+ \to \mathbb{R}^+$ with $\delta(0) = 0$ such that

$$\| \; u_\epsilon(t) - \hat{u}_\epsilon(t) \; \| \leq \delta(\epsilon) \quad , t \in [0, T_0]$$

for all $\epsilon > 0$, u_ϵ , $\hat{u}_\epsilon$ ϵ-approximate solutions of $(CP)_{u_o,f}$ on $[0, T_0]$.

COROLLARY 1 *Let $A \in \mathcal{A}(\omega)$, $u_0 \in \overline{D(A)}$, $f \in L^1_{loc}([0, T[; X)$, $\omega \in \mathbb{R}$. There exists a unique mild solution of $(CP)_{u_o,f}$ iff for all $0 < T_0 < T$, $\epsilon > 0$, there exists an ϵ-approximate solution of $(CP)_{u_o,f}$ on $[0, T_0]$.*

As an immediate consequence of the fact that if $A \in \mathcal{M}(\omega)$, then every discretization scheme of $(CP)_{u_o,f}$ is (uniquely) solvable for any $u_0 \in \overline{D(A)}$, $f \in L^1_{loc}([0,T[;X)$, we should also note as a further corollary:

COROLLARY 2 *Let $A \in \mathcal{M}(\omega)$, $\omega \in \mathbb{R}$. Then there exists a unique mild solution of $(CP)_{u_o,f}$ for all $u_0 \in \overline{D(A)}$, $f \in L^1_{loc}([0,T[;X)$.*

The condition of m-accretivity thus is sufficient to get existence of mild solutions for any given data $(u_0, f) \in \overline{D(A)} \times L^1(0,T;X)$, and we will see in one of the following subsections that this condition on the operator is rather optimal as well . However , to obtain existence only for some particular initial data , a much more refined analysis is possible . In particular, this remark applies to the homogeneous equation to which we will restrict our attention now.

1.5.3. The semigroup S^A

Note first that for $A \in \mathcal{A}(\omega)$, $\omega \in \mathbb{R}$, the semigroup S^A (as defined in 1.3) satisfies:
i) $D(S^A)$ is closed
ii) $\|S^A(t)x - S^A(t)\hat{x}\| \quad \leq e^{\omega t}\|x - \hat{x}\| \quad$ for all $x, \hat{x} \in D(S^A), t \geq 0$.

We recall further that, by definition, $D(S^A) = \overline{D(A)}$ iff $(CP)_{u_0}$ has a unique mild solution on $(0,\infty)$ for all $u_0 \in \overline{D(A)}$.

The following result due to [28] shows that in order to obtain existence of mild solutions of the homogeneous equation for all $u_0 \in \overline{D(A)}$, one may replace the (sufficient) range condition (RC) by the following "weak" range condition :

$$(WRC) \qquad \liminf_{\lambda \downarrow 0} \frac{1}{\lambda}\mathrm{dist}(x, R(I + \lambda A)) = 0 \quad \text{for all } x \in \overline{D(A)};$$

more precisely, we note :

PROPOSITION 1 *Let $A \in \mathcal{A}(\omega), \omega \in \mathbb{R}$ and assume (WRC) holds. Then $D(S^A) = \overline{D(A)}$.*

Note that condition (WRC) is properly weaker than (RC). One need only consider the following simple example :

Let $X = \mathbb{R}^2, D(A) = \{(x,y); x^2 + y^2 = 1\}, A(x,y) = (-y,x)$; then, obviously, $R(I + \lambda A) \cap \overline{D(A)} = \emptyset$ for all $\lambda > 0$, but it is immediately checked that (WRC) is fulfilled.

Moreover, in particular "smooth" cases (WRC) can be shown to be equivalent to a certain classical "tangency condition" which in turn is necessary for the existence of classical solutions of (CP) - the necessity of (WRC) for the existence of mild solutions of (CP) in this situation is then immediate (compare property 3) of 1.2.3.) :

PROPOSITION 2 *Let $A : D(A) \subset X \to X$ be continuous with $D(A)$ closed and $A \in \mathcal{A}(\omega), \omega \in \mathbb{R}$. The following assertions are equivalent :*
i) (WRC) holds
ii) The "subtangential condition" :

$$\liminf_{\lambda \downarrow 0} \frac{1}{\lambda} dist(x - \lambda Ax, D(A)) = 0 \quad \textit{holds for all} \ \ x \in D(A)$$

iii) $D(S^A) = D(A)$

We emphasize that in general, however, (WRC) is not necessary - it rather has to be replaced by a more complex, refined range condition, denoted by (CRC) below, which has been essentially developed in [34] (see also [9]) where also concrete examples of operators satisfying (CRC), but not (WRC) may be found. We state the precise result :

THEOREM 1 *Let* $A \in \mathcal{A}(\omega), \omega \in I\!R$ *then* $D(S^A) = \overline{D(A)}$ *iff*

$$(CRC) \qquad \begin{cases} \forall x \subset \overline{D(A)}, \epsilon > 0 \quad \exists \delta \subset]0, \epsilon], n \in I\!N \text{ and} \\ (x_i, y_i) \in A, h_i > 0 \text{ with } 2h_i\omega < 1 \text{ for } i = 1, \cdots, n \text{ such that} \\ \sum_i h_i = \delta \quad , \quad \sum_i \|x_i - x_{i-1} + h_i y_i\| \leq \epsilon\delta \end{cases}$$

is satisfied.

Let us now return to the relation between m-accretivity and the non-homogeneous evolution equation .

1.5.4. m-accretivity

Recalling continuation and closedness properties of mild solutions (cf 1.2), it is obvious that in order to get existence of mild solutions of $(CP)_{u_o,f}$ for all $u_o \in \overline{D(A)}$, $f \in L^1(O,T;X)$, it actually suffices to be able to solve the evolution equations for all constant right-hand sides , i.e. for $f \equiv y, \ y \in X$.

However , this in turn is equivalent to solve for all $y \in X$ the homogeneous evolution problem corresponding to the operator $A - y$ for all $u_0 \in D(A-y) \equiv D(A)$.

Obviously, if $A \in \mathcal{A}(\omega)$, then $A - y \in \mathcal{A}(\omega)$ for all y $\in X$, and by the results of the preceding subsection we know that , given some $y \in X$, then

$$\begin{cases} \dfrac{du}{dt} + (A-y)u \ni 0 \quad \text{on } (0,T) \\ \qquad u(0) = u_0 \end{cases}$$

has a unique mild solution for all $u_0 \in D(A)$ iff $(A - y)$ satisfies (CRC).

As a simple consequence of the preceding results, one can therefore note:

COROLLARY 1 *Let* $A \in \mathcal{A}(\omega), \omega \in I\!R, T > 0$. *Then the following are equivalent:*
i) $(CP)_{u_o,f}$ *has a unique mild solution for all* $u_0 \in \overline{D(A)}, f \in L^1(O,T;X)$.
ii) for all $y \in X$: (CRC) *holds for the operator* $A - y$

It is remarkable that actually these properties are essentially already equivalent to m-accretivity of the operator A (or , to be more precise, of $\bar{A}$, which is due to the fact that mild solutions do not "distinguish" between A and $\bar{A}$ (recall properties 1.2.3.)).

In other words, to get existence results for the inhomogeneous equation for any data $(u_0, f) \in \overline{D(A)} \times L^1(O,T;X)$, one has to check m-accretivity of the operator A .

This result has already been given in [5]:

THEOREM 1 *Let $A \in \mathcal{A}(\omega), 0 < T \leq \infty$. Then the following assertions are equivalent:*

i) $A + \omega I$ is m-accretive

ii) A is closed and for all $u_0 \in \overline{D(A)}, f \in L^1_{loc}([O,T[;X)$ there exists a unique mild solution of $(CP)_{u_o,f}$.

With regard to the importance of m-accretivity it is essential to have different characterizations of this property . In this direction , the following result is of particular interest :

THEOREM 2 *Let $\omega \in I\!R$. Then $A \in \mathcal{M}(\omega)$ iff $A \in \mathcal{A}(\omega), A$ is closed and*

$$(TC) \qquad \liminf_{\lambda \downarrow 0} \frac{1}{\lambda} \; dist\,(x + \lambda y, R(I + \lambda A)) = 0 \qquad \text{for all } y \in X, x \in \overline{D(A)}.$$

Observe that, in contrast to the corollary we deduced at the beginning of this subsection, (TC) amounts to the requirement that all translates $A - y,\ y \in X$, of the operator A do not only satisfy (CRC), but even (WRC).

The characterization of m-accretivity by means of (TC) is also particularly useful for applications in perturbation theory, as this property can be shown to be stable under a certain class of perturbations (more precisely, if A satisfies (TC) and $B : \overline{D(A)} \subset X \to X$ is continuous, then (TC) holds for $A+B$). This leads to the following simple conclusion :

PROPOSITION 1 *Let A be m-accretive, $B : \overline{D(A)} \subset X \to X$ continuous. If $A + B \in \mathcal{A}(\omega)$ for some $\omega \in I\!R$, then $A + B + \omega I$ is m-accretive. In particular, if $D(B) = X$ and B is accretive, then $A + B$ is m-accretive.*

1.6 Regularity of mild solutions

As we have already pointed out several times, though continuous by definition, mild solutions may not satisfy any additional regularity properties nor need they take their values in $D(A)$. In particular, in general they can not be interpreted as a solution of (CP) in a pointwise sense ; however they satisfy a number of differential inequalities in the weak sense (cf.1.5.1.).

Nevertheless, the question arises naturally whether under certain additional assumptions (for instance on the data u_0, f) one may obtain more regularity of mild solutions.

We recall in this direction the well-known fact that if $-A$ is the infinitesimal generator of a C_0-semigroup of bounded linear operators on X, then $S^A(.)u_0$ is even a classical solution of $(CP)_{u_0}$ for all $u_0 \in D(A)$. Likewise the question of regularity of a solution for the inhomogeneous equation may be easily answered by means of the classical Duhamel formula.

In the general (nonlinear) setting, the regularity question turns out to be much more delicate. The following two examples may give a first idea of the nature of the problem:

Example 1

Let $X = C[0,1]$, $D(A) = \{f \in C^1[0,1];\ f(1) = 0\}, A = -\dfrac{d}{dx}$. It is well-known that A is linear, m-accretive and the semigroup S^A generated by A on

$$D(S^A) = \overline{D(A)} = \{f \in C[0,1]; f(1) = 0\}$$

is given by

$$(S^A(t)f)(x) = \begin{cases} f(x+t) & , x+t \leq 1 \\ 0 & , x+t > 1 \end{cases}$$

However in sharp contrast to the classical linear case (i.e. densely defined operator A), we observe that if $f \in D(A), f'(1) \neq 0$, then $S^A(.)f$ is not even a strong solution of $(CP)_f$. Indeed, for all $t \in (0,1)$: $\frac{d}{dt}(S^A(t)f)(x) = \begin{cases} f'(x+t) & , x+t < 1 \\ 0 & , x+t < 1 \end{cases}$ and thus $\frac{d}{dt}(S^A(t)f) \notin C[0,1]$ i.e. $S^A(t)f \notin D(A)$ for any $t \in (0,1)$.

Example 2

Let $X = C_0(\mathbb{R}^+)$ and consider the semilinear operator $A + B$ in X where A is defined on $D(A) = \{f \in X; f' \in X\}$ by $Af = -f'$, and $Bf = f^+$ with $D(B) = X$.

Obviously, $-A$ is the infinitesimal generator of a C_0-semigroup of (linear) contractions on X, B is continuous, everywhere defined. Thus, by Proposition 1 of the preceding section, the (densely defined) operator $A + B$ is m-accretive in X.

Elementary calculations show that the semigroup S^{A+B} on X is given by

$$(S^{A+B}(t)u_0)(x) = \begin{cases} u_o(x+t) & , \text{ if } u_o(x+t) \leq 0 \\ e^{-t}u_0(x+t), & \text{ if } u_0(x+t) > 0 \end{cases}$$

So if $u_0 \in D(A) = D(A+B)$ changes sign at some $x_0 \in \mathbb{R}^+$ where $u_0'(x_0) \neq 0$, then for all $0 \leq t < x_0$, $S^{A+B}(t)u_0 \notin D(A+B)$ since the pointwise derivative of $S^{A+B}(t)u_0$ fails to be continuous at $x_0 - t$ .In particular, then $S^{A+B}(.)u_0$ is not a strong solution.

Note that, though not differentiable a.e. , the mild solution in each of the preceding examples satisfies a local Lipschitz condition. Thus, in particular, if the Banach space X satisfied the Radon - Nikodym property ((RN) for short), then these mild solutions would automatically be strong solutions. It is well-known that reflexive Banach spaces are (RN), whereas $C_0(\mathbb{R}^+)$ as well as $L^1(\Omega)$, are just the classical examples where this property fails.

As we will see in the following, the question whether a mild solution of (CP) is already a strong one or not, reduces essentially to the question of regularity - just as it is the case for solutions in the sense of distributions for example. In other words, if a mild solution turns out to be differentiable at some point, it will (under some general weak assumptions) automatically satisfy the differential inclusion.

One natural approach in the study of regularity thus consists in the determination of the set of initial data $(u_0, f) \in \overline{D(A)} \times L^1(0,T;X)$ for which the mild solution satisfies a local Lipschitz condition.

To this end we first have to introduce some notations :

DEFINITION 1 *For any operator A in X , $x \in X$, define*

$$| x |_A := \inf\{M \geq 0; \exists\ (x_n, y_n) \in A \text{ such that } x_n \to x \text{ in } X \text{and } \| y_n \| \leq M\},$$

and set $\widetilde{D}(A) := \{x \in X; | x |_A < \infty\}$, which will be called the "generalized" domain of A.

The "generalized domain" has been essentially introduced (under more restrictive assumptions however) in [20]; the definition given above as well as a detailed study of the generalized domain may be found in [9]. Let us further note that for the particular case of a generator of a C_0- semigroup of linear operators $-A$, the above definition coincides with the one of the set which is usually called the Favard class in classical linear theory (cf.[16]).

One immediately checks the following properties :

$$D(A) \subset \widetilde{D}(A) \subset \overline{D(A)},$$

$$|x|_A \leq \inf\{\ \|y\|\ \ ; y \in Ax\} \quad \text{for} \quad x \in D(A),$$

$$\widetilde{D}(A - y) = \widetilde{D}(A) \quad \text{for all } y \in X$$

The significance of the "generalized domain" already turns out in the following result :

THEOREM 1 *Let $A \in \mathcal{A}(\omega)$, $\omega \in \mathbb{R}$, , $u_0 \in \widetilde{D}(A)$, $f \in BV_{loc}([0, T[; X)$, u is a mild solution of $(CP)_{u_0,f}$. Then u is locally Lipschitz continuous on $[0, T[$ and for all $t \in [0, T[$, the following estimate holds :*

$$(*) \qquad \limsup_{h\downarrow 0} \frac{\| u(t+h) - u(t) \|}{h} \leq e^{\omega t} \ | u(0) |_{A-f(0+)} + \mathrm{Var}(f, [0, t]) + \omega \int_0^t e^{\omega(t-s)}\, \mathrm{Var}(f, [0, s])ds.$$

For the sake of completeness, let us recall that $BV(O, T; X)$ consists of the subspace of functions in $L^1(O, T; X)$ which are of bounded variation , i.e which satisfy

$$\mathrm{Var}(f, T) = \limsup_{h\downarrow 0} \int_0^{T-h} \frac{\|f(\tau + h) - f(\tau)\|}{h} d\tau < \infty.$$

For $t \in [0, T[$, we set

$$\mathrm{Var}(f, [0, t]) = \limsup_{h\downarrow 0} \int_0^t \frac{\|f(\tau + h) - f(\tau)\|}{h} d\tau.$$

Moreover, if $f \in L^1(0, T; X), t$ is a right Lebesgue point of f, then, as usual,

$$f(t+) := \lim_{h\downarrow 0} \frac{1}{h} \int_t^{t+h} f(\tau)d\tau.$$

Apart from results on global behaviour of mild solutions like the preceding one, it is also possible to characterize their local behaviour in some point. We need some more notations: first, for $u \in C([0,T[;X), t \in [0,T[$, we set

$$D^+u(t) := \cap_{\epsilon>0} \overline{conv}^{\sigma(X^{**},X^*)} \left\{\frac{u(t+h)-u(t)}{h} \; ; 0 < h < \epsilon\right\}.$$

As a motivation for this definition, we note that, in a general Banach space X, an estimate on the mild solution of the form (*) does not imply more than the existence of some $\sigma(X^{**},X^*)-$convergent filter of the differential quotients. However, u is $\sigma(X^{**},X^*)-$differentiable at $t \in [0,T[$ iff

$$D^+u(t) \quad \text{is a singleton and} \quad \limsup_{h\downarrow 0} \frac{\|u(t+h)-u(t)\|}{h} < \infty.$$

Moreover , for an operator A in X , we define

$$T_A := \{x \in X; \liminf_{\lambda\downarrow 0} \frac{1}{\lambda}\mathrm{dist}(x, R(I+\lambda A)) = 0\}.$$

The use of the set T_A of elements in X where "our familiar tangency conditions" hold allows to avoid a full strength of an $m-$ accretivity assumption in the following ("local") result (cf.[5],[9]).

THEOREM 2 *Let $A \in \mathcal{A}(\omega)$,$\omega \in \mathbb{R}, f \in L^1(0,T;X), u_0 \in \overline{D(A)}, t \in [0,T[$ right Lebesgue point of f, u a mild solution of $(CP)_{u_o,f}$ and assume that $u(t) \in T_{A-f(t+)}$. Then the following assertions are equivalent :*

i) $u(t) \in \hat{D}(A)$

ii) $\limsup_{h\downarrow 0} \frac{\|u(t+h)-u(t)\|}{h} < \infty$

iii) $D^+u(t) \neq \emptyset$

Moreover, then

$$\frac{\|u(t+h)-u(t)\|}{h} \longrightarrow |u(t)|_{A-f(t+)} \qquad \text{as } h \downarrow 0.$$

If, moreover, X is reflexive and strictly convex (resp. uniformly convex), then u is weakly (resp. strongly) differentiable from the right at t.

The essential question whether differentiable mild solutions satisfy automatically the equation , is solved in much greater generality by the next result (see also [5]) :

LEMMA 1 *Let $A \in \mathcal{A}(\omega), \omega \in \mathbb{R}, f \in L^1(0,T;X), u_0 \in \overline{D(A)}, t \in [0,T[$ a right Lebesgue point of f, u a mild solution of $(CP)_{u_0,f}$. Then for all $z \in D^+u(t)$, $(x,y) \in A$, and $w \in \mathcal{J}(u(t)-x)$ the following inequality holds :*

$$< w, z > \leq \; [u(t)-x, f(t+)-y] + \omega\|u(t)-x\|$$

Observe that this result shows that actually every pair of the form $(u(t), f(t+) - z)$, $z \in D^+u(t)$, extends the operator $A \in \mathcal{A}(\omega)$ as an ω-accretive operator in X^{**} . It is now obvious that if the mild solution u is differentiable at t , and moreover A can not be properly extended as an ω-accretive operator in X (in particular, for example, if $A \in \mathcal{M}(\omega)$), then u satisfies indeed the differential inclusion at t .

However, more can be said in general :

THEOREM 3 *Let $A \in \mathcal{A}(\omega), \omega \in \mathbb{R}$, and assume A is maximal in $\mathcal{A}(\omega)$ (i.e. $\forall B \in \mathcal{A}(\omega), B \supseteq A$ implies $B \equiv A$) . Assume further that $f \in L^1(0,T;X), u_0 \in \overline{D(A)}$, $t \in [0,T[$ a right Lebesgue point of f , and let u be a mild solution of $(CP)_{u_o,f}$. Then , if u is weakly differentiable from the right at t , we have*

$$\dot{u}_{wr}(t) + Au(t) \ni f(t+)$$

where $\dot{u}_{wr}(t)$ denotes the weak right derivate of u at t .

Let us also mention that in the homogeneous case the maximality assumption can be replaced by the range condition ; more precisely , we have (see [22]):

PROPOSITION 1 *Let $A \in \mathcal{A}(\omega), \omega \in \mathbb{R}$ and asume that A is closed and satisfies:*

$$(RC) \qquad R(I + \lambda A) \supseteq \overline{D(A)} \quad \text{for } \lambda > 0 \text{ small.}$$

Let u be a mild solution of $\dot{u} + Au \ni 0$ on $(0,T), 0 < T \leq \infty$. Then

i) If u is differentiable at $t \in]0,T[$, then $\dot{u}(t) + Au(t) \ni 0$
ii) If X is (RN) , then $(CP)_{u_o}$ has a unique strong solution for all $u_o \in \widetilde{D}(A)$.

We want to emphasize again that once regularity of the mild solution is shown, the problem whether the equation is fulfilled is completely solved. As to regularity itself, the most useful way for applications will be to verify first a Lipschitz type condition and to obtain differentiability by the Radom-Nikodym property of the space (if this holds). The regularity problem in a general Banach space , however , is still essentially open (cf. part 2 of the lecture) . Let us also underline that one may deduce differentiability by a different kind of arguments ; in particular, there exist clearly strong solutions which are not Lipschitz continuous .

To end this section , we want to collect the main useful consequences one may conclude from the preceding results under the additional assumption of m-accretivity which , as we know now , ensures existence and uiqueness of mild solutions as well as it guarantees the maximality needed (Theorem 3 / Proposition 1).

COROLLARY *Let $A \in \mathcal{M}(\omega)$, $\omega \in \mathbb{R}, f \in BV_{loc}([0,T[;X), u_o \in \widetilde{D}(A), u$ the mild solution of $(CP)_{u_o,f}$. Then :*

i) u is locally Lipschitz continuous on $[0,T[$; if X is (RN) , then u is a strong solution of $(CP)_{u_o,f}$.

ii) If X is reflexive and stricty convex (resp. uniformly convex), then u is weakly (resp. strongly) differentiable from the right at each $t \in [0, T[$ and

$$\dot{u}_{wr}(t) \in (A - f(t+))^0\, u(t) \quad \text{for all } t \in [0, T[$$

where for a set $E \subset X$:

$$E^0 := \{y \in E; \|y\| = min_{x \in E} \|x\|\}.$$

1.7 Continuous dependence and representation theorems

We want to conclude our "introduction" to the theory of nonlinear evolution equations by presenting some basic results precising the dependence of mild solutions of (CP) on the data u_0, f, A of the equation and, finally, also study to what extent well-known representation formulas from classical linear theory, like the Trotter product formula (cf.[33]), may be generalized to the nonlinear case.

The main result will, roughly, show that if a sequence of data $(u_{on}, f_n, A_n)_n$ converges in an appropriate sense to (u_0, f, A), then the corresponding mild solutions $(u_n)_n$ converge to a mild solution of $(CP)_{u_0,f}$.

In particular, this will allow to approximate mild solutions (under certain assumptions) by classical solutions of "nearby problems".

First of all, an appropriate concept of convergence of multivalued operators has to be introduced :

DEFINITION 1 *Let $(A_i)_i$ be a generalized sequence of operators in X. Then $\liminf_i A_i$ is defined to be the operator whose graph is the set :*

$$\{(x, y); \exists\, (x_i, y_i) \in A_i \text{ for all } i \text{ such that } (x_i, y_i) \longrightarrow (x, y) \text{ in } X \times X\}$$

The following result has been shown under some additional assumptions first in [15], and a proof of the form stated below may be found in [29]:

THEOREM 1 *Let $(A_i)_i$ be a generalized sequence in $\mathcal{A}(\omega), \omega \in \mathbb{R}$, $0 < T \leq \infty$; moreover, for any i, let $f_i \in L^1_{loc}([0, T[; X), u_{o,i} \in \overline{D(A_i)}$ and u_i be a mild solution of*

$$(CP_i) \qquad \begin{cases} \dfrac{du_i}{dt} + A_i u_i \ni f_i & \text{on } (0, T) \\ u_i(0) = u_{oi} \end{cases}$$

If $A \subseteq \liminf A_i$, $u_{0,i} \to u_0$ in $X, f_i \to f$ in $L^1_{loc}([0, T[; X)$ and u is a mild solution of $(CP)_{u_0,f}$ then $(u_i)_i$ converges to u uniformly on compact subsets of $[0, T[$.

The probably most useful way to approximate a given operator A in X is that by means of its so-called Yosida approximation A_λ defined for $\lambda > 0$ by$A_\lambda = \frac{1}{\lambda}(I - \mathcal{J}_\lambda^A)$. For any operator A , the inclusion $A \subseteq \liminf A_\lambda$ holds.

This kind of approximation, however, is of particular interest if $A \in \mathcal{A}(\omega), \omega \in I\!R$ and $R(I+\lambda A) \supseteq C \supseteq \overline{D(A)}$ for some closed convex set $C \subset X$ for all $\lambda > 0$ small enough. Indeed, under these assumptions, it is classical (cf. [19]) that for any $u_0 \in C$ there exists a unique classical solution $u_\lambda \equiv S^{A_\lambda}(.)u_0 \in C^1([0,\infty[;X)$ of

$$(CP_\lambda) \qquad \begin{cases} \dfrac{du_\lambda}{dt} + A_\lambda u_\lambda = 0 \quad \text{on} \quad (0,\infty) \\ u_\lambda(0) = u_0. \end{cases}$$

As an immediate consequence, we note :

COROLLARY 1 *Let $A \in \mathcal{A}(\omega), \omega \in I\!R$ and assume that $R(I+\lambda A) \supseteq \overline{conv}D(A)$ for all $\lambda > 0$ with $\lambda\omega < 1$. Then for all $u_0 \in \overline{D(A)}$, the mild solution $S^A(.)u_0$ of $(CP)_{u_0}$ may be locally uniformly approximated by the classical solutions $S^{A_\lambda}(.)u_0$ of (CP_λ).*

Apart from approximating the mild solution of (CP) by other mild solutions, one may also obtain explicit representation formulas in terms of the operator A, as, by the way, we have already seen in the case of the classical "exponential formula" due to Crandall-Liggett.

We recall this result at this place in a slightly generalized form :

THEOREM 2 *Let $A \in \mathcal{A}(\omega), \omega \in I\!R$, and $(\lambda_n)_n$ be some sequence such that $\lambda_n \downarrow 0$ and $R(I+\lambda_n A) \supseteq \overline{D(A)}$ for all n. Then*

$$S^A(t)x = \lim_{n\to\infty} (I+\lambda_n A)^{-[\frac{t}{\lambda_n}]}x$$

for all $x \in \overline{D(A)}$ uniformly on compact subsets of $I\!R^+$.

The exponential formula appears in fact as a particular case of the following general representation theorem due to [15], [29] :

THEOREM 3 *Let $(C_n)_n$ be a sequence of closed sets in X, $(\lambda_n)_n$ such that $\lambda_n \downarrow 0$, and for all n, $U_n : C_n \to C_n$ Lipschitz continuous with constant $(1+\omega\lambda_n)$ and $x_n \in C_n$. Let $A \subseteq \liminf_n \dfrac{I-U_n}{\lambda_n}$ and $x \in D(S^A)$. If $x_n \to x$, then*

$$S^A(t)x = \lim_n U_n{}^{[\frac{t}{\lambda_n}]}x_n$$

uniformly on compact subsets of $I\!R^+$.

As an immediate consequence of the preceding theorem one may obtain the kind of product formula we mentioned at the beginning.

COROLLARY *Let C_1, C_2 be closed subsets of X , $\omega_1, \omega_2 \geq 0$ and, for $0 < \rho < \rho_0$, let $F(\rho) : C_1 \to C_2$ be $(1+\omega_1\rho)$-Lipschitz continuous , $G(\rho) : C_2 \to C_1$ be $(1+\omega_2\rho)$-Lipschitz continuous. If the operators A, B satisfy*

$$(*) \qquad A + B \subseteq \liminf_{\rho\downarrow 0} \frac{I - F(\rho)G(\rho)}{\rho},$$

then, for all $x \in D(S^{A+B}) \subset C_1 \cap C_2$ *we have:*

$$S^{A+B}(t)x = \lim_n \left(F(\frac{t}{n})G(\frac{t}{n})\right)^n x = \lim_n \left(G(\frac{t}{n})F(\frac{t}{n})\right)^n x$$

uniformly on compact subsets of $\mathbb{R}^+$.

Remark

1) In the particular case where A, B satisfy $A \subseteq \liminf_{\rho\downarrow 0} \frac{I - F(\rho)}{\rho}$ and $Bx = \lim_{\rho\downarrow 0} \frac{x - G(\rho)x}{\rho}$ uniformly on compact subsets of C_2 (such that B is continuous), then the condition (*) of the corollary holds.

2) One may hope, having in mind the nice Lie-Trotter product formula for linear C_0-semigroups:

$$S^{A+B}(t) = \lim_n \left(S^A(\frac{t}{n})S^B(\frac{t}{n})\right)^n,$$

that, given some m-accretive operators A and B in X , by choosing $F(\rho) = S^A(\rho)$, and $G(\rho) = S^B(\rho)$ in the corollary, one could obtain an analogous formula in the nonlinear case.

We emphasize, however, that this is not possible in general. This is due to the particular peculiarity that, for an m-accretive (nonlinear) operator A , the inclusion $A \subset \liminf_{t\downarrow 0} \frac{I - S^A(t)}{t}$ does not hold in general (and thus, a fortiori, condition (*) is not satisfied in general); in other words, A can not be recovered by the semigroup S^A it has generated (in the sense of mild solutions) as its graph infinitesimal generator - the notion one should expect to be the appropriate extension of the notion of the classical infinitesimal generator of a linear C_0-semigroup (we will return to this "generation problem" in part 2 of the lecture).

2 OPEN PROBLEMS

In this part, we want to discuss some open problems of the theory of nonlinear evolution equations. Clearly, giving a complete survey would be beyond the scope of this lecture; we rather chose four important aspects of the theory and the most interesting questions arising in each range of ideas respectively. We begin by considering questions related to regularity of solutions of the evolution problem:

2.1. Different notions of solutions in the m-accretive case

We have already seen that, if A is m-accretive in X (or more generally, $A \in \mathcal{M}(\omega)$ for some $\omega \in \mathbb{R}$), then for all $f \in L^1(0,T;X)$ and all $u_0 \in \overline{D(A)}$ there exists a (unique) mild solution u of

$$(CP)_{u_0,f} \qquad \begin{cases} \dfrac{du}{dt} + Au \ni f & \text{on } (0,\infty) \\ u(0) = u_0. \end{cases}$$

It has also been pointed out that, if the Banach space X has the Radon-Nikodym property (RN) , then for certain initial data this mild solution turns out to be even a strong solution; more precisely, we have:

$$(*)\quad \begin{cases} \text{For all } u_0 \in \hat{D}(A) \text{ and } f \in BV(0,T;X), \text{ u is a strong solution of } (CP)_{u_0,f} \\ \qquad\qquad \text{iff u is the mild solution of } (CP)_{u_0,f}. \end{cases}$$

Property (*) in fact characterizes those Banach spaces which satisfy (RN) in the sense that, if X is not (RN), then there exists (for some equivalent norm on X) an m-accretive operator such that the only strong solutions of $(CP)_{u_0,f}$ are the constant ones. In particular, if f is not constant equal to some $y \in R(A)$, then $(CP)_{u_0,f}$ does not have a strong solution (cf.[26]).

Nevertheless, there are simple examples of operators for which (*) is satisfied even if X is not (RN):

Example 1:
If A is continuous with D(A) closed, then (*) holds.

Example 2:
Consider the problem

$$(P)\qquad \begin{cases} u_t = \Delta\Phi(u) + f & \text{on } (0,T)\times\Omega \\ \Phi(u) = 0 & \text{on } (0,T)\times\partial\Omega \\ u(0) = u_0 & \text{on } \Omega \end{cases}$$

where Ω is a bounded, open subset of $\mathbb{R}^N$, $\Phi : \mathbb{R} \to \mathbb{R}$ is continuous, nondecreasing, $T > 0$.

We may formulate (P) as an evolution problem in $X = L^1(\Omega)$ by setting

$$Au = -\Delta\Phi(u) \text{ on } D(A) = \{u \in L^1(\Omega); \Phi(u) \in W_0^{1,1}(\Omega), \Delta\Phi(u) \in L^1(\Omega)\}.$$

Then A is m-accretive and

$$\widehat{D}(A) = \{u \in L^1(\Omega); \Phi(u) \in W_0^{1,1}(\Omega), \Delta\Phi(u) \in \mathcal{M}_b(\Omega)\}.$$

Assume now that $\Phi \in C^1(\mathbb{R})$ and $\Phi' > 0$ a.e. on $\mathbb{R}$; then every essentially bounded mild solution of $(CP)_{u_0,f}$, with $u_0 \in \hat{D}(A), f \in BV(0,T;L^1(\Omega))$, is actually a strong solution (see [11]). Let β be a maximal monotone graph in $\mathbb{R}\times\mathbb{R}$ with $0 \in \beta(0)$ and define the operator B in $L^1(\Omega)$ by

$$w \in Bu \quad \text{iff} \quad u, w \in L^1(\Omega), w(x) \in \beta(u(x)) \;\; a.e.x.$$

Assuming $D(\beta) \subset R(\Phi)$, the operator $A + B$ is still m-accretive; the result claimed above shows that if $D(\beta)$ is bounded, then $A + B$ satisfies (*).

The above considerations motivate the following

PROBLEM 1 : How to characterize in a general Banach space X those m-accretive operators A that satisfy (*) ?

Even in particular cases , e. g. $X = L^1(\Omega)$, a formulation of abstract, merely sufficient conditions on the operator A to guarantee (*) would be of interest.

Problem 1 admits an anologous "semigroup version" corresponding to the homogeneous equation :

> **PROBLEM 1* : Characterization of those m-accretive operators for which the homogeneous equation $(CP)_{u_0}$ admits a strong solution for all $u_0 \in \widehat{D}(A)$.**

Note that this problem even seems to be still open in the classical linear setting where it is reduced to the question whether $S^A(t)(\widehat{D}(A)) \subseteq D(A)$ for all $t > 0$.

This fails to be true in general which is evident if one thinks of the translation semigroup in $X = L^1(\mathbb{R}^+)$ (resp. $C_0(\mathbb{R}^+)$) whose infinitesimal generator A defined on $D(A) = \{f \in X; f' \in X\}$ by $Au = -u'$, is known to have as generalized domain $BV(0,\infty)$ (resp. $Lip(\mathbb{R}^+)$, the set of Lipschitz continuous functions on $\mathbb{R}^+$).

In the linear case, we have the following result (cf. [25],[9]) :

$$\text{(1)} \qquad \begin{cases} \text{For all } u_0 \in D(A), f \in W^{1,1}(0,T;X) \text{ such that} \\ \text{the "compatibility condition" : } f(0) - Au_0 \cap \overline{D(A)} \neq \emptyset \text{ is satisfied} \\ \text{there exists a (unique) classical solution u of } (CP)_{u_0,f}. \end{cases}$$

A similar result, more precisely, the existence of strong solutions for initial data $u_0 \in D(A)$ and $f \in W^{1,1}(0,T;X)$ (now without compatibility condition) may also be established for a certain class of nonlinear operators, the so-called m-completely accretive operators (cf. [8] for the theory of completely accretive operators ; the essential idea for the existence of strong solutions has already been given in [31]).

In any of the above cases, however, we realize that the following property holds :

$$(**) \qquad \begin{cases} \{(u_0, f) \in \overline{D(A)} \times L^1(0,T;X); \exists! \text{ strong solution of } (CP)_{u_0,f}\} \\ \text{is dense in } \overline{D(A)} \times L^1(0,T;X) \end{cases}$$

(which, in fact, amounts to the property that every mild solution is a weak solution in the sense of [7]).

Note that property $(**)$ is actually weaker than property $(*)$. One needs only to consider once again the simple example due to [37] which we already considered in the first part of the lecture (see Example 2 in 1.6.):

Example 3 :
In $X = C_0(\mathbb{R}^+)$ we consider the operator A defined on $D(A) = \{u \in X; u' \in X\}$ by $Au = -u' + u^+$. We already know that there exist $u_0 \in D(A)$ such that $S^A(t)u_0 \notin D(A)$ for any $t > 0$; in particular, the mild solution $S^A(.)u_0$ fails to be a strong one and hence A does not verify property $(*)$.

However, being the sum of a generator of a C_0-semigroup of bounded linear operators and some continuous, everywhere defined operator, the operator A does satisfy property $(**)$ (cf. [10])

Consequently, the question that arises is the following

> **Problem 2 : How to characterize those m-accretive operators which satisfy (**) ?**

We emphasize that, just as in the case of (*) , also property (**) does not hold in general. This is harder to establish, at least in the brevity required here, but will become clear by the following discussion :

As already mentioned in part 1, if u is a strong solution of $(CP)_{u_0,f}$, then $v(t) := u(T-t)$ is a strong solution of the "time-reversed" problem

$$(2) \qquad \begin{cases} \frac{dv}{dt} + (-Av) \ni g \quad \text{on } (0,T) \\ v(0) = u(T) \end{cases}$$

where $g(t) := f(T-t)$, and therefore, a fortiori, a mild solution of (2).

Consequently one has that for an operator A satisfying (**) , the following property holds :

$(***)$ For all $f \in L^1(0,T;X), u_0 \in \overline{D(A)}$, if u is the mild solution of $(CP)_{u_0,f}$, then $v(t) := u(T-t)$ is a mild solution of (2)

or, in other words, mild solutions are reversible in time . As this is not the case in general (cf. part 1, and [10], for a concrete example), it follows that (***), and consequently (**), does not hold for every m-accretive operator and we are led to formulate

> **PROBLEM 3 : Characterization of those m-accretive operators that satisfy (***) ?**

We mention that, obviously, the same remark as that for Problem 1 applies to the other problems considered here in much the same way.

We further note that the properties (*) - (***) represent a certain graduation for parabolic up to hyperbolic nonlinear differential equations, and turn now to questions of a different nature, in fact, related to the "generation problem" for nonlinear semigroups:

2.2 A nonlinear version of the Hille-Yosida theorem

To begin with, we recall the classical Hille-Yosida theorem which may be divided into two parts and stated as follows :

1) Let A be a linear operator in X. Then the following properties are equivalent :

i) $\overline{D(S^A)} = X$ and A is closed

ii) $\overline{D(A)} = X$ and there exists an equivalent norm $|||.|||$ on X and $\omega \in \mathbb{R}$ such that $A \in \mathcal{M}(\omega)$.

2) Given some C_0 -semigroup $(S(t))_{t\geq 0}$ of bounded linear operators on X, there exists a unique linear operator A such that $S \equiv S^A$.

A first extension to the nonlinear case of the classical Hille-Yosida theorem has been given by the following result due to Komura in the case where X is a Hilbert space :

THEOREM (cf. [30])

1) Let A be a maximal monotone operator in the Hilbert space H . Then $\overline{D(A)}$ is a closed convex subset of H (and, trivially, $D(S_A) = \overline{D(A)}$) .

2) Given some closed convex set $C \subset H$ and a continuous semigroup of contractions $(S(t))_{t\geq 0}$ on C , then there exists a unique maximal monotone operator A in H such that $\overline{D(A)} = C$ and $S^A \equiv S$.

This result has been extended to uniformly smooth Banach spaces X (i.e. with uniformly convex dual X^*) by Baillon (cf. [3]), but it turns out to be false in general. The first problem that arises in general Banach spaces is the fact that, given any m-accretive operator in X, then $\overline{D(A)}$ fails to be convex in general.

However, if dim $X < \infty$, it is known that $C := \overline{D(A)}$ is a nonexpansive retract of X , i.e. there exists a contraction P of X onto C such that $P^2 = P$. The same holds for m-accretive operators in reflexive Banach spaces with uniformly Gâteaux differentiable norm and Fréchet differentiable dual norm (the setting, in which S.Reich presented his generalisation of the Hille-Yosida theorem for semigroups on nonexpansive retracts, cf. [35]) and, a fortiori, in Hilbert spaces (in fact, convex sets in Hilbert spaces are nonexpansive retracts) . However this is not true in the general case: if $X = L^\infty(\Omega)$, where Ω is a smooth bounded open set in $\mathbb{R}^N, C_0(\Omega)$ is not a nonexpansive retract of X, while it is $\overline{D(A)}$ where A is the m-accretive (linear) operator in X defined by $Au = -\Delta u$ on $D(A) = \{u \in C_0(\Omega); \Delta u \in X\}$.

In view of the above considerations we want to pose the following problems :

PROBLEM 4 : If A is m-accretive in a Banach space X such that spanD(A) is dense in X, is it true in general that $\overline{D(A)}$ is a nonexpansive retract of X ? If not, characterize the spaces X for which this is true.

PROBLEM 5 : Given a semigroup of contractions $(S(t))_{t\geq 0}$ on a nonexpansive retract C of X, does there exist an m-accretive operator A such that $\overline{D(A)} = C$ and $S^A \equiv S$?

We remark that, whereas the answer to Problems 4 and 5 is affirmative if X is of finite dimension, even in this case a complete extension of the Hille-Yosida theorem fails, as, in general, one does not have uniqueness of a "generator". More precisely, it was shown in [22] , that even in two dimensions an everywhere defined semigroup $S(t)$ may give rise to an infinity of different m-accretive operators A satisfying $\overline{D(A)} = X$ and $S^A \equiv S$.

But, in fact, the situation is even more complicated : by a result due to Kobayashi [28], we know that operators A which are merely quasi-accretive (which is actually a weaker notion than accretivity and is equivalent to the property : $2\|x_1 - x_2\| \leq \|x_1 - x_2 + \lambda y_1\| + \|x_2 - x_1 + \lambda y_2\|$ for all $\lambda > 0$, $(x_1, y_1), (x_2, y_2) \in A$) and satisfy, for example, the weak range condition, then S^A is a semigroup of contractions on $\overline{D(A)}$.

In this direction, one may note that the union of the infinity of accretive "generators" of the semigroup in the example of [22] is quasi-accretive, but possibly only by chance, as a union of accretive operators in general in no way needs to be quasi - accretive .

Anyway, with the intent to clear up the situation, already the answer to the following problem if X has finite dimension (without supposing X^* strictly convex) would be of great interest :

PROBLEM 6 : Definition of a class $\mathcal{C}$ of operators A with $\overline{D(A)} = X$ such that $A \to S^A$ is a bijection of $\mathcal{C}$ onto the set of continuous semigroups of contractions on X.

Let us recall at this place that in the linear case, in view of the possibility of renorming, the generation problem actually reduces to the accretive case which, as we have seen, is not the case in the nonlinear setting.

In this range of ideas, there arises naturally also the question if one may not replace the contraction property by a more general topological condition on the resolvents of A, just as in the linear case .

More precisely, we ask in

PROBLEM 7 : If $(I + \lambda A)^{-n}$ is uniformly Lipschitz continuous for all $\lambda > 0, n \in \mathbb{N}$, do we have $D(S^A) = \overline{D(A)}$?

Another characteristic feature of the classical linear case is the possibility to pass, by means of the Duhamel formula, from the semigroup $(S^A(t))_{t \geq 0}$, the solution operator of the homogeneous problem, to the general evolution problem, i.e. the application

$$(u_0, f) \in X \times L^1(0, T; X) \longrightarrow u(t) := S^A(t)_{u_0} + \int_0^t S^A(t-s)f(s)ds$$

(u is the mild solution of $(CP)_{u_0,f}$).

This gives rise to a different approach in the attempt to extend the Hille - Yosida theorem :

Given an m-accretive operator A in X, we denote by E^A the application

$$(x, y, t) \in \overline{D(A)} \times X \times \mathbb{R}^+ \longmapsto E^A(x, y, t)$$

where $E^A(x, y, .)$ is the mild solution of

$$(CP)_{x,y} \qquad \begin{cases} \dot{u} + Au \ni y \\ u(0) = x. \end{cases}$$

Then the following is true:

THEOREM (cf.[5]) *Given a closed subset C of X, the application $A \longrightarrow E^A$ is a bijection of the set $\{A; A$ m-accretive $, \overline{D(A)} = C\}$ onto the set of applications $E : C \times X \times \mathbb{R}^+ \to C$ satisfying*

i) $E(E(x, y, t), y, s) = E(x, y, t + s)$,

$E(x,y,0) = x$ *for all* $(x,y) \in C \times X, t, s \in \mathbb{R}^+$

ii) $t \in \mathbb{R}^+ \to E(x,y,t)$ *is continuous for all* $(x,y) \in C \times X$

iii) $\|E(x,y,t) - E(\hat{x},\hat{y},t)\| \leq \|x-\hat{x}\| + \int_0^t [(E(x,y,s) - E(\hat{x},\hat{y},s), y - \hat{y}]ds$ *for all* $(x,y,t),(\hat{x},\hat{y},t) \in C \times X \times \mathbb{R}^+$.

Although this theorem may be regarded as the complete extension of the Hille-Yosida theorem to the nonlinear case, the characterization of the map E in *iii)* of the theorem above is not satisfying, as the bracket $[.,.]$, though extremely useful in mathematical theory, does not admit a physical interpretation (just like it is the case with the norm). This gives rise to the final problem of this section

PROBLEM 8 : Is it possible to replace *iii)* **by the weaker property:**
iii)* $\|E(x,y,t) - E(\hat{x},\hat{y},t)\| \leq \|x-\hat{x}\| + t\|y-\hat{y}\|$
for all $(x,y,t),(\hat{x},\hat{y},t) \in C \times X \times \mathbb{R}^+$?

2.3. Linear phenomena

There is a number of characteristic linear phenomena that also appear in nonlinear theory. As for duality, for instance, we have the following result (*cf.*[12]) :

PROPOSITION *Let* A *be a subdifferential in the Hilbert space* H*, i.e.* $A = \partial\Psi$ *for some* $\Psi : H \to]-\infty,\infty]$ *proper, convex, lower semicontinuous ; let* $N : H \to [0,\infty]$ *be sublinear .*

Then the following assertions are equivalent :

i) A *is* N*-accretive, i.e.*

$$N(u-\hat{u}) \leq N(u-\hat{u}+\lambda(v-\hat{v})) \text{ for all } \lambda > 0, (u,v),(\hat{u},\hat{v}) \in A$$

ii) A *is* N^**-accretive, where*

$$N^*(x) := \sup \{(x,y) \; ; \; N(y) \leq 1\}$$

Phenomena of this type appear in many other cases :

Example 4

Let Ω be a bounded, open domain of $\mathbb{R}^N, \Phi : \mathbb{R} \to \mathbb{R}$ continuous, nondecreasing, $\Phi(0) = 0$ and consider the following problems :

$$(P) \qquad \begin{cases} u_t - \Delta\Phi(u) = 0 & \text{on } (0,\infty) \times \Omega \\ \Phi(u) = 0 & \text{on } (0,\infty) \times \partial\Omega \\ u(0,.) = u_0 & \text{on } \Omega \end{cases}$$

$$(P)^* \qquad \begin{cases} u_t + \Phi(-\Delta u) = 0 & \text{on } (0,\infty) \times \Omega \\ u = 0 & \text{on } (0,\infty) \times \partial\Omega \\ u(0,.) = u_0 & \text{on } \Omega \end{cases}$$

The problem (P), resp. $(P)^*$, gives rise to well-posed non-linear evolution problems in $L^1(\Omega)$, resp. $L^\infty(\Omega)$, in the sense that if we define the operator A in $L^1(\Omega)$ by $Au = -\Delta\Phi(u)$ on

$$D(A) = \{u \in L^1(\Omega); \Phi(u) \in W_0^{1,1}(\Omega), \Delta\Phi(u) \in L^1(\Omega)\}$$

and the operator A^* in $L^\infty(\Omega)$ by $A^*v = \Phi(-\Delta v)$ on $D(A^*) = \{v \in W^{1,\infty}(\Omega);\ \Delta v \in L^\infty(\Omega)\}$, these operators are m-accretive in $L^1(\Omega)$, resp. $L^\infty(\Omega)$, and hence there exist unique mild solutions for the corresponding evolution problems to (P), resp. $(P)^*$, for initial data $u_0 \in \overline{D(A)}$ (resp. $\overline{D(A^*)}$).

It is now natural to interpret $(P)^*$ as dual equation to (P) and thus the operator A^* in $L^\infty(\Omega)$ appears as the nonlinear transpose of the operator A in $L^1(\Omega)$.

This and other examples, like the "duality" between nonlinear scalar conservation laws and Hamilton-Jacobi equations, may be found in [18] . The general question, however, that arises in the theory of nonlinear operators is the following

> **PROBLEM 9 : Given (nonlinear) operators A in a Banach space X , A^* in the dual space X^* ; when can we interpret A^* as the transpose of the operator A ?**

In the case of a linear operator A, there exist equivalent characterizations for a continuous function to be a mild solution of $(CP)_{u_0,f}$; more precisely, the following result holds :

THEOREM (cf. [9] , [10]) *Let A be a linear operator in $X, f \in L^1(0,T;X)$ and $u \in C([0,T];X)$. Then the following assertions are equivalent :*

i) u is a mild solution of $\dot{u} + Au \ni f$ on $(0,T)$

ii) $\mathcal{U}(t) := \int_0^t u(s)ds$ is a classical solution of

$$\dot{\mathcal{U}} + \bar{A}\mathcal{U} \ni u(0) + \int_0^t f(s)ds \quad \text{on } [0,T]$$

iii) For all $(x^*, y^*) \in A^*$, the function $t \mapsto < x^*, u(t) >$ is absolutely continuous and

$$\frac{d}{dt} < x^*, u(t) > + < y^*, u(t) > = < x^*, f(t) > \quad a.e.\, t \in (0,T)$$

We remark that iii) characterizes mild solutions as weak solutions as they have been introduced first by Ball (cf. [4]) whereas ii) shows that the mild solution is in fact the classical solution of an "integrated problem". This property gave rise to a whole theory of so-called (k - times) integrated semigroups, $k \in \mathbb{N}$ (cf. [2] , [32]) .

It is now natural to ask if there exist analogous characterizations in the nonlinear case and hence we pose the following problems :

> **PROBLEM 10 : Is it possible to characterize the mild solution as in iii) of the above cited theorem using an appropriate nonlinear duality concept ?**

PROBLEM 11 : Is it possible to extend the characterization result ii) of the theorem to the nonlinear case ? In particular, what could be the corresponding nonlinear notion of a k - times integrated semigroup ?

2.4. Singular limits

In our final section of our survey on open problems in the theory of nonlinear evolution equations, we consider questions related to convergence properties of nonlinear semigroups and, more generally, continuous dependence of mild solutions on the given data u_0, f, A of the evolution problem .

We recall the following continuity result as a particular case of Theorem 1 in 1.7:

THEOREM *Let $(A_n)_n$ be a sequence of m - accretive operators in X, $\overline{D(A_n)} = X$ for all $n \in I\!N$, $T > 0$ and suppose that $A = \liminf_n A_n$ is m - accretive.*

Let $(u_0, f) \in X \times L^1(0,T;X)$ and, for all $n \in I\!N, u_n$ be the mild solution of

$$\begin{cases} \frac{du_n}{dt} + A_n u_n \ni f \\ u_n(0) = u_0 \end{cases}$$

If $u_0 \in \overline{D(A)}$, then $u_n \to u$ in $C([0,T];X)$, where u is the mild solution of $(CP)_{u_0,f}$.

We note that a typical example consists in approximating an m-accretive operator A by its Yosida approximations $A_\lambda = \frac{1}{\lambda}(I - (I + \lambda A)^{-1})$, which are m-accretive operators with $D(A_\lambda) = X$.

Now, the problem that arises is what happens in case the initial data u_0 do not belong to $\overline{D(A)}$.

In this direction we recall that in the Hilbert space setting for the special case of the Yosida approximation the following holds :

PROPOSITION (cf. [14]) *Let A be maximal monotone in H. Then for all $u_0 \in H$, $S^{A_\lambda}(.)u_0$ converges uniformly on compact subsets of $I\!R^+$ to $S^A(.)Pu_0$, where $P : H \to \overline{D(A)}$ denotes the projection operator onto the closed convex set $\overline{D(A)}$.*

However, even in a Hilbert space, the problem whether an anologous convergence result holds for a general approximation of A by some sequence $(A_n)_n$ seems to be still open . This problem has already been posed in [14] as Problem 16 and we restate it here in the semigroup version :

PROBLEM 12 : Let $(A_n)_n$ be a sequence of maximal monotone operators in the Hilbert space H with dense domain . Suppose $A = \liminf_n A_n$ is maximal monotone . Is it true that for all $u_0 \in H$, $S^{A_n}(.)u_0 \to S^A(.)P_{\overline{D(A)}}u_0$ as $n \to \infty$ uniformly on compact subsets of $I\!R^+$?

Clearly, the same question arises in the m-accretive case in a general Banach space X . There exist positive results in particular cases . One typical example is given by the so-called "mesa - problem" (cf. [17],[36],[6]).

Example 5 :
Let $u_0 \in L^1(\mathbb{R}^N), u_0 \geq 0, m > 1$. It is known (cf.[1]) that each of the problems

$$\begin{cases} \frac{\partial u_m}{\partial t} = \Delta {u_m}^m & \text{on } (0,\infty)\times \mathbb{R}^N \\ u_m(0,.) = u_0 & \text{on } \mathbb{R}^N \end{cases}$$

has a unique strong solution $u_m \in C([0,\infty[; L^1(\mathbb{R}^N)) \cap C([0,\infty[\times \mathbb{R}^N) \cap W^{1,1}_{loc}(0,\infty; L^1(\mathbb{R}^N))$.
Moreover, the following convergence result holds :

$$u_m \to \underline{u} \text{ in } C(]0,\infty[; L^1(\mathbb{R}^N)) \text{ as } m \to \infty$$

where $\underline{u} = u_0 + \Delta \underline{w}$ in $L^1(\mathbb{R}^N)$ and $\underline{w}$ is the solution of the variational inequality

$$\underline{w} \in L^1(\mathbb{R}^N),\ \Delta\underline{w} \in L^1(\mathbb{R}^N),$$
$$0 \leq u_0 + \Delta\underline{w} \leq 1,\ \underline{w} \geq 0,\ \underline{w}(u_0 + \Delta\underline{w} - 1) = 0 \ a.e.$$

This result may be interpreted in terms of nonlinear semigroups by defining for $m > 1$ operators A_m in $L^1(\mathbb{R}^N)$ by $A_m u = -\Delta|u|^{m-1}u$ on

$$D(A_m) = \{u \in L^1(\mathbb{R}^N) \cap L^m_{loc}(\mathbb{R}^N) \ ; \ \Delta|u|^{m-1}u \in L^1(\mathbb{R}^N)\}$$

Then A_m is m-accretive in $L^1(\mathbb{R}^N)$ for all $m > 1$ as well as the operator $A = \liminf_m A_m$ which is defined by

$$v \in Au \Longleftrightarrow u, v \in L^1(\mathbb{R}^N), \exists w \in L^1(\mathbb{R}^N) \text{ such that}$$
$$v = -\Delta w \text{ in } \mathcal{D}'(\mathbb{R}^N) \text{ and } u \in \text{sign } w \ a.e.$$

Let $u_0 \in L^1(\mathbb{R}^N), u_0 \geq 0$,and $\underline{u}$ as defined above. We have $\underline{u} = (I + \lambda A)^{-1} u_0$, $(I + \lambda A)^{-1}\underline{u} = \underline{u}$ for all $\lambda > 0$; hence, we may define

$$(*) \qquad P u_0 = \lim_{\lambda \to 0} (I + \lambda A)^{-1} u_0$$

and the following result is true :

$$(**) \qquad S^{A_m}(.)u_0 \to S^A(.)Pu_0 \text{ in } C(]0,\infty[; X) \text{ as } m \to \infty$$

with $X = L^1(\mathbb{R}^N)$.
We must note that the limit (*) still exists for any $u_0 \in L^1(\mathbb{R}^N)$, but it is still an open problem whether (**) holds for any $u_0 \in L^1(\mathbb{R}^N)$.

If A is an m-accretive operator in a Banach space X, it is not true in general that the limit (*) exists (in X) for any $u_0 \in X$, even in the linear case. As an example consider again $X = L^\infty(\Omega)$, where Ω is a smooth bounded open set in $\mathbb{R}^N$, and let A be defined by $Au = -\Delta u$ on $D(A) = \{u \in C_0(\Omega); \Delta u \in X\}$; for any $u_0 \in X, (I + \lambda A)^{-1}u_0 \to u_0$ in $L^1(\Omega)$ as $\lambda \to 0$, such that the limit (*) exists in X iff $u_0 \in C_0(\Omega)$.

A first problem, clearly related to Problem 4, is

> **PROBLEM 13 : If A is m-accretive in a Banach space X such that span$D(A)$ is dense in X, does the limit (*) exists in X for any $u_0 \in X$? If not, characterize the spaces X for which this is true.**

A second problem extending Problem 12 is

PROBLEM 14 : Suppose that $(A_n)_n, A = \liminf_n A_n$ are m-accretive in X and let $u_0 \in X\backslash\overline{D(A)}$ such that the limit (*) exists and $u_0 \in \overline{D(A_n)}$ for all n. Is it true that $S^{A_n}(.)u_0 \to S^A(.)Pu_0$ in $C(]0,\infty[;X)$ as $n \to \infty$?

Note that positive answers to Problems 13 and 14 hold for particular classes of operators. As an example one can see the results in [13] for the class of m-completely accretive operators in $L^1(\Omega)$ associated to $A + B$, where A is a fixed m-completely accretive operator in $L^1(\Omega)$ and B is an operator of the form

$$w \in Bu \Leftrightarrow w(x) \in \partial j(x, u(x)) \quad a.e.x,$$

with $j : \Omega \times \mathbb{R} \to [0, \infty]$ a normal convex functional.

REFERENCES

[1] Aronson,D.G., Bénilan,Ph.(1979). Régularité de l'équation des milieux poreux dans $\mathbb{R}^N$, C.R.A.S.Paris 288,103-105.

[2] Arendt,W.(1987). Resolvent positive operators, Proc. London Math. Soc. 54, 321-349.

[3] Baillon,J.B.(1978). Générateurs et semi-groupes dans les espaces de Banach uniformément lisses, J. Fct. An. 29, 199-213.

[4] Ball,J.M.(1977). Strongly continuous semigroups, weak solutions and the variation of constant formula, J. Am. Math. Soc., 370-374.

[5] Bénilan,Ph.(1972). Equations d'évolution dans un espace de Banach quelconque et applications, Thèse Orsay.

[6] Bénilan,Ph.,Boccardo,L.,Herrero,M.A.(1989). On the limit of solutions of $u_t = \Delta u^m$ as $m \to \infty$, Rend. Sem. Mat. Univ. Pol. Torino, Nonlinear PDE's.

[7] Bénilan,Ph.,Brézis,H.(1972). Solutions faibles d'équations d'évolution dans les espaces de Hilbert, Ann. de l'Inst. Fourier, XXII,2.

[8] Bénilan,Ph.,Crandall,M.(1991). Completely accretive operators, in *Semigroup theory and evolution equations*, Ph. Clément et al. (eds), Marcel Dekker Inc., 41-76.

[9] Bénilan,Ph.,Crandall,M.,Pazy,A. (book in preparation).*Evolution equations governed by accretive operators.*

[10] Bénilan,Ph.,Crandall,M.,Pazy,A.(1978). "Bonnes solutions" d'un probleme d'évolution semi-linéaire, C.R.A.S.Paris 306, Série I, 527-530.

[11] Bénilan,Ph.,Gariepy,R. Strong solutions for degenerate parabolic equations, in preparation.

[12] Bénilan,Ph.,Picard,C. Quelques aspects nonlinéaires du principe du maximum, in *Séminaire de théorie du Potentiel*, Lect.Notes in Math.713, Springer, 1-37.

[13] Bénilan,Ph.,Wittbold,P. Absorptions nonlinéaires, to appear, in J.Fct.An.

[14] Brézis,H. (1972). *Opérateurs maximaux monotones et semi-groupes de contractions dans les espaces de Hilbert*, North Holland.

[15] Brézis,H.,Pazy,A.(1972). Convergence and approximation of semigroups of nonlinear operators in Banach spaces, J.Fct.An.9, 63-74.

[16] Butzer,P.L.,Berens,H.(1967). *Semigroups of operators and approximation*, Springer-Verlag New York.

[17] Caffarelli,L.A.,Friedman,A.(1987). Asymptotic behaviour of solutions of $u_t = \Delta u^m$ as $m \to \infty$, Indiana Un.Math.J. 36,4, 711-718.

[18] Caselles,V.(1990). Duality and equations governed by accretive operators, Publ.-Math. de Besançon, 12.

[19] Crandall,M.(1970). Differential equation on convex sets, J.Math.Soc. Japan 22, 443-455.

[20] Crandall,M.(1973). A generalized domain for semigroup generators, Proc. Amer. Math. Soc. 37,2, 434-440.

[21] Crandall,M.,Evans,L.C.(1975). On the relation of the operator $\frac{\partial}{\partial s} + \frac{\partial}{\partial t}$ to evolution governed by accretive operators, Israel J.Math. 21, $N^0 4$, 261-278.

[22] Crandall,M.,Liggett,T.(1971). Generation of semigroups of nonlinear transformations on general Banach spaces, Amer.J.Math. 93, 265-298.

[23] Crandall,M.,Pazy,A.(1972). Nonlinear evolution equations in Banach spaces, Israel J.Math. 11, 57-94.

[24] Crandall,M.,Pazy,A.(1979). An approximation of integral functions by step functions with an application, Proc.Amer.Math.Soc. 76, 74-80.

[25] Da Prato,G.,Sinestrari,E.(1987). Differential operators with nondense domain, Ann.Sc.Norm.Sup. Pisa, 14, 285-344.

[26] Deville,R.(1989). Solutions fortes d'équations d'évolution et propriété de Radon-Nikodym, C.R.A.S.Paris 309, série I, 909-914.

[27] Evans,L.C.(1977). Nonlinear evolution equations in an arbitrary Banach space, Israel J Math. 26, 1-47.

[28] Kobayashi,Y.(1975). Difference approximation of Cauchy problems for quasi-dissipative operators and generation of nonlinear semigroups, J.Math.Soc. Japan, 27,4, 640-665.

[29] Kobayashi,Y.,Miyadera,I.(1978). Convergence and approximation of nonlinear semigroups, in *Functional Analysis and Numerical Analysis*, Japan-France Seminar Tokyo and Kyoto, 1976, H.Fujita ed. Japan Soc.for the Promotion of Science.

[30] Komura,Y.(1967). Nonlinear semigroup in Hilbert space, J.Math.Soc. Japan 19, 493-507.

[31] Le,C.H.(1977). Etude de la classe des opérateurs m-accrétifs dans $L^1(\Omega)$ et accrétifs dans $L^\infty(\Omega)$, Thèse de 3^0 cycle, Paris VI.

[32] Neubrander,F.(1988). Integrated semigroups and their applications to the abstract Cauchy problem, Pacific J.Math.185, 111-155.

[33] Pazy,A.(1983).*Semigroups of linear operators and applications to partial differential equations*, Springer.

[34] Pierre,M.(1976). Génération et perturbation des semigroupes de contractions nonlinéaires, Thèse de 3^0 cycle, Univ. Paris VI.

[35] Reich,S.(1980). Product formulas, nonlinear semigroups and accretive operators, J.Fct.An.36, 147-168.

[36] Sacks,P.E.(1989). A singular limit problem for the porous medium equation, J.Math.Anal.140,2, 456-466.

[37] Webb,G.F.(1972). Continuous nonlinear perturbations of linear accretive operators in Banach spaces, J.Fct.An.10, 191-203.

Vector-Valued Versions of Some Representation Theorems in Real Analysis

W. ARENDT Equipe de Mathématiques,CNRS - URA 741, Université de Franche-Comté, 25030 Besançon Cedex, France

1 INTRODUCTION

In the present paper we consider four classical theorems on real-valued functions involving the notion of positivity ; namely,

1. the representation theorem for convex functions ;
2. Riesz' representation theorem for positive functionals on $C[0,1]$;
3. Hausdorff's theorem on the moment problem ;
4. Bernstein's theorem characterizing completely monotonic functions.

We investigate the problem, whether these theorems still hold, if functions with values in an ordered Banach space X with normal cone are considered instead of real-valued functions. It turns out that any of these four theorems holds in X if and only if X has the *interpolation property* (i.e. for any two sequences $(x_n), (y_n)$ in X such that $x_n \leq x_{n+1} \leq y_{n+1} \leq y_n (n \in \mathbb{N})$ there exists $z \in X$ such that $x_n \leq z \leq y_n (n \in \mathbb{N})$). This property is more general than σ-order completeness and has been studied by Seever [Se] for spaces $C(K), K$ compact.

A vector-valued version of Riesz's and Hausdorff's theorem has been proved before by Schaefer [S1] in the case where the norm is order continuous adapting arguments of Hausdorff. In our proofs concerning the representation of convex functions (Section 2), Riesz' theorem and the moment problem (Section 3) we frequently use the corresponding scalar results in order to deduce vector-valued versions.

Our main interest concerns Laplace transforms. Here our approach is different. A well-known theorem due to Widder [Wi1] characterizes Laplace transforms of bounded real-valued functions. The vector-valued version of this theorem holds if and only if the space has the Radon-Nikodym property ([A2], [Z]). However, an integrated version of this theorem holds in every Banach space [A2]. We use this theorem to prove an integrated version of Bernstein's theorem ; then the desired characterization follows from Section 2. This approach, deducing Berstein's theorem from Widder's, is of independent interest ; the converse implication being well-known in the scalar case (see [Wi3]).

There exists a vector-valued version of Bernstein's theorem due to Bochner [Bo]. There σ-order complete partially ordered spaces are considered and integrals are understood in the order sense whereas we use norm convergence throughout. In a previous paper [A1], Bernstein's theorem was proved under more restrictive conditions (viz. order continuous norm), but the principal object was a resolvent positive operators, and order continuity of the norm was necessary in order to obtain other properties, and in particular, a strongly continuous integrated semigroup.

Bernstein's and Widder's theorem need special hypotheses on the space. This is not the case for all representation theorems. In the last section we show that the representation theorem for Laplace transforms of functions of bounded variation holds in arbitrary Banach spaces.

In the present article we consider Riemann-Stieltjes integrals instead of vector measures. One reason is that the Riemann-Stieltjes integral with respect to increasing functions is more general than the integral with respect to positive vector measures (see Section 3). Moreover, Stieltjes integrals are useful in applications since they allow the usual calculus (as integration by parts, etc.).

Acknowledgement : *The author is very grateful to Frank Räbiger for fruitful discussions and, in particular, valuable information about the interpolation property and F-spaces.*

2. CONVEX FUNCTIONS

Let $-\infty < a < b < \infty$ and let X be a Banach space.

Definition 2.1 A function $f : [a, b] \to X$ is of *bounded semivariation* if there exists a constant $M \geq 0$ such that

$$\left\|\sum_{j=1}^{n}(f(t_i) - f(s_i))\right\| \leq M \tag{2.1}$$

whenever $a \leq s_1 \leq t_1 \leq s_2 \leq t_2 \leq \cdots \leq s_n \leq t_n = b$. By $BSV([a, b]; X)$ we denote the space of all such functions.

Remark 2.2 a) A function $f : [a, b] \to X$ is of *bounded variation* if

$$\|f\|_{BV} := \sup \sum_{j=1}^{n} \|f(t_j - f(t_{j-1})\| < \infty, \tag{2.2}$$

where the supremum is taken over all subdivisions $a = t_0 \leq t_1 \leq \cdots \leq t_n = b$. It is easy to see that a scalar function is of bounded variation if and only if it is of bounded semivariation. From this and the uniform boundedness principle it follows that a function f is of bounded semivariation if and only if $\varphi \circ f$ is of bounded variation for all $\varphi \in X^*$ (cf. [HP ; 3.2.2]).

b) Here we deviate from the terminology in [HP] where bounded semivariation is called "bounded variation" and "strongly bounded variation" is used for condition (2.2). Our terminology is more usual today (cf. [DU]).

Let $f : [a, b] \to X$ be of bounded semivariation and let $g : [a, b] \to \mathbb{C}$ be continuous. Then the *Stieltjes integrals*

$$\int_a^b g(t)df(t) := \lim_{|\pi|\to 0} S_\pi(f, g) \quad \text{and} \tag{2.3}$$

$$\int_a^b f(t)dg(t) := \lim_{|\pi|\to 0} S_\pi(g, f) \tag{2.4}$$

exist in the norm topology. Here we let $S_\pi(f, g) = \sum_{j=1}^{n} g(s_i)(f(t_i) - f(t_{i-1}))$ for any π, a subdivision $a \leq t_0 \leq t_1 \leq \cdots \leq t_n \leq b$ with intermediate points $s_j \in [t_{j-1}, t_j]$, where $|\pi| = \max |t_j - t_{j-1}|$.

Moreover, one has

$$\langle \int_a^b g(t)df(t), \varphi\rangle = \int_a^b g(t)d\langle f(t), \varphi\rangle \quad \text{and}$$

$$\langle \int_a^b f(t)dg(t), \varphi\rangle = \int_a^b f(t)d\langle g(t), \varphi\rangle$$

for all $\varphi \in X^*$. All this follows from [HP, § 1 Sec. 3.3].

Note that in the case where $g(t) = t$, (2.4) is simply the Riemann integral. Thus any $f \in BSV([a, b] \; ; \; X)$ is integrable in the sense of Riemann.

In the sequel we assume that X is a real ordered Banach space with positive cone X_+. This means $X_+ \subset X$ is closed and $\mathbb{R}_+ X_+ \subset X_+$, $X_+ \cap (-X_+) = \{0\}$. Then, stipulating

$$x \leq y \iff y - x \in X_+,$$

defines an ordering on X.

We will assume throughout that the cone is normal, this means that all order intervals

$$[x, y] := \{z \in X; x \leq z \leq y\}$$

are norm-bounded, where $x, y \in X, x \leq y$. This is equivalent to the *dual cone* $X_+^* := \{\varphi \in X^* : \varphi(x) \geq 0 \text{ for all } x \in X_+\}$ being *generating* i.e. $X_+^* - X_+^* = X^*$. Every

Banach lattice and the hermitian part of a C^*-algebra satisfy these conditions. We refer to Schaefer [S2] or Batty-Robinson [BR] for details.

Lemma 2.3 *Let* $-\infty < a < b < \infty$ *and let* $f : [a, b] \to X$ *be increasing (i.e.* $f(s) \leq f(t)$ *for all* $a \leq s \leq t \leq b$*). Then* f *is of semibounded variation.*

Proof : Keeping the notation of Definition 2.1 we have

$$0 \leq \sum_{j=1}^{n} (f(t_j) - f(s_j)) \leq (f(t_1) - f(a)) + (f(t_2) - f(t_1)) + \\ + \cdots + (f(t_n) - f(t_{n-1})) \leq f(b) - f(a).$$

Since the cone is normal the claim follows. ◊

In particular, every increasing function $f : [a, b] \to X$ is integrable in the sense of Riemann.

Remark 2.4 Such a function need not be integrable in the sense of Bochner as the example $X = L^\infty(0, 1), f(t) = 1_{[0,t]}$ shows. In fact, f is not essentially separably valued.

Let $-\infty < a < b \leq \infty$ and let $f : [a, b) \to X_+$ be increasing. Let

$$F(t) = \int_0^t f(s)ds \qquad (a \leq t < b) \tag{2.5}$$

Then $F : [a, b) \to X_+$ is convex, i.e. $F(\lambda t + (1-\lambda)s) \leq \lambda F(t) + (1-\lambda)F(s)$ for the ordering in X for all $0 < \lambda < 1, s, t \in [a, b)$.

Remark 2.5 : Such properties are easily deduced from the real case since

$$\langle F(t), \varphi \rangle = \int_0^t \langle f(s), \varphi \rangle ds$$

for all $\varphi \in X_+^*$, and since the dual cone determines positivity ; i.e.

$$\langle x, \varphi \rangle \geq 0 \text{ for all } \varphi \in X_+^* \text{ implies } x \geq 0. \tag{2.6}$$

The purpose of this section is to investigate under which conditions on X every convex function $F : [a, b) \to X_+$ with $F(0) = 0$ is of the form (2.5).

Definition 2.6 We say that X has the *interpolation property* if given two sequences $(x_n), (y_n)$ in X such that

$$x_n \leq x_{n+1} \leq y_{n+1} \leq y_n \qquad (n \in \mathbb{N}),$$

there exists $z \in X$ such that $x_n \leq z \leq y_n$ $(n \in \mathbb{N})$.

Every σ-order complete Banach lattice and every von Neumann algebra has the interpolation property. However, the Banach lattice $C(\beta\mathbb{N}\backslash\mathbb{N})$ has the interpolation property but is not σ-order-complete.

More precisely, let K be compact. Then $C(K)$ has the interpolation property if and only if K is an *F-space* (i.e., if A, B are open disjoint F_σ-sets, then $\overline{A} \cap \overline{B} = \emptyset$), see Seever [Se]. Note that $C(K)$ is σ-order complete if and only if K is quasi-stonean (i.e., if $A \subset K$ is an open F_σ-set, then $\overline{A}$ is open), see [S3]. There exists a connected compact F-space. Moreover, every closed subspace of an F-space is an F-space (see [Se]).

The following are the main results of this section.

Theorem 2.7 *Assume that X has the interpolation property. Let $F : [a, b) \to X_+$ be convex such that $F(a) = 0$, where $-\infty < a < b \leq \infty$. Then there exists an increasing function $f : [a, b) \to X_+$ such that*

$$F(t) = \int_a^t f(s)ds \qquad (a \leq t < b). \tag{2.7}$$

Conversely, the following holds.

Theorem 2.8 *Let $-\infty < a < b < d < \infty$. Assume that for every convex function $F : [a, d] \to X_+$ with $F(a) = 0$ there exists an increasing function $f : [a, b) \to X_+$ such that (2.7) holds on $[a, b)$. Then X has the interpolation property.*

Proof of Theorem 2.7 : It follows from convexity that

$$\frac{1}{h}(F(s+h) - F(s)) \leq \frac{1}{k}(F(t+k) - F(t)) \tag{2.8}$$

whenever $a \leq s < s + h \leq t < t + k < b$. Let $f(a) = 0$. Let $a < t < b$. By the interpolation property there exists $f(t) \in X$ such that

$$\frac{1}{h}(F(s+h) - F(s)) \leq f(t) \leq \frac{1}{k}(F(t+k) - F(t)) \tag{2.9}$$

for all $s, h, k \in \mathbb{R}$ such that $a < s < s + h < t < t + k < b$.

It follows from (2.8) and (2.9) that f is increasing. Let $G(t) = \displaystyle\int_a^t f(s)ds$ $(t \in [a, b))$. We show that $G = F$. Let $a < t < b$. Let $a \leq t_0 < t_1 \cdots < t_n = t$ be a subdivision as above. Choosing $s_i = t_{i-1}$ and setting $h_i = (t_i - t_{i-1})$ we have

$$\sum_{i=1}^n f(s_i)(t_i - t_{i-1}) \leq \sum_{i=1}^n \frac{1}{h_i}(F(t_{i-1} + h_i) - F(t_{i_{-1}}))h_i$$

$$= \sum_{i=1}^n (F(t_i) - F(t_{i-1})) = F(t) - F(a) = F(t).$$

It follows that $G(t) \le F(t)$. Choosing $s_i = t_i$ one obtains

$$\sum_{i=1}^{n} f(s_i)(t_i - t_{i-1}) \ge \sum_{i=1}^{n} \frac{F(t_i - F(t_{i-1})}{t_i - t_{i-1}}(t_i - t_{i-1}) = F(t).$$

Consequently, $G(t) \ge F(t)$. Thus we have obtained the desired representation of F. $\diamond$

Proof of Theorem 2.8 : Let $x_n \le x_{n+1} \le y_{n+1} \le y_n$ $(n \in \mathbb{N})$. Let $a < c < b$ and let $a = t_1 < t_n < t_{n+1} < c < s_{n+1} < s_n < s_1 = d$ $(n \in \mathbb{N}, n \ge 2)$ such that $\lim_{n\to\infty} t_n = c$, $\lim_{n\to\infty} s_n = c$. Define $f : [a,d) \to X$ by

$$f(t) = \begin{cases} x_n & \text{if} \quad t \in [t_n, t_{n+1}), \\ y_n & \text{if} \quad t \in [s_{n+1}, s_n), \\ 0 & \text{if} \quad t = c. \end{cases}$$

Then f is measurable and $\int_a^d \|f(t)\|dt < \infty$. Thus f is Bochner integrable. Let $F_1 : [a,d] \to X_+$ be given by $F_1(t) = \int_a^t f(s)ds$. Then $F_1(t) \ge (t-a)x_1$ for all $t \in [a,b)$. Let $F(t) = F_1(t) - (t-a)x_1$. Then F is convex, $F(a) = 0$ and $F(t) \ge 0$ $(t \in [a,b])$. By assumption there exists $g : [a,b) \to X$ increasing such that $F(t) = \int_a^t g(s)ds$ $(t \in [a,b))$. Then

$$\frac{F(t-h) - F(t)}{-h} \le g(t) \le \frac{F(t+h) - F(t)}{h}$$

for all $t \in (a,b), h > 0$ small enough. It follows that $g(t) = F'(t)$ whenever F is differentiable in $t \in (a,b)$. Consequently $g(t) = F'(t)$ whenever $t \in (t_n, t_{n+1})$ or $t \in (s_{n+1}, s_n)$. Let $t'_n \in (t_n, t_{n+1}), s'_n \in (s_{n+1}, s_n)$. Then $x_n = g(t'_n) \le g(c) \le g(s'_n) = y_n$. Thus $z := g(c)$ interpolates. $\diamond$

Corollary 2.9 *Assume that X has the interpolation property. Let $-\infty \le a < c < b \le \infty$ and let $F : (a,b) \to X$ be convex. Then there exist $z \in X$ and $f : (a,b) \to X$ increasing such that*

$$F(t) := F(c) + (t-c)z + \int_c^t f(s)ds$$

for all $t \in (a,b)$.

Proof : We may assume that $c = 0$. It follows from convexity that

$$\frac{1}{t}(F(0) - F(-t)) \le \frac{1}{s}(F(s) - F(0))$$

for all $0 < s < b, 0 < t < -a$. Thus, by the interpolation property there exists $z \in X$ such that

$$\frac{1}{t}(F(0) - F(-t)) \le z \le \frac{1}{s}(F(s) - F(0))$$

for all $0 < t < -a,\quad 0 < s < b$. In particular, the function

$$G(t) = F(t) - F(0) - tz \qquad (t \in (a, b))$$

is positive, convex and $G(0) = 0$. By Theorem 2.7 there exist increasing functions $f_1 : [0, b) \to X_+$ and $f_2 : [0, -a) \to X_+$ such that $G(t) = \int_0^t f_1(s)ds$ for $t \in [0, b)$ and $G(t) = \int_0^t f_2(s)ds$ for $t \in [0, -a)$. Let $f(t) = f_1(t)$ for $t \in [0, b)$ and $f(t) = -f_2(-t)$ for $t \in (a, 0)$, then f is increasing and $G(t) = \int_0^t f(s)ds$ for all $t \in (a, b)$. ◊

3 Riesz' theorem and the Hausdorff moment problem

Let X be an ordered Banach space with normal cone. If $f : [0, 1] \to X$ is increasing, then

$$Th = \int_0^1 h(t)df(t) \qquad (h \in C[0, 1]) \tag{3.1}$$

defines a positive operator $T : C[0, 1] \to X$. Riesz' theorem asserts that every operator is of this form if $X = \mathbb{R}$.

A related problem is the *moment problem*.

Let

$$x_n = \int_0^1 t^n df(t) \qquad (n \in \mathbb{N}_0 := \mathbb{N} \cup \{0\}). \tag{3.2}$$

Then the sequence $x = (x_n)_{n\in\mathbb{N}}$ is *completely monotone* i.e.

$$(-\Delta)^k x \geq 0 \qquad \text{for all} \quad k \in \mathbb{N}_0. \tag{3.3}$$

Here $\Delta : X^{\mathbb{N}_0} \to X^{\mathbb{N}_0}$ is given by $\Delta x = (x_{n+1} - x_n)_{n\in\mathbb{N}_0}$ for all $x = (x_n)_{n\in\mathbb{N}_0} \in X^{\mathbb{N}_0}$; and we write $(x_n)_{n\in\mathbb{N}_0} \geq 0$ if $x_n \geq 0$ for all $n \in \mathbb{N}_0$.

Let $e_n(t) = t^n$ $(t \in [0, 1])$ for $n \in \mathbb{N}_0$. Then $(-\Delta)^k e = (e_n(e_0 - e_1)^k)_{n\in\mathbb{N}} \geq 0$ for all $k \in \mathbb{N}_0$.

It follows that a sequence $x = (x_n)_{n\in\mathbb{N}_0}$ given by (3.2) is completely monotone (in fact $((-\Delta)^k x)_n = \int_0^1 e_n(e_0 - e_1)^k df(t) \geq 0$). Hausdorff's classical theorem asserts that conversely, every completely monotone sequence has a representation (3.2) if $X = \mathbb{R}$.

Theorem 3.1 *The following assertions are equivalent*

(i) X has the interpolation property ;

(ii) for every positive $T \in C([0,1]$; $T \in \mathcal{L}(C[0,1];\ X)$ there exists $f : [0,1] \to X$ increasing such that (3.1) holds ;

(iii) for every completely monotone sequence $(x_n)_{n\in\mathbb{N}_0}$ there exists $f : [0,1] \to X$ increasing such that (3.2) holds.

Proof of (i) $\Longleftrightarrow$ (ii) : Let $T \in \mathcal{L}(C[0,1]\ ;\ X)$ be a positive operator. It follows from Riesz' classical theorem that there exists a unique increasing function $f_1 : [0,1] \to X^{**}$ such that $\varphi \circ f_1$ is normalized (in the sense of [Wi 2, Ch. I § 8]) for all $\varphi \in X^*_+$ and

$$Th = \int_0^1 h(t)df_1(t) \qquad (h \in C[0,1]). \tag{3.4}$$

Recall that X^* is again an ordered Banach space with normal cone. Let $F(t) = \int_0^t f_1(s)ds$, $0 \le t \le 1$. We claim that $F(t) \in X$ $(0 \le t \le 1)$. In fact, $f_1(1) = Te_0 \in X$ and $Te_n = \int_0^1 t^n df_1(t) = f_1(1) - n\int_0^1 t^{n-1}f_1(t)dt = Te_0 - n\int_0^1 t^{n-1}dF(t)$ $(n \in \mathbb{N})$. The function $F : [0,1] \to X^{**}$ is Lipschitz-continuous. Thus $Sh = \int_0^1 h(t)dF(t)$ defines a continuous operator $S \in \mathcal{L}(C[0,1]\ ;\ X^{**})$ such that $nSe_{n-1} = Te_0 - Te_n \in X$ for all $n \in \mathbb{N}$. Hence $Sh \in X$ for all $h \in C[0,1]$ since the polynomials are dense in $C[0,1]$. Denote by $q : X^{**} \to X^{**}/X$ the quotient mapping. Then $0 = q(Sh) = \int_0^1 hd(q\circ F)(t)$ $(h \in C[0,1])$. Since $q\circ F$ is Lipschitz continuous, it follows that $q\circ F \equiv 0$, i.e. $F(t) \in X$ for all $t \in [0,1]$.

Thus $F : [0,1] \to X_+$ is a convex function. Moreover

$$\frac{F(1) - F(t)}{1-t} \le f_1(1) = Te_0. \tag{3.5}$$

Now assume that X has the interpolation property. By Theorem 2.7 there exists $f : [0,1) \to X_+$ increasing such that $F(t) = \int_0^t f(s)ds$ $0 \le t \le 1$. Then $f(t) \le \frac{F(1)-F(t)}{1-t}$ for all $0 \le t < 1$. Set $f(1) = Te_0$, then $f : [0,1] \to X$ is increasing. Since $Te_n = f(1) - n\int_0^1 t^{n-1}f(t)dt = \int_0^1 t^n df(t)$ for all $n \in \mathbb{N}$, it follows that (3.1) holds. We have shown that (i) implies (ii).

Now let $F : [0,1] \to X_+$ be a convex function such that $F(0) = 0$. Assume that $x_0 \in X$ such that

$$\frac{F(1) - F(t)}{1-t} \le x_0 \qquad 0 \le t < 1. \tag{3.6}$$

Using the real representation of convex functions one obtains a unique normalized increasing function $f_1 : [0,1) \to X^{**}$ such that $F(t) = \int_0^t f_1(s)ds$ $(0 \le t < 1)$. It follows from (3.6) that $f_1(t) \le x_0$ $(0 \le t < 1)$. Letting $f_1(1) = x_0$, f_1 is increasing on $[0,1]$. Define $T \in \mathcal{L}(C[0,1], X^{**})$ by $Th = \int_0^1 h(t)df_1(t)$. Then $T \ge 0$ and $Te_0 = x_0$. Moreover, $Te_n = \int_0^1 t^n df_1(t) = f_1(t) - n\int_0^1 t^{n-1}f_1(t)dt = x_0 - n\int_0^1 t^{n-1}dF(t) \in X$. Hence $T \in \mathcal{L}(C[0,1], X)$. Now assume that (ii) holds. Then there exists $f : [0,1] \to X$ increasing such that (3.1) holds. Let $\widetilde{F}(t) = \int_0^t f(s)ds$. Then $x_0 - n\int_0^1 t^{n-1}d\widetilde{F}(t) = Te_n = x_0 - n\int_0^1 t^{n-1}dF(t)$ $(n \in \mathbb{N})$. It follows that $\int_0^1 t^n d\widetilde{F}(t) = \int_0^1 t^n dF(t)$ for all $n \in \mathbb{N}$ and this implies that $\widetilde{F}(t) = F(t)$. We have shown that (ii) implies that any convex function $F : [0,1] \to X_+$ satisfying $F(0) = 0$ and (3.6) is the antiderivative of an increasing function. By Theorem 2.8 this implies that X has the interpolation property. Thus the equivalence of (i) and (ii) is established.

Remark 3.2 The proof shows the following. Let $x_0 \in X_+$. The relation

$$Te_n = x_0 - n\int_0^1 t^{n-1}dF(t) \quad (n \in \mathbb{N}) \tag{3.7}$$

establishes a bijective correspondance between all positive operators $T \in \mathcal{L}(C[0,1]\,;\,X)$ satisfying $Te_0 = x_0$ and all convex functions $F : [0,1] \to X_+$ satisfying $F(0) = 0$ and (3.6).

The equivalence of (ii) and (iii) follows immediately from

Proposition 3.3 *A sequence $(x_n)_{n\in\mathbb{N}_0} \subset X$ is completely monotone if and only if there exists a positive operator $T \in \mathcal{L}(C[0,1]\,;\,X)$ such that $Te_n = x_n$ $(n \in \mathbb{N}_0$.)*

Proof: Assume that $Te_n = x_n$. Then by the arguments in the beginning of this section $((-\Delta)^k x)_n = T(e_n(e_0 - e_1)^k)$ $(n \in \mathbb{N}_0)$. This shows one implication. Conversely assume that $(x_n)_{n\in\mathbb{N}_0}$ is completely monotonic. Denote by P the polynomials in $C[0,1]$ and let $T : P \to X$ be the linear mapping such that $Te_n = x_n$ $(n \in \mathbb{N}_0)$. Then T is positive. In fact, let $0 \le h \in P$. Let $\varphi \in X_+^*$. It suffices to show that $\langle Th, \varphi\rangle \ge 0$.

By Hausdorff's classical theorem there exists $\alpha : [0,1] \to \mathbb{R}$ increasing such that $\langle x_n, \varphi\rangle = \int_0^1 t^n d\alpha(t)$ $(n \in \mathbb{N}_0)$. It follows that $(\varphi \circ T)e_n = \int_0^1 e_n d\alpha(t)$ $(n \in \mathbb{N}_0)$. Consequently, $\varphi \circ Th = \int_0^1 h(t)d\alpha(t) \ge 0$. Since T is positive and $Te_0 = x_0$ it follows that $0 \le h \le e_0$ implies $0 \le Th \le x_0$ for all $h \in P$. This implies that T has a unique positive extension to $C[0,1]$. $\diamondsuit$

Finally, we consider representation by vector measures.

Let Y be a Banach space. Denote by $\mathcal{B}[0,1]$ the σ-algebra of all Borel sets and let $\mu : \mathcal{B}[0,1] \to Y$ be a countably additive vector measure. Then

$$Th = \int_0^1 h(t)d\mu(t) \qquad (h \in C[0,1]) \tag{3.8}$$

defines a weakly compact operator $T : C[0,1] \to Y$, and every weakly compact operator is of this form (see [DU, VI.2 Theorem 5 p. 153]). Moreover, every operator $T : C[0,1] \to Y$ is weakly compact if and only if Y does not contain a closed subspace isomorphic to c_0. (In fact, one direction follows from [DU ; VI.2 Theorem 15, p. 159]. For the other observe that $T : C[0,1] \to c_0$, $Tf = \left(f(\frac{1}{n}) - f(0)\right)_{n \in \mathbb{N}}$ is not weakly compact (cf. [DU, p. 153]). If c_0 is isomorphic to a closed subspace of Y, then the topologies $\sigma(Y, Y^*)$ and $\sigma(c_0, c_0^*)$ coincide on that space by the Hahn-Banach theorem.)

The situation is different if we consider positive operators. Let X be an ordered Banach space with normal cone. We say that <u>X has order continuous norm</u> if every positive decreasing sequence converges (cf. [S3, II.5.10], [LT, Proposition 1,]).

Let $f : [0,1] \to X$ be increasing. Assume that X has order continuous norm. Then $f^*(t) = \lim_{h \downarrow 0} f(t+h)$ exists for all $t \in [0,1)$. Letting $f^*(1) = f(1)$, f^* is increasing and continuous on the right. It follows from the real-valued case that $\int_0^1 h(t)df(t) = \int_0^1 h(t)df^*(t)$ for all $h \in C[0,1]$. Thus, we obtain from Theorem 3.1

Proposition 3.4 *Assume that X has order continuous norm. Then for every positive operator $T : C[0,1] \to X$ there exists a unique right-continuous increasing mapping $f : [0,1] \to X$ such that (3.1) holds.*

Theorem 3.5 The following are equivalent

(i) X has order continuous norm ;

(ii) for every positive operator $T : C[0,1] \to X$ there exists a countably additive measure $\mu : \mathcal{B}[0,1] \to X$ such that (3.8) holds ;

(iii) every positive operator $T : C[0,1] \to X$ is weakly compact ;

(iv) for every increasing function $f : [0,1] \to X$ there exists a countably additive vector measure $\mu : \mathcal{B}[0,1] \to X$ such that $\mu([a,b)) = f(b) - f(a)$ for all $0 \le a < b$.

Note that c_0 has order continuous norm. Thus every positive operator from $C[0,1]$ into c_0 is weakly compact, but there exist non-weakly compact operators as seen above.

Proof : (i) $\Longrightarrow$ (ii). Let $T : C[0,1] \to X$ be positive and let f be the function of Proposition 3.4. Let $\mathcal{A}$ be the algebra of all finite unions of intervals of the form $(a,b], 0 \le a < b \le 1$. There exists a unique finitely additive vector measure ([DU, p. 1]) $\mu_0 : \mathcal{A} \to X$ such that $\mu_0((a,b]) = f(b) - f(a)$ whenever $0 \le a < b \le 1$.

It follows from real measure theory that μ_0 is weakly countably additive. We show that μ_0 is <u>strongly additive</u> ([DU, p. 7]). Let $A_n \in \mathcal{A}$ ($n \in \mathbb{N}$) such that $A_n \cap A_m = \emptyset$ if $n \neq m$. Then $0 \le \sum_{n=1}^{m} \mu_0(A_n) \le f(1) - f(0)$ for all $m \in \mathbb{N}$. It follows from (i) that $\sum_{n=1}^{\infty} \mu_0(A_n)$ converges. This proves the claim. Now it follows from the Carathéodory-Hahn-Kluvanek extension theorem ([DU, p. 27]) that there exists a unique countably additive extension μ of μ_0 to $\mathcal{B}[0,1]$. It follows from real measure theory that (3.8) holds.

(ii) $\Longleftrightarrow$ (iii) follows from [DU, VI.2 Theorem 5, p. 153].

(iii) $\Longrightarrow$ (i) Let $x_n \geq x_{n+1} \geq 0$ $(n \in \mathbb{N})$. Define $f : [0,1] \to X$ by $f(0) = 0, f(t) = x_n$ if $t \in (\frac{1}{n+1}, \frac{1}{n}]$ $(n \in \mathbb{N})$. Let $Th = \int_0^1 h(s)df(s)$ $(h \in C[0,1])$. Let $h_n \in C[0,1]$ such that $0 \leq h_n \leq \mathbf{1}$, $h_n(t) = 0$ if $t \leq \frac{1}{n+1}$ and $h_n(t) = 1$ if $t \geq \frac{1}{n}$.

Then $Th_n = (x_{n-1} - x_n) + (x_{n-2} - x_{n-1}) + \cdots + (x_1 - x_2) = (x_1 - x_n)$. Since T is weakly compact, it follows that the set $\{x_n : n \in \mathbb{N}\}$ is relatively weakly compact. Thus $(x_n)_{n\in\mathbb{N}}$ has a weakly convergent subsequence. Since $(x_n)_{n\in\mathbb{N}}$ is decreasing, it follows that $(x_n)_{n\in\mathbb{N}}$ converges weakly. Imbedding X into $C(B_+^*)$ by evaluation, where $B_+^* = \{x^* \in X_+ : \|x^*\| \leq 1\}$, one obtains by Dini's theorem that $(x_n)_{n\in\mathbb{N}}$ converges in norm.

(iv) $\Longrightarrow$ (i) This is the same as the proof (iii) $\Longrightarrow$ (i). In fact, let f be the function constructed in the proof of (iii) $\Longrightarrow$ (i). Let μ be the countably additive vector measure given by (iv). Then $Th = \int_0^1 h(t)d\mu(t)$ $(h \in C[0,1])$. Thus T is weakly compact by [DU, VI.2 Theorem 5, p. 153].

(i) $\Longrightarrow$ (iv) This is the same proof as (i) $\Longrightarrow$ (ii). $\diamondsuit$

We conclude this section by a Banach space version of the equivalence of (i) and (iv), which is a consequence of a theorem due to Bessaga-Pełczyński [DU, p. 22].

Proposition 3.6 *A Banach space Y does not contain a copy of c_0 if and only if for every $f \in BSV([0,1]\ ;\ X)$ there exists a countably additive measure μ such that $\mu([a,b)) = f(b) - f(a)$ whenever $0 \leq a < b \leq 1$.*

Proof : Let c be the space of all convergent sequences. Define $f : [0,1] \to c$ by $f(t) = (f_n(t))_{n\in\mathbb{N}}$ where $f_n(t) = 0$ if $t < 1 - \frac{1}{n}$ and $f_n(t) = 1$ if $t \geq 1 - \frac{1}{n}$. Then f is increasing and so $f \in BSV[0,1]$. Define $T : C[0,1] \to c$ by $Th = \int_0^1 h(t)df(t) = (h(1 - \frac{1}{n}))_{n\in\mathbb{N}}$. It is easy to see that T is not weakly compact. Thus f does not define *a* countably additive vector measure. Since c is isomorphic to c_0 this proves one direction. The other implication is proved as (i) $\Longrightarrow$ (ii) of Theorem 3.5. In fact, it follows from real measure theory that $\sum_{n=1}^{\infty} |\langle \mu_0(A_n), x^* \rangle| < \infty$ for all $x^* \in Y^*$. Now the Theorem of Bessaga-Pełczyński [DU, p. 22] implies that $\sum_{n=1}^{\infty} \mu_0(A_n)$ is unconditionnally convergent.
$\diamondsuit$

Problem 3.7 It would be interesting to characterize those operators T from $C[0,1]$ into a Banach space Y for which there exists a function $f \in BSV([0,1]\ ;\ Y)$ such that

$$Th = \int_0^1 h(t)df(t) \quad (h \in C[0,1]); \tag{3.9}$$

and to characterize those Banach spaces Y such that every operator $T \in \mathcal{L}(C[0,1]\ ;\ Y)$ is representable by a function $f \in BSV([0,1]\ ;\ Y)$ in the sense of (3.9).

4 Bernstein's theorem

Let X be an ordered Banach space with normal cone.

Let $f : [0,\infty) \to X$ an increasing function such that

$$r(\lambda) := \int_0^\infty e^{-\lambda t} df(t) := \lim_{b\to\infty} \int_0^b e^{-\lambda t} df(t) \tag{4.1}$$

exists for all $\lambda > 0$.

Remark 4.1 This is equivalent to $w(f) := \inf\{w : \sup_{t\geq 0} e^{-wt}\|f(t)\| < \infty\} \leq 0$ (see [HP ; Theorem 6.2.1 p. 215] or [ANS]).

In that case r is a completely monotone function on $(0,\infty)$ in the sense of the following definition.

Definition : A function $r : (0,\infty) \to X$ is *completely monotone* if r is infinitely differentiable and

$$(-1)^n r^{(n)}(\lambda) \geq 0 \qquad (\lambda > 0). \tag{4.2}$$

Bernstein's classical theorem [Wi 2, p. 160] or [Wi 3, p. 155] says that, if $X = \mathbb{R}$, then conversely, every completely monotone function has a representation (4.1).

Definition 4.2 We say that *Bernstein's theorem holds in* X, if for every completely monotone function $r \in C^\infty((0,\infty); X)$ there exists an increasing function $f : [0,\infty) \to X$ such that (4.1) holds for all $\lambda > 0, n \in \mathbb{N}_0$.

The main result of this section is the following.

Theorem 4.3 *Bernstein's theorem holds in X if and only if X has the interpolation property.*

Before proving Theorem 4.3 we establish a representation of completely monotone functions which holds without restrictions on the space (other than those indicated in the beginning of this section).

Theorem 4.4 *A function $r : (0,\infty) \to X$ is completely monotonic if and only if there exists a convex function $F : [0,\infty) \to X_+$ satisfying $F(0) = 0$ such that*

$$r(\lambda) = \lambda \int_0^\infty e^{-\lambda t} dF(t) \qquad (\lambda > 0). \tag{4.3}$$

Remark : The assumptions on F imply that F is increasing. (4.3) signifies more precisely : the improper Stieltjes integral converges for all $\lambda > 0$ and the identity (4.3) holds.

If (4.3) holds, then $r \in C^\infty((0,\infty)\,;\,X)$ and for all $x^* \in X^*_+$, $\langle r(\lambda), x^*\rangle = \int_0^\infty \lambda e^{-\lambda t} d\langle F(t), x^*\rangle = \int_0^\infty e^{-\lambda t} dg(t)$ where $g(t) = \frac{d}{dt}\langle F(t), x^*\rangle$ is increasing. Hence r is completely monotone. Thus one implication of Proposition 4.4 is obvious. For the other we use the following integrated version of Widder's theorem. In the following Y denotes a Banach space and we let $\mathrm{Lip}_0([0,\infty); Y) := \{F : [0,\infty) \to Y : F(0) = 0$, there exists $M \geq 0$ such that $\|F(t) - F(s)\| \leq M|t-s|$ for all $t, s \geq 0\}$.

Theorem 4.5 *Let $r \in C^\infty((0,\infty); Y)$. The following are equivalent :*

(i) $\sup\limits_{\substack{\lambda>0 \\ n\in\mathbb{N}_0}} \|\lambda^{n+1} r^{(n)}(\lambda)/n!\| < \infty$;

(ii) there exists $F \in \mathrm{Lip}_0([0,\infty)\,;\,Y)$ such that

$$r(\lambda) = \int_0^\infty e^{-\lambda t} dF(t) \qquad (\lambda > 0)$$

In that case one has

$$F(t) = \lim_{k\to\infty} F_k(t), \tag{4.4}$$

where $F_k(t) = G_k\left(\frac{k}{t}\right), G_k(\lambda) = \frac{(-1)^k}{k!}\lambda^{k+1}\left(\frac{r(\lambda)}{\lambda}\right)^{(k)} \quad (\lambda > 0, t > 0)$.

Theorem 4.5 has been proved in [A2] using Widder's classical theorem valid for $Y = \mathbb{R}$ [Wi 3, p. 157]. Then formula (4.4) follows from [HP, Theorem 6.3.5 p. 224]. A direct proof of Theorem 4.5 and (4.4) (based on Widder's arguments [Wi 1]) is given by Hennig, Neubrander [HN].

Lemma 4.6 *Let $r \in C^\infty((0,\infty)\,;\,Y)$. Then*

$$\frac{(-1)^n}{n!}\lambda^{n+1}\left(\frac{r(\lambda)}{\lambda}\right)^{(n)} = \sum_{m=0}^{n} \frac{(-1)^m}{m!}\lambda^m r^{(m)}(\lambda) \tag{4.5}$$

and

$$\left(\lambda^{k+n}\left(\frac{r(\lambda)}{\lambda^n}\right)^{(k)}\right)^{(n)} = \lambda^k r^{(k+n)}(\lambda) \tag{4.6}$$

for all $\lambda > 0, k, n \in \mathbb{N}_0$.

Proof : The first formula (4.5) is immediate. We show by induction over n that (4.6) holds for all $k \in \mathbb{N}_0$. It is obvious for $n = 0$.

Let $n = 1$. Then $\lambda^k r(\lambda)^{(k+1)} = \lambda^k\left(\lambda\frac{r(\lambda)}{\lambda}\right)^{(k+1)} = \lambda^k\{\lambda(\frac{r(\lambda)}{\lambda})^{(k+1)} + (k+1)(\frac{r(\lambda)}{\lambda})^k\} = \left(\lambda^{k+1}(\frac{r(\lambda)}{\lambda})^{(k)}\right)'$. Now assume that (4.6) holds for a fixed $n \in \mathbb{N}$ and all $k \in \mathbb{N}_0$. Then applying (4.6) to r' yields

$$\lambda^k r(\lambda)^{(k+n+1)} = \left(\lambda^{k+n}(r'(\lambda)/\lambda^n)^{(k)}\right)^{(n)}. \tag{4.7}$$

Observe that $[\lambda^{k+n+1}(r(\lambda)/\lambda^{n+1})^{(k)}]' = [\lambda^n.\lambda^{k+1}(r(\lambda)/\lambda^{n+1})^{(k)}]' =$
$n\lambda^{n-1}[\lambda^{k+1}(r(\lambda)/\lambda^{n+1})^{(k)}] + [\lambda^n[\lambda^{k+1}(r(\lambda)/\lambda^{n+1})^{(k)}]' =$
$n\lambda^{n-1}[\lambda^{k+1}(r(\lambda)/\lambda^{n+1})^{(k)}] + \lambda^n\lambda^k(r(\lambda)/\lambda^n)^{(k+1)}$ by applying (4.6) for $n = 1$ to the function $r(\lambda)/\lambda^n$ instead of r.
Hence $[\lambda^{k+n+1}(r(\lambda)/\lambda^{n+1})^{(k)}]' = n\lambda^{n+k}(r(\lambda/\lambda^{n+1})^{(k)} + \lambda^{n+k}(r'(\lambda)/\lambda^n -$
$nr(\lambda)/\lambda^{n+1})^{(k)} = \lambda^{n+k}(r'(\lambda)/\lambda^n)^{(k)}$. It follows from (4.7) that
$[\lambda^{k+n+1}(r(\lambda)/\lambda^{n+1})^{(k)}]^{(n+1)} = [\lambda^{n+k}(r'(\lambda)/\lambda^n)^{(k)}]^{(n)} = \lambda^k r(\lambda)^{(k+n+1)}$. ◊

It follows from (4.5) that $\frac{r(\lambda)}{\lambda}$ is completely monotone whenever $r(\lambda)$ is so.

Proposition 4.7 *Let $F \in \mathrm{Lip}_0([0,\infty)\ ;\ X)$ and let $r(\lambda) = \lambda \int_0^\infty e^{-\lambda t} dF(t)$ $(\lambda > 0)$. Then r is completely monotone if and only if $F(t) \geq 0$ $(t \geq 0)$ and F is convex.*

Proof : Assume that r is completely monotone. By (4.4) we have $F(t) = \lim_{k\to\infty} F_k(t)$ where $F_k(t) = G_k(k/t), G_k(\lambda) = \frac{(-1)^k}{k!}\lambda^{k+1}(r(\lambda)/\lambda^2)^{(k)}$. Since $r(\lambda)/\lambda^2$ is completely monotone, it follows that $F_k(t) \geq 0$. We show that F_k is convex, i.e. that $F_k''(t) = -\left(\frac{k}{t^2}G_k'(k/t)\right)' \geq 0$.

Let $H(\lambda) = -\lambda^2 k G_k'(k\lambda)$. Then $F_k''(t) = \frac{d}{dt}H(\frac{1}{t}) = -\frac{1}{t^2}H'(\frac{1}{t})$. Thus it suffices to show that $H'(\lambda) \leq 0$ or equivalently $+2\lambda k\ G_k'(k\lambda) + \lambda^2 k^2 G''(k\lambda) \geq 0$. Letting $\mu = k\lambda$ we have to show that $(\mu G_k(\mu))'' = 2G_k'(\mu) + \mu G_k''(\mu) \geq 0$ $(\mu > 0)$. This is true since by (4.6) for $n = 2$

$$(\mu G_k(\mu))'' = \frac{(-1)^k}{k!}\left(\mu^{k+2}(r(\mu)/\mu^2)^{(k)}\right)'' = \frac{(-1)^k}{k!}\mu^k r^{(k+2)}(\mu) \geq 0 \quad (\mu > 0).$$

This proves one implication. Conversely, if F is convex, then for all $x^* \in X_+^*$, $g(t) = \frac{d}{dt}\langle F(t), x^*\rangle$ is increasing. Thus $\langle r(\lambda), x^*\rangle = \int_0^\infty e^{-\lambda t} dg(t)$ is completely monotone. ◊

Next we prove the particular case of Proposition 4.4 where r has an extension to $[0,\infty)$.

Proposition 4.8 *Let $r \in C^\infty([0,\infty)\ ;\ X)$ be completely monotone. Then there exists a convex function $F : [0,\infty) \to X_+$ such that $F(0) = 0$ and*

$$r(\lambda) = \lambda \int_0^\infty e^{-\lambda t} dF(t) \qquad (\lambda > 0).$$

Proof : It follows from (4.5) that

$$p(\lambda) := \frac{(-1)^k}{k!}\lambda^{k+1}(r(\lambda)/\lambda)^k = \sum_{m=0}^{k} \frac{(-1)^m}{m!}\ \lambda^m r^{(m)}(\lambda) \geq 0 \quad (\lambda > 0).$$

Moreover, $\lim_{\lambda\downarrow 0} p(\lambda) = r(0)$. It follows from (4.6) for $n = 1$ that
$p'(\lambda) = \frac{(-1)^k}{k!}\lambda^k r^{(k+1)}(\lambda) \leq 0$. Hence $0 \leq p(\lambda) \leq r(0)$ for all $\lambda > 0$. Since the cone is

normal, condition (i) of Theorem 4.5 is satisfied and the claim follows from Theorem 4.5 and Proposition 4.7. $\diamondsuit$

Proof of Theorem 4.4 : Let $q(\lambda) = r(\lambda+1)$. Then $q \in C^\infty([0,\infty;X)$ is completely monotone.

By Proposition 4.8 there exists $G \in \operatorname{Lip}_0([0,\infty)\ ;\ X)$ positive and convex such that $q(\lambda) = \lambda \int_0^\infty e^{-\lambda t} dG(t)\ (\lambda > 0)$. Let $F(t) = \int_0^t (1-(t-s))e^s dG(s)$. Then F is positive and convex. In fact, let $\varphi \in X_+^*$. There exists an increasing positive function $g : [0,\infty) \to \mathbb{R}$ such that $\langle G(t),\varphi\rangle = \int_0^t g(s)ds$. Thus $\langle F(t),\varphi\rangle = \int_0^t e^s g(s)ds - \int_0^t (t-s)e^s g(s)ds = \int_0^t \{e^s g(s) - \int_0^s e^r g(r)dr\}ds = \int_0^t \int_0^s e^r dg(r)ds$ which shows that $\langle F(.),\varphi\rangle$ is convex.

For $\lambda > 0$, we have $\langle q(t),\varphi\rangle = \int_0^\infty e^{-\lambda t} dg(t)$. Hence for $\lambda > 1, \langle r(\lambda),\varphi\rangle = \langle q(\lambda - 1),\varphi\rangle = \int_0^\infty e^{-\lambda t}e^t dg(t) = \int_0^\infty e^{-\lambda t} df(t)$ where $f(t) = \int_0^t e^s dg(s) = e^t g(t) - \int_0^t e^s g(s)ds$. Since $\langle F(t),\varphi\rangle = \int_0^t f(s)ds$, it follows that $r(\lambda) = \lambda \int_0^\infty e^{-\lambda t} dF(t) \quad (\lambda > 1)$.

On the other hand $\int_0^\infty e^{-\lambda t} dF(t)$ converges for all $\lambda > 0$. In fact, let $\varphi \in X_+^*$. Since the abscissa of convergence of $\int_0^\infty e^{-\lambda t} d\langle F(t),\varphi\rangle$ is a singularity [Wi 3, p. 117], it follows that $\int_0^\infty e^{-\lambda t} d\langle F(t),\varphi\rangle$ converges for all $\lambda > 0$.

Consequently $\sup_{t\geq 0} e^{-wt}|\langle F(t),\varphi\rangle| < \infty$ for all $w > 0$ (by Remark 4.1).
It follows from the uniform boundedness principle that $\sup_{t\geq 0} e^{-wt}\|F(t)\| < \infty$ for all $w > 0$. Hence $\int_0^\infty e^{-\lambda t} dF(t)$ exists for Re $\lambda > 0$ (cf. [HP, Theorem 6.2.1]), is analytic and coincides with $r(\lambda)$ for $\lambda > 1$. Applying Proposition 4.8 to $r(\lambda+\delta)$ for $\delta > 0$ one sees that r is analytic on $(0,\infty)$. Hence $r(\lambda) = \int_0^\infty e^{-\lambda t} dF(t)$ for $\lambda > 0$. $\diamondsuit$

Proof of Theorem 4.3 : Assume that X has the interpolation property. Let $r \in C^\infty((0,\infty)\ ;\ X)$ be a completely monotone function. By Theorem 4.4 there exists a convex function $F : [0,\infty) \to X_+$ such that $F(0) = 0$ and $r(\lambda) = \lambda \int_0^\infty e^{-\lambda t} dF(t)\ (\lambda > 0)$.

By Theorem 2.7 there exists an increasing function $f : [0,\infty) \to X_+$ such that $F(t) = \int_0^t f(s)ds$. Thus $r(\lambda) = \int_0^\infty e^{-\lambda t} f(t)dt\ (\lambda > 0)$.

Conversely, assume that Bernstein's theorem holds. We show that X has the interpolation property. Let $F : [0,2] \to X_+$ be convex such that $F(0) = 0$. By Theorem 2.8, it suffices to show that there exists an increasing function $f : [0,1) \to X$ such that $F(t) = \int_0^t f(s)ds$ for $0 \le t < 1$.

Define $G : [0,\infty) \to X_+$ by $G(t) = F(t)$ for $0 \le t \le 1$, $G(t) = F(1) + (t-1)(G(2) - G(1))$ for $t > 1$. Then G is convex, $r(\lambda) = \lambda \int_0^\infty e^{-\lambda t} dG(t)$ exists for $\lambda > 0$ and defines a completely monotone function (by Proposition 4.4 or 4.7). By assumption there exists $f : [0,\infty) \to X_+$ increasing such that $r(\lambda) = \int_0^\infty e^{-\lambda t} df(t) = \lambda \int_0^\infty e^{-\lambda t} dH(t)$ where $H(t) = \int_0^t f(s)ds$. It follows from the uniqueness theorem that $H(t) = G(t)$ for all $t \ge 0$, in particular, $H(t) = F(t) \quad (0 \le t < 1)$. ◊

5 Laplace-Stieltjes transforms of functions of bounded variation

Whereas the preceding representation theorems required restrictive assumptions on the space the following characterization of Laplace-Stieltjes transforms of functions of bounded variation holds in all Banach spaces. Let $F : [0,\infty) \to X$. For $t > 0$ we define the variation of F on $[0,t]$ by

$$\operatorname{var}(F)\,|_0^t := \sup \sum_{i=1}^n \|F(t_i) - F(t_{i-1})\|,$$

where the supremum is taken over all finite sequences $0 \le t_0 < t_1 < \cdots < t_n \le t$. Moreover, we let

$$BV([0,\infty)\ ;\ X) := \{F : [0,\infty) \to X \ :\ \|F\|_{BV} := \sup_{t \ge 0} \operatorname{var}(F)\,|_0^t < \infty\}.$$

Example 5.1 : Let $f \in L^1((0,\infty)\ ;\ X)$ and $F(t) = \int_0^t f(s)ds$.

Then $F \in BV([0,\infty)\ ;\ X)$ and $\operatorname{var}(F)\,|_0^t = \int_0^t \|f(s)\|ds$.

In the proof of the following theorem we adapt arguments of [HP, Theorem 12.4.1], where resolvents are considered.

Theorem 5.2 *Let $r \in C^\infty((0,\infty)\ ;\ X)$ such that $\lim_{\lambda\to\infty} r(\lambda) = 0$ and let $M \ge 0$. The following are equivalent.*

(i) There exists $F \in BV([0,\infty)\ ;\ X)$ with $\|F\|_{BV} \le M$ such that $r(\lambda) = \int_0^\infty e^{-\lambda t} dF(t)\ (\lambda > 0)$;

(ii) $\int_0^\infty \|r^{(k)}(\lambda)\| \frac{\lambda^{k-1}}{(k-1)!} d\lambda \le M \quad (k \in \mathbb{N})$.

In the case where $X = \mathbb{R}$ this is [Wi 2, Theorem 12 a, p. 306].

Proof : (i) $\Rightarrow$ (ii). Let $G(t) = \text{ var } (F) \mid_0^t$. Then

$$\int_0^\infty \|r^{(k)}(\lambda)\| \frac{\lambda^{k-1}}{(k-1)!} d\lambda = \int_0^\infty \| \int_0^\infty (-t)^k e^{-\lambda t} dF(t) \| \frac{\lambda^{k-1}}{(k-1)!} d\lambda \le$$

$$\int_0^\infty \int_0^\infty t^k e^{-\lambda t} dG(t) \frac{\lambda^{k-1}}{(k-1)!} d\lambda = \int_0^\infty \int_0^\infty e^{-\lambda t} \frac{\lambda^{k-1}}{(k-1)!} d\lambda t^k dG(t) =$$

$$\int_0^\infty dG(t) = \|F\|_{BV} \le M.$$

(ii) $\Rightarrow$ (i). We use the corresponding theorem in the real case. Thus for every $\varphi \in X^*$ there exists a unique normalized function $F(.,\varphi) : [0,\infty) \to \mathbb{R}$ such that $\|F(.,\varphi)\|_{BV} \le M\|\varphi\|$ and

$$\langle r(\lambda), \varphi \rangle = \int_0^\infty e^{-\lambda t} dF(t,\varphi) \quad (\lambda > 0). \tag{5.1}$$

Assume that $\|\varphi\| \le 1$. Let $0 \le s \le t < \infty$. Denote by LIM a Banach limit on ℓ^∞. We define $G : [0,\infty) \to \mathbb{R}$ by

$$G(t) := \operatorname*{LIM}_{n\to\infty} \int_{n/t}^\infty \|r^{(n)}(\lambda)\| \; \lambda^{n-1}/(n-1)! d\lambda$$

for $t > 0$ and $G(0) = 0$. Then G is increasing and bounded by M. One has by [Wi 2, p. 299]

$$G(t) - G(s) = \operatorname*{LIM}_{n\to\infty} \int_{n/t}^{n/s} \|r^{(n)}(\lambda)\| \lambda^{n-1}/(n-1)! d\lambda$$

$$\ge \lim_{n\to\infty} \int_{n/t}^{n/s} |\langle r^{(n)}(\lambda), \varphi \rangle| \lambda^{n-1}/(n-1)! \; d\lambda = \text{ var } F(.,\varphi)|_s^t.$$

Hence

$$|F(t,\varphi) - F(s,\varphi)| \le G(t) - G(s). \tag{5.2}$$

In particular, $|F(t,\varphi)| \le G(t) \le M$ if $\|\varphi\| \le 1$. It follows from the uniqueness theorem that there exists $F(t) \in X^{**}$ such that $\langle F(t), \varphi \rangle = F(t,\varphi) \ (\varphi \in X^*, t \ge 0)$. By (5.2), $\|F(t) - F(s)\| \le G(t) - G(s), 0 \le s \le t$. Thus $F : [0,\infty) \to X^{**}$ is of bounded variation and $\|F\|_{BV} \le M$. It follows that $F(t+)$ and $F(t-)$ exist. Since $F(.,\varphi)$ is normalized for all $\varphi \in X^*$, it follows that F is normalized. Moreover, $r(\lambda) = \int_0^\infty e^{-\lambda t} dF(t)$ in X^{**} for all $\lambda > 0$ by (5.1). Let $q : X^{**} \to X^{**}/X$ be the quotient mapping. Then $0 = q(r(\lambda)) = \int_0^\infty e^{-\lambda t} dq(F(t))(\lambda > 0)$. It follows from the uniqueness theorem that $q \circ F \equiv 0$, i.e. $F(t) \in X \ (t \ge 0)$. $\Diamond$

REFERENCES

[A1] W. Arendt : Resolvent positive operators. Proc. London Math. Soc.**54** (1987), 321-349.

[A2] W. Arendt : Vector-valued Laplace transforms and Cauchy problems. Israel J. Math. **59** (1987), 327-352.

[ANS] W. Arendt, F. Neubrander, U. Schlotterbeck : Interpolation of semigroups and integrated semigroups. Semigroup Forum **45** (1992), 26-37.

[Bo] S. Bochner : Completely monotonic functions in partially ordered spaces. Duke Math. J. **9** (1942), 419-526.

[BR] C.J.K. Batty ; D.W. Robinson : Positive one-parameter semigroups on ordered Banach spaces. Acta Appl. Math. **2** (1984), 221-296.

[DU] J. Diestel, J.J. Uhl, Jr. : Vector Measures. Mathematical Surveys **15**, Amer. Math. Soc., Providence, Long Island, 1977.

[H] F. Hausdorff : Momentprobleme für ein endliches Intervall. Math Z. **16** (1923), 220-248.

[HN] B. Hennig ; F. Neubrander : On representations, inversions and approximations of Laplace transforms in Banach spaces. Applicable Analysis, to appear.

[HP] E. Hille ; R.S. Phillips : Functional Analysis and Semigroups. Amer. Math. Soc. Providence, Rhode Island 1957.

[LT] J. Lindenstrauss ; L. Tzafriri : Classical Banach Spaces II. Springer, Berlin 1979.

[S1] H.H. Schaefer : A generalized moment problem. Math. Ann. **146** (1962), 326-330.

[S2] H.H. Schaefer : Topological Vector Spaces. Springer, Berlin 1971.

[S3] H.H. Schaefer : Banach Lattices and Positive Operators. Springer, Berlin 1974.

[Se] G.L. Seever : Measures on F-spaces. Trans. Amer. Math. Soc. **133** (1968), 267-280.

[Wi1] D.V. Widder : The inversion of the Laplace integral and the related moment problem. Trans. Amer. Math. Soc. **36** (1934), 107-200.

[Wi2] D.V. Widder : The Laplace Transform. Princeton University Press 1946.

[Wi3] D.V. Widder : An Introduction to Transform Theory. Academic Press. New York 1971.

[Z] S. Zaidman : Sur un théorème de I. Miyadera concernant la représentation des fonctions vectorielles par des intégrales de Laplace. Tohoku Math. J. **12** (1960), 47-51.

Inhomogeneous Volterra Integrodifferential Equations for Hille–Yosida Operators

RAINER NAGEL Universität Tübingen, Mathematisches Institut, Auf der Morgenstelle 10, D-7400 Tübingen, Germany

EUGENIO SINESTRARI Universita di Roma "'La Sapienza"', Dipartimento di Matematica, P. Aldo Moro 2, I-00185 Roma, Italy

INTRODUCTION.

In this paper we study inhomogeneous Cauchy problems for abstract differential equations

(DE) $$\dot{x}(t) = Ax(t) + f(t), \quad t \geq 0, \quad x(0) = x_0,$$

and Volterra integrodifferential equations

(VE) $$\dot{x}(t) = Ax(t) + \int_0^t a(t-s)Ax(s)\,ds + f(t), \quad t \geq 0, \quad x(0) = x_0,$$

This work was done while R. N. was visiting professor at the University of Rome "La Sapienza" and it was partially supported by the project "Equazioni di evoluzione e applicazioni" of the Italian M.U.R.S.T..

where A is an unbounded linear operator with non-dense domain on some Banach space X and $a(\cdot)$ is a scalar function. Our approach can be described as follows: first we extend the operator A to the infinitesimal generator of a semigroup in a larger space X_{-1} (the so called extrapolation space of X with respect to A: see Section 1); then we show that the corresponding problem (DE) in X_{-1} is equivalent to an homogeneous one in the product space $X_{-1} \times L^1(\mathbb{R}_+, X_{-1})$ (see Corollary 2.4); this will enable us to obtain well posedness results for (DE) and (VE) in X by restriction and perturbation theorems for generators of semigroups. In this way we retrieve the DaPrato-Sinestrari theorem (see [DaP-S]) for (DE) and extend the results of Grimmer-Prüss (see [G-P]) for (VE) when A is not densely defined. Finally the method mentioned above allows us to study (DE) in a space intermediate between X and X_{-1} (the so called Favard class: see Section 3) which we characterize in an example where A is the first derivative in the space of continuous functions.

1 EXTRAPOLATION SPACES FOR HILLE-YOSIDA OPERATORS

Abstract extrapolation spaces have been introduced by several authors and used for various purposes (e.g., [Am], [DaP-G], [Na1], [vNe], [Wa]). We briefly sketch the simple construction from [Na] and start this section by assuming that

$$A_0 : D(A_0) \subseteq X_0 \to X_0$$

is the generator of a strongly continuous semigroup $(T_0(t))_{t\geq 0}$ of bounded linear operators on a Banach space $(X_0, \|\cdot\|)$. Using the rescaling procedure (see [Na2], A-I, 3.1) we assume, without loss of generality, that the growth bound (or type) of $(T_0(t))_{t\geq 0}$ is negative. This implies $0 \in \rho(A_0)$, i.e., $A_0^{-1} \in \mathcal{L}(X_0)$.

DEFINITION 1.1 *On X_0 we introduce a new norm by*

$$\|x\|_{-1} := \|A_0^{-1}x\|, \quad x \in X_0. \tag{1.1}$$

Then the completion of $Y_0 := (X_0, \|\cdot\|_{-1})$ will be called the extrapolation space of X_0 associated with A_0 and will be denoted by X_{-1}.

From $\|x\|_{-1} \leq \|A_0^{-1}\|_{\mathcal{L}(X_0)} \cdot \|x\|$ for each $x \in X_0$ we deduce that

$$X_0 \hookrightarrow X_{-1}, \tag{1.2}$$

with $X_0 = X_{-1}$ if and only if A_0 is bounded. Since $D(A_0)$ is $\|\cdot\|$-dense, hence $\|\cdot\|_{-1}$-dense in X, the extrapolation space X_{-1} is also obtained as the completion of $(D(A_0), \|\cdot\|_{-1})$.

The bounded operator A_0^{-1} commutes with each $T_0(t)$ and therefore

$$\|T_0(t)x\|_{-1} \leq \|T_0(t)\|_{\mathcal{L}(X_0)} \cdot \|x\|_{-1} \tag{1.3}$$

for each $t \geq 0$, $x \in X_0$. This shows that $T_0(t) : Y_0 \subseteq X_{-1} \to X_{-1}$ is bounded and can be extended continuously to the completion of Y_0.

DEFINITION 1.2 *For each $t \geq 0$ we denote by $T_{-1}(t) : X_{-1} \to X_{-1}$ the unique continuous extension of $T_0(t) : Y_0 \subseteq X_{-1} \to X_{-1}$.*

The extended operators clearly satisfy the semigroup law and, in fact, are strongly continuous on X_{-1}.

PROPOSITION 1.3 *The semigroup $(T_{-1}(t))_{t\geq 0}$ is strongly continuous on X_{-1} and, for each $t \geq 0$,*

$$\|T_{-1}(t)\|_{\mathcal{L}(X_{-1})} = \|T_0(t)\|_{\mathcal{L}(X_0)}. \tag{1.4}$$

Proof. From the density of Y_0 in X_{-1} and $D(A_0)$ in X_0 we obtain

$$\begin{aligned}
\|T_{-1}(t)\|_{\mathcal{L}(X_{-1})} &= sup\{\|T_{-1}(t)x\|_{-1} : x \in X_{-1}, \ \|x\|_{-1} \leq 1\} \\
&= sup\{\|T_0(t)x\|_{-1} : x \in Y_0, \ \|x\|_{-1} \leq 1\} \\
&= sup\{\|T_0(t)A_0^{-1}x\| : x \in X_0, \ \|A_0^{-1}x\| \leq 1\} \\
&= sup\{\|T_0(t)y\| : y \in D(A_0), \ \|y\| \leq 1\} \\
&= \|T_0(t)\|_{\mathcal{L}(X_0)}.
\end{aligned}$$

The strong continuity of $(T_{-1}(t))_{t\geq 0}$ follows easily. □

Next we identify the generator of $(T_{-1}(t))_{t\geq 0}$.

THEOREM 1.4 *Denote by $A_{-1} : D(A_{-1}) \subseteq X_{-1} \to X_{-1}$ the generator of the semigroup $(T_{-1}(t))_{t\geq 0}$. Then the following properties hold.*

(i) $D(A_{-1}) = X_0$.

(ii) *A_{-1} is the unique continuous extension of $A_0 : D(A_0) \subseteq X_0 \to X_{-1}$ to an isometry from X_0 to X_{-1}.*

(iii) *A_{-1} is invertible with $(A_{-1})^{-1} \in \mathcal{L}(X_{-1})$.*

Proof. For $x \in D(A_0)$ we have $T_{-1}(t)x = T(t)x \in D(A_0) \subset X_0$, for $t \geq 0$, and hence

$$\lim_{t\to\infty} \|\frac{1}{t}(T_{-1}(t)x - x) - A_0 x\|_{-1} = \lim_{t\to\infty} \|\frac{1}{t}(T_0(t)A_0^{-1}x - A_0^{-1}x) - x\| = 0.$$

This shows that A_{-1} extends A_0. But $D(A_0)$ is dense in X_{-1} and invariant for $(T_{-1}(t))_{t\geq 0}$, and hence, is a core for A_{-1}. This means that $D(A_{-1})$ is the closure of $D(A_0)$ with respect to the graph norm

$$\|x\|_{A_{-1}} := \|x\|_{-1} + \|A_{-1}x\|_{-1}, \quad x \in D(A_{-1}).$$

But, this norm is equivalent to $\|\cdot\|$ on $D(A_0)$ and so $D(A_{-1}) = X_0$.
The remaining assertions follow from the fact that $A_0 : D(A_0) \subseteq X_0 \to X_{-1}$ is an isometry with dense domain and range. □

The above construction will now be applied to (not necessarily densely defined) generators of Lipschitz continuous integrated semigroups. By [K-H], Theorem 2.4 these are characterized by the Hille-Yosida estimates.

DEFINITION 1.5 *A linear operator $A : D(A) \subseteq X \to X$ on a Banach space X is called a Hille-Yosida operator if there exists $w \in \mathbb{R}$ such that $(w, \infty) \subset \rho(A)$ and*

$$sup\{\|(\lambda - w)^n(\lambda - A)^{-n}\| : \lambda > w, \ n \in \mathbb{N}\} =: M < \infty. \tag{1.5}$$

If the constant w can be choosen smaller than zero then we say that A is of negative type.

It is well known ([H-P], Theorem 12.2.4) that to each Hille-Yosida operator corresponds a strongly continuous semigroup on the closure of its domain. More precisely:

PROPOSITION 1.6 *Let $A : D(A) \subseteq X \to X$ be a Hille-Yosida operator and set*

$$\begin{aligned} X_0 &:= \left(\overline{D(A)}, \|\cdot\|\right), \\ D(A_0) &:= \{x \in D(A) : Ax \in X_0\}, \\ A_0 x &:= Ax \quad for \quad x \in D(A_0). \end{aligned} \tag{1.6}$$

Then $A_0 : D(A_0) \subseteq X_0 \to X_0$ is the generator of a strongly continuous semigroup $(T_0(t))_{t\geq 0}$ on X_0. Moreover, if $\lambda \in \rho(A)$, then $\lambda \in \rho(A_0)$ and $(\lambda - A_0)^{-n}$ is the restriction of $(\lambda - A)^{-n}$ to X_0 for each $n \in \mathbb{N}$.

From this proposition it follows that, if A is a Hille-Yosida operator (of negative type), then the above definitions and results apply to A_0. Thus we obtain an extrapolation space X_{-1} containing X_0. We now show that our original space X lies between X_0 and X_{-1}.

THEOREM 1.7 *Let A be a Hille-Yosida operator of negative type. For the norm*

$$\|x\|_{-1} := \|A^{-1}x\|, \quad x \in X, \tag{1.7}$$

we have that X_0 (defined in (1.6)) is a dense subspace of $Y := (X, \|\cdot\|_{-1})$. Hence, the extrapolation space X_{-1} is also the completion of Y, hence $X \hookrightarrow X_{-1}$. Moreover, A_{-1} is an extension of A, hence $(A_{-1})^{-1}(X) = D(A)$.

Proof. For $x \in X$ we have $A^{-1}x \in D(A) \subseteq X_0$. Hence, for each $\epsilon > 0$, there exists $x_\epsilon \in D(A_0)$ satisfying $\|x_\epsilon - A^{-1}x\| < \epsilon$. As $\|x_\epsilon - A^{-1}x\| = \|Ax_\epsilon - x\|_{-1}$ and $Ax_\epsilon \in X_0$ this shows that X_0 is dense in Y and the completions of Y and of $(X_0, \|\cdot\|_{-1})$ coincide.
Since $A : D(A) \subseteq X \to X_{-1}$ is an isometry extending A_0 it coincides (by (1.6)) with the restriction of A_{-1} to $D(A)$. □

FINAL REMARKS 1.8 **1.** The mutual relation of the spaces considered above can be visualized by the following diagram, where we observe that in general the spaces X and $D(A)$ are not invariant subspaces for the semigroup $T_{-1}(t))_{t\geq 0}$.

$$
\begin{array}{ccc}
X_{-1} & \xrightarrow{T_{-1}(t)} & X_{-1} := (X_0, \|\cdot\|_{-1})^{\sim} \\
\uparrow & & \uparrow \\
X & & X \\
\uparrow & & \uparrow \\
X_0 & \xrightarrow{T_0(t)} & X_0 := \overline{D(A)} \\
\uparrow & & \uparrow \\
D(A) & & D(A) \\
\uparrow & & \uparrow \\
D(A_0) & \xrightarrow{T_1(t)} & D(A_0)
\end{array}
$$

2. In the following example an explicit characterization of the extrapolation space X_{-1} is easy. Take $X_0 := L^1(\mathbb{R}, dx)$ and a multiplication operator $A_0 f := q \cdot f$ for some measurable complex function q satisfying $\text{Re}q(s) \leq -1$ for a.e. $s \in \mathbb{R}$. The domain of A_0 is

$$D(A_0) = L^1(\mathbb{R}, |q| \cdot dx),$$

while

$$X_{-1} = L^1(\mathbb{R}, |q|^{-1} \cdot dx).$$

A more interesting example involving a differential operator is discussed at the end of Section 3.

2 INHOMOGENEOUS CAUCHY PROBLEMS AND SEMIGROUPS ON PRODUCT SPACES

In this section we assume $A : D(A) \subseteq X \to X$ to be a Hille-Yosida operator of negative type, hence, there is $M \geq 1$ and $w < 0$ such that

$$\|T_0(t)\|_{\mathcal{L}(X_0)} \leq M \cdot e^{wt}, \quad t \geq 0, \tag{2.1}$$

where $(T_0(t))_{t\geq 0}$ is the semigroup generated by A_0, i.e. the part of A in $X_0 := \overline{D(A)}$ (see Proposition 1.6). As in Section 1 we associate to A_0 its extension $A_{-1} : X_0 \to X_{-1}$ and are then interested in the following inhomogeneous Cauchy problem.

$$\begin{aligned} \dot{u}(t) &= A_{-1}u(t) + f(t), \quad t \geq 0 \\ u(0) &= u_0 . \end{aligned} \tag{2.2}$$

As A_{-1} generates a strongly continuous semigroup $(T_{-1}(t))_{t\geq 0}$ in X_{-1}, the Phillips theorem (see [Go], Theorem 1.3 and [DaP], Proposition III.3.17) provides conditions on f and u_0 such that differentiable solutions of (2.2) exist (see Corollary 2.4 below). It is our aim to investigate these solutions when f is an X-valued function.
We recall that the mild solution of (2.2) is given by

$$u(t) = T_{-1}(t)x + \int_0^t T_{-1}(t-s)f(s)\,ds \tag{2.3}$$

and we now prove some important properties of the integral term in (2.3).

PROPOSITION 2.1 *For $f \in L^1(\mathbb{R}_+, X)$ and $t \geq 0$ let*

$$(T_{-1} * f)(t) := \int_0^t T_{-1}(t-s)f(s)\,ds . \tag{2.4}$$

Then the following properties hold.

(i) $(T_{-1} * f)(t) \in X_0$.

(ii) $\|(T_{-1} * f)(t)\| \leq M_1 \|f\|_{L^1(]0,t[,X)}$, *where M_1 is independent of f and t.*

(iii) $\lim_{t\downarrow 0} \|(T_{-1} * f)(t)\| = 0$.

Proof. We will prove our result first for constant functions, then for step functions with compact support and finally obtain the assertion by density.
To this end, for each $t \geq 0$, we define $V(t) : X_{-1} \to X_{-1}$ by

$$V(t)x := \int_0^t T_{-1}(s)x\,ds, \quad x \in X_{-1} . \tag{2.5}$$

As $D(A_{-1}) = X_0$ it is well known (e.g., [Pa], Theorem 2.4) that

$$V(t)x \in X_0 \,. \tag{2.6}$$

Setting $y := (A_{-1})^{-1}x \in X_0$ we obtain

$$\frac{V(t)x}{t} = \frac{1}{t}\int_0^t A_{-1}T_{-1}(s)y\,ds = \frac{1}{t}(T_0(t)y - y)\,. \tag{2.7}$$

Take now $x \in X$, hence $y \in D(A)$ by Theorem 1.7. For $\lambda > w$ one has $(\lambda - w)R(\lambda, A_0)y = (\lambda - w)R(\lambda, A)y = y + R(\lambda, A)(A - w)y$, hence by (1.8)

$$\lim_{\lambda\to\infty} \|(\lambda - w)R(\lambda, A_0)y - y\| = 0\,.$$

This implies the estimate

$$\begin{aligned}
\|(T_0(t) - I)y\| &= \lim_{\lambda\to\infty} \|(T_0(t) - I)(\lambda - w)R(\lambda, A_0)y\| \\
&= \lim_{\lambda\to\infty} \|A_0 \int_0^t T_0(s)(\lambda - w)R(\lambda, A_0)y\,ds\| \\
&= \lim_{\lambda\to\infty} \|\int_0^t T_0(s)(\lambda - w)AR(\lambda, A)y\,ds\| \\
&\le M \cdot t \sup_{0\le s\le t} \|T_0(s)\|_{\mathcal{L}(X_0)} \cdot \|Ay\| \\
&\le t \cdot M^2\|x\|\,.
\end{aligned}$$

From this estimate and (2.7) we deduce that

$$\|V(t)x\| \le tM^2\|x\| \ \text{ for } x \in X\,. \tag{2.8}$$

Let $\mathcal{S}$ denote the space of all functions $\phi : \mathbb{R}_+ \to X$ such that $\phi := \sum_{i=0}^n x_i\alpha_i$ for some $n \in \mathbb{N}$, $x_0, \ldots, x_n \in X\backslash\{0\}$ and $0 = t_0 < t_1 < \ldots < t_{n+1}$ with $\alpha_i = \mathbb{I}_{[t_i, t_{i+1})}$.
Fix $t > 0$ and $\phi \in \mathcal{S}$. If the support of ϕ is contained in $[0, t]$ we obtain

$$\begin{aligned}
(T_{-1} * \phi)(t) &= \textstyle\sum_{i=0}^n \displaystyle\int_{t_i}^{t_{i+1}} T_{-1}(t - s)x_i\,ds \\
&= \textstyle\sum_{i=0}^n T_{-1}(t - t_{i+1}) \displaystyle\int_0^{t_{i+1}-t_i} T_{-1}(s)x_i\,ds \\
&= \textstyle\sum_{i=0}^n T_{-1}(t - t_{i+1})V(t_{i+1} - t_i)x_i\,.
\end{aligned}$$

From (2.6) and (2.8) we conclude that

$$(T_{-1} * \phi)(t) \in X_0 \tag{2.9}$$

and

$$\|(T_{-1} * \phi)(t)\| \le M^3\|\phi\|_{L^1(]0,t[,X)}\,. \tag{2.10}$$

If the support of ϕ is not contained in $[0, t]$, we set $\phi_1 := \phi \cdot \alpha$ where $\alpha = \mathbb{1}_{[0,t]}$. The above reasoning applied to ϕ_1 yields (2.9) and (2.10) because

$$(T_{-1} * \phi)(t) = (T_{-1} * \phi_1)(t) \in X_0$$

and

$$\|(T_{-1} * \phi)(t)\| = \|(T_{-1} * \phi_1)(t)\| \le M^3 \|\phi_1\|_{L^1(]0,t[,X)} \le M^3 \|\phi\|_{L^1(]0,t[,X)} \,.$$

For $\phi \in \mathcal{S}$ the assertion (iii) follows from (2.8) since $\|(T_{-1} * \phi)(t)\| = \|V(t)x_0\|$ on $[0, t_1)$. For the final step take $f \in L^1(\mathbb{R}_+, X)$ and a sequence $(\phi_n) \subset \mathcal{S}$ such that $\lim_{n\to\infty} \|\phi - \phi_n\|_{L^1(\mathbb{R}_+,X)} = 0$. The estimate (2.10) implies that $\lim_{n\to\infty} (T_{-1} * \phi_n)(t)$ exists in X_0 and coincides with $(T_{-1} * f)(t)$. This proves (i) and the remaining assertions (ii) and (iii) follow easily. □

The inhomogeneous Cauchy problem (2.2) will be substituted by a homogeneous one via a semigroup on a product space. Before introducing this semigroup we recall the well known characterization of the generator of the translation semigroup on vector valued L^1-spaces.

LEMMA 2.2 *The translation semigroup $(S(t))_{t\ge0}$ defined by*

$$(S(t)f)(s) := f(t+s)\,, \quad s, t \ge 0\,, \tag{2.11}$$

for $f \in L^1(\mathbb{R}_+, X_{-1})$ is strongly continuous and its generator B is given by

$$\begin{aligned} D(B) &= W^{1,1}(\mathbb{R}_+, X_{-1})\,, \\ Bf &= f'(s) \ \text{ for } f \in D(B)\,, \ 0 \le s \ \text{ a.e.}. \end{aligned} \tag{2.12}$$

From this lemma and from the fact that $(T_{-1}(t))_{t\ge0}$ is strongly continuous on X_{-1} one constructs a new semigroup on a product space.

PROPOSITION 2.3 *On the Banach space*

$$Z_{-1} := X_{-1} \times L^1(\mathbb{R}_+, X_{-1})\,, \tag{2.13}$$

with $\|\binom{x}{f}\|_{Z_{-1}} := \|x\|_{-1} + \|f\|_{L^1(\mathbb{R}_+,X_{-1})}$, define operators

$$G_{-1}(t)\begin{pmatrix} x \\ f \end{pmatrix} := \begin{pmatrix} T_{-1}(t)x + (T_{-1} * f)(t) \\ S(t)f \end{pmatrix} \tag{2.14}$$

for $\binom{x}{f} \in Z_{-1}$. Then $(G_{-1}(t))_{t\ge0}$ is a strongly continuous semigroup and its generator $\mathcal{A}_{-1}$:

$D(\mathcal{A}_{-1}) \subseteq Z_{-1} \to Z_{-1}$ is given by

$$\begin{aligned} D(\mathcal{A}_{-1}) &= X_0 \times W^{1,1}(\mathbb{R}_+, X_{-1}), \\ \mathcal{A}_{-1}\begin{pmatrix} x \\ f \end{pmatrix} &= \begin{pmatrix} A_{-1}x + f(0) \\ f' \end{pmatrix}. \end{aligned} \tag{2.15}$$

The proof consists in elementary calculations or can be deduced from the operator matrix setting studied in [Na3], Sect. 3.

In this paper we are interested in the semigroup $(G_{-1}(t))_{t\geq 0}$ since it yields the solutions of our inhomogeneous Cauchy problem (2.2). In fact, let

$$\begin{pmatrix} u(t) \\ v(t) \end{pmatrix} := G_{-1}(t)\begin{pmatrix} x \\ f \end{pmatrix} \in Z_{-1}.$$

Then its first coordinate $u(\cdot)$ is the unique mild solution (2.3) of (2.2) in the space X_{-1}. Since $G_{-1}(\cdot)\binom{x}{f}$ is differentiable on $\mathbb{R}_+$ if and only if $\binom{x}{f} \in D(\mathcal{A}_{-1})$ we immediately obtain Phillips theorem and the mentioned reduction of problem (DE) to an homogeneous one.

COROLLARY 2.4 *Let $\mathcal{U}_0 := \binom{x}{f} \in Z_{-1}$ such that $x \in X_0$ and $f \in W^{1,1}(\mathbb{R}_+, X_{-1})$. If $t \mapsto \mathcal{U}(t) := \binom{u(t)}{v(t)}$ is the solution of*

$$\begin{aligned} \dot{\mathcal{U}}(t) &= \mathcal{A}_{-1}\mathcal{U}(t), \quad t \geq 0, \\ \mathcal{U}(0) &= \mathcal{U}_0 \end{aligned} \tag{2.16}$$

(given by (2.14)) then $u(\cdot)$ is the unique differentiable solution of

$$\begin{aligned} \dot{u}(t) &= A_{-1}u(t) + f(t), \quad t \geq 0, \\ u(0) &= x. \end{aligned} \tag{2.17}$$

REMARK 2.5 Let us observe that the proofs of Proposition 2.3 and Corollary 2.4 use only the fact that G_{-1} is a semigroup on a Banach space X_{-1} and X_0 is the domain of its infinitesimal generator.

We now assume that f is an X-valued function and are interested in conditions on x and f in order to obtain a differentiable X-valued solution of (2.2). To this purpose we will restrict the semigroup $(G_{-1}(t))_{t\geq 0}$ to an invariant subspace of Z_{-1} and then apply the following lemma.

LEMMA 2.6 *Let $(T(t))_{t\geq 0}$ be a strongly continuous semigroup on the Banach space E having generator $(B, D(B))$. If E_1 is a Banach space such that $E_1 \hookrightarrow E$, $T(t)E_1 \subset E_1$ and the restrictions of $T(t)$ to E_1 form a strongly continuous semigroup $(T_|(t))_{t\geq 0}$ on E_1. Then its generator $C : D(C) \subset E_1 \to E_1$ is the 'part' of B in E_1, i.e.,*

$$\begin{aligned} D(C) &= \{x \in D(B) \cap E_1 : Bx \in E_1\} \\ Cx &= Bx \text{ for } x \in D(C). \end{aligned} \tag{2.18}$$

Proof. As $E_1 \hookrightarrow E$ we immediately have $D(C) \subset D := \{x \in D(B) \cap E_1 : Bx \in E_1\}$ and $Cx = Bx$ on $D(C)$. For the converse inclusion choose $0 < \lambda \in \rho(B) \cap \rho(C)$ sufficiently large and $y \in E_1$. Then $R(\lambda, C)y = \int_0^\infty e^{-\lambda s}T(s)y\,ds = R(\lambda, B)y$ and $x = R(\lambda, B)(\lambda - B)x = R(\lambda, C)(\lambda - C)x \in D(C)$ for $x \in D$. □

The explicit representation (2.14) of $G_{-1}(t)$ combined with Proposition 2.1, allows us to find a $(G_{-1}(t))_{t\geq 0}$-invariant subspace in Z_{-1} and to identify its generator.

THEOREM 2.7 *The Banach space*

$$Y := X_0 \times L^1(\mathbb{R}_+, X)$$

with norm $\|\binom{x}{f}\| := \|x\| + \|f\|_{L^1(\mathbb{R}_+,X)}$, satisfies $Y \hookrightarrow Z_{-1}$ and the restrictions $G(t)$ of $G_{-1}(t)$ form a strongly continuous semigroup on Y with generator $\mathcal{A} : D(\mathcal{A}) \subseteq Y \to Y$ given by

$$\begin{aligned} D(\mathcal{A}) &= \{\tbinom{x}{f} : x \in D(A),\ f \in W^{1,1}(\mathbb{R}_+, X),\ Ax + f(0) \in X_0\}, \\ \mathcal{A}\begin{pmatrix} x \\ f \end{pmatrix} &= \begin{pmatrix} Ax + f(0) \\ f' \end{pmatrix}. \end{aligned} \tag{2.19}$$

Proof. The space Y is invariant under $G_{-1}(t)$ by Proposition 2.1 (i). The strong continuity of $(G_{-1}(t))_{t\geq 0}$ on Y follows from (iii). Hence, we apply Lemma 2.6 and obtain $\binom{x}{f} \in D(\mathcal{A})$ if and only if $\binom{x}{f} \in D(\mathcal{A}_{-1}) \cap Y$ and $\mathcal{A}_{-1}\binom{x}{f} = \binom{A_{-1}x+f(0)}{f'} \in Y$. This is equivalent to $f \in W^{1,1}(\mathbb{R}_+, X)$, $x \in X_0$ and $A_{-1}x + f(0) \in X_0$. Since $f(0) \in X$ we must have $x \in D(A)$ by Theorem 1.7. □

As in Corollary 2.4 we can now take the first coordinate of $G(\cdot)\binom{x}{f}$ and obtain the X_0-valued solution of the inhomogeneous Cauchy problem (2.2). This solution is differentiable if (and only if) the initial vector $\binom{x}{f}$ is in the domain $D(\mathcal{A})$. Thus we have reproved the generalization of Phillips theorem due to [DaP-S], Theorem 2.3.

COROLLARY 2.8 *Let $A : D(A) \subseteq X \to X$ be a Hille-Yosida operator. For each $x \in D(A)$ and $f \in W^{1,1}(\mathbb{R}_+, X)$ such that*

$$Ax + f(0) \in \overline{D(A)},$$

the Cauchy problem

$$\begin{aligned} \dot{u}(t) &= Au(t) + f(t), \quad t \geq 0, \\ u(0) &= x \end{aligned} \tag{2.20}$$

has a unique solution $u \in C^1(\mathbb{R}_+, X)$. Moreover, there exists $M_1 > 0$ such that for $t \geq 0$

$$\begin{aligned} \|u(t)\| &\leq M_1(\|x\| + \textstyle\int_0^t \|f(s)\|\, ds), \\ \|\dot{u}(t)\| &\leq M_1(\|(Ax + f(0))\| + \textstyle\int_0^t \|f'(s)\|\, ds). \end{aligned} \tag{2.21}$$

Proof. It will be sufficient to prove the estimates (2.21). As $\mathcal{A}$ is the generator of the semigroup $(G(t))_{t\geq 0}$ in Y and $\binom{x}{f} \in D(\mathcal{A})$ there exists $M_1 > 0$ such that

$$\|G(t)\tbinom{x}{f}\|_Y \leq M_1\|\tbinom{x}{f}\|_Y \text{ and } \|\dot{G}(t)\tbinom{x}{f}\|_Y \leq M_1\|\mathcal{A}\tbinom{x}{f}\|_Y \text{for } t \geq 0 \quad .$$

This implies

$$\begin{aligned} \|u(t)\| &\leq M_1(\|x\| + \textstyle\int_0^{+\infty} \|f(s)\|\, ds), \\ \|\dot{u}(t)\| &\leq M_1(\|(Ax + f(0))\| + \textstyle\int_0^{+\infty} \|f'(s)\|\, ds) \text{ for } t \geq 0. \end{aligned} \tag{2.22}$$

Let us fix $t > 0$ and substitute f with $f_1 := f \cdot \alpha$ where α is the characteristic function of $[0, t]$. We obtain a solution v of (2.20) such that $v \in C^1(\mathbb{R}_+, X)$ and

$$\begin{aligned} \|v(\tau)\| &\leq M_1(\|x\| + \textstyle\int_0^t \|f(s)\|\, ds) \text{ and} \\ \|\dot{v}(\tau)\| &\leq M_1(\|(Ax + f(0))\| + \textstyle\int_0^t \|f'(s)\|\, ds) \text{ for } \tau \geq 0. \end{aligned} \tag{2.23}$$

But (2.14) shows that $v(\tau) = u(\tau)$ for $\tau \in [0, t]$ and so taking $\tau = t$ in (2.23) we deduce (2.21). □

3 FAVARD CLASS AND ITS EXTRAPOLATION SPACES

Let us recall the definition of the Favard class of a strongly continuous semigroup (see [B-B], Chapter 3):

DEFINITION 3.1 *Let $(T_0(t))_{t\geq 0}$ be a bounded strongly continuous semigroup on a Banach space X_0. Its Favard class is the Banach space*

$$F := \{x \in X_0 : \sup_{t>0} \frac{1}{t}\|T_0(t)x - x\| < \infty\} \tag{3.1}$$

with norm

$$\|x\|_F := \sup_{t>0} \frac{1}{t}\|T_0(t)x - x\|. \tag{3.2}$$

Obviously F is invariant under each $T_0(t)$ and it also is clear that $D(A_0) \subset F$ (with equality when X_0 is reflexive; see [B-B], Theorem 2.12) and similarly $X_0 \subset F_{-1}$ if F_{-1} is the Favard class of the extrapolated semigroup $(T_{-1}(t))_{t\geq 0}$ on X_{-1} (see Theorem 1.4). We now show that F_{-1} is indeed the extrapolation space of F.

PROPOSITION 3.2 *Let $A : D(A) \subseteq X \to X$ be a Hille-Yosida operator of negative type and denote by $(T_0(t))_{t\geq 0}$ the strongly continuous semigroup generated by the part A_0 of A in $X_0 := \overline{D(A)}$. If $(F_{-1}, \|\cdot\|_{F_{-1}})$ is the Favard class with respect to $(T_{-1}(t))_{t\geq 0}$, then the following properties hold.*

(i) $A_{-1}(F) = F_{-1}$.

(ii) $\|A_{-1}x\|_{F_{-1}} = \|x\|_F$ *for all* $x \in F$.

(iii) $D(A_0) \subseteq D(A) \subseteq F \hookrightarrow X_0 \subseteq X \subseteq F_{-1} \hookrightarrow X_{-1}$.

(iv) $T_{-1}(t)(F_{-1}) \subseteq F_{-1}$ *for all* $t \geq 0$.

Proof. By Theorem 1.4 the extended operator A_{-1} is an isometry between X_0 and X_{-1}. Therefore $\|T_0(t)x - x\| = \|A_{-1}(T_0(t)x - x)\|_{-1} = \|T_{-1}(t)A_{-1}x - A_{-1}x\|_{-1}$ for all $t > 0$ and $x \in X_0$. This proves (i) and (ii). In the proof of Proposition 2.1 we have shown that $D(A) \subseteq F$, which implies $X \subseteq F_{-1}$ since A_{-1} is an extension of A (Theorem 1.7). The remaining inclusions and assertion (iv) are immediate from the definitions. □

It is now possible to replace the space X by the larger space F_{-1} and still obtain an analogue to Proposition 2.1.

PROPOSITION 3.3 *Let $(A, D(A))$ be a Hille-Yosida operator of negative type on X and denote by F_{-1} the extrapolation space of its Favard class F. For each $f \in L^1(\mathbb{R}_+, F_{-1})$ and $t \geq 0$ we define*

$$(T_{-1} * f)(t) := \int_0^t T_{-1}(t-s)f(s)\,ds$$

and obtain for some constant $M_2 \geq 1$ and independent of f and t

(i) $(T_{-1} * f)(t) \in X_0$.

(ii) $\|(T_{-1} * f)(t)\| \leq M\|f\|_{L^1(]0,t[,F_{-1})}$.

(iii) $\lim_{t\downarrow 0} \|(T_{-1} * f)(t)\| = 0$.

Proof. Given $x \in F_{-1}$ and $t > 0$ we have that

$$V(t)x := \int_0^t T_{-1}(s)x\,ds \tag{3.3}$$

is contained in $X_0 = D(A_{-1})$. For $y := (A_{-1})^{-1}x \in F$ we obtain from Proposition 3.2.ii

$$\begin{aligned} \frac{\|V(t)x\|}{t} &= \|\frac{1}{t}A_{-1}\int_0^t T_{-1}(s)y\,ds\| = \|\frac{1}{t}A_0\int_0^t T_0(s)y\,ds\| = \|\frac{T_0(t)y-y}{t}\| \\ &\leq \|y\|_F = \|x\|_{F_{-1}}. \end{aligned} \tag{3.4}$$

This corresponds to the estimate (2.8) and we can proceed as in that proof. □

These properties allow to apply the ideas from Section 2 by first defining a strongly continuous semigroup on the product space $X_0 \times L^1(\mathbb{R}_+, F_{-1})$ and then using this semigroup to solve the inhomogeneous Cauchy problem for functions f taking values in F_{-1}. Since the proofs are analogous to those of Theorem 2.7 and Corollary 2.8 we only state the results.

THEOREM 3.4 *Under the above assumptions and with the notations from Proposition 2.3 we denote by $G_F(t)$ the restriction of $G_{-1}(t)$ to the Banach space*

$$Y_F := X_0 \times L^1(\mathbb{R}_+, F_{-1}). \tag{3.5}$$

Then $(G_F(t))_{t\geq 0}$ is a strongly continuous semigroup in Y_F with generator

$$\begin{aligned} D(\mathcal{A}_F) &= \{\tbinom{x}{f} : x \in F,\ f \in W^{1,1}(\mathbb{R}_+, F_{-1}),\ A_{-1}x + f(0) \in X_0\}, \\ \mathcal{A}_F\begin{pmatrix} x \\ f \end{pmatrix} &= \begin{pmatrix} A_{-1}x + f(0) \\ f' \end{pmatrix}. \end{aligned} \tag{3.6}$$

COROLLARY 3.5 *Let $A : D(A) \subseteq X \to X$ be a Hille-Yosida operator. For each $x \in F$ and $f \in W^{1,1}(\mathbb{R}_+, F_{-1})$ such that*

$$A_{-1}x + f(0) \in \overline{D(A)} \tag{3.7}$$

the Cauchy problem

$$\begin{aligned} \dot{u}(t) &= A_{-1}u(t) + f(t), \quad t \geq 0, \\ u(0) &= x \end{aligned} \tag{3.8}$$

has a unique solution $u \in \mathcal{C}^1(\mathbb{R}_+, X) \cap \mathcal{C}(\mathbb{R}_+, F)$.

Before closing this section we show in a simple but nontrivial example how the abstract extrapolation space F_{-1} can be identified with a concrete function space.

To this purpose consider the Banach space $X := \mathcal{C}[a,b]$ and the differential operator

$$\begin{aligned} D(A) &:= \{f \in \mathcal{C}^1[a,b] : f(a) = 0\} \\ Af &:= -f'. \end{aligned} \tag{3.9}$$

It is elementary to show that $\lambda \in \rho(A)$ for each $\lambda \in \mathbb{R}$ and

$$(R(\lambda, A)f)(x) = \int_a^x e^{-\lambda(x-y)} f(y)\, dy \tag{3.10}$$

for each $f \in X$, $x \in [a,b]$. Since $\|R(\lambda, A)\| \leq \frac{1}{\lambda}$ for $\lambda > 0$ we obtain that A is a Hille-Yosida operator. On the space

$$X_0 := \overline{D(A)} = \mathcal{C}_0[a,b] := \{f \in \mathcal{C}[a,b] : f(a) = 0\} \tag{3.11}$$

it is simple to verify that the part A_0 of A generates the nilpotent right translation semigroup

$$T(t)f(x) := \begin{cases} f(x-t) & \text{for } t \leq x-a \\ 0 & \text{for } t > x-a. \end{cases} \tag{3.12}$$

The Favard class corresponding to this translation semigroup is now easy to determine.

LEMMA 3.6 *In the above situation the Favard class F coincides with*

$$Lip_0[a,b] := \{f \in \mathcal{C}_0[a,b] : f \text{ Lipschitz continuous }\},$$

where the norm is given by

$$\|f\|_{Lip} := \sup_{a \leq x_1 < x_2 \leq b} \frac{|f(x_2) - f(x_1)|}{x_2 - x_1}.$$

It is well known that the functions $f \in \mathrm{Lip}_0[a,b]$ are a.e. differentiable yielding a derivative $f' \in L^\infty[a,b]$. We now show that our extrapolated operator A_{-1} coincides with this a.e. derivative. This proves that we can identify F_{-1} with $L^\infty[a,b]$. More precisely:

PROPOSITION 3.7 *There exists a linear isometry J from $L^\infty[a,b]$ onto F_{-1} and for $g \in Lip_0[a,b]$ we have*

$$A_{-1}g = Jg', \tag{3.13}$$

where g' is the a.e. derivative of g.

Proof. In the following we will write L^∞ and Lip_0 instead of $L^\infty[a,b]$ and $\mathrm{Lip}_0[a,b]$. For $\lambda \in \mathbb{R}$ and $f \in L^\infty$ let us define $R_\lambda f$ as

$$(R_\lambda f)(x) := \int_a^x e^{-\lambda(x-y)} f(y)\,dy\,, \quad x \in [a,b]\,.$$

Then the following properties hold:

$$R_0 \text{ is a linear isometry from } L^\infty \text{ onto } \mathrm{Lip}_0. \tag{3.14}$$

$$R_\lambda R_\mu = R_\mu R_\lambda \text{ for } \lambda, \mu \in \mathbb{R}\,. \tag{3.15}$$

$$R_\lambda|_{X_0} = R(\lambda, A_0) \text{ for } \lambda \in \mathbb{R}\,. \tag{3.16}$$

$$\lim_{\lambda\to\infty} \lambda R_\lambda f = f \text{ for } f \in X_0\,. \tag{3.17}$$

For $f \in L^\infty$ and $n \in \mathbb{N}$ we now define $f_n := nR_n f \in X_0$ and obtain from (3.10), (3.15) and (3.16) that

$$\begin{aligned} \|f_n - f_m\|_{-1} &= \|A_0^{-1}(nR_n - mR_m)f\| \\ &= \|R_0(nR_n - mR_m)f\| \\ &= \|(nR_n - mR_m)R_0 f\| \quad \text{for } n,m \in \mathbb{N}\,. \end{aligned}$$

Since $R_0 f \in X_0$ we conclude from (3.17) that $(f_n)_{n\in\mathbb{N}}$ converges in X_{-1} and hence

$$J : L^\infty \to X_{-1} \quad \text{with} \quad Jf := \|\cdot\|_{-1}\text{-} \lim_{n\to\infty} f_n \quad \text{for} \quad f \in L^\infty$$

is well defined.
Arguing as above we have

$$\|Jf\|_{-1} = \lim_{n\to\infty} \|f_n\|_{-1} = \lim_{n\to\infty} \|A_0^{-1} nR_n f\| = \lim_{n\to\infty} \|nR_n R_0 f\| = \|R_0 f\|\,,$$

which shows that J is injective. Moreover

$$A_{-1}^{-1}Jf = \lim_{n\to\infty} A_{-1}^{-1} f_n = \lim_{n\to\infty} A_0^{-1} f_n = \lim_{n\to\infty} nR_n R_0 f = R_0 f$$

and

$$Jf = A_{-1}R_0 f \quad \text{for every} \quad f \in L^\infty\,. \tag{3.18}$$

Since $A_{-1}(F) = F_{-1}$, from (3.14) and Lemma 3.6 we obtain

$$J(L^\infty) = A_{-1}(\mathrm{Lip}_0) = A_{-1}(F) = F_{-1}$$

and

$$\|f\|_{L^\infty} = \|R_0 f\|_{\mathrm{Lip}_0} = \|A_{-1}^{-1} J f\|_F = \|Jf\|_{F_{-1}} \text{ for } f \in L^\infty .$$

Finally if $g \in \mathrm{Lip}_0$ with a.e. derivative $g' \in L^\infty$ then $g = R_0 g'$ and by (3.18)

$$Jg' = A_{-1} R_0 g' = A_{-1} g .$$

This proves the assertion. □

4 VOLTERRA INTEGRODIFFERENTIAL EQUATIONS

In this section we still assume $A : D(A) \subset X \to X$ to be a Hille-Yosida operator on the Banach space X. To the inhomogeneous Cauchy problem (2.1) we now add an integral term and want to solve Volterra integrodifferential equations. We refer to [Pr] for a comprehensive study and restrict ourselves to problems of the form

$$\begin{aligned} \dot{u}(t) &= Au(t) + \int_0^t a(t-s) Au(s)\, ds + f(t) , \quad t \geq 0 , \\ u(0) &= u_0 . \end{aligned} \tag{4.1}$$

We assume the scalar valued function $a(\cdot)$ to be of bounded variation on $\mathbb{R}_+$, i.e., $a(\cdot)$ belongs to the space $BV(\mathbb{R}_+)$ of all functions $f : \mathbb{R}_+ \to \mathbb{C}$ satisfying

$$V_a^b(f) := sup \left\{ \sum_{k=0}^{m} |f(t_k) - f(t_{k-1})| : a = t_0 < \ldots < t_m = b \right\} < \infty \tag{4.2}$$

for all $0 \leq a < b$ and

$$V_0^\infty(f) := \lim_{b \to \infty} V_0^b(f) < \infty . \tag{4.3}$$

For functions with bounded variation the following property holds (see [Br], Lemma A.1).

LEMMA 4.1 *If $f \in BV(\mathbb{R}_+)$ then*

$$|f(t)| \leq f(0) + V_0^\infty(f) , \quad t \geq 0 , \tag{4.4}$$

and if $f \in BV(\mathbb{R}_+) \cap L^1(\mathbb{R}_+)$ *then*

$$\int_0^\infty |f(t+s) - f(s)|\, ds \le t \cdot V_0^\infty(f)\,, \quad t \ge 0\,. \tag{4.5}$$

In order to solve (4.1) for functions $a \in BV(\mathbb{R}_+)$ we use the semigroup approach due to R.K. Miller (see [D-S2]) and then combine the Desch-Schappacher perturbation theorem from [D-S1] with our construction of extrapolation spaces from Section 2 above.

To that purpose we recall from Theorem 2.7 the definitions of the Banach space

$$Y := X_0 \times L^1(\mathbb{R}_+, X)$$

and of the generator $(\mathcal{A}, D(\mathcal{A}))$. In addition we define new operators $C : D(A) \to L^1(\mathbb{R}_+, X)$ by

$$Cx := a(\cdot)Ax$$

for $x \in D(A)$ and $\mathcal{B} : D(\mathcal{A}) \to Y$ by

$$\mathcal{B}\begin{pmatrix} x \\ f \end{pmatrix} := \begin{pmatrix} 0 \\ Cx \end{pmatrix}$$

for $\binom{x}{f} \in D(\mathcal{A})$. To this operator $\mathcal{B}$ we want to apply the mentioned Desch-Schappacher theorem and therefore need the following result.

LEMMA 4.2 *Let* $(\mathcal{A}, D(\mathcal{A}))$ *and* $(\mathcal{B}, D(\mathcal{B}))$ *be the operators on* Y *defined above. Then* $\mathcal{B} \in \mathcal{L}(D(\mathcal{A}), F(\mathcal{A}))$ *where* $F(\mathcal{A})$ *is the Favard class of the semigroup* $(G(t))_{t\ge 0}$ *generated by* $\mathcal{A}$.

Proof. For $\binom{x}{f} \in D(\mathcal{A})$ and $t > 0$ we have

$$\frac{1}{t}\|(G(t) - I)\mathcal{B}\begin{pmatrix} x \\ f \end{pmatrix}\|_Y = \frac{1}{t}\|\int_0^t T_{-1}(t-s)a(s)Ax\, ds\| + \frac{1}{t}\int_0^\infty |a(t+s) - a(s)|\|Ax\|\, ds\,.$$

Since $Ax \in X$ we deduce from Proposition 2.1 (ii) and (4.4) that

$$\frac{1}{t}\|\int_0^\infty T_{-1}(t-s)a(s)Ax\, ds\| \le \frac{M_1}{t}\int_0^\infty \|a(s)Ax\|\, ds \le M^3\|Ax\|(|a(0)| + V_0^\infty(a))\,. \tag{4.6}$$

From (4.5) it follows that

$$\frac{1}{t}\int_0^\infty |a(t+s) - a(s)|\|Ax\|\, ds \le V_0^\infty(a)\|Ax\|\,. \tag{4.7}$$

Since $f \in W^{1,1}(\mathbb{R}_+, X)$ we have $f(0) = -\int_0^\infty f'(s)\, ds$ and hence,

$$\|Ax\| \le \|Ax + f(0)\| + \|f'\|_{L^1(\mathbb{R}_+,X)} = \|\mathcal{A}\begin{pmatrix} x \\ f \end{pmatrix}\|_Y\,. \tag{4.8}$$

Combining these estimates we obtain a constant $c > 0$ independent of $\binom{x}{f} \in D(\mathcal{A})$ such that

$$\sup_{t>0} \frac{1}{t} \|(G(t) - I)\mathcal{B}\binom{x}{f}\|_Y \leq c \cdot \|\mathcal{A}\binom{x}{f}\|_Y ,$$

i.e., $\mathcal{B}\binom{x}{f} \in F(\mathcal{A})$. □

This lemma permits the following conclusion from the Desch-Schappacher perturbation theorem (see [D-S1], Example 3).

PROPOSITION 4.3 *The operator* $\mathcal{A} + \mathcal{B} : D(\mathcal{A}) \subset Y \to Y$ *is the generator of a strongly continuous semigroup* $(H(t))_{t\geq 0}$ *on* Y.

It remains to show that this semigroup yields the solutions of problem (4.1). To that purpose we state the following relation between two semigroups.

LEMMA 4.4 *Let* A *and* B *be generators of two strongly continuous semigroups* $(T_A(t))_{t\geq 0}$ *and* $(T_B(t))_{t\geq 0}$ *and assume that* $D(A) = D(B)$. *Then*

$$T_A(t)x = T_B(t)x + \int_0^t T_B(t-s)(A-B)T_A(s)x\,ds \tag{4.9}$$

for all $x \in D(A)$.

Proof. Setting $u_A(t) := T_A(t)x$ and $u_B(t) := T_B(t)x$ we have

$$\begin{aligned} (u_A(t) - u_B(t))' &= B(u_A(t) - u_B(t)) + Au_A(t) - Bu_A(t) \\ u_A(0) - u_B(0) &= 0 , \end{aligned}$$

and the assertion follows (see [Pa], p. 106). □

Our main result solves Equation (4.1) for $a \in BV(\mathbb{R}_+) \cap L^1(\mathbb{R}_+)$ and therefore extends a result of Grimmer-Prüss [G-P] to Hille-Yosida operators.

THEOREM 4.5 *Let* $A : D(A) \subset X \to X$ *be a Hille-Yosida operator on a Banach space* X *and take* $a \in BV(\mathbb{R}_+) \cap L^1(\mathbb{R}_+)$. *Then the Cauchy problem for the Volterra integrodifferential equation*

$$\begin{aligned} \dot{u}(t) &= Au(t) + \int_0^t a(t-s)Au(s)\,ds + f(t) , \quad t \geq 0 , \\ u(0) &= u_0 \end{aligned} \tag{4.10}$$

has a unique solution $u \in \mathcal{C}^1(\mathbb{R}_+, X) \cap \mathcal{C}^0(\mathbb{R}_+, D(A))$ *for each* $u_0 \in D(A)$ *and* $f \in W^{1,1}(\mathbb{R}_+, X)$ *satisfying*

$$Au_0 + f(0) \in \overline{D(A)}. \tag{4.11}$$

Proof. The assumption (4.11) says that $\binom{u_0}{f} \in D(\mathcal{A}) = D(\mathcal{A} + \mathcal{B})$. Therefore $t \mapsto U(t) := H(t)\binom{u_0}{f}$ is the unique function $U \in \mathcal{C}^1(\mathbb{R}_+, Y) \cap \mathcal{C}^0(\mathbb{R}_+, D(\mathcal{A}))$ satisfying

$$\begin{aligned} \dot{U}(t) &= (\mathcal{A} + \mathcal{B})U(t), \quad t \geq 0, \\ U(0) &= \binom{u_0}{f}. \end{aligned} \tag{4.12}$$

Set $U(t) =: \binom{x(t)}{F(t)}$. Since the graph norm on $D(\mathcal{A})$ is equivalent to that of $D(A) \times W^{1,1}(\mathbb{R}_+)$ (use (4.8)) we obtain $x(\cdot) \in \mathcal{C}^1(\mathbb{R}_+, X) \cap \mathcal{C}(\mathbb{R}_+, D(A))$ and from (4.12)

$$\dot{x}(t) = Ax(t) + \dot{F}(t)(0). \tag{4.13}$$

Let us now apply Lemma 4.4 to $\mathcal{A}$ and $\mathcal{A} + \mathcal{B}$. The second component of Equation (4.9) in this situation can then be written in the space $L^1(\mathbb{R}_+, X)$ as

$$\dot{F}(t) = S(t)f + \int_0^t S(t-s)Cx\,ds = S(t)f + \int_0^t S(t-s)a(\cdot)Ax(s)\,ds.$$

Since $\dot{F}(t), S(t)f \in W^{1,1}(\mathbb{R}_+, X)$ we obtain

$$\dot{F}(t)(0) = f(t) + \int_0^t a(t-s)Ax(s)\,ds,$$

hence, (4.13) yields (4.10). □

REMARK 4.6 Note that (4.11) is a necessary condition for the existence of a solution $u \in \mathcal{C}^1(\mathbb{R}_+, X) \cap \mathcal{C}^0(\mathbb{R}_+, \underline{D(\mathcal{A})})$ of problem (4.10) because for such a solution we deduce that $Au_0 + f(0) = \dot{u}(0) \in \overline{D(A)}$.

REFERENCES

[**Am**] Amann, H.: *Parabolic evolution equations in interpolation and extrapolation spaces,* J. Functional Anal. **78** (1988), 233-270.

[Br] Brézis, H.: "Opérateurs maximaux monotones et semigroupes de contractions dans les spaces de Hilbert", North-Holland 1972.

[B-B] Butzer, P.L., Berens, H.: "Semigroups of Operators and Approximation", Springer-Verlag 1967.

[DaP] DaPrato, G.: "Applications croissantes et équations d'évolutions dans les espaces de Banach", Academic Press 1976.

[DaP-G] DaPrato, G., Grisvard, P.: *Maximal regularity for evolution equations by interpolation and extrapolation,* J. Functional Ana. **58** (1984), 107-124.

[DaP-S] DaPrato, G., Sinestrari, E.: *Differential operators with non dense domain,* Annali Scuola Norm. Sup. Pisa **14** (1987), 285-344.

[D-S1] Desch, W., Schappacher, W.: *On relatively bounded perturbations of linear C_0-semigroups,* Annali Scuola Norm. Sup. Pisa **11** (1984), 327-341.

[D-S2] Desch, W., Schappacher, W.: *A semigroup approach to integrodifferential equations in Banach spaces,* J. Integral. Eq. **10** (1985), 99-110.

[Go] Goldstein, J.: "Semigroups of Linear Operators and Applications", Oxford Univ. Press 1985.

[G-P] Grimmer, R., Prüss, J.: *On linear Volterra equations in Banach spaces,* Comp. & Math. with Appl. **11** (1985), 189-205.

[H-P] Hille, E., Phillips, R.S.: "Functional Analysis and Semigroups", Amer. Math. Soc. Colloq. Publ. **31** Providence R.I. 1957.

[K-H] Kellermann, H., Hieber, M.: *Integrated semigroups,* J. Functional Analysis **84** (1989), 160-180.

[Na1] Nagel, R.: *Sobolev spaces and semigroups,* Semesterbericht Funktionalanalysis, Sommersemester 1983, Tübingen.

[Na2] Nagel, R.(ed.): "One-parameter Semigroups of Positive Operators", Lect. Notes Math. **1184**, Springer-Verlag 1986.

[Na3] Nagel, R.: *Toward a "Matrix Theory" for unbounded operator matrices,* Math. Z. **201** (1989) 57-68.

[Pa] Pazy, A.: "Semigroups of Linear Operators and Applications to Partial Differential Equations", Springer-Verlag 1983.

[Pr] Prüss, J.: "Evolutionary Integral Equations and Applications", Birkhäuser (to appear)

[vNe] van Neerven, J.M.A.M.: *The Adjoint of a Semigroup of Linear Operators,* Lect. Notes Math. (to appear).

[Wa] Walther, Th.: *Abstrakte Sobolev-Räume und ihre Anwendung auf die Störungstheorie für Generatoren von C_0-Halbgruppen,* Dissertation, Tübingen 1986.

Nonlinear Almost Periodic Actions of Semigroups

VŨ QUÔC PHÓNG *) Department of Mathematics, The University of Iowa, Iowa City, IA 52242, U.S.A.

0 INTRODUCTION

Let $\varphi(t)$, $(t \in \mathbb{R})$, be an equicontinuous flow on a compact metric space X. Then the following two well known statements completely describe the structure of the flow. First, X can be decomposed into a union of pairwise disjoint closed subsets which are invariant with respect to $\varphi(t)$, and on each such invariant set the flow $\varphi(t)$ is minimal. Recall that a flow is called *minimal* if there does not exist a closed nonempty proper subset which is invariant with respect to all the mappings of the flow. Thus, the analysis of equicontinuous flows is reduced to that of minimal flows. The second statement is about this latter case. Namely, if there is a minimal equicontinuous flow $\varphi(t)$ on X, then X is homeomorphic to a compact abelian group, and the transformations of the flow correspond to multiplications in the group by elements of some one-parameter subgroup. Conversely, starting from a compact abelian group and a one-parameter subgroup of this group, one can define in a natural way a flow acting on the group which is equicontinuous. Thus, the theory of equicontinuous flows is reduced to the theory of compact abelian groups.

It is worthwhile to mention the connection between these facts and the theory of almost periodic functions (in Bohr's sense) on the real line. If $f(t)$ is an almost periodic function, then the closure of the family of translates of f in the topology of uniform convergence on the line forms a compact topological space X, and the translates in X form a flow,

*) On leave from Institute of Mathematics, P.O. Box 631, Bo Ho, Hanoi, Viet Nam

which is isometric (therefore, equicontinuous) in the uniform topology on X. One can show that this flow is minimal, so that X is a compact abelian group. The real line $\mathbb{R}$ is embedded into X as a dense subgroup of X, and one can say that f can be extended continuously to a continuous function on X. The theory of compact abelian groups then provides us, e.g., with a result such that f is a uniform limit of linear combinations of characters of X, i.e. f is a uniform limit of linear combinations of exponents on $\mathbb{R}$ (Bohr's Approximation Theorem).

These classical results remain true with some modifications when $\mathbf{R}$ is replaced by $\mathbf{R}_+$, and the compactness assumption on X is replaced by the assumption that every positive (semi-) trajectory is relatively compact. Namely, if $\{\mathbf{a}(t) : t \geq 0\}$ is an equicontinuous semiflow of mappings of X into itself such that all trajectories $\{\mathbf{a}(t)x : t \geq 0\}$, $x \in X$, are relatively compact, then the set $\omega(x)$ of ω-limit points of the trajectory of x is an invariant compact abelian group, such that the restriction of the semiflow to $\omega(x)$ can be extended to a compact minimal flow. Two ω-limit sets, $\omega(x)$ and $\omega(y)$, either coincide or are disjoint. Thus, the attractor Ω of the semiflow (i.e. the set of all ω-limit points) is a union of pairwise disjoint invariant compact abelian groups $\omega(x_\alpha)$. There is a retraction of X onto Ω, which commutes with all mappings of the semiflow. (See e.g. [D-S], [DeR], [H1]).

Now let S be a general topological semigroup and $\mathbf{a}$ be an action of S on a topological space X. This means that we are given a family $\{\mathbf{a}(s) : s \in S\}$ of mappings of X into itself, parameterized by elements of the semigroup S, such that the following conditions hold:

i) $\mathbf{a}(st) = \mathbf{a}(s)\mathbf{a}(t)$ *for all* $s, t \in S$;

ii) The mapping $s \mapsto \mathbf{a}(s)x$ *is continuous from* S *to* X, *for all* $x \in X$.

For convenience, we also assume throughout this paper that S has a unit element e, and

iii) $\mathbf{a}(e) = I_X$, where I_X denotes the identity mapping of X.

An action $\mathbf{a}$ of a semigroup S on X is said to be *equicontinuous*, if the family of mappings $\{\mathbf{a}(s) : s \in S\}$ is equicontinuous. An action is said to be *almost periodic* if it is equicontinuous and every trajectory $\gamma(x) \equiv \{\mathbf{a}(s)x : s \in S\}$ is relatively compact. In particular, equicontinuous actions on compact spaces are automatically almost periodic.

In this paper, we apply the theory of compact topological semigroups (see e.g. [D-G], [L3]) to the study of almost periodic actions of topological semigroups on general topological spaces. While we are giving exposition for the general case of arbitrary semigroups, we emphasize the most important case of actions of abelian semigroups, for which a full generalisation of the theory of one-parameter semigroups (i.e. semiflows) is obtained. As a basic example, we will have in mind the case of actions of the semigroup $\mathbf{R}_+^m$, which is equivalent to consideration of m commuting semiflows. In that case, a suitable and natural ordering on $\mathbf{R}_+^m$ is used to describe the asymptotic behavior of the action. Moreover, using this ordering, the structure theorems will be established for a class of *asymptotically almost periodic actions*, which is more general (for $m \geq 2$) than almost periodic actions as defined above. Our approach also sheds some new light on the asymptotic behavior of one-parameter nonlinear semigroups (Proposition 6.5).

1 COMPACTIFICATION

Let X be a Hausdorff topological space, X^X be the set of all mappings of X into itself. In X^X we consider the product topology, i.e., the topology of pointwise convergence. If X is compact, then X^X is a compact space, by Tikhonov's Theorem. Moreover, X^X is a (multiplicative) semigroup with unit. Although the elements in X^X are not, as a rule, continuous mappings, this semigroup turned out useful in consideration of actions on X. Generally speaking, the multiplication in X^X is not a continuous operation; however, it possesses some continuity properties. Namely, right multiplication $\mathbf{a} \mapsto \mathbf{ab}$ is continuous for all $\mathbf{b} \in X^X$, but left multiplication $\mathbf{a} \mapsto \mathbf{ba}$ is continuous only for continuous elements $\mathbf{b}$ in X^X. (Indeed, if $\{\mathbf{a}_i : i \in I\}$ is a net in X^X which converges (pointwise) to $\mathbf{a}$, then: i) the net $\{\mathbf{a}_i\mathbf{b} : i \in I\}$ converges to $\mathbf{ab}$, and ii) the net $\{\mathbf{ba}_i : i \in I\}$ converges to $\mathbf{ba}$, if $\mathbf{b}$ is continuous). In particular, if $\mathbf{a}$ is an action of a semigroup S on X, then left multiplication by elements of the action is continuous.

If $\mathbf{a}$ is an action of a semigroup S on X, then the set $\gamma(x) \equiv \{\mathbf{a}(s)x : s \in S\}$ is called *trajectory* of the point $x \in X$. Let $I(\mathbf{a}) = \{\mathbf{a}(s) : s \in S\}$ – the image of the action, and $\beta_{\mathbf{a}} = \overline{I(\mathbf{a})}$ (the closure in the introduced topology in X^X).

Proposition 1.1. *If every trajectory of the action* $\mathbf{a}$ *is relatively compact, then* $\beta_{\mathbf{a}}$ *is a compact space, and algebraically a semigroup. In addition, if the action* $\mathbf{a}$ *is equicontinuous, then* $\beta_{\mathbf{a}}$ *is a compact semigroup (i.e. the multiplication in* $\beta_{\mathbf{a}}$ *is continuous)*

Proof. It is straightforward to check that $\beta_{\mathbf{a}}$ is a semigroup. Its compactness (as a set) follows from a standard argument using Tikhonov's Theorem (cf [D-G]). If the action $\mathbf{a}$ is equicontinuous, then all the mappings from $\beta_{\mathbf{a}}$ are continuous, and therefore, they are equicontinuous. This in turn implies that the multiplication in $\beta_{\mathbf{a}}$ is continuous, so that $\beta_{\mathbf{a}}$ is a compact semigroup. □

The semigroup $\beta_{\mathbf{a}}$ is naturally called *Bohr's compactum* of the action $\mathbf{a}$. It is well known from the theory of compact semigroups that there exists a smallest two-sided ideal [1)] in $\beta_{\mathbf{a}}$, which will be denoted by $K_{\mathbf{a}}$. This object is called *Suskevic kernel of the action* $\mathbf{a}$. For convenience of the reader, we collect below the well known facts about compact semigroups which will be used in the sequel, and will be referred to as *the Structure Theorem of Compact Semigroups.*

Assume, that S is a compact semigroup. We denote by $\mathcal{L} = \{I_\alpha : \alpha \in \mathcal{A}\}$ the family of all minimal left ideals, and by $\mathcal{R} = \{J_\delta : \delta \in \mathcal{B}\}$ the family of all minimal right ideals of S. Then, $G_{\alpha\delta} = I_\alpha \cap J_\delta$ are compact groups for every α and δ, which are mutually isomorphic. Moreover, there exists the smallest two-sided ideal K in S, which has the following representation

$$K = \cup\{I_\alpha : \alpha \in \mathcal{A}\} = \cup\{J_\delta : \delta \in \mathcal{B}\} = \cup\{G_{\alpha\delta} : \alpha \in \mathcal{A}, \ \ \delta \in \mathcal{B}\}.$$

[1)] Recall that a subset I (resp. J) of a semigroup S is called left ideal (resp., right ideal), if $SI \subset I$ (resp., $JS \subset J$). It is called two-sided ideal, if it is simultaneously a left ideal and a right ideal (see [D-G], [L3])

In particular, every compact semigroup contains an idempotent (the units of $G_{\alpha\delta}$). Every minimal left ideal I has form $I = Se$, and likewise every minimal right ideal J has form $J = eS$, where $e \in K$ is some idempotent. For an idempotent $e \in K$, the set $G = eSe$ is a compact group with the unit e. The pair of cardinal numbers (l, r), where $l = \operatorname{card} \mathcal{A}$ and $r = \operatorname{card} \mathcal{B}$, is called *type* of the action.

The following conditions are equivalent:
i) the semigroup S is left-amenable (i.e. there is a left invariant mean on $C(S)$);
ii) it has type $(l, 1)$, i.e. minimal right ideal is unique; and
iii) for every two idempotents $e_1, e_2 \in K$, $e_1e_2 = e_2$.

Analogously, the following are also equivalent:
i') the semigroup S is right-amenable (i.e. there is a right invariant mean on $C(S)$),
ii') it has type $(1, r)$, i.e. minimal left ideal is unique, and
iii') for every two idempotents $e_1, e_2 \in K, e_1e_2 = e_1$.

The (compact) semigroup S is amenable (i.e. there exists a two-sided invariant mean on $C(S)$) if and only if its Suskevic kernel is a compact group, or if and only if there exists a unique idempotent in S. For more details, see [D-G].

2 ALMOST PERIODIC ACTIONS

Throughout this section, $\mathbf{a}$ is an almost periodic action of a topological semigroup S on a topological space X. By Proposition 1.1, one can define the Suskevic kernel $K_{\mathbf{a}}$ of this action. Since the action is continuous, the set $\{\mathbf{b}x : \mathbf{b} \in \beta_{\mathbf{a}}\}$ is closed and therefore coincides with the closure of the trajectory $\overline{\gamma(x)}$.

A point $x \in X$ is called *recurrent,* if $\overline{\gamma(x)} = \overline{\gamma(\mathbf{b}x)}$, for each $\mathbf{b}$ in $\beta_{\mathbf{a}}$ (equivalently, x is recurrent if $y \in \overline{\gamma(x)}$ implies $x \in \overline{\gamma(y)}$ and $\overline{\gamma(x)} = \overline{\gamma(y)}$). Denote by $X_r \equiv X_r(\mathbf{a})$ the set of recurrent points of $\mathbf{a}$.

Proposition 2.1. *Let x be a point in X. The following are equivalent:*
i) x is recurrent;
ii) For every $\mathbf{b} \in \beta_{\mathbf{a}}$ there exists $\mathbf{c} \in \beta_{\mathbf{a}}$, such that $\mathbf{cb}x = x$;
iii) There exists an idempotent element $\mathbf{e}$ in $K_{\mathbf{a}}$ such that $\mathbf{e}x = x$.

We note that idempotents in $K_{\mathbf{a}}$ are continuous retractions of X.

Proof. The equivalence of i) and ii) follows from the noted above equality $\overline{\gamma(x)} = \{\mathbf{b}x : \mathbf{b} \in \beta_{\mathbf{a}}\}$: $x \in X_r$ if and only if $\overline{\gamma(x)} = \overline{\gamma(y)}$ for each $y \in \gamma(x)$, i.e. $\overline{\gamma(x)} = \overline{\gamma(\mathbf{b}x)}$ for every $\mathbf{b} \in \beta_{\mathbf{a}}$, which is equivalent to ii).

Assume that ii) holds. Let $\mathbf{b} \in K_{\mathbf{a}}$ and $\mathbf{c} \in \beta_{\mathbf{a}}$ be such that $\mathbf{cb}x = x$. Since $\mathbf{cb} \in K_{\mathbf{a}}$, the set

$$L = \{\mathbf{c} \in \beta_{\mathbf{a}} : \mathbf{c}x = x\} \cap K_{\mathbf{a}}$$

is not empty. Clearly, L is a compact semigroup, and therefore there is an idempotent element $\mathbf{e} \in L$. Thus, ii) implies iii).

Finally, assume that there exists an idempotent element $\mathbf{e} \in K_{\mathbf{a}}$ such that $\mathbf{e}x = x$. The set $J = \{\mathbf{be} : \mathbf{b} \in \beta_{\mathbf{a}}\}$ is a minimal left ideal. By the Structure Theorem for Compact

Semigroups, for a given $\mathbf{b} \in \beta_\mathbf{a}$ there exists $\mathbf{c}$ in J such that $\mathbf{cbe} = \mathbf{e}$. Therefore, $\mathbf{cb}x = \mathbf{cbe}x = \mathbf{e}x = x$. Thus, iii) implies ii) and the proof is completed. □

Next we introduce a relation on X, which is called *proximality*. We say that x and y are *proximal*, $x \sim y$, if for every neighborhood V of the diagonal in $X \times X$, there exists $s \in S$ such that $(\mathbf{a}(s)x, \mathbf{a}(s)y) \in V$. In general, proximality is not an equivalence relation.

Proposition 2.2. *Let x and y be elements of X. The following are equivalent:*
i) $x \sim y$;
ii) There exists some $\mathbf{b} \in \beta_\mathbf{a}$ such that $\mathbf{b}x = \mathbf{b}y$;
iii) There exists an idempotent element $\mathbf{e} \in K$ such that $\mathbf{e}x = \mathbf{e}y$.

Proof. The only nontrivial implication which need to be verified is ii) $\Rightarrow$ iii). But this follows from the fact that the set $\{\mathbf{b} \in \beta_\mathbf{a} : \mathbf{b}x = \mathbf{b}y\}$ is a compact semigroup and hence contains an idempotent. □

As was noted, proximality is not in general an equivalence relation, but we will see later that if there is only one idempotent in $K_\mathbf{a}$ (which is exactly the case when $K_\mathbf{a}$ is a group), then proximality is an equivalence relation, and X_r is a maximal invariant set on which the action $\mathbf{a}$ can be extended to an action of a group.

A subset $A \subset X$ is said to be *completely invariant*, if $\mathbf{a}(s)A = A$ for all $s \in S$. Following the terminology of the representation theory, we say that a compact completely invariant subset A of X is *quasi-isometric*, if restrictions $\mathbf{a}(s)|A$, $s \in S$, are invertible. It is clear that if A is a compact quasi-isometric subset, then $\mathbf{a}(s)|A$ are contained in an equicontinuous group of mappings. A point $x \in X$ is called *quasi-isometric*, if $\overline{\gamma(x)}$ is a quasi-isometric subset.

Lemma 2.3. *If A is a quasi-isometric subset, then $A \subset X_r$.*

Proof. Since A is compact, and $\mathbf{a}(s)|A$ are invertible, it follows that $\beta_\mathbf{a}|A$ is contained in a compact group of equicontinuous mappings. Thus, $\beta_\mathbf{a}|A$ is a group and the inclusion $A \subset X_r$ follows from Proposition 2.1. □

Let $\Omega \equiv \Omega(\mathbf{a})$ be a subset of X which is the union of all compact quasi-isometric subsets of the action $\mathbf{a}$. Clearly, Ω is a completely invariant subset, on which all the mappings of the action $\mathbf{a}$ are invertible.

Proposition 2.4. *If G is a compact group of equicontinuous mappings of X into itself, then X can be decomposed as a union of pairwise disjoint compact quasi-isometric subsets.*

Proof. Since G is compact, the closures of orbits are compact subsets. From Proposition 2.1 it is easy to see that the relation on X defined by "$x \approx y \Leftrightarrow y \in \overline{\gamma(x)}$" is an equivalence relation, and the required conclusion follows. □

Lemma 2.5. *If an action* $\mathbf{a}$ *is left-amenable and* $\mathbf{e}$ *is an idempotent in* $K_{\mathbf{a}}$, *then* $\mathbf{e}(X) \subset \Omega$.

Proof. Left-amenability implies that $K_{\mathbf{a}}$ is the unique minimal right ideal in $\beta_{\mathbf{a}}$ and, moreover, the set

$$G = \{\mathbf{be} : \mathbf{b} \in \beta_{\mathbf{a}}\} = \{\mathbf{be} : \mathbf{b} \in K\} \tag{2.1}$$

is simultaneously a minimal left ideal and a compact group with the unit $\mathbf{e}$. By Proposition 2.4, $\mathbf{e}(X)$ can be decomposed into a union of pairwise disjoint compact subsets X_α which are quasi-isometric with respect to G. From (2.1) it follows that every subset X_α is invariant for mappings from $\beta_{\mathbf{a}}$. Therefore, $\mathbf{e}(X) \subset \Omega$. □

Now we are in a position to prove the following theorems, which are extensions of results of De Leeuw and Glicksberg [D-G] to nonlinear actions.

Theorem 2.6. *Let* $\mathbf{a}$ *be an almost periodic action of a topological semigroup* S *on a topological space* X. *The following are equivalent:*

i) The semigroup $\beta_{\mathbf{a}}$ *is right-amenable;*

ii) For every pair of idempotents $\mathbf{e}_1, \mathbf{e}_2 \in K$, *we have* $\mathbf{e}_1\mathbf{e}_2 = \mathbf{e}_1$;

iii) Proximality is an equivalence relation which is invariant in the sense that if $x \sim y$, *then* $\mathbf{a}(s)x \sim \mathbf{a}(s)y$, $(s \in S)$.

Proof. The equivalence of i) and ii) follows from the Structure Theorem of Compact Semigroups. Assume that ii) holds. If $x \sim y$ and $y \sim z$, then, by Proposition 2.2, there exist idempotents $\mathbf{e}_1$, $\mathbf{e}_2$ in $K_{\mathbf{a}}$ such that $\mathbf{e}_1 x = \mathbf{e}_1 y$, $\mathbf{e}_2 y = \mathbf{e}_2 z$. From ii) it follows that $\mathbf{e}_1\mathbf{e}_2 y = \mathbf{e}_1\mathbf{e}_2 z$, or $\mathbf{e}_1 y = \mathbf{e}_1 z$. Hence, $\mathbf{e}_1 x = \mathbf{e}_1 z$, so that "$\sim$" is an equivalence relation. We show that this equivalence relation is invariant with respect to all $\mathbf{a}(s)$, $s \in S$. Let $x \sim y$, i.e., by Proposition 2.2, $\mathbf{e}x = \mathbf{e}y$ for some idempotent $\mathbf{e}$ in $K_{\mathbf{a}}$. The set

$$G = \{\mathbf{eb} : \mathbf{b} \in \beta_{\mathbf{a}}\}$$

is simultaneously a minimal right ideal and a compact group with the unit $\mathbf{e}$. Therefore, $\mathbf{ea}(s)x = \mathbf{ea}(s)\mathbf{e}x = \mathbf{ea}(s)\mathbf{e}y = \mathbf{ea}(s)y$, hence $\mathbf{a}(s)x \sim \mathbf{a}(s)y$, and the implication ii)$\Rightarrow$ iii) is proved.

Conversely, assume that iii) holds. First we show that if $x \sim y$, $x = \mathbf{e}x$ and $y = \mathbf{e}y$, where $\mathbf{e}$ is an idempotent in $K_{\mathbf{a}}$, then $x = y$. Sine "$\sim$" is invariant, it follows from Proposition 2.2 that for every $\mathbf{b} \in \beta_{\mathbf{a}}$ there exists $\mathbf{c} \in \beta_{\mathbf{a}}$ such that $\mathbf{cbe}x = \mathbf{cbe}y$. The set $J = \{\mathbf{be} : \mathbf{b} \in \beta_{\mathbf{a}}\}$ is a minimal left ideal, hence, by the theory of compact semigroups, there exists $\mathbf{d} \in \beta_{\mathbf{a}}$ such that $\mathbf{dcbe} = \mathbf{e}$. Therefore, $\mathbf{dcbe}x = \mathbf{e}x = x = \mathbf{dcbe}y = \mathbf{e}y = y$, i.e. $x = y$.

Now let $\mathbf{e}_1$, $\mathbf{e}_2$ be idempotents in K, and $x \in X$. We have, by Proposition 2.2, $x \sim \mathbf{e}_1 x$, hence $\mathbf{e}_1 x \sim \mathbf{e}_1\mathbf{e}_2 x$. By the above proved it follows that $\mathbf{e}_1 x = \mathbf{e}_1\mathbf{e}_2 x$, i.e. $\mathbf{e}_1\mathbf{e}_2 = \mathbf{e}_1$, so iii) implies ii). □

Theorem 2.7. *Let* $\mathbf{a}$ *be an almost periodic action of a topological semigroup* S *on* X. *The following are equivalent:*

i) The semigroup $\beta_{\mathbf{a}}$ *is left-amenable;*

ii) For every pair of idempotents $\mathbf{e}_1, \mathbf{e}_2$ *in* K, *we have* $\mathbf{e}_1\mathbf{e}_2 = \mathbf{e}_2$;

iii) $X_r = \Omega$.

Proof. Again, the equivalence i)⇔ ii) follows from the theory of compact semigroups. Assume that i)–ii) hold. By Lemma 2.3, $\Omega \subset X_r$, and by Propositions 2.1 and 2.5, $X_r \subset \Omega$. Thus, $X_r = \Omega$.

Conversely, assume that iii) holds. By definition of Ω, if $x \in \Omega$, then x is contained in a compact quasi-isometric subset. Since for every idempotent $\mathbf{e}$ in $K_{\mathbf{a}}$, its restriction to this subset is invertible, we have $\mathbf{e}x = x$. Thus, elements in Ω are fixed points for idempotents in $K_{\mathbf{a}}$. In particular, if $\mathbf{e}_1$ and $\mathbf{e}_2$ are arbitrary idempotents in $K_{\mathbf{a}}$ and x is in X, then $\mathbf{e}_2 x \in X_r = \Omega$, hence $\mathbf{e}_1\mathbf{e}_2 x = \mathbf{e}_2 x$. This means that $\mathbf{e}_1\mathbf{e}_2 = \mathbf{e}_2$, i.e. ii) holds. □

From Theorems 2.6 and 2.7 we obtain the following result.

Theorem 2.8. *Let* $\mathbf{a}$ *be an almost periodic action of a topological semigroup* S *on* X. *The following are equivalent:*

i) $\beta_{\mathbf{a}}$ *is amenable;*

ii) $K_{\mathbf{a}}$ *is a compact group;*

iii) There exists a unique idempotent in $K_{\mathbf{a}}$;

iv) Proximality is an invariant equivalence relation and $X_r = \Omega$.

We show that if the semigroup S is itself left-amenable, (i.e., by definition, there exists a left invariant mean in the space $C(S)$ of bounded continuous functions on S), and if $\mathbf{a}$ is an almost periodic action of S on X, then $\mathbf{a}$ automatically is left-amenable. The same statement is valid also for right-amenable semigroups. We quote the following result which is contained in [D-G].

Proposition 2.9. *Let* S *and* S' *be topological semigroups such that* S' *is compact, and let* $\varphi : S \to S'$ *be a continuous homomorphism with dense range,* $\tilde{\varphi} : C(S') \to C(S)$ *be defined by* $(\tilde{\varphi} f)(s) = f(\varphi(s))$. *Then* $C(S')$ *has a left- (respectively, right-, two-sided) invariant mean if and only if* $\tilde{\varphi}(C(S'))$ *has a left- (resp., right-, two-sided) invariant mean.*

Suppose now that $\mathbf{a}$ is an almost periodic action of S. Let $\mathcal{S} = I(\mathbf{a})$, $\mathcal{S}' = \overline{\mathcal{S}}$, and i an imbedding from $\mathcal{S}$ to $\mathcal{S}'$. It is easy to see that the linear operator $\tilde{i}$ is surjective. Therefore, by Proposition 2.9, $C(\mathcal{S}')$ has a left- (right-, two-sided) invariant mean if and only if $C(\mathcal{S})$ has a left- (resp., right-, two-sided) invariant mean.

Assume that there is a left- (respectively, right-, two-sided) invariant mean φ on $C(S)$. For each f in $C(\mathcal{S})$ we put $\hat{\varphi}(f) \equiv \varphi(\hat{f})$, where $\hat{f}(s) \equiv f(\mathbf{a}(s))$, and obtain in this way a left (resp., right, two-sided) invariant mean on $C(\mathcal{S})$. Thus, we have proved the following statement, which shows that amenability properties are inherited from the semigroup S to the action $\mathbf{a}$. Such results are typical in [D-G] for semigroups of linear operators (cf. also [Gr]).

Proposition 2.10. *Assume that a semigroup* S *is left- (right-, two-sided) amenable and* $\mathbf{a}$ *is an almost periodic action of* S *on* X. *Then the action* $\mathbf{a}$ *is left- (resp., right-, two-sided) amenable.*

Assume that the action $\mathbf{a}$ is left-amenable. From Theorem 2.8 it follows that all idempotents in $\beta_{\mathbf{a}}$ have the same range, and X_r coincides with Ω. Moreover,

$$X_r = \cap\{\mathbf{b}(X) : \ \mathbf{b} \in \beta_{\mathbf{a}}\}. \tag{2.2}$$

Indeed, since for each idempotent $\mathbf{e}$ in $K_{\mathbf{a}}$ the subset

$$G = \{\mathbf{be} : \ \mathbf{b} \in \beta_{\mathbf{a}}\}$$

is simultaneously a minimal left ideal and a compact group with unit $\mathbf{e}$, for every $\mathbf{b} \in \beta_{\mathbf{a}}$ there is $\mathbf{c}$ in G such that $\mathbf{bec} = \mathbf{e}$. Let x be in X_r. Then $x = \mathbf{e}x = \mathbf{bec}x \in \mathbf{b}(X)$, for all $\mathbf{b} \in \beta_{\mathbf{a}}$, hence

$$X_r \subset \cap\{\mathbf{b}(X) : \ \mathbf{b} \in \beta_{\mathbf{a}}\}. \tag{2.3}$$

Since the inclusion inverse to (2.3) is obvious, we have shown that (2.2) holds. Thus, we obtain the following result which may be regarded as a noncommutative version of Poincare's Recurrence Theorem (cf. [L1], [Wa]).

Corollary 2.11. *Assume that the action* $\mathbf{a}$ *is left-amenable and* $x \in X$. *Then the following are equivalent:*

i) x *is a quasi-isometric point;*
ii) x *is a recurrent point;*
iii) x *belongs to the range of every mapping* $\mathbf{b}$ *in* $\beta_{\mathbf{a}}$.

In other words, $\Omega = X_r = \cap\{\mathbf{b}(X) : \ \mathbf{b} \in \beta_{\mathbf{a}}\}$. *Moreover, there are nonexpansive retractions* $\mathbf{e}$ *from* X *onto* Ω *such that* $\mathbf{ebe} = \mathbf{be}$ *(they are idempotents in* $K_{\mathbf{a}}$*).*

We remark that if x is in Ω, then from (2.2) it follows that $\overline{\gamma(x)} = \{\mathbf{g}x : \ \mathbf{g} \in G\}$. Therefore, the relation "$\sim$ " introduced above is an equivalence relation on Ω. Thus, we have the following corollary.

Corollary 2.12. *The relation "$\sim$" is an equivalence relation on* Ω *and decomposes* Ω *into a partition of pairwise disjoint minimal invariant subsets.*

If $\mathbf{a}$ is amenable, then $K_{\mathbf{a}}$ is a compact group and, therefore, there exists a unique idempotent – the unit of $K_{\mathbf{a}}$ – which we denote by $\mathbf{p}$. Obviously, $\mathbf{p}$ is a continuous retraction which commutes with all $\mathbf{a}(s)$ (hence with all elements of $\beta_{\mathbf{a}}$). The image of $\mathbf{p}$ coincides with Ω. Thus, the following holds.

Theorem 2.13. *Assume that* $\mathbf{a}$ *is an amenable almost periodic action of a topological semigroup* S *on* X. *Then there exists a continuous retraction* $\mathbf{p}$ *from* X *onto* Ω *such that* $\mathbf{a}$ *commutes with all* $\mathbf{b} \in \beta_{\mathbf{a}}$.

In this case the set Ω, which can be defined as either the maximal quasi-isometric subset, or the set of recurrent points, is called *support* of the action $\mathbf{a}$. The following corollary is a direct consequence of Theorem 2.13.

Corollary 2.14. *If X is connected, then Ω also is connected.*

For a point x in X we let $\omega(x) = \cap\{\overline{\mathbf{b}(\gamma(x))} : \mathbf{b} \in \beta_{\mathbf{a}}\}$. Then $\omega(x)$ is compact for every $x \in X$ and $\Omega = \cup\{\omega(x) : x \in X\}$. It is natural to call $\omega(x)$ the *local support of the action* $\mathbf{a}$ *at* x.

Corollary 2.15. *The binary relation $x \sim y \Leftrightarrow x \in \omega(y)$ is an equivalence on Ω, which decomposes Ω into a union of pairwise disjoint local supports.*

These equivalence classes are called *ergodic components* of the action $\mathbf{a}$. $\mathbf{a}$ is called *ergodic*, if it has a unique ergodic component. In other words, ergodicity of $\mathbf{a}$ means topological transitivity on the support Ω. The action $\mathbf{a}$ is called *primitive*, if Ω consists of a single point. Clearly, a primitive action is ergodic. (Cf. [D-S], [Go], [G-P], [H-T], [La], [Li], [L1], [Mil], [Pt], [Ta], [V2], [We]).

3 ABELIAN CASE

The case of abelian semigroups deserves a separate consideration because, firstly, in this case the class of almost periodic actions can be extended to an essentially more general and more natural class of *asymptotically almost periodic* actions, for which the previous results hold, and secondly, in the abelian case one can introduce and study the notion of *attractor* of the action which is important from the point of view of dynamical systems (cf. [Hl]).

The definition of asymptotically almost periodic actions is based on the following natural quasi-ordering of S: we define $s \leq t$ for arbitrary s, t in S, if there exists an element $u \in S$ such that $t = u + s$ (here, and in what follows, we denote "multiplication operation" in the abelian semigroup S additively). Thus S is a directed set with respect to this quasi-order, so we can speak about such notions as convergence through S, and so on (see [Ke]). A typical situation occurs when S is a subsemigroup of a locally compact abelian group G such that $S - S = G$. In this case $s \leq t$ if and only if $t - s \in S$ (cf. [R2, p. 193]). Examples: $S = \mathbb{R}^m_+$, $G = \mathbb{R}^m$, or $S = \mathbb{Z}^m_+$, $G = \mathbb{Z}^m$. If f is a function from R^m_+ to X then $\lim_S f(s_1, s_2, ..., s_m) = y$ means

$$\lim_{\min\{s_1,...,s_m\}\to\infty} f(s_1, ..., s_m) = y.$$

An action $\mathbf{a}$ of S on X is called *asymptotically almost periodic*, if it is equicontinuous and the net $\{\mathbf{a}(s)\}_{s\in S}$ is asymptotically relatively compact in the topology of pointwise convergence, i.e. for each subnet s_α in S such that $s_\alpha \to \infty$ (through semigroup S) there exists a subnet s_β such that $\lim_\beta \mathbf{a}(s_\beta)x$ exists for each x in X. It is clear that if S is an abelian semigroup and $\mathbf{a}$ is an almost periodic action of S on X, then $\mathbf{a}$ is asymptotically almost periodic. If $S = \mathbb{R}_+$ or $S = \mathbb{Z}_+$, and $\mathbf{a}$ is asymptotically almost periodic, then $\mathbf{a}$ is almost periodic, so that the definition of asymptotic almost periodicity does not give anything new for $\mathbb{R}_+$ and $\mathbb{Z}_+$. However, if $S = \mathbb{R}^m_+$, or $S = \mathbb{Z}^m_+$ with $m \geq 2$, then, as can be easily seen, the class of asymptotically almost periodic actions is essentially more general.

Assume that $\mathbf{a}$ is an asymptotically almost periodic action of an abelian semigroup S on X, and $\beta_{\mathbf{a}}$ is the closure of $\{\mathbf{a}(s) : s \in S\}$ in the topology of pointwise convergence. Then $\beta_{\mathbf{a}}$ is a topological abelian semigroup, which is not compact in general. However, the following holds.

Proposition 3.1. *There exists a Suskevic kernel (the smallest ideal) $K_{\mathbf{a}}$ in $\beta_{\mathbf{a}}$ which is a compact group.*

Proof. Let $K_{\mathbf{a}}$ be the set of all ω-limit points (in the topology of pointwise convergence) of the net $\{\mathbf{a}(s)\}$. Then $K_{\mathbf{a}}$ is nonempty and compact. If $\mathbf{v} \in K_{\mathbf{a}}$, i.e. there is a net $s_\alpha \to \infty$ in S such that $\mathbf{v} = \lim_\alpha \mathbf{a}(s_\alpha)$, then $\mathbf{a}(s)\mathbf{v} = \lim_\alpha \mathbf{a}(s + s_\alpha)$, so that $\mathbf{a}(s)\mathbf{v} \in K$. Hence $K_{\mathbf{a}}$ is an ideal.

It remains to show that $K_{\mathbf{a}}$ is a group (and then it will automatically imply that $K_{\mathbf{a}}$ is the smallest ideal), and for this it is enough to show that for every $\mathbf{v}, \mathbf{v}' \in K$, there exists $\mathbf{u} \in K_{\mathbf{a}}$ such that $\mathbf{v}' = \mathbf{u}\mathbf{v}$. Let $\mathbf{v} = \lim_\alpha \mathbf{a}(s_\alpha)$, $\mathbf{v}' = \lim_\beta \mathbf{a}(s_\beta)$, where $s_\alpha \to \infty$, $s_\beta \to \infty$. We choose for each α an element β_α such that $s'_\alpha = s_{\beta_\alpha} \geq s_\alpha + s_\alpha$. Hence, there exists a net t_α in S such that $t_\alpha \to \infty$ and $s'_\alpha = s_\alpha + t_\alpha$. Since $\mathbf{a}$ is asymptotically almost periodic, one can assume that $\mathbf{a}(t_\alpha)$ converges to some $\mathbf{u}$ in $K_{\mathbf{a}}$. Thus, $\mathbf{v}' = \mathbf{u}\mathbf{v}$. □

Proposition 3.1 enables one to extend results of Section 2 to asymptotically almost periodic actions of abelian semigroups. In particular, for each point $x \in X$ one can define the local support $\omega(x)$, which now can be written in the following form

$$\omega(x) = \cap\{\mathbf{a}(s)(\overline{\gamma(x)}) : s \in S\} = \cap\{\mathbf{a}(s)(\overline{\gamma(x)}) : s \geq t\}, \quad (\forall t \in S).$$

From this formula it follows that $\omega(x)$ consists of ω-limit points of the trajectory $\gamma(x)$, i.e. $y \in \omega(x)$ if and only if there is a net $s_\alpha \to \infty$ such that $y = \lim_\alpha \mathbf{a}(s_\alpha)x$. The set

$$\Omega = \cup_{x \in X}\omega(x)$$

is called in this case *attractor* of the action $\mathbf{a}$. It is not difficult to see that Ω is indeed an attractor in the sense that every trajectory converges to Ω through S.

Theorem 3.2. *On each subset $\omega(x)$ one can define a binary operation, which makes $\omega(x)$ a compact abelian group.*

Proof. We define a group operation in $\omega(x)$ by putting $x_1x_2 = (\mathbf{v}_1\mathbf{v}_2)x$, if $x_1 = \mathbf{v}_1x$, $x_2 = \mathbf{v}_2x$, $(\mathbf{v}_1, \mathbf{v}_2 \in K_{\mathbf{a}})$. This definition is correct, since if $x_1 = \mathbf{v}'_1x$, $x_2 = \mathbf{v}'_2x$, then $(\mathbf{v}_1\mathbf{v}_2)x = \mathbf{v}_1\mathbf{v}'_2x = \mathbf{v}'_2\mathbf{v}_1x = \mathbf{v}'_2\mathbf{v}'_1x = (\mathbf{v}'_1\mathbf{v}'_2)x$. Clearly, $\omega(x)$ is an abelian group with respect to this operation. This is a topological group (i.e. the operation and its inverse are continuous) thanks to equicontinuity of the family $\{\mathbf{v}|\omega(x) : \mathbf{v} \in K_{\mathbf{a}}\}$. Since the mapping $\pi : K_{\mathbf{a}} \to \omega(x)$ defined by $\pi\mathbf{v} = \mathbf{v}x$, $(\mathbf{v} \in K_{\mathbf{a}})$, is continuous and surjective, $\omega(x)$ is a compact group. □

Corollary 3.3. (cf. [G-P], [Si2]). *The attractor Ω is a union of pairwise disjoint compact abelian groups.*

It is well known that every connected and locally connected finite-dimensional compact abelian group is a torus (see e.g. [Po]). Therefore, the following corollary holds.

Corollary 3.4. *Let X be finite dimensional and* $\mathbf{a}$ *be an asymptotically almost periodic action of an abelian semigroup S on X such that the closure of every trajectory of* $\mathbf{a}$ *is connected and locally connected. Then the attractor Ω is a union of fixed points and pairwise disjoint invariant tori.*

Corollary 3.4 points out a common principle of occurrence of invariant tori in classical dynamical systems (see e.g. [S-N]).

A nonempty closed invariant subset $M \subset X$ is called *local attractor* of the action $\mathbf{a}$, if every point of M is an ω-limit point of some trajectory $\gamma(x)$, and if M is an attracting set for some of its neighborhoods V, i.e. all ω-limit points of trajectories $\gamma(y)$, $y \in V$, are in M. It is easy to see from this definition that a set M is a local attractor if and only if it can be represented in the following form

$$M = \cup\{\omega(x) : x \in V\}.$$

In particular, if M is a single point, say x_0, then M is a local attractor means that it is an attractive fixed point for $\mathbf{a}$, i.e. all trajectories of points from a neighborhood of x_0 converge to x_0.

Theorem 3.5. *Assume that X is connected and M is a local attractor. Then $M = \Omega$.*

Proof. Assume that M attracts some of its neighborhoods V. Consider the domain of attraction to M, i.e. the following subset $Y = \{y \in X : \omega(y) \subset M\}$. Since $M \subset \Omega$ and $\omega(x) = \omega(\mathbf{a}(s)x)$ $(x \in \Omega, s \in S)$, it follows that Y is a completely invariant subset for $\mathbf{a}$ (i.e. Y and its complement are invariant sets). Therefore

$$\Omega = (\Omega \cap M) \cup (\Omega \cap (X \setminus Y)) = M \cap (\Omega \cap (X \setminus Y)).$$

Hence, Ω is a disjoint union of closed subsets. By Corollary 2.9, Ω is connected, it follows that one of these sets is empty, and one can easily see that $\Omega \cap (X \setminus Y) = \emptyset$. This implies $X \setminus Y = \emptyset$, i.e. $Y = X$, so $M = \Omega$. □

Corollary 3.6. *If X is connected and there exists an attractive fixed point, then all trajectories converge to this point (i.e. the action* $\mathbf{a}$ *is primitive).*

If $\mathbf{a}$ is an action of an abelian semigroup such that all trajectories $\{\mathbf{a}(s)x\}$ converge to the same fixed point x_0 for all $x \in X$, then $\mathbf{a}$ is also called *asymptotically stable*. Corollaries 3.3, 3.4, 3.6 and Theorem 3.5 for dissipative actions on compact metric spaces were obtained earlier by Lyubich [L1].

4 REPRESENTATIONS

If X is a topological vector space and $\mathbf{a}$ is an action of a semigroup S on X such that every $\mathbf{a}(s), s \in S$, is a continuous linear operator, then $\mathbf{a}$ is called *representation* of S [2]. For almost periodic representations the results of Sections 2 and 3 remain true.

[2] For representations, we will use the notation T instead of $\mathbf{a}$

More precisely, let T be an almost periodic representation of a topological semigroup S on a locally convex space X, X_r the set of recurrent vectors of T, Ω the closed subspace spanned by invariant subspaces on which T is equivalent to a unitary representation, and $X_0 = \{x \in X : 0 \in \overline{\gamma(x)}\}$.

Theorem 4.1. *The following are equivalent*

i) The representation T *is left-amenable;*

ii) For every pair of projections E_1, E_2 *in* K_T *we have* $E_1E_2 = E_2$*;*

iii) $X_r = \Omega$.

The following are also equivalent:

iv) The representation T *is right-amenable;*

v) For every pair of projections E_1, E_2 *in* K_T *we have* $E_1E_2 = E_1$*;*

vi) X_0 *is an invariant subspace.*

Theorem 4.1 shows, in particular, that if T is amenable, then X_0 and $X_r = \Omega$ are closed invariant subspaces, and X has a decomposition into direct sum $X = X_0 \oplus \Omega$. If X is a Banach space, de Leeuw and Glicksberg [D-G] and Lyubich [L3] have shown, using the classical Peter-Weil Theory, that Ω is spanned by *finite dimensional* invariant subspaces X_λ, such that $T|X_\lambda$ is equivalent to an irreducible unitary representation for each λ. We generalise these results to almost periodic representations on suitable locally convex spaces. To be short, we refer the reader to [D-G] and [L3] for details in the Banach space case.

Let U_λ be a (finite dimensional) irreducible unitary representation of a compact group G, χ_λ be its character, and n_λ the multiplicity of the representation U_λ. In order to apply the Peter-Weil Theory, we need to consider the Fourier transform of the function $\pi_\lambda = n_\lambda\chi_\lambda$ with respect to the representation T, i.e. the following integral

$$P_\lambda x = n_\lambda \int_G \chi_\lambda(g)T(g^{-1})x dg, \quad x \in X.$$

In order to make possible the integration within the space X, we have to impose a standard condition on X. Namely, X must satisfy the so-called *Krein-Smul'jan condition*: if $\Lambda \subset X$ is a compact set, then its closed convex hull also is compact. For details about the theory of vector integration the reader is referred to [Bo]. We also note that the Krein-Smul'jan condition was considered in the context of representation theory on locally convex spaces in [C-D]. Assume from now on that the space X satisfies the Krein-Smul'jan condition. Then, applying results of Sections 2 and 3 and the Peter-Weil theory we obtain the following result.

Theorem 4.2. *Let* T *be an amenable almost periodic representation of a semigroup* S *on a locally convex space* X. *Then* $X = X_0 \oplus \Omega$, *where* X_0 *and* Ω *are closed invariant subspaces,* Ω *is spanned by finite dimensional subspaces on which* T *is equivalent to an irreducible unitary representation. In particular, if* S *is abelian and* T *is asymptotically almost periodic, then* Ω *is spanned by eigenvectors corresponding to unitary eigenvalues of* T.

We consider some examples which will be needed later.

First, assume that Y is a Banach space and $C(X,Y)$ is the space of all bounded continuous functions defined on X and with values in Y. Endowed with the compact-open topology, $C(X,Y)$ is a locally convex space. Moreover, a standard argument involving Ascoli's Theorem shows that $C(X,Y)$ satisfies the Krein-Smul'jan condition. If $\mathbf{a}$ is an almost periodic action of a topological semigroup S on X, then $\mathbf{a}$ induces in a natural way a representation $T_{\mathbf{a}}$ of S by linear operators on $C(X,Y)$ by the following formula

$$(T_{\mathbf{a}}(s)f)(x) = f(\mathbf{a}(s)x), \quad s \in S, x \in X, f \in C(X,Y).$$

One can show easily that $T_{\mathbf{a}}$ is left- (resp., right- , two-sided) amenable if and only if $\mathbf{a}$ is left- (resp., right- , two-sided) amenable. If S is abelian and $\mathbf{a}$ is asymptotically almost periodic, then so is $T_{\mathbf{a}}$.

Now let X be a Banach space. Recall that X is *strictly convex*, if from $\|x+y\| = \|x\| + \|y\|$, $(x, y \in X)$, it follows that x and y are linearly dependent. A projection P in X is said to be *orthogonal*, if $\|P\| = 1$.

Lemma 4.3. *Let X be a strictly convex Banach space. Then every orthogonal projection in X is uniquely determined by its kernel.*

Proof. We show that $\mathrm{Im}P = B$ where we put $B = \{x \in X : \|x+y\| \geq \|x\| \;\; \forall y \in \mathrm{Ker}P\}$. Since $\|P\| = 1$, the inclusion $\mathrm{Im}\ P \subset B$ is obvious. It is also obvious that B is linear and closed. Let $x \in B$. We write $x = x_1 + x_2$ with $x_1 \in \mathrm{Im}P$, $x_2 \in \mathrm{Ker}P$. Then $\|x_1\| \leq \|x_1 + x_2\| \leq \|x_1 + x_2 + y\|$ for all y in $\mathrm{Ker}P$. Taking $y = -x_2$, we have $\|x_1 + x_2\| = \|x_1\|$. Since $x_1 - x_2$ also is in B, we have by the same argument that $\|x_1 - x_2\| = \|x_1\|$, which is possible only if $x_2 = 0$, thanks to strict convexity of X. □

Proposition 4.4. *If T is an almost periodic contraction representation of a semigroup S in a strictly convex Banach space X, then T is left-amenable.*

Proof. In the compact semigroup β_T there exists a minimal left ideal $\mathcal{I}$, and there is an idempotent (projection) E in $\mathcal{I}$. We show that there is exactly one idempotent in $\mathcal{I}$. Since $\mathcal{I}E$ is a left ideal, it follows $\mathcal{I}E = \mathcal{I}$. If F is another idempotent in $\mathcal{I}$, then there is $Q \in \mathcal{I}$ such that $QE = F$, therefore $FE = QEE = QE = F$, i.e. $FE = F$. In particular, $\mathrm{Ker}E = \mathrm{Ker}F$, and Lemma 4.3 shows that $E = F$. Thus, there is exactly one idempotent in $\mathcal{I}$, which implies by the Structure Theorem of Compact Semigroups that the minimal right ideal is unique, so β_T has type $(l, 1)$, hence T is left-amenable. □

A Banach space X is said to be *smooth*, if for each $x \in X$, $\|x\| = 1$, there exists a unique functional x^* such that $\|x^*\| = x^*(x) = 1$. It is not difficult to verify that X is smooth if and only if, for every x, $\|x\| = 1$, the set $\{y \in X : \|x + ty\| \geq \|x\| \;\; \forall t \in \mathbb{C}\}$ is a hyperplane. Recall that if X^* is strictly convex, then X is smooth. The converse holds for finite dimensional spaces, but fails in general for infinite dimensional Banach spaces (see [Di]).

Lemma 4.5. *If X is a smooth Banach space, then every orthogonal projection in X is uniquely determined by its range.*

Proof. Let P be an orthogonal projection, $B = \{y \in X : \|x + ty\| \geq \|x\|, \; \forall x \in \mathrm{Im}P, \; t \in \mathbb{C}\}$. Since $\|P\| = 1$, it follows $\mathrm{Ker}P \subset B$. Conversely, let y be in B. We write $y = y_1 + y_2$, where $y_1 \in \text{Im}P$, $y_2 \in \text{Ker}P$. Since y and y_2 are in B, and B is a hyperplane, it follows that y_1 also is in B, which is possible only if $y_1 = 0$, i.e. $y \in \text{Ker}P$. □

Proposition 4.6. *If T is an almost periodic contraction representation of a semigroup S in a smooth Banach space, then T is right-amenable.*

Proof. Let $\mathcal{J}$ be a minimal right ideal in β_T, and E be an idempotent in $\mathcal{J}$. By the same reason as in the proof of Theorem 4.3, $E\mathcal{J} = \mathcal{J}$, and if F is another idempotent in $\mathcal{J}$, then there is $Q \in \mathcal{J}$ such that $EQ = F$, hence $EF = EEQ = EQ = F$. In particular, $\text{Im}E = \text{Im}F$, which implies $E = F$ by virtue of Lemma 4.5. Therefore, the idempotent in $\mathcal{J}$ is unique, so that, again by the Structure Theorem of Compact Semigroups, the minimal left ideal in β_T is unique, or the representation T is right-amenable. □

From Propositions 4.4 and 4.6 it follows that if X is a strictly convex and smooth Banach space (for instance,if X is uniformly convex), then every representation of a topological semigroup S by linear contractions on X is amenable (cf. [D-G]).

5 FIXED POINTS

A point x in X is called *fixed point* of the action $\mathbf{a}$, if $\mathbf{a}(s)x = x$ $(s \in S)$. Let $F(\mathbf{a})$ denote the set of fixed points of $\mathbf{a}$. The existence of fixed points of almost periodic actions can be established by averaging the trajectories with respect to the Haar measure defined on the Suskevic kernel $K_\mathbf{a}$, or on a suitable subgroup of $K_\mathbf{a}$.

Suppose that the action $\mathbf{a}$ is left-amenable. Then for each idempotent $\mathbf{e}$ in $K_\mathbf{a}$ the set

$$G = \{\mathbf{be} : \ \mathbf{b} \in \beta_\mathbf{a}\} \tag{5.1}$$

is simultaneously a minimal left ideal and a compact group with unit $\mathbf{e}$. Let $d\mathbf{g}$ be the Haar measure on G and define a mapping $\mathcal{P} : X \to X$ by

$$\mathcal{P}x = \int_G \mathbf{g}x d\mathbf{g}. \tag{5.2}$$

Assume further that X is a closed convex subset in a Banach space and that the action $\mathbf{a}$ is *affine*, i.e. $\mathbf{a}(s)[\lambda x + (1-\lambda)y] = \lambda \mathbf{a}(s)x + (1-\lambda)\mathbf{a}(s)y$, $(x, y \in X, s \in S, 0 \le \lambda \le 1)$. Then, thanks to affineness of $\mathbf{a}(s)$,

$$\begin{aligned}\mathbf{a}(s)\mathcal{P}x = \mathbf{a}(s)\int_G \mathbf{g}x d\mathbf{g} &= \int_G \mathbf{a}(s)\mathbf{g}x d\mathbf{g} \\ = \int_G (\mathbf{a}(s)\mathbf{e})\mathbf{g}x d\mathbf{g} &= \int_G \mathbf{g}'x d\mathbf{g}'.\end{aligned}$$

Therefore, $\mathbf{a}(s)\mathcal{P}x = \mathcal{P}x$, $(s \in S)$, so we have proved the following theorem.

Theorem 5.1. *If* $\mathbf{a}$ *is a left-amenable affine action, then* $F(\mathbf{a}) \neq \emptyset$. *Moreover, the mapping* $\mathcal{P}$ *defined by (5.2) is a continuous retraction from* X *onto* $F(\mathbf{a})$.

Assume now that $\mathbf{a}$ is right-amenable. Then for each idempotent $\mathbf{e}$ the set

$$G = \{\mathbf{eb} : \mathbf{b} \in \beta_{\mathbf{a}}\} \tag{5.3}$$

is simultaneously a minimal right ideal and a compact group with unit $\mathbf{e}$. Again, let $d\mathbf{g}$ denote the Haar measure on G, and $\mathcal{P}$ be a mapping from X into itself defined by

$$\mathcal{P}x = \int_G \mathbf{g}x d\mathbf{g}. \tag{5.4}$$

From (5.4) it follows that

$$\begin{aligned} \mathcal{P}\mathbf{a}(s)x &= \int_G \mathbf{g}\mathbf{a}(s)x d\mathbf{g} = \int_G \mathbf{g}(\mathbf{e}\mathbf{a}(s))x d\mathbf{g} \\ &= \int G\mathbf{g}'x d\mathbf{g}' = \mathcal{P}x, \end{aligned}$$

i.e. $\mathcal{P}\mathbf{a}(s) = \mathcal{P}(s \in S)$. Thus, the following holds.

Theorem 5.2. *If* $\mathbf{a}$ *is right-amenable and* $\mathcal{P}$ *is given by (5.4), then* $\mathcal{P}\mathbf{a}(s)x = \mathcal{P}x$, $(x \in X, s \in S)$.

From Theorem 5.2 one can also derive some results on fixed points. For this, we have to put some more conditions on X and $\mathbf{a}$. Namely, we assume that X is a closed convex subset in a Banach space and the action $\mathbf{a}$ is affine. We say, that $\mathbf{a}$ *satisfies condition* (A), if the following holds :

For each x_1, x_2 *in* X *and* s *in* S *and for each neighborhood* V *of zero in the Banach space, the exists* $s' \in S$ *such that* $\mathbf{a}(s's)x_1 - x_1 \in V$ *and* $\mathbf{a}(s's)x_2 - x_2 \in V$.

Corollary 5.3. *If the action* $\mathbf{a}$ *is right-amenable, affine, and satisfies condition* (A), *then the set* $F(\mathbf{a})$ *is non-empty, closed, convex and the mapping* $\mathcal{P}$ *defined by (5.4) is a retraction from* X *onto* $F(\mathbf{a})$.

Proof. Let $\mathbf{b}_1$, $\mathbf{b}_2$ be arbitrary elements in $\beta_{\mathbf{a}}$ and $x \in X$. We put $x_1 = \mathbf{b}_1\mathcal{P}x$, $x_2 = \mathbf{b}_2\mathcal{P}x$. By condition (A), for each neighborhood of zero V there exists $\mathbf{b}'$ in $\beta_{\mathbf{a}}$ such that

$$\mathbf{b}'\mathbf{e}\mathbf{b}_1\mathcal{P}x - \mathbf{b}_1\mathcal{P}x \in V, \quad \mathbf{b}'\mathbf{e}\mathbf{b}_2\mathcal{P}x - \mathbf{b}_2\mathcal{P}x \in V. \tag{5.5}$$

Since $\mathbf{eb}_1 \in G$, affineness and (5.4) imply that $\mathbf{eb}_1\mathcal{P}x = \mathcal{P}x$, $\mathbf{eb}_2\mathcal{P}x = x$. Therefore, by virtue of (5.5) we have

$$\mathbf{b}'\mathcal{P}x - \mathbf{b}_1\mathcal{P}x \in V, \quad \mathbf{b}'\mathcal{P}x - \mathbf{b}_2\mathcal{P}x \in V,$$

which implies that $\mathbf{b}_1\mathcal{P}x - \mathbf{b}_2\mathcal{P}x \in V + V$. Since V is an arbitrary neighborhood of zero, the latter means that $\mathbf{b}_1\mathcal{P}x = \mathbf{b}_2\mathcal{P}x$, so that $\mathcal{P}x$ is a fixed point. That $\mathcal{P}$ is a retraction follows from its construction. □

We note that Corollary 5.3 is a generalization of Kakutani's Fixed Point Theorem (see e.g. [R1]).

We say that the action $\mathbf{a}$ *satisfies condition (B)*, if for an arbitrary neighborhood of zero V, $x \in X$, $s, s' \in S$, there exists $s'' \in S$ such that

$$\mathbf{a}(ss'')x - \mathbf{a}(s's)x \in V. \tag{5.6}$$

Corollary 5.4. *If the action* $\mathbf{a}$ *is right-amenable, affine, and satisfies condition* (B)*, then* $F(\mathbf{a})$ *is non-empty, closed, convex, and the mapping* $\mathcal{P}$ *defined by (5.4) is a retraction from* X *onto* $F(\mathbf{a})$.

Proof. Let $\mathbf{b} \in G$, $\mathbf{b}' \in \beta_{\mathbf{a}}$, $x \in X$ be arbitrary. From (5.6) it follows that there exists $\mathbf{b}'' \in \beta_{\mathbf{a}}$ such that $\mathbf{b}\mathbf{b}''\mathcal{P}x - \mathbf{b}'\mathbf{b}\mathcal{P}x \in V$. Since G is a right ideal, $\mathbf{b}\mathbf{b}'' \in G$. Thanks to affineness, we have

$$\mathbf{g}\mathcal{P}x = \mathcal{P}x \quad (\mathbf{g} \in G, \quad x \in X),$$

which implies $\mathbf{b}\mathbf{b}''\mathcal{P}x = \mathcal{P}x$. Therefore $\mathbf{b}\mathbf{b}''\mathcal{P}x - \mathbf{b}'\mathbf{b}\mathcal{P}x = \mathcal{P}x - \mathbf{b}'\mathcal{P}x \in V$, for all $x \in X$. Hence $\mathbf{b}'\mathcal{P}x = \mathcal{P}x$ for all $\mathbf{b}'$. $\square$

Corollary 5.4 is a version of the Markov-Kakutani Fixed Point Theorem, where commutativity is replaced by the less restrictive condition (5.6) (cf. [Yo]).

For an amenable action $\mathbf{a}$, all its fixed poins are contained in Ω. The existence of fixed points in the amenable case can be established by averaging the trajectories with respect to the Haar measure on the Suskevic kernel, as is shown in the following corollary.

Corollary 5.5. *Let* X *be a closed convex subset of a Banach space* E *and* $\mathbf{a}$ *be an amenable almost periodic action of* S *on* X*, such that the restriction of* $\mathbf{a}$ *to* Ω *is affine. Then the set* $F(\mathbf{a})$ *of fixed points is non-empty, closed, convex, and the mapping* $\mathcal{P}$ *defined by*

$$\mathcal{P}x = \int_{K_{\mathbf{a}}} \mathbf{v}x d\mathbf{v}$$

is a continuous affine retraction from X *onto* $F(\mathbf{a})$.

Let X be a convex subset in a Banach space E. We say, that the action $\mathbf{a}$ of S on X is of *non-contractive type*, if for arbitrary x, y in X, $x \neq y$, we have $0 \notin \overline{\{\mathbf{a}(s)x - \mathbf{a}(s)y : s \in S\}}$. This definition is a version of one due to Ryll-Nardzewskii [Ry].

Proposition 5.6. *If* $\mathbf{a}$ *is an almost periodic action of non-contractive type, then* $K_{\mathbf{a}}$ *is a compact group (so that* $\mathbf{a}$ *is amenable).*

Proof. Let $\mathbf{e}$ be an idempotent in $\beta_{\mathbf{a}}$. From the condition it follows that if $x \neq y$, then $\mathbf{e}x \neq \mathbf{e}y$. Since $\mathbf{e}\mathbf{e}x = \mathbf{e}x$, it follows that $\mathbf{e}x = x$ for every $x \in X$, so that $\mathbf{e}$ is an identity mapping of X onto itself, i.e. there is a unique idempotent in $\beta_{\mathbf{a}}$, which implies that $K_{\mathbf{a}}$ is a compact group. $\square$

Corollary 5.7. *Let* $\mathbf{a}$ *be an almost periodic affine action of non-contractive type. Then for every* $x \in X$ *there exists a unique fixed point which belongs to the closed convex hull of the trajectory* $\gamma(x)$. *Moreover,* $\overline{\gamma(x)}$ *is a compact group, and* $\mathcal{P}x$ *is obtained by averaging points on* $\overline{\gamma(x)}$ *with respect to the Haar measure on this orbit.*

Affine actions arise, e.g., under consideration of restrictions of bounded linear representations on closed convex invariant subsets. Thus, there is a possibility of a synthesis of the preceeding results with the theory of almost periodic representations (see Section 4). In this way, one can obtain some essential extensions of the known results of L. Alaoglu, G. Birkhoff, D. Day, and others.

Theorem 5.8. *Let* T *be a weakly almost periodic representation of an arbitrary semigroup* S *by contractions on a stricly convex Banach space* E *and let* X *be a closed convex subset which is invariant with respect to* T. *Let* $\mathbf{a}$ *be an action of* S *on* X *which is the restriction of* T *to* X. *Then the set* $F(\mathbf{a})$ *of fixed points of* $\mathbf{a}$ *is non-empty, closed, convex, and there exists a continuous affine retraction from* X *onto* $F(\mathbf{a})$.

Proof. By Mazur's Theorem, X is weakly closed. The action $\mathbf{a}$ is almost periodic (in the topology in X which is induced by the weak topology). Since E is strictly convex, the representation T is left-amenable. From this it follows that the action $\mathbf{a}$ also is left-amenable, by Proposition 4.3, and the conclusion follows from Theorem 5.1. □

The following is a direct consequence of Theorem 5.8.

Corollary 5.9. *If* T *is a representation of* S *by contractions on a strictly convex, reflexive, Banach space and* X *is a closed convex invariant subset, then the set* $F(\mathbf{a})$ *of fixed points of* $\mathbf{a}$ *is non-empty, closed, convex, and there exists a continuous affine retraction from* X *onto* $F(\mathbf{a})$.

In particular, Corollary 5.9 holds when E is a uniformly convex Banach space (L. Alaoglu- G. Birkhoff's Theorem).

A representation T is called *doubly bounded*, if there are $c > 0$, $C > 0$ such that $c\|x\| \le \|T(s)x\| \le C\|x\| \quad (s \in S,\ x \in E)$.

Theorem 5.10. *Let* T *be a doubly bounded representation of a semigroup* S *on a Banach space* E, *and let* X *be a weakly compact convex invariant subset. Assume that* $\mathbf{a}$ *is an action on* X *which is a restriction of* T *to* X. *Then the set* $F(\mathbf{a})$ *of fixed points of* $\mathbf{a}$ *is non-empty, closed, convex, and there exists a continuous affine retraction from* X *onto* $F(\mathbf{a})$.

The proof of Theorem 5.10 follows from Proposition 5.6. By the same reasoning one obtains the following result.

Theorem 5.11. *Let* T *be a bounded representation of a left-amenable semigroup* S *in a Banach space* E, *and let* X *be a weakly compact convex invariant set. Then the conclusion of Theorem 5.10 holds.*

For various results on fixed points of a nonexpansive mapping the reader is referred to [Bd], [Br], [D-M], [Ki] and references therein.

6 DISSIPATIVE ACTIONS

An action $\mathbf{a}$ of a semigroup S on a metric space is called *dissipative*, if every mapping $\mathbf{a}(s)$, $(s \in S)$, is non-expansive, i.e. $d(\mathbf{a}(s)x, \mathbf{a}(s)y) \leq d(x, y)$, $(x, y \in X,\ s \in S)$, where d denotes the metric on X. It is clear that every dissipative action is equicontinuous, so that a dissipative action is almost periodic if and only if every trajectory is relatively compact. On the other hand, given an almost periodic action $\mathbf{a}$ on a compact metric space X, one can introduce an equivalent metric in which $\mathbf{a}$ is dissipative.

For dissipative almost periodic actions the previous results remain true with the following concretizations. The maximal quasi-isometric subset is in this case the maximal invariant subset on which the action $\mathbf{a}$ is isometric (i.e. all mappings $\mathbf{a}(s)|\Omega$ are isometries). If $\mathbf{a}$ is left-amenable, then Ω coincides with the set of recurrent points and there is a nonexpansive retraction from X onto Ω (moreover, every idempotent in $K_{\mathbf{a}}$ is such a retraction). In the abelian case, every asymptotically almost periodic action has a Suskevic kernel $K_{\mathbf{a}}$ which is a compact abelian group, Ω coincides with the attractor of the action, and there is a non-expansive retraction from X onto Ω (namely, the unit of $K_{\mathbf{a}}$ is the retraction). If, moreover, X is a closed convex subset of a strictly convex Banach space E and $\mathbf{a}$ is a dissipative almost periodic action on X, then the condition of affineness of $\mathbf{a}$ in Theorem 5.1 and Corollaries 5.3–5.4. is superfluous. This can be achieved through the following two lemmas.

Lemma 6.1. (cf. [D-S], [L1]). *If X is a convex subset of a strictly convex Banach space E and $\mathbf{e}$ is a non-expansive retraction of X, then the image of $\mathbf{e}$ is convex.*

Proof. Let Ω denote the image of X under $\mathbf{e}$ and x, y be arbitrary points in Ω. Consider the image of the interval $[x, y] = \{\lambda x + (1-\lambda)y,\ 0 \leq \lambda \leq 1\}$ under action of $\mathbf{e}$. It is a rectifiable curve in X with endpoints at x and y. Since $\mathbf{e}$ is nonexpansive, it follows that the length of this curve is less than or equal to $\|y - x\|$. Let $z = \lambda x + (1-\lambda)y$ be an arbitrary point in $[x, y]$. Then $\|x - \mathbf{e}z\| + \|\mathbf{e}z - y\| \leq \|x - z\| + \|z - y\| = \|x - [\lambda x + (1-\lambda)y]\| + \|\lambda x + (1-\lambda)y - y\| = \|x - y\|$. Since E is strictly convex, it follows that $\mathbf{e}z \in [x, y]$ and $\|\mathbf{e}z - \mathbf{e}x\| = \|z - x\|$. But then $\mathbf{e}z = \lambda \mathbf{e}x + (1-\lambda)\mathbf{e}y = \lambda x + (1-\lambda)y = z$, i.e. $z \in \Omega$. □

Lemma 6.2. (see [B-M]). *Every isometry of a convex compact set is affine.*

Let $\mathbf{a}$ be a left-amenable, dissipative, almost periodic action of a semigroup S on X and let G be the group defined by (5.1). Then for every $\mathbf{u}$ in G the restriction $\mathbf{u}|\Omega$, being an isometry of a convex compact set, is affine by Lemma 6.2. Thus, the following theorem holds.

Theorem 6.3. *Assume that $\mathbf{a}$ is a left-amenable, dissipative, almost periodic action of a semigroup S on a closed, convex subset of a strictly convex Banach space E. Then the set $F(\mathbf{a})$ of fixed points of $\mathbf{a}$ is nonempty, closed, convex, and there exists a nonexpansive affine retraction $\mathcal{P}$ from X onto $F(\mathbf{a})$ which is defined by (5.2).*

REMARK. The condition of strict convexity of E is essential for Ω to be convex, as is shown by the following example, (see [L1]).

Consider $\mathbb{R}^2$ with the norm $\|(\psi,\zeta)\| = \max(|\psi|,|\zeta|)$. Let $\zeta = f(\psi)$, $(0 \le \psi \le 1)$ be an arbitrary nonlinear function satisfying the condition $|f(\psi_1)-f(\psi_2)| \le |\psi_1-\psi_2|$, $f(0)=1$. Put $P(\psi,\zeta) = (\psi, f(\psi))$, $(0 \le \psi,\zeta \le 1)$. Then $P^2 = P$, P is non-expansive, and the set of fixed points of P coincides with the graph of f. This set is the attractor of the trivial action of the semigroup $S = \{P\}$.

The following result is dual to Theorem 6.3 and follows from Corollaries 5.3 - 5.4.

Theorem 6.4. *Let* $\mathbf{a}$ *be a right-amenable, dissipative, almost periodic action of a semigroup* S *on a closed convex subset* X *of a strictly convex Banach space* E, G *be a group defined by (5.3), and* $\mathcal{P}$ *a mapping from* X *into itself defined by (5.4). Then* $\mathcal{P}$ *is a non-expansive affine retraction such that* $\mathcal{P}\mathbf{a}(s)x = \mathcal{P}x$ *for all* $s \in S$, $x \in X$. *If, in addition,* $\mathbf{a}$ *satisfies condition* (A) *or* (B), *then* $\mathcal{P}$ *is a retraction from* X *onto* $F(\mathbf{a})$.

As a new application, we prove the following theorem on stability of one-parameter semigroups of nonlinear contractions (see [L-V], [A-B] and [B-V] for stronger results of this type for linear semigroups).

Let X be a closed convex subset of a real strictly convex Banach space E and $\mathbf{a}(t)$, $t \ge 0$, be a one-parameter semigroup of nonlinear contractions on X, with generator A. We assume that every trajectory of $\mathbf{a}(t)$ is bounded (since $\mathbf{a}(t)$ is dissipative, this is equivalent to boundedness of some trajectory). We say that the semigroup $\mathbf{a}(t)$ is *asymptotically stable* if there is a point x_0 such that $\mathbf{a}(t)x$ converges to x_0, as $t \to \infty$, for every $x \in X$. In other words, $\mathbf{a}(t)$ is asymptotically stable if and only if all solutions of the differential equation

$$x'(t) \in Ax(t), \quad t \ge 0,$$

converge to the same point x_0 (see e.g. [Cle]).

Proposition 6.5. *Assume that* X *is a convex subset of* $\mathbb{R}^n$ *with strictly convex norm,* A *is a generator of a nonlinear contraction semigroup* $\mathbf{a}(t)$, $t \ge 0$, *on* X. *Then the following are equivalent:*

(i) The semigroup $\mathbf{a}(t)$ *is asymptotically stable;*

(ii) There are no x_1, x_2, y_1, y_2 *in* X *and real numbers* α,β *and* λ, $\alpha > 0, \beta > 0$ *such that* $\alpha(Ax_1 - Ax_2) = -\lambda\beta(y_1 - y_2)$, *and* $\beta(Ay_1 - Ay_2) = \lambda\alpha(x_1 - x_2)$.

Proof. Let Ω be the attractor of $\mathbf{a}$ and L be the linear manifold which is spanned by Ω. Since X is strictly convex, Ω is convex by Lemma 6.1 (and Theorem 2.13), so L coincides with the set of vectors of the form $z = \alpha(x-y)$, $x, y \in \Omega, \alpha > 0$. It is easy to see that $\mathbf{a}(t)$ is asymptotically stable if and only if Ω consists of a single point x_0, or equivalently, $L = \{0\}$. Define, as in [D-S] or [L2],

$$T(t)z = \alpha(\mathbf{a}(t)x - \mathbf{a}(t)y), \quad (t \ge 0). \tag{6.1}$$

Since $\mathbf{a}(t)|\Omega$ is affine, this definition is correct, and $T(t) : L \to L$ is a linear isometric operator, and it is also easy to see that $T(t)$ is strongly continuous, so $T(t)$ is an isometric C_0-semigroup on L, and since L is finite dimensional, $T(t)$ extends to an isometric group.

Now we take the complexification $\widetilde{L} = L + iL$ of L and extend $T(t)$ to $\widetilde{T}(t)$ on $\widetilde{L}$ in a natural way. It is clear that the space $\widetilde{L}$ is equal to $\{0\}$ if and only if there do not exist $\lambda \in \mathbb{R}$ and $z = x + iy \in \widetilde{L}$ such that

$$\widetilde{T}(t)z = e^{i\lambda t}z, \quad \forall t \geq 0. \tag{6.2}$$

We can write $x = \alpha(x_1 - x_2)$, $y = \beta(y_1 - y_2)$, for some x_1, x_2, y_1, y_2 in Ω and $\alpha > 0$, $\beta > 0$. Then from (6.1) and (6.2) it follows

$$\alpha(\mathbf{a}(t)x_1 - \mathbf{a}(t)x_2) = \alpha(\cos\lambda t)(x_1 - x_2) - \beta(\sin\lambda t)(y_1 - y_2),$$
$$\beta(\mathbf{a}(t)y_1 - \mathbf{a}(t)y_2) = \alpha(\sin\lambda t)(x_1 - x_2) + \beta(\cos\lambda t)(y_1 - y_2), \quad \forall t \geq 0.$$

The conclusion of Proposition 6.5 now follows from general properties of contraction semigroups (see e.g [Cle]). □

For related and further results on the asymptotic behavior of contraction semigroups (or iterates of a non-expansive mapping) see [Br], [D-S], [G-P], [Hr2], [Hi], [H-T], [I-R1-2], [L1-2], [Miy], [N-R], [Pz1-4], [R-S], [Si], [V1-2] .

7 ERGODIC THEOREMS

Let $\mathbf{a}$ be an amenable almost periodic (or asymptotically almost periodic, if S is abelian) action of a semigroup S on a topological space X. Let Y be a Banach space and $C(X,Y)$ be the space of bounded continuous functions from X to Y, endowed with the compact-open topology.

With the action $\mathbf{a}$ we associate a representation $T \equiv T_{\mathbf{a}}$ of S by continuous linear operators on $C(X,Y)$, as defined in Section 4. Recall that T is an amenable almost periodic (resp., asymptotically almost periodic) representation. Let K_T denote the Suskevic kernel of T.

By Corollary 5.5, the operator

$$Q = \int_{K_T} VfdV, \quad (dV \text{ is the Haar measure on } K_T),$$

is a linear projection onto the subspace of functions invariant with respect to T.

Let $\mathcal{B}$ be another Banach space and ϕ be a continuous linear operator from $C(X,Y)$ to $\mathcal{B}$, which is invariant with respect to $\mathbf{a}$, i.e. $\phi(f) = \phi(f \circ \mathbf{a}(s))$ $(s \in S)$. Then it is easy to see that the following simple relation holds

$$\phi(f) = \phi(Qf), (f \in C(X,Y)). \tag{7.1}$$

We show how one can derive from (7.1) ergodic theorems for actions of some suitable abelian semigroups. For the case of one-parameter semigroups, the reader is referred to [G-P], [R-S] (see also [Kr] and references therein). We also note that multi-parameter ergodic theorems may have important applications in the asymptotic behavior of solutions of multi-dimensional differential equations (cf. [Sh]).

Let S be a subsemigroup of a locally compact abelian group G such that S is measurable with respect to the Haar measure on G and with non-empty interior S°. Let S^* be the space of continuous bounded characters of S; thus S^* consists of the non-zero continuous bounded homomorphisms of S into the multiplicative semigroup $\mathbb{C}$. Let

$$S_u^* = \{\chi \in S^* : |\chi(s)| = 1 \text{ for all } s \in S\}.$$

Note that S satisfies Følner's condition: there is a net (Ω_α) of compact subsets of S such that

$$\lim_\alpha \frac{|(\Omega_\alpha + t) \bigtriangleup \Omega_\alpha|}{|\Omega_\alpha|} = 0 \tag{7.2}$$

uniformly for t in compact subsets of S, where $|\Omega|$ is the Haar measure of Ω and $\bigtriangleup$ denotes the symmetric difference. The locally compact abelian semigroup S° satisfies the condition (see [Gr], [H-R]) and it follows easily that S satisfies it. Moreover, if S is σ-compact, then the net may be chosen to be a sequence. Since unitary characters of compact abelian groups are mutually orthogonal (see e.g. [H-R], [L3]), a standard argument using Bohr's compactification shows that the net Ω_α satisfies the following condition

$$\lim_\alpha \frac{1}{|\Omega_\alpha|} \int_{\Omega_\alpha} \chi(s) ds = 0 \quad \text{for all } \chi \in S_u^*, \quad \chi \neq 1. \tag{7.3}$$

We fix an element $x \in X$ and let

$$\phi_x(f) \equiv \lim_\alpha \frac{1}{|\Omega_\alpha|} \int_{\Omega_\alpha} f(\mathbf{a}(s)x) ds. \tag{7.4}$$

We show that the limit in (7.4) exists for every function f in $C(X,Y)$. Let $C(X,Y) = \mathcal{B}_0 \oplus \mathcal{B}_1$ be the decomposition as in Theorem 4.2. First assume that $f \in \mathcal{B}_0$. Then $T(s)f$ converges to zero uniformly on every compact, hence for a given $\epsilon > 0$, there exists $t \in S$ such that $\|f(\mathbf{a}(s)x)\| < \epsilon$ if $s \geq t$. From this and (7.2) it follows

$$\lim_\alpha \frac{1}{|\Omega_\alpha|} \int_{\Omega_\alpha} f(\mathbf{a}(s)x) ds = \lim_\alpha \frac{1}{|\Omega_\alpha|} \int_{\Omega_\alpha \setminus (\Omega_\alpha + t)} f(\mathbf{a}(s)x) ds +$$
$$\lim_\alpha \frac{1}{|\Omega_\alpha|} \int_{\Omega_\alpha \cap (\Omega_\alpha + t)} f(\mathbf{a}(s)x) ds \leq \frac{|\Omega_\alpha \setminus (\Omega_\alpha + t)|}{|\Omega_\alpha|} \|f\| + \epsilon.$$

Therefore,

$$\lim_\alpha \frac{1}{|\Omega_\alpha|} \int_{\Omega_\alpha} f(\mathbf{a}(s)x) ds = 0 \quad (f \in \mathcal{B}_0).$$

Next, let $f \in \mathcal{B}_1$ be an eigenfunction corresponding to an eigenvalue $\chi \neq 1$. Then

$$\lim_\alpha \frac{1}{|\Omega_\alpha|} \int_{\Omega_\alpha} f(\mathbf{a}(s)x) ds = \lim_\alpha \frac{1}{|\Omega_\alpha|} \int_{\Omega_\alpha} \chi(s) f(x) ds = 0,$$

by virtue of (7.3). Finally, if f is an invariant function, then

$$\phi_x(f) = \lim_\alpha \frac{1}{|\Omega_\alpha|} \int_{\Omega_\alpha} f(x)ds = f(x). \tag{7.5}$$

Therefore, $\phi(f)$ is correctly defined for all f in $C(X,Y)$. From (7.1) and (7.5) it follows that

$$\phi_x(f) = \phi_x(Qf) = (Qf)(x) = \int_{K_T} (Vf)(x)dV = \int_K f(\mathbf{v}x)d\mathbf{v}.$$

Note that if we make a substitution $\psi = \mathbf{v}x$, then by virtue of Theorem 3.2 we have

$$\phi_x(f) = \int_K f(\mathbf{v}x)d\mathbf{v} = \int_{\omega(x)} f(\psi)d\psi,$$

where $d\psi$ is the Haar measure on $\omega(x)$. Thus, we have proved the following theorem.

Theorem 7.1. *Let* $\mathbf{a}$ *be an asymptotically almost periodic action of* S *and* Ω_α *be a Følner's net. Then for every* f *in* $C(X,Y)$

$$\lim_\alpha \frac{1}{|\Omega_\alpha|} \int_{\Omega_\alpha} f(\mathbf{a}(s)x)ds = \int_{\omega(x)} f(\psi)d\psi,$$

uniformly on compact sets.

REFERENCES

[A-B] W. Arendt and C.J.K. Batty, *Tauberian theorems and stability of C_0-semigroups*, Trans. Amer. Math. Soc. **306** (1988), 837–852.

[B-V] C.J.K. Batty and Vu Quoc Phong, *Stability of strongly continuous representations of abelian semigroups*, Math. Z. **209** (1992), 75–88.

[Bo] N. Bourbaki, *Eléments de mathématique. XXIX Livre VI. "Intégration. Vector intégration. Mesure de Haar. Convolution et représentations"*, Hermann, Paris, 1963.

[B-M] M.S. Brodskii and D.P. Mil'man, *On center of a convex set*, Dokl. Akad. Nauk SSSR **49(5)** (1948), 837–840.

[Bd] F.E. Browder, *Fixed point theory and nonlinear problems*, Bull. Amer. Math. Soc. **9** (1983), 1–39.

[Br] R. Bruck, *Asymptotic behavior of nonexpansive mappings*, Contemp. Math. **18** (1983), 1–47.

[C-D] F. Combes and C. Delaroches, *Représentations de groupes localement compactes et applications aux algèbres d'opérateurs*, Asterisque **55**.

[Cle] Ph. Clément et al., *One-Parameter Semigroups*, North-Holland, Amsterdam, 1987.

[D-S] C.M. Dafermos and M. Slemrod, *Asymptotic behavior of nonlinear contraction semigroups*, J. Funct. Anal. **13** (1973), 97–106.

[D-M] F.S. De Blasi and J. Myjak, *Generic properties of contraction semigroups and fixed points of nonexpansive operators*, Proc. Amer. Math. Soc. **77** (1979), 341–347.

[D-G] K. de Leeuw and I. Glicksberg, *Applications of almost periodic compactifications*, Acta Math. **105** (1961), 63–97.

[DeR] G. Della Riccia, *Equicontinuous Semiflows (One-Parameter Semi-Groups) on Locally Compact or Complete Metric Spaces*, Math. Systems Theory 4 (1970), 29–34.

[Di] J. Diestel, *Geometry of Banach spaces - Selected topics, Lecture Notes in Math.* **485**, Springer, Berlin, 1975.

[Go] W.H. Gottschalk, *Almost periodic points with respect to transformation semi-groups*, Ann. of Math. **47** (1946), 762–766.

[Gr] F.P. Greenleaf, *Invariant Means on Topological Groups*, Van Nostrand, New York, 1969.

[G-P] S. Gutman and A. Pazy, *An ergodic theorem for semigroups of contractions*, Proc. Amer. Math. Soc. **88** (1983), 254–256.

[Hl] J.K. Hale, *Asymptotic Behavior of Dissipative Systems*, Amer. Math. Soc., Providence, R.I., 1988.

[H1] A. Haraux, *Nonlinear Evolution Equations - Global Behavior of Solutions, Lecture Notes in Math.*, **841**, Springer-Verlag, New York, 1981.

[H2] ———, *Asymptotic behavior for two-dimensional, quasi-autonomous, almost periodic evolution equations*, J. Diff. Equ. **66** (1987), 62–70.

[H-R] E. Hewitt and K.A. Ross, *Abstract Harmonic Analysis*, Springer, Berlin, 1979.

[Hi] N. Hirano, *Nonlinear ergodic theorems and weak convergence theorems*, J. Math. Soc. Japan **34** (1982), 35–46.

[H-T] N. Hirano and W. Takahashi, *Nonlinear ergodic theorems for an amenable semigroup of nonexpansive mappings in a Banach space*, Pacific J. Math. **112** (1984), 333–346.

[I-R1] M.M. Israel and S. Reich, *A note on nonlinear semigroups with large set of fixed points*, Global Analysis - Analysis on Manifolds, Teubner-Texte zur Math., Leipzig (1983), 154–157.

[I-R2] ———, *Extension and selection problems for nonlinear semigroups in Banach spaces*, Math. jap. **28** (1983), 1–8.

[Ke] J.L. Kelley, *General Topology*, Van Nostrand, New York, 1955.

[Ki] W. Kirk, *Fixed point theory for nonexpansive mappings*, Contemp. Math. **18** (1983), 121–140.

[Kr] U. Krengel, *Ergodic Theorems*, De Gruyter, Berlin, 1985.

[La] A.T. Lau, *Semigroups of nonexpansive mappings in Hilbert space*, J. Math. Anal. Appl. **105** (1985), 514–522.

[Li] Lim Teck-Cheong, *Asymptotic center and non-expansive mappings in conjugate Banach spaces*, Pacific J. Math. **90** (1980), 135–143.

[L1] Yu.I. Lyubich, *Dissipative actions and almost periodic representation of abelian semigroups*, Ukr. Math. J. **40** (1988), 70–74.

[L2] ———, *On trajectory convergence of dissipative flows in Banach spaces*, Integral Equations and Operator Theory **13** (1990), 138–144.

[L3] ———, *Introduction to the Theory of Banach Representations of Groups*, Birkhäuser, Basel, 1988.

[L-V] Yu.I. Lyubich and Vu Quoc Phong, *Asymptotic stability of differential equations in Banach spaces*, Studia Math. **88** (1988), 37–42.

[Mil] P. Milnes, *Left mean ergodicity, fixed points, and invariant means*, J. Math. Anal. Appl. **65** (1978), 32–43.

[Miy] I. Miyadera, *On the infinitesimal generators and the asymptotic behavior of nonlinear contraction semigroups*, Proc. Japan Acad. **A58** (1982), 1-4.

[N-S] V.V. Nemytskii and V.V. Stepanov, *Qualitative Theory of Differential Equations*, Princeton Univ. Press, Princeton, NJ, 1960.

[N-R] O. Nevanlinna and S. Reich, *Strong convergence of contraction semigroups and of iterative methods for accretive operators*, Israel J. Math. **32** (1979), 44–58.

[Pt] A.L.T. Paterson, *Amenability and locally compact semigroups*, Math. Scand. **42** (1982), 271-288.

[Pz1] A. Pazy, *On the asymptotic behavior of semigroups of nonlinear contractions in Hilbert spaces*, J. Funct. Anal. **27** (1978), 292–307.

[Pz2] ———, *The asymptotic behavior of semigroups of nonlinear contractions having a large set of fixed points*, Proc. Royal Soc. Edinburgh **A80** (1978), 261–271.

[Pz3] ———, *Remarks on nonlinear ergodic theory in Hilbert space*, Nonlinear Anal.: Theory, Methods Appl. **3** (1979), 863–871.

[Pz4] ———, *Semigroups of nonlinear contractions and their asymptotic behavior*, In: Nonlinear analysis and mechanics, Heriot-Watt Symp. (R.J. Knops, Ed.), Pitman Research Notes Math. **30** (1979)), 36–134.

[Po] L. Pontrjagin, *Topological groups*, Princeton University Press, Princeton, NJ, 1946.

[Re] S. Reich, *A note on the mean ergodic theorem for nonlinear semigroups*, J. Math. Anal. and Appl. **91** (1983), 547–551.

[R-S] S. Roehrig and R. Sine, *The structure of ω-limit sets of nonexpansive maps*, Proc. Amer. Math. Soc. **81** (1981), 398–400.

[R1] W. Rudin, *Functional Analysis*, McGraw Hill, New York, 1973.

[R2] ———-, *Fourier Analysis on Groups*, Interscience, New York, 1962.

[R-S] W.M. Ruess and W.H. Summers, *Ergodic theorems for semigroups of operators*, Proc. Amer. Math. Soc. **114** (1992), 423–432.

[Ry] C. Ryll-Nardzewski, *On fixed points of semigroups of endomorphisms of linear spaces*, Proceedings of the Fifth Berkeley Symposium on Math. Statistics and Prob., v. II, Berkeley (1966).

[Sh] B.A. Shcherbakov, *Exponential stability of solutions of multidimensional differential equations*, Differential Equations **26** (1990), 998–1002.

[Si1] R. Sine, *Structure and asymptotic behavior of abstract dynamical systems*, Nonlinear Anal.: Theory Methods and Appl. **2** (1978), 119–127.

[Si2] ———-, *Recurrence of nonexpansive mappings in Banach spaces*, Contemp. Math. **18** (1983), 175–200.

[Ta] W. Takahashi, *A nonlinear ergodic theorem for an amenable semigroup of nonexpansive mappings in a Hilbert space*, Proc. Amer. Math. Soc. **81** (1981), 253–256.

[V1] Vu Quoc Phong, *Dissipative almost periodic actions of semigroups*, Ukr. Math. J. **40** (1988), 110–113.

[V2] —————, *Theory of almost periodic actions of topological semigroups and its applications*, Doctor Thesis, Kiev, 1987.

[Wa] P. Walter, *An Introduction to Ergodic Theory*, Springer, New York, 1982.

[We] G.F. Webb, *Compactness of bounded trajectories of dynamical systems in infinite dimensional spaces*, Proc. Royal Soc. Edinburgh **A84** (1979), 19–33.

[Yo] K. Yosida, *Functional Analysis*, Springer-Verlag, New York, 1968.

Stability of the L_p-Spectrum of Schrödinger Operators with Form-Small Negative Part of the Potential

Gabriele Schreieck and Jürgen Voigt

Fachbereich Mathematik der Universität Oldenburg
Ammerländer Heerstraße 114-118
W-2900 Oldenburg, Germany

INTRODUCTION

In [2] it was shown that $\sigma((H + V)_p)$ is p-independent, where $(H + V)_p = -\frac{1}{2}\Delta + V$ is a Schrödinger operator in $L_p(\mathbb{R}^d)$, $1 \le p \le \infty$, for potentials $V = V_+ - V_-$, with V_+ admissible and V_- in the Kato class. If the negative part V_- is no longer assumed to belong to the Kato class but is only assumed to be $(-\frac{1}{2}\Delta)$–form small, then $H + V$ can be defined as an operator in L_p only for $p_0 \le p \le p_0'$, with suitable $p_0 > 1$. It is the purpose of this paper to show that the spectrum of these realizations in L_p is the same for $p_0 < p < p_0'$. In the proof given in [2], the use of L_1 and L_∞ is essential; therefore part of the proof requires a new method.

More generally, we treat the case of form perturbations, where the perturbation consists of a form small distributional part as in [3], and a part defined by a (possibly large) positive measure (as in [9]).

We mention that Sturm [12] has shown p-independence of the L_p-spectrum for Laplace-Beltrami operators on suitable Riemannian manifolds.

In Section 1 we fix our notation, define the kind of perturbation we treat, and recall and derive the basic facts and inequalities needed in the sequel.

In Section 2 one part of the spectral inclusion is shown by the known method; we only have to derive L_p - L_q-smoothing, which is done by using Stein interpolation.

In Section 3 it is shown that the L_2-resolvent set is contained in the L_p-resolvent set. This is shown by proving that the L_2-resolvent is continuous in L_p. In order to achieve this the operators are also considered in suitable weighted L_p-spaces.

1. NOTATIONS AND BASIC FACTS

By $\mathfrak{h}$ we shall denote the classical Dirichlet form in $L_2(\mathbb{R}^d)$,

$$D(\mathfrak{h}) = W_2^1(\mathbb{R}^d),$$

$$\mathfrak{h}[u,v] = \frac{1}{2} \int \nabla u \; \overline{\nabla v} \; dx.$$

The corresponding selfadjoint operator will be denoted by H $(= -\frac{1}{2}\Delta)$.

The perturbation we shall consider will consist of two parts. The first part μ_o is the "distributional part" introduced in [3]: μ_o is symmetric and $\mathfrak{h}$-small, i.e., $D(\mu_o) = D(\mathfrak{h})$, and there exist $0 < \beta < 1$, $b \geq 0$ such that

$$|\mu_o[u]| \leq \beta \, \mathfrak{h}[u] + b \, \|u\|_2^2 \qquad (u \in D(\mathfrak{h})); \tag{1.1}$$

and moreover there exists a distribution τ on $C_c^\infty(\mathbb{R}^d)$ such that

$$\mu_o[u,v] = \tau(u\bar{v}) \qquad (u,\, v \in C_c^\infty(\mathbb{R}^d)).$$

The other part of the perturbation is a measure μ_+; $\mu_+ : \mathfrak{B} \longrightarrow [0,\infty]$ a σ-additive set function defined on the Borel sets $\mathfrak{B}$ of $\mathbb{R}^d$. Additionally we require the property $\mu_+(B) = 0$ for each set $B \in \mathfrak{B}$ with $\mathrm{cap}(B) = 0$, where cap denotes capacity,

$$\mathrm{cap}(B) := \inf \left\{ \|f\|_{2,1}^2 ;\; f \in W_2^1(\mathbb{R}^d),\; f \geq 1 \text{ on a neighbourhood of } B \right\}.$$

This implies that the form $\mathfrak{h} + \mu_+$, where the form μ_+ is defined by

$$D(\mu_+) := \left\{ u \in W_2^1(\mathbb{R}^d) \,;\; u^\sim \in \mathcal{L}_2(\mu_+) \right\}$$

($u^\sim$ denoting a quasi-continuous version of u),

$$\mu_+[u,v] := \int u^\sim \overline{v^\sim}\, d\mu_+ \qquad (u,v \in D(\mu_+)),$$

is a closed (symmetric, positive) form, in general not densely defined; cf. [9].

It further follows that $\mathfrak{h} + \mu_0 + \mu_+$ is a closed semibounded symmetric form. The space $\overline{D(\mathfrak{h} + \mu_0 + \mu_+)}^{L_2(\mathbb{R}^d)}$ is of the form $L_2(Y)$, for a measurable set $Y \subset \mathbb{R}^d$, and $\mathfrak{h} + \mu_0 + \mu_+$ generates a selfadjoint operator $H_\mu = H_{\mu_0 + \mu_+}$ in $L_2(Y)$.

We summarize the fundamental facts which will be needed later in the following remarks.

1.1. Remarks. (a) Denoting $D := D(\mathfrak{h})$, the form μ_0 can be interpreted as a mapping $\mu_0 : D \to D^*$, $u \mapsto \mu_0[u,\cdot]$ (D^* the space of conjugate linear continuous functionals on D). Let $\rho \in C_c^\infty(\mathbb{R}^d)_+$, $\int \rho\, dx = 1$, $\rho_n(x) := n^d \rho(nx)$, $V_n := \rho_n * \tau$ ($\in L_\infty(\mathbb{R}^d)$). It was shown in [3] that then

$$\left|\int |u|^2 V_n\, dx\right| \le \beta\, \mathfrak{h}[u] + b\, \|u\|_2^2 \qquad (u \in D(\mathfrak{h})) \tag{1.2}$$

for all $n \in \mathbb{N}$, with β, b from (1.1), and that $V_n \to \mu_0$ strongly in $L(D,D^*)$. This implies strong resolvent convergence

$$H_{V_n + \mu_+} \to H_{\mu_0 + \mu_+},$$

(cf. [3], [10 ; Appendix A]), and strong convergence of the semigroups,

$$e^{-tH_{V_n + \mu_+}} \to e^{-tH_{\mu_0 + \mu_+}}$$

for $t \ge 0$, uniformly for t in bounded sets (by the Trotter convergence theorem; cf. [6; Sec. 3.4]).

(b) The distribution τ is real; therefore the V_n are real, and the operators $e^{-t(H + V_n)}$, $e^{-tH_{\mu_0}}$ are positivity preserving for all $t \ge 0$; cf. [3]. From [10; Appendix B] (or [5]) it follows that

$$0 \le e^{-tH_{\mu_0 + \mu_+}} f \le e^{-tH_{\mu_0}} f$$

for all $f \in L_2(Y)_+$, $t \ge 0$. Therefore, all norm estimates derived for $e^{-tH_{\mu_0}}$ carry over to $e^{-tH_{\mu_0+\mu_+}}$.

(c) Let $V \in L_\infty(\mathbb{R}^d)$,

$$-\int |u|^2\, V\, dx \le \beta\, \mathfrak{h}[u] + b\, \|u\|_2^2 \qquad (u \in D(\mathfrak{h})) \tag{1.3}$$

Let $p_0 := \frac{2}{1+\sqrt{1-\beta}}$ (then $p_0' = \frac{2}{1-\sqrt{1-\beta}}$). Since $(e^{-tH};\ t \ge 0)$ acts also as a C_0-semigroup on $L_p(\mathbb{R}^d)$, for $1 \le p < \infty$, it follows that the same is true for $(e^{-t(H+V)};\ t \ge 0)$. It was shown in [1], [4] that

$$\| e^{-t(H+V)} \|_{p,p} \le e^{bt} \qquad (t \ge 0),$$

for all $p \in [p_0, p_0']$, where $\|\cdot\|_{p,p}$ denotes the operator norm in $L(L_p(\mathbb{R}^d))$. (In the references quoted above, V is assumed to be nonpositive. It is obvious from the proofs that the asserted inequality depends only on the estimate (1.3).) Applying this to V_n from (a) and letting $n \to \infty$, the estimate carries over to the limit,

$$\| e^{-tH_{\mu_0}} \|_{p,p} \le e^{bt} \qquad (t \ge 0),$$

and, by (b),

$$\| e^{-tH_{\mu_0+\mu_+}} \|_{p,p} \le e^{bt} \qquad (t \ge 0).$$

It is easy to see (cf. [13]) that, for $p \in [p_0, p_0']$, $(e^{-tH_{\mu_0+\mu_+}};\ t \ge 0)$ is also a C_0-semigroup on $L_p(Y)$; its generator will be denoted by $-H_{\mu_0+\mu_+,p} = -H_{\mu,p}$. (Note $H_\mu = H_{\mu,2}$.)

After these preparations we are able to state the main result of this paper.

1.2. Theorem. *With the notations introduced so far, there is equality*

$$\sigma(H_{\mu,p}) = \sigma(H_{\mu,2})$$

for all $p \in (p_0, p_0')$.

The proof will consist in showing the two inclusions, in the following sections.

1.3. Remarks. (a) We assumed $\beta > 0$ in inequality (1.1). Inequality (1.1) with $\beta = 0$ would lead to $\mu_0 = V$ with $V \in L_\infty(\mathbb{R}^d)$, and to $p_0 = 1$. An adequate assumption for treating the case $p_0 = 1$, however, is to assume that $-\mu_0$ is a measure in a suitable Kato class. We refer to [10] for the definition and for properties of the corresponding semigroups. We note that, combining the estimates given in [10] with the methods in [2] shows the result stated in Theorem 1.2 for this case as well. Our proof given below would also yield this result.

(b) In [3; Theorem 9.1] it is also shown that a form small perturbation gives rise to a semigroup acting in L_p for p in a certain interval $(\tilde{p}_0, \tilde{p}_0')$, with $\tilde{p}_0 = \frac{2}{2-\beta}$. It is easy to see that

$$p_0 = \frac{2}{1 + \sqrt{1-\beta}} < \tilde{p}_0$$

for $0 < \beta < 1$.

(c) We mention that perturbation by μ_+ includes perturbed situations which are far away from the unperturbed. For instance, if $\Omega \subset \mathbb{R}^d$ is open, then μ_+ can be chosen such that H_{μ_+} is the Dirichlet Laplacian in $L_2(\Omega)$; see, e.g., [11]. We note, however, that for this situation, even if it is not covered by the hypotheses made in [2] the proof given there applies to show $\sigma(H_{\mu_+,p}) = \sigma(H_{\mu_+,2})$ for all $p \in [1,\infty]$.

2. $\sigma(H_{\mu,2}) \subset \sigma(H_{\mu,p})$

2.1. Proposition. *Let* μ_0, μ_+ *etc. be as in Section 1. Then* $\rho(H_{\mu,p}) \subset \rho(H_{\mu,2})$ *for all* $p \in (p_0, p_0')$, *and*

$$(H_{\mu,p} - z)^{-1} \mid L_p \cap L_2 = (H_{\mu,2} - z)^{-1} \mid L_p \cap L_2 \qquad (z \in \rho(H_{\mu,p})).$$

The proof is achieved in the same way as in [2], as soon as one shows that, for $p_0 < p < q < p_0'$, $t > 0$, the operator $e^{-tH_{\mu,p}}$ maps L_p into L_q (p-q-smoothing). In Section 3 we shall need a quantitative version of this fact as stated in the following result.

2.2. Theorem. *Let* $p_0 < p_1 < 2$. *Then there exist constants* C, ω *only depending on* β, b, p_1 *such that*

$$\|e^{-tH_{\mu,p}}\|_{p,q} \le C\, t^{-\gamma}\, e^{\omega t}$$

for all $t > 0$, $p_1 \le p \le q \le p_1'$, *with* $\gamma = \frac{d}{2}(\frac{1}{p} - \frac{1}{q})$.
(Here $\|\cdot\|_{p,q}$ denotes the norm in $L(L_p, L_q)$.)

Proof. The proof uses known techniques; it is in particular very similar to the corresponding proof in [10]. For this reason we sketch only the main steps.

(i) Because of Remarks 1.1 it is sufficient to treat the case $\mu_+ = 0$, $\mu_0 = V$, $V \in L_\infty(\mathbb{R}^d)$.

(ii) Let $c > 1$ be such that $\frac{2}{1 + \sqrt{1 - c\beta}} < p_1$. Then $c\beta < 1$,

$$-\int |u|^2 \, cV dx \le c\beta \, \mathfrak{h}[u] + cb \|u\|_2^2 \qquad (u \in D(\mathfrak{h})),$$

and Remark 1.1 (c) implies

$$\|e^{-t(H + cV)}\|_{p_1, p_1} \le e^{cbt} \qquad (t \ge 0).$$

Further, there exist constants $C \ge 1$, $\omega \ge cb$, such that

$$\|e^{-tH}\|_{p_1, p_1'} \le C\, t^{-\frac{d}{2}(\frac{1}{p_1} - \frac{1}{p_1'})} e^{\omega t} \qquad (t > 0).$$

(Note that e^{-tH} is convolution by the heat kernel.) The Stein interpolation theorem (cf. [8; p. 69] applied to the operator function $z \mapsto e^{-t(H + zcV)}$, implies

$$\|e^{-t(H + V)}\|_{p_1, q} \le C\, t^{-\frac{d}{2}(\frac{1}{p_1} - \frac{1}{q})} e^{\omega t},$$

where $\frac{1}{q} = \frac{1 - \frac{1}{c}}{p_1'} + \frac{\frac{1}{c}}{p_1}$.

(iii) From (ii) one obtains the assertion of the theorem by duality, Riesz-Thorin interpolation and the semigroup property. ■

3. $\sigma(H_{\mu,p}) \subset \sigma(H_{\mu,2})$

3.1. Proposition. *Let μ_0, μ_+ etc. be as in Section 1. Then $\rho(H_{\mu,2}) \subset \rho(H_{\mu,p})$ for all $p \in (p_0, p_0')$.*

The idea of the proof is to show that, for $z \in \rho(H_{\mu,2})$, the L_2-resolvent $(z - H_{\mu,2})^{-1}$ is also continuous in L_p, and therefore z should belong to the resolvent set of $H_{\mu,p}$. In order to reason in this way, a uniformity of the L_p- bounds on compact sets is required.

The main additional tool needed in our proof, in comparison with the proof given in [2], is the result we state next. We use the notation

$$L_{\infty,c} := \left\{ f \in L_\infty(\mathbb{R}^d);\ \operatorname{supp} f \text{ compact} \right\}.$$

3.2. Proposition. *Let* $1 \le p \le q \le \infty$, $\varepsilon > 0$. *Then there exists* $C \ge 0$ *with the following property. For each linear operator*

$$A : L_{\infty,c} \longrightarrow L_{1,loc}$$

satisfying

$$\| e^{\xi\cdot x} A e^{-\xi\cdot x} \|_{p,q} \le 1 \quad \textit{for all } \xi \in \mathbb{R}^d \textit{ with } |\xi| \le \varepsilon$$

one has

$$\|A\|_{r,r} \le C \quad \textit{for all } r \in [p,q].$$

Expressed differently: For each linear operator $A : L_{\infty,c} \longrightarrow L_{1,loc}$ *one has*

$$\| A \|_{r,r} \le C \sup_{|\xi| \le \varepsilon} \| e^{\xi\cdot x} A e^{-\xi\cdot x} \|_{p,q} \quad \textit{for } p \le r \le q.$$

(By "$e^{\xi\cdot x}$" we denote the operator of multiplication with the function $x \mapsto e^{\xi\cdot x}$.)

Proof. Assume that A satisfies the assumption. It is sufficient to show $\|A\|_{p,p} \le C$, $\|A\|_{q,q} \le C$ (and then apply Riesz-Thorin interpolation).

We subdivide $\mathbb{R}^d$ into cubes of unit side length as follows. For $j \in \mathbb{Z}^d$ define

$$Q_j := \left\{ x \in \mathbb{R}^d;\ |x-j|_\infty \le \tfrac{1}{2} \right\}.$$

Let $k,j \in \mathbb{Z}^d$, $f \in L_\infty(\mathbb{R}^d)$, supp $f \subset Q_j$. Then, choosing $\xi = \frac{\varepsilon}{|k-j|}(k-j)$, one estimates

$$\|Af\|_{Q_k,q} = \| e^{-\xi\cdot x} e^{\xi\cdot x} A e^{-\xi\cdot x} e^{\xi\cdot x} f \|_{Q_k,q} \le c\, e^{-\xi\cdot k} \| e^{\xi\cdot x} A e^{-\xi\cdot x} (e^{\xi\cdot x} f) \|_{Q_k,q}$$

(where $\sup_{x \in Q_k} e^{\xi\cdot k} e^{-\xi\cdot x} = \sup_{|y|_\infty \le \frac{1}{2}} e^{\xi\cdot y} \le e^{\frac{1}{2}\varepsilon\sqrt{d}} =: c$)

$$\le c\, e^{-\xi\cdot k} \| e^{\xi\cdot x} f \|_{Q_j,p} \le c^2 e^{-\xi\cdot(k-j)} \|f\|_{Q_j,p} = c^2\, e^{-\varepsilon|k-j|} \|f\|_{Q_j,p}.$$

Now we can estimate for arbitrary $f \in L_{\infty,c}$:

$$\|Af\|_p^p = \sum_{k \in \mathbb{Z}^d} \|Af\|_{Q_k,p}^p \le \sum_k \|Af\|_{Q_k,q}^p \le \sum_k \Big(\sum_j \|A(\chi_{Q_j} f)\|_{Q_k,q}\Big)^p$$

$$\le c^{2p} \sum_k \Big(\sum_j e^{-\varepsilon|k-j|} \|f\|_{Q_j,p}\Big)^p \le c^{2p}\, \|(e^{-\varepsilon|k|})_{k \in \mathbb{Z}^d}\|_{\ell_1(\mathbb{Z}^d)}^p \cdot \|f\|_p^p.$$

Similarly,

$$\|Af\|_q^q \le c^{2q} \, \|(e^{-\varepsilon|k|})_k\|^q_{\ell_1(\mathbb{Z}^d)} \|f\|_q^q .$$

Therefore, the assertion holds with $C = c^2 \|(e^{-\varepsilon|k|})_k\|_{\ell_1(\mathbb{Z}^d)}$. ■

3.3. Corollary. *Let* $1 \le p \le q \le \infty$, $0 < \varepsilon' < \varepsilon''$. *Then there exists* $C \ge 0$ *such that for each linear operator*

$$A : L_{\infty,c} \longrightarrow L_{1,loc}$$

satisfying

$$\|e^{\xi\cdot x} A e^{-\xi\cdot x}\|_{p,q} \le 1 \textit{ for all } \xi \in \mathbb{R}^d \textit{ with } |\xi| \le \varepsilon''$$

one has

$$\|e^{\xi\cdot x} A e^{-\xi\cdot x}\|_{r,r} \le C$$

for $p \le r \le q$, $|\xi| \le \varepsilon'$.

Proof. Choose C according to Proposition 3.2, with $\varepsilon = \varepsilon'' - \varepsilon'$. ■

In the following remark we summarize further preparations for the proof of Proposition 3.1.

3.4. Remarks. (a) Let $\varepsilon > 0$. Then there exist constants C, ω such that

$$\|e^{\xi\cdot x} e^{-tH} e^{-\xi\cdot x}\|_{p,q} \le Ct^{-\gamma} e^{\omega t}$$

for all $t > 0$, $1 \le p \le q \le \infty$, $\xi \in \mathbb{R}^d$ with $|\xi| \le \varepsilon$, where $\gamma := \frac{d}{2}(\frac{1}{p} - \frac{1}{q})$.

For the proof see [7; Lemma B.6.1], [2; Lemma 3.4].

(b) Let μ_0, μ_+ etc. be as in Section 1. Let $1 < c < \frac{1}{\beta}$, $p_c := \frac{2}{1 + \sqrt{1 - c\beta}}$, $\frac{1}{c} + \frac{1}{c'} = 1$, and let $p_c \le p \le q \le p_c'$. Then, for all $\xi \in \mathbb{R}^d$,

$$\|e^{\xi\cdot x} e^{-tH_\mu} e^{-\xi\cdot x}\|_{p,q} \le \|e^{-tH_{c\mu_0}}\|_{p,q}^{1/c} \; \|e^{c'\xi\cdot x} e^{-tH} e^{-c'\xi\cdot x}\|_{p,q}^{1/c'} .$$

As previously, it is sufficient to treat the case $\mu_+ = 0$, $\mu_0 = V$, with $V \in L_\infty$. In this case the stated inequality follows from the Stein interpolation theorem, applied to the function

$$F(z) := w^{(1-z)c'} e^{-t(H+zcV)} w^{-(1-z)c'},$$

where $w(x) = e^{\xi \cdot x}$. Alternatively, the inequality can be obtained by Hölder's inequality in function space; cf. [7; Lemma B.6.2(b)], [2; Lemma 3.6].

(c) Combining the statements of Theorem 2.2 and (a), (b) above one obtains: Let $p_0 < p_1 < 2$, $\varepsilon > 0$. Then there exist constants C, ω such that

$$\| e^{\xi \cdot x} e^{-tH_\mu} e^{-\xi \cdot x} \|_{p,q} \le C t^{-\gamma} e^{\omega t}$$

for all $t > 0$, $p_1 \le p \le q \le p_1'$, $\xi \in \mathbb{R}^d$ with $|\xi| \le \varepsilon$, where $\gamma = \frac{d}{2}(\frac{1}{p} - \frac{1}{q})$.

Note that, in the application of Theorem 2.2 to the estimate in (b), one has to replace μ by $c\mu_0$, where c is chosen sufficiently close to 1.

(d) Let μ_0, μ_+ etc. be as in Section 1. Let $p_0 < p_1 < 2$, $\varepsilon > 0$. Then there exist constants C, ω such that

$$\| e^{\xi \cdot x} (H_\mu - w)^{-1} e^{-\xi \cdot x} \|_{p,q} \le C\left(\frac{1}{1-\gamma} + \frac{1}{-w-\omega}\right)$$

for all $w \in \mathbb{R}$ with $w < -\omega$, $p_0 \le p \le q \le p_0'$ with $\gamma := \frac{d}{2}(\frac{1}{p} - \frac{1}{q}) < 1$, $|\xi| \le \varepsilon$. Further $(-\infty, -\omega) \subset \rho(H_{\mu,p})$ for $p_0 \le p \le p_0'$, and

$$(H_{\mu,p} - w)^{-1} = (H_\mu - w)^{-1}$$

on $L_p(Y) \cap L_2(Y)$, for $w < -\omega$.

The proof consists in integrating the inequality in (c) after multiplying by e^{wt}; cf. [7; Theorem B.6.3], [2; Proposition 3.7].

Proof of Proposition 3.1. Let $p_0 < p_1 < 2$. It is sufficient to prove the assertion for all $p \in [p_1, 2]$. By symmetry it then follows for all $p \in [p_1, p_1']$. According to Remark 3.4 (d) we find w $(< -\omega)$, C such that

$$\| e^{\xi \cdot x} (H_\mu - w)^{-1} e^{-\xi \cdot x} \|_{p,q} \le C$$

whenever $p_1 \le p \le q \le 2$, $\frac{d}{2}(\frac{1}{p} - \frac{1}{q}) \le \frac{1}{2}$, $|\xi| \le 1$.

Let $K \subset \rho(H_{\mu,2})$ be compact and connected, $K = \overline{\mathring{K}}$, $w \in \mathring{K}$. Then there exist $\varepsilon > 0$ and a constant C' such that $K \subset \rho(e^{\xi \cdot x} H_{\mu,2} e^{-\xi \cdot x})$ for $|\xi| \le \varepsilon$, and

$$\| e^{\xi \cdot x} (H_{\mu,2} - z)^{-1} e^{-\xi \cdot x} \| = \| (e^{\xi \cdot x} H_{\mu,2} e^{-\xi \cdot x} - z)^{-1} \| \le C' \quad (|\xi| \le \varepsilon,\ z \in K).$$

This is shown with the same proof as in [2; Proposition 3.8] or in [12; Lemma 1]. (Note that the perturbed operator

$$e^{\xi \cdot x} H_{\mu,2} e^{-\xi \cdot x} = H_\mu + \xi \cdot \nabla - \tfrac{1}{2} \xi \cdot \xi$$

is not symmetric). Without restriction $\varepsilon \le 1$. Now the resolvent equation

$$(H_{\mu,2} - z)^{-1} = (I + (z - w)(H_{\mu,2} - z)^{-1})(H_{\mu,2} - w)^{-1}$$

implies

$$\| e^{\xi\cdot x}(H_{\mu,2} - z)^{-1} e^{-\xi\cdot x}\|_{p,2} \le (1 + C'\sup_{z\in K}|z-w|)C$$

for $z \in K$, $p_1 \le p \le 2$, $\frac{d}{2}(\frac{1}{p} - \frac{1}{2}) \le \frac{1}{2}$, $|\xi| \le \varepsilon$. Applying Corollary 3.3 we obtain a constant C'' such that

$$\| e^{\xi\cdot x}(H_{\mu,2} - z)^{-1} e^{-\xi\cdot x}\|_{p,p} \le C''$$

for $z \in K$, $p_1 \le p \le 2$, $\frac{d}{2}(\frac{1}{p} - \frac{1}{2}) \le \frac{1}{2}$, $|\xi| \le \frac{\varepsilon}{2}$. (The operator $(H_{\mu,2} - z)^{-1}$ on $L_2(Y)$ can be considered as an operator from $L_{\infty,c}$ to $L_{1,loc}$ in the obvious way.)

Let $\tilde{p} := \min\left\{p \in [p_1, 2];\ \frac{d}{2}(\frac{1}{p} - \frac{1}{2}) \le \frac{1}{2}\right\}$. If $\tilde{p} > p_1$ we use again the resolvent equation and obtain

$$\| e^{\xi\cdot x}(H_{\mu,2} - z)^{-1} e^{-\xi\cdot x}\|_{p,\tilde{p}} \le (1 + C''\sup_{z\in K}|z-w|)C$$

for $z \in K$, $p_1 \le p \le \tilde{p}$, $\frac{d}{2}(\frac{1}{p} - \frac{1}{\tilde{p}}) \le \frac{1}{2}$, $|\xi| \le \frac{\varepsilon}{2}$. Corollary 3.3 implies

$$\| e^{\xi\cdot x}(H_{\mu,2} - z)^{-1} e^{-\xi\cdot x}\|_{p,p} \le C'''$$

for $z \in K$, p as above, $|\xi| \le \frac{\varepsilon}{4}$, and some C'''.

Repeating this procedure, if necessary, we obtain a constant C_1 and $\varepsilon_1 > 0$ such that

$$\| e^{\xi\cdot x}(H_{\mu,2} - z)^{-1} e^{-\xi\cdot x}\|_{p,p} \le C_1$$

for $z \in K$, $p_1 \le p \le 2$, $|\xi| \le \varepsilon_1$. Using this estimate for $\xi = 0$ and exploiting the fact that $(H_{\mu,2} - w)^{-1} = (H_{\mu,p} - w)^{-1}$ on $L_p \cap L_2$ we obtain $K \subset \rho(H_{\mu,p})$. ∎

3.5. Remark. We comment shortly on the differences between our proof and the proof given in [2]. The proof in [2] uses the Dunford-Pettis theorem, and therefore suitable powers of the resolvents are taken mapping L_1 to L_∞. (The resolvents themselves have local singularities and do not have bounded integral kernels.) In the present more general context the space L_1 cannot be used. Corollary 3.3 replaces the use of the Dunford-Pettis theorem.

REFERENCES

1. A.G. Belyi and Yu.A. Semenov: L^p-theory of Schrödinger semigroups II. Sibirskii Matematicheskii Zhurnal **31**, 16-26 (1990) (russian). Translation: Siberian Mathematical Journal **31**, 540-549 (1990).

2. R. Hempel and J. Voigt: The spectrum of a Schrödinger operator in $L_p(\mathbb{R}^\nu)$ is p-independent. Commun. Math. Phys. **104**, 243-250 (1986).

3. I.W. Herbst and A.D. Sloan: Perturbation of translation invariant positivity preserving semigroups on $L^2(\mathbb{R}^N)$. Trans. Amer. Math. Soc. **236**, 325-360 (1978).

4. V.A. Liskevich and Yu.A. Semenov: Some inequalities for submarkovian generators and their applications to the perturbation theory. Preprint 1992.

5. E.M. Ouhabaz: Propriétés d'ordre et de contractivité des semigroupes avec applications aux opérateurs elliptiques. Thèse de doctorat, Besançon, 1992.

6. A. Pazy: Semigroups of linear operators and applications to partial differential equations. Springer-Verlag, Berlin, 1983.

7. B. Simon: Schrödinger semigroups. Bull. Amer. Math. Soc., New Ser., **7**, 447-526 (1982).

8. E.M. Stein: Topics in harmonic analysis related to the Littlewood-Paley theory. Princeton Univ. Press, Princeton, N.J., 1970.

9. P. Stollmann: Smooth measures for regular Dirichlet forms. Proc. Amer. Math. Soc., to appear.

10. P. Stollmann and J. Voigt: Perturbation of Dirichlet forms by measures. Preprint 1992.

11. K.-Th. Sturm: Schrödinger operators with arbitrary nonnegative potentials. Preprint 1992. In "Operator Calculus and Spectral Theory", Proc. Lambrecht 1991, Demuth, Gramsch, Schulze eds., Birkhäuser, 1992, p. 291-306.

12. K.-Th. Sturm: On the L^p-spectrum of Laplace-Beltrami operators. Preprint 1992.

13. J. Voigt: One-parameter semigroups acting simultaneously on different L_p-spaces. Preprint 1992.

Singularity Theory and Semilinear Elliptic Equations

T. RUNST Mathematical Institute, Friedrich-Schiller-University Jena, UHH, 17.OG, D(O)- 6900 Jena, Germany

1 INTRODUCTION

At first we consider the following nonlinear elliptic equation

$$\begin{aligned} -\Delta\, u(x) - f(u(x)) &= h(x), \quad x \in \Omega \\ u(x) &= 0, \qquad x \in \partial\Omega, \end{aligned} \tag{1}$$

where $\Omega \subset \mathbf{R}^1$ is a bounded and smooth domain with boundary $\partial\Omega$, h is a given function and f is a sufficiently smooth real-valued function on $\mathbf{R}^1$ which generates the nonlinearity. In the last years, many mathematicians were interested in existence and multiplicity results of solutions of equation (1) in dependence on the kind of the nonlinearity f and on the given function h. We denote, as usual, by $0 < \lambda_1 < \lambda_2 \leq \lambda_3 \leq ..$ the sequence of eigenvalues of

$$\begin{aligned} -\Delta\, v &= \lambda\, v\, in\, \Omega \\ v &= 0 \quad on\, \partial\Omega. \end{aligned} \tag{2}$$

It is known that our problem (1) admits multiple solutions in dependence on the behaviour between the values of the first derivative f' and the eigenvalues of $-\Delta$ under Dirichlet conditions if h belongs to the classical Hölder space $C^\alpha(\Omega)$, $0 < \alpha < 1$ fixed. First note that if the range of f' contains no eigenvalue λ_k, then (1) is uniquely solvable in $C^{2+\alpha}(\Omega)$, see C.L.Dolph (1949). The first result, where the range of f' contains only the eigenvalue λ_1 was proved by A.Ambrosetti, G.Prodi (1972) in „ On the inversion of some differentiable mappings with singularities between Banach spaces". They showed that the conditions for $f \in C^2(\mathbf{R}^1)$, $0 < f'(-\infty) < \lambda_1 < f'(+\infty) < \lambda_2$ and $f'' > 0$ on $\mathbf{R}^1$ imply the existence of a smooth hypersurface $M_1 \subset C^\alpha(\Omega)$ of codimension 1 which disconnects $C^\alpha(\Omega)$ in two components M_0 and M_2 with the property that (1) has exactly i solutions if h belongs to M_i, $i = 0, 1, 2$. For the proof they used the local behaviour of the corresponding Fredholm mapping $\Phi : C_0^{2+\alpha}(\Omega) \to C^\alpha(\Omega)$ defined by $\Phi(u) = -\Delta u - f(u)$. In that case the origin is a singular point of $\Phi'(u)[v] = -\Delta v - f'(u)v$, $\Phi' : C_0^{2+\alpha}(\Omega) \to C^\alpha(\Omega)$. Using tools from singularity theory the behaviour of Φ can be completely described by means of the so-called „normal form" corresponding to such a singularity. The above assumptions on f imply the existence of a fold point at 0. A corresponding result within the framework of Besov and Triebel- Lizorkin spaces was proved by M.Geisler, T.Runst (1991). A.Manes, A.M. Micheletti (1973) replaced the condition $0 < f'(-\infty) < \lambda_1$ by $-\infty < f'(-\infty) < \lambda_1$. The aim of the following survey is to describe how one can use results of the singularity theory in quasi-Banach and Banach spaces, respectively, for the investigation of the solution structure. As above mentioned that approach was already taken by A.Ambrosetti and G. Prodi. In recent years, the connection between singularity theory in Banach spaces and multiplicity results has been developped further by M.S.Berger, P.T.Church, J.G.Timourian, V.Cafagna, G.Tarantello, B.Ruf,.... The singularity theory was elaborated for Banach spaces. For general quasi-Banach spaces some striking tools fail: not locally convex, the quasi-norm is not necessarily continuous, the dual space may be empty, ... However, if there are „enough" continuous linear functionals over the quasi-Banach spaces some results of the theory can be carried over. We refer to M.Geisler, T.Runst (1991).

2 DUAL-RICH QUASI-BANACH SPACES

Let X be a quasi-Banach space, i.e. the topology is generated by a quasi-norm $\| \cdot \|$ satisfying the usual conditions of a norm except the triangle inequality which is replaced by $\| x + y \| \le K(\| x \| + \| y \|)$ with some $K \ge 1$. In the following, we describe the concept of dual-rich quasi-Banach spaces. We explain „how many" continuous functionals a space should have to become a „good" space in our considerations.

DEFINITION 1 *A quasi-Banach space X is said to be dual-rich if for all $0 \ne x \in X$ there exists a continuous linear functional $x^\star$ with $x^\star(x) = 1$.*

Remark 1 : Clearly every Banach space is dual-rich; on the other hand, $L_p(\mathbf{R}^1)$ with $0 < p < 1$ is not dual-rich because its dual is trivial.
Remark 2 : In dual-rich quasi-Banach spaces one can define the factorization with respect to spaces of finite dimension and finite codimension, respectively; i.e., these factor spaces are isomorphic to a closed subspace of the dual-rich quasi-Banach space itself.

For our purpose Besov spaces $B^s_{p,q}(\mathbf{R}^n)$ and Triebel-Lizorkin spaces $F^s_{p,q}(\mathbf{R}^n)$ are of special interest. We start by listing some basic material. Let $\mathcal{S} = \mathcal{S}(\mathbf{R}^n)$ denote the Schwartz space of rapidly decreasing functions on the euclidean n-space $\mathbf{R}^n$ and let $\mathcal{S}' = \mathcal{S}'(\mathbf{R}^n)$ be its topological dual. Then one can find functions ϕ_0, $\phi \in \mathcal{S}(\mathbf{R}^n)$ such that

$$supp\ \phi_0 \subset \{x \mid |x| \le 2\}\ , \quad supp\ \phi \subset \{x \mid 1 \le |x| \le 4\}$$

$$0 \le \phi_0(x)\,, \phi(x) \le 1, \ \phi_j(x) := \phi(2^{-j+1}x), \ \sum_{j=0}^{\infty} \phi_j(x) = 1.$$

We call these a dyadic resolution of unity.

DEFINITION 2 *For $-\infty < s < \infty$, $0 < p, q \le \infty$ we define*
$B^s_{p,q}(\mathbf{R}^n) = \{f \in \mathcal{S}'(\mathbf{R}^n) :$

$$\|f|B^s_{p,q}(\mathbf{R}^n)\| = \left(\sum_{j=0}^{\infty} 2^{jsq}\Big(\int_{\mathbf{R}^n} | \mathcal{F}^{-1}\phi_j\mathcal{F}f(x) |^p \, dx\Big)^{\frac{q}{p}}\right)^{\frac{1}{q}} < \infty\}$$

and

$F^s_{p,q}(\mathbf{R}^n) = \{f \in \mathcal{S}'(\mathbf{R}^n) :$

$$\|f|F^s_{p,q}(\mathbf{R}^n)\| = \left(\int_{\mathbf{R}^n} (\sum_{j=0}^{\infty} 2^{jsq} \mid \mathcal{F}^{-1}\phi_j \mathcal{F} f(x) \mid^q)^{\frac{p}{q}} dx\right)^{\frac{1}{p}} < \infty\}$$

(with obvious modification for $p = \infty$ or $q = \infty$).

For $F^s_{p,q}(\mathbf{R}^n)$ we always assume that $p < \infty$. Here $\mathcal{F}$ stands for Fourier transform in $\mathcal{S}'$ and $\mathcal{F}^{-1}$ for its inverse; $F^s_{p,q}(\mathbf{R}^n)$ and $B^s_{p,q}(\mathbf{R}^n)$ are Banach spaces for $p, q \geq 1$ and quasi-Banach spaces otherwise.

LEMMA 1 *For $-\infty < s < \infty$, $0 < p \leq \infty$, $F^s_{p,q}(\mathbf{R}^n)$ and $B^s_{p,q}(\mathbf{R}^n)$ are dual-rich.*

Proof: We know that $\mathcal{S} \subset B^s_{p,q}(\mathbf{R}^n) \subset \mathcal{S}'$ and $\mathcal{S} \subset F^s_{p,q}(\mathbf{R}^n) \subset \mathcal{S}'$, where the embeddings are continuous. If $\psi \in \mathcal{S}$, this means that

$$\mid f(\psi) \mid \leq \|f \mid B^s_{p,q}(\mathbf{R}^n)\| \, P(\psi).$$

Thereby c does not depend on f and ψ, and P stands for a proper seminorm in the $\mathcal{S}$-topology. From this we conclude that $\mathcal{S} \subset (B^s_{p,q}(\mathbf{R}^n))'$, where the prime denotes duality. If $f(\psi) = 0$ for all $\psi \in \mathcal{S}$ then $f = 0$ in $B^s_{p,q}(\mathbf{R}^n)$. Consequently for all $f \neq 0$ we find $\psi \in \mathcal{S}$ with $f(\psi) = 1$. The same arguments for $F^s_{p,q}(\mathbf{R}^n)$ finish the proof.

Let us remark that the two scales of spaces $B^s_{p,q}(\mathbf{R}^n)$ and $F^s_{p,q}(\mathbf{R}^n)$ cover many well-known function spaces. We give some examples; for details we refer to H.Triebel (1983). Let $s > 0$; then $B^s_{\infty,\infty}(\mathbf{R}^n) = \mathcal{C}^s(\mathbf{R}^n)$, a Zygmund space. If s is not an integer then $B^s_{\infty,\infty}(\mathbf{R}^n) = C^s(\mathbf{R}^n)$, a Hölder space. Let $1 < p < \infty$, $-\infty < s < \infty$, then $F^s_{p,2}(\mathbf{R}^n) = H^s_p(\mathbf{R}^n)$, a Bessel-potential space. Let $1 \leq p < \infty$, $1 \leq q \leq \infty$, $s > 0$, then $B^s_{p,q}(\mathbf{R}^n) = \Lambda^s_{p,q}(\mathbf{R}^n)$, the classical Besov space.
For dealing with boundary value problems it is useful to define Besov and Triebel-Lizorkin spaces on domains. Let Ω be a bounded C^∞- domain in $\mathbf{R}^n$ with boundary $\partial\Omega$. Then one can introduce the spaces $B^s_{p,q}(\partial\Omega)$ and $F^s_{p,q}(\partial\Omega)$ by standard procedure via local charts, see H.Triebel (1983). The spaces $B^s_{p,q}(\Omega)$ and $F^s_{p,q}(\Omega)$ are defined as usually by the restriction method, see next definition. If $g \in \mathcal{S}'$ then the restriction of g to Ω is an element of $\mathcal{D}'(\Omega)$ which will be denoted by $g \mid \Omega$. We recall that $\mathcal{D}'(\Omega)$ is the collection of all complex-valued distributions on Ω.

DEFINITION 3 *(i) Let* $0 < p \leq \infty$, $0 < q \leq \infty$ *and* $-\infty < s < \infty$. *Then*

$$B^s_{p,q}(\Omega) = \{f \in \mathcal{D}'(\Omega),\ \exists g \in B^s_{p,q}(\mathbf{R}^n)\ with\ g \mid \Omega = f\}$$

$$\| f \mid B^s_{p,q}(\Omega) \| = \inf \| g \mid B^s_{p,q}(\mathbf{R}^n) \|,$$

where the infimum is taken over all $g \in B^s_{p,q}(\mathbf{R}^n)$ *with* $g \mid \Omega = f$.

(ii) Let $0 < p < \infty$, $0 < q \leq \infty$ *and* $-\infty < s < \infty$. *Then*

$$F^s_{p,q}(\Omega) = \{f \in \mathcal{D}'(\Omega),\ \exists g \in F^s_{p,q}(\mathbf{R}^n)\ with\ g \mid \Omega = f\}$$

$$\| f \mid F^s_{p,q}(\Omega) \| = \inf \|g \mid F^s_{p,q}(\mathbf{R}^n) \|,$$

where the infimum is taken over all $g \in F^s_{p,q}(\mathbf{R}^n)$ *in the above sense.*

As above the following embeddings are continuous:

$$\mathcal{D}(\Omega) \subset B^s_{p,q}(\Omega) \subset \mathcal{D}'(\Omega),$$

$$\mathcal{D}(\Omega) \subset F^s_{p,q}(\Omega) \subset \mathcal{D}'(\Omega),$$

This yields an analogue to Lemma 1.

LEMMA 2 *Let* $-\infty < s < \infty$, $0 < p, q \leq \infty$, Ω *be a bounded domain with smooth boundary. Then the spaces* $F^s_{p,q}(\Omega)$ *and* $B^s_{p,q}(\Omega)$ *are dual-rich.*

It is well-known that for

$$0 < p, q \leq \infty,\ s > s_0 = \frac{1}{p} + \max\left(0,\ (n-1)\,(\frac{1}{p} - 1)\right)$$

the spaces $B^s_{p,q}(\Omega)$ and $F^s_{p,q}(\Omega)$ (here $p \neq \infty$) admits traces on the boundary $\partial\Omega$. These boundary values can be characterized in a similar fashion, namely as $B^{s-\frac{1}{p}}_{p,q}(\partial\Omega)$ and $F^{s-\frac{1}{p}}_{p,p}(\partial\Omega)$, respectively. For the subspaces with vanishing trace in the above sense we write $B^s_{p,q,0}(\Omega)$ and $F^s_{p,q,0}(\Omega)$, respectively.

Let Δ denote the Laplacian. If $0 < p, q \leq \infty$ and $s > s_0$ then Δ yields an isomorphic map from $B^s_{p,q,0}(\Omega)$ onto $B^{s-2}_{p,q}(\Omega)$. If $0 < p < \infty$ a corresponding result holds in $F^s_{p,q,0}(\Omega)$.

DEFINITION 4 *Let X be a quasi-Banach space. A set* $M \in X$ *is called a* C^k *-manifold of codimension 1 if for all* $m \in M$ *one can find a neighbourhood U and a* C^k *-map* $F : U \to \mathbf{R}$ *such that*

$$(i)\ F'(u) \neq 0\,,\ (ii)\ M \cap U = \{x \in U, F(x) = 0\} \tag{3}$$

Now we can formulate the generalization of the above mentioned Ambrosetti- Prodi result (see M.Geisler, T.Runst (1991).

THEOREM 1 *Let $\Omega \subset \mathbf{R}^n$ be a bounded connected domain with smooth boundary $\partial\Omega$. Let $0 < p \leq \infty$, $0 < q \leq \infty$ and $s > \frac{n}{p}$. Assume that the real-valued smooth function f fulfils the following conditions*

$$f(0) = 0\,,\, 0 < f'(-\infty) < \lambda_1 < f'(+\infty) < \lambda_2\,,\, f'' > 0$$

where λ_1, λ_2 have the usual meaning. Then for the solutions of the boundary value problem

$$\begin{aligned} -\Delta\, u(x) - f(u(x)) &= h(x) \quad \text{in } \Omega \\ u(x) &= 0 \qquad \text{on } \partial\Omega \end{aligned} \tag{4}$$

the following assertions are valid:

(i) There exists a closed connected C^1-manifold M_1 of codimension 1 in $B^{s-2}_{p,q}(\Omega)$ such that $B^{s-2}_{p,q}(\Omega) \setminus M_1$ consists of two connected components M_0 and M_2 with the following properties

(a) if $h \in M_0$ then (4) has no solution in $B^s_{p,q,0}(\Omega)$,

(b) if $h \in M_1$ then (4) has exactly one solution in $B^s_{p,q,0}(\Omega)$,

(c) if $h \in M_2$ then (4) has exactly two solutions in $B^s_{p,q,0}(\Omega)$.

(ii) Let $p < \infty$ then a corresponding result holds for $F^s_{p,q,0}(\Omega)$, too.

We remark that $B^s_{p,q}(\Omega) \subset C(\Omega)$ if $s > \frac{n}{p}$ or $s = \frac{n}{p}$ and $0 < q \leq 1$. A corresponding result holds within the framework of Triebel-Lizorkin spaces: We have $F^s_{p,q}(\Omega) \subset C(\Omega)$ if $s > \frac{n}{p}$ or $s = \frac{n}{p}$ and $0 < p \leq 1$.

3 THE SINGULAR SET OF NONLINEAR OPERATORS

At first we give a short survey on singularity theory in dual-rich quasi-Banach spaces. We assume that $E \subset F$ are dual-rich quasi-Banach spaces. A smooth nonlinear mapping

$$\Phi : E \to F$$

can be studied by properties of its Fréchet derivative

$$\Phi'(u) : E \to F \quad at\, u \in E.$$

Let $E \subset F$ be dual-rich quasi-Banach spaces and U a connected open subset in E. A mapping $\Phi \in C^1(U, F)$ is called a (nonlinear) Fredholm operator if the Fréchet derivative of Φ , $\Phi'(u)$, is a linear Fredholm map belonging to $\mathcal{L}\ (E\ , \ F)$ for each $u \in U$. In this case the index of Φ, ind Φ, is defined by

$$ind\, \Phi\, (u) \; := \; ind\, \Phi'(u) \; = \; dim\, ker\, \Phi'(u) \; - \; dim\, coker\, \Phi'(u) \;\; for \;\; all\, u \; \in \; U\, .$$

It is known that the index of Φ is independent of $u \in U$. In the following we suppose that Φ is a (nonlinear) Fredholm operator of index 0, with

$$dim\, ker\, \Phi'(u) \leq 1 \quad for \;\; all \;\; u \in E.$$

DEFINITION 5 *(i) A point $u \in E$ is a singular point of Φ if the equation*

$$\Phi'(u)[v] = 0$$

has a nontrivial solution $v \in E$. Otherwise, u is a regular point of Φ. We denote by S_Φ the set of all singular points of Φ.
(ii) A point $y \in F$ is said to be a critical point if $y \in \Phi(S_\Phi)$.

The importance of the singular set for understanding the solution structure becomes clear from the following result.

LEMMA 3 *Let E, F be dual-rich quasi-Banach spaces and let $\Phi : E \to F$ be a proper C^1 - mapping (i.e. for every compact $K \subset F$ the preimage $\Phi^{-1}(K) \subset E$ is compact). Then the following assertions hold.*
(i) If S_Φ is empty then Φ is a global homeomorphism of E onto F.
(ii) For $y \in F \setminus \Phi(S_\Phi)$ (i.e. y is not a critical point) the cardinality of $\Phi^{-1}(y)$

$$N(y) \;\; = \;\; \# \, \{u \in E, \; \Phi(u) \;\; = \;\; y\}$$

is finite and locally constant and consequently constant on every connected component of $F \setminus \Phi(S_\Phi)$.

The basic tools for the classification of the singularities are the following two lemmas. We remark that the singularities investigated here belong to infinite-dimensional quasi-Banach spaces. But by our assumption with respect to Φ these mappings can be handled like their finite-dimensional analogeous. The first result is called the local representation lemma.

LEMMA 4 *Let E, F be dual-rich quasi-Banach spaces, $E \subset F, \Phi : E \to F$ as above. Let $\tilde{u} \in E$. We suppose that*

$$dim\ ker\ \Phi'(\tilde{u}) \ = \ dim\ coker\ \Phi'(\tilde{u}) \ = \ 1 \ at\ \tilde{u} \in \ E.$$

Let $\Phi'(\tilde{u})[v] = 0$ for $0 \neq v \in E$. Then there exist neighbourhoods $U(\tilde{u}) \subset E$, $V(\Phi(U)) \subset F$, a quasi-Banach space X and smooth diffeomorphisms

$$\alpha : U \to \alpha(U) \subset \mathbf{R}^1 \times X$$

$$\beta : V \to \beta(V) \subset \mathbf{R}^1 \times X,$$

such that

$$\Psi = \beta \circ \Phi \circ \alpha^{-1} : \alpha(U) \subset \mathbf{R}^1 \times X \to \beta(V) \subset \mathbf{R}^1 \times X$$

has the form

$$\Psi\ (t, x) = (f(t, x); x).$$

Proof: This lemma is due to B.Ruf (1990b) for a special case in Banach spaces. The proof for dual-rich quasi-Banach spaces is the same as in B.Ruf (1990b) because the splitting procedures do work in this case. Let $v = v(\tilde{u})$ be a nontrivial solution of $\Phi'(\tilde{u})[v] = 0$. Let $P_0 : F \to [v]$, $P_1 = I - P_0$ and $P_1 F = F_1$. Now we put $F_1 := X$ and u = tv + y, where $y \in P_1 E$. Hence there exists a $\tilde{t}$ such that $\tilde{u} = \tilde{t}v + \tilde{y}$. For fixed $\tilde{t}$ the operator $P_1\Phi(\tilde{t}v + \cdot) : P_1 E \to X$ is locally invertible. For $\tilde{x} = P_1\Phi(\tilde{u})$, we denote by $y(t, x)$ the unique solution of

$$P_1\Phi(tv + y) = x \in U(\tilde{x}) \subset X\ .$$

Using the implicit function theorem with respect to quasi-Banach spaces, we obtain that $y(t, x)$ depends smoothly on (t, x) for (t; x) near $(\tilde{t};\ \tilde{x})$. Let α^{-1} be the local diffeomorphism

$$\begin{aligned} \alpha^{-1} : U(\tilde{t};\ \tilde{x}) \subset \mathbf{R}^1 \times X &\to [v] \oplus P_1 E \\ (t; x) &\to tv + y(t, x). \end{aligned}$$

Then we get

$$\Phi \circ \alpha^{-1}\,(t,x) \;=\; P_0\Phi(tv \;+\; y(t,x)) \;+\; x\,.$$

Denoting the natural identification by β,

$$\begin{aligned} \beta \;:\; [v] \oplus X \;&\rightarrow\; \mathbf{R}^1 \times X \\ tv \;+\; x \;&\rightarrow\; (t;x), \end{aligned}$$

then

$$\Psi \;=\; \beta \circ \Phi \circ \alpha^{-1} \;:\; \mathbf{R}^1 \times X \;\rightarrow\; \mathbf{R}^1 \times X$$

has the desired form.

Remark 3: We have proved that $f(t,x)\cdot v \;=\; P_0\Phi(tv \;+\; y(t,x))$.
Furthermore, we apply the Malgrange preparation theorem, see B. Ruf (1990b), P. Michor (1980).

LEMMA 5 *Let X be a Banach space, U an open neighbourhood of the origin (0;0) of $\mathbf{R}^1 \times X$ and $f : U \rightarrow \mathbf{R}^1$ a smooth function such that*

$$f(0;0) \;=\; \frac{\partial}{\partial t}\, f(0;0) \;=\; \ldots \;=\; \frac{\partial^{k-1}}{\partial t^{k-1}}\, f(0;0) \;=\; 0\;,\; \frac{\partial^{k}}{\partial t^{k}}\, f(0;0) \;\neq\; 0$$

Then there exist smooth functions

$$a_j \;.\; U \;\rightarrow\; \mathbf{R}^1\;,\; j \;=\; 0,\ldots,k-1$$

vanishing at (0;0) such that

$$t^k \;=\; \sum_{j=0}^{k-1} a_j(f(t,z),z)\cdot t^j$$

Up to now we are not able to carry over this result to quasi-Banach spaces. Therefore we have to restricted our next considerations to Banach spaces. But it should be also possible to extend the last lemma to dual-rich quasi-Banach spaces.

4 CLASSIFICATION OF SINGULARITIES

In what follows, we consider the nonlinear operator

$$\Phi_\lambda \;:\; E \;\rightarrow\; F\,,$$

generated by the semilinear elliptic boundary value problem with Neumann condition

$$\begin{aligned} -\Delta u - \lambda u + f(u) &= h \ in\, \Omega \\ \frac{\partial u}{\partial n} &= 0 \ on\ \partial\Omega, \end{aligned} \tag{5}$$

where $\Omega \subset \mathbf{R}^n$ is a bounded and smooth domain, $\lambda \in \mathbf{R}^1$ is a parameter, f is a sufficiently smooth real-valued function on $\mathbf{R}^1$ with $f'(t) \geq f'(0) = 0$ for all $t \in \mathbf{R}^1$ and $h \in C^\alpha(\Omega)$, with $0 < \alpha < 1$ fixed, is a given function. Furthermore, we put

$$E = \{ u \in C^{2+\alpha}(\Omega), \frac{\partial u}{\partial n} = 0 \ on\ \partial\Omega\}, \ F = C^\alpha(\Omega). \tag{6}$$

Let $0 = \lambda_1 < \lambda_2 \leq \lambda_3 \leq ...$ denote the eigenvalues of the Laplacian on Ω under Neumann condition. In our example we suppose that

$$\Phi_\lambda(u) = -\Delta u - \lambda u + f(u), \ \Phi_\lambda : E \to F$$

is a proper mapping (it holds, for instance, when $f(t) = h(t)t^3$, where h is a bounded and smooth function). We get

$$\Phi_\lambda{}'(u)[v] = -\Delta v - \lambda v + f'(u)v, \ \Phi_\lambda{}'(u) : E \to F.$$

By the linear theory of the Laplacian we obtain that $\Phi_\lambda{}'(u)$ is a Fredholm operator of index 0. Furthermore, let $\mu_1^\lambda(u) < \mu_2^\lambda(u) \leq \mu_3^\lambda(u) \leq ....$ denote the eigenvalues of $\Phi_\lambda{}'(u)$:

$$\Phi_\lambda{}'(u)[v_i^\lambda(u)] = (-\Delta - \lambda + f'(u))\, v_i^\lambda(u),$$

where $\{v_i^\lambda(u)\}_{i=1}^\infty$ denotes a corresponding orthonormal system of eigenfunctions (with respect to the Hilbert space $L_2(\Omega)$). We remark that the first eigenvalue is simple and the following results are known :

(i) Equation (5) has a unique solution in E if h belongs to F and $\lambda < 0$: The linearization of Φ_λ, i.e.,

$$\Phi_\lambda{}'(u)[v] = -\Delta v - \lambda v + f'(u)v$$

is regular at every $u \in E$ (see Lemma 3).

(ii) If $\lambda > \lambda_1 = 0$ then Φ_λ has singular points, i.e., there exist nontrivial functions $v \in E$ with

$$\begin{aligned} -\Delta v - \lambda v + f'(u)v &= 0 \ in\, \Omega \\ \frac{\partial v}{\partial n} &= 0 \ on\ \partial\Omega. \end{aligned} \tag{7}$$

Remark 4: Applying Lemma 3 we obtain results within the framework of corresponding spaces of Besov-Triebel-Lizorkin type. For the sake of simplicity we formulate the results in Hölder spaces, only. The most of the following assertions may be found in the paper by B. Ruf (1990a).

At first we describe the singular set S_Φ^λ .

THEOREM 2 *For* $\lambda_1 < \lambda < \lambda_2$ *one has*

$$S_\Phi^\lambda = \{ u \in E, \mu_1^\lambda(u) = 0 \} \tag{8}$$

and S_Φ^λ *is a smooth manifold of codimension 1 in E. Furthermore, we have*

$$\mu_2^\lambda \geq \lambda_2 - \lambda \textit{ for every } u \in E. \tag{9}$$

Proof: Assertion (8) follows from (9) since for $\lambda < \lambda_2$ we get $\mu_2^\lambda > 0$ for all $u \in E$. The inequality in (9) one obtains by the variational characterization of the eigenvalues and the properties of f.

Remark 5: Furthermore, one can consider the case $\lambda_{k-1} < \lambda < \lambda_k$, where λ_{k-1} is a simple eigenvalue, too. Then $S_\Phi^\lambda = \{u \in E, \mu_{k-1}^\lambda(u) = 0\}$. By the hypothesis on λ one gets *dim ker* $\Phi_\lambda{}'(u) = 1$ for $u \in S_\Phi^\lambda$.

Now we begin with the classification of the singularities using the methods by B.Ruf (1990b). He showed that the kinds of singularities can completely described by derivatives of the function $\mu^\lambda(u)$. Here we only consider the case $0 = \lambda_1 < \lambda < \lambda_2$. With regard to the last theorem we use the following notations: $S := S_\Phi^\lambda$, $\mu(u) := \mu_1^\lambda(u)$, $v(u) := v_1^\lambda(u)$ *and* $\Phi := \Phi_\lambda$ for fixed λ .

DEFINITION 6 *(a) A point* $u \in S$ *is 1-transverse, if there exists a* $w \in E$ *such that* $\mu'(u)[w] \neq 0$. *We define* $S_1 = \{ u \in S, u \, is \, 1 - transverse\}$.

(b) Let $u \in S_1$. *Then u is a fold point if*

$$\mu'(u)[v(u)] \neq 0. \tag{10}$$

This is equivalent to say that $v(u) \notin T_u S$, *where* $T_u S$ *denotes the tangent space to S at the point u.*

Remark 6: Because of $\lambda_1 < \lambda < \lambda_2$ we get that $S = S_1$ holds. Applying Lemma 4 and Lemma 5 we obtain the normal form for folds.

THEOREM 3 *If u is a fold point of Φ, then there exist neighbourhoods $U(u) \subset E$, $V(\Phi(u)) \subset F$, a Banach space X and diffeomorphisms $\alpha : U \to \alpha(U) \subset \mathbf{R}^1 \times X$, $\beta : V \to \beta(V) \subset \mathbf{R}^1 \times X$ such that $\beta \circ \Phi \circ \alpha^{-1}(t,x) = \Phi_F(t,x)$ where $\Phi_F : \mathbf{R}^1 \times X \to \mathbf{R}^1 \times X$ is given by $(t;x) \to (t^2;x)$.*

Examples: (i) If $\Phi = \Phi_\lambda = -\Delta - \lambda + f$ then condition $\mu'(u)[v(u)] \neq 0$ is equivalent to $\int_\Omega f''(u)v^3(u)d\omega \neq 0$, ie. if $f(u) = u^3$ one gets $\int_\Omega uv^3(u)d\omega \neq 0$.
(ii) In the above mentioned paper by A.Ambrosetti, G.Prodi (1972) the conditions of the function f imply that the origin is a global fold point.
(iii) Further results concerning fold points may be found in the paper by V.Cafagna, G.Tarantello (1987).
In a similar way, one can introduce higher singularities.

DEFINITION 7 *(a) Let $u \in S_1$ be not a fold point, i.e. $\mu'(u)[v(u)] = 0$. We say that u is 1-1-transverse, if there exists a $w \in T_uS_1$ such that*

$$(\mu'(u)[v(u)])'[w] \neq 0\,. \tag{11}$$

Let $S_{1,1} = \{u \in S_1\,, u \, is\, 1-1-transverse\}$.
(b) A point $u \in S_{1,1}$ is said to be a cusp point, if

$$\mu_{vv}(u) = (\mu'(u)[v(u)])'[v(u)] \neq 0. \tag{12}$$

Here we use the notation $\mu_{vw} = (\mu'(u)[v(u)])'[w]$.

This is equivalent to say that v(u) is in T_uS_1 and does not belong to the tangent space of $S_{1,1}$. By B.Ruf (1990b) the following result concerning the normal form for cusps was proved by application of Lemma 4 and Lemma 5.

THEOREM 4 *If $u \in S$ is a cusp point for Φ, then there are neighbourhoods $U(u) \subset E$, $V(\Phi(u)) \subset F$, a Banach space X, and diffeomorphisms $\alpha : U \to \alpha(U) \subset \mathbf{R}^2 \times X$, $\beta : V \to \beta(V) \subset \mathbf{R}^2 \times X$ such that $\beta \circ \Phi \circ \alpha^{-1}(s,t,x) = \Phi_C(s,t,x)$ where $\Phi_C : \mathbf{R}^2 \times X \to \mathbf{R}^2 \times X$ is given by $(s,t;x) \to (s^3 - st, t; x)$.*

Examples: (i) If $\Phi = \Phi_\lambda = -\Delta - \lambda + f$ then the cusp condition (12) is equivalent to

$$C_\lambda(u) := \int_\Omega f'''(u)v^4(u)d\omega + 3\int_\Omega f''(u)v^2(u)v_v(u)d\omega \neq 0\,, \tag{13}$$

where

$$v_v(u) = (-\Delta - \lambda + f'(u))^{-1}\left(-f''(u)v^2(u)\right) \; and \; \frac{\partial v_v}{\partial n} = 0 \; on \; \partial\Omega\,. \tag{14}$$

Here we used that $f''(u)v^2(u)$ is normal to v(u).

(ii) B.Ruf (1990b) considered the case $f(u) = u^3$.

(iii) Further results concerning cusp singularities may be found in the paper by V.Cafagna, G.Tarantello (1987).

In the same way one can introduce higher singularities.

DEFINITION 8 *(a) Let* $u \in S_{1,1}$ *be not a cusp point,i.e.* $\mu_{vv}(u) = 0$. *We say that* u *is 1-1-1-transverse, if there exists a* $w \in T_uS_{1,1}$ *such that*

$$\mu_{vvw}(u) \neq 0\,. \tag{15}$$

Let $S_{1,1,1} = \{u \in S_{1,1};\, u \,is\, 1-1-1-transverse\}$.

(b) A point $u \in S_{1,1,1}$ *is a swallow tail singularity, if*

$$\mu_{vvv}(u) \neq 0\,. \tag{16}$$

Applying Lemma 4 and Lemma 5 one obtain the normal form for swallow tails, see B.Ruf (1990b).

THEOREM 5 *If* $u \in S$ *is a swallow tail singularity for* Φ, *then there exist neighbourhoods* $U(u) \subset E$, $V(\Phi(u)) \subset F$, *a Banach space* X, *and local diffeomorphisms* $\alpha : U \to \alpha(U) \subset \mathbf{R}^3 \times X$ *and* $\beta : V \to \beta(V) \subset \mathbf{R}^3 \times X$ *such that* $\beta \circ \Phi \circ \alpha^{-1}(r,s,t,x) = \Phi_S(r,s,t,x)$ *where* $\Phi_S : \mathbf{R}^3 \times X \to \mathbf{R}^3 \times X$ *is given by* $(r,s,t;x) \to (t^4 + st^2 + rt, s, r; x)$.

DEFINITION 9 *(a) Let* $u \in S_{1,1,1}$ *be not a swallow tail,i.e.* $\mu_{vvv}(u) = 0$. *We say that* u *is 1-1-1-1-transverse, if there exists a* $w \in T_uS_{1,1,1}$ *such that*

$$\mu_{vvvw}(u) \neq 0. \tag{17}$$

Let $S_{1,1,1,1} = \{u \in S_{1,1,1};\ u\, is\, 1-1-1-1-transverse\}$.

(b) A point $u \in S_{1,1,1,1}$ *is a butterfly singularity, if*

$$\mu_{vvvvv}(u) \neq 0. \tag{18}$$

THEOREM 6 *If* $u \in S$ *is a butterfly singularity, then there exist neighbourhoods* $U(u) \subset E, V(\Phi(u)) \subset F$, *a Banach space* X, *and local diffeomorpisms* $\alpha : U \to \alpha(U) \subset \mathbf{R}^4 \times X$, $\beta : V \to \beta(V) \subset \mathbf{R}^4 \times X$ *such that* $\beta \circ \Phi \circ \alpha^{-1}(q,r,s,t,x) = \Phi_B(q,r,s,t,x)$ *where* $\Phi_B : \mathbf{R}^4 \times X \to \mathbf{R}^4 \times X$ *is given by* $(q,r,s,t;x) \to (t^5 + st^3 + rt^2 + qt, s, r, q; x)$

5 CONCLUDING REMARKS

As above showed (see Theorem 2) the singular set S_Φ^λ of our semilinear boundary value problem

$$\begin{aligned} -\Delta u - \lambda u + f(u) &= h \ in\ \Omega \\ \frac{\partial u}{\partial n} &= 0 \ on\ \partial\Omega \end{aligned} \tag{19}$$

is defined by

$$S_\Phi^\lambda = \{u \in E,\ \mu_1^\lambda(u) = 0\} \tag{20}$$

if $0 = \lambda_1 < \lambda < \lambda_2$.

In the case when $\lambda_1 < \lambda < \frac{\lambda_2}{7}$ holds B.Ruf (1990a) proved that $C_\lambda(u)$ is positive for all $u \in S_\Phi^\lambda$ (see (13)). It means that the singular set contains only fold and cusp points. A similar result (for a different range of λ) holds also if $f(t) = t^3$ is replaced by $f(t) = h(t)t^3$, where h is a bounded C^3 ($\mathbf{R}^1$)-function. Up to now it is a open problem whether corresponding results hold also for more general nonlinearities f. Furthermore, B.Ruf (1990b) proved that for certain domains the singular set contains also higher singularities (swallow tail and butterfly points) when $\frac{\lambda_2}{7} < \lambda < \lambda_2$. To prove it B.Ruf (1990b) used a step function $z(x) \in B_{p,p}^{\frac{1}{p}}(0,1) \cap L_\infty(0,1)$ which does not belong to the classical spaces $C^{2+\alpha}(0,1)$ and $W_2^1(0,1)$, respectively. This shows that it is useful to try to deal with problems of the above type not only, say, in Hölder and Sobolev spaces, but in more general function spaces. It was the aim of this survey to describe such an approach with respect to the two scales of function spaces of Besov and Triebel- Lizorkin type. For further results and open questions in Hölder spaces we refer also to B.Ruf (1990b).

6 REFERENCES

Ambrosetti, A., and Prodi, G. (1972). On the inversion of some differentiable mappings with singularities between Banach spaccs, Ann. Mat. Pura Appl., 93: pp. 231 246.

Cafagna, V., and Tarantello, G. (1987). Multiple solutions for some semilinear elliptic equations, Math. Ann., 276: pp. 643-656.

Dolph, C. L. (1949). Nonlinear integral equations of the Hammerstein-type, Trans. AMS, 66: pp. 289-307.

Geisler, M., and Runst, T. (1991). On a superlinear Ambrosetti-Prodi problem in Besov and Triebel-Lizorkin spaces, J. London Math. Soc., 43: pp. 324-336.

Manes, A., and Micheletti, A.M. (1973). Un 'estensione della teoria variazionale classica degli autovalori per operatori elliptic del secondo ordine, Boll. Un. Mat. Ital., 7: pp. 285-301.

Michor, P. (1980). The division theorem on Banach spaces, Österr. Ak. Wiss. II-189-1-3: pp. 1-18.

Ruf, B. (1990a). Singularity theory and the geometry of a nonlinear elliptic equation, Ann. Sc. Norm. Sup. Pisa, C. Sc. 17: pp. 1-33.

Ruf, B. (1990b). Singularity theory and forced secondary bifurcation, preprint Dip. Mat. "F. Enriques", Milano, Quarderno n. 48/1990: pp. 1-32.

Triebel, H. (1983). Theory of function spaces, Birkhäuser Verl., Boston, Stuttgart; and Akad. Verlagsges. Geest & Portig, Leipzig

Random Approximation in Numerical Analysis

STEFAN HEINRICH Fachbereich Informatik, Universität Kaiserslautern
Erwin-Schrödinger-Str., D-W-6750 Kaiserslautern, Germany

1 Introduction

One of the most intensively disputed questions of computational mathematics is the following: What is the use of Monte Carlo methods, i. e. can it be of help to involve chance, randomness into numerical processes, and if yes, in which situations is this advisable? Due to the widespread application of Monte Carlo in physical simulation, there is a lot of empirical evidence, and based on this, some intuitive feeling about the problem. However, such impressions are often subjective and depend on the concrete example being computed. This brings up the need for a mathematical basis and a theoretical understanding of the efficiency of randomized algorithms. The first results in this direction were obtained at the end of the fifties, while an intensive wider research started only during the last decade. This is an achievement of information-based complexity theory [TWW88], a new branch of complexity theory dealing with problems of (numerical) analysis. This theory is concerned with the minimal computational effort needed to solve an analytic task, that is, it studies the intrinsic difficulty of continuous (as opposed to discrete,combinatorial) problems.

Information-based complexity provides the notions to give a precise mathematical meaning to the efficiency issues both for deterministic and stochastic methods. The complexity of basic numerical problems can be determined using tools of this theory. This way some comparison between the deterministic and the stochastic setting becomes possible, and some basic orientation for using this or that method can be derived, providing theoretical foundation for practical experience.

The aim of this paper is twofold. In the first part (sections 2 - 6) I want to give a survey on recent developments of Monte Carlo complexity. This will include techniques to derive sharp lower bounds as well as the construction of concrete numerical methods which attain these optimal bounds. The field covered here lies at the frontiers of several disciplines, among them theoretical computer science, numerical analysis, probability theory, approximation theory and to a large extent functional analysis. I want to stress the latter aspect and show how new techniques from Banach space and operator theory can be applied to Monte Carlo complexity.

In the second part I want to present new results - the solution to a problem concering the Monte Carlo complexity of Fredholm integral equations. This will demonstrate in detail the general approach outlined in part one. We develop a new, fast algorithm - it is a combination of Monte Carlo methods with the Galerkin technique, an approach which seems to be new to this field. The basis functions used for the Galerkin discretization are orthogonal splines of minimal smoothness. They lead to an implementable procedure of minimal computational cost.

The paper is organized as follows. In section 2, the main notions of information-based complexity theory are explained. We cover both the deterministic and the stochastic setting in detail, also for the sake of later comparisons. Some relations to s-number theory are presented in section 3. The role of the average case in proofs of lower bounds for Monte Carlo methods is explained in Section 4. In the following three sections, we analyse the complexity of basic numerical problems: Section 5 deals with numerical integration and contains classical results on the complexity of Monte Carlo quadrature, together with an outline of the numerical methods involved. The deterministic counterpart will be mentioned, as well. Section 6 is concerned with random approximation of smooth functions. After recalling well-known estimates on n-widths, which cover the deterministic case, we present recent results on the Monte Carlo setting and sketch the functional analytic background of them. The rest of the paper is devoted to the Monte Carlo solution of Fredholm integral equations. In section 7 we formulate the setting, review some prevoius results and state the new theorems. The proof of the new results, Theorems 7.3 and 7.4, is given in sections 8 - 12. In section 8 we explain the algorithm, which consists of a deterministic and a stochastic part. We also introduce the required spline system. In section 9 we prove the convergence of the deterministic part. Section 10 is devoted to the analysis of the stochastic part. Section 11 completes the proof of the upper bounds. The final section 12 contains the proof of the lower bound.

There are some standard references which may be consulted about the background of the material presented: For information-based complexity theory we refer to Traub, Wasilkowski and Woźniakowski [TWW88], for the basics of Monte Carlo to Ermakov [Erm71] and for s-numbers and operator theory to Pietsch [Pie78], [Pie87] and König[Kön86].

2 Basics of numerical complexity theory

Let us first have a short glance at the history of this field. The origins go back to Kolmogorov's work on foundations of information theory, n-widths and their relation to the quality of numerical methods (see [Kol87]). This was followed by investigations on optimality of numerical methods due to Sard [Sar49], Nikolskij [Nik50], Kiefer [Kie53], and later on, to Bakhvalov [Bak59], Babuška and Sobolev [BS65]. For the seventies we mention the work of Babenko [Bab76] and the study of optimal recovery by Micchelli and Rivlin [MR77]. Nowadays the subject is represented by Traub, Woźniakowski [TW80], [TWW88] dealing with information-based complexity, Smale [Sma85], studying the efficiency of algorithms of analysis, and many others. We refer to [TWW88] for bibliography and more historical details.

Now we shall introduce basic concepts of numerical complexity theory, following essentially the approach of Traub, Wasilkowski and Woźniakowski [TWW88]. With each numerical problem in consideration, we shall associate the so-called solution operator. Let S be a continuous (possibly nonlinear) mapping from a closed bounded subset X_0 of a Banach space X into a Banach space Y. Here X_0 is interpreted as the set of problem elements, that is, the collection of data for which we want to solve the given numerical problem (i. e. sets of integrants, right-hand sides of equations etc.). S is the operator mapping the data to the exact solution of the problem. X and Y are usually function spaces. How this setting appears in concrete cases can be seen in Sections 5 - 7.

Next we want to consider numerical approximations to this operator. For this sake, we fix a non-empty set of continuous linear functionals $\Lambda \subset X^*$, where X^* denotes the dual space of X. We shall call Λ the set of admissible functionals. First we define the deterministic setting (see [TWW88]). A mapping $A : X_0 \to Y$ is called a deterministic method, if there is an $n \in \mathbf{N}$, an operator $N : X_0 \to \mathbf{R}^n$ of the form

$$Nx = (\lambda_1(x), \dots, \lambda_n(x)) \qquad (x \in X),$$

with $\lambda_1, \dots, \lambda_n \in \Lambda$, and an (arbitrary) mapping $\varphi : N(X_0) \to Y$ such that $A = \varphi \circ N$. The mapping N is called the information operator, φ is called the algorithm, and A is the resulting approximation. What is the background of this definition? The data $x \in X_0$ are usually elements of an infinite dimensional function space, and consequently, a computer cannot work with them directly. Hence one has to obtain certain parameters, certain information about the data, which will be fed into the computational process. The n-tuple $Nx \in \mathbf{R}^n$ represents these parameters. The collection Λ consists of those functionals, which are accessible to us, i. e. whose values at x can be involved into the computational process. One can think of a subroutine (or, as common in theoretical computer science, of an oracle) supplying e. g. function values at given points or Fourier coefficients with given numbers. After having obtained the information, the actual computational process takes place, that is, arithmetic operations are performed involving the components of Nx. This process is symbolized by the mapping φ, sending the information Nx to the outcome of the computation, i. e. to the approximation $\varphi(Nx)$ of the true solution Sx. Although formally separated, the process of obtaining information parameters and the arithmetic computation may, of course, intertwine. Let $\mathcal{A}_n(X_0, Y, \Lambda)$, or shortly $\mathcal{A}_n(\Lambda)$, be the set of all deterministic methods possessing a representation as above with this given n. The error of the approximation $A \in \mathcal{A}_n(\Lambda)$ to S is defined as

$$e(S, A) = \sup_{x \in X_0} \|S(x) - A(x)\|. \tag{1}$$

Here $+\infty$ is admitted as a possible value. The n-th minimal deterministic error is defined by

$$e_n^{det}(S, \Lambda) = \inf_{A \in \mathcal{A}_{n-1}(\Lambda)} e(S, A).$$

The shift in the index to $n - 1$ is convenient for the application of s-number theory. To give a meaning to this definition also for $n = 1$, we let $\mathcal{A}_0(X_0, Y, \Lambda)$ be the set of all constant mappings, i. e. $A \in \mathcal{A}_0(X_0, Y, \Lambda)$ iff there is a $y_0 \in Y$ such that $A(x) = y_0$ for all $x \in X_0$. This corresponds to $N = 0$, the zero information operator.

Thus, $e_n^{det}(S, \Lambda)$ is the best possible precision that can be obtained by any approximation to S using less than n information functionals from Λ. Since in a real numerical program each subroutine call or function evaluation takes a certain minimal time t_0, the quantity introduced above plays the following role as a lower bound: No deterministic program with running time below $(n - 1)t_0$ can produce an approximation to S better than $e_n^{det}(S, \Lambda)$. Although certainly only a rough bound, for many concrete numerical problems there are implementable algorithms whose error, obtained in time proportional to n, matches e_n^{det}. (The numerical problems we discuss below belong to this group, with the possible exception of some cases of approximation of smooth functions in section 6.)

So in many instances, e_n^{det} stands for the order of the error attainable in time n (or with n arithmetic operations). On the other hand, when there is a discrepancy in the behaviour of e_n^{det} and of known concrete algorithms, it is a challenge to understand if there are implementable methods of the same quality as abstract ones. For further discussion of complexity and models of computation we refer to [TWW88] and [HK91].

Next we pass to the setting of stochastic, or Monte Carlo methods. As compared to deterministic procedures, they involve one more feature: A random number generator can be called and the numbers it produces can be used for the computation. That means, the whole method, and hence also the approximate result, depends on chance, or, as we formalize it, on a random parameter.

Now we give the abstract definition, following [TWW88] again. A Monte Carlo method is a pair $M = ((\Omega, \Sigma, \mu), (A_\omega)_{\omega \in \Omega})$, where (Ω, Σ, μ) is a probability space, i. e. Ω is a nonempty set, Σ a σ-algebra on Ω, and μ a (σ-additive) probability measure on (Ω, Σ). The family A_ω is supposed to have the following properties: There is a Σ measurable function $n(\omega)$ from Ω to $\mathbf{N}$ such that for all $\omega \in \Omega$, $A_\omega \in \mathcal{A}_{n(\omega)}(\Lambda, X_0, Y)$, and the mapping

$$(x, \omega) \rightarrow A_\omega(x)$$

is $(\mathcal{B}(X_0) \times \Sigma, \mathcal{B}(Y))$ measurable. Here $\mathcal{B}(X_0)$ stands for the σ-algebra of Borel subsets of X_0. Given $n \in \mathbf{N}$, we let $\mathcal{M}_n(X_0, Y, \Lambda)$, or $\mathcal{M}_n(\Lambda)$ denote the class of all Monte Carlo methods for which there is a function $n(\omega)$ as above with

$$\int_\Omega n(\omega) d\mu(\omega) \leq n.$$

The error of a Monte Carlo method is defined as

$$e(S, M) = \sup_{x \in X_0} \int_\Omega \|S(x) - A_\omega(x)\| d\mu(\omega). \tag{2}$$

That is, we consider the expected error over the realizations of ω, for each $x \in X_0$, and then the uniform deviation over X_0. This is the usual approach in Monte Carlo theory. Now we define the n-th minimal Monte Carlo error as

$$e_n^{MC}(S, \Lambda) = \inf_{M \in \mathcal{M}_{n-1}(\Lambda)} e(S, M).$$

Thus, $e_n^{MC}(S, \Lambda)$ describes the best possible (expected) error of a randomized approximation to S which uses (on the average) not more than $n - 1$ information functionals from Λ. The remarks about complexity issues made for deterministic methods pertain accordingly to the Monte Carlo setting.

3 Relations to s-number theory

Now we shall discuss some general functional analytic aspects of minimal errors. Throughout this section we shall assume that $X_0 = B_X$ is the unit ball of X and $S \in L(X, Y)$ is a bounded linear operator. So the solution operator is, strictly speaking, the restriction of S to B_X, but there is no need to make a notational difference. We start with the deterministic setting. The quantities $e_n^{det}(S, \Lambda)$ turn out to be closely related to the Gelfand numbers, which have been studied for a long time in the theory of linear operators on Banach spaces ([Pie78],

[Kön86]). Let us generalize the standard definition in the following way: Let Λ be, as above, a non-empty subset of the dual space X^*. The n-th Λ-Gelfand number of $T \in L(X, Y)$ is defined as

$$c_n(T, \Lambda) = \inf \|T|_E\|, \tag{3}$$

where the infimum is extended over all subspaces E of X of the form

$$E = \{x \in X : \lambda_1(x) = \lambda_2(x) = \ldots = \lambda_{n-1}(x) = 0\}$$

for some $\lambda_1, \ldots, \lambda_{n-1} \in \Lambda$. Clearly, if $\Lambda = X^*$, then $c_n(T, X^*)$ are just the usual Gelfand numbers $c_n(T)$.

Proposition 3.1 *Assume that $X_0 = B_X$ and $S \in L(X, Y)$. Then for $n \in \mathbf{N}$*

$$c_n(S, \Lambda) \leq e_n^{det}(S, \Lambda) \leq 2c_n(S, \Lambda).$$

Proof: To verify the lower estimate, let $A \in \mathcal{A}_{n-1}(X_0, Y, \Lambda)$, $A = \varphi \circ N$, with $Nx = (\lambda_1(x), \ldots, \lambda_{n-1}(x))$ $(x \in B_X)$. Then

$$\begin{aligned} \sup_{x \in B_X} \|Sx - \varphi(Nx)\| &\geq \sup_{\substack{x \in B_X \\ \lambda_1(x) = \ldots = \lambda_{n-1}(x) = 0}} \|Sx - \varphi(0)\| \\ &\geq \sup_{\substack{x \in B_X \\ \lambda_1(x) = \ldots = \lambda_{n-1}(x) = 0}} \|Sx\| \geq c_n(S, \Lambda). \end{aligned}$$

Here we used central symmetry of B_X and the triangle inequality. Passing to the infimum over N and φ, we arrive at the required inequality. To check the upper estimate, we fix a $\delta > 0$ and choose $\lambda_1, \ldots, \lambda_{n-1} \in \Lambda$ in such a way that

$$\|S|_E\| \leq c_n(S, \Lambda) + \delta$$

with $E = \{x \in X : \lambda_i(x) = 0, \quad i = 1, \ldots, n-1\}$. We put $Nx = (\lambda_1(x), \ldots, \lambda_{n-1}(x))$ and define for $z \in N(B_X)$, $z = (z_i)_{i=1}^{n-1}$

$$F_z = \{Sx : x \in B_X, \quad \lambda_i(x) = z_i, \quad i = 1, \ldots, n-1\}.$$

For $z \in N(B_X)$ we let $\varphi(z)$ be any δ-center of F_z, that is, any element of Y satisfying

$$\sup_{y \in F_z} \|y - \varphi(z)\| \leq \mathrm{rad} F_z + \delta,$$

where the radius of F_z is defined as

$$\mathrm{rad} F_z = \inf_{u \in Y} \sup_{y \in F_z} \|y - u\|.$$

We shall also use the diameter of F_z given by

$$\mathrm{diam} F_z = \sup_{y_1, y_2 \in F_z} \|y_1 - y_2\|.$$

With φ defined as above we have

$$\sup_{x \in B_X} \|Sx - \varphi(Nx)\| = \sup_{z \in N(B_X)} \sup_{y \in F_z} \|y - \varphi(z)\| \leq \sup_{z \in N(B_X)} \mathrm{rad} F_z + \delta. \tag{4}$$

Furthermore, if $y_1, y_2 \in F_z$, then $(y_1 - y_2)/2 \in F_0$, where, by definition,

$$F_0 = \{Sx : x \in B_X,\quad \lambda_i(x) = 0,\quad i = 1, \ldots, n-1\}.$$

Consequently,

$$\text{rad}F_z \le \text{diam}F_z \le 2\text{rad}F_0 = 2\|S|_E\| \le 2c_n(S, \Lambda) + 2\delta.$$

Combining this with (4) and letting $\delta \to 0$ yields the desired estimate.

This argument was essentially taken from [TW80], Th. 6.1 and [TWW88], Th. 5.4.1. No counterpart of (3) is known for the randomized setting, i. e. there is no obvoius way of defining in analogy to (3) "randomized" Gelfand numbers. On the other hand, taking into account how close $e_n^{det}(S, \Lambda)$ and $c_n(S, \Lambda)$ are, one might simply view $e_n^{MC}(S, \Lambda)$ as something like randomized Gelfand numbers.

The two kinds of sets Λ we shall encounter in concrete cases below are either the set of all function evaluations (i. e. δ-functionals), or else $\Lambda = X^*$. This latter case usually stands for the situation that besides point functionals also others, like e. g. Fourier coefficients or scalar products with certain basis splines, are available. Then instead of specifying the analysis to each particular situation, one tries to find a lower bound for all possible $\lambda \in X^*$. We shall have a closer look at this case $\Lambda = X^*$. For simplicity, we write

$$e_n^{det}(S, X^*) = e_n^{det}(S)$$

provided $X_0 = B_X$. Similarly,

$$e_n^{MC}(S, X^*) = e_n^{MC}(S).$$

It turns out that these quantities possess an important property - they are pseudo-s-numbers. What this means is explained in the statement of the following result.

Proposition 3.2 *The quantities e_n^{det} and e_n^{MC} are additive pseudo-s-numbers, i. e. for all Banach spaces X, Y, F, G, for all operators $T, U \in L(X, Y)$, $V \in L(F, X)$, $W \in L(Y, G)$ and natural numbers $m, n \in \mathbf{N}$, for $s_n = e_n^{det}$ or $s_n = e_n^{MC}$ the following holds:*

(i) $\|T\| = s_1(T) \ge s_2(T) \ge \ldots,$

(ii) $s_{m+n-1}(T + U) \le s_m(T) + s_n(U)$, *and*

(iii) $s_n(WTV) \le \|W\| s_n(T) \|V\|.$

The proof of this result is straightforward, and we omit it therefore. Although elementary, Proposition 3.2 is crucial for handling the new quantities, e_n^{MC}. Indeed, the presence of these properties provides access to a variety of tools from the theory of operators between Banach spaces, such as Mayorov's discretization technique for Sobolev spaces and a powerful machinery for handling the resulting problems on operators between finite dimensional ℓ_p spaces. In connection with information-based complexity, this aspect was systematically exploited by Mathé [Mat90], [Mat91] and Heinrich [Hei92]. In section 6 we discuss some of these results. Let us mention that e_n^{det} is, in fact, an s-number sequence, which means that, in addition to the properties above, we also have

$$e_n^{det}(Id_{\ell_2^n}) = 1.$$

Here $Id_{\ell_2^n}$ stands for the identity on the space ℓ_2^n. On the other hand, it can be shown, using techniques described in the next sections, that

$$e_n^{MC}(Id_{\ell_2^n}) \asymp n^{-1/2}.$$

So e_n^{MC} is not even equivalent to an s-number sequence. The notation $a_n \asymp b_n$ means that there are constants $c_1, c_2 > 0$ and $n_0 \in \mathbf{N}$ such that for all $n \geq n_0$, $c_1 a_n \leq b_n \leq c_2 a_n$.

Let us finally establish a relation between $e_n^{det}(S, \Lambda)$ and $e_n^{MC}(S, \Lambda)$.

Proposition 3.3 *If $X_0 = B_X$ and $S \in L(X, Y)$, then*

$$e_n^{MC}(S, \Lambda) \leq 2e_n^{det}(S, \Lambda).$$

The idea of the proof is natural: We take a deterministic method $A = \varphi \circ N$ and view it as a Monte Carlo method by simply assigning to it the trivial one-point probability space. There is, however, one technical detail. According to the definition of a Monte Carlo method, $\varphi \circ N$ has to be $(\mathcal{B}(X_0), \mathcal{B}(Y))$ measurable. So what we will actually do is to replace an arbitray φ by a continuous - and hence measurable one. This idea is taken from [Mat90].

Proof: Fix $\delta > 0$ and let $\lambda_1, \ldots, \lambda_{n-1} \in \Lambda$ be such that

$$\|S|_E\| \leq c_n(S, \Lambda) + \delta,$$

again with

$$E = \{x \in X : \lambda_1(x) = \ldots = \lambda_{n-1}(x) = 0\}.$$

Let $N : X \to \mathbf{R}^{n-1}$ be given by

$$Nx = (\lambda_1(x), \ldots, \lambda_{n-1}(x)).$$

Without loss of generality we can assume that N is onto. Now we use a result of Michael on continuous selections, [Mic56], Proposition 7.2. There is a continuous mapping $\psi : \mathbf{R}^{n-1} \to X$ such that

$$N \circ \psi = Id_{\mathbf{R}^{n-1}} \tag{5}$$

and for all $z \in \mathbf{R}^{n-1}$,

$$\|\psi(z)\| \leq (1 + \delta) \inf\{\|x\| : Nx = z\}. \tag{6}$$

Put $\tilde{\varphi} = S \circ \psi$. Then $\tilde{\varphi}$ is continuous, and hence, $\varphi \circ N$ is Borel measurable. Moreover, by (5), $x - \psi(Nx) \in E$, so we get

$$\begin{aligned}
\sup_{x \in B_X} \|Sx - \tilde{\varphi}(Nx)\| &= \sup_{x \in B_X} \|S(x - \psi(Nx))\| \\
&\leq c_n(S, \Lambda) \sup_{x \in B_X} \|x - \psi(Nx)\| \\
&\leq (2 + \delta) c_n(S, \Lambda),
\end{aligned}$$

where the last inequality is a consequence of (6). Using Proposition 3.1 and letting $\delta \to 0$, we arrive at the required result.

4 Lower bounds for Monte Carlo through average case analysis

The analysis of average case efficiency of numerical methods is an interesting topic in itself. Early work started in the 50's ([Sul59], [Sul60]), while a broader development took place during the last decade. It turns out that average case analysis is also closely tied to the study of Monte Carlo methods. In fact, it is the basic mean of proving lower bounds. This idea was first used by Bakhvalov [Bak59], to establish lower estimates for Monte Carlo integration. Later, the idea was taken up again by Traub, Wasilkowski, Woźniakowski [TWW88] and Novak [Nov88]. Recently, general structural results in this direction were obtained by Mathé [Mat91], [Mat92] and Heinrich [Hei92]. The aim of this section is to present the basic relation between the Monte Carlo and average case setting.

Let again X, Y be Banach spaces, let $X_0 \subset X$ be a closed bounded subset and $S : X_0 \to Y$ a continuous mapping. Now we assume furthermore, that we are given a probability measure ν on $(X_0, \mathcal{B}(X_0))$. Let $\mathcal{A}_n^{mes}(X_0, Y, \Lambda)$, or $\mathcal{A}_n^{mes}(\Lambda)$ denote the subset of those $A \in \mathcal{A}_n(X_0, Y, \Lambda)$ which are $(\mathcal{B}(X_0), \mathcal{B}(Y))$ measurable. For $A \in \mathcal{A}_n^{mes}(X_0, Y, \Lambda)$ we define the average error with respect to the measure ν as

$$e(S, A, \nu) = \int_{X_0} \|S(x) - A(x)\| d\nu(x).$$

Comparing with the error $e(S, A)$ defined above (1), we see that the supremum over all possible $x \in X_0$ (the so-called "worst case") is replaced by the integral over X_0 with respect to the distribution ν (the "average case").

Let us point out the principal difference to Monte Carlo methods. As introduced in section 2, a Monte Carlo method consists of a random family of deterministic methods (A_ω), and in order to define the error, we integrate over this family of methods. In the average case, we have one fixed deterministic method A, while the problem elements are random, i. e. we integrate over X_0. So both approaches are conceptually completely different. Inspite of this, there are surprisingly close connections between both settings. To present them we define the minimal average error with respect to ν by

$$e_n^{avg}(S, \Lambda, \nu) = \inf_{A \in \mathcal{A}_{n-1}^{mes}(\Lambda)} e(S, A, \nu).$$

Proposition 4.1 *For every probability measure ν on X_0 and for all $n \in \mathbf{N}$,*

$$e_n^{MC}(S, \Lambda) \geq \frac{1}{2} e_{2n-1}^{avg}(S, \Lambda, \nu).$$

Proof: Let $M \in \mathcal{M}_{n-1}(X_0, Y, \Lambda)$, $M = ((\Omega, \Sigma, \mu), (A_\omega)_{\omega \in \Omega})$ with $A_\omega \in \mathcal{A}_{n(\omega)}(\Lambda)$ and

$$\int_\Omega n(\omega) d\mu(\omega) \leq n - 1. \tag{7}$$

Define the set $\Omega_0 \in \Sigma$ by

$$\Omega_0 = \{\omega \in \Omega : n(\omega) \leq 2n - 2\}.$$

It follows from (7) that

$$\mu(\Omega_0) \geq \frac{1}{2}.$$

So, for each $\omega \in \Omega_0$, we have $A_\omega \in \mathcal{A}^{mes}_{2n-2}(\Lambda)$, the measurability following from the definition of Monte Carlo methods. Using Fubini's Theorem we get

$$\begin{aligned}
e(S,M) &= \sup_{x\in X_0}\int_\Omega \|S(x)-A_\omega(x)\|d\mu(\omega) \\
&\geq \int_{X_0}\int_\Omega \|S(x)-A_\omega(x)\|d\mu(\omega)d\nu(x) \\
&= \int_\Omega\int_{X_0} \|S(x)-A_\omega(x)\|d\nu(x)d\mu(\omega) \\
&\geq \int_{\Omega_0}\int_{X_0} \|S(x)-A_\omega(x)\|d\nu(x)d\mu(\omega) \\
&\geq \mu(\Omega_0)\inf_{\omega\in\Omega_0}\int_{X_0} \|S(x)-A_\omega(x)\|d\nu(x) \\
&\geq \frac{1}{2}e^{avg}_{2n-1}(S,\Lambda,\nu).
\end{aligned}$$

Thus, in order to prove lower bounds for Monte Carlo methods, one has to design an appropriate measure and then estimate the minimal average error from below. Surprisingly enough, this problem is much better suited to the application of functional analytic and approximation theoretic tools than the task of directly deriving lower bounds for Monte Carlo methods.

In principle, one could take the supremum over all possible measures ν as a lower bound. Here an interesting general open question is involved: To which extent is such kind of supremum not only a lower bound, but close or even equal to the minimal Monte Carlo error? Packel [Pac87] gave a positive answer for some discrete situations. Partial answers for continuous problems are contained in recent work by Mathé [Mat92], who uses functional analytic tools, in particular Ky Fan's Lemma.

5 Numerical integration

We begin our treatment of concrete numerical problems with classical results about quadrature. To put the integration problem into the setting of section 2, we let $d, r \in \mathbf{N}$, and $1 \leq p \leq \infty$. $G = [0,1]^d$ is the d-dimensional unit cube, $X = W^r_p(G)$ the Sobolev space (see [Tri78]), $X_0 = B_X$ and $Y = \mathbf{R}$. We want to approximate the integral of functions from the unit ball of $W^r_p(G)$, so the solution operator $S^I : X_0 \to \mathbf{R}$ is defined as

$$S^I f = \int_G f(t)dt \qquad (f \in B_{W^r_p(G)}).$$

We assume that we can use function values for the computational process, that is, the set of admissible functionals will be

$$\Lambda = \Delta := \{\delta_t : t \in G\},$$

where, as usual, $\delta_t(f) = f(t)$. We assume that $r/d > 1/p$, so that by the Sobolev embedding theorem [Tri78], $W^r_p(G)$ is embedded into $C(G)$, the space of continuous functions on G. Hence the delta functionals are well-defined and belong to X^*.

Now we are ready to present the estimate of the minimal error of deterministic approximate integration. It is a classical result of numerical quadrature theory. The first approach to the lower bound is due to Sard [Sar49] and Nikolskij [Nik50].

Theorem 5.1 *Let $r/d > 1/p$. Then*

$$e_n^{det}(S^I, \Delta) \asymp n^{-r/d}.$$

Although we are mainly concerned with the Monte Carlo setting, I want to include a short proof of this result, because it shows some basic arguments in a very simple situation and gives us the opportunity of distinguishing the specifics of the Monte Carlo setting. The upper bound will be derived from the following by now well-known approximation theoretic result, which we shall repeatedly use later on . It can be found e. g. in Ciarlet [Cia78], Ch. 3.1.

Proposition 5.2 *Let $1 \le p, q \le \infty$ and $r/d > 1/p - 1/q$. Then there is a constant $c > 0$ such that for all $n \in \mathbf{N}$ there exist $t_1, \dots, t_n \in G$, $h_1, \dots, h_n \in C(G)$ with the property that, if $P_n : C(G) \to C(G)$ is defined by*

$$(P_n f)(t_i) = \sum_{i=1}^{n} f(t_i) h_i(t),$$

then for all $f \in W_p^r(G)$

$$\|f - P_n f\|_{L_q} \le c \begin{cases} n^{-r/d+1/p-1/q} \|f\|_{W_p^r} & (p < q) \\ n^{-r/d} \|f\|_{W_p^r} & (p \ge q). \end{cases}$$

Let us briefly comment on how P_n looks like. For $d = 1$ one could, e. g., use Lagrange interpolation of degree $(r-1)$ on intervals $[\frac{k}{m}, \frac{k+1}{m}]$ for some m and $k = 0, \dots, m-1$, for $d > 1$ tensor products of the one-dimensional interpolation operator, where m has to be chosen in such a way that the total number of nodes does not exceed n. (See [HK91], where this is carried through in detail). Other approximations by piecewise polynomials may serve this purpose, as well. A general procedure to obtain such operators as well as proofs of the approximation rates are contained in [Cia78]. Let us emphasize that in all these cases the points t_i as well as the functions h_i are explicitly given and easy to compute. Moreover the functions h_i have small supports in the sense that for each i the support of h_i intersects the support of at most c_1 other h_j, where c_1 is independent of n.

Proof of Theorem 5.1 The upper bound is a direct consequence of 5.2: For $f \in W_p^r(G)$ we put

$$Q_n f = \int_G (P_n f)(t) dt = \sum_{i=1}^{n} f(t_i) \int_G h_i(t) dt.$$

So we arrive at a quadrature formula of the form

$$Q_n f = \sum_{i=1}^{n} w_i f(t_i),$$

and we put $Nf = (f(t_1), \dots, f(t_n))$ and $\varphi(z_1, \dots, z_n) = \sum_{i=1}^{n} w_i z_i$. We get

$$e(S^I, \varphi \circ N) = \sup_{f \in X_0} |S^I f - \varphi(Nf)|$$

$$= \sup_{f \in X_0} |S^I f - Q_n f|$$

$$\le \|(Id - P_n) : W_p^r(G) \to L_1(G)\| \|\text{Int} : L_1(G) \to \mathbf{R}\| \le c n^{-r/d}.$$

Here Id denotes the identity mapping, and Int the integration functional.

In the proof of the lower bound we can assume without loss of generality that n is of the form $n = m^d$. We subdivide $G = [0,1]^d$ into $\tilde{n} = (2m)^d$ closed cubes $(G_i)_{i=1}^{\tilde{n}}$ of sidelength $1/(2m)$ with mutually disjoint interior. Let g_i denote the "left lower" angle of G_i, or precisely, the point of G_i with minimal coordinate in each direction. Let now ψ be a function with $\psi \in C^\infty(\mathbf{R}^d)$, $\psi \geq 0$, $\operatorname{supp} \psi \subseteq G^0$, G^0 the interior of G, $\int_G \psi(t)dt = 1$, and define

$$\psi_i(t) = \psi(2m(t - g_i)).$$

So ψ is a smooth "bump" function, and the ψ_i are obtained by shrinking G to sidelength $1/(2m)$ and shifting to G_i. It is easily checked that

$$\int_G \psi_i(t)dt = (2m)^{-d} = 2^{-d}n^{-1} \tag{8}$$

and that there are constants $c_1, c_2 > 0$, independent of n such that for all $\alpha_1, \ldots, \alpha_{\tilde{n}} \subset \mathbf{R}$

$$c_1 n^{r/d-1/p} \|(\alpha_i)\|_{\ell_p^{\tilde{n}}} \leq \|\sum_{i=1}^{\tilde{n}} \alpha_i \psi_i\|_{W_p^r} \leq c_2 n^{r/d-1/p} \|(\alpha_i)\|_{\ell_p^{\tilde{n}}}. \tag{9}$$

Now assume that we are given any method $A = \varphi o N \in \mathcal{A}_n(\Delta)$, with $Nf = (f(t_1)), \ldots, f(t_n))$ for some $t_1, \ldots, t_n \in G$. We put

$$J = \{j : 1 \leq j \leq \tilde{n}, \ \ \{t_1, \ldots, t_n\} \cap G_j^0 = \emptyset\}. \tag{10}$$

Cleary,

$$|J| \geq \tilde{n} - n = (2^d - 1)n. \tag{11}$$

We let

$$f_0 = \frac{\sum_{j \in J} \psi_j}{\|\sum_{j \in J} \psi_j\|_{W_p^r(G)}}. \tag{12}$$

Hence $f_0 \in X_0$. Moreover, $Nf_0 = 0$, while (8), (9) and (11) give

$$\int_G f_0(t)dt \geq c_3 n^{-r/d},$$

where $c_3 > 0$ is another constant independent of n, which can be calculated from (9).

Finally we get

$$\sup_{f \in X_0} |S^I f - \varphi(Nf)| \geq \max_{\sigma = \pm 1} |S^I(\sigma f_0) - \varphi(0)| \geq |S^I f_0| \geq c_3 n^{-r/d}.$$

Next we present the stochastic case, which is due to Bakhvalov [Bak59], [Bak62]. Related results for weighted integration were obtained by Novak [Nov88].

Theorem 5.3 *Let $r/d > 1/p$. Then*

$$e_n^{MC}(S^I, \Delta) \asymp \begin{cases} n^{-r/d-1/2}, & \text{if } 2 \leq p \leq \infty \\ n^{-r/d-1+1/p}, & \text{if } 1 \leq p < 2. \end{cases}$$

Before we prove it, let us make a few remarks. When comparing the rates in the deterministic and Monte Carlo case, one can see that for the integration problem Monte Carlo provides (except for the case $p = 1$) a faster rate of convergence than deterministic quadrature. (However, we should be aware of different error criteria applied in both cases - so the comparison can only be rough, heuristical.) An important instance about the exponent of improvement $-1/2$ (or $-1 + 1/p$ for $1 \le p < 2$) is its independence of dimension d, in contrast to the deterministic rate $-r/d$. This reflects quite realistically the role of Monte Carlo integration in practice. Let us look at the case $p = \infty$. For small dimensions d (and moderate smoothness), the improvement is negligible, and deterministic quadrature rules will perform satisfactorily. On the other hand, for large d, the deterministic contribution almost vanishes, and only the Monte Carlo contribution leads to reasonable precision. This can be observed in practical computation, where for $d \ge 6$, deterministic rules usually break down and Monte Carlo is the preferred choice.

In order to prove theorem 5.3, we first analyse the standard Monte Carlo quadrature. It is defined as follows: Let $(\xi_i)_{i=1}^n$ be independent, G-valued, uniformly distributed over G random variables on some probability space (Ω, Σ, μ). For $f \in C(G)$, we put

$$Q_\omega f = \frac{1}{n}\sum_{i=1}^{n} f(\xi_i(\omega)) \qquad (\omega \in \Omega). \tag{13}$$

Then we have

$$\mathbf{E}Q_\omega f = \mathrm{Int}(f).$$

First we derive the following estimate.

Proposition 5.4 *Let $1 \le p \le \infty$. Then for all $f \in C(G)$*

$$\mathbf{E}|\mathrm{Int}(f) - Q_\omega f| \le \begin{cases} n^{-1/2}\|f\|_{L_p(G)} & \text{if } \ 2 \le p \le \infty \\ c_p n^{1-1/p}\|f\|_{L_p(G)} & \text{if } \ 1 \le p < 2. \end{cases},$$

where $c_p = 2^{2/p-1}$.

Proof: First we check the case $p = 2$ which will imply the result also for $2 < p \le \infty$.

$$\begin{aligned} (\mathbf{E}|\mathrm{Int}(f) - \tfrac{1}{n}\textstyle\sum_{i=1}^n f(\xi_i)|)^2 &\le \mathbf{E}(\mathrm{Int}(f) - \tfrac{1}{n}\textstyle\sum_{i=1}^n f(\xi_i))^2 \\ = \tfrac{1}{n^2}\mathbf{E}(\textstyle\sum_{i=1}^n(\mathrm{Int}(f) - f(\xi_i)))^2 &= \tfrac{1}{n^2}\mathbf{E}\textstyle\sum_{i=1}^n(\mathrm{Int}(f) - f(\xi_i))^2 \\ = \tfrac{1}{n}\mathbf{E}(\mathrm{Int}(f) - f(\xi_1))^2 &= \tfrac{\mathrm{Var}(f)}{n}. \end{aligned} \tag{14}$$

Here we used the fact that $\mathbf{E}f(\xi_i) = \mathrm{Int}(f)$ and the independence of ξ_i, which implied

$$\mathbf{E}\sum_{\substack{i,j=1\\ i\ne j}}^{n}(\mathrm{Int}(f) - f(\xi_i))(\mathrm{Int}(f) - f(\xi_j)) = 0.$$

The variance of f is defined as

$$\mathrm{Var}(f) = \|f - \mathrm{Int}(f)\|_{L_2(G)}^2 = \int_G (f(t) - \mathrm{Int}(f))^2 dt.$$

It is readily verified that

$$\mathrm{Var}(f) = \int_G f(t)^2 dt - (\int_G f(t)dt)^2 \le \|f\|_{L_2(G)}^2. \tag{15}$$

Inequalities (14) and (15) prove the case $p = 2$. For $f \in C(G)$ we trivially have

$$\mathbf{E}|\mathrm{Int}(f) - \frac{1}{n}\sum_{i=1}^{n} f(\xi_i)| \le |\mathrm{Int}(f)| + \frac{1}{n}\sum_{i=1}^{n} \mathbf{E}|f(\xi_i)| \le 2\|f\|_{L_1(G)}.$$

Hence, if we denote

$$Tf = \mathrm{Int}(f) - \frac{1}{n}\sum_{i=1}^{n} f(\xi_i),$$

T is a linear operator from $C(G)$ into $L_1(\Omega, \mu)$. By the above, we have

$$\|Tf\|_{L_1(\Omega,\mu)} \le 2\|f\|_{L_1(G)}$$

and

$$\|Tf\|_{L_1(\Omega,\mu)} \le n^{-1/2}\|f\|_{L_2(G)}.$$

Real interpolation between the spaces $L_1(G)$ and $L_2(G)$ provides the desired result for $1 < p < 2$ (see [Tri78]):

$$\|Tf\|_{L_1(\Omega,\mu)} \le 2^{1-\theta}(n^{-1/2})^{\theta}\|f\|_{L_p(G)}$$

where θ is such that $1/p = (1 - \theta) + \theta/2$, i. e. $\theta = 2 - 2/p$. This proves Proposition 5.4.

Proof of Theorem 5.3: To show the upper bound, let (Ω, Σ, μ), $(\xi_i)_{i=1}^{n}$ and Q_ω $(\omega \in \Omega)$ be defined as above. Let $f \in W_p^r(G)$, $\|f\|_{W_p^r(G)} \le 1$, i. e. $f \in X_0$. We put, for $\omega \in \Omega$,

$$\begin{aligned} A_\omega(f) &= Q_\omega(f - P_n f) + \mathrm{Int}(P_n f) \\ &= Q_\omega(f - P_n f) + Q_n f. \end{aligned} \tag{16}$$

It is easily seen that $M = ((\Omega, \Sigma, \mu), (A_\omega)_{\omega\in\Omega})$ is a Monte Carlo method in the sense of section 2, where the required measurability follows from the fact that the mapping

$$(f, z) \mapsto f(z) - (P_n f)(z)$$

from $C(G) \times G$ into $\mathbf{R}$ is continuous. Obviously, $M \in \mathcal{M}_{2n}(\Delta)$. By Proposition 5.2

$$\|f - P_n f\|_{L_p(G)} \le cn^{-r/d} \qquad (f \in X_0).$$

Using Proposition 5.4, we get for $f \in X_0$

$$\begin{aligned} &\mathbf{E}|S^I f - A_\omega(f)| = \mathbf{E}|\mathrm{Int}(f) - Q_\omega(f - P_n f) - \mathrm{Int}(P_n f)| \\ = \;& \mathbf{E}|\mathrm{Int}(f - P_n f) - Q_\omega(f - P_n f)| \le n^{-\alpha_p}\|f - P_n f\|_{L_p(G)} \\ \le \;& c_1 n^{-\alpha_p - r/d}, \end{aligned}$$

where $\alpha_p = 1/2$ for $2 \le p \le \infty$, $\alpha_p = 1 - 1/p$ for $1 \le p < 2$, and $c_1 > 0$ is a constant not depending on n. In the proof of the lower bound we shall distinguish two cases. First let $2 \le p \le \infty$. Then it suffices to consider $p = \infty$. We shall define a probability measure on $X_0 = B_{W_\infty^r(G)}$. For this purpose we assume again $n = m^d$ and let $\tilde{n} = 2^d n, (G_i)_{i=1}^{\tilde{n}}$ and $(\psi_i)_{i=1}^{\tilde{n}}$ be as in the proof of Theorem 5.1. Furthermore, let $(\varepsilon_i)_{i=1}^{\tilde{n}}$ be a sequence of independent, $\{-1, 1\}$-valued random variables on some probability space (Ω', Σ', ν') with

$$\nu'\{\varepsilon_i = 1\} = \nu'\{\varepsilon_i = -1\} = 1/2 \qquad (i = 1, \ldots, \tilde{n}), \tag{17}$$

in other words, the (ε_i) are Bernoulli. Let

$$\tilde{\psi}_i = \psi_i / \|\psi_i\|_{W^r_\infty(G)} \qquad (i = 1, \ldots, \tilde{n})$$

and put

$$\nu = \text{dist}\left(\sum_{i=1}^{\tilde{n}} \varepsilon_i \tilde{\psi}_i\right), \tag{18}$$

where dist means the distribution of the X_0-valued random variable. Now we shall use Proposition 4.1. Let $A \in \mathcal{A}_n^{mes}(X_0, \mathbf{R}, \Delta)$, $A = \varphi \circ N$, $Nf = (f(t_1), \ldots, f(t_n))$. Define J as above by

$$J = \{j : 1 \le j \le \tilde{n},\ \{t_1, \ldots, t_n\} \cap G_j^0 = \emptyset\}.$$

Then $N\tilde{\psi}_j = 0$ for $j \in J$, and we can estimate

$$\begin{aligned} & \int_{X_0} |S^I f - \varphi(Nf)| d\nu(f) \\ = \;& \mathbf{E}_{\nu'} |S^I(\textstyle\sum_{i=1}^{\tilde{n}} \varepsilon_i \tilde{\psi}_i) - \varphi(N \sum_{i=1}^{\tilde{n}} \varepsilon_i \tilde{\psi}_i)| \\ = \;& \mathbf{E}_{\nu'} |\textstyle\sum_{i \in J} \varepsilon_i S^I \tilde{\psi}_i + \sum_{i \notin J} \varepsilon_i S^I \tilde{\psi}_i - \varphi(N \sum_{i \notin J} \varepsilon_i \tilde{\psi}_i)|. \end{aligned} \tag{19}$$

The distribution of $(\varepsilon_i)_{i=1}^{\tilde{n}}$ does not change if we replace ε_i by $-\varepsilon_i$ for $i \in J$ (and leave it unchanged for $i \notin J$). Consequently we can continue (19) as follows

$$\begin{aligned} = \;& \mathbf{E}_{\nu'} | - \textstyle\sum_{i \in J} \varepsilon_i S^I \tilde{\psi}_i + \sum_{i \notin J} \varepsilon_i S^i \tilde{\psi}_i - \varphi(N \sum_{i \notin J} \varepsilon_i \tilde{\psi}_i)| \\ \ge \;& \mathbf{E}_{\nu'} |\textstyle\sum_{i \in J} \varepsilon_i S^I \tilde{\psi}_i| \ge c_1 (\sum_{i \in J} |S^I \tilde{\psi}_i|^2)^{1/2} \\ \ge \;& c_1 |J|^{1/2} |S^I \tilde{\psi}_1| \ge c_2 n^{-r/d - 1/2}. \end{aligned}$$

Here we used Khintchine's inequality [Kön86], 1.c.3, relations (8),(9), (11), and c_1, c_2 denote constants independent of n (we use the same symbols for possibly different constants). This completes the proof of the case $2 \le p \le \infty$. Now let $1 \le p < 2$ and let ν be the equidistribution on the set

$$\{\pm \tilde{\psi}_i : \quad i = 1, \ldots, \tilde{n}\},$$

where this time,

$$\tilde{\psi}_i = \psi_i / \|\psi_i\|_{W^r_p(G)}.$$

With $A = \varphi \circ N$ and J as above, we get

$$\begin{aligned} & \int_{X_0} |S^I f - \varphi(Nf)| d\nu(f) = \tfrac{1}{2\tilde{n}} \textstyle\sum_{i=1}^{\tilde{n}} \sum_{\sigma = \pm 1} |\sigma S^I \tilde{\psi}_i - \varphi(N(\sigma \tilde{\psi}_i))| \\ \ge \;& \tfrac{1}{2\tilde{n}} \textstyle\sum_{i \in J} \sum_{\sigma = \pm 1} |\sigma S^I \tilde{\psi}_i - \varphi(0)| \\ \ge \;& \tfrac{1}{2\tilde{n}} \textstyle\sum_{i \in J} |S^I \tilde{\psi}_i| \ge c n^{-r/d + 1/p - 1}, \end{aligned}$$

using (8), (9) and (11) again. This proves theorem 5.3.

Note that the idea of the upper bound proof (see (16)) was to apply the standard Monte Carlo (13) to $f - P_n f$ instead of f and then add a correction term, $\text{Int}(P_n f)$. Using standard

Monte Carlo for $f - P_n f$ means that, according to Proposition 5.4, and in particular to (14), the smaller variance of $f - P_n f$ instead of that of f enters the estimate. This is a simple case of a technique widely used in Monte Carlo, called variance reduction. Let us also note that the method underlying (16) is easily translated into an implementable computer program. It uses dn calls of a random generator providing the uniform distribution on $[0,1]$, $2n$ function value calls, and $O(n)$ arithmetic operations. The latter statement still needs some explanation, namely the expression

$$Q_\omega(P_n f) = \sum_{i=1}^{n}\sum_{j=1}^{n} f(t_j)h_j(\xi_i). \tag{20}$$

By the remarks after Proposition 5.2,

$$\max_i |\{j : \mathrm{supp}(h_i) \cap \mathrm{supp}(h_j)\} \neq \emptyset| \leq c,$$

with c independent of n. Hence only $O(n)$ terms in the above sum (20) are non-zero. Finally, let us mention that for the integration problem, the choice $\Lambda = X^*$ makes no sense, since the solution operator $S^I = \mathrm{Int}$ itself belongs to X^*.

6 Approximation of functions

The next problem we consider is the numerical approximation of smooth functions. We let $X = W_p^r(G)$ be as in the previous section, where $d, r \in \mathbf{N}$, $1 \leq p \leq \infty$, $G = [0,1]^d$, and put $X_0 = B_X$. Furthermore, let $1 \leq q \leq \infty$, $Y = L_q(G)$, and assume that $r/d > 1/p - 1/q$. Let

$$S^E = Id : W_p^r(G) \to L_q(G)$$

be the identical embedding (which exists and is continuous by the Sobolev embedding theorem [Tri78]). For the approximation problem it makes sense to consider $\Lambda = \Delta$ as well as $\Lambda = X^*$, and we shall see that there is a considerable difference between the results. First we treat the (classical) case $\Lambda = \Delta$, which means that we want to approximate Sobolev smooth functions when knowing their values at a certain number of points, which can be chosen in the most favorable way. When $\Lambda = \Delta$, we shall assume that $r/d > 1/p$, so that the δ-functionals belong to X^*. The deterministic case of the following result belongs to classical approximation theory. The fact that in this situation Monte Carlo does not bring improvement, was observed in [TWW88], [Nov88] and [Mat91].

Theorem 6.1 *Let $r/d > 1/p$. Then*

$$e_n^{det}(S^E, \Delta) \asymp e_n^{MC}(S^E, \Delta) \asymp \begin{cases} n^{-r/d}, & q \leq p \\ n^{-r/d+1/p-1/q}, & q > p. \end{cases}$$

Proof: Let us first sketch the deterministic case. The upper bound is a direct consequence of Proposition 5.2. The lower bound can be handled similarly to the case of integration. If $q \leq p$ we define J and f_0 by (10) and (12) and get analogously, from (9)

$$e(S^E, \varphi \circ N) \geq \|S^E f_0\| \geq cn^{-r/d}.$$

For $q > p$ we simply choose a $j \in J$ and put

$$f_0 = \frac{\psi_j}{\|\psi_j\|_{W_p^r(G)}}.$$

Then

$$\|S^E f_0\| \geq cn^{-r/d+1/p-1/q}.$$

In the stochastic case, the upper bound is clear, by Propostion 3.3. The proof of the lower bound is similar to that for stochastic integration. If $q \leq p$, it suffices to consider $p = \infty, \quad q = 1$. Define ν by (17) and (18). Then an analogous argument yields (without using Khintchine's inequality)

$$e(S^E, \varphi \circ N, \nu) \geq \mathbf{E}_{\nu'} \|\sum_{i \in J} \varepsilon_i S^E \tilde{\psi}_i\|_{L_1(G)} \geq cn^{-r/d}.$$

If $q > p$, then we let ν be as in the integration case $1 \leq p < 2$. Here we obtain

$$e(S^E, \varphi \circ N, \nu) \geq \frac{1}{2\tilde{n}} \sum_{i \in J} \|S^E \tilde{\psi}_i\| \geq cn^{-r/d+1/p-1/q}.$$

Now we turn to the case $\Lambda = X^*$. Because of Proposition 3.1 the deterministic case reduces directly to the Gelfand numbers of Sobolev embeddings. For some relations between p and q, the following result belongs to classical approximation theory, while for others the order of the Gelfand numbers was a longstanding problem, which was settled in 1977 by Kashin [Kas77]. Let us state the result without proof (cf. also [Pin85])

Theorem 6.2 *Let $r/d > 1/p - 1/q$. Then*

$$e_n^{det}(S^E) \asymp c_n(S^E) \asymp \begin{cases} n^{-r/d} & if \quad 1 \leq q \leq p \leq \infty \\ & or \quad 1 \leq p \leq q \leq 2 \\ n^{-r/d+1/2-1/q} & if \quad 1 \leq p \leq 2 \leq q \leq \infty \\ n^{-r/d+1/p-1/q} & if \quad 2 \leq p \leq q \leq \infty. \end{cases}$$

The Monte Carlo aspect of this problem has not been considered in approximation theory. It arose in complexity theory and remained open for some time. Recently, it was settled by Mathé [Mat91] and Heinrich [Hei92]. I want to discuss the proof of this result in some detail because of the functional analytic techniques involved.

Theorem 6.3 *Assume $r > d$. Then*

$$e_n^{MC}(S^E) \asymp \begin{cases} n^{-r/d} & if \quad 1 \leq p, q < \infty \quad or \quad p = q = \infty \\ n^{-r/d}(\log n)^{1/2} & if \quad 1 \leq p < \infty, \quad q = \infty \end{cases}$$

and

$$n^{-r/d}(\log n)^{-1/2} \prec e_n^{MC}(S^E) \prec n^{-r/d} \qquad \text{if} \qquad p = \infty, \quad 1 \leq q < \infty.$$

Note that in the case $p = \infty, \quad 1 \leq q < \infty$, the precise order is not known. It is also unknown if the restrictive assumption $r > d$ can be relaxed. In the sequel, we shall only discuss the case $1 \leq p, q < \infty$, in order to avoid technicalities. The full proof is contained in [Hei92].

For the proof of the upper estimate we introduce the following notion. For $T \in L(X,Y)$ and $n \in \mathbf{N}$ define the n-th Monte Carlo approximation number of T by

$$a_n^{MC}(T) = \sup_{x \in B_X} \int_\Omega \|Tx - U_\omega x\| d\mu(\omega)$$

where the infimum is taken over all probability spaces (Ω, Σ, μ) and measurable mappings $\omega \to U_\omega$ from (Ω, Σ) to $(L(X,Y), \mathcal{B}(L(X,Y)))$ with rank $U_\omega \leq n-1$ for all $\omega \in \Omega$. Obviously

$$e_n^{MC}(T) \leq a_n^{MC}(T) \leq a_n(T),$$

where $a_n(T)$ are the usual approximation numbers (cf. [Pie78]). The quantities e_n^{MC} are pseudo-s-numbers. The following proposition is essentially due to Gordon, König, Schütt [GKS87]. They used it to estimate the usual approximation numbers. Mathé [Mat91] observed that the proof can be modified to cover the Monte Carlo approximation numbers, as well. Let us introduce the following notation: For $(x_i)_{i=1}^n \subset X$ and $(x_i^*)_{i=1}^n \subset X^*$ we write

$$h((x_i)) = \mathbf{E}\|\sum_{i=1}^n \gamma_i x_i\|,$$

where $\gamma_1, \ldots, \gamma_n$ is a sequence of independent standard Gaussian (i. e. $N(0,1)$) random variables, and

$$\varepsilon_2((x_i^*)) = \sup_{x \in B_X} (\sum_{i=1}^n (x, x_i^*)^2)^{1/2}.$$

Proposition 6.4 *Let X and Y be Banach spaces, $x_i^* \in X^*$,- $y_i \in Y$, $i = 1, \ldots, n$, and $T = \sum_{i=1}^n x_i^* \otimes y_i$. Then for $1 \leq k \leq n$,*

$$a_{k+1}^{MC}(T) \leq 2k^{-1/2} \varepsilon_2((x_i^*)) h((y_i)).$$

Proof: We can assume without loss of generality that the $(y_i)_{i=1}^n$ are independent. Let $(y_i^*)_{i=1}^n \subset Y^*$ be a biorthogonal sequence to $(y_i)_{i=1}^n$. Let $(\gamma_{ij})_{i,j=1}^n$ be independent standard Gaussian random variables on a probability space (Ω, Σ, μ) and define for $\omega \in \Omega$, $G_\omega \in L(X,Y)$, $P, H_\omega \in L(Y)$ by

$$\begin{aligned} G_\omega &= k^{-1/2} \textstyle\sum_{i,j=1}^n \gamma_{ij}(\omega) x_i^* \otimes y_j, \\ P &= \textstyle\sum_{j=1}^k y_j^* \otimes y_j \\ H_\omega &= k^{-1/2} \textstyle\sum_{\ell,j=1}^n \gamma_{\ell j} y_j^* \otimes y_\ell. \end{aligned}$$

Then clearly

$$a_{k+1}^{MC}(T) \leq \sup_{x \in B_X} \mathbf{E}\|Tx - H_\omega P G_\omega x\|.$$

Observe that

$$\mathbf{E}(H_\omega P G_\omega) = k^{-1/2} \mathbf{E}(H_\omega \textstyle\sum_{j=1}^k \sum_{i=1}^n \gamma_{ij} x_i^* \otimes y_j)$$

$$= k^{-1} \mathbf{E}(\textstyle\sum_{i,\ell=1}^n \sum_{j=1}^k \gamma_{\ell j} \gamma_{ij} x_i^* \otimes y_\ell) = \sum_{i=1}^n x_i^* \otimes y_i = T.$$

Now, if $G_{\omega'}$ and $H_{\omega'}$ are (independent) copies of G_ω and H_ω, on some other probability space(Ω', Σ', μ'), then

$$\mathbf{E}_{\omega'}(H_{\omega'}PG_\omega) = \mathbf{E}_{\omega'}(H_\omega PG_{\omega'}) = 0,$$

and, by invariance properties of Gaussian measures, the sequence

$$\left\{\frac{\gamma'_{ij}+\gamma_{ij}}{\sqrt{2}}, \frac{\gamma'_{ij}-\gamma_{ij}}{\sqrt{2}}\right\}_{i,j=1}^n$$

has the same distribution as $(\gamma'_{ij}, \gamma_{ij})_{i,j=1}^n$, where γ'_{ij} are the corresponding copies of γ_{ij} on (Ω', Σ', μ'). Now we get

$$\begin{aligned} \mathbf{E}\|Tx - H_\omega PG_\omega x\| &= \mathbf{E}_\omega\|\mathbf{E}_{\omega'}(H_{\omega'}PG_{\omega'}x) - H_\omega PG_\omega x\| \\ &= \mathbf{E}_\omega\|\mathbf{E}_{\omega'}(H_{\omega'}PG_{\omega'} - H_{\omega'}PG_\omega + H_\omega PG_{\omega'} - H_\omega PG_\omega)x\| \\ &\le 2\mathbf{E}_\omega\mathbf{E}_{\omega'}\|\tfrac{H_{\omega'}+H_\omega}{\sqrt{2}}P\tfrac{G_{\omega'}-G_\omega}{\sqrt{2}}x\| = 2\mathbf{E}_\omega\mathbf{E}_{\omega'}\|H_{\omega'}PG_\omega x\|. \end{aligned}$$

Let $(\gamma_i)_{i=1}^n$ and $(\gamma'_i)_{i=1}^n$ be sequences of independent standard Gaussian random variables, independent also of (γ_{ij}) and (γ'_{ij}). Then we have

$$\sum_{i=1}^n \gamma_{ij}(x, x_i^*) \sim \|(x, x_i^*)_{i=1}^n\|_{\ell_2^n}\gamma_j$$

where $\sim$ denotes equal distribution, hence

$$PG_\omega x \sim k^{-1/2}\|(x, x_i^*)\|_{\ell_2^n}\sum_{j=1}^k \gamma_j y_j$$

and

$$\begin{aligned} H_{\omega'}PG_\omega x &\sim k^{-1}\|(x, x_i^*)\|_{\ell_2^n}\textstyle\sum_{\ell=1}^n\sum_{j=1}^k \gamma'_{\ell j}\gamma_j y_\ell \\ &\sim k^{-1}\|(x, x_i^*)\|_{\ell_2^n}\|(\gamma_j)_{j=1}^k\|_{\ell_2^k}\textstyle\sum_{\ell=1}^n \gamma'_\ell y_\ell. \end{aligned}$$

Integrating, we get

$$\begin{aligned} \mathbf{E}_{\omega'}\mathbf{E}_\omega\|H_{\omega'}PG_\omega x\| &= k^{-1}\|(x, x_i^*)\|_{\ell_2^n}\mathbf{E}_\omega\|(\gamma_j)_{j=1}^k\|_{\ell_2^k}\mathbf{E}_{\omega'}\|\textstyle\sum_{\ell=1}^n \gamma'_\ell y_\ell\| \\ &\le k^{-1/2}\varepsilon_2((x_i^*))h((y_i))\|x\|, \end{aligned}$$

which proves the proposition.

If $(e_i)_{i=1}^n$ is the unit vector basis, considered in ℓ_q^n $(1 \le q < \infty)$, then

$$h((e_i)) \asymp n^{1/q}$$

(this is well-known, see, e.g. [Pis89], Lemma 4.14).

Corollary 6.5 *For $2 \le q < \infty$ there is a constant $c > 0$ such that for all $n \in \mathbf{N}$, and $1 \le k \le n$,*

$$a_k^{MC}(Id : \ell_2^n \to \ell_q^n) \le ck^{-1/2}n^{1/q}.$$

The following multiplicativity property turns out to be useful for the upper estimate.

Proposition 6.6 *Let X, Y, Z be Banach spaces, $S \in L(X,Y), T \in L(Y,Z)$ and let $m, n \in$ **N**. Then*

$$e_{m+n-1}^{MC}(TS) \leq a_n^{MC}(T) e_m^{MC}(S).$$

The proof consists in a direct application of the definitions and uses the linearity involved into a_n^{MC}. Combining 6.5 and 6.6 with the following result of Kashin [Kas77], Garnayev and Gluskin [GG84]:

$$c_k(Id : \ell_1^n \to \ell_2^n) \leq ck^{-1/2}(\log(1+n/k))^{1/2},$$

we get

Corollary 6.7 *Let $2 \leq q < \infty$, then there is a constant $c > 0$ such that for $1 \leq k \leq n$,*

$$e_k^{MC}(Id : \ell_1^n \to \ell_q^n) \leq ck^{-1}n^{1/q}(\log(1+n/k))^{1/2}.$$

The rest of the upper bound proof is an application of Mayorov's discretization technique ([May75], see also [Kön86] for an exposition). Now we pass to the proof of the lower estimate. When $\Lambda = X^*$, no discrete measures ν can give a lower estimate since, as it is easily checked, $e_n^{\text{avg}}(S, X^*, \nu) = 0$ for all $n > 1$ [Mat90], Cor. 19. Therefore we shall consider Gaussian measures. In information-based complexity theory, these measures were studied for the purposes of average case analysis, see [TWW88]. Here essentially Hilbert space methods were applied. A general Banach space approach using γ-summing operators and related approximation quantities was introduces into this field by the author [Hei90a], [Hei90b], [Hei91]. We refer to these papers for detailed exposition of such techniques. For the subsequent analysis, we introduce a further notion. Given $S \in L(X,Y)$, a Borel measure ν on X and $n \in$ **N**, we define the n-th approximation number of S with respect to ν as

$$a_n(S,\nu) = \inf_{\substack{T \in L(X,Y) \\ rank T < n}} \int_X \|Sx - Tx\| d\nu(x).$$

The basis for the lower estimate is the following result from [Hei92].

Proposition 6.8 *There is a constant $c > 0$ such that for all Banach spaces X, Y, operators $S \in L(X,Y)$ and centered Gaussian measures ν on X, the following holds:*

$$e_n^{MC}(S) \geq \frac{a_{2n-1}(S,\nu)}{\int_X \|x\| d\nu(x)}.$$

Sketch of proof: By Proposition 4.1 we can estimate $e_n^{MC}(S)$ from below by the $(2n-1)$-st minimal average error of S with respect to the restriction of ν to B_X. Hence, for each $N \in L(X, \mathbf{R}^{2n-2})$ and Borel measurable $\varphi : \mathbf{R}^{2n-2} \to Y$ we need a lower bound on

$$\int_{B_X} \|Sx - \varphi(Nx)\| d\nu(x). \tag{21}$$

Let $X_1 = \text{Ker}N$. Since ν is Gaussian, there exists a subspace $X_2 \subset X$, $\text{dim}X_2 \leq 2n-2$ such that $X = X_1 \oplus X_2$ and $\nu = \nu_1 \times \nu_2$, where ν_1 and ν_2 are centered Gaussian measures on X_1

and X_2, respectively. It follows that ν is invariant under the mapping $x_1 + x_2 \to -x_1 + x_2$ $(x_1 \in X_1, x_2 \in X_2)$. Hence we can estimate (21) as follows

$$\geq \int_{\frac{1}{2}B_{X_1} + \frac{1}{2}B_{X_2}} \| \pm Sx_1 + Sx_2 - \varphi(Nx_2)\| d\nu(x)$$

$$\geq \int_{\frac{1}{2}B_{X_1} + \frac{1}{2}B_{X_2}} \|S(I-P)x\| d\nu(x) \tag{22}$$

where P is the projection from X onto X_2 along X_1. The expression (22) could be bounded from below by $a_{2n-1}(S,\nu)$ if the domain of integration were X. The passage from the subset $\frac{1}{2}B_{X_1} + \frac{1}{2}B_{X_2}$ to X is accomplished using the result of Maurey and Pisier ([Pis89], Th. 2.1 and remark on p. 180): For all $t > 1$

$$\nu\{x \in X : \|x\| > t\mathbf{E}_\nu\|x\|\} \leq \exp(-(t-1)^2/\pi).$$

Here the γ-summing norm enters the argument. For details we refer to [Hei92] and to section 12 below, Lemma 12.1.

To conclude the proof of Theorem 6.3, we apply Mayorov's discretization technique again, together with the following finite dimensional estimates. Let ν be the standard Gaussian measure on $\mathbf{R}^{4n}$. Then for $1 \leq p, q < \infty$

$$e_n^{MC}(Id : \ell_p^{4n} \to \ell_q^{4n}) \succ \frac{a_{2n}(Id : \ell_p^{4n} \to \ell_q^{4n}, \nu)}{\mathbf{E}_\nu\|x\|_p} \asymp \frac{\mathbf{E}_\nu\|x\|_q}{\mathbf{E}_\nu\|x\|_p} \asymp n^{1/q-1/p}.$$

When we compare Theorems 6.1 - 6.3, we see that both in the deterministic and stochastic cases, the use of linear functionals brings advantages over the use of function values only. For large d, these advantages can be considerable. Comparing deterministic and Monte Carlo methods for fixed Λ, it turns out that no gain is reached for function values, but when linear functionals are available, Monte Carlo can perform still better than deterministic approximation (compare Theorems 6.2 and 6.3, e. g. for $2 \leq p \leq q < \infty$). Speaking, though, of implementable algorithms, in contrast to sections 5 and 7, here the proofs of the estimates of e_n^{det} and e_n^{MC} do not lead to arithmetic procedures of cost (running time) proportional to n. To find suitable, implementable algorithms providing the optimal rates is still an open, challenging problem.

7 Numerical solution of Fredholm integral equations

Here we shall be concerned with the following problem. Fix $t_0 \in G = [0,1]^d$ and consider the Fredholm integral equation of the second kind

$$u(s) - \int_G k(s,t)u(t)dt = f(s) \tag{23}$$

in the space $C(G)$. Given functions $f \in C(G)$ and $k \in C(G^2)$, we want to find the value $u(t_0)$ of the solution function u in the given point t_0. To make the setting precise, we put

$$X = C(G^2) \oplus_\infty C(G).$$

Let $r \in \mathbf{N}$ and let C^r denote the space of r times continuously differentiable functions with the usual norm (maximum of the sup norms of all partial derivatives up to order r). We fix $\kappa_1, \kappa_3 > 0$, $0 < \kappa_2 < 1$, put

$$X_0 = \{(k,f) \in X : \|k\|_{C^r} \leq \kappa_1, \|k\|_C \leq \kappa_2, \|f\|_{C^r} \leq \kappa_3\}$$

and define the solution operator $S^F : X_0 \to \mathbf{R}$ by

$$S^F(k,f) = u(t_0) = \left\langle (I - T_k)^{-1} f, \delta_{t_0} \right\rangle . \tag{24}$$

Here I is the identity on $C(G)$ and T_k is the integral operator acting in $C(G)$ as

$$(T_k g)(s) = \int_G k(s,t) g(t) dt \qquad (g \in L_\infty(G)).$$

It follows from the definition of X_0 that $\|T_k\| \leq \kappa_2$, hence $I - T_k$ is invertible, and thus, the solution operator is well-defined. Observe that, contrasting the previous sections, the solution operator S^F is nonlinear (in k). Note also that we do not seek the full solution function u, but only the value in a given point. The full problem can also be considered, although in connection with Monte Carlo analysis it is usually not. The delta functional δ_{t_0} could be replaced by other linear functionals as e. g. scalar products with $L_1(G)$ functions, without changing the results presented below.

As in the prevoius sections we shall first consider the case when function values are available. Here we have

$$\Lambda = \Delta = \{\delta_{(s,t)} : (s,t) \in G^2\} \cup \{\delta_t : t \in G\}$$

and the admissible information operators take (up to permutation of the components) the form

$$N(k,f) = (k(s_1,t_1), \ldots, k(s_p,t_p), f(t_{p+1}) \ldots, f(t_n)).$$

The following is the classical result of Emelyanov and Ilin [EI67] which marked the beginning of the deterministic complexity analysis for integral equations.

Theorem 7.1 $e_n^{det}(S^F, \Delta) \asymp n^{-\frac{r}{2d}}$.

We shall not give a proof here but rather comment on some details. The lower bound proof is of functional-analytic nature. The idea is to choose a subset $X_1 \subset X_0$ in such a way that, on one hand, the full complexity is still preserved and on the other hand, the restriction of S^F to X_1 becomes more handy (in particular linear in k). After that, similar arguments as in the proof of Theorem 5.1 will work. For the proof of the upper bound Emelyanov and Ilin [EI67] constructed a Nystöm multigrid iteration.

Let us now pass to the stochastic setting. The solution of Fredholm integral equations has interested Monte Carlo researchers since the fifties. The methods developed are basically modifications of the classical von Neumann-Ulam scheme, which I would like to describe in view of its role in the later discussion. It starts from a sequence $(\xi_i)_{i=1}^\infty$ of independent copies of the equidistribution on G. Furthermore, an $\mathbf{N}$-valued random variable τ is needed, independent of the sequence (ξ_i), with the distribution

$$\text{prob}\{\tau = j\} = (1-\theta)\theta^{j-1},$$

where θ is some constant, $0 < \theta < 1$. The sequence

$$(\xi_1, \xi_2, \ldots, \xi_{\tau-1})$$

can be viewed as an absorbing Markov chain with absorption probability $1 - \theta$. Using this we define a new random variable, which depends on k and f, by setting

$$\eta(k,f) = \begin{cases} (1-\theta)^{-1} f(t_0) & \text{if} \quad \tau = 1 \\ (1-\theta)^{-1}\theta^{-(\tau-1)} k(t_0, \xi_{\tau-1}) k(\xi_{\tau-1}, \xi_{\tau-2}) \ldots k(\xi_2, \xi_1) f(\xi_1) \\ \text{if} \quad \tau > 1. \end{cases}$$

It is not difficult to check that, if $\theta > \kappa_2$,

$$\mathbf{E}\eta(k,f) = f(t_0) + \sum_{j=1}^{\infty} (T_k^j f)(t_0) = u(t_0).$$

The final approximation to $u(t_0)$ is obtained by producing n independent realizations of η and averaging,

$$u(t_0) \approx \zeta(k,f) = \frac{1}{n} \sum_{i=1}^{n} \eta_i(k,f).$$

This is an implementable procedure, we need, on the average, $O(n)$ calls of a random generator providing the equidistribution on $[0,1]$, and $O(n)$ arithmetic operations. The error of this procedure (as well as of modifications given in the literature) is $O(n^{-1/2})$, see e. g. [Erm71]. Combining this with the deterministic result, we get

$$e_n^{MC}(S^F, \Delta) \prec \min(n^{-r/(2d)}, n^{-1/2}),$$

as a first upper estimate of the complexity. No better upper bounds nor lower bounds were known till the problem was recently settled by Heinrich and Mathé [HM92]:

Theorem 7.2 $e_n^{MC}(S^F, \Delta) \asymp n^{-r/(2d)-1/2}$.

Thus we see that both rates - the optimal deterministic and the classical Monte Carlo rate - multiply to yield the optimal Monte Carlo rate. Let us discuss some aspects of the proof. Lower bounds are derived on the basis of Proposition 4.1, i. e. by constructing a suitable measure and then analyzing the average case. The measure is arranged in such a way that its support is a set of the type X_1 mentioned above, i. e. the restriction of the solution operator to X_1 is simpler than on the original X_0.

The proof of the upper bound consists of the construction of a concrete algorithm. It is again a variance reduction technique. The idea is the following: First we use an optimal deterministic method to produce an approximation to the solution function. Then we use this kowledge for a preconditioning of the von Neumann-Ulam procedure, i. e. we apply this method to modified data. These data are obtained in the following way. Let v be an approximate solution of (23) obtained using $O(n)$ operations. The optimal rate one can reach is

$$\|u - v\|_{C(G)} \asymp n^{-\frac{r}{2d}},$$

see [EI67]. Now we construct an *approximate* equation whose *exact* solution is v. For this sake, let $\widetilde{P}_n : C(G^2) \to C(G^2)$ be the projection defined in Proposition 5.2, just with $2d$ and G^2 instead of d and G. Put

$$h = \widetilde{P}_n k$$

and define

$$g = v - T_k v.$$

It can easily be checked that

$$\|k - h\|_{C(G^2)} \asymp n^{-\frac{r}{2d}}$$

and

$$\|f - g\|_{C(G)} \asymp n^{-\frac{r}{2d}}.$$

Now we define a new random variable by

$$\tilde{\eta}(k, f) = v(t_0) + \eta(k, f) - \eta(h, g).$$

Obviously,

$$\mathbf{E}\tilde{\eta}(k, f) = u(t_0).$$

It is shown in [HM92] that the square root of the variance of $\tilde{\eta}$ is of the order $n^{-\frac{r}{2d}}$, thus, together with the standard Monte Carlo contribution $n^{-1/2}$ arising from averaging n independent realizations of $\tilde{\eta}$, the optimal rate is obtained. A detailed discussion in [HM92] shows that all steps of the described procedure can be arranged in such a way, that an implementable numerical algorithm emerges, which requires on the average $O(n)$ random generator calls, $O(n)$ function and kernel values, and $O(n)$ arithmetic operations. It seems that this new method might also be of numerical interest.

Finally, let us discuss the case when linear functionals can be used. Here we shall consider a more general setting: $X = L_\infty(G^2) \oplus_\infty L_\infty(G)$, X_0 as above, and

$$S_\chi^F(k, f) = \left\langle (I - T_k)^{-1} f, \chi \right\rangle \tag{25}$$

with $\chi \in L_\infty(G)^*$ an arbitrary continuous linear functional. (This includes δ_{t_0}, if we take χ formally to be any continuous extension of $\delta_{t_0} \in C(G)^*$ to $L_\infty(G)$.) We have $\Lambda = X^* = L_\infty(G^2)^* \oplus_1 L_\infty(G)^*$. It is readily checked that, by at most doubling the number of functionals used, one can consider only information operators of the form

$$N(k, f) = ((k, h_1), \ldots, (k, h_p), (f, g_{p+1}), \ldots, (f, g_n)).$$

with $h_1, \ldots, h_p \in L_\infty(G^2)^*$ and $g_{p+1}, \ldots, g_n \in L_\infty(G)^*$. Pereverzev [Per89] analyzed the complexity of obtaining the full solution to (23) using linear functionals. He proved that with n functionals, the order $n^{-r/d}$ can be obtained, up to a logarithmic factor. Comparing this to Theorem 7.1 shows that the use of linear functionals (Fourier coefficients, as it turns out) can provide a significantly better rate of precision than the use of function values only. Pereverzev's result provides, of course, also an upper bound to the complexity of computing only a single functional of the solution. It turns out, however, that this bound is not optimal. In the subsequent sections we shall prove

Theorem 7.3 *For each $q > 2$ there is a constant $c > 0$ such that for all $\chi \in L_\infty(G)^*$ with $\|\chi\| \leq 1$ and for all $n \in \mathbf{N}$*

$$e_n^{\det}(S_\chi^F) \leq c(\log(n+1))^q n^{-\frac{3r}{2d}}.$$

It is not known if - even only up to logarithmic factors - the estimate of Theorem 7.3 is sharp, i. e. can also serve as a lower bound. This is tied to a difficult problem concerning Gelfand numbers of tensor product operators [Hei93]. Let us now look at the Monte Carlo case. After the solution of the case of function value information in [HM92], it remained open to determine the Monte Carlo complexity for $\Lambda = X^*$. The previous methods did not extend to this case. The following result settles this problem (up to logarithmic factors). It will also be proves in the sections below.

Theorem 7.4 *For each $q > 3.5$ there are constants $c_1, c_2 > 0$ such that for all $n \in \mathbf{N}$,*

$$c_1(\log(n+1))^{-3/2} n^{-\frac{3r}{2d}-\frac{1}{2}} \le \sup_{\substack{\chi \in L_\infty(G)^* \\ \|\chi\| \le 1}} e_n^{MC}(S_\chi^F, X^*) \le c_2(\log(n+1))^q n^{-\frac{3r}{2d}-\frac{1}{2}}.$$

8 The algorithm

In this chapter we start the proof of the upper bounds in 7.3 and 7.4 by constructing an algorithm which later on will be shown to provide the optimal rates. The algorithm connects Galerkin methods with the Monte Carlo approach, a combination which seems to be new in the literature.

We begin with describing the spline basis that will be used for the Galerkin discretization. It will correspond to a sequence of partitions of G. We let, for $\ell \ge 0$,

$$J_\ell = \{1, 2, 3, \dots, 2^\ell\}^d,$$

for $i = (i_1, \dots, i_d) \in J_\ell$,

$$((i-1)2^{-\ell}, i2^{-\ell}) = \Pi_{m=1}^d ((i_m - 1)2^{-\ell}, i_m 2^{-\ell})$$

and define the partition π_ℓ of G by

$$\pi_\ell = \{((i-1)2^{-\ell}, i2^{-\ell}) : \ i \in J_\ell\}.$$

The space of splines corresponding to π_ℓ is defined as $\mathcal{S}^r(\pi_\ell) = \{f \in L_\infty(G) : f|_D$ is a polynomial of degree $< r$, for all $D \in \pi_\ell\}$. Let us add, for definiteness, that the degree of the monomial $x_1^{\alpha_1} \dots x_d^{\alpha_d}$ is $\alpha_1 + \alpha_2 + \dots + \alpha_d$, i. e. we use the sum degree. $\mathcal{S}_\ell^r$ is the maximal space of splines of degree $r - 1$, with no correlation in the nodes (faces), i. e. no continuity or smoothness conditions imposed. Let

$$L = \text{closure } \Big(\bigcup_{\ell \ge 0} \mathcal{S}^r(\pi_\ell)\Big)$$

where the closure is taken in $L_\infty(G)$. Next we shall construct an orthonormal (with respect to $L_2(G)$) basis of L. Let $w_1, \dots, w_M$ be any orthonormal basis in the space of polynomials $\mathcal{S}^r(\pi_0)$. Extend this basis to an orthonormal basis of $\mathcal{S}^r(\pi_1)$ by adding $w_{M+1}, \dots, w_{M'}, (M' = 2^d M)$. By repeating this process on higher levels we shall go on constructing the basis. For this purpose we need the contraction operators $Q_{\ell i} : L_\infty(G) \to L_\infty(G), \ell = 0, 1, 2, \dots, i \in J_\ell$, given by

$$(Q_{\ell i} x)(t) = \begin{cases} x(2^\ell t - (i-1)) & \text{for } t \in ((i-1)2^{-\ell}, i2^{-\ell}) \\ 0 & \text{otherwise,} \end{cases}$$

and the extension operators $R_{\ell i} : L_\infty(G) \to L_\infty(G)$

$$(R_{\ell i}x)(t) = x((i-1)2^{-\ell} + t2^{-\ell}).$$

Now we define z_{0ij} for $i \in J_0, j = 1, \dots, M$, and $z_{\ell ij}$ for $\ell \geq 1, i \in J_{\ell-1}, j = 1, 2, \dots, (M'-M)$ by setting

$$\begin{aligned} z_{0ij} &= w_j \\ z_{\ell ij} &= 2^{(\ell-1)d/2} Q_{\ell-1,i} w_{M+j} \quad (\ell \geq 1). \end{aligned}$$

It is elementary to verify that this system is orthonormal and that there are constants $c_1, c_2 > 0$ such that for all $\ell \geq 0$ and all tuples of reals a_{ij},

$$c_1 2^{\ell d/2} \max|a_{ij}| \leq \| \sum_{i,j} a_{ij} z_{\ell ij} \|_{L_\infty} \leq c_2 2^{\ell d/2} \max|a_{ij}|. \tag{26}$$

Let P_ℓ be the orthogonal projection onto $\mathcal{S}^r(\pi_\ell)$ in $L_2(G)$. Then

$$P_\ell = \sum_{0 \leq m \leq \ell} \sum_{i,j} z_{mij} \otimes z_{mij}, \tag{27}$$

but, on the other hand, it can also be represented as

$$P_\ell = \sum_{i \in J_\ell} Q_{\ell i} P_0 R_{\ell i}.$$

This latter representation implies that

$$\sup_{\ell \geq 0} \|P_\ell : L_\infty(G) \to L_\infty(G)\| < \infty. \tag{28}$$

It follows from (26) - (28) that $(z_{\ell ij})$ is a basis in L.
We also have to consider the approximation of smooth functions by this basis. It is well-known that there is a constant $c > 0$ such that for $\nu \in \{0, 1, \dots, r\}$ and $x \in C^\nu(G)$

$$d(x, \mathcal{S}^r(\pi_\ell)) \leq c 2^{-\ell\nu} \|x\|_{C^\nu},$$

where d denotes the distance in $L_\infty(G)$. Together with (28) this gives

$$\|P_\ell x - P_{\ell-1} x\|_{L_\infty} \leq c 2^{-\ell\nu} \|x\|_{C^\nu}.$$

(In this and the following sections often the same letter will be used to denote different constants.) Using (26), we get from the relation above

$$|(x, z_{\ell ij})| \leq c 2^{-(\nu + \frac{d}{2})\ell} \|x\|_{C^\nu}. \tag{29}$$

Let us now introduce a more convenient notation by letting

$$(z_{\ell ij}) = (e_m)_{m=1}^\infty$$

be the lexicographical order. We consider e_m as an element of $L_\infty(G)$, and let e_m^* be the same function as e_m, just considered as a functional, $e_m^* \in L_\infty(G)^*$. Then (26) and (28) imply

$$c_1 m^{1/2} \max_{m \leq i < 2m} |a_i| \leq \| \sum_{i=m}^{2m-1} a_i e_i \|_{L_\infty} \leq c_2 m^{1/2} \max_{m \leq i < 2m} |a_i| \tag{30}$$

while (29) transforms into

$$|(x, e_m^*)| \le cm^{-\nu/d-1/2}\|x\|_{C^\nu}, \tag{31}$$

for all $x \in C^\nu(G)$. For further simplification of notation we set for $f \in L_\infty(G)$

$$f(i) = (f, e_i^*)$$

and for $k \in L_\infty(G^2)$

$$k(i,j) = \langle T_k e_j, e_i^* \rangle = \int_G \int_G k(s,t) e_i(s) e_j(t) ds dt.$$

In a way similar to the derivation of (29), (31) we get for $k \in C^r(G)$

$$|k(m,n)| \le c(mn)^{-1/2} \max(m,n)^{-r/d} \|k\|_{C^r}. \tag{32}$$

The system e_i is an orthogonal spline basis sequence, similar to the systems constructed by Ciesielski [Cie75] and Ciesielski, Figiel [CF83a], [CF83b], just simpler, because no smoothness is required. It provides the same rates of approximation, but is computationally easier to handle, since the Schmidt orthogonalization procedure has to be carried out only for a fixed initial segment of the sequence. This reduces the complexity to the amount needed for our purposes.

The algorithm we present will depend on the parameters $n \in \mathbf{N}, 0 < \theta < 1$ and $\sigma > 1$. The number n determines the size of the discretization, and it will turn out that the number of arithmetic operations will be of that magnitude. The parameters θ and σ control the stochastic part. We start with the deterministic part. Here it is our goal to produce approximations v, h, g to u, k, f respectively so that the equation

$$v - T_h v = g$$

is fulfilled exactly. This is done as follows. We define $k_0 \in L_\infty(G^2)$ by

$$k_0(i,j) = \begin{cases} k(i,j) & \text{if } \max(i,j) \le n^{1/3} \\ 0 & \text{otherwise.} \end{cases} \tag{33}$$

We also define $f_0 \in L_\infty(G)$ by

$$f_0(i) = \begin{cases} f(i) & \text{if } i \le n^{3/4} \\ 0 & \text{otherwise.} \end{cases} \tag{34}$$

Note that the systems (e_i) and $(e_i \otimes e_j)$ are total on $L_\infty(G)$ and $L_\infty(G^2)$ respectively, so that k_0 and f_0 are uniquely determined. Define $H_n \subset \mathbf{N} \times \mathbf{N}$ by

$$H_n = \{(i,j) : ij\max(i,j) \le n^{3/2}\} \tag{35}$$

and $h \in L_\infty(G^2)$ by

$$h(i,j) = \begin{cases} k(i,j) & \text{if } (i,j) \in H_n \\ 0 & \text{otherwise.} \end{cases}$$

Now the following iteration is performed. Put $v_0 = 0$ and determine $v_\ell \in L_\infty(G)$ for $= 1, 2, \ldots, \ell_0$ from

$$(I - T_{k_0})(v_\ell - v_{\ell-1}) = f_0 - (I - T_h)v_{\ell-1}. \tag{36}$$

or equivalently

$$(I - T_{k_0})v_\ell = f_0 + (T_h - T_{k_0})v_{\ell-1}. \tag{37}$$

To obtain the optimal rate, it suffices to take $\ell_0 = 5$. We set $v = v_{\ell_0}$ and compute g as

$$v - T_h v = g \tag{38}$$

To accomplish the algorithm for the deterministic approximation of (25) (and thus for the proof of Theorem 7.3), we put

$$A_n(k, f) = \langle f, \chi\rangle + \langle v, T_k^* \chi\rangle . \tag{39}$$

Now we come to the stochastic part of the algorithm. As in section 7, we let τ be an $\mathbf{N}$-valued random variable with

$$\mu\{\tau = j\} = (1-\theta)\theta^{j-1}$$

Let $(\xi_\ell)_{\ell=1}^\infty$ be a sequence of independent $\mathbf{N}$-valued random variables (independent also of τ) with

$$\text{prob}\{\xi_\ell = i\} = p_i = (\log(i+1))^{1-\sigma} - (\log(i+2))^{1-\sigma}, \tag{40}$$

where log is the logarithm to the base 2, so that $\sum p_i = 1$. Again, the sequence

$$(\xi_1, \xi_2, \ldots, \xi_\tau)$$

can be considered as a Markov chain. The following alternative, but equivalent view is convenient for our purposes. Let Ω be the set of all sequences of natural number of the form

$$\omega = (i_1, i_2, \ldots, i_m) \tag{41}$$

$(m = 1, 2, \ldots)$ and associate to $\omega \in \Omega$ the discrete probability

$$\mu\{\omega\} = (1-\theta)\theta^{m-1} p_{i_1} \ldots p_{i_m}.$$

Then (Ω, μ) is the probability space on which our algorithm is based. Let us introduce the following notation:

$$\theta_m = \theta^{m-1}(1-\theta)$$

and

$$k(\chi, j) = \langle T_k e_j, \chi\rangle .$$

Now let $\omega \in \Omega$ be as in (41). We shall define a random variable η on (Ω, μ) as follows. If $m = 1$, we put

$$\eta(\omega, k, f) = (\theta_1 p_{i_1})^{-1} k(\chi, i_1)(f(i_1) - g(i_1)), \tag{42}$$

and if $m > 1$, we set

$$\begin{aligned}\eta(k, f, \omega) = (\theta_m p_{i_1} \ldots p_{i_m})^{-1} k(\chi, i_1)[&k(i_1, i_2) \ldots k(i_{m-1}, i_m) f(i_m) \\ &- h(i_1, i_2) \ldots h(i_{m-1}, i_m) g(i_m)].\end{aligned} \tag{43}$$

To obtain the final approximation, we set $(\Omega_n, \mu_n) = (\Omega, \mu)^n$ and for $\omega \in \Omega_n, \omega = (\omega_i)_{i=1}^n$

$$\zeta_n(\omega, k, f) = \langle f, \chi\rangle + \langle T_k v, \chi\rangle + \frac{1}{n}\sum_{i=1}^{n} \eta(\omega_i, k, f) \tag{44}$$

This concludes the description of the algorithm. Let us comment on the numerical implementation and thus on the arithmetic cost of this procedure. In accordance to our assumption on linear information we suppose that the following numbers are available: $\langle f, \chi\rangle$, $f(i), k(i,j), k(\chi, j)$. This can be imagined as follows : We have a subroutine which provides, within constant time, for given imput i or (i, j) the requested number. This, of course, depends on the structure of the a priori given χ. Now think of χ as being δ_{t_0} or $e^*_{i_0}$ for some $t_0 \in G$ or $i_0 \in \mathbf{N}$. Then we have to be able to compute scalar products of f and k with the spline basis functions constructed in the beginning of this section. But due to their simple structure this reduces to computing scalar products with polynomials of degree $< r$, on at most 2^d subcubes of G, per coefficient. So the assumption on linear information may be interpreted in this case as the assumption that these scalar products can be computed very quickly, e. g. by automatic integration. Let us now look at the number of informations and arithmetic operations the algorithm requires.
First observe that $|H_n| \asymp n$, which can be verified by elementary estimates. Put $n_1 = [n^{1/3}]$ and $n_2 = [n^{3/4}]$. Since $f_0(i) = 0$ for $i > n_2$ and $h(i, j) = 0$ for $\max(i, j) > n_2$, it follows from (36) that $v_\ell(i) = 0$ for all $i > n_2$. For the remaining coefficients (37) yields the following system

$$v_\ell(i) - \sum_{j=1}^{n_1} k(i,j)v_\ell(j) = f(i) + \sum_{\{j:(i,j)\in H_n, j>n_1\}} k(i,j)v_{\ell-1}(j) \tag{45}$$

if $i \le n_1$, and

$$v_\ell(i) = f(i) + \sum_{\{j:(i,j)\in H_n\}} k(i,j)v_{\ell-1}(j) \tag{46}$$

if $n_1 < i \le n_2$. Due to the number of elements in H_n the right-hand sides of (45) and (46) can be computed in a total of $O(n)$ operations. Solving the system (45) by Gaussian elimination can be accomplished in $O(n_1^3) = O(n)$ operations. The computation of $g(i)$, $i = 1, \ldots, n_2$ from $v(i)$, on the basis of (38) also needs only $O(n)$ operations, again because of the structure of H_n. Note that $g(i) = 0$ for all $i > n_2$. Finally, to produce $A_n(k, f)$, we just have to compute

$$\langle v, T_k^*\chi\rangle = \sum_{j=1}^{n_2} k(\chi, j)v(j).$$

The number of informations, i. e. $f(i)'s, k(i,j)'s$ and $k(\chi, j)'s$, required so far, is $O(n)$, as well. The unique solvability of (45) will be verified later on.
Now we pass to the stochastic part. We assume that the computer provides us with realizations of a sequence of independent random variables, each uniformly distributed on $[0, 1]$. (In practice, we get only an approximation to this - pseudo-random numbers, but we will not discuss this issue here.) These uniform distributions can be transformed into the required variables τ and $\xi_1, \ldots, \xi_\tau$ by standard methods (see, e. g. [Erm71]).
Since $\mathbf{E}\tau < \infty$, we need, on the average, $O(n)$ random number calls, informations and arithmetic operations to carry out the stochastic part.

9 Convergence of the deterministic part

In this and the following section we study convergence rates provided by the algorithm described before. For the analysis of the deterministic part we need some preparations. We shall study the sequence spaces of coefficients associated with the spline basis system of section 8. We introduce the following notation. Let $s, q \in \mathbf{R}$ and define for $a = (a_i)_{i=1}^{\infty} \in \mathbf{R}^{\mathbf{N}}$

$$\alpha_q^s(a) = \sup_i (i^{s+1/2}(\log(i+1))^q |a_i|)$$

and

$$\mathcal{A}_q^s = \left\{a \in \mathbf{R}^N : \alpha_q^s(a) < \infty\right\}.$$

For $b = (b_{ij})_{i,j=1}^{\infty} \in \mathbf{R}^{\mathbf{N}\times\mathbf{N}}$ we set

$$\beta_q^s(b) = \sup_{i,j}((ij)^{1/2} \max(i,j)^s (\log(i+1)\log(j+1))^q |b_{ij}|)$$

and

$$\mathcal{B}_q^s = \{b \in \mathbf{R}^{\mathbf{N}\times\mathbf{N}} : \beta_q^s(b) < \infty\}.$$

Observe first that

$$\mathcal{A}_p^s \subseteq \mathcal{A}_q^t \quad , \quad \mathcal{B}_p^s \subseteq \mathcal{B}_q^t$$

whenever $s > t$ or $s = t$ and $p > q$. Given $b \in \mathbf{R}^{\mathbf{N}\times\mathbf{N}}$, we denote by T_b the operator acting in $\mathbf{R}^{\mathbf{N}}$ defined formally by

$$T_b a = \sum_j b_{ij} a_j.$$

(We consider this expression to be defined if the right-hand side converges absolutely for all i.) In the sequel we list some elementary properties of the sequence spaces and operators. The following result can be checked using standard rules of series summation. We omit the proof.

Lemma 9.1 *Let $\rho > 0$, $q > 1$, $t \in \mathbf{R}$. Then there is a constant $c > 0$ such that for $b \in \mathcal{B}_0^{\rho}$*

$$c\beta_0^{\rho}(b) \geq \begin{cases} \|T_b : \mathcal{A}_0^t \to \mathcal{A}_0^{\rho}\| & if \quad 0 < t < \rho \\ \|T_b : \mathcal{A}_0^t \to \mathcal{A}_0^{t+\rho}\| & if \quad -\rho < t < 0 \\ \|T_b : \mathcal{A}_q^{-\rho} \to \mathcal{A}_{q-1}^0\|. \end{cases}$$

Furthermore, for $b \in \mathcal{B}_q^0$

$$\|T_b : \mathcal{A}_0 \to \mathcal{A}_0\| \leq c\beta_q^0(b).$$

Given $a \in \mathbf{R}^{\mathbf{N}}$, $b \in \mathbf{R}^{\mathbf{N}\times\mathbf{N}}$, $m \in \mathbf{N}$, we define $P_m a \in \mathbf{R}^{\mathbf{N}}$ by

$$(P_m a)_i = \begin{cases} a_i & \text{if} \quad i \leq m \\ 0 & \text{otherwise,} \end{cases}$$

$Q_m b \in \mathbf{R}^{\mathbf{N}\times\mathbf{N}}$ by

$$(Q_m b)_{ij} = \begin{cases} b_{ij} & \text{if} \quad \max(i,j) \leq m \\ 0 & \text{otherwise,} \end{cases}$$

and $\Pi_m b \in \mathbf{R}^{\mathbf{N}\times\mathbf{N}}$ by

$$(\Pi_m b)_{ij} = \begin{cases} b_{ij} & \text{if } \; ij\max(i,j) \le m \\ 0 & \text{otherwise.} \end{cases}$$

Lemma 9.2 *Let* $\rho > 0$, $q \in \mathbf{R}$. *Then there are constants* $c_1, c_2, c_3 > 0$ *such that for all* $a \in \mathcal{A}_0^\rho$ *and* $b \in \mathcal{B}_0^\rho$, $m \in \mathbf{N}$

$$\begin{aligned} \alpha_q^{-\rho}(a - P_m a) &\le c_1(\log(m+1))m^{-2\rho}\alpha_0^\rho(a), \\ \beta_q^0(b - Q_m b) &\le c_2(\log(m+1))^{2q}m^{-\rho}\beta_0^\rho(b), \end{aligned}$$

and

$$\|T_{b-\Pi_m b} : \mathcal{A}_0^\rho \to \mathcal{A}_q^{-\rho}\| \le c(\log(m+1))^q m^{-\rho}\beta_0^\rho(b).$$

Proof: The first two statements are immediately seen. We shall verify the third one. Let $C_m = \{(i,j) : ij\max(i,j) \le m\}$. We have

$$\{j : (i,j) \notin C_m\} = \{m/i^2 < j \le i\} \cup \{j > \max(i, (m/i)^{1/2})\}.$$

Moreover

$$\sum_{m/i^2<j\le i} (ij)^{-1/2} i^{-\rho} j^{-\rho-1/2} \le \begin{cases} ci^{-\rho-1/2} & \text{for } i > m^{1/2} \\ ci^{\rho-1/2}m^{-\rho} & \text{for } m^{1/3} < i \le m^{1/2} \\ 0 & \text{for } i \le m^{1/3}. \end{cases}$$

Similary,

$$\sum_{j>\max(i,(m/i)^{1/2})} (ij)^{-1/2} j^{-\rho} j^{-\rho-1/2} \le \begin{cases} ci^{-2\rho-1/2} & \text{for } \; i > m^{1/3} \\ ci^{\rho-1/2}m^{-\rho} & \text{for } \; i \le m^{1/3}. \end{cases}$$

To proof the statement, we have to estimate

$$\begin{aligned} &\max_i i^{-\rho+1/2}(\log(i+1))^q \sum_{j:(i,j)\notin C_m} (ij)^{-1/2}\max(i,j)^{-\rho} j^{-\rho-1/2} \\ &\le c(\log m)^q m^{-\rho}. \end{aligned}$$

Lemma 9.3 *Let* $\rho > 0, q \in \mathbf{R}$ *and consider three cases: 1.* $t = -\rho$, $q > 1$, *2.* $t = 0$, $q \ge 0$, *3.* $t = \rho$, $q = 0$. *Then for all reals* $\kappa_1 > 0$ *and* $\kappa_2 > 1$ *there are* $c_1, c_2 > 0$, *and* $m_0 \in \mathbf{N}$ *such that for each* $b \in \mathcal{B}_0^\rho$ *with* $\beta_0^\rho(b) \le \kappa_1$ *and* $\|(I - T_b)^{-1} : \mathcal{A}_0^0 \to \mathcal{A}_0^0\| \le \kappa_2$ *it follows that*

$$\|(I - T_b)^{-1} : \mathcal{A}_q^t \to \mathcal{A}_q^t\| \le c_1$$

and for all $m \ge m_0$,

$$\|(I - T_{Q_m b})^{-1} : \mathcal{A}_q^t \to \mathcal{A}_q^t\| \le c_2.$$

When we write an expression like $\|(I-T)^{-1} : X \to X\| \le c$, we usually understand it as shorthand for "the operator $(I - T)$ is invertible in X and its inverse satisfies ...".

Proof: Assume first $t = \rho$, $q = 0$. We have

$$(I - T_b)^{-1} = I + T_b + T_b^2(I - T_b)^{-1}.$$

Because of Lemma 9.1,

$$\begin{aligned} \|T_b : \mathcal{A}_0^\rho \to \mathcal{A}_0^\rho\| &\leq c\kappa_1 \\ \|T_b : \mathcal{A}_0^0 \to \mathcal{A}_0^{\rho/2}\| &\leq c\kappa_1 \\ \|T_b : \mathcal{A}_0^{\rho/2} \to \mathcal{A}_0^\rho\| &\leq c\kappa_1. \end{aligned}$$

It follows that

$$\|(I - T_b)^{-1} : \mathcal{A}_0^\rho \to \mathcal{A}_0^\rho\| \leq c(1 + \kappa_1 + \kappa_1^2\kappa_2).$$

The case $t = 0$, $q > 0$ is proved analogously. If $t = -\rho$, $q > 1$, we argue as follows. $(I - T_b)^{-1} = I + T_b + (I - T_b)^{-1}T_b^2$. Again by 9.1, we have

$$\begin{aligned} \|T_b : \mathcal{A}_q^{-\rho} \to \mathcal{A}_0^{-\rho/2}\| &\leq c\kappa_1 \\ \|T_b : \mathcal{A}_0^{-\rho/2} \to \mathcal{A}_0^0\| &\leq c\kappa_1. \end{aligned}$$

Hence

$$\|(I - T_b)^{-1} : \mathcal{A}_q^{-\rho} \to \mathcal{A}_q^{-\rho}\| \leq c(1 + \kappa_1 + \kappa_1^2\kappa_2).$$

To prove the second assertion, we use Lemmas 9.1 and 9.2, which give, for some $p > 1$

$$\|T_b - T_{Q_m b} : \mathcal{A}_0^0 \to \mathcal{A}_0^0\| \leq c\beta_p^0(b - Q_m b) \leq c(\log(m+1))^{2p} m^{-\rho}\kappa_1 \leq (2\kappa_2)^{-1}.$$

for $m \geq m_0$ with m_0 properly chosen, depending only on c, κ_1 and κ_2. It follows that

$$\|(I - T_{Q_m b})^{-1} : \mathcal{A}_0^0 \to \mathcal{A}_0^0\| \leq 2\kappa_2.$$

Since we have, of course, also

$$\beta_0^\rho(Q_m b) \leq \beta_0^\rho(b) \leq \kappa_1,$$

we can apply the already proven first assertion to derive rest of the statement.

In our subsequent analysis of the algorithm we shall identify a function $f \in L_\infty(G)$ with the sequence of coefficients $(f(i))_{i=1}^\infty \in \mathbf{R}^\mathbf{N}$ and $k \in L_\infty(G^2)$ with $(k(i,j))_{i,j=1}^\infty \in \mathbf{R}^{\mathbf{N}\times\mathbf{N}}$. Moreover, we identify T_k with the operator from $\mathbf{R}^\mathbf{N}$ to $\mathbf{R}^\mathbf{N}$ generated by $(k(i,j))$. In this sense we get from (30) and (31)

$$\mathcal{A}_q^0 \subset L_\infty(G) \subset \mathcal{A}_0^0 \tag{47}$$

for all $q > 1$. Similarly

$$\mathcal{B}_q^0 \subset L_\infty(G^2) \subset \mathcal{B}_0^0. \tag{48}$$

From now on we set

$$\rho = r/d.$$

By (31)

$$C^r(G) \subset \mathcal{A}_0^\rho \tag{49}$$

and, due to (32)

$$C^r(G^2) \subset \mathcal{B}_0^\rho. \tag{50}$$

Now we define a new class X_1 larger than X_0 (provided the parameters are chosen properly, see below). Let $\kappa_1, \kappa_3 > 0,\ \kappa_2 > 1$ and define

$$\begin{aligned} X_1 &= \{(k,f):\ k \in \mathcal{B}_0^\rho,\ \beta_0^\rho(k) \le \kappa_1, \\ &\quad \|(I-T_k)^{-1} : \mathcal{A}_0^0 \to \mathcal{A}_0^0\| \le \kappa_2,\ f \in \mathcal{A}_0^\rho,\ \alpha_0^\rho(f) \le \kappa_3\} \end{aligned} \tag{51}$$

By the canonical identifications, (47) and (48), we have $k \in L_\infty(G^2)$ and $f \in L_\infty(G)$ whenever $(k,f) \in X_1$. Because of (47), each $\chi \in L_\infty(G)^*$ acts continuously on $\mathcal{A}_q^0$. For $(k,f) \in X_1$, the equation

$$(I-T_k)u = f$$

has a unique solution in $\mathcal{A}_0^0$. But Lemma 9.3 gives for $t=\rho,\ q=0$ and for $t=-\rho,\ q>1$,

$$\|(I-T_k)^{-1} : \mathcal{A}_q^t \to \mathcal{A}_q^t\| \le c. \tag{52}$$

Consequently, $u \in \mathcal{A}_0^\rho$ and hence we can extend S_χ^F to X_1:

$$S_\chi^F(k,f) = \left\langle (I-T_k)^{-1}f, \chi \right\rangle.$$

Since the algorithm of section 8 uses only the coefficients $f(i), \langle f,\chi\rangle, k(i,j)$ and $k(\chi,j)$, which are well-defined also for $(k,f) \in X_1$, we can analyze this algorithm on the larger class X_1. Let $A_n(k,f)$ be the deterministic approximation defined in (39). In this section we shall prove

Theorem 9.4 *For each $q>2$ there are constants $c>0$ and $n_0 \in \mathbf{N}$ such that for any $\chi \in L_\infty(G)^*$ with $\|\chi\| \le 1$ and all $n \ge n_0$*

$$\sup_{(k,f)\in X_1} \|S_\chi^F(k,f) - A_n(k,f)\| \le c(\log(n+1))^q n^{-\frac{3r}{2d}}.$$

Let us first discuss how 7.3 follows from 9.4. It is easy to verify that for all $\kappa_1, \kappa_3 > 0$, $0 < \kappa_2 < 1$, there are $\kappa_1', \kappa_3' > 0, \kappa_2' > 1$ such that

$$X_0(\kappa_1,\kappa_2,\kappa_3) \subseteq X_1(\kappa_1',\kappa_2',\kappa_3').$$

Indeed, if $(k,f) \in X_0(\kappa_1,\kappa_2,\kappa_3)$, then $\beta_0^\rho(k) \le c\kappa_1,\ \alpha_0^\rho(f) \le c\kappa_2$, because of (31) and (32). Moreover we have

$$\|(I-T_k)^{-1} : L_\infty(G) \to L_\infty(G)\| \le (1-\kappa_2)^{-1}.$$

By Lemma 9.1, for any $q>1$,

$$\|T_k : \mathcal{A}_0^0 \to \mathcal{A}_q^0\| \le c,$$

thus, from (47)

$$\|T_k : \mathcal{A}_0^0 \to L_\infty(G)\| \le c.$$

Taking into account that

$$(I-T_k)^{-1} = I + (I-T_k)^{-1}T_k$$

we conclude that

$$\|(I-T_k)^{-1} : \mathcal{A}_0^0 \to \mathcal{A}_0^0\| \le 1 + c(1-\kappa_2)^{-1}.$$

This observation and the remarks on the number of information and arithmetic operations needed for the algorithm show that 9.4 implies Theorem 7.3.

Proof of Theorem 9.4: Let $(k, f) \in X_1$. If k_0 is defined by (33) and if $n^{1/3} \geq m_0$, then Lemma 9.3 gives

$$\|(I - T_{k_0})^{-1} : \mathcal{A}_q^t \to \mathcal{A}_q^t\| \leq c, \tag{53}$$

for $t = \rho,\ q = 0$, or $t = 0,\ q = 0$, or $t = -\rho,\ q > 1$. First we shall analyze the precision of the iteration algorithm. We have

$$(I - T_k)u = f,$$

and we let $u_0 \in \mathcal{A}_0^0$ be the solution of

$$(I - T_k)u_0 = f_0, \tag{54}$$

where here and below we shall use the notation of section 8 developed in connection with the algorithm. From (52), we get

$$\alpha_0^\rho(u_0) \leq c. \tag{55}$$

Fix $q > 2$. Then, by Lemma 9.2

$$\alpha_q^{-\rho}(f - f_0) \leq c(\log(n+1))^q (n^{3/4})^{-2\rho} = c(\log(n+1))^q n^{-3\rho/2}.$$

Consequently, by (52)

$$\begin{aligned} \alpha_q^{-\rho}(u - v) &\leq \alpha_q^{-\rho}(u - u_0) + \alpha_q^{-\rho}(u_0 - v) \\ &\leq c\alpha_q^{-\rho}(f - f_0) + \alpha_q^{-\rho}(u_0 - v) \\ &\leq c(\log(n+1))^q n^{-3\rho/2} + \alpha_q^{-\rho}(u_0 - v). \end{aligned} \tag{56}$$

To estimate the last summand, we recall that by (37) for $\ell = 1, \ldots, \ell_0$

$$(I - T_{k_0})v_\ell = f_0 + (T_h - T_{k_0})v_{\ell-1},$$

with $v_0 = 0$. Hence, combining this with (54), we get

$$\begin{aligned} &(I - T_{k_0})(u_0 - v_\ell) \\ =\ &(I - T_k)u_0 + (T_k - T_h)u_0 + (T_h - T_{k_0})u_0 - f_0 - (T_h - T_{k_0})v_{\ell-1} \\ =\ &(T_h - T_{k_0})(u_0 - v_{\ell-1}) + (T_k - T_h)u_0. \end{aligned}$$

Setting

$$(I - T_{k_0})^{-1}(T_h - T_{k_0}) = Y \tag{57}$$

$$(I - T_{k_0})^{-1}(T_k - T_h) = Z, \tag{58}$$

it follows that

$$u_0 - v_\ell = Y(u_0 - v_{\ell-1}) + Zu_0 \quad , \quad (\ell = 1, \ldots, \ell_0)$$

and hence for $v = v_{\ell_0}$, $\ell_0 = 5$

$$u_0 - v = Y^5 u_0 + \sum_{i=0}^{4} Y^i Z u_0. \tag{59}$$

Lemma 9.1 and 9.2 yield

$$\begin{aligned} &\|T_h - T_{k_0} :\ \mathcal{A}_0^0 \to \mathcal{A}_0^0\| \leq c\beta_q^0(h - k_0) \\ &\leq c(\log(n+1))^{2q} n^{-\rho/3} \beta_0^\rho(h) \leq c(\log(n+1))^{2q} n^{-\rho/3}. \end{aligned}$$

Together with (53), we obtain

$$\|Y : \mathcal{A}_0^0 \to \mathcal{A}_0^0\| \le c(\log(n+1))^{2q} n^{-\rho/3}. \tag{60}$$

Since $\beta_0^\rho(h - k_0) \le c$, we also get from Lemma 9.1

$$\|T_h - T_{k_0} : \mathcal{A}_q^{-\rho} \to \mathcal{A}_q^{-\rho}\| \le c,$$

and, with (53),

$$\|Y : \mathcal{A}_q^{-\rho} \to \mathcal{A}_q^{-\rho}\| \le c. \tag{61}$$

Finally, we derive from (53) and Lemma 9.2

$$\begin{aligned} \|Z : \mathcal{A}_0^\rho \to \mathcal{A}_q^{-\rho}\| &\le c\|T_k - T_h : \mathcal{A}_0^\rho \to \mathcal{A}_q^{-\rho}\| \\ &\le c(\log(n+1))^q (n^{3/2})^{-\rho} \beta_0^\rho(k) \le c(\log(n+1))^q n^{-3\rho/2}. \end{aligned} \tag{62}$$

Now we combine (55), (59) - (62) and obtain

$$\begin{aligned} \alpha_q^{-\rho}(u_0 - v) &\le c\alpha_0^0(Y^5 u_0) + \sum_{i=0}^{4} \alpha_q^{-\rho}(Y^i Z u_0) \\ &\le c(\log(n+1))^{10q} n^{-5\rho/3} + c\|Z : \mathcal{A}_0^\rho \to \mathcal{A}_q^{-\rho}\| \alpha_0^\rho(u_0) \\ &\le c(\log(n+1))^q n^{-3\rho/2}. \end{aligned}$$

Joining this with (56) yields

$$\alpha_q^{-\rho}(u - v) \le c(\log(n+1))^q n^{-3\rho/2}. \tag{63}$$

Finally we consider

$$\begin{aligned} S_\chi^F(k,f) - A_n(k,f) &= \langle (I - T_k)^{-1} f, \chi\rangle - A_n(k,f) \\ &= \langle f, \chi\rangle + \langle T_k(I - T_k)^{-1} f, \chi\rangle - \langle f, \chi\rangle - \langle v, T_k^* \chi\rangle \\ &= \langle T_k(u - v), \chi\rangle . \end{aligned}$$

Using (47) and Lemma 9.1, we get for $q > 2$

$$\begin{aligned} |\langle T_k(u - v), \chi\rangle| &\le c\alpha_{q-1}^0(T_k(u - v)) \|\chi\|_{L_\infty^*} \\ &\le c\beta_0^\rho(k) \alpha_q^{-\rho}(u - v) \le c(\log(n+1))^q n^{-3\rho/2}. \end{aligned}$$

This concludes the proof of Theorem 9.4.

For the analysis of the stochastic part we want to extract an estimate of $\alpha_q^{-\rho}(f - g)$ from the above proof, where we recall that

$$g = (I - T_h)v.$$

We have

$$\begin{aligned} f - g &= (I - T_k)u - (I - T_h)v \\ &= (I - T_h)(u - v) - (T_k - T_h)u, \end{aligned}$$

thus, by Lemmas 9.2, 9.3 and relation (63),

$$\begin{aligned} \alpha_q^{-\rho}(f - g) &\le \|I - T_h : \mathcal{A}_q^{-\rho} \to \mathcal{A}_q^{-\rho}\| \alpha_q^{-\rho}(u - v) \\ &+ \|T_k - T_h : \mathcal{A}_0^\rho \to \mathcal{A}_q^{-\rho}\| \alpha_0^\rho(u) \\ &\le c(\log(n+1))^q n^{-3\rho/2}. \end{aligned} \tag{64}$$

for $q > 2$.

10 Convergence of the stochastic part

The analysis carried out in this section will be restricted to a class X_2 slightly smaller than X_1. It will be discussed in section 11 how to extend the results to X_1. Let $0 < \delta < 1$ and let X_2 be the class of all $(k, f) \in X_1$ such that

$$|k(i,j)| \le \delta (p_i p_j)^{1/2} \tag{65}$$

for $i, j \in \mathbf{N}$ with p_i as in (40). Recall that the algorithm developed in section 8 depended on θ, σ and n and produced an approximation $\zeta_n(\omega, k, f)$ to $S_\chi^F(k, f) = \langle (I - T_k)^{-1} f, \chi \rangle$ (see (44)). Let $\mathbf{E}$ denote the expectation with respect to the probability space (Ω_n, μ_n) defined there.

Theorem 10.1 *Assume that $\delta < \theta < 1$ and $\sigma > 1$. Then there are constants $c > 0$ and $n_0 \in \mathbf{N}$ such that for any $\chi \in L_\infty(G)^*$ with $\|\chi\| \le 1$ and all $n \ge n_0$*

$$\sup_{(k,f)\in X_2} \mathbf{E}|S_\chi^F(k,f) - \zeta_n(\omega,k,f)| \le c(\log n)^{7\sigma/2} n^{-\frac{3r}{2d} - \frac{1}{2}}.$$

Proof: There are constants $c > 0$ such for any $(k, f) \in X_2,\ i, j \in \mathbf{N}$,

$$\begin{aligned} |k(i,j)| &\le c(ij)^{-1/2}(\max(i,j))^{-\rho} && (66)\\ |f(i)| &\le ci^{-\rho-1/2} && (67)\\ |k(\chi,j)| &\le c(\log(j+1))^{\sigma} j^{-\rho-1/2}. && (68) \end{aligned}$$

To derive the third relation, observe that

$$\begin{aligned} &|k(\chi,j)| = |\langle T_k e_j, \chi\rangle| \le c\alpha_\sigma^0(T_k e_j)\\ &\le c\max_i \left\{(\log(i+1))^\sigma j^{-1/2}(\max(i,j))^{-\rho}\right\}\\ &\le c(\log(j+1))^\sigma j^{-\rho-1/2}. \end{aligned}$$

Note that by (40),

$$p_i \asymp i^{-1}(\log(i+1))^{-\sigma}. \tag{69}$$

First we shall estimate the variance of η. In fact, we shall estimate the maximum of η on Ω. The following notation will be convenient for the coming estimates: Let

$$\begin{aligned} K_i &= p_i^{-1/2} k(\chi, i) && (70)\\ K_{ij} &= (p_i p_j)^{-1/2} k(i,j) && (71)\\ H_{ij} &= (p_i p_j)^{-1/2} h(i,j) && (72)\\ F_i &= p_i^{-1/2} f(i) && (73)\\ G_i &= p_i^{-1/2} g(i). && (74) \end{aligned}$$

Fix $\omega = (i_1, \ldots, i_m)$. Then we have

$$\begin{aligned} \eta(\omega,k,f) &= \theta_m^{-1} K_{i_1}(K_{i_1 i_2}\ldots K_{i_{m-1} i_m} F_{i_m} - H_{i_1 i_2}\ldots H_{i_{m-1} i_m} G_{i_m})\\ &= \theta_m^{-1}(\mathcal{W}_1 + \mathcal{W}_2). \end{aligned}$$

Here $\mathcal{W}_1 = 0$ if $m = 1$, and

$$\mathcal{W}_1 = K_{i_1}(K_{i_1 i_2} \dots K_{i_{m-1} i_m} - H_{i_1 i_2} \dots H_{i_{m-1} i_m})F_{i_m}$$

for $m > 1$. Furthermore, for all $m \geq 1$,

$$\mathcal{W}_2 = K_{i_1} H_{i_1 i_2} \dots H_{i_{m-1} i_m}(F_{i_m} - G_{i_m}).$$

It is readily verified that for $m > 1$.

$$\mathcal{W}_1 = \sum_{\ell=1}^{m-1} \mathcal{W}_{1\ell},$$

with

$$\mathcal{W}_{1\ell} = K_{i_1} H_{i_1 i_2} \dots H_{i_{\ell-1} i_\ell}(K_{i_\ell i_{\ell+1}} - H_{i_\ell i_{\ell+1}})K_{i_{\ell+1} i_{\ell+2}} \dots K_{i_{m-1} i_m} F_{i_m}. \tag{75}$$

First we estimate $\mathcal{W}_{1\ell}$. By (69), (70), (73), we have

$$|K_i| \leq c \quad , \quad |F_i| \leq c, \tag{76}$$

and by (65)

$$|K_{ij}| \leq \delta \quad , \quad |H_{ij}| \leq \delta. \tag{77}$$

The middle part of (75), that is

$$M = H_{i_{\ell-1} i_\ell}(K_{i_\ell i_{\ell+1}} - H_{i_\ell i_{\ell+1}})K_{i_{\ell+1} i_{\ell+2}}$$

will be estimated next. Note that $H_{i_{\ell-1} i_\ell}$ has to be replaced by K_{i_1} if $\ell = 1$, and $K_{i_{\ell+1} i_{\ell+2}}$ by F_{i_m} if $\ell = m - 1$. From (66), (69), (71) and (72) we get

$$|H_{i_{\ell-1} i_\ell}| \leq c(\log(i_\ell + 1))^\sigma i_\ell^{-\rho} \tag{78}$$

$$|K_{i_{\ell+1} i_{\ell+2}}| \leq c(\log(i_{\ell+1} + 1))^\sigma i_{\ell+1}^{-\rho} \tag{79}$$

and from (67) and (68)

$$|K_{i_1}| \leq c(\log(i_1 + 1))^{3\sigma/2} i_1^{-\rho} \tag{80}$$

$$|F_{i_m}| \leq c(\log(i_m + 1))^{\sigma/2} i_m^{-\rho}. \tag{81}$$

Finally,

$$K_{i_\ell i_{\ell+1}} - H_{i_\ell i_{\ell+1}} = 0$$

whenever $i_\ell i_{\ell+1} \max(i_\ell, i_{\ell+1}) \leq n^{3/2}$, while in the opposite case

$$|K_{i_\ell i_{\ell+1}} - H_{i_\ell i_{\ell+1}}| = |K_{i_\ell}|$$

$$\leq c(\log(i_\ell + 1)\log(i_{\ell+1} + 1))^{\sigma/2}(\max(i_\ell, i_{\ell+1}))^{-\rho}.$$

Joining the estimates obtained, and taking into account $i_\ell i_{\ell+1} \max(i_\ell, i_{\ell+1}) > n^{3/2}$, we get

$$|M| \leq c(\log(n + 1))^{7\sigma/2} n^{-3\rho/2}.$$

Inserting this as well as (76) and (77) into (75), we get

$$|\mathcal{W}_1| \leq c(m - 1)\delta^{m-4}(\log(n + 1))^{7\sigma/2} n^{-3\rho/2}. \tag{82}$$

For the estimate of $\mathcal{W}_2$ we consider the term

$$V = H_{i_{m-1}i_m}(F_{i_m} - G_{i_m}),$$

where the first factor has to be replaced by K_{i_1} if $m = 1$. Now we use the estimate (64) obtained at the end of the previous section. It gives for $q = 2\sigma$,

$$|f(i_m) - g(i_m)| \leq c(\log(n+1))^{2\sigma} n^{-3\rho/2} (\log(i_m+1))^{-2\sigma} i_m^{\rho-1/2},$$

consequently, using (78) and (80),

$$\begin{aligned} |V| &\leq c(\log(i_m+1))^{3\sigma/2} i_m^{-\rho} \cdot i_m^{1/2} (\log(i_m+1))^{\sigma/2} |f(i_m) - g(i_m)| \\ &\leq c(\log(n+1))^{2\sigma} n^{-3\rho/2}. \end{aligned}$$

We get,

$$|\mathcal{W}_2| \leq c\delta^{m-2} (\log(n+1))^{2\sigma} n^{-3\rho/2}. \tag{83}$$

Recalling that $\theta_m = \theta^{m-1}(1-\theta)$, we see that for θ satisfying $\delta < \theta < 1$,

$$|\eta(\omega, k, f)| \leq c(\log(n+1))^{7\sigma/2} n^{-\frac{3r}{2d}} \tag{84}$$

with a constant $c > 0$ independent of ω, k and f. Next we study the expectation of η. Note that, since (e_i) is a basis of $L \subset L_\infty(G)$, we have for all $x \in L$ and $x^* \in L^*$,

$$\langle x, x^* \rangle = \left\langle \sum_{i=1}^{\infty} \langle x, e_i^* \rangle \, e_i, x^* \right\rangle = \sum_{i=1}^{\infty} \langle x, e_i^* \rangle \langle e_i, x^* \rangle . \tag{85}$$

Here the convergence could - in general - be conditional. However, due to the requirements imposed on k and f, all series considered below converge absolutely. This is easily derived from the estimates in the first part of this section, but we omit details. Consequently, we can change the order of summation as needed. We have

$$\begin{aligned} \mathbf{E}\eta(\omega, k, f) = \sum_{m=1}^{\infty} \sum_{i_1,\ldots,i_m=1}^{\infty} & k(\chi, i_1)(k(i_1, i_2) \ldots k(i_{m-1}, i_m) f(i_m) \\ & - h(i_1, i_2) \ldots h(i_{m-1}, i_m) g(i_m)) \end{aligned}$$

On the basis of (85) we get

$$\begin{aligned} \sum_{i_1=1}^{\infty} k(\chi, i_1) k(i_1, i_2) &= \sum_{i_1=1}^{\infty} \langle e_{i_1}, T_k^* \chi \rangle \left\langle T_k e_{i_2}, e_{i_1}^* \right\rangle \\ = \langle T_k e_{i_2}, T_k^* \chi \rangle &= \left\langle e_{i_2}, T_k^{*2} \chi \right\rangle . \end{aligned}$$

In the same way, we obtain

$$\begin{aligned} \sum_{i_2=1}^{\infty} \left\langle e_{i_2}, T_k^{*2} \chi \right\rangle k(i_2, i_3) &= \sum_{i_2=1}^{\infty} \left\langle e_{i_2}, T_k^{*2} \chi \right\rangle \left\langle T_k e_{i_3}, e_{i_2}^* \right\rangle \\ = \left\langle T_k e_{i_3}, T_k^{*2} \chi \right\rangle &= \left\langle e_{i_3}, T_k^{*3} \chi \right\rangle . \end{aligned}$$

Continuing, we arrive at

$$\begin{aligned}
\mathbf{E}\eta(\omega,k,f) &= \sum_{m=1}^{\infty}(\langle f, T_k^{*m}\chi\rangle - \left\langle g, T_h^{*(m-1)}T_k^*\chi\right\rangle) \\
&= \sum_{m=1}^{\infty}\left\langle T_k^{m-1}f - T_h^{m-1}g, T_k^*\chi\right\rangle \\
&= \left\langle (I-T_k)^{-1}f - (I-T_h)^{-1}g, T_k^*\chi\right\rangle \\
&= \langle u-v, T_k^*\chi\rangle = \langle T_k(u-v),\chi\rangle .
\end{aligned}$$

Recalling the definition (44) of $\zeta_n(\omega,k,f)$, we get for $\omega = (\omega_1,\ldots,\omega_n) \in \Omega_n$ and $\eta_i = \eta(\omega_i,k,f)$,

$$\begin{aligned}
\zeta_n(\omega,k,f) - S_\chi^F(k,f) &= \langle f,\chi\rangle + \langle T_k v,\chi\rangle + \frac{1}{n}\sum_{i=1}^{n}\eta_i - \left\langle (I-T_k)^{-1}f,\chi\right\rangle \\
&= \frac{1}{n}\sum_{i=1}^{n}\eta_i - \left\langle T_k(I-T_k)^{-1}f,\chi\right\rangle + \langle T_k v,\chi\rangle \\
&= \frac{1}{n}\sum_{i=1}^{n}\eta_i - \langle T_k(u-v),\chi\rangle = \frac{1}{n}\sum_{i=1}^{n}(\eta_i - \mathbf{E}\eta_i).
\end{aligned}$$

It follows that

$$\mathbf{E}\zeta_n(\omega,k,f) = S_\chi^F(k,f)$$

and

$$\begin{aligned}
&(\mathbf{E}|\zeta_n(\omega,k,f) - S_\chi^F(k,f)|)^2 \\
\le\ & \mathbf{E}(\zeta_n(\omega,k,f) - S_\chi^F(k,f))^2 = \frac{1}{n^2}\mathbf{E}\sum_{i=1}^{n}(\eta_i - \mathbf{E}\eta_i)^2 \\
=\ & \frac{1}{n}\mathbf{E}(\eta - \mathbf{E}\eta)^2 \le \frac{1}{n}\mathbf{E}\eta^2 \le \frac{1}{n}\sup|\eta(\omega,k,f)|^2
\end{aligned}$$

Now (84) implies the statement of the theorem.

11 Preconditioning

In this section we briefly discuss a preconditioning process which allows us to handle the larger class X_1 without spoiling the efficiency estimates. Fix an $m \in \mathbf{N}$. For $(k,f) \in X_1$ we define w, k' and f' formally by

$$w(i,j) = \begin{cases} k(i,j) & \text{if } \max(i,j) \le m \\ 0 & \text{otherwise,} \end{cases} \tag{86}$$

$$T_{k'} = (I-T_w)^{-1}(T_k - T_w), \tag{87}$$

and

$$f' = (I-T_w)^{-1}f. \tag{88}$$

Clearly, (87) is equivalent to setting for $t \in G$,

$$k'(\cdot, t) = (I - T_w)^{-1}(k(\cdot, t) - w(\cdot, t)), \tag{89}$$

or, in terms of spline basis coefficients

$$k'(\cdot, j) = (I - T_w)^{-1}(k(\cdot, j) - w(\cdot, j)). \tag{90}$$

Next we check that this preconditioning is well-defined and can be used to reduce problems from X_1 to problems from X_2. Aspects of numerical realization will be discussed later on in this section.

Proposition 11.1 *Let $\kappa_1, \kappa_3 > 0$, $\kappa_2 > 1$, $0 < \delta < 1$, $\sigma > 1$. Then there exist $m \in \mathbf{N}$, $\kappa_1', \kappa_3' > 0$, $\kappa_2' > 1$, such that for all $(k, f) \in X_1(\kappa_1, \kappa_2, \kappa_3)$, the preconditioning process (86) - (88) is well-defined and $(k', f') \in X_2(\kappa_1', \kappa_2', \kappa_3', \sigma, \delta)$.*

Proof: By Lemma 9.3, there is an $m_0 \in \mathbf{N}$ such that for $m > m_0$ and $(k, f) \in X_1$,

$$\|(I - T_w)^{-1} : \mathcal{A}_0^0 \to \mathcal{A}_0^0\| \le c \tag{91}$$

$$\|(I - T_w)^{-1} : \mathcal{A}_0^\rho \to \mathcal{A}_0^\rho\| \le c, \tag{92}$$

and

$$\|(I - T_w)^{-1} : \mathcal{A}_{\sigma/2}^0 \to \mathcal{A}_{\sigma/2}^0\| \le c. \tag{93}$$

Relation (91) shows that (k', f') is well-defined. (92) implies

$$\alpha_0^\rho(f') \le c\kappa_3.$$

From (87) we get

$$I - T_{k'} = I - (I - T_w)^{-1}(T_k - T_w) = (I - T_w)^{-1}(I - T_k), \tag{94}$$

which yields, combined with Lemma 9.1,

$$\|(I - T_{k'})^{-1} : \mathcal{A}_0^0 \to \mathcal{A}_0^0\| \le c\kappa_2.$$

By the definition of β_0^ρ, we have for $b \in \mathbf{R}^{\mathbf{N}\times\mathbf{N}}$

$$\beta_0^\rho(b) = \sup_j \max(j^{1/2}\alpha_0^\rho(b_j), j^{\rho+1/2}\alpha_0^0(b_j)),$$

with $b_j = (b_{ij})_{i=1}^\infty$. Consequently, (90), (91) and (92) give

$$\alpha_0^\rho(k'(\cdot, j)) \le c\alpha_0^\rho(k(\cdot, j) - w(\cdot, j))$$
$$\alpha_0^0(k'(\cdot, j)) \le c\alpha_0^0(k(\cdot, j) - w(\cdot, j)),$$

and hence

$$\beta_0^\rho(k') \le c\beta_0^\rho(k - w) \le c\kappa_1.$$

Arguing similarly, we have for $b \in \mathbf{R}^{\mathbf{N}\times\mathbf{N}}$,

$$\beta^0_{\sigma/2}(b) = \sup_j j^{1/2}(\log(j+1))^{\sigma/2}\alpha^0_{\sigma/2}(b_j),$$

hence, using (93) and Lemma 9.2,

$$\sup_{i,j}(p_i p_j)^{-1/2}|k'(i,j)| \le c\beta^0_{\sigma/2}(k') \le c\beta^0_{\sigma/2}(k-w) \le c(\log(m+1))^\sigma m^{-\rho} \le \delta,$$

if m is properly chosen, depending only on c, σ, ρ and δ. This proves the proposition.

The solution of the integral equation with data (k, f) is equivalent to the solution of that with (k', f'). Indeed, by (94)

$$\begin{aligned} S^F_\chi(k', f') &= \left\langle (I - T_{k'})^{-1} f', \chi \right\rangle \\ &= \left\langle (I - T_k)^{-1}(I - T_w)(I - T_w)^{-1} f, \chi \right\rangle \\ &= \left\langle (I - T_k)^{-1} f, \chi \right\rangle = S^F_\chi(k, f). \end{aligned}$$

If we have to handle $(k, f) \in X_1$, the numerical process goes as follows. We start with the preconditioning described above and obtain (k', f'). Then the algorithm of section 8 (deterministic and stochastic part) is applied to (k', f'). As we mentioned in section 8, for this we need to have $\langle f', \chi \rangle$, $f'(i)$, $k'(\chi, j)$ and $k'(i, j)$ at our disposal. So the preconditioning is actually a procedure of obtaining these coefficients. To describe it in more detail, let us first consider the task of determining $a' = (a'_i)_{i=1}^\infty \in \mathcal{A}^0_0$ from the equation

$$(I - T_w)a' = a,$$

where $a \in \mathcal{A}^0_0$ is given and w is as defined above. This reduces to the system

$$a'_i - \sum_{\ell=1}^m k(i,\ell)a'_\ell = a_i$$

for $i \le m$, and $a'_i = a_i$ for $i > m$. Hence we get for $i \le m$

$$a'_i = \sum_{\ell=1}^m b_{i\ell} a_\ell, \tag{95}$$

where $B = (b_{i\ell})_{i,\ell=1}^m$ is the inverse of the matrix $(\delta_{i\ell} - k(i,\ell))$. The matrix B can be precomputed in $O(1)$ operations (note that m is constant, i. e. does not depend on the size parameter n of the algorithm). Consequently, we get

$$\begin{aligned} f'(i) &= \sum_{\ell=1}^m b_{i\ell} f(\ell) \\ k'(i,j) &= \sum_{\ell=1}^m b_{i\ell}(k(\ell,j) - w(\ell,j)) \end{aligned}$$

for $i \le m$, $j \in \mathbf{N}$ and $f'(i) = f(i)$, $k'(i,j) = k(i,j)$ for $i > m$, $j \in \mathbf{N}$. The definition of w implies that

$$k'(i,j) = 0$$

for $i \le m$ and $j \le m$. For $i \le m$ and $j > m$ we get

$$k'(i,j) = \sum_{\ell=1}^{m} b_{i\ell} k(\ell, j).$$

Equation (95) can also be used to determine $\langle f', \chi \rangle$ and $k'(\chi, j)$ from $\langle f, \chi \rangle$ and $k(\chi, j)$. Indeed, from (95) we get

$$a'_i = a_i + \sum_{\ell=1}^{m} b'_{i\ell} a_\ell$$

with $b'_{i\ell} = b_{i\ell} - \delta_{i\ell}$, thus

$$a' = a + \sum_{i,\ell=1}^{m} b'_{i\ell} a_\ell e_i.$$

Consequently,

$$\langle f', \chi \rangle = \langle f, \chi \rangle + \sum_{i,\ell=1}^{m} b'_{i\ell} f(\ell) \langle e_i, \chi \rangle$$

and

$$\begin{aligned} k'(\chi, j) &= \langle k'(\cdot, j), \chi \rangle \\ &= \langle k(\cdot, j) - w(\cdot, j), \chi \rangle + \sum_{i,\ell=1}^{m} b'_{i\ell}(k(\ell, j) - w(\ell, j)) \langle e_i, \chi \rangle . \end{aligned}$$

It follows that for $j \le m$

$$k'(\chi, j) = k(\chi, j) - \langle w(\cdot, j), \chi \rangle = k(\chi, j) - \sum_{i=1}^{m} k(i, j) \langle e_i, \chi \rangle ,$$

while for $j > m$

$$k'(\chi, j) = k(\chi, j) + \sum_{i,\ell=1}^{m} b'_{i\ell} k(\ell, j) \langle e_i, \chi \rangle .$$

Thus, we have seen that each of the values $\langle f', \chi \rangle$, $f'(i), k'(\chi, j), k'(i, j)$ can be computed in $O(1)$ operations from $O(1)$ of the original information $\langle f, \chi \rangle$, $f(i), k(\chi, j), k(i, j)$. Since χ is a fixed part of the problem, the numbers $\langle e_i, \chi \rangle$ can be precomputed before any input (k, f) is processed, so that we consider them as constants. Hence the extended algorithm for the larger class X_1 possesses the same order of required information and arithmetic operations as the one given in section 8. This together with Theorem 10.1 and the fact that $X_0 \subseteq X_1$ proves the upper bound in Theorem 7.4.

12 The lower bound

We shall give here the proof of the lower estimate of Theorem 7.4. Due to Proposition 4.1, we have to construct a suitable measure on X_0 and analyze the average case error. Fix

$t_0 \in G$ and let χ be any extension of $\delta_{t_0} \in C(G)^*$ to $L_\infty(G)$ with $\|\chi\| = 1$. Then we have for $(k, f) \in X_0$,

$$S_\chi^F(k, f) = ((I - T_k)^{-1} f)(t_0).$$

First we decompose G into three parts, depending on the position of t_0. Let $t_0 = (t_{01}, \dots, t_{0d})$. We shall assume that $1/2 \le t_{01} \le 1$. The construction can be adapted to the other situation $0 \le t_{01} < 1/2$ in an obvious way. We define

$$\begin{aligned} G_1 &= [0, 1/6] \times [0, 1]^{d-1} \\ G_2 &= [1/6, 1/3] \times [0, 1]^{d-1} \\ G_3 &= [1/3, 1] \times [0, 1]^{d-1}. \end{aligned}$$

Let $n \in \mathbf{N}$ be of the form $n = \ell^d$, $\ell \in \mathbf{N}$ and let ψ_{i1} $(i = 1, \dots, n)$ be constructed in the same way as the ψ_i in the proof of Theorem 5.1, just with G replaced by G_1, $\tilde{n}$ by n, and instead of $\int_G \psi(t)dt = 1$ we assume $\int_{G_1} \psi(t)^2 dt = 1$. Similarly, we let ψ_{i2} be constructed with supports in G_2. Finally, let $\lambda \in C^r(G)$ be a function with supp $\lambda \subseteq G_3$ and $\lambda(t_0) = 1$. There are constants $c_1, c_2, c_3 > 0$ such that for $\alpha_i \in \mathbf{R}$, $i = 1, \dots, n$, $j = 1, 2$

$$c_1 n^{r/d} \|(\alpha_i)\|_{\ell_\infty^n} \le \|\sum_{i=1}^n \alpha_i \psi_{ij}\|_{C^r} \le c_2 n^{r/d} \|(\alpha_i)\|_{\ell_\infty^n} \tag{96}$$

and

$$\|\lambda\|_{C^r} \le c_3. \tag{97}$$

The constants depend neither on n nor on t_0. Let $E = \ell_2^n$, with the unit vector basis $(e_1, \dots, e_n)$ and let $L = \ell_2^{n^2}$, with the unit vector basis $(e_{ij})_{i,j=1}^n$. Define the mappings $J_1 : E \to C^r(G)$ by

$$J_1 e_i = \theta_{n1} \psi_{i1} \tag{98}$$

$J_2 : E \to C^r(G)^*$ by

$$J_2 e_i = \theta_{n2} \psi_{i2} \tag{99}$$

(here we consider ψ_{i2} as an element of the dual of $C^r(G)$, in the canonical way), $\Phi_1 : L \to C^r(G^2)$ by

$$(\Phi_1 e_{ij})(s, t) = \beta_n \psi_{i2}(s) \psi_{j1}(t), \tag{100}$$

and $\Phi_2 : E \to C^r(G^2)$ by

$$(\Phi_2 e_j)(s, t) = \theta_{n2} \lambda(s) \psi_{j2}(t). \tag{101}$$

The real numbers θ_{n1}, θ_{n2} and β_n will be chosen below. In order to define measures, we need some more notation. Let H be a finite dimensional real Hilbert space and let γ_H denote the canonical Gaussian measure on H, i. e. the measure with density

$$(2\pi)^{-\dim H/2} e^{-\|h\|^2/2} \qquad (h \in H).$$

Let J be a linear operator from H into a Banach space X. Then we let $\gamma_H \circ J^{-1} = \mu$ be the measure on X induced by γ_H under the action of J. Furthermore, we let

$$E_\gamma(J) = \int_H \|Jh\| d\gamma_H(h).$$

Clearly, we have

$$E_\gamma(J) = \int_X \|x\| d\mu(x).$$

We shall use the following result ([Hei92], Lemma 1).

Lemma 12.1 *There is a constant $c' > 0$ such that for all finite dimensional Hilbert spaces H, and for all operators $J \in L(H,X)$, $T \in L(X,Y)$, where X and Y are Banach spaces, the following holds with $\mu = \gamma_H \circ J^{-1}$,*

$$\int_{\|x\|>c'E_\gamma(J)} \|Tx\| d\mu(x) \leq \frac{1}{2} \int_X \|Tx\| d\mu(x). \tag{102}$$

Let $\kappa_1, \kappa_2, \kappa_3$ be the parameters determining the class X_0. Now we choose $\theta_{n1}, \theta_{n2}, \beta_n$ in such a way that

$$E_\gamma(J_1) = \kappa_3/c' \tag{103}$$

and, with $\kappa = \min(\kappa_1, \kappa_2)$,

$$E_\gamma(\Phi_1) = E_\gamma(\Phi_2) = \kappa/(4c'). \tag{104}$$

Since, by (96)

$$E_\gamma(J_1) \asymp \theta_{n1} n^{r/d} E_\gamma(Id : \ell_2^n \to \ell_\infty^n) \asymp \theta_{n1} n^{r/d} (\log n)^{1/2},$$

it follows that

$$\theta_{n1} \asymp n^{-r/d} (\log n)^{-1/2}. \tag{105}$$

Similarly, we get

$$\theta_{n2} \asymp \beta_n \asymp n^{-r/d} (\log n)^{-1/2}, \tag{106}$$

where we used (96) and (97). Next we define measures with the help of these operators. For the sake of brevity we put $K = C^r(G^2)$, $F = C^r(G)$, $W = C^r(G)^*$, and define measures

$$\nu_i = \gamma_E \circ J_i^{-1} \qquad (i = 1, 2)$$

on F and W, respectively, and

$$\mu_i = \gamma_L \circ \Phi_i^{-1} \qquad (i = 1, 2)$$

on K. Furthermore, we let

$$\rho = \mu_1 \times \mu_2 \times \nu_1$$

on $K^2 \times F$, and define a mapping

$$\Psi : K^2 \times F \to K \times F$$

by

$$\Psi(k, h, f) = (k + h, f).$$

We put

$$\tilde{\sigma} = \rho \circ \Psi^{-1},$$

and for $B \subseteq X_0$ a Borel set

$$\sigma(B) = \frac{\tilde{\sigma}(B)}{\tilde{\sigma}(X_0)}.$$

Then σ is a Borel probability measure on X_0. We shall use Proposition 4.1 and estimate

$$e^{\mathrm{avg}}(S_X^F, A, \sigma),$$

where it suffices to consider A of the form $A = \varphi \circ N$, $N = N_1 \times N_2$, $N_1 \in L(K, \mathbf{R}^{n_1})$, $N_2 \in L(F, \mathbf{R}^{n_2})$, $\varphi : \mathbf{R}^{n_1+n_2} \to \mathbf{R}$ is Borel measurable, and

$$n_1 + n_2 \leq m.$$

Set

$$\begin{aligned} K' &= \mathrm{Ker}\ N_1 \\ L' &= \Phi_1^{-1}(K') = \mathrm{Ker}\ N_1\Phi_1. \end{aligned}$$

Let $L'' = (L')^\perp$ be the orthogonal complement of L' in L. Since K' is of finite codimension, L'' is finite dimensional, and Φ_1 is an injection, we can find a complement K'' of K' in K with $\Phi_1(L'') \subseteq K''$. Define measures on K' and K'', respectively,

$$\begin{aligned} \mu_1' &= (\Phi_1|_{L'})^{-1}\gamma_{L'} \\ \mu_1'' &= (\Phi_1|_{L''})^{-1}\gamma_{L''}. \end{aligned}$$

Because of $\gamma_L = \gamma_{L'} \times \gamma_{L''}$, we also have $\mu_1 = \mu_1' \times \mu_1''$. Let

$$K_0' = \{k \in K' : \|k\|_{C^r} \leq \kappa/4\} \tag{107}$$

$$K_0'' = \{k \in K'' : \|k\|_{C^r} \leq \kappa/4\} \tag{108}$$

$$K_0 = \{k \in K : \|k\|_{C^r} \leq \kappa/2\} \tag{109}$$

$$F_0 = \{f \in F : \|f\|_{C^r} \leq \kappa_3\}. \tag{110}$$

It follows that

$$\begin{gathered} K_0' + K_0'' \subseteq K_0 \\ (K_0 + K_0, F_0) \subseteq X_0. \end{gathered}$$

Now we estimate

$$e^{\mathrm{avg}}(S_\chi^F, A, \sigma) = \int_{X_0} |S_\chi^F(k,f) - \varphi(N(k,f))| d\sigma(k,f)$$

$$\geq \int_{K_0\times K_0\times F_0} |S_\chi^F(k+h,f) - \varphi(N_1(k+h), N_2 f)| d\mu_1(k) d\mu_2(h) d\nu_1(f).$$

Using the symmetry of μ_1', we derive

$$\begin{aligned}
&\int_{K_0} |S_\chi^F(k+h,f) - \varphi(N_1(k+h), N_2 f)| d\mu_1(k) \\
\geq\ & \int_{K_0'\times K_0''} |S_\chi^F(k'+k''+h,f) - \varphi(N_1(k''+h), N_2 f)| d\mu_1'(k') d\mu_1''(k'') \\
=\ & \frac{1}{2}\sum_{\tau=\pm 1} \int_{K_0'\times K_0''} |S_\chi^F(\tau k'+k''+h,f) - \varphi(N_1(k''+h), N_2 f) d\mu_1'(k') d\mu_1''(k'') \\
\geq\ & \frac{1}{2}\int_{K_0'\times K_0''} |S_\chi^F(k'+k''+h,f) - S_\chi^F(-k'+k''+h,f)| d\mu_1'(k') d\mu_1''(k'').
\end{aligned}$$

Next obeserve that (99) - (101) and the construction of ψ_{ij} and λ imply for $f \in \mathrm{Im}\ J_1, k \in \mathrm{Im}\ \Phi_1, h \in \mathrm{Im}\ \Phi_2$,

$$\begin{aligned}
\mathrm{supp}\ f &\subseteq G_1 \\
\mathrm{supp}\ k &\subseteq G_2 \times G_1 \\
\mathrm{supp}\ h &\subseteq G_3 \times G_2.
\end{aligned}$$

Consequently,

$$\begin{aligned}
&T_k^2 = T_h^2 = T_k T_h = 0, \\
&T_h f = 0 \\
&\langle T_k f, \delta_{t_0}\rangle = 0
\end{aligned}$$

It follows that

$$(T_{k+h})^j = 0$$

for $j \geq 3$. Hence

$$\begin{aligned}
&S_\chi^F(k+h,f) = \left\langle (I - T_{k+h})^{-1} f, \delta_{t_0}\right\rangle \\
=\ & \left\langle f + T_{k+h} f + T_{k+h}^2 f, \delta_{t_0}\right\rangle = \langle T_h T_k f, \delta_{t_0}\rangle.
\end{aligned}$$

Consequently, if $k = k' + k''$, $k' \in K'$, $k'' \in K''$, then

$$S_\chi^F(k'+k''+h,f) - S_\chi^F(-k'+k''+h,f) = 2\langle T_h T_{k'} f, \delta_{t_0}\rangle,$$

which gives

$$e^{\mathrm{avg}}(S_\chi^F, A, \sigma) \geq \int_{K_0'\times K_0\times F_0} |\langle T_h T_{k'} f, \delta_{t_0}\rangle| d\mu_1'(k') d\mu_2(h) d\nu_1(f).$$

Using Lemma 12.1, (103), (104), (107) - (110), and the fact that $E_\gamma(\Phi_1|_{L'}) \le E_\gamma(\Phi_1)$ we can continue this inequality as follows,

$$\begin{aligned}
&\ge c\int_{K'\times K\times F} |\langle T_h T_{k'} f, \delta_{t_0}\rangle|\, d\mu_1'(k')d\mu_2(h)d\nu_1(f) \\
&= c\int_{K'\times K\times F} |\langle T_{k'} f, T_h^*\delta_{t_0}\rangle|\, d\mu_1'(k')d\mu_2(h)d\nu_1(f) \\
&= c\int_{K'\times W\times F} |\langle T_{k'} f, w\rangle|\, d\mu_1'(k')d\nu_2(w)d\nu_1(f).
\end{aligned} \tag{111}$$

Here we used the fact that for $\xi = (\xi_j)_{j=1}^n \in E$, $h = \Phi_2\xi$ and $g \in F$ we have by (99) and (101)

$$\langle T_h g, \delta_{t_0}\rangle = \left\langle g, \theta_{n2}\sum_{j=1}^n \xi_j\psi_{j2}\right\rangle \lambda(t_0) = \langle g, J_2\xi\rangle\,.$$

Now let $k' = \Phi_1 a$, $a \in L'$, $a = (a_{ij})_{i,j=1}^n$, let $f = J_1 x$, $w = J_2 y$, $x, y \in E$, $x = (x_j)_{j=1}^n$, $y = (y_i)_{i=1}^n$. Then

$$\begin{aligned}
T_{k'} &= \beta_n \sum_{i,j=1}^n a_{ij}\psi_{j1}\otimes\psi_{i2} \\
f &= J_1 x = \theta_{n1}\sum_{j=1}^n x_j\psi_{j1} \\
w &= J_2 y = \theta_{n2}\sum_{i=1}^n y_i\psi_{i2},
\end{aligned}$$

and consequently

$$\begin{aligned}
\langle T_{k'} f, w\rangle &= \beta_n\theta_{n1}\theta_{n2}\sum_{i,j=1}^n a_{ij}x_j y_i(\psi_{j1},\psi_{j1})(\psi_{i2},\psi_{i2}) \\
&= (4n)^{-2}\beta_n\theta_{n1}\theta_{n2}\sum_{i,j=1}^n a_{ij}x_j y_i.
\end{aligned}$$

Since L and E are finite dimensional, we can rewrite (111) as

$$\begin{aligned}
&= \int_{L'\times E\times E} |\langle T_{\Phi_1 a}J_1 x, J_2 y\rangle|\, d\gamma_{L'}(a)d\gamma_E(x)d\gamma_E(y) \\
&= (4n)^{-2}\beta_n\theta_{n1}\theta_{n2}\int_{L'\times E\times E} \Big|\sum_{i,j=1}^n a_{ij}x_j y_i\Big| d\gamma_{L'}(a)d\gamma_E(x)d\gamma_E(y).
\end{aligned} \tag{112}$$

Using well-known facts about the standard Gaussian measure in $\mathbf{R}^n$, we get

$$\int_{E\times E} \Big|\sum_{i,j=1}^n a_{ij}x_j y_i\Big| d\gamma_E(x)d\gamma_E(y) \ge c\int_E \left(\sum_{i=1}^n \Big|\sum_{j=1}^n a_{ij}x_j\Big|^2\right)^{1/2} d\gamma_E(x)$$

$$= c\int_E \|ax\|_E d\gamma_E(x) \ge c\left(\sum_{i,j=1}^n a_{ij}^2\right)^{1/2} = c\|a\|_L$$

with the constants $c > 0$ not depending on n and a. Inserting (105) and (106), we can continue (112) as

$$\begin{aligned} &\geq cn^{-3r/d-2}(\log(n+1))^{-3/2}\int_{L'}\|a\|d\gamma_{L'}(a) \\ &\geq cn^{-3r/d-2}(\log(n+1))^{-3/2}(n^2-m)^{1/2} \\ &\geq c(\log(n+1))^{-3/2}n^{-3r/d-1}, \end{aligned}$$

if $m \leq n^2/2$. The lower estimate of Theorem 7.4 follows.

References

[Bab76] K. I. Babenko. Estimating the quality of computational algorithms. *Computer Methods in Appl. Mech. and Engineering*, **7**: 47 – 73 (part I), 135 – 152 (part II), 1976.

[Bak59] N. S. Bakhvalov. On approximate calculation of integrals (in Russian). *Vestnik Moskow. Gos. Univ., Ser Mat., Mekh. Astronom, Fiz. Khim*, **4**: 3 – 18, 1959.

[Bak62] N. S. Bakhvalov. On the convergence of indeterministic integration processes within the functional classes $W_p^{(\ell)}$ (in Russian). *Teor. Veroyatn. i Prim.*, **7**: 238, 1962.

[BS65] I. Babuška and S. L. Sobolev. Optimization of numerical methods (in Russian). *Appl. Mat.*, **10**: 96 – 130, 1965.

[CF83a] Z. Ciesielski and T. Figiel. Spline bases in classical function spaces on compact C^∞ manifolds, part I. *Studia Math.*, **76**: 1 – 58, 1983.

[CF83b] Z. Ciesielski and T. Figiel. Spline bases in classical function spaces on compact C^∞ manifolds, part II. *Studia Math.*, **76**: 95 – 136, 1983.

[Cia78] P. G. Ciarlet. *The Finite Element Method for Elliptic Problems*. North-Holland, 1978.

[Cie75] Z. Ciesielski. Constructive function theory and spline systems. *Studia Math.*, **53**: 277 – 302, 1975.

[EI67] K. V. Emelyanov and A. M. Ilin. On the number of arithmetic operations, necessary for the approximate solution of Fredholm integral equations of the second kind (in Russian). *Zh. Vychisl. Mat. i Mat, Fiz.*, **7**: 905 – 910, 1967.

[Erm71] S. M. Ermakov. *The Monte Carlo Method and Related Problems* (in Russian). Nauka, Moscow, 1971.

[GG84] A. Yu. Garnayev and E. D. Gluskin. On the diameters of the Euclidean ball. *Dokl. Akad. Nauk SSSR*, **227**: pages 1 048 – 1 052, 1984.

[GKS87] Y. Gordon, H. König, and C. Schütt. Geometric and probabilistic estimates for entropy and approximation numbers of operators. *J. Approx. Th.*, **49**: 219 – 239, 1987.

[Hei90a] S. Heinrich. Invertibility of random Fredholm operators. *Stochastic Anal. Appl.*, **8**: 1 – 59, 1990.

[Hei90b] S. Heinrich. Probabilistic complexity analysis for linear problems in bounded domains. *J. Complexity*, **6**: 231 – 255, 1990.

[Hei91] S. Heinrich. Probabilistic analysis of numerical methods for integral equations. *J. Integral Equations Appl.*, **3**: 289 – 319, 1991.

[Hei92] S. Heinrich. Lower bounds for the complexity of Monte Carlo function approximation. To appear in *J. Complexity*, 1992.

[Hei93] S. Heinrich. Complexity of integral equations and relations to s-numbers. Submitted to J. *Complexity*, 1993.

[HK91] S. Heinrich and J.-D. Kern. Parallel information-based complexity. *J. Complexity*, **7**: 339 – 370, 1991.

[HM92] S. Heinrich and P. Mathé. The Monte Carlo complexity of Fredholm integral equations. To appear in *Math. of Computation*, 1992.

[Kas77] B. S. Kashin. Diameters of some finite dimensional sets and classes of smooth functions (in Russian). *Izv. Akad. Nauk SSSR*, **41**: 334 – 351, 1977.

[Kie53] J. Kiefer. Sequential minimax search for a maximum. *Proc. Amer. Math. Soc.*, **4**: 502 – 505, 1953.

[Kol87] A. N. Kolmogorov. *Information Theory and Theory of Algorithms, Collected Works*. Nauka, Moscow, 1987.

[Kön86] H. König. *Eigenvalue Distribution of Compact Operators*. Birkhäuser, Basel, 1986.

[Mat90] P. Mathé. s-numbers in information-based complexity. *J. Complexity*, **6**: 41 – 66, 1990.

[Mat91] P. Mathé. Random approximation of Sobolev embeddings. *J. Complexity*, **7**: 261 – 281, 1991.

[Mat92] P. Mathé. A minimax principle for the optimal error of Monte Carlo methods. To appear in *Constructive Approx.*, 1992.

[May75] V. E. Mayorov. Discretization of the problem of diameters (in Russian). *Uspehi Mat. Nauk*, **30**: 179 – 180, 1975.

[Mic56] E. Michael. Continuous selections I. *Ann. of Math.*, **63**: 361 – 382, 1956.

[MR77] C. A. Micchelli and T. J. Rivlin. A survey of optimal recovery. In C. A. Micchelli and T. J. Rivlin, editors, *Optimal Estimation in Approximation Theory*, pages 1 – 54. Plenum Press, New York, 1977.

[Nik50] S. M. Nikolskij. On the problem of approximation estimate by quadrature formulas (in Russian). *Usp. Mat. Nauk*, **5**: 165 – 177, 1950.

[Nov88] E. Novak. *Deterministic and Stochastic Error Bounds in Numerical Analysis.* Springer: Berlin, Heidelberg, New York, London, Paris and Tokio, 1988.

[Pac87] E. W. Packel. The algorithm designer versus nature: A game-theoretic approach to information-based complexity. *J. Complexity*, **3**, 1987.

[Per89] S. V. Pereverzev. On the complexity of the problem of finding solutions of Fredholm equations of the second kind with differentiable kernels (in Russian). *Ukrain. Mat. Sh.*, **41**: 1422 - 1425, 1989.

[Pie78] A. Pietsch. *Operator Ideals.* Deutscher Verlag der Wissenschaften, Berlin and North Holland, Amsterdam, 1978.

[Pie87] A. Pietsch. *Eigenvalues and s-numbers.* Geest and Portig, Leipzig and Cambridge University Press, Cambridge, 1987.

[Pin85] A. Pinkus. *n-Widths in Approximation Theory.* Springer: Berlin, Heidelberg and New York, 1985.

[Pis89] G. Pisier. *The Volume of Convex Bodies and Banach Space Geometry.* Cambridge University Press, Cambridge, 1989.

[Sar49] A. Sard. Best approximate integration formulas; best approximation formulas. *Amer. J. Math.*, **71**: 80 - 91, 1949.

[Sma85] S. Smale. On the efficiency of algorithms of analysis. *Bull. Amer. Math. Soc.*, **13**: 87 - 121, 1985.

[Sul59] A. V. Suldin. Wiener measure and its applications to approximation methods, I (in Russian). *Izv. Vyssh. Ucheb. Zaved. Mat.*, **13**: 145 - 158, 1959.

[Sul60] A. V. Suldin. Wiener measure and its application to approximation methods, II (in Russian). *Izv. Vyssh. Ucheb. Zaved. Mat.*, **18**: 165 - 179, 1960.

[Tri78] H. Triebel. *Interpolation Theory, Function Spaces, Differential Operators.* Deutscher Verlag der Wissenschaften, Berlin, 1978.

[TW80] J. F. Traub and H. Woźniakowski. *A General Theory of Optimal Algorithms.* Academic Press, New York, 1980.

[TWW88] J. F. Traub, G. W. Wasilkowski, and H. Woźniakowski. *Information-Based Complexity.* Academic Press, New York, 1988.

Best Constants for the Norm of the Complexification of Operators Between L_p-Spaces

ANDREAS DEFANT, Fachbereich Mathematik, Carl v. Ossietzky Universität, 2900 Oldenburg, Germany

Abstract.

Denote the natural complexification of a linear operator

$$T : L_q^{\mathbb{R}}(\mu_1) \longrightarrow L_p^{\mathbb{R}}(\mu_2)$$

(real valued functions) by

$$T^{\mathbb{C}} : L_q^{\mathbb{C}}(\mu_1) \longrightarrow L_p^{\mathbb{C}}(\mu_2)$$

(complex valued functions, $1 \le p, q \le \infty$). For $1 \le p \le q \le 2$ (and dually for $2 \le p \le q \le \infty$) we calculate the best constant $k_{q,p}$ such that $\|T^{\mathbb{C}}\| \le k_{q,p}\|T\|$ for all T - this improves some results of [1], [2]. Example: $k_{2,1} = \frac{\pi}{\sqrt{8}}$.

1. The result.

For $1 \le q \le \infty$ and an arbitrary measure space (Ω, μ) let $L_q^{\mathbb{K}}(\mu)$ be the space of all $\mathbb{K}$-valued integrable functions, $\mathbb{K} = \mathbb{R}$ or $\mathbb{C}$. By a result of Figiel, Iwaniec and Pełzcyński [3] the norm of a (linear and continuous) operator $T : L_q^{\mathbb{R}}(\mu_1) \longrightarrow L_p^{\mathbb{R}}(\mu_2)$ and the norm of its complexification

$$T^{\mathbb{C}} : L_q^{\mathbb{C}}(\mu_1) \longrightarrow L_p^{\mathbb{C}}(\mu_2),\ T(u + iv) := Tu + iTv$$

coincide whenever $1 \le q \le p \le \infty$, but in general these norms are different for

$1 \le p < q \le \infty$ as easy examples show. For $1 \le p, q \le \infty$ let $k_{q,p}$ be the best constant such that

$$\|T^{\mathbb{C}}\| \le k_{q,p}\|T\| \quad \text{for all} \quad T : L_q^{\mathbb{R}}(\mu_1) \longrightarrow L_p^{\mathbb{R}}(\mu_2) \ .$$

It is a highly non-trivial fact due to Krivine [4], [5] that

$$k_{\infty,1} = \sqrt{2} \ .$$

By use of a formula for $k_{q,p}$ in terms of L_p-factorable and absolutely p-summing norms, an upper estimate of $k_{q,p}$ for $1 \le p < q \le 2$ is given in [1], [2] - moreover, for $q = 2$ this estimate is even best possible. We reprove this estimate by an easy direct argument and show that it is best possible for all $1 \le p < q \le 2$:

Theorem. *For* $1 \le p \le q \le 2$

$$k_{q,p} = \sqrt{\pi}^{\,1/p - 1/q} \left[\frac{\Gamma\left(\frac{q+1}{2}\right)}{\Gamma\left(\frac{q+2}{2}\right)} \right]^{1/q} \left[\frac{\Gamma\left(\frac{p+2}{2}\right)}{\Gamma\left(\frac{p+1}{2}\right)} \right]^{1/p},$$

where Γ *as usual stands for the Gamma function.*

By the obvious duality $k_{q,p} = k_{p',q'}$ this also gives the best constants for the case $2 \le p \le q \le \infty$.

For more information on the constants $k_{q,p}$ — in particular, their reformulation in terms of certain (tensor) norms of the function $\cos(x - y)$ as an element of $C[-\pi, \pi] \otimes C[-\pi, \pi] \subset C([-\pi, \pi]^2)$ as well as their relation with the n-dimensional Grothendieck constant — see [1], [2] and [5].

2. A tensor product formulation.

For any Banach space E the tensor product $L_p(\mu) \otimes E$ can be considered as a subspace of the Banach space $L_p(\mu, E)$ of Bochner integrable functions via the canonical mapping

$$L_p(\mu) \times E \longrightarrow L_p(\mu, E), \ (f, x) \rightsquigarrow f \cdot x \ .$$

The norm which $L_p(\mu, E)$ induces on $L_p(\mu) \otimes E$ is denoted by Δ_p:

$$L_p(\mu) \otimes_{\Delta_p} E \hookrightarrow L_p(\mu, E) \text{ isometrically.}$$

An obvious consideration shows:

Remark. *For any operator* $T : L_q(\mu_1) \to L_p(\mu_2)$

$$\|T^{\mathbb{C}}\| = \left\|T \otimes id_{\mathbb{R}_2^2} : L_q^{\mathbb{R}}(\mu_1) \otimes_{\Delta_q} \mathbb{R}_2^2 \to L_p^{\mathbb{R}}(\mu_2) \otimes_{\Delta_p} \mathbb{R}_2^2\right\| ,$$

here $id_{\mathbb{R}_2^2}$ *denotes the identity on the 2-dimensional real Hilbert space* $\mathbb{R}_2^2$.

The transposed norm Δ_p^t on $E \otimes L_p(\mu)$ is defined by the natural transposition map

$$E \otimes_{\Delta_p^t} L_p(\mu) := L_p(\mu) \otimes_{\Delta_p} E$$

$$x \otimes f \rightsquigarrow f \otimes x .$$

We will need two simple facts:

(1) $L_q(\mu_1) \otimes_{\Delta_q} L_q(\mu_2) = L_q(\mu_1) \otimes_{\Delta_q^t} L_q(\mu_2)$ holds isometrically.

(2) For any operator $T : E \to F$ between Banach spaces, every probability space (Ω, μ) and $1 \le p \le q \le \infty$

$$\|T\| = \left\|I \otimes T : L_q(\mu) \otimes_{\Delta_q} E \to L_p(\mu) \otimes_{\Delta_p} F\right\|$$

$$= \left\|T \otimes I : E \otimes_{\Delta_q^t} L_q(\mu) \to F \otimes_{\Delta_p^t} L_p(\mu)\right\| ,$$

where $I : L_q \hookrightarrow L_p$ denotes the canonical embedding. If $q = p$ and T is an isometry (into), then $I \otimes T$ and $T \otimes I$ are isometries as well.

Statement (1) is a consequence of Fubini's theorem and (2) follows from the obvious fact that

$$L_q(\mu, E) \to L_p(\mu, F)$$

$$f \rightsquigarrow T \circ f$$

has norm $\|T\|$ (resp. is an isometry).

For a detailed study of the norm Δ_p and its various relations to the metric theory of tensor norms on tensor products of Banach spaces see [1], [2].

3. The upper estimate.

This estimate is based on the following elementary

Lemma. *For* $x \in \mathbb{R}^n$ *and* $1 \le p < \infty$

$$\|x\|_2 = \frac{c_p^{(n)}}{c_p^{(1)}} \Big(\int_{S_{n-1}} |\langle x, w \rangle|^p \lambda_{n-1}(dw) \Big)^{1/p},$$

where S_{n-1} *is the* $(n-1)$*-dimensional sphere in* $\mathbb{R}_2^n$, λ_{n-1} *the normalized Lebesgue measure on* S_{n-1} *and*

$$c_p^{(n)} := \Big(\int_{\mathbb{R}^n} \|w\|_2^p \gamma_n(dw) \Big)^{1/p} = \sqrt{2} \left[\frac{\Gamma\big(\frac{n+p}{2}\big)}{\Gamma\big(\frac{n}{2}\big)} \right]^{1/p}$$

the p - th *moment of the Gauss measure* γ_n *on* $\mathbb{R}^n$.

For a proof see e. g. [1] or [6].

Clearly, this result can be formulated in terms of an embedding: For $1 \le p, q < \infty$ and $n \in \mathbb{N}$ let

$$I_{p,q}^{(n)} : \mathbb{R}_2^n \to L_q(S_{n-1}, \lambda_{n-1}),\ I_{p,q}^{(n)}(x) := \frac{c_p^{(n)}}{c_p^{(1)}} \langle \cdot, x \rangle ;$$

then $I_{p,p}^{(n)}$ is an isometry (into) and

$$\big\| I_{p,q}^{(n)} \big\| = \frac{c_q^{(1)}}{c_q^{(n)}} \frac{c_p^{(n)}}{c_p^{(1)}} .$$

Since for any operator $T : L_q^{\mathbb{R}}(\mu_1) \to L_p^{\mathbb{R}}(\mu_2)$ the diagramm

$$\begin{array}{ccc} L_q^{\mathbb{R}}(\mu_1) \otimes_{\Delta_q} \mathbb{R}_2^2 & \xrightarrow{T \otimes id_{\mathbb{R}_2^2}} & L_p^{\mathbb{R}}(\mu_2) \otimes_{\Delta_p} \mathbb{R}_2^2 \\ \Big\downarrow id \otimes I_{p,q}^{(2)} & & \Big\downarrow id \otimes I_{p,p}^{(2)} \\ L_q^{\mathbb{R}}(\mu_1) \otimes_{\Delta_q^t} L_q(\lambda_1) & \xrightarrow{T \otimes I} & L_p^{\mathbb{R}}(\mu_2) \otimes_{\Delta_p^t} L_p(\lambda_1) \end{array}$$

commutes, by the remark and (1), (2) of the preceding section

$$\|T^{\mathbb{C}}\| = \big\|T \otimes id_{\mathbb{R}_2^2}\big\|$$

$$\le \big\|I_{p,q}^{(2)}\big\| \, \|T\| = \frac{c_q^{(1)}}{c_q^{(2)}} \frac{c_p^{(2)}}{c_p^{(1)}} \|T\| .$$

Hence an easy calculation gives the upper estimate of the theorem. □

Of course, the proof also could have been written as a chain of inequalities instead of a chain of maps.
Finally, we note that the assumption $1 \le p \le q \le 2$ was not used so far, hence the upper estimate for $k_{q,p}$ holds for all $1 \le p \le q \le \infty$.

4. Absolutely summing and mixing norms.

In order to show that this estimate is best possible we recall some facts from the theory of absolutely summing operators. See [1] and [6] for more information and the proofs.

The absolutely p-summing norm of an operator $T : E \to F$, where E and F are two Banach spaces and $1 \le p < \infty$, is defined by

$$\mathbf{P}_p(T) := \sup\Big(\sum_1^n \|Tx_k\|^p\Big)^{1/p} \in [0, \infty] ,$$

the supremum taken over all $x_1, \ldots, x_n \in E$ such that

$$\sup_{\|x'\|\le 1}\Big(\sum_1^n |x'(x_k)|^p\Big)^{1/p} \le 1 .$$

By a result of Gordon

$$\mathbf{P}_p\big(id_{\mathbb{R}_2^n}\big) = \frac{c_p^{(n)}}{c_p^{(1)}} .$$

For $1 \le p \le q < \infty$ the (q, p)-mixing norm of $T : E \to F$ is given by

$$\mathbf{M}_{q,p}(T) := \sup \mathbf{P}_p(ST) \in [0, \infty] ,$$

where the supremum is taken over all $S \in \mathcal{L}(F, G)$ such that $\mathbf{P}_q(S) \le 1$. It was shown by Maurey that this norm has the following discrete formulation:

$$\mathbf{M}_{q,p}(T) := \sup\bigg(\sum_{\ell=1}^m\Big(\sum_{k=1}^n |y_k'(Tx_\ell)|^q\Big)^{p/q}\bigg)^{1/p} ,$$

the supremum taken over all $x_1, \ldots, x_m \in E$ and $y_1', \ldots, y_n' \in F$ such that

$$\sup_{\|x'\|\le 1}\Big(\sum_{\ell=1}^m |x'(x_\ell)|^p\Big)^{1/p} \le 1 \quad \text{and} \quad \Big(\sum_{k=1}^n \|y_k'\|^q\Big)^{1/q} \le 1 .$$

5. The lower estimate.

So-called Lévy measures will be used: For $1 \le p < q \le 2$ there is a constant $c_{p,q}$ such that for every n there is a probability measure μ_q^n on $\mathbb{R}^n$ satisfying

$$\|x\|_q = \frac{1}{c_{q,p}} \Big(\int_{\mathbb{R}^n} |< x, w >|^p \mu_q^n(dw) \Big)^{1/p}$$

for all $x \in \mathbb{R}^n$. See e. g. [1] or [6] for the details of this non-trivial fact. We will consider the following isometries

$$A_n : \mathbb{R}_q^n \hookrightarrow L_p^{\mathbb{R}}(\mathbb{R}^n, \mu_q^n), \quad A_n(x) := \frac{1}{c_{q,p}} < \cdot, x >$$

and show that the norms of their complexifications $A_n^{\mathbb{C}}$ satisfy

$$\frac{P_p(id_{\mathbb{R}_2^2})}{P_q(id_{\mathbb{R}_2^2})} \le M_{q,p}(id_{\mathbb{R}_2^2}) \le \sup_n \|A_n^{\mathbb{C}}\| \le k_{q,p} ,$$

which gives the conclusion by what was said before. The first and the last inequalities are clear by the definitions, the remaining one uses Maurey's discrete formulation of the mixing norm:

For $x_1, \ldots, x_m \in \mathbb{R}^2$ and $y_1, \ldots, y_n \in \mathbb{R}^2$

$$\Big(\sum_{\ell=1}^m \Big(\sum_{k=1}^n |< y_k, x_\ell >|^q \Big)^{p/q} \Big)^{1/p}$$

$$= \frac{1}{c_{q,p}} \Big(\sum_{\ell=1}^m \int_{\mathbb{R}^n} \Big| \sum_{k=1}^n w_k < y_k, x_\ell > \Big|^p \mu_q^n(dw) \Big)^{1/p}$$

$$= \frac{1}{c_{q,p}} \Big(\int_{\mathbb{R}^n} \sum_{\ell=1}^m \Big|< \sum_{k=1}^n w_k y_k, x_\ell > \Big|^p \mu_q^n(dw) \Big)^{1/p}$$

$$= \frac{1}{c_{q,p}} \Big(\int_{\mathbb{R}^n} \Big\| \sum_{k=1}^n w_k y_k \Big\|_2^p \sum_{\ell=1}^m \Big|< \Big\| \sum_{k=1}^n w_k y_k \Big\|_2^{-1} \sum_{k=1}^n w_k y_k, x_\ell > \Big|^p \mu_q^n(dw) \Big)^{1/p}$$

$$\le \frac{1}{c_{q,p}} \Big(\int_{\mathbb{R}_n} \Big\| \sum_{k=1}^n w_k y_k \Big\|_2^p \mu_q^n(dw) \Big)^{1/p} \Big(\sup_{\|y\|_2 \le 1} \sum_{\ell=1}^m |< y, x_\ell >|^p \Big)^{1/p}$$

$$\le \Big\| A_n \otimes id_{\mathbb{R}_2^2} : \mathbb{R}_q^n \otimes_{\Delta_q} \mathbb{R}_2^2 \to L_p(\mu_q^n) \otimes_{\Delta_q} \mathbb{R}_2^2 \Big\|$$

$$\cdot \sup_{\|y\|_2 \le 1} \Big(\sum_{\ell=1}^m |< y, x_\ell >|^p \Big)^{1/p} \cdot \Big(\sum_{k=1}^n \|y_k\|_2^q \Big)^{1/q}$$

$$\le \sup_j \|A_j^{\mathbb{C}}\| \cdot \sup_{\|y\|_2 \le 1} \Big(\sum_{\ell=1}^m |< y, x_\ell >|^p \Big)^{1/p} \cdot \Big(\sum_{k=1}^n \|y_k\|_2^q \Big)^{1/q} .$$

This completes the proof of the theorem in 1. □

6. Remarks.

Finally, we remark that our proof also gives the exact value of the n-dimensional complexification constant $k_{q,p}^{(n)}$ whenever $1 \le p < q \le 2$:

$$k_{q,p}^{(n)} := \sup \| T \otimes id : L_q(\mu_1) \otimes_{\Delta_q} \mathbb{R}_2^n \to L_p(\mu_2) \otimes_{\Delta_p} \mathbb{R}_2^n \| = \frac{c_q^{(1)}}{c_q^{(n)}} \frac{c_p^{(n)}}{c_p^{(1)}},$$

the supremum taken over all measures μ_1, μ_2 and $\|T\| \le 1$.

The exact value of $k_{q,p}$ for $1 \le p < 2 < q \le \infty$ and $(p, q) \neq (1, \infty)$ remains open.

Conjecture: For $1 < p < 2$

$$k_{p',p} = \begin{cases} 2^{1/p - 1/2} & \text{for } 1 \le p \le p_0 \\ \frac{1}{\sqrt{2}} \left[\frac{\sqrt{\pi}}{\Gamma\left(\frac{p+1}{2}\right)} \right]^{1/p} & \text{for } p_0 \le p < 2 \end{cases}$$

where $p_0 \approx 1,84742$ is the solution of $\Gamma\left(\frac{p+1}{2}\right) = \frac{\sqrt{\pi}}{2}$ in $]1, 2[$.
As shown by Haagerup these constants apear as best constants in Khintchine's inequality (see e. g. [1]).

References

[1] Defant, A. - K. Floret: Tensor norms and operator ideals; North-Holland Math. Studies 176, 1993

[2] Defant, A. - K. Floret: Continuity of tensor product operators between spaces of Bochner integrable functions; Progress in Functional Analysis, Proc. Peñiscola Meeting 1990 (eds.: Bierstedt, Bonet, Horvath, Maestre), North-Holland Math. Studies 170 (1992) 367-381

[3] Figiel, T. - T. Iwaniec - A. Pełczyński: Computing norms of some operators in L^p-spaces; Studia Math. 79 (1984) 227-274

[4] Krivine, J. L.: Sur la complexification des opèrateurs de L^∞ dans L^1; C. R. A. S. Paris 284 (1977) 377-379

[5] Krivine, J. L.: Constantes de Grothendieck et fonctions de type positif sur les sphères; Advances in Math. 31 (1979) 16-30

[6] Pietsch, A.: Operator ideals; Deutscher Verlag der Wiss., 1978 and North-Holland, 1980

Antisymmetric Tensor Products of Nuclear Operators on Banach Spaces

STEFAN GEISS, Mathematisches Institut, Friedrich-Schiller-Universität Jena
D-O-6900 Jena, Germany

1 INTRODUCTION

Let $\mathcal{L}(X, Y)$ be the Banach space of all linear and continuous operators from a Banach space X into a Banach space Y, and let B_X be the closed unit ball of X. An operator $T \in \mathcal{L}(X, Y)$ is nuclear provided that there are sequences $\{a_n\}_{n=1}^{\infty} \subseteq B_{X'}$, $\{y_n\}_{n=1}^{\infty} \subseteq B_Y$ and $\lambda_1 \geq \lambda_2 \geq \ldots \geq 0$ with $\sum \lambda_n < \infty$ and $T = \sum \lambda_n a_n \otimes y_n$. Setting

$$\nu(T) := \inf \left\{ \sum_n \lambda_n : T = \sum_n \lambda_n a_n \otimes y_n \text{ as above} \right\},$$

$$N(X, Y) := \{T \in \mathcal{L}(X, Y) : T \text{ is nuclear}\}$$

we obtain a Banach operator ideal $[N, \nu]$ (see Pietsch (1980)). The non-increasing sequence of the eigenvalues of a nuclear operator T on a complex Banach space X, where every eigenvalue is counted according to its multiplicity, is denoted by $\{\lambda_n(T)\}_{n=1}^{\infty}$. The asymptotic behaviour of the eigenvalues of an operator $T \in N(X, X)$ is closely related to the geometry of the

The author is supported by the DFG (Ko 962/3-1).

underlying Banach space X. We describe this geometry with the help of the Grothendieck numbers. The n–th Grothendieck number of an operator $T \in \mathcal{L}(X,Y)$ is defined as

$$\Gamma_n(T) := \sup\left\{ |\det(\langle Tx_i, b_j\rangle)_{i,j=1}^n|^{\frac{1}{n}} : x_i \in B_X,\ b_j \in B_{Y'} \right\}.$$

In the case $T = I_X$ (I_X is the identity of X) we set $\Gamma_n(X) := \Gamma_n(I_X)$. The quantity $\Gamma_n(X)$ can be understood as the maximal distance of the n–dimensional subspaces E of X to the Hilbert space ℓ_2^n if we approximate the unit ball B_E with the ellipsoids of minimal and maximal volume (see Geiss (1992)(Corollary 4.4)). For convenience we will say that $X \in \Gamma_\alpha$ $(0 \le \alpha \le \frac{1}{2})$ if there is a constant $c > 0$ such that

$$\Gamma_n(X) \le cn^\alpha \qquad \text{for} \qquad n = 1, 2, \ldots$$

The scale of classes Γ_α is well investigated. For example, Γ_0 is the class of weak Hilbert spaces and $\Gamma_{\frac{1}{2}}$ is the class of all Banach spaces. For detailed information see Geiss (1990:a, 1992), Pietsch (1991), Pisier (1988) and the references there. In Geiss (1990:a)(Proposition 3.5) it is shown that

$$X \in \Gamma_\alpha \qquad \text{if and only if} \qquad |\lambda_n(T)| \le cn^{\alpha-1}\nu(T)$$

for all $T \in N(X,X)$ and some constant $c > 0$ depending only on X. In section 2 of this paper we improve one direction of this result by showing that

$$\lim_n n^{1-\alpha}|\lambda_n(T)| = 0$$

for all $T \in N(X,X)$ whenever $X \in \Gamma_\alpha$; see Corollary 2.9. We can also replace the eigenvalues by the Weyl numbers, and obtain in Theorem 2.8

$$X \in \Gamma_\alpha \qquad \text{if and only if} \qquad \lim_n n^{1-\alpha}x_n(T) = 0$$

for all Banach spaces Y and all $T \in N(Y,X)$. Consequently, if $X \in \Gamma_\alpha$ then there does not exist any operator $T \in N(X,X)$ or $T \in N(Y,X)$ such that

$$|\lambda_n(T)| \sim n^{\alpha-1} \qquad \text{or} \qquad x_n(T) \sim n^{\alpha-1}.$$

Let us recall that, for an operator $T \in \mathcal{L}(X,Y)$, the n-th approximation number $a_n(T)$ and the n-th Weyl number $x_n(T)$ are given by

$$a_n(T) := \inf\{\|T - L\| : L \in \mathcal{L}(X,Y),\ rank(L) < n\},$$

$$x_n(T) := \sup\{a_n(Tu) : u \in \mathcal{L}(\ell_2, X),\ \|u\| = 1\}.$$

The technique presented here is a more precise variant of Geiss (1990:a). The new aspect is the concept of the antisymmetric tensor product of a Radon measure which naturally arises from the consideration of antisymmetric tensor products of nuclear operators. Using this concept Lemma 2.4 yields the additional null-sequence (for instance $\sup_n n^{1-\alpha}|\lambda_n(T)| < \infty$ turns into $\lim_n n^{1-\alpha}|\lambda_n(T)| = 0$).

Let us first introduce the antisymmetric ε–tensor product $\Lambda_\varepsilon^n X$ of a Banach space X as the closure of

$$\Lambda^n X := \left\{ s = \sum_{j=1}^N x_{1j} \wedge \ldots \wedge x_{nj} : \quad x_{ij} \in X,\ N = 1, 2, \ldots \right\} \subseteq \bigotimes^n X$$

in the ε–tensor product $\bigotimes_\varepsilon^n X$. As usual $x_1 \wedge \ldots \wedge x_n := \sum_{\pi \in S_n} sgn\pi \; x_{\pi(1)} \otimes \cdots \otimes x_{\pi(n)}$ (see Geiss (1992)). The antisymmetric ε–tensor product of an operator $T \in \mathcal{L}(X,Y)$ is naturally given by

$$(\Lambda_\varepsilon^n T)(x_1 \wedge \ldots \wedge x_n) := Tx_1 \wedge \ldots \wedge Tx_n.$$

It is clear that $\Lambda_\varepsilon^n T \in \mathcal{L}(\Lambda_\varepsilon^n X, \Lambda_\varepsilon^n Y)$. A probability measure μ on the σ–algebra $\mathcal{B}(X)$ generated by the norm-open subsets of X is called Radon if for all $\varepsilon > 0$ there is a norm-compact subset $K \subseteq X$ with $\mu(K) \geq 1 - \varepsilon$. The set of all Radon probability measures μ on $\mathcal{B}(X)$ with $\|\mu\|_p := (\int \|x\|^p d\mu(x))^{\frac{1}{p}} < \infty$ $(1 \leq p < \infty)$ is denoted by $\mathcal{R}_p(X)$. For $\mu \in \mathcal{R}_p(X)$ we define the antisymmetric product $\Lambda_\varepsilon^n \mu \in \mathcal{R}_p(\Lambda_\varepsilon^n X)$ as the image measure of μ with respect to the map

$$\Phi : X \times \ldots \times X \to \Lambda_\varepsilon^n X \qquad \text{given by} \qquad \Phi(x_1, \ldots, x_n) := x_1 \wedge \ldots \wedge x_n.$$

We get $\|\Lambda_\varepsilon^n \mu\|_p = (\int \ldots \int \|x_1 \wedge \ldots \wedge x_n\|^p d\mu(x_1) \ldots d\mu(x_n))^{1/p}$.

The next statement is the key for this paper and has its origin in Grothendieck (1956). To formulate this statement we define for $a_1, \ldots, a_n \in X'$ the functional $\phi_{a_1,\cdots,a_n} \in (\Lambda_\varepsilon^n X)'$ by

$$\langle \sum_{j=1}^N x_{1j} \wedge \ldots \wedge x_{nj}, \phi_{a_1,\cdots,a_n} \rangle := \sum_{j=1}^N \det(\langle x_{ij}, a_k \rangle)_{i,k=1}^n.$$

PROPOSITION 1.1 *Let $T \in N(Y,X)$ with $T = \sum_1^\infty \lambda_i b_i \otimes x_i$ where $\{b_i\}_1^\infty \subset B_{Y'}$, $\{x_i\}_1^\infty \subset B_X$, $\lambda_1 \geq \lambda_2 \geq \ldots \geq 0$ and $0 < \lambda := \sum \lambda_i < \infty$. Then*

1. $\Lambda_\varepsilon^n T = \frac{1}{n!} \sum_{i_1,\ldots,i_n} \lambda_{i_1} \cdots \lambda_{i_n} \phi_{b_{i_1},\ldots,b_{i_n}} \otimes (x_{i_1} \wedge \ldots \wedge x_{i_n})$,
2. $\nu(\Lambda_\varepsilon^n T) \leq \frac{\lambda^n}{n!} \|\Lambda_\varepsilon^n \mu\|_1$, *where* $\mu = \frac{1}{\lambda} \sum_1^\infty \lambda_i \delta_{x_i} \in \mathcal{R}_1(X)$.

Consequently, for all $\theta > 0$ there exists a measure $\mu_\theta \in \mathcal{R}_1(X)$ such that

$$\nu(\Lambda_\varepsilon^n T)^{\frac{1}{n}} \leq (1+\theta)\, n!^{-\frac{1}{n}} \, \|\Lambda_\varepsilon^n \mu\|_1^{\frac{1}{n}} \, \nu(T) \leq (1+\theta)\, n!^{-\frac{1}{n}} \, \Gamma_n(X)\, \nu(T).$$

Proof. (1) Assuming $a_1, \ldots, a_n \in X'$ and $s = \sum_{l=1}^N y_{1l} \wedge \ldots \wedge y_{nl}$ the proof of Geiss (1990:a) (Lemma 2.2) gives

$$\begin{aligned} \langle (\Lambda_\varepsilon^n T)\, s, \phi_{a_1,\ldots,a_n} \rangle &= \sum_{l=1}^N \sum_{i_1 < \ldots < i_n} \lambda_{i_1} \cdots \lambda_{i_n} \det(\langle y_{kl}, b_{i_j} \rangle)_{j,k} \det(\langle x_{i_j}, a_k \rangle)_{j,k} \\ &= \frac{1}{n!} \sum_{i_1,\ldots,i_n} \lambda_{i_1} \cdots \lambda_{i_n} \langle s, \phi_{b_{i_1},\ldots,b_{i_n}} \rangle \langle x_{i_1} \wedge \ldots \wedge x_{i_n}, \phi_{a_1,\ldots,a_n} \rangle. \end{aligned}$$

(2) Since $\|\phi_{b_{i_1},\ldots,b_{i_n}}\|_{(\Lambda_\varepsilon^n Y)'} \leq \|b_{i_1}\| \cdots \|b_{i_n}\| \leq 1$ we have

$$\nu(\Lambda_\varepsilon^n T) \leq \frac{1}{n!} \sum_{i_1,\ldots,i_n} \lambda_{i_1} \cdots \lambda_{i_n} \|x_{i_1} \wedge \ldots \wedge x_{i_n}\| = \frac{\lambda^n}{n!} \|\Lambda_\varepsilon^n \mu\|_1.$$

□

In section 3 we show that the antisymmetric tensor products of Radon measures on the Hlibert space ℓ_2^n are related to the Gauss-inequality. We give precise estimates of the quanities $\|\Lambda_\varepsilon^n \mu\|_p$.

2 THE CLASSES Γ_α

The basic results of this section are

PROPOSITION 2.1 *Let $1 \le p < \infty$ and let X be an infinite-dimensional Banach space. Then*

$$\lim_n \frac{\|\Lambda_\varepsilon^n \mu\|_p^{\frac{1}{n}}}{\Gamma_n(X)} = 0 \quad \text{for all} \quad \mu \in \mathcal{R}_p(X).$$

PROPOSITION 2.2 *Let $1 \le p < \infty$, $0 \le \alpha \le \frac{1}{2}$ and let X be a Banach space. Then the following assertions are equivalent.*

1. $X \in \Gamma_\alpha$.

2. $\lim_n n^{-\alpha}\|\Lambda_\varepsilon^n \mu\|_p^{\frac{1}{n}} = 0$ *for all* $\mu \in \mathcal{R}_p(X)$.

3. $\sup_n n^{-\alpha}\|\Lambda_\varepsilon^n \mu\|_p^{\frac{1}{n}} < \infty$ *for all* $\mu \in \mathcal{R}_p(X)$.

In order to prove these statements we introduce for $x_1, ..., x_n \in X$

$$d(x_n|x_{n-1}, ..., x_1) := \inf\{\|x_n - y\| : y \in span\{x_1, ..., x_{n-1}\}\}$$

and

$$v_n(x_1, ..., x_n) := d(x_n|x_{n-1}, ..., x_1)d(x_{n-1}|x_{n-2}, ..., x_1)...d(x_2|x_1)\|x_1\|.$$

Furthermore, two lemmata are needed. The first one can be verified in a straightforward way.

LEMMA 2.3 *Let X be a Banach space and let $x_1, ..., x_n \in X$. Then*

1. $v_n(x_1, ..., x_n) \le \|x_{i_1}\| \cdots \|x_{i_k}\| v_{n-k}(x_{j_1}, ..., x_{j_{n-k}})$ *for* $\{i_1, ..., i_k\} \cup \{j_1, ..., j_{n-k}\} = \{1, ..., n\}$,

2. $\|x_1 \wedge ... \wedge x_n\|_{\Lambda_\varepsilon^n X} \le \Gamma_n(X)^n \, v_n(x_1, ..., x_n)$.

LEMMA 2.4 *Let $1 \le p < \infty$ and $\mu \in \mathcal{R}_p(X)$. Then*

$$\lim_n \left(\int ... \int v_n(x_1, ..., x_n)^p d\mu(x_1)...d\mu(x_n)\right)^{1/pn} = 0.$$

Proof. Let $K \subset X$ be compact and $K^c := X \setminus K$. Setting $a_p := \int_X \|x\|^p d\mu(x)$ and $b_p := \int_{K^c} \|x\|^p d\mu(x)$ we obtain from 2.3 (1)

$$\left(\int \dots \int v_n(x_1,\dots,x_n)^p d\mu(x_1)\dots d\mu(x_n)\right)^{1/pn}$$

$$= \left(\sum_{k=0}^{n} \sum_{\substack{k \text{ times } K \\ (n-k) \text{ times } K^c}} \int \dots \int v_n(x_1,\dots,x_n)^p d\mu(x_1)\dots d\mu(x_n) \right)^{1/pn}$$

$$\leq \left(\sum_{k=0}^{n} \binom{n}{k} b_p^{n-k} \int_K \dots \int_K v_k(x_1,\dots,x_k)^p d\mu(x_1)\dots d\mu(x_k) \right)^{1/pn}.$$

Let $0 < m < n$, $d(K) := \sup\{\|x\| : x \in K\}$, and

$$\varepsilon_m(K) := \inf\left\{\delta > 0 : K \subseteq \cup_1^m \{x_i + \delta B_X\} \text{ for some } x_1,\dots,x_m \in X\right\},$$

then, for $k \geq m$,

$$\int_K \dots \int_K v_k^p d\mu \dots d\mu \leq d(K)^{pm} \left(2\varepsilon_m(K)\right)^{p(k-m)}.$$

Consequently (we can assume that $d(K) > 0$ and $\varepsilon_m(K) > 0$),

$$\left(\int \dots \int v_n^p d\mu \dots d\mu\right)^{1/pn}$$

$$\leq b_p^{1/p} \left(\sum_{k=0}^{m} \binom{n}{k} b_p^{-k} a_p^k \right)^{1/pn} + \left(\sum_{k=m+1}^{n} \binom{n}{k} b_p^{n-k} d(K)^{pm} \left(2\varepsilon_m(K)\right)^{p(k-m)} \right)^{1/pn}$$

$$\leq b_p^{1/p} \left(\sum_{k=0}^{m} \binom{n}{k} b_p^{-k} a_p^k \right)^{1/pn} + d(K)^{m/n} \left(2\varepsilon_m(K)\right)^{-m/n} \left(\sum_{k=0}^{n} \binom{n}{k} b_p^{n-k} \left(2\varepsilon_m(K)\right)^{pk} \right)^{1/pn}.$$

For fixed m we can continue to

$$\limsup_n \left(\int \dots \int v_n^p d\mu \dots d\mu\right)^{1/pn} \leq b_p^{1/p} + \left(b_p + \left(2\varepsilon_m(K)\right)^p\right)^{1/p}.$$

Since $\varepsilon_m(K) \to_m 0$ this gives

$$\limsup_n \left(\int \dots \int v_n^p d\mu \dots d\mu\right)^{1/pn} \leq 2b_p^{1/p}.$$

Finally, we take an increasing sequence $\{K_l\}_{l=1}^{\infty}$ of compact subsets of X (μ is Radon) with $\mu(K_l) \geq 1 - \frac{1}{l}$ such that $\lim_l \int_{X \setminus K_l} \|x\|^p d\mu(x) = 0$. □

Now 2.1 immediately follows from 2.3 and 2.4.

Proof of Proposition 2.2. (1) ⇒ (2) is a consequence of 2.1. We consider (3) ⇒ (1). Let $T \in \mathcal{N}(\ell_2, X)$ with $T = \sum_1^\infty \lambda_i b_i \otimes x_i \neq 0$, $\{b_i\}_1^\infty \subseteq B_{\ell_2}$, $\{x_i\}_1^\infty \subseteq B_X$, $\lambda_1 \geq \lambda_2 \geq \dots \geq 0$, and $0 < \lambda = \sum \lambda_i \leq 2\nu(T)$. Setting $\mu := \frac{1}{\lambda}\sum_j \lambda_j \delta_{x_j} \in \mathcal{R}_p(X)$ and using 1.1 we get for $y_1, ..., y_n \in B_{\ell_2}$

$$\begin{aligned} \|Ty_1 \wedge ... \wedge Ty_n\|_{\Lambda_\varepsilon^n X} &\leq \frac{1}{n!} \sum_{i_1,\cdots,i_n} \lambda_{i_1} \cdots \lambda_{i_n} |\det(\langle y_k, b_{i_l}\rangle)_{k,l}| \, \|x_{i_1} \wedge \ldots \wedge x_{i_n}\|_{\Lambda_\varepsilon^n X} \\ &\leq \frac{(2v(T))^n}{n!} \|\Lambda_\varepsilon^n \mu\|_p. \end{aligned}$$

Hence $\sup_n n^{1-\alpha}\Gamma_n(T) < \infty$ whenever $T \in \mathcal{N}(\ell_2, X)$. By Pietsch (1991) (1.11,1.13,2.11) and Pietsch (1987) (2.1.8) the quantity $\|T\|_\alpha := \sup_n n^{1-\alpha}\Gamma_n(T) < \infty$ defines a quasi-norm and a corresponding quasi-Banach operator ideal. By standard arguments (Pietsch (1980) (7.2.2); note that $\ell_2(\ell_2) \cong \ell_2$) there is a constant $c(X) > 0$ such that for all $T \in \mathcal{N}(\ell_2, X)$

$$\sup_n n^{1-\alpha}\Gamma_n(T) \leq c(X)\nu(T).$$

Assuming $x_1, ..., x_n \in B_X$ and setting $T := \sum_1^n e_i \otimes x_i \in \mathcal{N}(\ell_2, X)$ we finally obtain

$$\|x_1 \wedge ... \wedge x_n\|_{\Lambda_\varepsilon^n X}^{\frac{1}{n}} = \Gamma_n(T) \leq c(X)n^{\alpha-1}\nu(T) \leq c(X)n^\alpha$$

which proves (1). □

The operator versions of 2.1 and 2.2 are as follows.

THEOREM 2.5 *Let X be an infinite-dimensional Banach space. Then*

$$\lim_n n \frac{\nu(\Lambda_\varepsilon^n T)^{\frac{1}{n}}}{\Gamma_n(X)} = 0 \qquad \textit{for all } Y \textit{ and } T \in N(Y, X).$$

Proof. Apply 1.1 and 2.1. □

THEOREM 2.6 *Let $0 \leq \alpha \leq \frac{1}{2}$ and let X be a Banach space. Then the following assertions are equivalent.*

1. $X \in \Gamma_\alpha$.
2. $\lim_n n^{1-\alpha}\nu(\Lambda_\varepsilon^n T)^{\frac{1}{n}} = 0$ *for all Y and $T \in N(Y, X)$.*
3. $\sup_n n^{1-\alpha}\nu(\Lambda_\varepsilon^n T)^{\frac{1}{n}} < \infty$ *for all Y and $T \in N(Y, X)$.*

Proof. (1) ⇒ (2) follows from 2.5. (3) ⇒ (1) Since

$$\Gamma_n(T) \leq \Gamma_n(Y)\|\Lambda_\varepsilon^n T\|^{1/n} \leq \Gamma_n(Y)\nu(\Lambda_\varepsilon^n T)^{1/n}$$

for $T \in \mathcal{L}(Y, X)$ we obatin $\sup_n n^{1-\alpha}\Gamma_n(T) < \infty$ for all $T \in N(\ell_2, X)$. Using the arguments given in the proof of 2.2 we arrive at (1). □

In the context of the Weyl numbers the above results read as follows.

THEOREM 2.7 *Let X be infinite-dimensional. Then*

$$\lim_n n \frac{x_n(T)}{\Gamma_n(X)} = 0 \qquad \textit{for all } Y \textit{ and } T \in N(Y, X).$$

Proof. Let $T = \sum_1^\infty \lambda_i b_i \otimes x_i$ with $\{b_i\} \subset B_{Y'}$, $\{x_i\} \subset B_X$, $\lambda_1 \geq \lambda_2 \geq \ldots \geq 0$ and $\lambda := \sum \lambda_i \leq 2\nu(T)$. Via $\mu := \frac{1}{\lambda} \sum \lambda_i \delta_{x_i}$ Proposition 1.1 implies for all subspaces $E \subseteq X$

$$\nu(\Lambda_\varepsilon^n(T|_E))^{1/n} \leq 2\, n!^{-1/n}\, \nu(T) \|\Lambda_\varepsilon^n \mu\|_1^{1/n}.$$

Since $\Gamma_n(T|_E) \leq \Gamma_n(E)\nu(\Lambda_\varepsilon^n(T|_E))^{1/n}$ we obtain

$$\sup_{dim(E)=n} \frac{\Gamma_n(T|_E)}{\Gamma_n(E)} \leq \frac{2e}{n} \|\Lambda_\varepsilon^n \mu\|_1^{1/n} \nu(T).$$

Geiss (1990:b)(Theorem 2.1) gives $x_n(T) \leq \frac{2e}{n}\|\Lambda_\varepsilon^n \mu\|_1^{1/n}$. Finally, 2.1 yields the desired result. □

THEOREM 2.8 *Let $0 \leq \alpha \leq \frac{1}{2}$ and let X be a Banach space. Then the following assertions are equivalent.*

1. $X \in \Gamma_\alpha$.
2. $\lim_n n^{1-\alpha} x_n(T) = 0$ *for all Y and $T \in N(Y, X)$.*
3. $\sup_n n^{1-\alpha} x_n(T) < \infty$ *for all Y and $T \in N(Y, X)$.*

Proof.(1) $\Rightarrow$ (2) is clear from 2.7. (3) $\Rightarrow$ (1) can be proved as in 2.2. □

Finally, $|\lambda_1(T) \cdots \lambda_n(T)| \leq \|\Lambda_\varepsilon^n T\| \leq \nu(\Lambda_\varepsilon^n T)$ and 2.5 imply

COROLLARY 2.9 *Let X be a complex and infinite-dimensional Banach space and let $T \in N(X, X)$. Then*

$$\lim_n n \frac{|\lambda_1(T) \cdots \lambda_n(T)|^{1/n}}{\Gamma_n(X)} = 0.$$

3 THE HILBERT SPACE SITUATION

This section is devoted to a generalization of the classical Gauss-inequality (cf. Beckenbach and Bellmann (1961) (p.11)). This inequality states that

$$k! \sum_{i_1 < \cdots < i_k} \lambda_{i_1} \cdots \lambda_{i_k} \leq \frac{n \cdots (n-k+1)}{n^k}$$

for all non-negative reals $\lambda_1, \ldots, \lambda_n$ with $\sum \lambda_i = 1$ and will be extended to a measure theoretical version in Corollary 3.6. Necessary and sufficient conditions for the equality are stated.

For simplicity we will only treat real Banach spaces. In the previous section we have considered the ε -tensor product $\Lambda^n_\varepsilon X$ of X. In the case $X = H$ it is more natural to take the 2-tensor product $\Lambda^n_2 H$ which is the closure of $\Lambda^n H$ with respect to the scalar product

$$\langle s,t\rangle := \sum_{i,j} det((\langle x_{ki}, y_{lj}\rangle)^n_{k,l=1},$$

where $s = \sum_i x_{1i} \wedge ... \wedge x_{ni}$ and $t = \sum_j y_{1j} \wedge ... \wedge y_{nj}$. It is easy to see that for 'elementary' antisymmetric tensors $s = x_1 \wedge ... \wedge x_n$

$$\|x_1 \wedge ... \wedge x_n\|_{\Lambda^n_\varepsilon H} = \|x_1 \wedge \cdots \wedge x_n\|_{\Lambda^n_2 H} = v_n(x_1, ..., x_n)$$

such that we simply use $\|x_1 \wedge ... \wedge x_n\|$ and

$$\begin{aligned} \|\Lambda^n \mu\|_p &= \left(\int_H \cdots \int_H \|x_1 \wedge ... \wedge x_n\|^p d\mu(x_1)...d\mu(x_n)\right)^{1/p} \\ &= \left(\int_H \cdots \int_H |\det(< x_i, x_j >)^n_{i,j=1}|^{p/2} d\mu(x_1)...d\mu(x_n)\right)^{1/p}. \end{aligned}$$

Throughout in this section Δ^n_k stands for $\{(i_1, ..., i_k) : 1 \le i_1 < ... < i_k \le n\}$. Furthermore, it turns out to be reasonable to restrict our considerations on Radon measures μ on $\mathcal{B}(\ell^n_2)$ with $\mu(B_{\ell^n_2}) = 1$. We denote this set of measures by $\mathcal{R}(B_{\ell^n_2})$. For technical reason a slightly different kind of the Grothendieck numbers is required. Let $T \in \mathcal{L}(X,Y)$ and μ be a normalized regular Borel measure on the unit ball $B_{Y'}$ equipped with the $\sigma(Y',Y)$ topology. Then

$$\Gamma_n(T;\mu) := \sup\left\{\int_{B_{Y'}} \cdots \int_{B_{Y'}} |\det(\langle Tx_i, b_j\rangle)^n_{i,j=1}|^2 d\mu(b_1)...d\mu(b_n)\right\}^{\frac{1}{2n}}$$

where the supremum is taken over all $x_1, \dots, x_n \in B_X$.

The next lemma summarizes the facts which are needed in the sequel.

LEMMA 3.1 *(Geiss (1990:a) (Proposition 1.3), (1992) (Lemma 2.1))* *Let $T \in \mathcal{L}(X,Y)$. Then*

1. *$n!^{\frac{1}{2n}}\,\Gamma_n(JT) = \Gamma_n(T;\mu)$, where μ is a regular normalized Borel measure on $B_{Y'}$ and $J : Y \to L_2(B_{Y'};\mu)$ is the canonical embedding,*
2. *$\Gamma_n(T) = (a_1(T) \cdots a_n(T))^{\frac{1}{n}}$ if X and Y are Hilbert spaces.*

The following multiplication formula is the basic result of this section.

THEOREM 3.2 *Let $\mu \in \mathcal{R}(B_{\ell^n_2})$, $1 \le k, l \le n$ and $k + l \le n$. Then*

$$\|\Lambda^{k+l}\mu\|_2 \le \left(\frac{(n-k)!(n-l)!}{n!(n-k-l)!}\right)^{\frac{1}{2}} \|\Lambda^k \mu\|_2 \cdot \|\Lambda^l \mu\|_2.$$

The proof follows immediately from the following two facts.

LEMMA 3.3 *Let $\{\lambda_i\}$ be a sequence of non-negative reals and $1 \le k, l \le n$ with $k + l \le n$. Then*

$$(k+l)! \sum_{\Delta^n_{k+l}} \lambda_{j_1} \cdots \lambda_{j_{k+l}} \le \frac{(n-k)!(n-l)!}{n!(n-k-l)!} \left(k! \sum_{\Delta^n_k} \lambda_{j_1} \cdots \lambda_{j_k} \right) \left(l! \sum_{\Delta^n_l} \lambda_{j_1} \cdots \lambda_{j_l} \right).$$

Proof. Let $p_k := \frac{(n-k)!k!}{n!} \sum_{\Delta^n_k} \lambda_{j_1} \cdots \lambda_{j_k}$ and $p_0 := 1$. Using $p_{k-1}p_{k+1} \le p_k^2$ for $k = 1, ..., n$ from Beckenbach and Bellmann (1961) (p.11) and assuming $p_{k+l-1} > 0$ we obtain

$$\frac{p_{k+l}}{p_k} = \frac{p_{k+1}}{p_k} \frac{p_{k+2}}{p_{k+1}} \cdots \frac{p_{k+l}}{p_{k+l-1}} \le \frac{p_k}{p_{k-1}} \frac{p_{k+1}}{p_k} \cdots \frac{p_{k+l-1}}{p_{k+l-2}} = \frac{p_{k+l-1}}{p_{k-1}}$$

and

$$\frac{p_{k+l}}{p_k} \le \frac{p_{k+l-1}}{p_{k-1}} \le \frac{p_l}{p_0} = p_l.$$

Hence $p_{k+l} \le p_k p_l$ for all $0 \le k, l \le n$ (if $p_i = 0$ then $p_j = 0$ for all $j > i$), which implies the desired inequality. □

LEMMA 3.4 *Let $\mu \in \mathcal{R}(B_{\ell_2^n})$ and $1 \le k \le n$. Then*

$$\int_{B_{\ell_2^n}} \cdots \int_{B_{\ell_2^n}} \|x_1 \wedge ... \wedge x_k\|^2 d\mu(x_1)...d\mu(x_k) = k! \sum_{\Delta^n_k} a_{j_1}(J)^2 \cdots a_{j_k}(J)^2,$$

where $J : \ell_2^n \to L_2(B_{\ell_2^n}, \mu)$ is canonical.

Proof. Let $J = \sum_1^n a_i(J) a_i \otimes y_i$ be a Schmidt-representation such that $\{a_{j_1} \wedge \cdots \wedge a_{j_k}\}_{\Delta^n_k}$ is a basis of $\Lambda_2^k \ell_2^n$. Then it follows from 3.1 that

$$\begin{aligned}
&\int_{B_{\ell_2^n}} \cdots \int_{B_{\ell_2^n}} \|x_1 \wedge ... \wedge x_k\|^2 d\mu(x_1)...d\mu(x_k) \\
= &\sum_{i_1 < \cdots < i_k} \int_{B_{\ell_2^n}} \cdots \int_{B_{\ell_2^n}} |\langle a_{i_1} \wedge ... \wedge a_{i_k}, x_1 \wedge ... \wedge x_k \rangle|^2 \, d\mu(x_1)...d\mu(x_k) \\
= &\sum_{i_1 < \cdots < i_k} \Gamma_k \left(I|_{[a_{i_1}, ..., a_{i_k}]}; \mu \right)^{2k} \\
= &\sum_{i_1 < \cdots < i_k} k! \Gamma_k \left(JI|_{[a_{i_1}, ..., a_{i_k}]} \right)^{2k} \\
= &\sum_{i_1 < \cdots < i_k} k! a_{i_1}(J)^2 \cdots a_{i_k}(J)^2,
\end{aligned}$$

where $[a_{i_1}, \ldots, a_{i_k}]$ stands for the span of $a_{i_1}, \ldots, a_{i_k}$ and $I := I_{\ell_2^n}$. □

Directly from 3.2 we will deduce in 3.6 the generalized Gauss-inequality. To give necessary and sufficient conditions for the equality we introduce

$$\mathcal{R}^{st}(B_{\ell_2^n}) := \left\{\mu \in \mathcal{R}(B_{\ell_2^n}) : \frac{1}{n}\|a\|_2^2 = \int |\langle a, x\rangle|^2 d\mu(x) \text{ for all } a \in \ell_2^n\right\},$$

$\mathcal{R}^{\delta}(B_{\ell_2^n}) := \{ \mu \in \mathcal{R}(B_{\ell_2^n}) : \mu = \sum_1^n (p_i\delta_{x_i} + q_i\delta_{-x_i})$ for some non-negative $p_1, \ldots, p_n$, $q_1, \ldots, q_n$ with $1/n = p_i + q_i$ for $i = 1, \ldots, n$ and some orthonormal system $\{x_1, \ldots, x_n\} \subseteq B_{\ell_2^n}\}$.

The symbol 'st' stands for 'standard covariance matrix' and 'δ' stands for 'Dirac measure'. It is clear that $\mathcal{R}^{\delta}(B_{\ell_2^n}) \subset \mathcal{R}^{st}(B_{\ell_2^n})$. Furthermore we need

LEMMA 3.5 *Let $\mu \in \mathcal{R}^{st}(B_{\ell_2^n})$ and $2 \le k \le n$. Then $\mu \in \mathcal{R}^{\delta}(B_{\ell_2^n})$ if and only if*

$$\mu \times \cdots \times \mu \{(x_1, ..., x_k) : \|x_1 \wedge ... \wedge x_k\| \in \{0, 1\}\} = 1.$$

Proof. We only prove the 'if-part' since the other one is trivial. Recall that $x \in \mathrm{supp}(\mu)$ if $\mu(U_\varepsilon(x)) > 0$ for all $\varepsilon > 0$ where $U_\varepsilon(x) := \{y : \|x - y\| < \varepsilon\}$. Assuming $x_1, \ldots, x_k \in \mathrm{supp}(\mu)$ such that $0 < \delta := \|x_1 \wedge \ldots \wedge x_k\| < 1$ we find an $\varepsilon > 0$ with $0 < \frac{\delta}{2} \le \|y_1 \wedge \ldots \wedge y_k\| \le \delta + \frac{1-\delta}{2} < 1$ for all $y_i \in U_\varepsilon(x_i)$. Since $\mu \times \cdots \times \mu(U_\varepsilon(x_1) \times \cdots \times U_\varepsilon(x_k)) > 0$ this would imply $\mu \times \cdots \times \mu \{(x_1, ..., x_k) : \|x_1 \wedge ... \wedge x_k\| \in \{0, 1\}\} < 1$. Consequently, $\|x_1 \wedge \ldots \wedge x_k\| \in \{0, 1\}$ whenever $x_1, \ldots, x_k \in \mathrm{supp}(\mu) \subseteq B_{\ell_2^n}$. Furthermore, $\mu \in \mathcal{R}^{st}(B_{\ell_2^n})$ gives $\mathrm{span}\{\mathrm{supp}(\mu)\} = \ell_2^n$. Now let $x, y \in \mathrm{supp}(\mu)$ with $\|x \wedge y\| > 0$. Then one can find $z_1, \ldots z_{k-2} \in \mathrm{supp}(\mu)$ with $\|x \wedge y \wedge z_1 \wedge ... \wedge z_{k-2}\| > 0$. But then $\|x \wedge y \wedge z_1 \wedge ... \wedge z_{k-2}\| = 1$ and $x \perp y$. Consequently,

$$\mathrm{supp}(\mu) \subseteq \{0, x_1, -x_1, \ldots, x_n, -x_n\}$$

for some orthonormal system $\{x_1, \ldots, x_n\}$. Finally, using again $\mu \in \mathcal{R}^{st}(B_{\ell_2^n})$ implies the desired representation. □

COROLLARY 3.6 *Let $\mu \in \mathcal{R}(B_{\ell_2^n})$, $2 \le k \le n$, and $2 \le p < \infty$ Then the following holds.*

1. $\|\Lambda^k \mu\|_p \le \left(\frac{n \ldots (n-k+1)}{n^k}\right)^{1/p}$.
2. *If $p = 2$ then the equality holds if and only if $\mu \in \mathcal{R}^{st}(B_{\ell_2^n})$.*
3. *If $p > 2$ then the equality holds if and only if $\mu \in \mathcal{R}^{\delta}(B_{\ell_2^n})$.*

Proof. (1) Using 3.2 and taking into account that $\|x_1 \wedge \ldots \wedge x_k\| \le 1$ whenever $x_1, ..., x_k \in B_{\ell_2^n}$ we get

$$\|\Lambda^k \mu\|_p \le \left(\int \cdots \int \|x_1 \wedge \ldots \wedge x_k\|^2 d\mu(x_1) ... d\mu(x_k)\right)^{1/p} \le \left(\frac{n \ldots (n-k+1)}{n^k}\right)^{1/p}.$$

(2) Assuming equality we obtain from 3.2

$$\begin{aligned}&\int\int \|x\wedge y\|^2 d\mu(x)d\mu(y)\\ \geq\ & \frac{n}{n-2}\cdots\frac{n}{n-k+1}\int\ldots\int \|x_1\wedge\ldots\wedge x_k\|^2 d\mu(x_1)\ldots d\mu(x_k)\\ =\ & \frac{n-1}{n}.\end{aligned}$$

According to 3.4 we conclude for the canonical embedding $J : \ell_2^n \to L_2(B_{\ell_2^n}, \mu)$

$$\frac{n-1}{n}\left(\sum a_j(J)^2\right)^2 \geq \left(\sum a_j(J)^2\right)^2 - \sum a_j(J)^4 = \sum_{i\neq j} a_i(J)^2 a_j(J)^2 \geq \frac{n-1}{n}.$$

Consequently, $1 \leq (\sum a_j(J)^2)^{\frac{1}{2}} \leq 1$ (Pietsch (1987) (1.5.2,2.11.28)) and $\sum a_j(J)^4 = \frac{1}{n}$. Hence

$$a_1(J) = \ldots = a_n(J) = n^{-\frac{1}{2}}$$

which implies that $n^{\frac{1}{2}} J$ is an isometric embedding and $\mu \in \mathcal{R}^{st}(B_{\ell_2^n})$. Now let us assume that $\mu \in \mathcal{R}^{st}(B_{\ell_2^n})$. If $H \subseteq l_2^n$ is an m-dimensional subspace with an orthonormal basis $\{a_1, ..., a_m\}$ then it is easy to see that

$$\int \|P_H x\|^2 d\mu(x) = \sum_{i=1}^{m} \int |\langle a_i, x\rangle|^2 \mu(x) = \frac{m}{n},$$

whereas $P_H \in \mathcal{L}(\ell_2^n, \ell_2^n)$ is the orthogonal projection onto H. Consequently,

$$\begin{aligned}\|\Lambda^k \mu\|_2^2 &= \int\ldots\int \|x_1\wedge\ldots\wedge x_k\|^2 d\mu(x_1)\ldots d\mu(x_k)\\ &= \int\ldots\int \|x_1\wedge\ldots\wedge x_{k-1}\|^2 \|P_{\{x_1,\ldots,x_{k-1}\}}(x_k)\|^2 d\mu(x_1)\ldots d\mu(x_k)\\ &= \frac{n-k+1}{n}\int\ldots\int \|x_1\wedge\ldots\wedge x_{k-1}\|^2 d\mu(x_1)\ldots d\mu(x_{k-1})\end{aligned}$$

where $P_{\{x_1,\ldots,x_{k-1}\}}$ is the orthogonal projection onto $span\{x_1,\ldots,x_{k-1}\}^\perp$ (we may assume that $x_1,\ldots,x_{k-1}$ are linearly independent). If we continue in this way then the desired equality follows.

(3) Again we first assume equality. Then from (1) and

$$\begin{aligned}\frac{n\cdots(n-k+1)}{n^k} &= \int\ldots\int \|x_1\wedge\ldots\wedge x_k\|^p d\mu(x_1)\ldots d\mu(x_k)\\ &\leq \int\ldots\int \|x_1\wedge\ldots\wedge x_k\|^2 d\mu(x_1)\ldots d\mu(x_k)\end{aligned}$$

we get that $\mu \in \mathcal{R}^{st}(B_{\ell_2^n})$ and $\mu\times\cdots\times\mu\,\{(x_1, ..., x_k) : \|x_1\wedge ... \wedge x_k\| \in \{0,1\}\} = 1$. Applying 3.5 gives $\mu \in \mathcal{R}^{\delta}(B_{\ell_2^n})$. Finally, if $\mu \in \mathcal{R}^{\delta}(B_{\ell_2^n})$ then

$$\begin{aligned}\|\Lambda^k \mu\|_p^p &= \int\ldots\int \|x_1\wedge\ldots\wedge x_k\|^p d\mu(x_1)\ldots d\mu(x_k)\\ &= \int\ldots\int \|x_1\wedge\ldots\wedge x_k\|^2 d\mu(x_1)\ldots d\mu(x_k)\\ &= \frac{n\cdots(n-k+1)}{n^k}.\end{aligned}$$

□

For completness we state the much more easier counterpart of 3.6 which can be easily proved in the same way.

PROPOSITION 3.7 *Let* $\mu \in \mathcal{R}^{st}(B_{\ell_2^n})$, $2 \leq k \leq n$, *and* $1 \leq p \leq 2$. *Then the following holds.*

1. $\|\Lambda^k \mu\|_p \geq \left(\frac{n \ldots (n-k+1)}{n^k}\right)^{1/p}$.

2. *The equality holds if and only if* $\mu \in \mathcal{R}^{\delta}(B_{\ell_2^n})$.

References

Beckenbach, R., Bellmann, R. (1961). *Inequalities,* Springer.

Geiss, S., (1990:a). Grothendieck numbers of linear and continuous operators on Banach spaces, *Math. Nachr., 148:* 65-79.

Geiss, S., (1990:b). Grothendieck numbers and volume ratios of operators on Banach spaces, *Forum Math., 2:* 323-340.

Geiss, S., (1992). Antisymmetric tensor products of absolutely p-summing operators, *J. Appr. Theory, 68:* 223-246.

Grothendieck, A., (1956). La théorie de Fredholm, *Bull. Soc. Math. France, 84:* 319-384.

Pietsch, A., (1980). *Operator Ideals,* North-Holland, Amsterdam.

Pietsch, A., (1987). *Eigenvalues and s-numbers,* Cambridge Univ. Press, Cambridge.

Pietsch, A., (1991). Eigenvalue distributions and geometry of Banach spaces, *Math. Nachr., 150:* 41-81.

Pisier, G., (1988). " Weak Hilbert spaces," Proc. London Math. Soc (3), 56, pp. 547-579.

Absolutely Summing Composition Operators

HANS JARCHOW MATHEMATISCHES INSTITUT DER UNIVERSITÄT ZÜRICH
CH 8001 ZÜRICH, SWITZERLAND

INTRODUCTION

One of the possibilities to compare function theoretic properties of analytic functions and functional analytic properties of linear operators is through "change of variable" type formulae. The corresponding operators are known as *composition operators*; they have been intensively studied on various classical function spaces, in particular Hilbert function spaces. In this paper we continue investigating composition operators within the framework of general Hardy spaces on the open unit disk.

We shall concentrate on two closely related properties, *absolute summability* and *order boundedness*, with special emphasis on the question of coincidence within our setting. Whereas order boundedness of a composition operator can be characterized in terms of a rather simple integrability condition involving the generating function, nothing comparable seems to exist in the literature regarding absolute summability in general Hardy spaces. The first few partial results concerning the problem of coincidence of these two concepts for composition operators appear in Shapiro-Taylor [12], Hunziker-Jarchow [5] and Jarchow [6].

We plan to discuss related questions in some detail in this note. We shall prove that, whenever the parameters of the underlying Hardy spaces are chosen from an appropriate range, order bounded and absolutely summing composition operators are indeed one and the same. But we shall also provide examples showing that, at least for special parameters from outside this range, absolutely summing composition operators are available which fail to be order bounded; apparently, these are the first examples of this kind. Interesting enough, our operators even take their values in a Hilbert space.

PRELIMINARIES

We will work with classical Hardy spaces

$$H^p \quad (0 < p \le \infty) \ .$$

H^p consists of all analytic functions f on the open unit disk

$$D := \{z \in \mathbf{C} : |z| < 1\}$$

in the complex plane which satisfy

$$\|f\|_p := \sup_{r<1} \Big(\int_{-\pi}^{\pi} |f(re^{it})|^p \frac{dt}{2\pi} \Big)^{1/p} < \infty$$

if $0 < p < \infty$, resp. which are bounded when $p = \infty$:

$$\|f\|_\infty := \sup_{z \in D} |f(z)| < \infty \ .$$

With respect to $\|\cdot\|_p$, H^p is a Banach space when $1 \le p \le \infty$, and a p-Banach space when $0 < p < 1$.

Fatou's Theorem asserts that, no matter how we choose $0 < p \le \infty$ and $f \in H^p$, the limit $\tilde{f}(e^{it}) := \lim_{r \nearrow 1} f(re^{it})$ exists for almost all points e^{it} on the unit circle ∂D; in this way, an element $\tilde{f} \in L^p(m)$ is generated, m being the normalized Lebesgue measure on ∂D. If $0 < p < \infty$, then $\tilde{f}$ is also the $L^p(m)$-limit of the functions $t \mapsto f(re^{it})$ as $r \nearrow 1$; if $p = \infty$ such a statement remains only valid if the norm topology is replaced by the weak$*$topology of $L^\infty(m)$. In any case, $f \mapsto \tilde{f}$ defines an isometric isomorphism of H^p into $L^p(m)$; the range is the closure of the polynomials in the metric topology if $0 < p < \infty$, resp. in the weak$*$topology if $p = \infty$.

In the sequel, we shall identify H^p with its image in $L^p(m)$ and no longer distinguish formally between f and $\tilde{f}$ for H^p functions f.

We refer to Duren [4] for this and for further results on H^p spaces to be used in the sequel. As for Banach spaces, we shall employ standard terminology and notation; our main reference for ideals of Banach space operators is Pietsch [9].

For notational convenience, let us denote by

$$\Phi$$

the set of all analytic functions $\varphi : D \to \mathbf{C}$ such that $\varphi(D) \subset D$. In other words, Φ is obtained from the unit ball of H^∞ by just deleting the constant functions generated by the elements of ∂D.

Take any $\varphi \in \Phi$. It is a well known consequence of *Littlewood's Subordination Principle* (Duren [2], p.10) that, regardless of how we select $0 < p \le \infty$,

$$f \mapsto f \circ \varphi$$

defines a bounded linear operator

$$C_\varphi : H^p \to H^p ,$$

the so-called *composition operator* induced by φ.

It is clear that $C_\varphi : H^\infty \to H^\infty$ has norm one. If $0 < p < \infty$, then

$$\sup_{z\in D} \frac{1-|z|^2}{1-|\varphi(z)|^2} \leq \|C_\varphi : H^p \to H^p\|^p = \|C_\varphi : H^1 \to H^1\| \leq \frac{1+|\varphi(0)|}{1-|\varphi(0)|} ;$$

in particular, $C_\varphi : H^p \to H^p$ has norm one if and only if $\varphi(0) = 0$. To determine the precise value of $\|C_\varphi\|$ if $\varphi(0) \neq 0$ is one out of several open basic questions on composition operators. Compare e.g. with Cowen [1] and Hunziker [4].

In the sequel, p will always be a positive real number. We shall need to work with composition operators $C_\varphi : H^p \to H^p$ which have the property that, for some $1 < \beta < \infty$, $C_\varphi(B_{H^p})$ is even contained in $H^{\beta p}$. It turns out that this property does depend on φ and β but *not* on p, so it is justified to label such composition operators as *β-bounded.* There are examples of composition operators which have very special properties (e.g., as operators $H^2 \to H^2$, they can be in the Hilbert-Schmidt class) but which fail to be β-bounded for every $\beta > 1$; see again [4] and [5].

We shall need a little terminology on operators.

Recall that a Banach space operator $u : X \to Y$ is *(absolutely) p-summing* $(1 \leq p < \infty)$ if it takes weak ℓ^p sequences $(x_n)_n$ in X into strong ℓ^p sequences $(ux_n)_n$ in Y. In other words, we require $\sum_{n=1}^\infty \|ux_n\|^p$ to be finite whenever $\sum_{n=1}^\infty |\langle x^*, x_n\rangle|^p$ converges for each x^* in X^*, the dual of X. The operator $u : X \to Y$ is p-summing if and only if there is a probability measure μ such that u admits a factorization

$$u : X \xrightarrow{w} Z_\infty \xrightarrow{j} Z_p \xrightarrow{v} Y ,$$

where Z_∞ is a subspace of $L^\infty(\mu)$, Z_p is Z_∞'s closure in $L^p(\mu)$, j is induced by the formal identity $j_p : L^\infty(\mu) \to L^p(\mu)$, and v and w are suitably chosen operators. The operator u is called *p-integral* if even a factorization

$$k_Y u : X \xrightarrow{w} L^\infty(\mu) \xrightarrow{j_p} L_p(\mu) \xrightarrow{v} Y^{**}$$

is available, k_Y being the canonical embedding $Y \hookrightarrow Y^{**}$. Still a smaller class consists of all *p-nuclear* operators: u is p-nuclear if it factors

$$u : X \xrightarrow{w} \ell^\infty \xrightarrow{\Delta} \ell^p \xrightarrow{v} Y$$

where Δ is a diagonal operator induced by a scalar ℓ^p-sequence. We refer to Pietsch's monograph [9] for more on these classes of operators.

We shall work with order boundedness only within the framework of L^p-spaces (for $0 < p < \infty$). Let X be a Banach space, μ a measure (≥ 0), and p a positive real number. We say that an operator $u : X \to L^p(\mu)$ is *order bounded* if it maps B_X, the unit ball of X, into an order interval of $L^p(\mu)$: in other words, we require the existence of a non-negative function $h \in L^p(\mu)$ such that $|ux| \leq h$ μ-a.e. for each $x \in B_X$.

A classical result due to Grothendieck [3] states that $u : X \to L_1(\mu)$ is order bounded if and only if it is 1-integral. Back to Schwartz [11] and Kwapień [7] goes the observation that if $1 < p < \infty$, then $u : X \to L_p(\mu)$ is p-integral whenever it is order bounded, and it is order bounded whenever u^* is p-summing. The converse is known to fail in both cases.

ORDER BOUNDED COMPOSITION OPERATORS

Fix now $0 < \beta < \infty$; unlike in the case of β-bounded operators, we do *not* exclude the case $\beta < 1$! We are going to consider composition operators $C_\varphi : H^p \to H^{\beta p}$ which are order bounded when considered as operators with values in $L^{\beta p}(m)$. It is known (see [4] or [5]) that this is again a property which does *not* depend on the particular choice of p but only on φ and β; therefore it is appropriate to use the label *β-order bounded composition operator* to describe the situation.

To begin, let us recall the main characterizations of β-order bounded composition operators from [4], or [5]:

THEOREM 1: *Let $0 < \beta < \infty$ be given. The following statements about $\varphi \in \Phi$ are equivalent:*

(i) *C_φ is β-order bounded.*

(ii) $\left(1 - |\varphi|^2\right)^{-1} \in L^\beta(m)$.

(iii) *For some (and then all) $0 < r < \infty$, the sequence $(\|\varphi^n\|_1)_{n=0}^\infty$ belongs to the Lorentz sequence space $\ell^{(r/\beta, r)}$.*

Notice that $(\|\varphi_n\|_r)_n$ is always a decreasing sequence (for any r).

It follows that if $m(|\varphi| = 1) > 0$ then C_φ cannot be β-order bounded for any $\beta > 0$. In particular, no inner function φ can ever generate a β-order bounded composition operator, though every composition operator is of course continuous as a map $H^p \to H^{\beta p}$ for any p when $0 < \beta \leq 1$. However, when $\beta > 1$ it is already much less trivial to find functions $\varphi \in \Phi$ such that C_φ is β-bounded but not β-order bounded ; examples can be found in Hunziker [4].

The case $\beta = 1$ of the preceding theorem was naturally the first to be investigated. J.H.Shapiro and P.D.Taylor noted in [12] that 1-order boundedness of C_φ is equivalent to saying that $(\|\varphi^n\|_2)_n$ belongs ℓ_2, a property classically known to characterize when $C_\varphi : H^2 \to H^2$ is a Hilbert-Schmidt operator : after all, the monomials z^n , $n = 0, 1, 2, ...$ form an orthonormal basis in H^2. More generally, the author proved in [6] that

- if $1 \leq \beta < \infty$, then C_φ is β-order bounded if and only if it admits a factorization $C_\varphi : H^{2/\beta} \xrightarrow{v} \ell_2 \xrightarrow{u} H^2$ where v is a bounded operator and u is a Hilbert-Schmidt operator ,

and that

- if $0 < \beta \leq 1$, then C_φ is β-order bounded if and only if it admits a factorization $C_\varphi : H^2 \xrightarrow{v} \ell_2 \xrightarrow{u} H^{2\beta}$ where now u is bounded and v is a Hilbert-Schmidt operator .

RELATIONS TO ABSOLUTELY SUMMING OPERATORS

Let X be any Banach space, μ any measure and $u : X \longrightarrow L^p(\mu)$ any operator. As we have mentioned before, if u is order bounded then it is p-summing and even p-integral, and if u's adjoint u^* is p-summing, then u is order bounded.

We are going to investigate the interplay between these properties within the framework of composition operators. The simplest case occurs when $\beta = 1$. Here the Hausdorff-Young Theorem (Duren [2], p.94) leads to the following:

PROPOSITION 2: *The following statements about $\varphi \in \Phi$ are equivalent:*

(i) C_φ *is 1-order bounded.*

(ii) $C_\varphi : H^p \longrightarrow H^p$ *is p-nuclear for some (and then all)* $2 \leq p < \infty$.

(iii) $C_\varphi : H^p \longrightarrow H^p$ *is p-summing for some (and then all)* $2 \leq p < \infty$.

(ii') $C_\varphi^* : (H^p)^* \longrightarrow (H^p)^*$ *is p-nuclear for some (and then all)* $1 \leq p \leq 2$.

(iii') $C_\varphi^* : (H^p)^* \longrightarrow (H^p)^*$ *is p-summing for some (and then all)* $1 \leq p \leq 2$.

Note that "p-integral" could be added free of charge.

The equivalence of (i) and (iii) and the nuclearity on H^1 of 1-order bounded composition operators was already noted in Shapiro-Taylor [12].

Proof. (i)$\Rightarrow$(ii): Let $2 \leq p < \infty$. We know that order bounded operators with values in L^p are p-integral. But we are dealing with a reflexive domain situation so that, by a result of Persson [8], our operator must even be p-nuclear.

(ii)$\Rightarrow$(iii) is automatic. To derive (i) from (iii), just observe (as in [12]) that, thanks to the Hausdorff-Young Theorem and since $p \geq 2$, the monomials z^n form a weak ℓ^r sequence in H^p.

We know already that (iii')$\Rightarrow$(i) is generally true. To prove (i)$\Rightarrow$(iii'), let us agree to denote by μ_n the functional which associates with each (appropriate) f its n-th Fourier coefficient $\hat{f}(n)$. Notice that we may write, in the sense of pointwise convergence,

$$C_\varphi = \sum_{n=0}^{\infty} \mu_n \otimes \varphi^n .$$

Since $1 \leq p \leq 2$ we can again invoke the Hausdorff-Young Theorem to see that $(\mu_n)_n$ is a weak ℓ^{p^*} sequence in $(H^p)^*$. On the other hand, our hypothesis (i) yields that $\left(\|\varphi^n\|\right)_n$ is a member of ℓ_p, and so (ii) follows suit. The remaining implications are valid for trivial reasons, so we are done. QED

A number of possibilities exist to express β-order boundedness of composition operators by looking at its action on various other function spaces (Bloch spaces, weighted Bergman spaces, etc.). We plan to discuss these topics in a systematic fashion elsewhere. Here we confine ourselves to just briefly pointing out one such possibility which is particularly simple.

For $0 < p \leq \infty$ let

$$A^p$$

be the space of all analytic functions $f : D \to \mathbf{C}$ whose Fourier coefficients $\hat{f}(n)$ $(n = 0, 1, 2, \ldots)$ form a sequence in ℓ^p. In the canonical fashion, this is a (p-) Banach space – just another model of ℓ^p.

PROPOSITION 3: *Let $0 < \beta < \infty$ and $\varphi \in \Phi$ be given. The following statements about the composition operator C_φ are equivalent:*

(i) *C_φ is β-order bounded.*

(ii) *For each $1 \leq p < \infty$, $f \mapsto f \circ \varphi$ defines an order bounded operator $A^{p^*} \to L^{\beta p}(m)$.*

(iii) *For some $1 \leq p < \infty$, $f \mapsto f \circ \varphi$ defines an order bounded operator $A^{p^*} \to L^{\beta p}(m)$.*

Proof. This is indeed straightforward: to say that, for some $1 \leq p < \infty$, $C_\varphi : A^{p^*} \to L^{\beta p}(m)$ exists as an order bounded operator means to require the existence of some $h \geq 0$ in $L_{\beta p}(m)$ such that $|C_\varphi f| \leq h$ m-a.e. for each $f \in B_{A^{p^*}}$, i.e., such that $|\sum_n a_n \varphi^n| \leq h$ m-a.e. for each $(a_n) \in B_{\ell^{p^*}}$. But this is equivalent to asking for $\left(\sum_n |\varphi^n|^p\right)^{1/p} \leq h$ m-a.e, i.e. for $\left(\sum_n |\varphi^n|^p\right)^{1/p} \in L^{\beta p}(m)$, which in turn happens if and only if $(1 - |\varphi|^2)^{-1} \in L^\beta(m)$. QED

In case $1 \leq p \leq 2$, (iii)$\Rightarrow$(i) can also be deduced from the fact that, by the Hausdorff-Young Theorem, H^p embeds contractively into A^{p^*}. Notice that in this case the above statement takes the form of a special factorization result.

Similarly, A^{p^*} embeds contractively into H^p when $2 \leq p < \infty$; this can be used to take care of part of the implication (i)$\Rightarrow$(ii).

Recall that an L^p-valued operator with a p-summing adjoint must be order bounded. Therefore, for operators $u : L^{p^*}(\mu) \to L^p(\nu)$, order boundedness is equivalent to u being p-summing, as well as to u^* being p-summing. Thus, by application of Proposition 3:

COROLLARY 4: *A composition operator C_φ is 1-order bounded if and only if for some (and then every) $1 \leq p < \infty$, C_φ is p-summing as an operator $A^{p^*} \to H^p$, if and only if for some (and then every) $1 \leq p < \infty$, C_φ^* is p-summing as an operator $(H^p)^* \to A^p$.*

As to be expected, the situation becomes more involved when we pass to β-order boundedness for arbitrary $0 < \beta < \infty$. The argument used in the proof of the preceding corollary can of course be applied to show that β-order boundedness of a composition operator C_φ is equivalent to C_φ, or C_φ^*, being $(1+\beta)$-summing as an operator from $H^{1+1/\beta}$ into $H^{1+\beta}$, cf. [6]. But otherwise not much seems to be known. In what follows, we are going to provide some additional information on the subject. We start by extending the corresponding part of Proposition 2:

THEOREM 5: *Let $1 \leq \beta < \infty$ and $\varphi \in \Phi$ be given. The following statements are equivalent:*

(i) *C_φ is β-order bounded.*

(ii) *For some (and then all) $p \in [\beta^{-1}, 1+\beta^{-1}]$, C_φ^* is βp-summing.*

(iii) *For some (and then all) $p \in [\beta^{-1}, 1+\beta^{-1}]$, C_φ^* is βp-nuclear.*

(iv) *For some (and then all) $p \in [\beta^{-1}, 1+\beta^{-1}]$, C_φ admits a factorization*

$$C_\varphi : H^p \xrightarrow{v} \ell^{(\beta p)^*} \xrightarrow{u} H^{\beta p}$$

where v is bounded and u is βp-nuclear.

Proof. The results used before also yield that, in the situation of (iv), βp-nuclearity for u and u^* are equivalent. Therefore the implications (iv)$\Rightarrow$(iii) $\Rightarrow$(ii)$\Rightarrow$(i) are automatic. The work is in proving that (i) implies (iv).

We use the hypothesis of (i) in the form $\sum_n (n+1)^{\beta-1}\|\varphi^n\|_{\beta p}^{\beta p} < \infty$. It follows that

$$u : \ell^{(\beta p)^*} \to H^{\beta p} : (\xi_n) \mapsto \sum_n (n+1)^{(1/p)-(1/\beta p)} \xi_n \varphi^n$$

is well defined (we have $\beta p \geq 1$). Since u^* is clearly βp-nuclear all we must check is that

$$v : H^p \to \ell^{(\beta p)^*} : f \mapsto \left(\frac{1}{(n+1)^{(1/p)-(1/\beta p)}} \cdot \hat{f}(n) \right)_n$$

is well defined and bounded.

Since $p \leq 2$, we can apply an inequality by Hardy and Littlewood (cf. Duren [2], p.96) to find a constant $c > 0$ such that

(1)
$$\Big(\sum_{n=0}^{\infty} \frac{|\hat{f}(n)|^p}{(n+1)^{2-p}} \Big)^{1/p} \leq c \cdot \|f\|_p \quad \forall f \in H^p .$$

Let us treat the case $p \geq 1$ first. We can use the Hausdorff-Young Theorem to obtain that

(2)
$$\Big(\sum_{n=0}^{\infty} |\hat{f}(n)|^{p^*} \Big)^{1/p^*} \leq \|f\|_p \quad \forall f \in H^p .$$

Put $\alpha := (2-p)(\beta p - 1)\cdot(\beta-1)^{-1}$ and note that $\alpha \geq 1$ since $\beta(p-1) \leq 1$ and $p \geq 1$ and that $\alpha^* = (2-p)(\beta p - 1)\cdot[(p-1)(1-\beta(p-1))]^{-1}$. By our assumptions, $\theta := (\beta-1)(\beta(2-p))^{-1}$ is in $[0,1]$, and we have, for any $f \in B_{H^p}$,

$$\begin{aligned}
\sum_n \Big(\frac{|\hat{f}(n)|}{(n+1)^{(1/p)-(1/\beta p)}} \Big)^{(\beta p)^*} &= \sum_n \frac{|\hat{f}(n)|^{\beta p/(\beta p - 1)}}{(n+1)^{(\beta-1)/(\beta p-1)}} \\
&\leq \Big(\sum_n \frac{|\hat{f}(n)|^{\beta p \theta \alpha/(\beta p-1)}}{(n+1)^{(\beta-1)\alpha/(\beta p - 1)}} \Big)^{1/\alpha} \Big(\sum_n |\hat{f}(n)|^{\beta p (1-\theta)\alpha^*/(\beta p-1)} \Big)^{1/\alpha^*} \\
&= \Big(\sum_n \frac{|\hat{f}(n)|^p}{(n+1)^{2-p}} \Big)^{1/\alpha} \Big(\sum_n |\hat{f}(n)|^{p^*} \Big)^{1/\alpha^*} \quad \leq \quad c^{p/\alpha}
\end{aligned}$$

by (1) and (2). This settles the case $p \geq 1$.

Suppose next that $1/\beta \leq p < 1$. Another result by Hardy and Littlewood (Duren [2], p.98) informs us that there is a constant $C > 0$ such that

$$\sup_n |\hat{f}(n)| \leq C \cdot (n+1)^{(1/p)-1} \cdot \|f\|_p \quad \forall f \in H^p .$$

Note that $\beta p/(\beta p - 1) \geq p$ and put $\gamma := C^{(\beta p/(\beta p-1))-p}$. We have, again for all $f \in B_{H^p}$,

$$\begin{aligned}\sum_n \Big(\frac{|\hat{f}(n)|}{(n+1)^{(1/p)-(1/\beta p)}} \Big)^{(\beta p)^*} &= \sum_n \frac{|\hat{f}(n)|^{\beta p/(\beta p-1)}}{(n+1)^{(\beta-1)/(\beta p-1)}} \\ &= \sum_n |\hat{f}(n)|^p \cdot \frac{|\hat{f}(n)|^{(\beta p/(\beta p-1))-p}}{(n+1)^{(\beta-1)/(\beta p-1)}} \\ &\leq \gamma \cdot \sum_n \frac{|\hat{f}(n)|^p}{(n+1)^{2-p}} \quad \leq \quad \gamma \cdot c^p ,\end{aligned}$$

and this takes care of the second case. QED

For the time being, it is not clear what happens if the values of β and p are not covered by the hypothesis of the preceding theorem. All what we can say is that outside of the described range there are parameters for which order boundedness and absolute summability are still equivalent, and there are parameters where coincidence breaks down. It is interesting to note that in both cases one of the involved spaces is the Hilbert space H^2.

PROPOSITION 6: *Suppose that $\beta \geq \frac{1}{2}$. For C_φ to be a β-order bounded composition operator it is necessary and sufficient that C_φ is 2β-summing as an operator $H^2 \to H^{2\beta}$.*

Proof. In fact, by a result of S.Kwapień [7], if $C_\varphi : H^2 \to H^{2\beta}$ is q-summing for some q, then its adjoint C_φ^* is s-summing for any $0 < s < \infty$, see also Pietsch [10], 1.3.15. In particular, C_φ^* is 2β-summing, and so C_φ must be β-order bounded. QED

We see again that if $\beta \leq 1$, then $C_\varphi : H^2 \to H^{2\beta}$ contains a Hilbert-Schmidt factor if and only if it is β-order bounded, as was reported earlier.

It is conceivable that the range of parameters in Theorem 5 can be extended. However, as our next two results show, there are some (as yet unknown) restrictions which cannot be removed.

PROPOSITION 7: *For every $1 < \beta \leq 2$, there are functions $\varphi \in \Phi$ such that C_φ exists as an absolutely 1-summing operator $H^{2/\beta} \to H^2$, but fails to be β-order bounded.*

Proof. For notational convenience set $p := 2/\beta$. Then $1 \leq p < 2$ by our hypothesis, and $p^* = 2/(2-\beta)$.

Apply the Hardy-Littlewood Inequality one more time to see that

$$v : H^p \to \ell^p : f \to \Big(\frac{\hat{f}(n)}{(n+1)^{(1/p)-(1/p^*)}} \Big)_n$$

is a well defined bounded operator.

Now, if $(\|\varphi^n\|_2)$ belongs to the Lorentz sequence space $\ell^{(p,p^*)}$, then

$$u : \ell^p \to H^2 : (\alpha_n) \mapsto \sum_n \alpha_n (n+1)^{(1/p)-(1/p^*)} \varphi^n$$

is also well defined and bounded. It factors

$$u : \ell^p \stackrel{u_2}{\to} \ell^1 \stackrel{u_1}{\to} H^2$$

where the u_k's are given by

$$u_1((\xi_n)_n)) = \sum_n \xi_n \|\varphi^n\|_2^{-1} \varphi^n$$

and

$$u_2((\xi_n)_n) = \left(n^{(1/p)-(1/p^*)} \|\varphi^n\|_2 \xi_n\right) ,$$

respectively (u^* is p^*-nuclear!). An appeal to Grothendieck's Theorem (cf. [9] 22.4.4) shows that u_1, and so $C_\varphi = uv$, must be 1-summing,.

But now our requirement $(\|\varphi^n\|_2) \in \ell^{(p,p^*)}$ is equivalent to $(\|\varphi^n\|_1)$ being a member of $\ell^{(1/\beta,1/(2-\beta))}$. Since $\beta \geq 1$, we are dealing with a sequence space which is strictly bigger than $\ell^{(1/\beta,1)}$.

There are indeed functions $\varphi \in \Phi$ such that $(\|\varphi^n\|_1)$ belongs to $\ell^{(1/\beta,1/(2-\beta))}$ but *not* to $\ell^{(1/\beta,1)}$; this was shown in [3] by H.Hunziker in the course of his characterization of sequences (a_n) of non-negative real numbers for which there is a $\varphi \in \Phi$ satisfying $a_n = \|\varphi^n\|$ for each n.
QED

We finish by a related observation involving the spaces A^p introduced in Proposition 3 in order to characterize β-order boundedness of composition operators. First notice that, however we choose $1 \leq p < \infty$ and $0 < \beta < \infty$, for C_φ to exist as a bounded linear map $A^{p^*} \to L^{\beta p}(m)$, it is necessary and sufficient that (φ^n) is a weak ℓ^p sequence in $H^{\beta p}$. This is certainly the case if we require $(\|\varphi_n\|_{\beta p})_n \in \ell^p$, a condition which is clearly equivalent to $(\|\varphi^n\|_1)_n$ being a member of $\ell^{1/\beta}$. As before, this implies that C_φ^* is p-nuclear. As long as $\beta \geq 1$, it follows that C_φ^* is βp-summing, and so C_φ is β-order bounded, by Proposition 3.

But p-nuclearity of C_φ^* also provides us with a corresponding canonical factorization $A^{p^*} \to \ell^1 \to H^{\beta p}$. So we may argue as before and get:

PROPOSITION 8: *If $0 < \beta < 1$, then each $\varphi \in \Phi$ with $(\|\varphi^n\|_1)_n \in \ell^{1/\beta}$ induces a 2-summing composition operator $C_\varphi : A^{2/(2-\beta)} \to H^2$. It fails to be β-order bounded unless $(\|\varphi^n\|_1)_n$ belongs to $\ell^{(1/\beta,1)}$.*

It follows from the preceding two propositions that one cannot hope for one single condition to characterize βp-summability of composition operators $C_\varphi : H^p \to H^{\beta p}$ simultaneously for all p's, as it was the case with β-boundedness and β-order boundedness.

REFERENCES

[1] C.C.Cowen, Composition operators on Hilbert spaces of analytic functions: a status report. *Proc. AMS 1988 Summer Res. Inst. on Operator Theory and Operator Algebras* 131-145 (1990).

[2] P.L.Duren, Theory of H^p Spaces, *Academic Press* (1970).

[3] A.Grothendieck, Produits tensoriels topologiques ets espaces nucléaires. *Memoirs of the Amer. Math. Soc. 16* (1955).

[4] H.Hunziker, Kompositionsoperatoren auf klassischen Hardyräumen. *Thesis, Universität Zürich* (1989).

[5] H.Hunziker and H.Jarchow, Composition operators which improve integrability. *Math. Nachr. 152*: 83-99 (1991).

[6] H.Jarchow, Some factorization properties of composition operators. *Progress in Functional Analysis (Elsevier - North-Holland)*: 405-413 (1992).

[7] S.Kwapień, On a theorem of L.Schwartz and its applications to absolutely summing operators. *Studia Math. 38*: 193-201 (1970).

[8] A.Persson, On some properties of p-nuclear and p-integral operators. *Studia Math. 33*: 213-222 (1969).

[9] A.Pietsch, Operator Ideals. *VEB Deutscher Verlag der Wissenschaften* (1978) and *North-Holland* (1980).

[10] A.Pietsch, Eigenvalues and s-Numbers.*Cambridge University Press* (1987).

[11] L.Schwartz, Applications p-radonifiantes et theéorème de dualité. *Studia Math. 38*: 203-213 (1970).

[12] J.Shapiro and P.D.Taylor, Compact, nuclear, and Hilbert-Schmidt composition operators on H^2. *Indiana Univ. Math. Journ. 23*: 471-496 (1973).

Factorization of Operator Ideals and Bounded Sets in Tensor Products of Locally Convex Spaces

H. JUNEK Department of Mathematics, University of Potsdam, Germany

1 INTRODUCTION

In this paper we study (non injective) ideals of operators between locally convex spaces and we prove a factorization and uniqueness theorem for certain operator ideal components mapping DF-spaces into F-spaces. A representation of the bounded subsets is given. The application to certain tensor products of Fréchet spaces leads to the characterization of the bounded sets in $\alpha-$ tensor products, which includes the case of $\pi-$ tensor products. The problem of the characterization of the bounded subsets of the $\pi-$tensor product of F-spaces, known as Grothendieck's "problème des topologies" was answered in the negative by J. Taskinen [13], and after that several papers appeared to get positive results in particular cases. The method presented in this paper allows to derive several of the known results using the tool of operator ideals. This includes the cases of (FG)-spaces [2], and of Fréchet-Montel spaces of $\mathcal{L}_\infty-$ type which were treated by A. Defant and M. Maestre [5] using tensor product methods.

2 OPERATOR IDEALS ON LOCALLY CONVEX SPACES AND IDEAL TOPOLOGIES

A *quasi-normed ideal* $(\mathcal{A}, \alpha)$ of operators in the sense of A. Pietsch [12] is a subclass $\mathcal{A}$ of the class $\mathcal{L}$ of all linear bounded operators between Banach spaces with the components $\mathcal{A}(X, Y) \subseteq \mathcal{L}(X, Y)$ such that the following conditions are satisfied:

(I1) *For all Banach spaces X, Y the set $\mathcal{A}(X, Y)$ is a linear subspace of $\mathcal{L}(X, Y)$ containing the set $\mathcal{F}(X, Y)$ of all finite rank operators.*

(I2) *The product STR of any operators $T \in \mathcal{A}(X, Y), R \in \mathcal{L}(X_0, X)$ and $S \in \mathcal{L}(Y, Y_0)$ belongs to $\mathcal{A}(X_0, Y_0)$.*

(I3) *On each component $\mathcal{A}(X, Y)$ there is some complete quasi-norm α satisfying $\alpha(x' \otimes y) = \|x'\|\|y\|$ and $\alpha(STR) \leq \|S\|\|R\|\alpha(T)$ for all operators R, S, T as in (I2).*

Let E be any locally convex space. As usually, we denote by $\mathcal{U}(E)$ the set of all absolutely convex closed zero-neighborhoods and by $\mathcal{B}(E)$ the set of all absolutely convex closed bounded subsets of E. For $A \in \mathcal{B}(E)$ and $U \in \mathcal{U}(E)$ let E_A and E_U denote the Banach spaces associated to the balls, and if E is quasi-complete, there are canonical linear maps $C_A : E_A \to E$ and $C_U : E \to E_U$.

Given any operator ideal $\mathcal{A}$ of operators on Banach spaces there is a plenty of possibilities to extend it to the category of locally convex spaces. In [12] six typical extensions are given. Among them there are a smallest (strong extension) and a largest (weak) extension denoted by $\mathcal{A}^s$ and $\mathcal{A}^w$, respectively. The components of these ideals are defined by the following conditions: Let E and F be any quasi-complete locally convex spaces and let $T \in \mathcal{L}(E, F)$ be any operator. Then we put

$T \in \mathcal{A}^w(E, F)$ if for all Banach spaces X, Y and for all operators $R \in \mathcal{L}(X, E)$ and $S \in \mathcal{L}(F, Y)$ the product STR belongs to $\mathcal{A}(X, Y)$, and

$T \in \mathcal{A}^s(E, F)$ if there is some factorization

$$\begin{array}{ccc} E & \xrightarrow{\quad T \quad} & F \\ {\scriptstyle R_0}\big\downarrow & & \big\uparrow{\scriptstyle S_0} \\ Y_0 & \xrightarrow{\quad T_0 \quad} & X_0 \end{array}$$

with some $R_0 \in \mathcal{L}(E, X_0)$, $S_0 \in \mathcal{L}(Y_0, F)$ and $T_0 \in \mathcal{A}(X_0, Y_0)$.

Obviously, the Banach spaces X, Y, X_0 and Y_0 appearing in the above definitions can be choosen to be the canonical Banach spaces E_A, F_U, E_V and F_B for some $A \in \mathcal{B}(E)$, $U \in \mathcal{U}(F)$, $V \in \mathcal{U}(E)$ and $B \in \mathcal{B}(F)$, respectively.

In generally, the ideal components $\mathcal{A}^w(E, F)$ and $\mathcal{A}^s(E, F)$ are rather different. For instance, if $E = F$ is choosen to be a Banach space X equipped with the weak topology $\sigma(X, X^*)$ then we get $\mathcal{F}^w(E, E) = \mathcal{L}(E, E)$ and $\mathcal{F}^s(E, E) = \mathcal{F}(E, E)$, where $\mathcal{F}$ is the ideal of finite rank operators. In contrast, if E is a DF-space and if F is an F-spaces, then the coincidence of $\mathcal{A}^w(E, F)$ and $\mathcal{A}^s(E, F)$ could be shown by the author in [8] for the case of injective and surjective quasi-normed operator ideals $\mathcal{A}$. But several interesting ideals, included the ideal of nuclear operators, are neither surjective nor injective. In this paper we are interested in such ideals.

If $(\mathcal{A}, \alpha)$ is a quasi-normed ideal of operators in Banach spaces, then $\mathcal{A}^w$ and $\mathcal{A}^s$ admit natural topologies. Let us suppose that E and F are complete. The space $\mathcal{A}^w(E, F)$ admits the embedding

$$\mathcal{A}^w(E,F) \subseteq \operatorname*{proj}_{A \in \mathcal{B}(E), U \in \mathcal{U}(F)} \mathcal{A}(E_A, F_U)\ ,$$

and the induced topology τ_α on $\mathcal{A}^w(E,F)$ is given by the seminorms

$$p^\alpha_{A,U}(T) = \alpha(C_U T C_A) \quad \text{for} \quad T \in \mathcal{A}^w(E,F), \quad A \in \mathcal{B}(E), \ U \in \mathcal{U}(F)\ .$$

Since $\mathcal{A}^s(E,F)$ coincides by definition with the following inductive limit

$$\mathcal{A}^s(E,F) = \operatorname*{ind}_{V \in \mathcal{U}(E),\ B \in \mathcal{B}(F)} \mathcal{A}(E_V, F_B)\ ,$$

we can consider on $\mathcal{A}^s(E,F)$ the associated locally convex inductive topology.

Let E be any complete DF-space and let F be any Fréchet space. Since every linear operator T from E into F which is bounded on bounded sets is even continuous ([8, 6.2.1]), we have $\mathcal{L}(E,F) = \operatorname*{proj}_{A,U} \mathcal{L}(E_A, F_U)$ in this case. This implies

$$\mathcal{A}^w(E,F) = \operatorname*{proj}_{A,U} \mathcal{A}(E_A, F_U)\ .$$

Therefore, $(\mathcal{A}^w(E,F), \tau_\alpha)$ is a complete metrizable space in this case. We will use this result later on.

3 FACTORIZATION OF OPERATORS MAPPING DF-SPACES INTO F-SPACES

In this section we prove the coincidence of the ideal component $\mathcal{A}^w(E,F)$ and $\mathcal{A}^s(E,F)$ for reflexive DF-spaces E and F-spaces F satisfying some additional conditions, but we don't assume the injectivity or surjectivity of $\mathcal{A}$. Furthermore, we will obtain a characterization of the τ_α-bounded subsets of $\mathcal{A}^w(E,F) = \mathcal{A}^s(E,F)$. The mentioned additional conditions concern the geometry of the spaces E and F and they will be stated now. We recall the definition of Grothendieck-$\mathcal{A}$-spaces (c.f.[12]). Let $\mathcal{A}$ be any ideal of operators on Banach spaces. A locally convex space F is called to be an $\mathcal{A}$-*space*, and we write $F \in sp(\mathcal{A})$, if for every closed convex neighborhood $U \in \mathcal{U}(F)$ there is some $V \in \mathcal{U}(F)$ contained in U such that the canonical mapping $C_{VU} : \ F_V \to F_U$ belongs to $\mathcal{A}(F_V, F_U)$. In this paper the $\mathcal{L}_2-$ and the $\mathcal{L}_\infty-$spaces are of special interest. The ideals $\mathcal{L}_p$ of p-factorable operators $(1 \le p \le \infty)$ consist of all operators $T: \ X \to Y$ which admit a factorization

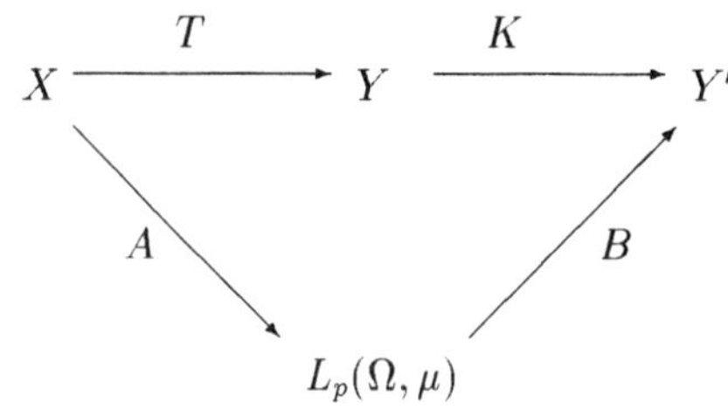

for some measure space (Ω, μ). Here, $K: Y \to Y''$ stands for the canonical embedding of Y in its second dual. For more details we refer to [12]. The 2-factorable operators admit obviously even a factorization

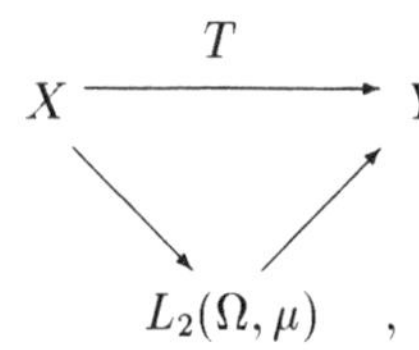

,

since each subspace in L_2 is complemented. Therefore, the class $sp(\mathcal{L}_2)$ coincides with the class of all Hilbertizable locally convex spaces. The spaces in $sp(\mathcal{L}_\infty)$ were intensively studied by R. Hollstein [10] as *ε−spaces.*

A locally convex space F is called to be a *co $\mathcal{A}$ − space* and we write $F \in co(\mathcal{A})$, if for every closed convex bounded subset $A \in \mathcal{B}(F)$ there is some $B \in \mathcal{B}(F)$ containing A such that the canonical map $C_{AB}: F_A \to F_B$ belongs to $\mathcal{A}(F_A, F_B)$.

With this notations we have:

PROPOSITION 3.1.

(a) *Let F be any Fréchet-space. Then F belongs to $sp(\mathcal{L}_2)$ iff it belongs to $co(\mathcal{L}_2)$.*

(b) *Let F be any Fréchet-Montel space. Then F belongs to $sp(\mathcal{L}_\infty)$ iff it belongs to $co(\mathcal{L}_\infty)$.*

Proof: The first assertion was proved by the author in [8,6.5.6]. Let us turn to (b). It was shown in [8,6.4.2] that every metrizable co($\mathcal{L}_\infty$)−space is an $\mathcal{L}_\infty$−space. Since the ideal $\mathcal{L}_\infty$ is not injective, the result in [8,6.2.5] cannot be used to prove the converse direction. But it was shown by R. Hollstein [10] and by A. Defant and M. Maestre in [5, 4.Corollary] that strong duals F'_b of Fréchet-Montel spaces contained in $sp(\mathcal{L}_\infty)$ belong to $sp(\mathcal{L}_1)$. This means that for each $A \in \mathcal{B}(F)$ there is some $B \in \mathcal{B}(F)$ containing A such that the canonical map $C_{A^0B^0}: F'_{A^0} \to F'_{B^0}$ belongs to $sp(\mathcal{L}_1)$. Dualizing the assoziated factorization of the map $F'_{A^0} \to F'_{B^0} \to (F'_{B^0})''$ we get a factorization

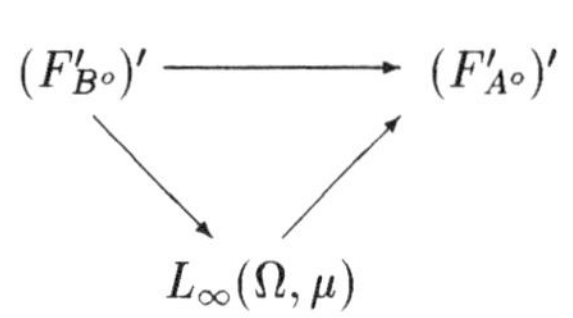

Since F is reflexive, we may assume $(F'_{A^0})' = F_A$ and $(F'_{B^0})' = F_B$ by [8,4.3.10], and this shows $F \in co(\mathcal{L}_\infty)$.

As a third class of Fréchet-spaces we will consider the decomposable (FG)-spaces. This class was introduced by J. Bonet, J.C. Diaz and J. Taskinen in [2] as a class solving the BB-problem in the positive.

DEFINITION. *A Fréchet space F is called to be a decomposable (FG)-space if there is an increasing fundamental sequence of seminorms $(\|\cdot\|_k)_{k\in\mathbb{N}}$ such that for every sequence of numbers $0 < c_k \leq 1, k \geq 1$, there is some sequence $(P_k)_{k\in\mathbb{N}} \subseteq \mathcal{L}(F,F)$ of operators satisfying the following conditions:*

(FG1) $x = \sum_{k\in\mathbb{N}} P_k(x)$ *for all* $x \in F$.

(FG2) $\|P_k x\|_{k-1} \leq c_k \|x\|_k$ *for all* $x \in F$ *and all* $k \geq 2$.

(FG3) *For each* $j \geq k$ *there is some constant* $\lambda_{jk} \geq 1$ *such that* $\|P_k(x)\|_j \leq \lambda_{jk}\|x\|_k$ *for all* $x \in F$.

(FG4) $P_k\, P_j = \delta_{jk}\, P_k$ *for each* $k, j \in \mathbb{N}$.

Together with (FG4) the condition (FG3) means that the seminorm $\|\cdot\|_k$ induces the topology of F on $P_k(F)$. For Fréchet-Montel spaces this implies $\dim P_k(F) < \infty$. Permanence properties and examples of decomposable (FG)-spaces have been presented in [2]. Among these examples there are all Banach space-valued Köthe echelon spaces $\lambda_p(A, (X_n)_{n\in\mathbb{N}})$ of order $1 \leq p < \infty$ where $(X_n)_{n\in\mathbb{N}}$ is any sequence of Banach spaces, and all Fréchet-Schwartz spaces with a finite dimensional decomposition and a continuous norm ([2,§3] and [1]). Now, we can state the following result:

PROPOSITION 3.2. *Let $(\mathcal{A}, \alpha)$ be any quasi-normed operator ideal, let X be any Banach space and let F be any Fréchet space. Suppose that one of the following conditions is satisfied:*

(i) *F belongs to $sp(\mathcal{L}_2)$.*

(ii) *F is a Fréchet-Montel space belonging to $sp(\mathcal{L}_\infty)$.*

(iii) *F is a decomposable (FG)-space.*

Then we have

(j) $\mathcal{A}^s(X,F) = \mathcal{A}^w(X,F)$.

(jj) *For each τ_α-bounded subset $\mathcal{M} \subseteq \mathcal{A}^w(X,F)$ there is some $B \in \mathcal{B}(F)$ such that each $T \in \mathcal{M}$ admits a factorization through the unit ball of $\mathcal{A}(X, F_B)$.*

We will use the same construction in all cases (i) - (iii), but before we should state the following easy lemma which is a direct consequence of the completeness of $(\mathcal{A}, \alpha)$ (cf [8, Lemma 6.2.4]).

LEMMA 3.1. *Let $(\mathcal{A}, \alpha)$ be any quasi-normed operator ideal, let X and Y_n be any Banach spaces and let $1 \leq p \leq \infty$ be any fixed number. For any sequence of numbers $K_n > 0$ there is some sequence of numbers $c_n > 0$ such that for each sequence of operators $T_n \in \mathcal{A}(X, Y_n)$ with $\alpha(T_n) \leq K_n$ the operator*

$$T:\ X \to l_p((Y_n),(c_n)) \quad \textit{defined by} \quad Tx = ((T_n x))$$

belongs to $\mathcal{A}(X, l_p((Y_n),(c_n)))$ and satisfies $\alpha(T) \leq 1$.

Proof of Proposition 3.2: It suffices to prove the assertion (jj). Let $\mathcal{M}$ be any τ_α−bounded subsets of $\mathcal{A}^w(X, F)$. We choose some countable decreasing zero-neighborhood basis (U_n) in $\mathcal{U}(F)$. Let us denote the spaces F_{U_n} simply by F_n and the canonical maps $C_{U_n}:\ F \to F_n$ by C_n. Under the assumption (i) we may suppose that each of the spaces F_n is a Hilbert space. By the boundedness of $\mathcal{M}$ with respect to τ_α, all numbers

$$K_n = \sup\{\alpha(C_n T):\ T \in \mathcal{M}\}$$

are finite. Using the Lemma, there are weights $c_n > 0$ such that the mappings

$$\hat{T}:\ X \to l_2((F_n),(c_n)) \ \text{ given by } \quad \hat{T}x = \sum_{n=1}^{\infty} C_n T x$$

belong to $\mathcal{A}(X, l_2((F_n),(c_n)))$ and satisfy $\alpha(\hat{T}) \leq 1$ for each $T \in \mathcal{M}$.
Define

$$B = \left\{ y \in F:\ p_B^2(y) = \sum_{n=1}^{\infty} \|C_n y\|_n^2 \cdot c_n < \infty \right\}.$$

This set belongs to $\mathcal{B}(F)$ and the map

$$J:\ F_B \to l_2((F_n),(c_n))$$

defined by $Jx = (C_1 x, C_2 x, \ldots)$ is a metric injection. Furthermore, for all $x \in X$ we have

$$\begin{aligned} p_B^2(Tx) &= \sum_{n=1}^{\infty} \|C_n T x\|_n^2\, c_n = \|JTx\|^2_{l^2((F_n),(c_n))} = \|\hat{T}x\|^2_{l^2((F_n),(c_n))} \\ &\leq \|\hat{T}:\ X \to l^2((F_n),(c_n))\|^2 \cdot \|x\|^2 \leq \alpha(\hat{T})^2 \cdot \|x\|^2 \leq \|x\|^2 . \end{aligned}$$

Therefore, the map $\hat{T} = JT$ admits a linear continuous factorization through F_B:

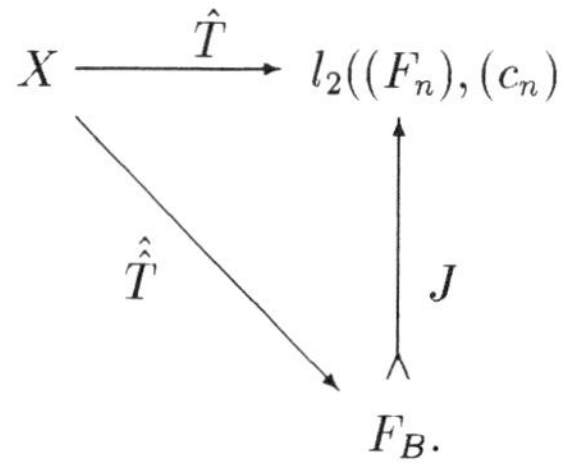

Now we can show (j) and (jj). Under the assumption (i) the space F_B is a Hilbert space. If P denotes the orthoprojection from $l_2((F_n),(c_n))$ onto F_B, we get $\hat{\hat{T}} = P\hat{T}$ and $\alpha(\hat{\hat{T}}) \leq \|P\|\ \alpha(\hat{T}) = \alpha(\hat{T}) \leq 1$. Under the assumption (ii) we will use the Proposition 3.1b. Therefore, there is some $C \in \mathcal{B}(F)$ containing B such that the canonical map C_{BC} satisfies $C_{BC} \in \mathcal{L}_\infty(F_B, F_C)$. Since F is reflexive as a Fréchet-Montel space, F_C can supposed to be reflexive, too ([Ju, 6.5.5]). This shows the existence of some diagram

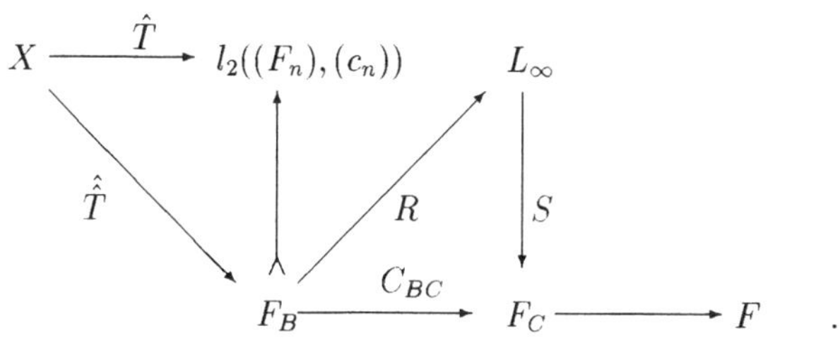

Since L_∞ has the metric extension property ([12, C.3]), there is an extension $R_1 : l_2((F_n),(c_n)) \to L_\infty$ of R. Since $C_{BC}\hat{\hat{T}} = SR\hat{\hat{T}} = SR_1\hat{T}$, we get $C_{BC}\hat{\hat{T}} \in \mathcal{A}(X, F_C)$ and

$$\alpha(C_{BC}\hat{\hat{T}}) \leq \|S\|\|R_1\|\alpha(\hat{T}) \leq \|S\|\|R_1\| \ .$$

The factorization $T = C_C \cdot C_{BC}\,\hat{\hat{T}}$ is obvious. This shows (j) with C instead of B. If we replace C by the multiple $D = \|S\|\|R_1\|C$, then we get the wanted factorization of each $T \in \mathcal{M}$ through the unit ball of $\mathcal{A}(X, F_D)$.

Finally, we consider the assumption (iii). Instead of $l_2((F_n),(c_n))$ we use the space $l_1((F_n),(c_n))$ with some other constants c_n given by the Lemma 3.1 and define as above

$$\hat{T} : \ X \to l_1((F_n),(c_n)) \quad \text{by} \quad \hat{T}x = \sum_{n=1}^{\infty} C_n T x$$

Next we choose a sequence of mappings (P_k) adapted to (c_k) and satisfying (FG1) – (FG4). We are going to show that a mapping

$$P : \ l_1((F_n),(c_n)) \to F$$

can be defined by $P((C_k x_k)) = \sum_{k=1}^{\infty} P_k x_k$. To prove the convergence of the series $\sum P_k x_k$, the correctness of the definition and the continuity of this map let any seminorm $\|\cdot\|_j$ be given. Then we have by (FG2) and (FG3) the estimation

$$\begin{aligned}
\|\sum_k P_k x_k\|_j &\le \sum_{k\le j} \|P_k x_k\|_j + \sum_{k>j} \|P_k x_k\|_j \\
&\le \sum_{k\le j} \|P_k x_k\|_j + \sum_{k>j} \|P_k x_k\|_{k-1} \\
&\le \sum_{k\le j} \|P_k x_k\|_j + \sum_{k>j} c_k \|x_k\|_k \\
&\le \sum_{k\le j} \lambda_{jk} \|x_k\|_k + \sum_{k>j} c_k \|x_k\|_k \\
&\le \Lambda_j \sum_{k=1}^{\infty} c_k \|x_k\|_k = \Lambda_j \|(x_k)\|_{l_1((F_n),(c_n))}
\end{aligned}$$

for some large constant Λ_j. Now, for each $T \in \mathcal{M}$ we get the factorization

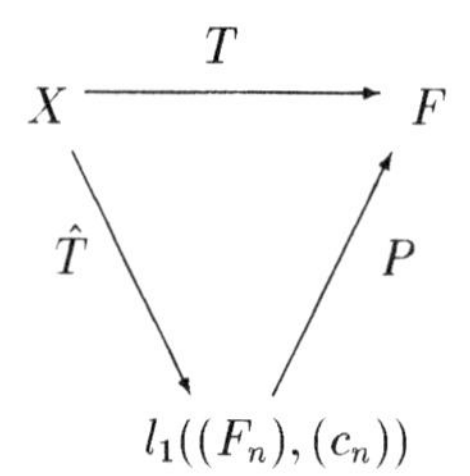

,

since

$$P\hat{T}x = P((C_k Tx)) = \sum_k P_k Tx = Tx$$

by (FG1). If we define B to be the image of the unit ball of $l_1((F_n),(c_n))$ under the map P, we get the desired factorization of $\mathcal{M}$ through the unit ball of $\mathcal{A}(X, F_B)$. Thus we are done. ■

Using adjoint operators we obtain a dual version of Proposition 3.2 for DF-spaces. For sake of simplicity we state this only for reflexive spaces and for symmetric operator ideals. As in [12] an operator ideal $(\mathcal{A}, \alpha)$ is called *symmetric*, if $T \in \mathcal{A}(X, Y)$ implies $T' \in \mathcal{A}(Y', X')$ and $\alpha(T') \le \alpha(T)$.

PROPOSITION 3.3. *Let $(\mathcal{A}, \alpha)$ be any symmetric quasi-normed operator ideal, let Y be any reflexive Banach space and let E be any reflexive Fréchet space satisfying one of the conditions (i) - (iii) of Proposition 3.2. Then we have*

(j) $\mathcal{A}^s(E'_b, Y) = \mathcal{A}^w(E'_b, Y)$.

(jj) *For each τ_α−bounded subset $\mathcal{M} \subseteq \mathcal{A}^w(E'_b, Y)$ there is some set $B \in \mathcal{B}(E)$ such that each $T \in \mathcal{M}$ admits a factorization through the unit ball of $\mathcal{A}(E'_{B^0}, Y)$.*

Proof: It is sufficient to prove (jj). Let $\mathcal{M} \subseteq \mathcal{A}^w(E'_b, Y)$ be any τ_α−bounded set. We consider the set $\mathcal{M}' = \{T' :\ Y' \to E \,|\, T \in \mathcal{M}\}$ and we will show that $\mathcal{M}'$ is a τ_α−bounded subset of $\mathcal{A}^w(Y', E)$. Indeed, for each $U \in \mathcal{U}(E)$ we have $C_U T' = (T\, C_{U^0})'$, and the symmetry of $(\mathcal{A}, \alpha)$ implies now

$$\alpha(C_U T') = \alpha((T\, C_{U^0})') \leq \alpha(T\, C_{U^0}) \leq K_U$$

for some number K_U depending on U and $\mathcal{M}$ by the τ_α−boundedness of $\mathcal{M}$. So we can apply Proposition 3.2 to the set $\mathcal{M}'$ and by 3.2(jj) there is some $B \in \mathcal{B}(E)$ such that each $T' \in \mathcal{M}'$ admits a factorization

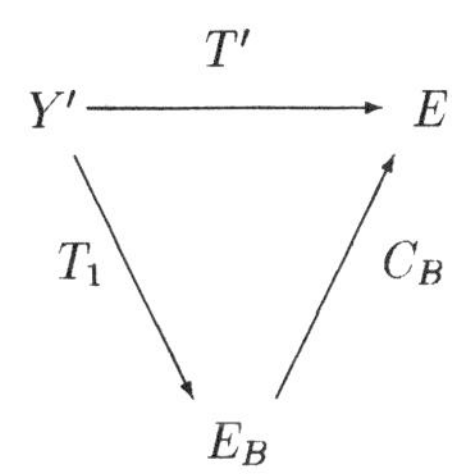

for some $T_1 \in \mathcal{A}(Y, E_B)$ with $\alpha(T_1) \leq 1$. Taking the adjoint, we get a factorization

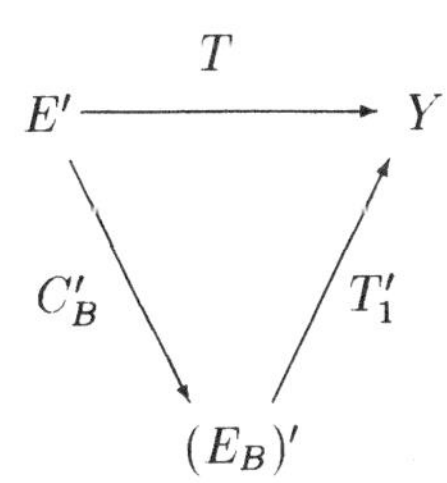

with $T'_1 \in \mathcal{A}((E_B)', Y)$ and $\alpha(T'_1) \leq \alpha(T_1) \leq 1$. Since E is reflexive, we may assume the reflexivity of E_B and this implies $(E_B)' = E'_{B^0}$ ([8,6.5.5]). ∎

Now, we are ready to show the main result of this section.

THEOREM. *Suppose that E and F are any reflexive Fréchet-spaces. Let $(\mathcal{A}, \alpha)$ be any quasi-normed, symmetric operator ideal. Suppose that one of the following conditions is satisfied:*

(i) *E and F belong to $sp(\mathcal{L}_2)$.*

(ii) *E and F are Fréchet-Montel spaces belonging to $sp(\mathcal{L}_\infty)$.*

(iii) *E and F are decomposable (FG)-spaces.*

Then we have

(j) $\mathcal{A}^s(E'_b, F) = \mathcal{A}^w(E'_b, F)$.

(jj) *For each τ_α−bounded subset $\mathcal{M} \subseteq \mathcal{A}^w(E'_b, F)$ there are some $B \in \mathcal{B}(E)$ and some $A \in \mathcal{B}(F)$ such that each $T \in \mathcal{M}$ admits a factorization through the unit ball of $\mathcal{A}(E'_{B^0}, F_A)$.*

In other words, the equation

$$\underset{A,\,B}{ind}\ \mathcal{A}(E'_{A^\circ}, F_B) \;=\; \underset{U,V}{proj}\ \mathcal{A}(E'_U, F_V)\ .$$

holds algebraically, and both spaces have the same bounded subsets with respect to their topologies.

Proof: Let (U_k) be any fundamental system of zero-neighborhoods of F with canonical mappings $C_k : F \to F_{U_k}$ such that the spaces F_{U_k} are reflexive. We put

$$\mathcal{M}_k = \{C_k T :\ E'_b \to F \to F_{U_k} \ \mid\ T \in \mathcal{M}\}.$$

By the assumption on $\mathcal{M}$, the sets $\mathcal{M}_k$ are τ_α−bounded subsets of $\mathcal{A}^w(E'_b, F_{U_k})$ for each $k \in \mathbb{N}$. By Proposition 3.3 there are bounded subsets $A_k \in \mathcal{B}(E)$ such that the operators contained in $\mathcal{M}_k$ can be factorized through the unit ball of $\mathcal{A}(E_{A_k^0}, F_{U_k})$. Since E is metrizable, there is some bounded set $A \in \mathcal{B}(E)$ absorbing each A_k. Hence the canonical mappings $C_{A^0 A_k^0} :\ E'_{A^0} \to E'_{A_k^0}$ exist and the sets $\mathcal{M}_k$ factorize through some α−bounded subset of $\mathcal{A}(E'_{A^0}, F_{U_k})$ for each $k \in \mathbb{N}$. This proves by definition that $\mathcal{M}$ factorizes through some τ_α−bounded subset $\mathcal{N}$ of $\mathcal{A}^w(E'_{A^0}, F)$. Now, we can apply Proposition 3.2, and this yields the existence of some bounded set $B \in \mathcal{B}(F)$ and a factorization of $\mathcal{N}$ through the unit ball of $\mathcal{A}(E'_{A^0}, F_B)$. ■

We close this section with the application to the operator ideals $\mathcal{K}_{p,q}$. These ideals were introduced by A. Pietsch [12]. The associated tensor norms were firstly defined by J.T. Lapresté and extensively studied in [3].

Let $1 \le p,\ q \le \infty$ with $1/p + 1/q \ge 1$ be given and define $r > 0$ by $1 + 1/r = 1/p + 1/q$. An operator $T \in \mathcal{L}(X, Y)$ is called to be a *(p,q)-compact operator* and we write $T \in \mathcal{K}_{p,q}(X, Y)$ in this case, if T admits a representation

$$T = \sum_{i=1}^{\infty} \lambda_i\, a_i \otimes y_i$$

for some sequences $(\lambda_i) \in l_r$, $(a_i) \subseteq E'$ and $(y_i) \subseteq F$ such that the number

$$\alpha_{p,q}(T) = \inf \|(\lambda_i)\|_r \cdot \mathbf{w}_{q'}((a_i)) \cdot \mathbf{w}_{p'}((y_i))$$

is finite. Here, the infimum is taken over all representations of T. The numbers $\mathbf{w}_{q'}((a_i))$ and $\mathbf{w}_{p'}((y_i))$ are defined by

$$\begin{array}{rcll} \mathbf{w}_{q'}((a_i)) & = & \sup\{\|(\langle x, a_i\rangle)\|_{q'} : \ \|x\| \le 1\} & \text{and} \\ \mathbf{w}_{p'}((y_i)) & = & \sup\{\|(\langle y_i, b\rangle)\|_{p'} : \ \|b\| \le 1\} \ , & \end{array}$$

and they are called weakly q'–summing and weakly p'–summing norms, respectively. With this definition, $(\mathcal{K}_{p,q}, \alpha_{p,q})$ is a normed operator ideal in the category of Banach spaces ([12, 18.4.1]).

Obviously, an operator $T \in \mathcal{L}(X, Y)$ belongs to $\mathcal{K}_{p,q}(X, Y)$, if there is a factorization

$$\begin{array}{ccc} X & \xrightarrow{\quad T \quad} & Y \\ S \big\downarrow & & \big\uparrow R \\ l_p & \xrightarrow{\quad T_0 \quad} & l_{q'} \end{array}$$

with some diagonal operator $T_0((\xi_i)) = (\lambda_i\, \xi_i)$ and some scalar sequence $(\lambda_i) \in l_r$ where r is as above. The ideal of nuclear operators is included as the special case, we have $\mathcal{N}(X, Y) = \mathcal{K}_{1,1}(X, Y)$, and the nuclear norm ν coincides with $\alpha_{1,1}$. Since $1/p + 1/q \ge 1$ is to be assumed, we see that $T \in \mathcal{K}_{p,q}(X, Y)$ implies $T' \in \mathcal{K}_{q,p}(Y', X')$ and $\alpha_{q,p}(T') \le \alpha_{p,q}(T)$. Hence, the ideals $\mathcal{K}_{p,p}$ are symmetric for $1 \le p \le 2$, and Proposition 3.2 and Theorem 3.3 are applicable. For $p \ne q$ we have at least $T'' \in \mathcal{K}_{p,q}(X'', Y'')$, but a careful reading of the proof of Proposition 3.2 shows that the assertion holds true even for the ideals $\mathcal{K}_{p,q}$ because of the estimation $\alpha_{p,q}(T'') \le \alpha_{q,p}(T') \le \alpha_{p,q}(T)$. This leads to the following result:

COROLLARY. *Theorem 3.3 holds true for E and F as in 3.3 and for $(\mathcal{K}_{p,q}, \alpha_{p,q})$ for all $1/p + 1/q \ge 1$.*

4 BOUNDED SUBSETS IN TENSOR PRODUCTS OF FRÉCHET SPACES

We shall apply the results of the preceding section to tensor products of Fréchet spaces. Under some assumptions on the Fréchet spaces E and F this will lead to a representation of the complete tensor product $E\tilde{\otimes}_{\alpha_{p,q}}F$ as the space $\mathcal{K}^s_{p,q}(E'_b, F) = \mathcal{K}^w_{p,q}(E'_b, F)$ of (p, q)–compact operators and to the characterization of the bounded subsets in $E\tilde{\otimes}_{\alpha_{p,q}}F$. Especially, this applies to the case $\pi = \alpha_{1,1}$ and gives an alternative proof to the results concerning the BB-property contained in [5].

We introduce now the $\alpha_{p,q}$–tensor product which is associated to the ideal norm $\alpha_{p,q}$. For normed spaces X, Y and numbers $p, q \ge 1$ with $1/p + 1/q \ge 1$ we define $X\tilde{\otimes}_{\alpha_{p,q}}Y$ as the completion of the algebraic tensor product $X \otimes Y$ with respect to the finitely generated tensor norm

$$\alpha_{p,q}(z) = \inf\{\|(\lambda_i)\|_r \ \mathbf{w}_{q'}(x_i)\mathbf{w}_p(y_i) : \ z = \sum_{i=1}^{n} \lambda_i\, x_i \otimes y_i\} \ ,$$

where r is determined by $1/r = 1/p + 1/q - 1$. The $\alpha_{p,q}$–tensor product of two locally convex spaces E and F is then defined by

$$E\tilde{\otimes}_{\alpha_{p,q}}F = \operatorname*{proj}_{U \in \mathcal{U}(E), V \in \mathcal{U}(F)} E_U \tilde{\otimes}_{\alpha_{p,q}} F_V .$$

The tensor norm $\alpha_{1,1}$ is nothing else the π−tensor norm, and we have $E\tilde{\otimes}_{\alpha_{1,1}}F = E\tilde{\otimes}_\pi F$ in this case.

Now we use the following representation of the $\alpha_{p,q}$−tensor product of Banach spaces by operator ideals which is a special case of [3, 7.1 COROLLARY] and a generalization of the well known representation of the π−tensor product of Banach spaces as a space of nuclear operators:

PROPOSITION 4.1 *Let X and Y be any Banach spaces such that X' or Y have the approximation property. Then we have*

$$X'\tilde{\otimes}_{\alpha_{p,q}}Y = \mathcal{K}_{p,q}(X, Y)$$

isometrically, where the isomorphism is an extension of the canonical embedding $\chi : X' \otimes Y \to \mathcal{L}(X, Y)$ given by $\chi(\sum x_i' \otimes y_i) = \sum\langle\cdot, x_i'\rangle\ y_i$.

We will not repeat the proof here but only mention that the approximation property has to be used twice in this proof: It is used to get the injectivity of the map $X'\tilde{\otimes}_\pi Y$ into $\mathcal{L}(X, Y)$ and to get the isometry $\alpha_{p,q}(z) = \alpha_{p,q}(T_z)$ for $z = \sum_{i=1}^{n} x_i' \otimes y_i \in X' \otimes Y$ and $T_z(x) = \sum_{i=1}^{n}\langle x, x_i'\rangle y_i$.

Now we can prove the main result of this section:

THEOREM 4.2. *Let E and F be any Fréchet-spaces satisfying one of the following conditions*

(i) *E and F belong to $sp(\mathcal{L}_2)$.*

(ii) *E and F are Fréchet-Montel spaces belonging to $sp(\mathcal{L}_\infty)$.*

(iii) *E and F are decomposable (FG) and Fréchet-Montel spaces.*

Then we have

(j) $E\tilde{\otimes}_{\alpha_{p,q}}F = (\mathcal{K}^s_{p,q}(E_b', F), \tau_{\alpha_{p,q}}) = (\mathcal{K}^w_{p,q}(E_b', F), \tau_{\alpha_{p,q}})$ *isometrically .*
In particularly, the elements $z \in E\tilde{\otimes}_{\alpha_{p,q}}F$ admit a representation

$$z = \sum_{i=1}^{\infty} \lambda_j x_j \otimes y_j$$

for some zero sequences (x_j), (y_j) and some $(\lambda_j) \in l_r$ such that

$$\|(\lambda_j)\|_r \cdot \mathbf{w}_{q',A}((x_j)) \cdot \mathbf{w}_{p',B}((y_j)) < \infty$$

for some $A \in \mathcal{B}(E)$, $B \in \mathcal{B}(F)$, *where*

$$\mathbf{w}_{q',A}((x_j)) = \sup_{x' \in A^0} \left(\sum_j |\langle x_j, x' \rangle|^{q'} \right)^{1/q'}$$

(jj) *A subset* $\mathcal{M}$ *of* $E\tilde{\otimes}_{\alpha_{p,q}}F$ *is bounded iff it is contained in a canonical way in some space* $E_A\tilde{\otimes}_{\alpha_{p,q}}F_B$ *with* $A \in \mathcal{B}(E)$ *and* $B \in \mathcal{B}(F)$ *as a bounded subset. In particularly, bounded subsets of* $E\tilde{\otimes}_\pi F$ *are contained in the closed absolutely convex hull of some set* $A \otimes B$.

Proof: Under the assumptions (i) and (iii) the spaces E and F admit fundamental systems $\mathcal{U}_0(E)$ and $\mathcal{U}_0(F)$ of neighborhoods such that the spaces E_U and F_V have the approximation property for $U \in \mathcal{U}_0(E)$ and $V \in \mathcal{U}_0(F)$: For Hilbertizable spaces this is obvious and for (FG)-spaces of Montel type the projections P_j are of finite rank, as it was mentioned above. Thus we can apply the Proposition 4.1, and this yields

$$E_U\tilde{\otimes}_{\alpha_{p,q}}F_V = \mathcal{K}_{p,q}(E'_{U^0}, F_V) .$$

Therefore, we have topologically

$$\mathcal{K}^w_{p,q}(E'_b, F) = \operatorname*{proj}_{U,V} \mathcal{K}_{p,q}(E'_{U^0}, F_V) = \operatorname*{proj}_{U,V} E_U\tilde{\otimes}_{\alpha_{p,q}}F_V = E\tilde{\otimes}_{\alpha_{p,q}}F ,$$

and finally $\mathcal{K}^s_{p,q}(E'_b, F) = E'_b\tilde{\otimes}_{\alpha_{p,q}}F$ by Proposition 3.3. This Proposition implies also the statement (jj). Under the assumption (ii) the spaces E and F are at least representable as projective limits of $L_\infty(\mu)-$spaces, and this allows the repetition of the above proof with a slight modification and one more use of Proposition 3.1b. ∎

A pair (E, F) of locally convex spaces satisfying (jj) for $\pi = \alpha_{1,1}$ is said to have the BB-*property*. The question wether all pairs of Fréchet-Montel spaces have the BB-property was stated by Grothendieck as the "problème des topologies" and answered in the negative by J. Taskinen in [13]. Recently it was recovered that the BB-property is of importance in both, in the theory of algebras of unbounded operators the describe some natural topologies [11] and in the theory of holomorphic functions on Fréchet-Montel spaces to describe the coincidence of certain topologies [7]. This caused a great number of papers studying particular cases for an affirmative answer to the BB-property. In a very recent paper by A. Defant and M. Maestre [5] the conditions (i)-(iii) of Theorem 4.2 were used to get a positive answer to this problem in the case of the $\pi-$tensor product. The condition (ii) appears there for the first time. The method they used was the technique of tensor products, especially the so-called desintegration theorem.

Acknowledgements. I am very grateful to Andreas Braunß for reading the manuscript and to Andreas Defant for valuable remarks concerning the final part of this paper.

REFERENCES

[1] J.Bonet and J.C.Diaz, The problem of topologies of Grothendieck and the class of Fréchet T-spaces, (to appear in *Math. Nachrichten*)

[2] J.Bonet, J.C.Diaz and J. Taskinen, Tensor stable Fréchet and (DF)-spaces, (to appear).

[3] A.Defant and K.Floret, Aspects of the metric theory of tensor products and operator ideals, *Note di Mathematica 8*(1988).

[4] A.Defant and K.Floret, Topological tensor products and the approximation property of locally convex spaces, *Bull. Soc. R. Liège 58,1*(1989), 29-51.

[5] A.Defant and M.Maestre, Holomorphic functions and (BB)-property on Fréchet-Montel spaces, (to appear).

[6] S.Dineen, Holomorphic functions on Fréchet-Montel spaces, *J. Math. Anal. Appl. 163(2)*(1992), 581-587.

[7] S.Dineen, Holomorphic functions and the BB-property, (to appear).

[8] H.Junek, Locally convex spaces and operator ideals, *Teubner Text zur Math., Vol 56*, Leipzig 1983.

[9] H.Junek, Maximal Op*-algebras on DF-domains, *Zeitschr. Anal. Anw. 9(5)*(1990), 403-414.

[10] R.Hollstein, Inductive limits and ε-tensor products, *J. Reine Angew. Math. 319*(1980)38-62.

[11] K.D.Kürsten, Bounded sets in projective tensor products of hilbertizable locally convex spaces, *Studia Math. 99(3)*(1991), 185-198.

[12] A.Pietsch, Operator ideals, *Akademie- Verlag*, Berlin 1978.

[13] J.Taskinen, The projective tensor product of Frèchet-Montel spaces, *Studia Math. 91*(1988), 17-30.

Real Interpolation of Hardy Spaces on the Disk and on the Bidisc

S.V.KISLIAKOV Steklov Math. Institute, St. Petersburg branch, Fontanka 27, 191011, St. Petersburg, Russia

S U M M A R Y A very short proof of the result of P.Jones on the K-closedness of the couple $(H^1(\mathbf{T}), H^\infty(\mathbf{T}))$ is presented. It is shown also that the real interpolation method applied to the couple $(H^1(\mathbf{T}^2), H^\infty(\mathbf{T}^2))$ gives what is expected.

This paper is an extended version of the lecture given by the author to the functional analysis meeting in Essen in November 1991. We begin with recalling some standard definitions. Let (X, Y) be a couple of quasi-Banach spaces. The word "couple" means that the both spaces are embedded into a greater topological vector space, so that, in particular, one can consider sums of vectors lying in different spaces. A convenient tool to study the decompositions $w=x+y$ with $x\in X$, $y\in Y$ from the quantitative viewpoint is provided by the so called K-functional. For $w\in X+Y$ and $t>0$ the K-functional of (w, t) with respect to the couple (X, Y) is defined by

$$K(w, t; X, Y)=\inf\left\{\|x\|+t\|y\|:\ x\in X,\ y\in Y,\ x+y=w\right\}.$$

The real interpolation spaces $(X, Y)_{\theta,r}$, $0<r\leq\infty$, $0<\theta<1$ are defined in terms of this functional. Namely, $(X, Y)_{\theta,r}=\left\{w\in X+Y:\ \|w\|_{\theta,r}<\infty\right\}$, where

$$\|w\|_{\theta,r}=\left(\int_0^\infty (t^{-\theta}K(w,\ t;\ X,\ Y))^r t^{-1}dt\right)^{1/r},\quad w\in X+Y$$

(for $r=\infty$ the usual modification is understood). We refer the reader to [BL] for the essentials on the real interpolation method. Here we recall only the well-known formula for the Lorentz spaces

$$(L^{p,s},\ L^{q,t})_{\theta,r}=L^{u,r},\quad \text{where } p\neq q,\ 1/u=(1-\theta)/p+\theta/q$$

So, $(L^p,\ L^q)_{\theta,u}=L^u$ if $1/u=(1-\theta)/p+\theta/q$.

Astonishingly, it was not until 1981 that the complete analogue of the last formula for the Hardy spaces H^p on the unit circle was proved. This was done by P.Jones. See his papers [J1 - J3], which in fact contain much more information (in particular, about the complex interpolation method as well). We recall that $H^p=H^p(\mathbf{T})$ is the closed linear span in $L^p(\mathbf{T})$ of the monomials z^j, j=0, 1, 2, ... (if $p=\infty$, the closure is taken in the weak* topology of L^∞).

Actually, Jones proved a result stronger than the H^p-version of the above formula. In the terminology introduced much later (cf. [P]) this result says that the couple $(H^p,\ H^q)$ is K-closed with respect to the couple $(L^p,\ L^q)$. In general, let (X, Y) be a couple of quasi-Banach spaces and let A, B be subspaces of X and Y. The smaller couple (A, B) is said to be K-closed with respect to the bigger one (X, Y) if there is a constant C such that for every $w\in A+B$ and every $t>0$ we have

$$K(w,\ t;\ A,\ B)\leqslant CK(w,\ t;\ X,\ Y).$$

(The reverse inequality with C=1 is always true). Clearly this estimate enables us to use the following simple recipe to identify the space $(A,\ B)_{\theta,r}$: apply the same functor $(\cdot,\ \cdot)_{\theta,r}$ to the bigger couple (X, Y) and intersect the result with A+B. In the case of H^p-spaces this means that K-closedness does infer the analogue of the L^p result quoted above.

It is worthwhile to restate the property of being K-closed without mentioning K-functionals. It simply means that whenever $w\in A+B$ is decomposed as $w=x+y$ with $x\in X$, $y\in Y$, we can find another decomposition $w=a+b$ with $a\in A$, $b\in B$ and $\|a\|\leqslant C\|x\|$, $\|b\|\leqslant C\|y\|$, C being independent of the vectors involved.

In the case of H^p spaces it is very easy to see that for $1<p,q<\infty$ the couple $(H^p,\ H^q)$ is K-closed with respect to $(L^p,\ L^q)$. Indeed, it is well-known that the orthogonal projection $\mathbf{P}$ of L^2 onto H^2 (called the Riesz projection) is bounded on L^p for $1<p<\infty$, and hence projects L^p onto H^p. Thus, given a decomposition $g+h=f\in H^p+H^q$ with $g\in L^p$, $h\in L^q$, we can simply apply $\mathbf{P}$ to obtain $f=\mathbf{P}g+\mathbf{P}h$, the summands being analytic (i.e. lying in some H^r each). Moreover, their norms in H^p and H^q are controlled by the norms of g and h, respectively, in L^p and L^q. Of course, this argument fails if the above condition on p and q is violated. However, if it is violated in the way that at least one of these numbers becomes $\leqslant 1$, but they remain finite, it is not difficult to reduce the problem to the case already considered by means of an argument based on factorization of H^p functions.

So the unique case that seems to present trouble is that of $\max(p, q)=\infty$. To the moment of Jones's work on the subject this had been, moreover, the unique unknown case. The method employed by Jones to consider this case was not elementary in the sense that a serious use was made (via some subtle estimates of solutions of the $\bar\partial$ equation) of the representation of H^p as a space of functions analytic in the unit disc rather than a space of functions on the circle, as defined above. It should be noted that the result on K-closedness of the couple (H^p, H^∞) with respect to the couple (L^p, L^∞) was not stated in [J1 - J3] explicitly, though in fact proved.

In the recent time some new proofs of the result just mentioned have appeared. See, e.g., the papers of Pisier [P] and Bourgain [B2]. One does not need to leave the unit circle in the course of these proofs, and so they are of more elementary nature. Also, they can be applied in many other situations (the arguments in [P] work, for example, for H^p with values in Schatten classes, whereas the arguments in [B2] apply to Hardy spaces on the ball of $\mathbf{C}^n$).

We are going to present here yet another very short and elementary proof of Jones's theorem. The author has a strong feeling that the argument should have been known since very long ago, but probably not in the interpolation context. More tricky variants of the same argument were used in [K3, K4, K-X1, K-X2] to embrace the case of weighted H^p-spaces on the circle, but the extremely simple form the argument takes when there is no weight was not given there explicitly. I am indebted to Gilles Pisier who convinced me that this might be of interest.

THEOREM 1 Let $0<p<\infty$. Then the couple (H^p, H^∞) is K-closed with respect to the couple (L^p, L^∞).

Proof: Let $f\in H^p+H^\infty=H^p$ be represented as $f=g+h$ with $g\in L^p$, $h\in L^\infty$. We have to produce another decomposition $f=g'+h'$ with g', h' analytic and of roughly the same norms as g, h. The desired functions will be found in the form $g'=(1-G)f$, $h'=Gf$, where G is some function in H^∞. Denote $t=\|h\|_\infty$.

Let us consider first the case p=2. Then we proceed as follows. Define $u=\max(1, t^{-1}|g|)$ and $G=(u+i\mathbf{H}u)^{-1}$, where **H** is the harmonic conjugation operator. Clearly $|G|\leqslant u^{-1}\leqslant 1$, and so $G\in H^\infty$ (since G is analytic). Thus the functions g' and

h' defined as above are analytic, and it only remains to estimate their norms. We have $|h'| = |G||f| \leq |G||g| + |G||h| \leq u^{-1}|g| + t \leq 2t$, by the definition of u. Let us show that the L^2 norm of g' is dominated by the L^2 norm of g (this will finish the proof). We have $g'=(1-G)g+(1-G)h$, and the first summand is already of correct order, because $|1-G| \leq 2$. Now

$$\|(1-G)h\|_2 \leq t\|1-G\|_2,$$

and, since $\mathbf{H}1=0$,

$$1-G=\frac{(1-u)+i\mathbf{H}(1-u)}{u+i\mathbf{H}u}.$$

Hence $|1-G| \leq u-1+|\mathbf{H}(u-1)|$ (recall that $u \geq 1$). It follows by the L^2 continuity of $\mathbf{H}$ that

$$\|1-G\|_2 \leq \|u-1\|_2+\|\mathbf{H}(u-1)\|_2 \leq 2\|u-1\|_2 \leq 2t^{-1}\|g\|_2$$

(the last estimate is valid in view of the pointwise inequality $0 \leq u-1 \leq t^{-1}|g|$). This completes the proof in the case p=2.

We note at once that the same reasoning applies whenever $1<p<\infty$ (because then $\mathbf{H}$ is bounded on L^p), but it clearly fails if p=1. The following modification works for p=1 (in fact for $p>\frac{1}{2}$). Define this time $u=(\max(1, t^{-1}|g|))^{\frac{1}{2}}$, and let F be an auxiliary H^∞-function defined as G above, but with this new u: $F=(u+i\mathbf{H}u)^{-1}$. Then let $G=1-(1-F^2)^2$. Clearly $|G| \leq \text{const}\,|F|^2 \leq \text{const}\ u^{-2}$, and hence $|h'| \leq \text{const}\,|g|\,u^{-2}+Ct \leq C't$, as before. It remains to majorize the L^1-norm of $(1-G)f$, that amounts to estimating the L^1-norm of 1-G. But $|1-G| \leq 2|1-F|^2$, whence $\|1-G\|_1 \leq 2(\|u-1\|_2+\|\mathbf{H}(u-1)\|_2)^2$, and so again the L^2 boundedness of $\mathbf{H}$ applies. The remaining calculations are left to the reader, as well as the case $p \leq \frac{1}{2}$.

We pass now to a discussion of related multidimensional problems, restricting ourselves to the Hardy classes in the complex balls and the polydiscs. As it has already been mentioned, satisfactory information for the case of the ball can be found in [B2]. It is proved there (among other things) that for $1<p<\infty$ the couples $(H^1(B_n), H^p(B_n))$ and $(H^p(B_n), H^\infty(B_n))$ are K-closed with respect to the couples of the corresponding L^q-spaces against the surface measure on the sphere in $\mathbf{C}^n$.

The situation with the polydiscs is less clear. Since we are going to discuss it in more detail, we recall the definition of H^p-spaces. As in the case of dimension 1 we prefer to deal with the boundary values of analytic functions on the frame of the polydisc, i. e. on the n-dimensional torus $\mathbf{T}^n$. The space $H^p(\mathbf{T}^n)$ is the closed linear span in $L^p(\mathbf{T}^n)$ (in the weak* topology if $p=\infty$) of the monomials $z_1^{k_1}\ldots z_n^{k_n}$ with $k_1, \ldots, k_n$ non-negative integers. We do not consider here the values of p strictly smaller than 1. If $1<p<\infty$, then the multiple Riesz projection $\mathbf{P}\otimes\mathbf{P}\otimes\ldots\otimes\mathbf{P}$ acts on L^p. So if p and q are both strictly between 1 and ∞, then the couple $(H^p(\mathbf{T}^n), H^q(\mathbf{T}^n))$ is K-closed with respect to the couple $(L^p(\mathbf{T}^n), L^q(\mathbf{T}^n))$, for trivial reasons. The same is true (but is no longer simple for $n>1$) if $p=1$ and q is finite, see the papers of Xu [X] and Müller [M].

It is unknown whether the couple $(H^p(\mathbf{T}^n), H^\infty(\mathbf{T}^n))$ is K-closed with respect to the couple $(L^p(\mathbf{T}^n), L^\infty(\mathbf{T}^n))$ for $n\geq 2$. But for n=2 it turns out to be possible to prove the weaker result mentioned in the summary which we are going to discuss now. Let us restrict ourselves to the following precise statement.

THEOREM 2 Let $1\leq p<r<\infty$ and let θ be defined by the equation $1/r=(1-\theta)/p+\theta/\infty$. Then $(H^p(\mathbf{T}^2), H^\infty(\mathbf{T}^2))_{\theta,r}=H^r(\mathbf{T}^2)$.

We note at once that the method described below does not apply if n>2. The scheme of the proof will be as follows. The statement will be reduced, with the help of some general results on the real interpolation method, to two concrete facts. One of them is the theorem of Müller and Xu mentioned above; it is needed only to include the case p=1. The main case p>1 will be reduced to a multiplicative inequality for quotient norms (LEMMA 1 below). This lemma goes back to [K1], though it was stated there in a slightly weaker form. The proper statement can be found in [B1] and [K2]. Since that time, however, it has become clear that the reason for which this inequality is true lies again in K-closedness of some Hardy type spaces (see LEMMA 2 below). To emphasise this (and also for the sake of completeness), we include a proof of LEMMA 1 based on LEMMA 2. The reader will easily recognize in the proof of LEMMA 2 the main idea of the above proof of THEOREM 1.

Let us start doing all that. First of all, we check that it is sufficient to prove THEOREM 2 for p>1 only. Indeed, suppose this has been done. Fix $1<p<q<\infty$. The supposition just made means that the scale $H^r(\mathbf{T}^2)$ admits a "nice" (i.e., L^r-like) interpolation when r ranges over the interval $[p, \infty)$. The result of Müller and Xu mentioned above implies that the same is true when r ranges over the interval $[1, q]$. Since these intervals overlap, we can, by a general result of Wolff (cf. [W]), "glue together" these two scales to obtain that the statement of THEOREM 2 holds for p=1 as well.

Note that this reasoning shows in fact that it is sufficient to prove THEOREM 2 for p=2 only. In the sequel we restrict ourselves to this case, though the argument will work for any $p>1$ without serious changes.

So let $2<r<\infty$. We need to check that $H^r(\mathbf{T}^2)=(H^2(\mathbf{T}^2), H^\infty(\mathbf{T}^2))_{\theta,r}$ if $1/r=(1-\theta)/2$. The interpolation space on the right clearly embeds into $(L^2(\mathbf{T}^2), L^\infty(\mathbf{T}^2))_{\theta,r}=L^r(\mathbf{T}^2)$; on the other hand, it consists of analytic functions. Thus this space is contained in $H^r(\mathbf{T}^2)$. To prove the reverse inclusion it is sufficient to verify that

$$H^r(\mathbf{T}^2)\subset(H^2(\mathbf{T}^2), H^\infty(\mathbf{T}^2))_{\theta,\infty} \quad . \tag{1}$$

Indeed, let $2<v<r<w<\infty$, and define s ant t by $1/v=(1-s)/2$ and $1/w=(1-t)/2$. It follows from (1) that

$$H^v(\mathbf{T}^2)\subset(H^2(\mathbf{T}^2), H^\infty(\mathbf{T}^2))_{s,\infty}, \quad H^w(\mathbf{T}^2)\subset(H^2(\mathbf{T}^2), H^\infty(\mathbf{T}^2))_{t,\infty} \quad .$$

Now define σ by $1/r=(1-\sigma)/v+\sigma/w$. Since $(1-\sigma)s+\sigma t=\theta$, we have by reiteration (cf. [BL, Theorem 3.5.3])

$$(H^v(\mathbf{T}^2), H^w(\mathbf{T}^2))_{\sigma,r}\subset (H^2(\mathbf{T}^2), H^\infty(\mathbf{T}^2))_{\theta,r}.$$

But the space on the left is just $H^r(\mathbf{T}^2)$. Indeed, v and w are strictly between 1 and ∞, hence by the boundedness of the double Riesz projection we can interpolate "as between L^p-spaces".

To prove (1) we first pass to the dual statement. For $1<p\leq\infty$, $H^p(\mathbf{T}^2)$ is a dual space. Its predual, that we denote by X_q $(1/p+1/q=1)$, can be identified with the quotient space of $L^q(\mathbf{T}^2)$ by its subspace

$$Y_q=\overline{\text{span}}\{z_1{}^k z_2{}^n: (k, n)\in\mathbf{Z}^2\setminus(\mathbf{Z}_-)^2\}$$

($\mathbf{Z}_-$ is the set of strictly negative integers). Denoting the norm in X_q by $[\ ]_q$, we agree that for a function h in L^q the symbol $[h]_q$ means the norm of the element of X_q generated by h.

Now we are ready to state the dual version of (1) (of course, s below is connected with r in (1) by the formula $1/s+1/r=1$). This duality is well-known (see [BL, Theorem 3.7.1, Theorem 3.5.2(b), and statement (b) after Definition 3.5.1], but anyway it can easily be established directly.

LEMMA 1 Let $1<s<2$ and let σ satisfy $1/s=(1-\sigma)/1+\sigma/2$. Then for every square summable function h on the torus $\mathbf{T}$ we have $[h]_s\leq[h]_1^{1-\sigma}[h]_2^{\sigma}$.

So it remains to prove LEMMA 1. It has already been mentioned that this lemma is a consequence of a certain K-closedness result that we are going to state now. A function on the circle **T** will be called analytic if it belongs to H^s for some s>0. We denote by $\underline{H}^{p,q}$ the subspace of the Lorentz space $L^{p,q}(\mathbf{T}^2)$ consisting of the functions analytic in the first variable (more precisely, this means that if we fix any value of the second variable off an exceptional set of measure 0, the resulting function in the first variable must be analytic). We write $\underline{H}^p$ for $\underline{H}^{p,p}$.

LEMMA 2 The couple $(\underline{H}^{1,\infty}, \underline{H}^2)$ is K-closed with respect to the couple $(L^{1,\infty}(\mathbf{T}^2), L^2(\mathbf{T}^2))$.

We first derive LEMMA 1 from LEMMA 2 and then prove LEMMA 2 itself. To do the first part of this job recall that the Riesz projection **P** is an operator of weak type (1, 1), i.e. it is bounded as an operator from L^1 to $L^{1,\infty}$. Let us denote by I the identity operator on functions in one variable and consider the operator $I\otimes\mathbf{P}$ on functions on the torus (this is nothing but the operator **P** acting in the second variable). It follows that $I\otimes\mathbf{P}$ is a bounded operator from $L^1(\mathbf{T}^2)$ to $L^{1,\infty}(\mathbf{T}^2)$. To see this just integrate in z over the unit circle the corresponding inequality for **P**:

$$t\left|\left\{|\mathbf{P}f(z,\ \cdot)|>t\right\}\right|\leqslant C\|f(z,\ \cdot)\|_{L^1(\mathbf{T})}.$$

Now we make the crucial observation that whenever $f\in Y_1$ (Y_1 has been defined in the course of the above discussion of preduals of Hardy spaces on the torus), the function $(I\otimes(I-\mathbf{P}))f$ is analytic in the first variable. Hence $I\otimes(I-\mathbf{P})$ maps Y_1 into $\underline{H}^{1,\infty}$. Clearly it also maps Y_2 into $\underline{H}^2$, by the L^2-continuity of **P**.

We are ready to derive LEMMA 1 from LEMMA 2. Let h be in $L^2(\mathbf{T}^2)$. Denote $a=[h]_1$, $b=[h]_2$. We can assume without loss of generality that $\|h\|_2\leqslant 2b$. There exists also a function g in $L^1(\mathbf{T}^2)$ such that $\|g\|_1\leqslant 2a$ and $g-h\in Y_1$.

If we could ensure g=h, LEMMA 1 would follow, because we would be able to write $[h]_s\leqslant\|h\|_s\leqslant\|h\|_1^{1-\sigma}\|h\|_2^{\sigma}$, by the Hölder inequality. So we are going to modify g and h to make them coincide, trying to retain the estimates for their norms. This will turn out to be possible if we pass from the L^1-norm to the weaker $L^{1,\infty}$-norm.

Set $F=(I\otimes(I-\mathbf{P}))(g-h)$, $G=(I\otimes(I-\mathbf{P}))g$, $H=(I\otimes(I-\mathbf{P}))h$, so that F=G-H. Clearly H defines the same class modulo Y_2 as h. By the continuity properties of $I\otimes(I-\mathbf{P})$ we conclude that $\|G\|_{1,\infty}\leqslant Ca$, $\|H\|_2\leqslant 2b$ for some numerical constant C. On the other hand, $F\in\underline{H}^{1,\infty}$, because $g-h\in Y_1$. By LEMMA 2 we can find $G'\in\underline{H}^{1,\infty}$ and $H'\in\underline{H}^2$ such that F=G'-H' and $\|G'\|_{1,\infty}\leqslant ca$, $\|H'\|_2\leqslant cb$ (c is another numerical constant). Now let h'=H-H'. Clearly h' is in the same class modulo Y_2 as h, so that, in particular, $[h]_s \leq \text{const}\|h'\|_s$. It follows from the definition of h' that $\|h'\|_2 \leq \|H\|_2+\|H'\|_2 \leq \text{const } b$. Also, since F=G-H=G'-H', we have G-G'=H-H', and so $\|h'\|_{1,\infty} \leq \|G-G'\|_{1,\infty} \leq \text{const } a$. By the Hölder type inequality involving Lorentz norms (which is well-known and easy to prove) we obtain

$$\|h'\|_s \leq \text{const } \|h'\|_{1,\infty}^{1-\sigma}\|h'\|_2^{\sigma} \leq \text{const } a^{1-\sigma}b^{\sigma}.$$

This completes the proof of LEMMA 1.

It remains to prove LEMMA 2. We note at once that an argument similar to what follows was used in [K-X2] to treat K-closedness problem for Hardy spaces of vector valued functions on the unit circle. In many concrete cases these spaces of vector valued functions can be considered as spaces of functions in two variables, but we emphasize that the space $\underline{H}^{1,\infty}$ from LEMMA 2 is n o t among them. Moreover, LEMMA 2 is in a sense one dimensional: the second variable almost does not interfere, though we must retain it all the time.

Consider the operator $\mathbf{H}\otimes I$ (the harmonic conjugation in the first variable). It can easily be checked (see the discussion of the operator $I\otimes\mathbf{P}$ above) that this operator is bounded on $L^p(\mathbf{T}^2)$ for $1<p<\infty$ and acts from $L^1(\mathbf{T}^2)$ into $L^{1,\infty}(\mathbf{T}^2)$. By interpolation, it is bounded also on the space $L^{p,q}(\mathbf{T}^2)$, $1<p<\infty$, $1\leq q\leq\infty$. (We shall need the case p=2, q=∞).

Now let a function $f\in\underline{H}^{1,\infty}$ be decomposed as f=g+h with $g\in L^{1,\infty}(\mathbf{T}^2)$ and $h\in L^2(\mathbf{T}^2)$. To prove LEMMA 2 we need to find another decomposition f=g'+h', where g' and h' are analytic in the first variable and have roughly the same norms as g and h. As in the proof of THEOREM 1, the desired decomposition will be found in the form g'=(1-G)f, h'=Gf, where G is a certain function in $\underline{H}^\infty$. First we need to construct a majorant for h with nice properties (examining carefully the both proofs, the reader will see that in the proof of THEOREM 1 the role of this majorant was played by the constant function everywhere equal to $\|h\|_\infty$).

SUBLEMMA There exists a function $B\in L^2(\mathbf{T}^2)$ bounded away from zero and verifying the following properties: (i) $B\geq|h|$; (ii) $\|B\|_2\leq c\|h\|_2$; (iii) for $b=B^{1/2}$ we have $|(\mathbf{H}\otimes I)b|\leq cb$. Here c is a numerical constant.

Proof: Take a small positive number t and set $b_0=(|h|+t)^{\frac{1}{2}}$ (t is needed only to ensure that B will be bounded away from 0). Then define by induction $b_{n+1}=|(\mathbf{H}\otimes I)b_n|$, $n\geq0$. Denoting by s the norm of $\mathbf{H}\otimes I$ as an operator on L^4, we obtain

$$\|b_n\|_4\leq s\|b_{n-1}\|_4\leq\ldots\leq s^n\|b_0\|_4=s^n\|\,|h|+t\|_2^{2}.$$

Hence the function

$$b=b_0+(2s)^{-1}b_1+(2s)^{-2}b_2+\dots$$

is well defined (the series converges in L^4). It is easily seen that $B=b^2$ enjoys the desired properties if t is small enough.

Now we define $u=(\max(1,\ b^{-1}|g|))^{\frac{1}{2}}$ and

$$F=(b+i(\mathbf{H}\otimes I)b)/(bu+i(\mathbf{H}\otimes I)(bu)),\quad G=1-(1-F^2)^2.$$

Since b and u are bounded away from 0, it follows that F is analytic in the first variable. By (iii) in SUBLEMMA, $F\leqslant(1+c)b/bu\leqslant(1+c)/u$, whence, in particular, $|F|\leqslant 1+c$, $|G|\leqslant$ const. So F and G are in $\underline{H}^\infty$.

Now define $g'=(1-G)f$, $h'=Gf$. We have to estimate the norms of these functions in the corresponding spaces. Since $|G|\leqslant \text{const}\,|F|^2\leqslant Cu^{-2}$, we obtain

$$|h'|\leqslant Cu^{-2}|g|+\text{const}\,|h|\leqslant C'B,$$

by (i) and the definition of u. Hence $\|h'\|_2\leqslant C\|h\|_2$ by (ii).

It remains to check that $\|g'\|_{1,\infty}\leqslant C\|g\|_{1,\infty}$. Since $g'=(1-G)g+(1-G)h$ and the first summand is already of desired order, we only have to estimate the second one. But after elementary calculations we obtain

$$|1-G|\leqslant 2|1-F|^2\leqslant 2[(b(u-1)+|(\mathbf{H}\otimes I)(b(u-1))|)/bu]^2\leqslant C[(u-1)^2+|(\mathbf{H}\otimes I)(b(u-1))|^2/B].$$

Consequently, by (i)

$$\|(1-G)h\|_{1,\infty}\leqslant c\|(1-G)B\|_{1,\infty}\leqslant$$

$$C''(\|B(u-1)^2\|_{1,\infty}+\|[(\mathbf{H}\otimes I)(b(u-1))]^2\|_{1,\infty}).$$

Now note that for an arbitrary function v we have $\|v^2\|_{1,\infty}=\|v\|_{2,\infty}$. Since $\mathbf{H}\otimes I$ acts on $L^{2,\infty}$, it follows that the second summand in the last line of the above inequality is dominated by the first one. Since $B(u-1)\leqslant|g|$, we are done.

REFERENCES

[BL] Bergh, J. and Löfström, J. (1976). Interpolation spaces, Gründlerhen der Mathematischen Wissenschaften, 226, Springer-Verlag, Berlin, etc.

[B1] Bourgain, J. (1984). Bilinear forms on H and bounded bianalytic functions, Trans. Amer. Math. Soc., 286: 313-337.

[B2] Bourgain, J. (1991). Some consequences of Pisier's approach to interpolation, to appear.

[J1] Jones, P.W. (1981). "Interpolation between Hardy spaces", Conference in Harmonic Analysis in Honor of Antony Zygmund, part II, Chicago, pp. 437 -451.

[J2] Jones, P.W. (1983). L^∞ estimates of the $\bar\partial$ problem in a half-plane, Acta Math., 150: 138-152

[J3] Jones, P.W. (1984). "On interpolation between H^1 and H^∞", Interpolation spaces and allied topics in analysis, Proceedings of the conference held in Lund, Sweden, August 29 - September 1, 1983, Lecture Notes in Math., Springer-Verlag, 1070: 143-151.

[K1] Kisliakov, S.V. (1981). Fourier coefficients of boundary values of functions analytic in the disc and in the bidisc, Proceedings of V.A.Steklov Institute, 130: 77-94 (Russian).

[K2] Kisliakov, S.V. (1988). Fourier coefficients of continuous functions and a class of multipliers, Ann. Inst. Fourier, 38: 147-183.

[K3] Kisliakov, S.V. (1989). Truncating functions in weighted H^p and two theorems of J.Bourgain, Uppsala Univ. Dept. of Math., report 1989:10.

[K4] Kisliakov, S.V. (1991). Absolutely summing operators on the disc algebra, Algebra & Analysis, 3, no 4: 1-77 (Russian).

[K-X1] Kisliakov, S.V., Xu Quanhua (1991). Sur l'interpolation des espaces H^p. C. R. Acad. Sci. Paris, sér.1, 312: 249-254.

[K-X2] Kisliakov, S.V., Xu Quanhua (1991). Interpolation of weighted and vector-valued Hardy spaces, Pub. IRMA, Lille, 25: no 2.

[M] Müller, P. (1990). Holomorphic martingales and interpolation between Hardy spaces, Journ. d'Analyse Math., to appear.

[P] Pisier, G. (1990). Interpolation between H^p spaces and non-commutative generalizations, to appear.

[W] Wolff, T.W. (1982). A note on interpolation spaces, Lecture Notes in Math., Springer Verlag, 908: 199-204.

[X] Xu, Quanhua (1991). Some properties of the quotient space $L^1(\mathbf{T}^d)/H^1(\mathbf{T}^d)$, to appear.

Vector-Valued Lagrange Interpolation and Mean Convergence of Hermite Series

HERMANN KÖNIG[1] Mathematisches Seminar der Universität Kiel, Ludewig-Meyn-Str. 4, D-2300 Kiel 1, Germany.

Abstract

Let X be a Banach space and $1 \leq p < \infty$. We prove interpolation inequalities of Marcinkiewicz-Zygmund type for X-valued polynomials g of degree $\leq n$ on $\mathbb{R}$,

$$c_p \left(\sum_{i=1}^{n+1} \mu_i \| g(t_i) e^{-\frac{t_i^2}{2}} \|^p \right)^{\frac{1}{p}} \leq \left(\int_{\mathbb{R}} \| g(t) e^{-\frac{t^2}{2}} \|^p dt \right)^{\frac{1}{p}} \leq d_p \left(\sum_{i=1}^{n+1} \mu_i \| g(t_i) e^{-\frac{t_i^2}{2}} \|^p \right)^{\frac{1}{p}},$$

where $(t_i)_1^{n+1}$ are the zeros of the Hermite polynomial H_{n+1} and $(\mu_i)_1^{n+1}$ are suitable weights. The validity of the right inequality requires $1 < p < 4$ and X being an UMD-space. This implies a mean convergence theorem for the Lagrange interpolation polynomials of continuous functions on $\mathbb{R}$ taken at the zeros of the Hermite polynomials. In the scalar case, this improves a result of Nevai [N]. Moreover, we give vector-valued extensions of the mean convergence results of Askey-Wainger [AW] in the case of Hermite expansions.

[1]Supported by the Brazilian-German CNP$_q$ - GMD agreement and by DFG-grant Ko 962/3-1

1 Introduction and results

Marcinkiewicz and Zygmund [Z, chap.X] proved interpolation inequalities for trigonometric polynomials g of degree n of the form

$$1/3\left(\sum_{j=1}^{2n+1}|g(x_j)|^p/(2n+1)\right)^{1/p}\leq\left(\int_0^{2\pi}|g(x)|^p dx\right)^{1/p}\leq c_p\left(\sum_{j=1}^{2n+1}|g(x_j)|^p/(2n+1)\right)^{1/p}$$

where $x_j=\frac{\pi j}{2n+1}$, $1<p<\infty$ and $c_p>0$ depends on p only. The left inequality is true for $p=1,\infty$ as well. A similar result holds for ordinary polynomials with weight function $(1-x^2)^{-1/2}$, the x_j's being replaced by the zeros of the Tchebychev polynomials, or, more generally, for Jacobi polynomial weights $(1-t)^{\alpha}(1+t)^{\beta}$ and corresponding zeros under suitable restrictions on p, provided that $1/(2n+1)$ is replaced by the weight sequence of the corresponding Gaussian quadrature formula, which is natural if one considers the case $p=2$. See Askey [A], König-Nielsen [KN].

We prove an analogue of this type of inequality in the case of the Hermite polynomials H_n, orthogonal with respect to $e^{-t^2}dt$ on $\mathbb{R}$. Let

$$h_n:=\pi^{-1/4}(2^n n!)^{-1/2}H_n,\quad \mathcal{H}_n(t):=h_n(t)e^{-t^2/2}$$

denote the L_2-normalized Hermite polynomials and Hermite functions, respectively. Thus

$$\int_{\mathbb{R}} h_n(t)h_n(t)e^{-t^2}dt=\int_{\mathbb{R}}\mathcal{H}_n(t)\mathcal{H}_m(t)dt,$$

see Szegö [Sz,5.5]. Let $t_1>\ldots>t_{n+1}$ denote the zeros of H_{n+1} and $\lambda_1,\ldots,\lambda_{n+1}$ the corresponding Gaussian quadrature weights. Thus, for any polynomial q of degree $\leq 2n+1$

$$\int_{\mathbb{R}} q(t)e^{-t^2}dt=\sum_{j=1}^{n+1}\lambda_j q(t_j). \tag{1}$$

The Lagrange functions l_j, $l_j(t):=h_{n+1}(t)/\left(h'_{n+1}(t_j)(t-t_j)\right)$, satisfy $l_j(t_i)=\delta_{ij}$. The weights λ_j can be calculated from

$$\lambda_j=2/h'_{n+1}(t_j)^2=1/\left(nh_n(t_j)^2\right)=\int_{\mathbb{R}}|l_j(t)|^2e^{-t^2}dt \tag{2}$$

$j=1,\ldots,n+1$, see Szegö [Sz, chap. 3.4, 5.5, 15.3]. Let $\mu_j:=\lambda_j e^{t_j^2}$. If necessary, we indicate the dependence on h_{n+1} by superscript, $t_j=t_j^{n+1},\lambda_j=\lambda_j^{n+1},\mu_j=\mu_j^{n+1}$. Recall that $t_1 \leq \sqrt{2n+3}$.

Let X be a Banach space and $L_p(\mathbb{R}; X)$ denote the space of (classes of) p-th power Bochner-integrable functions $f : \mathbb{R} \longrightarrow X$ with norm $\|f\|_p = (\int \mathbb{R} \|f(t)\|^p \, dt)^{1/p}$. Choose $1 < p < \infty$. A Banach space X is an UMD-space provided that the Hilbert-transform on $\mathbb{R}$

$$Hf(t) := p.v. \int_{\mathbb{R}} \frac{f(s)}{t-s} ds, \qquad f \in L_p(\mathbb{R}; X)$$

defines a bounded linear operator $H : L_p(\mathbb{R}; X) \longrightarrow L_p(\mathbb{R}; X)$. It is well-known that this holds for some $1 < p < \infty$ if and only if it holds for all $1 < p < \infty$, and the property is thus independent of $1 < p < \infty$, see J.Schwartz [S]. All $L_q(\mu)$-spaces where $1 < q < \infty$ or all reflexive Orlicz spaces are UMD-spaces, see Fernandez-Garcia [FG].

Given $n \in \mathbb{N}$ and a Banach space X, the X-valued polynomials of degree $\leq n$ will be denoted by $\Pi_n(X) = \{\sum_{j=0}^{n} x_j t^j | x_j \in X\}$. We let $\Pi_n = \Pi_n(\mathcal{K})$ for $\mathcal{K} \in \{\mathbb{R}, \mathbb{C}\}$. Recall that (t_j) were the zeros of H_{n+1}. The Marcinkiewicz–Zygmund interpolation inequalities for the Hermite polynomials then state:

THEOREM 1. *Let X be a Banach space and $1 \leq p \leq \infty$.*

(a) *The following are equivalent:*

(1) *There is $c_p > 0$ such that for all $n \in \mathbb{N}$ and for all $q \in \Pi_n(X)$*

$$\left(\int_{\mathbb{R}} \left\| q(t) e^{-t^2/2} \right\|^p dt \right)^{1/p} \leq c_p \left(\sum_{j=1}^{n+1} \mu_j \left\| q(t_j) e^{-t_j^2/2} \right\|^p \right)^{1/p} \tag{3}$$

(2) *X is an UMD-space and $1 < p < 4$.*

(b) (1) *Let $0 < \delta < 1$. Then there is $c_\delta > 0$ such that for all $n \in \mathbb{N}$, $N := 2n+3$ and $q \in \Pi_m(X)$ where $m \leq 2n$,*

$$\left(\sum_{|t_j| \leq \delta\sqrt{N}} \mu_j \left\| q(t_j) e^{-t_j^2/2} \right\|^p \right)^{1/p} \leq c_\delta \left(\int_{\mathbb{R}} \left\| q(t) e^{-t^2/2} \right\|^p dt \right)^{1/p} \tag{4}$$

(2) *If X is an UMD-space and $4/3 < p \leq \infty$, there is $c_p > 0$ such that for all n and $q \in \Pi_n(X)$*

$$\left(\sum_{j=1}^{n+1} \mu_j \left\| q(t_j) e^{-t_j^2/2} \right\|^p\right)^{1/p} \leq c_p \left(\int_{\mathbb{R}} \left\| q(t) e^{-t^2/2} \right\|^p dt\right)^{1/p} \tag{5}$$

REMARKS. (i) As indicated, the μ_j's and t_j's depend on n as well. It seems likely that (5) holds for all $1 \leq p \leq \infty$ and all Banach spaces as in the Jacobi case; in (4) the terms involving the zeros with largest absolute value are omitted on the left.

(ii) The papers of Pollard [P], Askey-Wainger [AW] and Nevai [N] strongly suggest the choice of the weight function $e^{-p/2t^2}$ instead of e^{-t^2} for $p \neq 2$ if positive results are to be expected.

Inequalities of type (3) imply mean convergence results for interpolating polynomials. Given a continuous function $f : \mathbb{R} \longrightarrow X$, $I_n f := \sum_{j=1}^{n+1} f(t_j) l_j$ is the interpolating polynomial at the zeros of the Hermite polynomial H_{n+1}, $I_n f(t_j) = f(t_j)$.

Let $L_p\left(\mathbb{R}, e^{-p/2t^2}; X\right) = \{f : \mathbb{R} \longrightarrow X | g \in L_p(\mathbb{R}; X) \text{ where } g(t) = f(t) e^{-t^2/2}\}$.

THEOREM 2. *Let X be an UMD-space and $1 < p < 4$. Let $\alpha > 1/p$ and $f : \mathbb{R} \longrightarrow X$ be be a continuous function satisfying*

$$\lim_{|t| \to \infty} \|f(t)\|_X (1 + |t|)^\alpha e^{-t^2/2} = 0. \tag{6}$$

Then the interpolating polynomials $I_n f$ at the zeros of the Hermite polynomials H_{n+1} converge to f in the $L_p\left(\mathbb{R}, e^{-p/2t^2}; X\right)$-norm,

$$\left(\int_{\mathbb{R}} \left\| (f(t) - I_n f(t)) e^{-t^2/2} \right\|^p dt\right)^{1/p} \longrightarrow 0. \tag{7}$$

The same statement is false, in general, if $p > 4$.

REMARKS. (i) In the scalar case $X = \mathcal{K}$, this improves the result of Nevai [N] where $\alpha = 1$ is assumed. Nevai's result, however, holds for any $1 < p < \infty$.

(ii) Condition (5) clearly guarantees that $f \in L_p\left(\mathbb{R}, e^{-p/2t^2}; X\right)$. Under the weaker assumption that $f \in C(\mathbb{R}; X) \cap L_p\left(\mathbb{R}; e^{-p/2t^2}; X\right)$ convergence (7) does not hold, in general, as we show below.

Askey-Wainger [AW] prove their mean convergence result for the expansions of the L_p-functions on $\mathbb{R}$ into Hermite functions for $4/3 < p < 4$. Their proof generalizes to the vector-valued setting if X is an UMD-space. The necessity of the UMD condition of the following result is a consequence of theorem 1. Given $f \in L_p(\mathbb{R}; X)$, we let $a_j = \int_{\mathbb{R}} f(t)\mathcal{H}_j(t)\,dt$, $P_n f = \sum_{j=0}^{n} a_j \mathcal{H}_j$. We have:

THEOREM 3. *Let X be a Banach space and $1 \leq p \leq \infty$. The following are equivalent:*

(1) *For all $f \in L_p(\mathbb{R}; X)$, $P_n f \longrightarrow f$ in L_p-norm, i.e.*

$$\int_{\mathbb{R}} \|f(t) - P_n f(t)\|^p \, dt \longrightarrow 0.$$

(2) *X is an UMD-space and $4/3 < p < 4$.*

Work on this paper started during a visit of the University of Campinas. The author greatfully acknowledges the hospitality of the collegues there, in particular D.L.Fernandez and K.Floret.

2 Hermite asymptotics, zeros and quadrature weights

We need some estimates for the quadrature weights and the zeros of the Hermite polynomials. The zeros $t_1 > \ldots > t_{n+1}$ of H_{n+1} are in in the interval $\left(-\sqrt{N}, \sqrt{N}\right)$ where $N := 2n+3$; $t_{\left[\frac{n+2}{2}\right]}$ is the one closest to 0 and, more precisely, $\sqrt{N} - t_1 = \mathcal{O}\left(1/\sqrt[6]{N}\right)$, Szegö [Sz, 6.32]. Given sequences $(a_n)_{n\in\mathbb{N}}, (b_n)_{n\in\mathbb{N}}$ and $c > 0$, we write $a_n \underset{c}{\lesssim} b_n$ if $a_n \leq c\, b_n$ holds for all $n \in \mathbb{N}$. We write $a_n \underset{c}{\sim} b_n$ if $a_n \underset{c}{\lesssim} b_n$ and $b_n \underset{c}{\lesssim} a_n$. A similar notation will be used for real functions. Define $\Phi : [0,1] \longrightarrow \mathbb{R}$ by

$$2/3\ \Phi(x)^{3/2} = \int_x^1 \left(1 - s^2\right)^{1/2} ds, \qquad x \in [0,1]$$

One easily checks that $\Phi(x) \underset{c}{\sim} (1 - x^2)$ with $c = 2^{2/3}$ and that

$$|\Phi'(x)|^{-1/2} = \left(\Phi(x) \,/\, \left(1 - x^2\right)\right)^{1/4} \underset{d}{\sim} 1$$

with $d = 2^{1/6}$. Skovgaard's asymptotic formula for H_{n+1} yields that for $0 \leq t \leq \sqrt{N} - {}^{1/6}\!\sqrt{N}$

$$\mathcal{H}_{n+1}(t) = c_n / n^{1/12}\, |\Phi'\left(t/\sqrt{N}\right)|^{-1/2} \left[Ai\left(-N^{2/3}\,\Phi\left(\frac{t}{\sqrt{N}}\right)\right) + \mathcal{O}\left(\frac{1}{n^{7/6}\,\Phi\left(t/\sqrt{N}\right)^{1/4}}\right)\right]$$

where $\lim_n c_n = 2^{5/12}$, cf. Askey-Wainger [AW, p.700] (there is a misprint, it should be H_n instead of $\mathcal{H}_n$). Using the relation between Airy and Bessel functions

$$Ai(-z) = \sqrt{z}/3 \left(J_{1/3}(\zeta) + J_{-1/3}(\zeta)\right) = \frac{1}{\sqrt{\pi}} \frac{1}{\sqrt[4]{z}} \cos(\zeta - \pi/4) \left(1 + \mathcal{O}\left(\frac{1}{\zeta}\right)\right)$$

for large $\zeta := 2/3\, z^{3/2}$

$$\mathcal{H}_{n+1}(t) = \frac{g_n(t)}{n^{\frac{1}{8}} \left(\sqrt{N} - t\right)^{\frac{1}{4}}} \cos\left(\frac{2}{3} N\,\Phi\left(\frac{t}{\sqrt{N}}\right)^{\frac{3}{2}} - \frac{\pi}{4}\right) + \mathcal{O}\left(\frac{1}{n^{\frac{9}{8}} \left(\sqrt{N} - t\right)^{\frac{1}{4}}}\right) \tag{8}$$

where $g_n(t) \underset{c}{\sim} 1$ for some $c > 0$ independent of n and $t \le \sqrt{N} - {}^{1/6}\!\sqrt{N}$. We note that $N\,\Phi\left(t/\sqrt{N}\right)^{3/2} \sim n^{1/4}\left(\sqrt{N} - t\right)^{3/2}$.

Define $k_{n,j}$:=sup$\{|\mathcal{H}_{n+1}(t)| : t \in [t_{j+1}, t_j]\}$ for $n \in \mathbb{N}, j = 1, \dots, n$. By (8), there is $c > 0$ independent of n and j such that

$$k_{n,j} \underset{c}{\sim} n^{-1/8} \left(\sqrt{N} - |t_j|\right)^{-1/4}. \tag{9}$$

LEMMA 1. *There is a constant $c > 0$ such that for all $n \in \mathbb{N}$ and $j = 1, \dots, n$*

$$(t_j - t_{j+1}) \underset{c}{\sim} \mu_j = \frac{2}{n\mathcal{H}_n(t_j)^2} = \frac{1}{\mathcal{H}'_{n+1}(t_j)^2} \underset{c}{\sim} \frac{1}{n^{1/4}} \frac{1}{\sqrt{\sqrt{N} - |t_j|}}$$

The sequence $(\mu_j)_{j=1,\dots,[(n+2)/2]}$ is decreasing in j.

PROOF. The equalities follow from (2) and $\mu_j = \lambda_j e^{t_j^2}$. By symmetry we may assume $t_j \ge 0$, $j \le \left[\frac{n+2}{2}\right]$. $\mathcal{H}_{n+1}$ satisfies the differential equation

$$\mathcal{H}''_{n+1}(t) + \left(N - t^2\right) \mathcal{H}_{n+1} = 0, \qquad N = 2n + 3, \tag{10}$$

Szegö [Sz, 5.5]. Let $f(t) := \mathcal{H}'_{n+1}(t)^2 + (N - t^2)\,\mathcal{H}_{n+1}(t)^2$. Then $f'(t) = -2t\mathcal{H}_{n+1}(t)^2 \le 0$ for $t \ge 0$, i.e. f is decreasing. Hence $\mu_j = 1/\mathcal{H}'_{n+1}(t_j)^2 \ge 1/\mathcal{H}'_{n+1}(t_{j+1})^2 = \mu_{j+1}$, using $\mathcal{H}_{n+1}(t_i) = 0$. Denote $\nu_{n,j} := \sup\{|\mathcal{H}'_{n+1}(t)| : t \in [t_{j+1}, t_j]\}$. In view of (10), $\mathcal{H}_{n+1}$ is concave or convex in (t_{j+1}, t_j), depending on whether $\mathcal{H}_{n+1}$ is positive or negative there. Thus

$$\nu_{n,j} = \max\{|\mathcal{H}'_{n+1}(t_{j+1})|, |\mathcal{H}'_{n+1}(t_j)|\} = |\mathcal{H}'_{n+1}(t_{j+1})|.$$

Let $\bar{t} \in (t_{j+1}, t_j)$ be such that $\mathcal{H}'_{n+1}(\bar{t}) = 0$, i.e. $\kappa_{n,j} = |\mathcal{H}_{n+1}(\bar{t})|$. Using that f is decreasing, we find

$$\nu_{n,j}^2 = \mathcal{H}'_{n+1}(t_{j+1})^2 \geq \left(N - \bar{t}^2\right)\kappa_{n,j}^2 \geq \mathcal{H}'_{n+1}(t_j)^2 = \nu_{n,j-1}^2,$$

and using (9) and $N - \bar{t}^2 \sim \sqrt{n}\left(\sqrt{N} - t_j\right)$,

$$\nu_{n,j} \underset{c}{\sim} \kappa_{n,j} \cdot \sqrt[4]{n}\sqrt{\sqrt{N} - t_j} \underset{c}{\sim} n^{1/8}\left(\sqrt{N} - t_j\right)^{1/4}.$$

Hence $\kappa_{n,j} \cdot \nu_{n,j} \underset{d}{\sim} 1$ with c,d independent of n and $j = 1, \ldots, [(n+2)/2]$. Since $\mu_{j+1} = 1/\nu_{n,j}^2$ $(j \geq 2)$, the right estimate $\mu_j \sim n^{-1/4}\left(\sqrt{N} - t_j\right)^{-1/2}$ follows.

The mean value theorem, applied to $\mathcal{H}_{n+1}$ in (t_{j+1}, t_j), yields

$$t_j - t_{j+1} \geq \frac{\kappa_{n,j}}{\nu_{n,j}} = \frac{\kappa_{n,j} \cdot \nu_{n,j}}{\nu_{n,j}^2} \underset{d}{\sim} \frac{1}{\nu_{n,j}^2} = \mu_{j+1} \sim \mu_j. \tag{11}$$

For $t \in [-t_j, t_j]$, $N - t^2 \geq N - t_j^2$. Comparing the differential equation (10) in the interval $[-t_j, t_j]$ with

$$\mathcal{K}''(t) + \left(N - t_j^2\right)\mathcal{K}(t) = 0, \qquad t \in [-t_j, t_j],$$

Sturm's comparison principle yealds the converse to (11),

$$t_j - t_{j+1} \lesssim \frac{\pi}{\sqrt{N - t_j^2}} \lesssim \frac{\pi}{\sqrt[4]{n}\sqrt{\sqrt{N} - t_j}} \underset{c}{\sim} \mu_j.$$

□

More precise information on the constants involved can be found from (8), analyzing the zeros of the cosine-term. We do not need this. We note however, that for all j with $|t_j| < \delta\sqrt{N}$ with fixed $0 < \delta < 1$, one has $\mu_j \underset{c_\delta}{\sim} n^{-1/2}$, cf. Nevai [N]. In contrast to this, $\mu_1 \underset{c}{\sim} n^{-1/6}$. It is in the range in between that lemma 1 is of importance. As a corollary, we get

LEMMA 2. *Let $1 \leq p \leq 2$. Then there is $c > 0$ such that for all $n \in \mathbb{N}$*

$$\sum_{j=1}^{n+1} \mu_j^p \leq c n^{1-p/2} \quad \textit{(bounded for } p = 2\textit{)}$$

Proof. For $p=1$ see Nevai [N]. Let $\alpha := p-1$. Then by lemma 1

$$\sum_{j=1}^{n+1} \mu_j^p \underset{c_1}{\sim} \sum_{j=1}^{n} (t_j - t_{j+1})\, \mu_j^\alpha \underset{c_2}{\sim} \sum_{j=1}^{n} (t_j - t_{j+1}) \frac{1}{n^{\alpha/4}} \frac{1}{\left(\sqrt{N} - |t_j|\right)^{\alpha/2}}.$$

Since $1/\left(\sqrt{N} - t\right)$ is monotone for $t \geq 0$ and $(t_j - t_{j+1})$ is monotone in j (for the positive t_j's), the last Riemann sum can be replaced by an integral

$$\sum_{j=1}^{n+1} \mu_j^p \underset{c_3}{\sim} \frac{1}{n^{\alpha/4}} \int_{-\sqrt{N}}^{\sqrt{N}} \frac{dt}{\left(\sqrt{N} - |t|\right)^{\alpha/2}} \underset{c_4}{\sim} n^{(1-\alpha)/2} = n^{1-p/2}.$$

□

3 The interpolation inequalities

To prove theorem 1, we need two well-known facts about continuity in L_p, cf. Benedek-Murphy-Panzone [BMP] or Pollard [P].

Lemma 3. *Let X be a Banach space, $1 \leq p \leq \infty$, (Ω, μ) be a measure space and $k, r : \Omega^2 \longrightarrow \mathcal{K}$ be measurable such that*

$$\sup_u \int_\Omega |k(u,v)|\, |r(u,v)|^{p'}\, d\mu(v) \leq M, \quad \sup_v \int_\Omega |k(u,v)|\, |r(u,v)|^{-p}\, d\mu(u) \leq M \qquad (12)$$

Then $T_k f(u) := \int_\Omega k(u,v) f(v)\, d\mu(v)$ defines an operator $T_k : L_p(\Omega, \mu; X) \longrightarrow L_p(\Omega, \mu; X)$ of norm $\leq M$. Here $1/p + 1/p' = 1$.

This follows from an application of Hölder's inequality. One consequence is

Lemma 4. *Let X be a Banach space, $1 \leq p \leq \infty, b \in \mathbb{R}$ and $k : \mathbb{R}^2 \longrightarrow \mathbb{R}$ be defined by*

$$k(u,v) := \left| \left| \frac{u}{v} \right|^b - 1 \right| |u - v|^{-1}.$$

Then T_k is bounded as a map $T_k : L_p(\mathbb{R}; X) \longrightarrow L_p(\mathbb{R}; X)$ provided that $-1/p < b < 1 - 1/p$ (actually if and only if).

For the convenience of the reader, here is a sketch of the proof (cf. [BMP]):

Take $r(u,v) := |u/v|^{1/pp'}$. To check (12), substitute $v/u = t$ to find

$$\sup_{u(\neq 0)} \int_{\mathbb{R}} |k(u,v)\, | r(u,v)^{p'}\, dv = \int_{\mathbb{R}} |t^{-b} - 1|\, |t|^{-1/p} / |t-1|\, dt.$$

This is finite: integrability at 0 is assured since $b < 1 - 1/p$, integrability at $\pm\infty$ since $b > -1/p$. Note that for $t \longrightarrow 1$, the integrand tends to $|b|$. The second condition in (12) is checked similarly. □

Instead of using this simple lemma 4 below one could also apply the general theory of weighted singular integral operators with weights in the Muckenhaupt class A_p, see. Garcia-Cuerva, Rubio de Francia [GR, ch. IV].

PROOF OF THEOREM 1.

a) (2) ⇒ (1) We prove inequality (3) if X is an UMD-space and $1 < p < 4$. Let $q \in \Pi_n(X)$ and put $y_j := q(t_j)\, e^{-t_j^2/2} / \left(n^{1/8}\mathcal{H}'_{n+1}(t_j)\right)$. Then q concides with its interpolating polynomial

$$q(t) = I_n q(t) = \sum_{j=1}^{n+1} q(t_j)\, l_j(t) = \sum_{j=1}^{n+1} q(t_j) \frac{h_{n+1}(t)}{h'_{n+1}(t_j)(t - t_j)}$$

and we have to estimate

$$J := \left(\int_{\mathbb{R}} \left\| q(t)\, e^{-t^2/2} \right\|^p dt \right)^{1/p} = \left(\int_{\mathbb{R}} \left\| n^{1/8}\mathcal{H}_{n+1}(t) \sum_{j=1}^{n+1} \frac{y_j}{t - t_j} \right\|^p dt \right)^{1/p} \tag{13}$$

from above. Let $I_j := (t_j, t_{j-1})$, $|I_j| = (t_{j-1} - t_j)$ and χ_j be the characteristic function of I_j, for $j = 1, \ldots, n+1$ (with $t_0 := \sqrt{N}, N = 2n + 3$). The proof uses essentially that $1/(t - t_j)$ is close enough to the Hilbert transform of $-\chi_j/|I_j|$ at t which is

$$H\left(-\frac{\chi_j}{|I_j|}\right)(t) = \frac{1}{|I_j|} \ln \left| \frac{t - t_{j-1}}{t - t_j} \right| = \frac{1}{|I_j|} \ln \left| 1 - \frac{|I_j|}{t - t_j} \right|.$$

We claim that for all $j = 1, \ldots, n+1$

$$n^{1/8}|\mathcal{H}_{n+1}(t)| \left| \frac{1}{t - t_j} + H\left(\frac{\chi_j}{|I_j|}\right)(t) \right| \leq f_j(t), \quad t \in \mathbb{R} \tag{14}$$

where

$$f_j(t) = c_1 \min\left(\frac{1}{|I_j|}, \frac{|I_j|}{(t - t_j)^2} \right) |\sqrt{N} - |t||^{-1/4}.$$

By Skovgaard's asymptotic formula [AW, p.700] ((8) for $|t| < \sqrt{N}$)

$$n^{1/8}|\mathcal{H}_{n+1}(t)| \leq c_2|\sqrt{N} - |t||^{-1/4}, \quad t \in \mathbb{R}.$$

Thus, for $|t-t_j|>2|I_j|$, (14) follows from $|x-\ln(1+x)|\le x^2$ for $|x|\le 1/2$, i.e.

$$\left|\frac{1}{t-t_j}+H\left(\frac{\chi_j}{|I_j|}\right)(t)\right|\le\frac{|I_j|}{(t-t_j)^2}$$

For $|t-t_j|\le 2|I_j|$, one uses that $\mathcal{H}_{n+1}$ has zeros at t_j to get (14). We remark that for $j=1$, there is a singularity of the $H\left(\chi_1/|I_1|\right)$-term at $t_0=\sqrt{N}$ (where $\mathcal{H}_{n+1}$ has no zero to compensate); in this case use $|\ln|x||\le 4|x|^{-1/4}$ for $|x|\le 1$ and $|I_1|\sim n^{-1/6}$, $|\mathcal{H}_{n+1}(t)|\lesssim n^{-1/12}$ [Sz, 6.32], [AW] to find

$$n^{1/8}|\mathcal{H}_{n+1}(t)|\left|H\left(\frac{\chi_1}{|I_1|}\right)(t)\right|\le c_3/\left(|I_1||\sqrt{N}-|t|\,|^{1/4}\right)$$

for $|t-t_1|<2|I_1|$. Now (13) and (14) imply $J\le J_1+J_2$ where

$$J_1 = c_2\left(\int_{\mathbb{R}}\left\|H\left(\sum_{j=1}^{n+1}y_j\chi_j/|I_j|\right)(t)\right\|^p|\sqrt{N}-|t|\,|^{-p/4}dt\right)^{1/p}$$

$$J_2 = \left(\int_{\mathbb{R}}\left(\sum_{j=1}^{n+1}\|y_j\|\,f_j(t)\right)^p dt\right)^{1/p}.$$

We estimate the "main term" J_1 and the "error-term" J_2 seperately.

By lemma 4, the kernel $\frac{1}{|t-s|}\left||\frac{s}{t}|^{1/4}-1\right|$ defines a bounded operator in $L_p(\mathbb{R};X)$ for $1<p<4$. Since the Hilbert transform with kernel $1/(t-s)$ is bounded in $L_p(\mathbb{R};X)$ by assumption, so is the weighted Hilbert transform with kernel $|s/t|^{1/4}/(t-s)$. Replacing s and t by $\sqrt{N}-s$ and $\sqrt{N}-t$, we find that for $g\in L_p(\mathbb{R};X)$

$$\left(\int_{\mathbb{R}}\left\|\int_{\mathbb{R}}\frac{g(s)}{t-s}\left|\frac{\sqrt{N}-s}{\sqrt{N}-t}\right|^{1/4}ds\right\|^p dt\right)^{1/p}\le c_p\left(\int_{\mathbb{R}}\|g(s)\|^p ds\right)^{1/p},$$

c_p independent of g and N. Replacing t by $(-t)$ and g by g^-, $g^-(s)=g(-s)$, and then putting $f(s)=|\sqrt{N}-|s|\,|^{1/4}g(s)$, we find

$$\left(\int_{\mathbb{R}}\left\|\int_{\mathbb{R}}\frac{f(s)}{t-s}ds\right\|^p\left(\sqrt{N}-|t|\right)^{-p/4}dt\right)^{1/p}\le 2\,c_p\left(\int_{\mathbb{R}}\|f(s)\|^p|\sqrt{N}-|s|\,|^{-p/4}ds\right)^{1/p}.$$

Take $f:=\sum_{j=1}^{n+1}y_j\chi_j/|I_j|$ to estimate J_1

$$J_1 \le c_4\left(\int_{\mathbb{R}}\left\|\sum_{j=1}^{n+1}y_j\chi_j(s)/|I_j|\right\|^p|\sqrt{N}-|s|\,|^{-p/4}ds\right)^{1/p},\qquad c_4=2\,c_p\,c_2,$$

$$= \quad c_4 \left[\sum_{j=1}^{n+1} \|y_j\|^p / |I_j|^p \left(\int_{t_j}^{t_{j-1}} |\sqrt{N} - |s|\,|^{-p/4} ds \right) \right]^{1/p}$$

$$\underset{c_5}{\sim} \quad \left[\sum_{j=1}^{n+1} \|y_j\|^p / |I_j|^{p-1} \left(\sqrt{N} - |t_j| \right)^{-p/4} \right]^{1/p}$$

$$= \quad \left(\sum_{j=1}^{n+1} \left\| q(t_j)\, e^{-t_j^2/2} \right\|^p / \left(n^{p/8} |\mathcal{H}_{n+1}(t_j)|^p |I_j|^{p-1} \left(\sqrt{N} - |t_j| \right)^{p/4} \right) \right)^{1/p}$$

$$\sim \quad \left(\sum_{j=1}^{n+1} \mu_j \left\| q(t_i)\, e^{-t_j^2/2} \right\|^p \right)^{1/p}$$

using lemma 1: $n^{1/8} |\mathcal{H}'_{n+1}(t_j)|\, |I_j| \left(\sqrt{N} - |t_j| \right)^{1/4} \sim 1$, $|I_j| \sim \mu_j$.

To estimate J_2, we note that there is c_6 such that for all $t \in \left(\ell 2\sqrt{N}, (\ell+1) 2\sqrt{N} \right)$ and $\ell \in \mathbb{N}$: $f_j(t) \le c_4 \ell^{-9/4} f_j\left(t - 2\ell\sqrt{N} \right)$. Moreover for $t \in \left(\sqrt{N}, 2\sqrt{N} \right)$ and $s := 2\sqrt{N} - t \in \left(0, \sqrt{N} \right)$ one finds $f_j(t) \le f_j(s)$. Similar statements hold for $t < -\sqrt{N}$. Since $\sum_\ell \ell^{-9/4} < \infty$, this implies that there is c_7 such that

$$J_2 \quad \le \quad c_7 \left(\int_{-2\sqrt{N}}^{2\sqrt{N}} \left(\sum_{j=1}^{n+1} \|y_j\|\, f_j(t) \right)^p dt \right)^{1/p} \le 2c_7 \left(\int_{-\sqrt{N}}^{\sqrt{N}} \left(\sum_{j=1}^{n+1} \|y_j\|\, f_j(t) \right)^p dt \right)^{1/p}$$

$$\underset{c_8}{\sim} \quad \left\{ \sum_{k=1}^{n+1} |I_k| \left(\sum_{\substack{j=1 \\ j \ne k}}^{n+1} \|y_j\|\, |I_j| / (t_k - t_j)^2 + \|y_k\| / |I_k| \right)^p \left(\sqrt{N} - |t_k| \right)^{-p/4} \right\}^{1/p}$$

$$\le \quad \left(\sum_{k=1}^{n+1} \left(\sum_{\substack{j=1 \\ j \ne k}}^{n+1} \frac{|I_j| |I_k|^{1/p}}{(t_k - t_j)^2} \frac{\|y_j\|}{\left(\sqrt{N} - |t_k| \right)^{1/4}} \right)^p \right)^{1/p} +$$

$$+ \left(\sum_{k=1}^{n+1} \|y_k\|^p / |I_k|^{p-1} \cdot \left(\sqrt{N} - |t_k| \right)^{-p/4} \right)^{1/p}$$

$$=: \quad J_{21} + J_{22}.$$

The step involving c_8 is by discretization, decomposing $\left(-\sqrt{N}, \sqrt{N} \right)$ into the intervals I_j and using the definition of f_j. The term J_{22} is estimated as before; using the definition of y_j and lemma 1, we find

$$J_{21} \sim \left(\sum_{k=1}^{n+1} | \sum_{j=1}^{n+1} a_{kj} \left(\mu_j^{1/p} \left\| g(t_j)\, e^{-t_j^2/2} \right\| \right) |^p \right)^{1/p}$$

where $a_{kj} = \mu_j^{3/2-1/p} \mu_k^{1/2+1/p} / (t_k - t_j)^2$ for $j \ne k$ and $a_{kk} = 0$. To bound J_{21} by $M \left(\sum_{j=1}^{n+1} \mu_j \left\| g(t_j)\, e^{-t_j^2/2} \right\|^p \right)^{1/p}$, we have to show that $A = (a_{kj})_{k,j=1}^{n+1}$ has norm $\le M$ as a map from ℓ_p^{n+1} to ℓ_p^{n+1}, M being independent of $n \in \mathbb{N}$. To do so, use lemma 3 with $\Omega = \{1, \ldots, n+1\}, \mu\{j\} = 1, r_{kj} = (\mu_j/\mu_k)^{1/pp'}$. Calculation shows that the two conditions in (12) reduce to one condition , namely

$$\sup_{j=1,\ldots,n+1} \sum_{\substack{k=1 \\ k\neq j}}^{n+1} \mu_k^{3/2} \mu_j^{1/2} / (t_k - t_j)^2 \leq M.$$

We check this using lemma 1: $\mu_k \sim |t_k - t_{k+1}| \leq |t_k - t_j|$ for $k \neq j$. Thus we may replace discrete sums by integrals to find with constants independent of j and n

$$\sum_{\substack{k=1 \\ k\neq j}}^{n+1} \mu_k^{3/2} \mu_j^{1/2} / (t_k - t_j)^2 \leq c_9 \sqrt{\mu_j} \sum_{k\neq j} \mu_k / |t_k - t_j|^{3/2} \leq c_{10} \sqrt{\mu_j} \int\limits_{|t-t_j|\geq \mu_j} |t - t_j|^{-3/2} dt = 4c_{10} =: M.$$

The estimates for J_1 and J_{21}, J_{22} together yield inequality (3).

(b) (2) This statement is a dualization of the inequality just proved: Let $4/3 < p < \infty$ and X be an UMD-space. For any $q \in \Pi_n(X)$ there are functionals $\xi_j \in X^*$ with $\left(\sum_{j=1}^{n+1} \mu_j \left\| \xi_j e^{-t_j^2/2} \right\|_{X^*}^{p'} \right)^{1/p'} = 1$ such that

$$\left(\sum_{j=1}^{n+1} \mu_j \left\| q(t_j) e^{-t_j^2/2} \right\|^p \right)^{1/p} = \sum_{j=1}^{n+1} \mu_j \langle q(t_j), \xi_j \rangle e^{-t_j^2} = \sum_{j=1}^{n+1} \lambda_j \langle q(t_j), \xi_j \rangle =: I.$$

Let $r := \sum_{j=1}^{n+1} \xi_j \ell_j \in \Pi_n(X^*)$. Then $\langle q, r \rangle \in \Pi_{2n}$ is integrated exactly by Gaussian quadrature, and Hölder's inequality yields

$$\begin{aligned} I &= \int\limits_{\mathbb{R}} \langle q(t), r(t) \rangle e^{-t^2} dt \\ &\leq \left(\int\limits_{\mathbb{R}} \left\| q(t) e^{-t^2/2} \right\|^p dt \right)^{1/p} \left(\int\limits_{\mathbb{R}} \left\| r(t) e^{-t^2/2} \right\|_{X^*}^{p'} dt \right)^{1/p'} \end{aligned}$$

Since X^* is an UMD-space as well and $1 < p' < 4$, we have by (3)

$$\left(\int\limits_{\mathbb{R}} \left\| r(t) e^{-t^2/2} \right\|_{X^*}^{p'} dt \right)^{1/p'} \leq c_{p'} \left(\sum_{j=1}^{n+1} \mu_j \left\| \xi_j e^{-t_j^2/2} \right\|_{X^*}^{p'} \right)^{1/p'} = c_{p'}$$

(b) (1) We denote by $\mathcal{K}_j(t,s) = \sum_{i=0}^{j} \mathcal{H}_i(t) \mathcal{H}_i(s)$, $\mathcal{K}^m(t,s) = \frac{1}{m} \sum_{j=0}^{m-1} \mathcal{K}_j(t,s)$ the kernels of the orthogonal projection P_j onto $\Pi_j \cdot e^{-t^2/2} \subseteq L_2(\mathbb{R})$ and the first Césaro mean operator σ_m in $L_2(\mathbb{R})$, respectively. It was shown by Freud [F1] and independently by Poiani [Po] that

$$\sup_{m\in\mathbb{N}} \sup_{s\in\mathbb{R}} \int\limits_{\mathbb{R}} |\mathcal{K}^m(t,s)| dt \leq M \tag{15}$$

Let $0 < \delta < 1$. By Nevai [N,p.265] there is $c_\delta \geq 1$ such that for all $q \in \Pi_{4n}$

$$\sum_{|t_j| \leq \delta\sqrt{N}} \mu_j |q(t_j) e^{-t_j^2/2}| \leq c_\delta \int_{\mathbb{R}} |q(t) e^{-t^2/2}| dt.$$

We apply this to $q(t) e^{-t^2/2} = \mathcal{K}^m(t,s)$ for fixed $s \in \mathbb{R}$ and $m \leq 4n$ to get

$$\sup_{m \leq 4n} \sup_{s \in \mathbb{R}} \sum_{|t_j| \leq \delta\sqrt{N}} \mu_j |\mathcal{K}^m(t_j, s)| \leq c_\delta \cdot M \tag{16}$$

From (15) and (16), we find for $m \leq 4n$ and $f \in L_p(\mathbb{R}, X)$, $p = 1$ or ∞,

$$\sum_{|t_j| \leq \delta\sqrt{N}} \mu_j \|\sigma_m f(t_j)\| \leq \sup_s \sum_{|t_j| \leq \delta\sqrt{N}} \mu_j |\mathcal{K}^m(t_j, s)| \cdot \|f\|_{L_1(X)} \leq c_\delta M \|f\|_{L_1(X)}$$

$$\sup_{|t_j| \leq \delta\sqrt{N}} \|\sigma_m f(t_j)\| \leq \sup_j \int_{\mathbb{R}} |\mathcal{K}^m(t_j, s)| ds \cdot \|f\|_{L_\infty(X)} \leq M \|f\|_{L_\infty(X)} \cdot$$

Hence, by the Riesz-Thorin interpolation theorem, for $f \in L_p(X)$, $m \leq 4n$

$$\left(\sum_{|t_j| \leq \delta\sqrt{N}} \mu_j \|\sigma_m f(t_j)\|^p \right)^{1/p} \leq c_\delta M \cdot \left(\int_{\mathbb{R}} \|f(t)\|^p dt \right)^{1/p}. \tag{17}$$

Replacing $c_\delta M$ by $3c_\delta M$, we may substitute the operator $v_{2n} := 2\sigma_{4n} - \sigma_{2n} = \frac{1}{2n} \sum_{j=2n}^{4n-1} P_j$ for σ_m in (17). However, $V_{2n} f = f$ for functions of the form $f(t) = q(t) e^{-t^2/2}$ where $q \in \Pi_{2n}(X)$. Thus for any $q \in \Pi_{2n}(X)$

$$\left(\sum_{|t_j| \leq \delta\sqrt{N}} \mu_j \left\| q(t) e^{-t^2/2} \right\|^p \right)^{1/p} \leq 3c_\delta M \left(\int_{\mathbb{R}} \left\| q(t) e^{-t^2/2} \right\|^p dt \right)^{1/p}$$

This ends the proof of (b) of theorem 1.

Remark. It is likely that (5) holds for all p and X; however, this proof does not work: for $m = (1+\epsilon) n$ with $\epsilon > 0$

$$\mu_1 |\mathcal{K}^m(t_1, t_1)| \underset{c\epsilon}{\sim} n^{-1/6} \cdot n^{1/2} = n^{1/3}$$

tends to ∞ with n, and thus (16) does not hold if the sum is extended over all $j = 1, \ldots, n+1$. On the other hand,

$$\sup_{m \leq n} \sup_{s \in \mathbb{R}} \sum_{j=1}^{n+1} \mu_j |\mathcal{K}^m(t_j, s)| \leq M \tag{18}$$

is correct; Freud's rather elegant proof of (15) in [F1] may be modified to yield (18) in the discrete case. The reason why this only works for $m \leq n$ is that the biorthogonality relations

$$\delta_{kl} = \int_{\mathbb{R}} h_k(t)\, h_l(t)\, e^{-t^2} dt = \sum_{j=1}^{n+1} \lambda_j h_k(t_j)\, h_l(t_j)$$

are used which in the discrete case only hold if $k + l \leq 2n+1$, i.e. essentially $k, l \leq n$ is satisfied. Instead of V_{2n} as above, one might take $\epsilon > 0$ and $V_\epsilon := \frac{1}{\epsilon}\sigma_n - \frac{1-\epsilon}{\epsilon}\sigma_{(1-\epsilon)n}$ (assuming WLOG that $\epsilon n \in \mathbb{N}$) to find an inequality of type (5) for polynomials of degree $\leq (1-\epsilon)\, n$ and of type (3) by dualization without assumptions on X and p. Thus one finds the

PROPOSITION. *Let $1 \leq p \leq \infty$, X be a Banach space and $0 < \epsilon < 1$. Then there is $c_\epsilon > 0$ such that for all $m, n \in \mathbb{N}$ with $m \leq (1-\epsilon)\, n$ and all $q \in \Pi_m(X)$*

$$c_\epsilon^{-1} \left(\sum_{i=1}^{n+1} \mu_i \left\| q(t_i)\, e^{-t_i^2/2} \right\|^p \right)^{1/p} \leq \left(\int_{\mathbb{R}} \left\| q(t)\, e^{-t^2/2} \right\|^p dt \right)^{1/p} \leq c_\epsilon \left(\sum_{i=1}^{n+1} \mu_i \left\| q(t_i)\, e^{-t_i^2/2} \right\|^p \right)^{1/p}$$

This should be compared with theorem 1. We now return to the one part of theorem 1 still to be proved.

(a) (1) ⇒ (2) Assume (1) holds. We have to show that necessarily $1 < p < 4$ holds and that X is an UMD-space. Choose $q = h_n$ in inequality (3). Since by (2), $|\mathcal{H}_n(t_j)| = \sqrt{2/(n\mu_j)}$, the right side in (3) reads

$$\left(\sum_{j=1}^{n+1} \mu_j |\mathcal{H}_n(t_j)|^p \right)^{1/p} = \sqrt{\frac{2}{n}} \left(\sum_{j=1}^{n+1} \mu_j^{1-p/2} \right)^{1/p} \leq \sqrt{\frac{2}{n}} \left(\sum_{j=1}^{n+1} \mu_j \right)^{1/p} \sup_j \mu_j^{-1/2}$$
$$\sim \quad n^{-1/2} n^{1/2p} n^{1/4} = n^{1/2p - 1/4}$$

by using lemma 1 again. For the left side, we find using only the asymptotic behavior of $\mathcal{H}_n$ near its maximum [AW]

$$\|\mathcal{H}_n\|_p \gtrsim n^{-1/6p - 1/12}.$$

Hence (3) requires $-\frac{1}{6p} - \frac{1}{12} \leq \frac{1}{2p} - \frac{1}{4}$, i.e. $p \leq 4$. For $p = 4$, a slightly more careful use of the formulas in [AW] shows

$$\|\mathcal{H}_n\|_4 \sim n^{-1/8} (\log n)^{1/4}$$

which is larger than $n^{-1/8}$ and excludes $p = 4$ as well.

We now show that (3) implies that the Hilbert matrix $A = (1/(i-j+1/2))_{i,j\in\mathbb{N}}$ defines a bounded operator $A : l_p(X) \longrightarrow l_p(X)$. Since A is not bounded in l_1, this excludes $p = 1$. A well-known approximation and scaling argument shows that the boundedness of A in $l_p(X)$ is equivalent to the boundedness of the Hilbert transform in $L_p(\mathbb{R}; X)$, i.e. X is an UMD-space. In this sense, A is a discrete version of the Hilbert transform. For $n \in \mathbb{N}$, we need the zeros $\left(t_j^{n+1}\right)$ of H_{n+1} and (t_i^n) of H_n. Let $J := \{j \mid |t_j^{n+1}| \le 1\}$, $I := \{i \mid |t_i^n| \le 1\}$. Take any system $(x_j)_{j\in J} \subseteq X$ and define $q \in \Pi_n$ by

$$q(t) := \sum_{j\in J} \left(\mu_j^{n+1}\right)^{-1/p} x_j e^{\left(t_j^{n+1}\right)^2/2} l_j^{n+1}(t),$$

where $l_i^{n+1} \in \Pi_n$, $l_i^{n+1}\left(t_k^{n+1}\right) = \delta_{jk}$ for $i, k = 1, \dots, n+1$. Note that $q(t_k) = 0$ for $k \notin J$. Thus using assumption (3) and inequality (4) of (b)(1) of theorem 1 – but with μ_i^n, t_i^n instead of μ_j^{n+1}, t_j^{n+1} – we find

$$\begin{aligned}\sum_{i\in I} \mu_i^n \left\| q(t_i^n)\, e^{-(t_i^n)^2/2} \right\|^p &\le c_1 \int_{\mathbb{R}} \left\| q(t)\, e^{-t^2/2} \right\|^p dt \\ &\le c_2 \sum_{j\in J} \mu_j^{n+1} \left\| q\left(t_j^{n+1}\right) e^{-\left(t_j^{n+1}\right)^2/2} \right\|^p,\end{aligned}$$

i.e.

$$\sum_{i\in I} \left\| \sum_{j\in J} a_{ij} x_j \right\|^p \le c_2 \sum_{j\in J} \|x_j\|^p .$$

where

$$\begin{aligned} a_{ij} &:= \left(\mu_i^n/\mu_j^{n+1}\right)^{1/p} l_j^{n+1}(t_i^n)\, e^{\left(t_j^{n+1}\right)^2/2} e^{-(t_i^n)^2/2} \\ &= \left(\mu_i^n/\mu_j^{n+1}\right)^{1/p} \mathcal{H}_{n+1}(t_i^n) / \left[\mathcal{H}_{n+1}'\left(t_j^{n+1}\right)\left(t_i^n - t_j^{n+1}\right)\right] \quad \text{for} \quad i \in I, j \in J.\end{aligned}$$

By (2), $1/|\mathcal{H}_{n+1}'\left(t_j^{n+1}\right)| = \sqrt{\mu_j^{n+1}/2}$. Using the recursive formulas of the H_n's, see Szegö [Sz, 5.5], $\mathcal{H}_n(t_i^n) = 0$, and again (2), one finds that

$$\begin{aligned} |\mathcal{H}_{n+1}(t_i^n)| &= \left| -\sqrt{\frac{n}{n+1}}\mathcal{H}_{n-1}(t_i^n) \right| = \frac{\sqrt{n/(n^2-1)}}{\sqrt{\mu_i^n}}, \\ a_{ij} &= \epsilon_i \delta_j \left(\frac{\mu_i^n}{\mu_j^{n+1}}\right)^{1/p-1/2} \frac{\sqrt{n/(2(n^2-1))}}{t_i^n - t_j^{n+1}},\end{aligned}$$

$\epsilon_i = \text{sgn}\ \mathcal{H}_{n+1}(t_i^n), \delta_j = \text{sgn}\ \mathcal{H}'_{n+1}\left(t_j^{n+1}\right)$. By lemma 1, $\mu_i^n \underset{c_3}{\sim} \frac{1}{n} \underset{c_3}{\sim} \mu_j^{n+1}$ for $i \in I$, $j \in J$. Hence $B = (b_{ij})$ with $b_{ij} = \frac{1}{\sqrt{N}(t_i^n - t_j^{n+1})}$, $N = 2n+3$ defines a map $B : l_p^{|J|}(X) \longrightarrow l_p^{|I|}(X)$ of norm $\le b_2 b_3$ bounded independently of $n \in \mathbb{N}$. Near zero, the asymptotic formula with $N = 2n + 3$

$$\mathcal{H}_{n+1}(t) = \frac{1}{\sqrt{\pi}}\left(\frac{2}{n}\right)^{1/4}\left[\cos\left(\sqrt{N}\,t - n\frac{\pi}{2}\right) + \frac{t^3}{6\sqrt{N}}\sin\left(\sqrt{N}t - n\frac{\pi}{2}\right) + \mathcal{O}\left(\frac{1}{n}\right)\right],$$

for the Hermite functions holds, [Sz, 8.22.6]. The zeros of $\mathcal{H}_{n+1}$ in $[-1, 1]$ may be determined from the cos-term up to an error of $\mathcal{O}\left(\frac{1}{n}\right)$, since $\left(\sqrt{N}t_j^{n+1} - n\frac{\pi}{2}\right)$ is determined up to $\mathcal{O}\left(\frac{1}{\sqrt{N}}\right)$. The zeros of $\mathcal{H}_{n+1}$ separate those of $\mathcal{H}_n$; for $\mathcal{H}_n$, $N = 2n + 3$ is replaced by $\tilde{N} = 2n + 1$ with (again) $\sqrt{N} - \sqrt{\tilde{N}} = \mathcal{O}\left(\frac{1}{\sqrt{N}}\right)$. Thus the difference $t_j^n - t_j^{n+1}$ is $\frac{\pi}{2\sqrt{N}} + \mathcal{O}\left(\frac{1}{n}\right)$ and

$$\left|\left(t_i^n - t_j^{n+1}\right) - \frac{\pi(i - j + 1/2)}{\sqrt{N}}\right| \le c_4/n; \qquad i \in I,\ j \in J$$

and thus

$$\left|b_{ij} - \frac{1}{\pi(i - j + 1/2)}\right| \le c_5 \frac{1}{(i - j + 1/2)^2} =: c_{ij}; \qquad i \in I,\ j \in J$$

where c_4, c_5 are independent of n, $i \in I$ and $j \in J$. Since $C = (c_{ij})_{i,j \in \mathbb{N}}$ is bounded as a map $l_p(X) \longrightarrow l_p(X)$, we conclude that the finite Hilbert matrix $\left(\frac{1}{i-j+1/2}\right)_{i \in I, j \in J}$: $l_p^{|J|}(X) \longrightarrow l_p^{|I|}(X)$ has norm independent of $n \in \mathbb{N}$, i.e. of I, J. Note that for $n \longrightarrow \infty$ $|I| \sim |J| \sim \sqrt{n} \longrightarrow \infty$. Thus A is continuous in $l_p(X)$. This ends the proof of theorem 1. □

REMARKS. (a) The proof shows that X has to be an UMD-space already if (3) only holds for those $q \in \Pi_n(X)$ with $q(t_j) = 0$ for all $|t_j| > 1$. (b) Theorem 1 shows that $\left(\int_{\mathbb{R}} |q(t)\, e^{-t^2/2}|^p dt\right)^{1/p}$ is essentially determined by the values of q in $\left(-\sqrt{N}, \sqrt{N}\right)$, see also Freud [F2]. This corresponds to the fact that, if $|q(\bar{t})|e^{-\bar{t}^2/2} = \max_{t \in \mathbb{R}} |q(t)|e^{-t^2/2}$ for $q \in \Pi_n$, then $|\bar{t}| \le \sqrt{N}$, as can be shown using Gaussian quadrature techniques. (c) For an UMD-space and $4/3 < p < 4$, the subspace $\Pi_n(X)$ of $L_p\left(\mathbb{R}, e^{-t^2/2p}; X\right)$ is uniformly isomorphic to $l_p^{n+1}(X)$, the maps

$$\begin{aligned} J_n : \Pi_n(X) &\longrightarrow l_p^{n+1}(X), \\ q &\longrightarrow \left(\mu_j^{1/p} e^{-t_j^2/2} q(t_j)\right)_{j=1}^{n+1} \end{aligned}$$

satisfy $\sup_n \|J_n\| \ \|J_n^{-1}\| \leq c_p < \infty$, interpolation essentially yields the Banach-Mazur distance. This probably also holds for $1 < p \leq 4/3$.

4 Mean convergence of interpolating polynomials and expansions

We now give the PROOF OF THEOREM 2 :

Let X be an UMD-space, $1 < p < 4$ and $\alpha > 1/p$. Define the norm

$$|||g||| := \sup_{t\in\mathbb{R}} \|g(t)\|_X (1+|t|)^\alpha e^{-t^2/2}$$

for those $g \in C(\mathbb{R};X)$ where this is finite. Take $f \in C(\mathbb{R};X)$ satisfying (6). Then $|||f||| < \infty$, and moreover, f can be approximated by polynomials $q_n \in \Pi_n(X)$ in $|||\cdot|||$, $|||f - q_n||| \longrightarrow 0$ ((6) allows the restriction to a finite interval where this clearly is possible).

Let $\|g\|_p := \left(\int_{\mathbb{R}} \|g(t)\|^p e^{-t^2/2p} dt\right)^{1/p}$. Since $q_n - I_n f \in \Pi_n(X)$ and $I_n f(t_j) = f(t_j)$, theorem 1 yields

$$\begin{aligned} \|f - I_n f\|_p &\leq \|f - q_n\|_p + \|q_n - I_n f\|_p \\ &\leq \left(\int_{\mathbb{R}} \frac{dt}{(1+|t|)^{\alpha p}}\right)^{1/p} |||f - q_n||| + \left(\sum_{j-1}^{n+1} \mu_j \left\|(q_n(t_j) - f(t_j)) e^{-t_j^2/2}\right\|^p\right)^{1/p} \end{aligned}$$

The integral is finite ($\alpha p > 1$), thus the first term tends to zero. The second term approaches zero just as well: using again $\mu_j \sim (t_j - t_{j+1})$, see lemma 1, we find

$$\begin{aligned} \left(\sum_{j=1}^{n+1} \mu_j \left\|(f(t_j) - q_n(t_j)) e^{-t_j^2/2}\right\|^p\right)^{1/p} &\leq |||f - q_n||| \cdot \left(\sum_{j=1}^{n+1} \mu_j/(1+|t_j|)^{\alpha p}\right)^{1/p} \\ &\leq c|||f - q_n||| \cdot \left(\int_{\mathbb{R}} dt/(1+|t|)^{\alpha p}\right)^{1/p}. \end{aligned}$$

Hence $\|f - I_n f\|_p \longrightarrow 0$.

For $p > 4$ the same statement does not hold, as we will show now. The example is an extension and modification of Nevai's [N]. Assume $p > 4$ and choose α with $1/p < \alpha < 1/4$. Consider the Banach spaces

$$C_0 := \left(\{f : \mathbb{R} \to \mathbb{R} \mid f \text{ continuous}, |f(t)|(1+|t|)^\alpha e^{-t^2/2} \to 0 \text{ for } |t| \longrightarrow \infty\}, |||\cdot|||\right)$$

$$L_p := \left\{ f : \mathbb{R} \to \mathbb{R} \mid \|f\|_p = \left(\int_{\mathbb{R}} \left(|f(t)| e^{-t^2/2} \right) dt \right)^{1/p} < \infty \right\}.$$

If theorem 2 would hold for this α and p, the interpolating operators $I_n : C_0 \longrightarrow L_p$, $f \longmapsto I_n f$ would be uniformly bounded by the Banach-Steinhaus theorem, $\sup_{n \in \mathbb{N}} \|I_n : C_0 \longrightarrow L_p\| = M < \infty$.

Let $\epsilon_j :=\text{sgn}\, \mathcal{H}'_{n+1}(t_j)$, $J := \{j \mid t_j \leq 0\}$ and define

$$f(t_j) := \left\{ \begin{array}{ccc} \epsilon_j (1 + |t_j|)^{-\alpha} e^{t_j^2/2} & : & j \in J \\ 0 & : & j \notin J \end{array} \right\} .$$

This obviously can be extended to define a continuous function $f : \mathbb{R} \longrightarrow \mathbb{R}$ with $|||f||| = 1$. We show that $I_n f$ is bounded from below by a suitable multiple of h_{n+1}, for $0 \leq t \leq \sqrt{N}$: Then $0 \leq t - t_j \leq 2\sqrt{N}$ for $j \in J$. Using (2), $\mu_j \geq c_1/\sqrt{n}$ (lemma 1) and $|J| \sim n/2$, we find

$$\begin{aligned} |I_n f(t)| e^{-t^2/2} &= \sum_{j \in J} \frac{|\mathcal{H}_{n+1}(t)|}{|\mathcal{H}'_{n+1}(t_j)| (1 + |t_j|)^{\alpha} (t - t_j)} \geq c_2 \sum_{j \in J} \sqrt{\mu_j} n^{-(1+\alpha)/2} |\mathcal{H}_{n+1}(t)| \\ &\geq c_3 n^{-1/4-\alpha/2} |\mathcal{H}_{n+1}(t)|, \qquad 0 \leq t \leq \sqrt{N}. \end{aligned}$$

Using Skovgaard's formula (8), $p > 4$ and $\alpha < 1/4$, we get for all $n \in \mathbb{N}$

$$\begin{aligned} M &\geq \|I_n\| \, |||f||| \geq \|I_n f\|_p \geq c_3 \, n^{1/4-\alpha/2} \left(\int_0^{\sqrt{N}} |\mathcal{H}_{n+1}(t)|^p dt \right)^{1/p} \\ &\sim n^{1/4-\alpha/2} n^{-1/12-1/6p} = n^{1/6-1/6p-\alpha/2} \geq n^{1/24-1/6p}. \end{aligned}$$

Thus $1/24 - 1/6p \leq 0$, $p \leq 4$, contradicting our assumption $p > 4$.
This proves theorem 2. □

In a similar way we may prove that the assumption

$$f \in C(\mathbb{R}) \cap L_p \left(\mathbb{R}; e^{-t^2/2} \right) =: Y$$

does not suffice, in general, to prove the convergence $I_n f \longrightarrow f$ in L_p-norm (Remark (ii) after theorem 2):

Assume this would hold. Introduce the norm

$$\|f\|_0 := \max\left(\sup_{t\in\mathbb{R}} |f(t)| e^{-t^2/2}, \left(\int_{\mathbb{R}} \left(|f(t)| e^{-t^2/2}\right)^p dt\right)^{1/p}\right), \qquad f \in Y$$

on Y. Then $(Y, \|\cdot\|_0)$ is a Banach space and by Banach-Steinhaus we would have $\|I_n f\|_p \le M\|f\|_0$. Define $f \in Y$ as above, with $\alpha = 0$, and $\|f\|_0 = 1$. By the above estimates (for all $1 \le p < \infty, n \in \mathbb{N}$)

$$M \ge \|I_n f\|_p \ge c_4 n^{1/4} \left(\int_0^{\sqrt{N}} |\mathcal{H}_{n+1}(t)|^p dt\right)^{1/p}.$$

The right side is of order $n^{1/6p'}$ for $p > 4$, $n^{1/8}(\log n)^{1/4}$ for $p = 4$ and $n^{1/2p}$ for $p < 4$. In any of these cases, this contradicts the uniform boundedness by M.

□

It remains to give the PROOF OF THEOREM 3 :

Recall that we put $a_j = \int_{\mathbb{R}} f(t)\mathcal{H}_j(t)\,dt$, $P_n f = \sum_{j=0}^{n} a_j \mathcal{H}_j$ for $f \in L_p(\mathbb{R}; X)$.

(2) ⇒ (1) If X is an UMD-space and $4/3 < p < 4, P_n f \longrightarrow f$ in $L_p(\mathbb{R}; X)$. The scalar proof of Askey-Wainger $X = \mathbb{R}$ [AW] directy generalizes to the X-valued case, by using the boundedness of the Hilbert-transform and of the weighted Hilbert-transforms with kernels $\frac{1}{t-s}\left|\frac{t}{s}\right|^{\pm 1/4}$ also in the vector valued case of $L_p(\mathbb{R}; X)$ for $4/3 < p < 4$ (the latter folows using lemma 4).

(1) ⇒ (2) We assume that $P_n f \longrightarrow f$ for all $f \in L_p(\mathbb{R}; X)$. Already in the scalar case, $4/3 < p < 4$ is necessary [AW]. By the Banach-Steinhaus theorem,

$$\sup_{n\in\mathbb{N}} \|P_n : L_p(\mathbb{R}; X) \longrightarrow L_p(\mathbb{R}; X)\| = M < \infty.$$

Using this, we can dualize inequality (4). Let $0 < \delta < 1$. We claim that there is a_δ such that for all $q \in \Pi_n(X)$ with $q(t_j) = 0$ for $|t_j| > \delta\sqrt{N}$, one has

$$\left(\int_{\mathbb{R}} \left\|q(t)\, e^{-t^2/2}\right\|^p dt\right)^{1/p} \le a_\delta \left(\sum_{j=1}^{n+1} \mu_j \left\|q(t_j)\, e^{-t_j^2/2}\right\|^p\right)^{1/p} \tag{19}$$

By remark (a) at the end of section 3, this will imply that X has to be an UMD-space. For such q, let $f(t) := q(t)\, e^{-t^2/2}$. Since P_n projects onto $\Pi_n(X) \cdot e^{-t^2/2}$, $P_n f = f$. Let

$\epsilon > 0$. Then there is $g \in L_p(\mathbb{R}; X^*)$, $\int_{\mathbb{R}} \|g(t)\|_{X^*}^{p'} = 1$, such that

$$(1-\epsilon)\left(\int_{\mathbb{R}} \left\|q(t)\,e^{-t^2/2}\right\|_X^p dt\right)^{1/p} = (1-\epsilon)\left(\int_{\mathbb{R}} \|f(t)\|_X^p\, dt\right)^{1/p} \leq \int_{\mathbb{R}} \langle g(t), f(t)\rangle dt$$
$$= \int_{\mathbb{R}} \langle g(t), P_n f(t)\rangle dt = \int_{\mathbb{R}} \langle P_n g(t), f(t)\rangle dt =: I$$

Since $P_n g(t)$ is of the form $r(t)\,e^{-t^2/2}$ for some $r \in \Pi_n(X^*)$, $\langle P_n g(t), f(t)\rangle = \langle r(t), q(t)\rangle e^{-t^2}$ is integrated exactly by Gaussian quadrature. Using $q(t_j) = 0$ for $|t_j| > \delta\sqrt{N}$, inequality (4) for X^* (being UMD as well) and $4/3 < p' < 4$, we find

$$\begin{aligned} I &= \sum_{|t_j| \leq \delta\sqrt{N}} \lambda_j \langle r(t_j), q(t_j)\rangle \\ &= \sum_{|t_j| \leq \delta\sqrt{N}} \mu_j \langle P_n g(t_j), f(t_j)\rangle \\ &\leq \left(\sum_{|t_j| \leq \delta\sqrt{N}} \mu_j \|P_n g(t_j)\|_{X^*}^{p'}\right)^{1/p'} \left(\sum_j \mu_j \|f(t_j)\|^p\right)^{1/p} \\ &\leq c_\delta \left(\int_{\mathbb{R}} \|P_n g(t)\|^{p'} dt\right)^{1/p'} \left(\sum_j \mu_j \left\|q(t_j)\,e^{-t_j^2/2}\right\|^p\right)^{1/p} \\ &\leq c_\delta M \left(\sum_j \mu_j \left\|q(t_j)\,e^{-t_j^2/2}\right\|^p\right)^{1/p} \end{aligned}$$

which gives (19). In the last step $\|P_n = P_n^* : L_{p'}(\mathbb{R}; X^*) \longrightarrow L_{p'}(\mathbb{R}; X^*)\| \leq M$ was used. This ends the proof of theorem 3. □

REFERENCES.

[A] R. Askey; Mean convergence of orthogonal series and Lagrange interpolation, Acta Math. Sci. Hungar. 23(1972), 71-85.

[AW] R. Askey, S. Wainger; Mean convergence of expansions in Laguerre and Hermite series, Amer. J. Math. 87(1965), 695-708.

[BMP] A.I. Benedek, E.R. Murphy, R. Panzone; Cuestions del analisis de Fourier, Notas de Algebra y Analisis 5(1974), Univ. Bahia Blanca, Argentinia.

[FG] D.L. Fernandez, J.B. Garcia; Interpolation of Orlicz-valued function spaces and U.M.D. property, Studia Math. 99(1991), 23-39.

[F1] G. Freud; Über die $(C, 1)$- Summen der Entwicklungen nach orthogonalen Polynomen, Acta Math. Sci. Hungar. 14(1963), 197-208.

[F2] G. Freud; On an inequality of Markov type, Soviet Math. Dokl. 12(1971), 570-573.

[GR] J. Garcia-Cuerva, J.L. Rubio de Francia; Weighted norm inequalities and related topies, North Holland, 1985.

[KN] H. König, N.J. Nielsen; Vector-valued L_p-convergence of orthogonal series and Lagrange interpolation, preprint, 1992.

[N] P.G. Navai; Mean convergence of Lagrange interpolation II, J. Appr. Th. 30(1980), 263-276.

[P] H. Pollard; Mean convergence of orthogonal series II, Transact. AMS 63(1948), 355-367.

[PO] E.L. Poiani; Mean Cesaro summability of Laguerre and Hermite series, Transactions AMS 173(1972), 1-31.

[S] J. Schwartz; A remark on inequalities of Calderon-Zygmund type for vector-valued functions, Comm. Pures Appl. Math. 14(1961), 785-799.

[Sz] G. Szegö; Orthogonal polynomials, AMS, Providence, 1959.

[Z] A. Zygmund; Trigonometric series, Cambridge Univ. Press, 1968.

Orthonormal Trigonometric Systems and Banach Space Geometry

ALBRECHT PIETSCH* and JÖRG WENZEL† Mathematisches Institut, Friedrich–Schiller–Universität Jena, Germany

Introduction

The main goal of functional analysis is to provide powerful tools for a unified treatment of differential and integral equations, integral transformations, expansions and approximations of functions, and various other topics. A basic idea consists in extending classical results about real or complex functions to operators acting between topological linear spaces. Another important goal of functional analysis is the classification of objects such as spaces and operators. Luckily enough, both of these goals are closely connected with each other.

A significant trend in Banach space theory is the search for numerical parameters which can be used to quantify special properties. For example, everybody would agree that Hilbert spaces are the most beautiful among all Banach spaces. Thus it is of importance to decide whether a given Banach space X can be renormed to become a Hilbert space. In the affirmative case, X is called *Hilbertian*. If such a renorming does not exist, then we may ask for a measure of non–Hilbertness. To what extent is the sequence space l_3 nearer to the Hilbert space l_2 than l_{1892}?

We now exemplify the preceding statements by asking whether Bessel's inequality holds also for functions f with values in a Banach space X. Do we have

$$\text{(b)} \qquad \left(\sum_{k=1}^{n} \left\| \frac{1}{2\pi} \int_{-\pi}^{+\pi} f(t) \exp(-ikt)\, dt \right\|^2 \right)^{1/2} \leq \left(\frac{1}{2\pi} \int_{-\pi}^{+\pi} \|f(t)\|^2\, dt \right)^{1/2} ?$$

*supported by DFG (Ko 962/3–1)

†supported by Studienstiftung des dt. Volkes and by DFG (Ko 962/3–1)

As observed by S. Bochner in 1933, this is not so in general; see [Boc33]. Later on, it became clear that the validity of (b), even if it is true for some fixed $n > 1$ only, characterizes Hilbert spaces isometrically; see [AMI86, p. 51].

An isomorphic counterpart of this criterion was established by S. Kwapień in 1972. He showed that X is Hilbertian if and only if there exists a constant $c \geq 1$ such that

$$\text{(B)} \qquad \left(\sum_{k=1}^{n} \left\| \frac{1}{2\pi} \int_{-\pi}^{+\pi} f(t) \exp(-ikt)\, dt \right\|^2 \right)^{1/2} \leq c \left(\frac{1}{2\pi} \int_{-\pi}^{+\pi} \|f(t)\|^2\, dt \right)^{1/2}$$

for all square integrable X-valued functions f and $n \in \mathbb{N}$. Now it is only a little step to fix n and ask for the smallest constant $c \geq 1$ such that (B) holds in a given Banach space X. Call, for a moment, this quantity $\varphi_n(X)$. Then we always have $\varphi_n(X) \leq n^{1/2}$. In particular,

$$\varphi_n(l_r) \asymp n^{|1/r-1/2|} \quad \text{for } 1 \leq r \leq \infty.$$

That observation suggests to associate with every exponent $0 \leq \varrho \leq 1/2$ the class F_ϱ consisting of all Banach spaces X for which the sequence $(\varphi_n(X))$ is dominated by a positive multiple of (n^ϱ). Then

$$l_r \in \mathsf{F}_\varrho \setminus \mathsf{F}_{\varrho-\varepsilon}$$

for $\varrho = |1/r - 1/2| \geq \varepsilon > 0$. Thus F_ϱ increases strictly with ϱ, and we have obtained a useful classification of Banach spaces. Remember that, by Kwapień's criterion, F_0 is the class of all Hilbertian Banach spaces.

There are various other ways to define quantities similar to $\varphi_n(X)$. For instance, given orthonormal systems $\mathcal{A}_n = (a_1, \ldots, a_n)$ and $\mathcal{B}_n = (b_1, \ldots, b_n)$ in Hilbert spaces $L_2(M, \mu)$ and $L_2(N, \nu)$, respectively, we can look for the smallest constant $c \geq 1$ such that

$$\left(\int_N \left\| \sum_{k=1}^{n} x_k b_k(t) \right\|^2 d\nu(t) \right)^{1/2} \leq c \left(\int_M \left\| \sum_{k=1}^{n} x_k a_k(s) \right\|^2 d\mu(s) \right)^{1/2},$$

where $x_1, \ldots, x_n$ range over a Banach space X. An obvious modification allows us to extend these concepts even to (bounded linear) operators acting from a Banach space X into a Banach space Y.

Of course, one may consider concrete orthonormal systems $\mathcal{A}_n$ and $\mathcal{B}_n$ consisting of Haar functions, Walsh functions, Jacobi polynomials, Rademacher functions, Gauss random variables, etc. In this survey, however, we restrict our attention to trigonometric systems only.

Concerning the lengthy proofs we refer the reader to a book, which is being prepared [P–W9?]; see also [Wen93] .

Finally, we express our hope that the general view presented in this paper yields more insights even in such well-known concepts like B-convexity, superreflexivity, the vector-valued Fourier transform and the vector-valued Hilbert transform.

1 Notation and basic concepts

Throughout, X and Y denote *Banach spaces* over $\mathbb{K}$ (synonym of the real field $\mathbb{R}$ or the complex field $\mathbb{C}$). We write L for the *class of all Banach spaces.* The *closed unit ball* of X is defined by $U_X := \{x \in X : \|x\| \leq 1\}$.

The *dual Banach space* X^* consists of all (bounded linear) *functionals* $x^* : X \to \mathbb{K}$. The value of x^* at $x \in X$ is denoted by $\langle x, x^* \rangle$, and we let

$$\|x^*\| := \sup\{|\langle x, x^* \rangle| : x \in U_X\}.$$

In the complex case, the linear structure of X^* is defined in such a way that $\langle x, x^* \rangle$ becomes conjugate linear in x^*, i. e.

$$\langle x, \alpha_1 x_1^* + \alpha_2 x_2^* \rangle = \overline{\alpha}_1 \langle x, x_1^* \rangle + \overline{\alpha}_2 \langle x, x_2^* \rangle$$

for all $x \in X$, $x_1^*, x_2^* \in X^*$ and $\alpha_1, \alpha_2 \in \mathbb{C}$. This unconventional definition is motivated by the wish to have analogy with the scalar products in Hilbert spaces.

Throughout, T denotes a (bounded linear) *operator* from X into Y. Its norm is given by

$$\|T\| := \sup\{\|Tx\| : x \in U_X\}.$$

The Banach space of all operators from X into Y is denoted by $\mathfrak{L}(X, Y)$. Moreover, $\mathfrak{L}$ stands for the *class of all operators* acting between arbitrary Banach spaces. This means that

$$\mathfrak{L} = \bigcup_{X,Y} \mathfrak{L}(X, Y),$$

where X and Y range over L.

For $T \in \mathfrak{L}(X, Y)$, the *dual operator* $T^* \in \mathfrak{L}(Y^*, X^*)$ is defined by

$$\langle x, T^* y^* \rangle = \langle Tx, y^* \rangle \quad \text{whenever } x \in X \text{ and } y^* \in Y^*.$$

Note that $(\alpha T)^* = \bar{\alpha} T^*$, in the complex case.

Let (M, μ) be a σ–finite measure space, and let $1 \leq p < \infty$. For every Banach space X, we denote by $[L_p(M, \mu), X]$ the Banach space of all strongly measurable functions $f : M \to X$ for which

$$\|f|L_p\| := \left(\int_M \|f(s)\|^p \, d\mu(s) \right)^{1/p}$$

is finite. If X is the scalar field $\mathbb{K}$, we simply write $L_p(M, \mu)$ instead of $[L_p(M, \mu), \mathbb{K}]$.

2 Ideal norms

An *ideal norm* $\boldsymbol{\alpha}$ is a function which assigns to *every* operator T between arbitrary Banach spaces a non-negative number $\boldsymbol{\alpha}(T)$ such that the following conditions hold:

(1) $\boldsymbol{\alpha}(S+T) \leq \boldsymbol{\alpha}(S) + \boldsymbol{\alpha}(T)$ for all $S, T \in \mathfrak{L}(X,Y)$.

(2) $\boldsymbol{\alpha}(BTA) \leq \|B\|\boldsymbol{\alpha}(T)\|A\|$ for all $A \in \mathfrak{L}(X_0, X)$, $T \in \mathfrak{L}(X,Y)$, $B \in \mathfrak{L}(Y, Y_0)$.

(3) $\boldsymbol{\alpha}(T) = 0$ implies $T = O$.

Sometimes it is necessary to indicate that the operator T acts from X into Y. In this case, we use the more precise notation $\boldsymbol{\alpha}(T : X \to Y)$.

Note that

$$\boldsymbol{\alpha}(\lambda T) = |\lambda|\boldsymbol{\alpha}(T) \quad \text{for all } T \in \mathfrak{L}(X,Y) \text{ and } \lambda \in \mathbb{K}.$$

To simplify matters, we use the symbol $\boldsymbol{\alpha}(X)$ instead of $\boldsymbol{\alpha}(I_X)$, where I_X denotes the identity map of the Banach space X.

Let $\boldsymbol{\alpha}$ be an ideal norm, and let $c > 0$ be a constant. We write

$$\boldsymbol{\alpha} \leq c \quad \text{if} \quad \boldsymbol{\alpha}(X) \leq c \quad \text{for all } X \in \mathsf{L}.$$

Then it follows from $\boldsymbol{\alpha}(T) \leq \|T\|\, \boldsymbol{\alpha}(X)$ that

$$\boldsymbol{\alpha}(T) \leq c\,\|T\| \quad \text{for all } T \in \mathfrak{L}.$$

Every ideal norm is equivalent to the operator norm. In particular, we have

$$\boldsymbol{\alpha}(I_{\mathbb{K}})\,\|T\| \leq \boldsymbol{\alpha}(T) \quad \textit{for all } T \in \mathfrak{L},$$

where $I_{\mathbb{K}}$ denotes the identity map of the 1–dimensional Banach space (scalar field).

Let $(\boldsymbol{\alpha}_n)$ and $(\boldsymbol{\beta}_n)$ be sequences of ideal norms. We write $\boldsymbol{\alpha}_n \asymp \boldsymbol{\beta}_n$ if there exists a constant $c > 0$ such that

$$\boldsymbol{\alpha}_n(T) \leq c\boldsymbol{\beta}_n(T) \quad \text{and} \quad \boldsymbol{\beta}_n(T) \leq c\boldsymbol{\alpha}_n(T) \quad \text{for all } T \in \mathfrak{L} \text{ and } n \in \mathbb{N}.$$

Then $\boldsymbol{\alpha}_n$ and $\boldsymbol{\beta}_n$ are said to be *uniformly equivalent.* That property is basic for the following considerations.

With every ideal norm $\boldsymbol{\alpha}$ we associate the *dual ideal norm* $\boldsymbol{\alpha}^*$ defined by

$$\boldsymbol{\alpha}^*(T) := \boldsymbol{\alpha}(T^*) \quad \text{for all } T \in \mathfrak{L}.$$

An ideal norm $\boldsymbol{\alpha}$ is called *symmetric* if $\boldsymbol{\alpha}^* = \boldsymbol{\alpha}$.
An ideal norm $\boldsymbol{\alpha}$ is said to be *injective* if

$$\boldsymbol{\alpha}(JT) = \boldsymbol{\alpha}(T) \quad \text{for all } T \in \mathfrak{L}(X,Y) \text{ and any metric injection } J \in \mathfrak{L}(Y, Y_0).$$

An ideal norm $\boldsymbol{\alpha}$ is said to be *surjective* if

$$\boldsymbol{\alpha}(TQ) = \boldsymbol{\alpha}(T) \quad \text{for all } T \in \mathfrak{L}(X,Y) \text{ and any metric surjection } Q \in \mathfrak{L}(X_0, X).$$

Here, by definiton, $J \in \mathfrak{L}(Y, Y_0)$ is a metric injection if $\|Jy\| = \|y\|$ for all $y \in Y$. A metric surjection $Q \in \mathfrak{L}(X_0, X)$ maps the open unit ball of X_0 onto the open unit ball of X; see [PIE78, B.3.2 and B.3.6].

If an ideal norm $\boldsymbol{\alpha}$ is injective (surjective), then $\boldsymbol{\alpha}^*$ is surjective (injective).

3 Ideal norms associated with orthonormal systems

In the following

$$\mathcal{A}_n = (a_1, \ldots, a_n) \quad \text{and} \quad \mathcal{B}_n = (b_1, \ldots, b_n)$$

denote arbitrary orthonormal systems in the Hilbert spaces $L_2(M, \mu)$ and $L_2(N, \nu)$, respectively.

For $x_1, \ldots, x_n \in X$, we write

$$\|(x_k)|\mathcal{A}_n\| := \left(\int\limits_M \left\| \sum_{k=1}^n x_k a_k(s) \right\|^2 d\mu(s) \right)^{1/2}.$$

This expression yields a norm on the n-th Cartesian power of X.

For $T \in \mathfrak{L}(X, Y)$, we denote by $\boldsymbol{\pi}(T|\mathcal{B}_n, \mathcal{A}_n)$ the smallest constant $c \geq 0$ such that

$$\text{(P)} \qquad \left(\int\limits_N \left\| \sum_{k=1}^n T x_k b_k(t) \right\|^2 d\nu(t) \right)^{1/2} \leq c \left(\int\limits_M \left\| \sum_{k=1}^n x_k a_k(s) \right\|^2 d\mu(s) \right)^{1/2}$$

whenever $x_1, \ldots, x_n \in X$.

More concisely, the preceding inequality reads as follows

$$\text{(P)} \qquad \|(Tx_k)|\mathcal{B}_n\| \leq c \, \|(x_k)|\mathcal{A}_n\|.$$

We refer to

$$\boldsymbol{\pi}(\mathcal{B}_n, \mathcal{A}_n) : T \longrightarrow \boldsymbol{\pi}(T|\mathcal{B}_n, \mathcal{A}_n)$$

as the *Parseval ideal norm* associated with the orthonormal systems $\mathcal{A}_n$ and $\mathcal{B}_n$.

We denote by $\boldsymbol{\delta}(T|\mathcal{B}_n, \mathcal{A}_n)$ the smallest constant $c \geq 0$ such that

$$\text{(D)} \qquad \left(\int\limits_N \left\| \sum_{k=1}^n T \langle f, a_k \rangle b_k(t) \right\|^2 d\nu(t) \right)^{1/2} \leq c \left(\int\limits_M \|f(s)\|^2 \, d\mu(s) \right)^{1/2}$$

whenever $f \in [L_2(M, \mu), X]$. Here

$$\langle f, a_k \rangle := \int\limits_N f(s) \, \overline{a_k(s)} \, d\mu(s)$$

denotes the k-th *Fourier coefficient* of f with respect to $\mathcal{A}_n$.

More concisely, the preceding inequality reads as follows

$$\text{(D)} \qquad \|(T\langle f, a_k \rangle) \, |\mathcal{B}_n\| \leq c \, \|f|L_2\|.$$

We refer to

$$\boldsymbol{\delta}(\mathcal{B}_n, \mathcal{A}_n) : T \longrightarrow \boldsymbol{\delta}(T|\mathcal{B}_n, \mathcal{A}_n)$$

as the *Dirichlet ideal norm* associated with the orthonormal systems $\mathcal{A}_n$ and $\mathcal{B}_n$.

The Dirichlet ideal norms enjoy a remarkable duality property, namely

$$\boldsymbol{\delta}(\mathcal{B}_n, \mathcal{A}_n) = \boldsymbol{\delta}^*(\mathcal{A}_n, \mathcal{B}_n).$$

The ideal norm $\boldsymbol{\delta}(\mathcal{B}_n, \mathcal{A}_n)$ is injective and surjective, while $\boldsymbol{\pi}(\mathcal{B}_n, \mathcal{A}_n)$ is, in general, injective only.

Dirichlet ideal norms behave well with respect to (complex) interpolation.

Let (X_0, X_1) and (Y_0, Y_1) be interpolation couples [B–L76, p. 25], *and assume that $T \in \mathfrak{L}(X_0 + X_1, Y_0 + Y_1)$ transforms X_0 into Y_0 and X_1 into Y_1. Then*

$$\boldsymbol{\delta}(T\colon [X_0, X_1]_\theta \to [Y_0, Y_1]_\theta | \mathcal{B}_n, \mathcal{A}_n) \le \boldsymbol{\delta}(T\colon X_0 \to Y_0 | \mathcal{B}_n, \mathcal{A}_n)^{1-\theta} \boldsymbol{\delta}(T\colon X_1 \to Y_1 | \mathcal{B}_n, \mathcal{A}_n)^{\theta}.$$

The following estimate holds

$$\|T\| \le \boldsymbol{\pi}(T|\mathcal{B}_n, \mathcal{A}_n) \le \boldsymbol{\delta}(T|\mathcal{B}_n, \mathcal{A}_n) \le n^{1/2}\|T\| \quad \text{for all } T \in \mathfrak{L}.$$

Let u_k be the characteristic function of the intervall $[k-1, k)$. Then

$$\mathcal{U}_n := (u_1, \dots, u_n)$$

is an orthonormal system in $L_2(\mathbb{R})$, where $\mathbb{R}$ is equipped with the Lebesgue measure.

Sometimes it is useful to replace $\mathcal{U}_n := (u_1, \dots, u_n)$ by the discrete analogue

$$\mathcal{U}_n^\circ := (u_1^{(n)}, \dots, u_n^{(n)})$$

which is formed by the unit vectors in l_2^n.

In both cases, we have

$$\|(x_k)|\mathcal{U}_n\| = \|(x_k)|\mathcal{U}_n^\circ\| = \|(x_k)|l_2^n\| = \left(\sum_{k=1}^n \|x_k\|^2\right)^{1/2}.$$

We now consider some specializations of the Parseval ideal norms defined above. Let $\mathcal{A}_n = (a_1, \dots, a_n)$ be any orthonormal system in a Hilbert space $L_2(M, \mu)$, and let $T \in \mathfrak{L}(X, Y)$.

$\boldsymbol{\pi}(T|\mathcal{A}_n, \mathcal{U}_n)$ is the smallest constant $c \ge 0$ such that

$$\text{(T)} \qquad \left(\int_M \left\|\sum_{k=1}^n T x_k a_k(s)\right\|^2 d\mu(s)\right)^{1/2} \le c \left(\sum_{k=1}^n \|x_k\|^2\right)^{1/2}$$

whenever $x_1, \dots, x_n \in X$.

$\boldsymbol{\pi}(T|\mathcal{U}_n, \mathcal{A}_n)$ is the smallest constant $c \ge 0$ such that

$$\text{(C)} \qquad \left(\sum_{k=1}^n \|T x_k\|^2\right)^{1/2} \le c \left(\int_M \left\|\sum_{k=1}^n x_k a_k(s)\right\|^2 d\mu(s)\right)^{1/2}$$

whenever $x_1, \dots, x_n \in X$.

More concisely, the preceding inequalities read as follows:

$$\text{(T)} \qquad \|(Tx_k)|\mathcal{A}_n\| \le c\,\|(x_k)|l_2^n\|,$$
$$\text{(C)} \qquad \|(Tx_k)|l_2^m\| \le c\,\|(x_k)|\mathcal{A}_n\|.$$

Next, we fix some terminology, which goes back to [M–P76].

The map

$$\boldsymbol{\pi}(\mathcal{A}_n, \mathfrak{U}_n) : T \longrightarrow \boldsymbol{\pi}(T|\mathcal{A}_n, \mathfrak{U}_n)$$

is called the *type ideal norm* associated with the orthonormal system $\mathcal{A}_n$.

The map

$$\boldsymbol{\pi}(\mathfrak{U}_n, \mathcal{A}_n) : T \longrightarrow \boldsymbol{\pi}(T|\mathfrak{U}_n, \mathcal{A}_n)$$

is called the *cotype ideal norm* associated with the orthonormal system $\mathcal{A}_n$.

It follows from $\boldsymbol{\delta}(X|\mathfrak{U}_n, \mathfrak{U}_n) = 1$ for all $X \in \mathsf{L}$ that

$$\boldsymbol{\delta}(\mathcal{A}_n, \mathfrak{U}_n) = \boldsymbol{\pi}(\mathcal{A}_n, \mathfrak{U}_n) \quad \text{and} \quad \boldsymbol{\delta}(\mathfrak{U}_n, \mathcal{A}_n) = \boldsymbol{\pi}^*(\mathcal{A}_n, \mathfrak{U}_n).$$

Therefore it is not necessary to consider the Dirichlet ideal norms $\boldsymbol{\delta}(\mathcal{A}_n, \mathfrak{U}_n)$ and $\boldsymbol{\delta}(\mathfrak{U}_n, \mathcal{A}_n)$.

Next, we compare the type and the cotype ideal norms. It easily follows that

$$\boldsymbol{\pi}(\mathfrak{U}_n, \mathcal{A}_n) \le \boldsymbol{\pi}^*(\mathcal{A}_n, \mathfrak{U}_n) \le \boldsymbol{\pi}(\mathfrak{U}_n, \mathcal{A}_n)\boldsymbol{\delta}(\mathcal{A}_n, \mathcal{A}_n)$$

and

$$\boldsymbol{\pi}^*(\mathfrak{U}_n, \mathcal{A}_n) \le \boldsymbol{\pi}(\mathcal{A}_n, \mathfrak{U}_n) \le \boldsymbol{\delta}(\mathcal{A}_n, \mathcal{A}_n)\boldsymbol{\pi}^*(\mathfrak{U}_n, \mathcal{A}_n).$$

The ideal norm $\boldsymbol{\pi}(\mathcal{A}_n, \mathfrak{U}_n)$ is injective and surjective, while $\boldsymbol{\pi}(\mathfrak{U}_n, \mathcal{A}_n)$ is, in general, injective only.

4 Orthonormal trigonometric systems

We write

$$e_k(t) := \exp(ikt) \quad \text{for } k \in \mathbb{Z},$$

$$c_k(t) := \sqrt{2}\cos kt \quad \text{and} \quad s_k(t) := \sqrt{2}\sin kt \quad \text{for } k \in \mathbb{N}.$$

Note that

$$\mathcal{E}_n := (e_1, \dots, e_n), \quad \mathcal{C}_n := (c_1, \dots, c_n) \quad \text{and} \quad \mathcal{S}_n := (s_1, \dots, s_n)$$

are orthonormal systems in $L_2(-\pi, +\pi)$ equipped with the scalar product

$$\langle f, g\rangle = \frac{1}{2\pi}\int_{-\pi}^{+\pi} f(t)\overline{g(t)}\,dt.$$

We have

$$\|(x_k)|\mathcal{C}_n\| = \left(\frac{2}{\pi}\int_0^\pi \left\|\sum_{k=1}^n x_k \cos kt\right\|^2 dt\right)^{1/2}$$

and

$$\|(x_k)|\mathcal{S}_n\| = \left(\frac{2}{\pi}\int_0^\pi \left\|\sum_{k=1}^n x_k \sin kt\right\|^2 dt\right)^{1/2}.$$

Next, discrete versions of the above trigonometric systems will be introduced. Letting

$$e_k^{(n)} := (\varepsilon_{1k}^{(n)}, \dots, \varepsilon_{nk}^{(n)}),$$

where

$$\varepsilon_{hk}^{(n)} := n^{-1/2} \exp\left(\tfrac{2\pi i}{n} hk\right),$$

we define the orthonormal system

$$\mathcal{E}_n^\circ := (e_1^{(n)}, \dots, e_n^{(n)})$$

in l_2^n.

Moreover, letting

$$c_k^{(n)} := (\gamma_{-nk}^{(n)}, \dots, \gamma_{0k}^{(n)}, \dots, \gamma_{+nk}^{(n)}) \quad \text{and} \quad s_k^{(n)} := (\sigma_{-nk}^{(n)}, \dots, \sigma_{0k}^{(n)}, \dots, \sigma_{+nk}^{(n)}),$$

where

$$\gamma_{hk}^{(n)} := \sqrt{\tfrac{2}{2n+1}} \cos \tfrac{2\pi}{2n+1} hk \quad \text{and} \quad \sigma_{hk}^{(n)} := \sqrt{\tfrac{2}{2n+1}} \sin \tfrac{2\pi}{2n+1} hk,$$

we obtain the orthonormal systems

$$\mathcal{C}_n^\circ := (c_1^{(n)}, \dots, c_n^{(n)}) \quad \text{and} \quad \mathcal{S}_n^\circ := (s_1^{(n)}, \dots, s_n^{(n)})$$

in the $(2n+1)$–dimensional Hilbert space $l_2(I_n)$ with $I_n := \{-n, \dots, 0, \dots, +n\}$. It turns out that

$$\|(x_k)|\mathcal{C}_n^\circ\| = \left(\frac{2}{2n+1}\left\|\sum_{k=1}^n x_k\right\|^2 + \frac{4}{2n+1}\sum_{h=1}^n \left\|\sum_{k=1}^n x_k \cos \tfrac{2\pi}{2n+1} hk\right\|^2\right)^{1/2}$$

and

$$\|(x_k)|\mathcal{S}_n^\circ\| = \left(\frac{4}{2n+1}\sum_{h=1}^n \left\|\sum_{k=1}^n x_k \sin \tfrac{2\pi}{2n+1} hk\right\|^2\right)^{1/2}.$$

We are now in a position to specialize the Parseval and Dirichlet ideal norms and, in particular, the type and cotype ideal norms to the case of the trigonometric systems. Note that the quantity $\varphi_n(X)$, defined in the introduction, is nothing but $\boldsymbol{\delta}(X|\mathcal{U}_n, \mathcal{E}_n)$.

5 The tensor product technique

We now describe a basic tool for proving the main results of this paper, which will be formulated in the next section.

Let $\mathcal{A}_n = (a_1, \dots, a_n)$ and $\mathcal{B}_n = (b_1, \dots, b_n)$ be orthonormal systems in the Hilbert spaces $L_2(M, \mu)$ and $L_2(N, \nu)$, respectively. We denote by $\mathcal{A}_n \otimes \mathcal{B}_n$ the orthonormal system in $L_2(M \times N, \mu \times \nu)$ formed by the functions $a_k \otimes b_k : (s,t) \to a_k(s)b_k(t)$ with $k = 1, \dots, n$. Note that

$$\|(x_k)|\mathcal{A}_n \otimes \mathcal{B}_n\| = \left(\int\limits_M \int\limits_N \left\| \sum_{k=1}^{n} x_k a_k(s) b_k(t) \right\|^2 d\mu(s)\, d\nu(t) \right)^{1/2}.$$

In particular, we have

$$\|(x_k)|\mathcal{A}_n \otimes \mathcal{U}_n\| = \|(x_k)|\mathcal{U}_n\|$$

for every orthonormal system $\mathcal{A}_n$.

If $\mathcal{F}_n = (f_1, \dots, f_n)$ *is another orthonormal system, then*

$$\boldsymbol{\pi}(\mathcal{B}_n \otimes \mathcal{F}_n, \mathcal{A}_n \otimes \mathcal{F}_n) \le \boldsymbol{\pi}(\mathcal{B}_n, \mathcal{A}_n).$$

The following results hold; see section 7:

$$\begin{array}{rclcrcl}
\boldsymbol{\pi}(\mathcal{E}_n \otimes \mathcal{E}_n, \mathcal{E}_n) & \le & 1, & \quad & \boldsymbol{\pi}(\mathcal{E}_n \otimes \mathcal{E}_n^\circ, \mathcal{E}_n) & \le & 1, \\
\boldsymbol{\pi}(\mathcal{C}_n \otimes \mathcal{C}_n, \mathcal{C}_n) & \le & \sqrt{2}, & & \boldsymbol{\pi}(\mathcal{C}_n \otimes \mathcal{C}_n^\circ, \mathcal{C}_n) & \le & \sqrt{2}, \\
\boldsymbol{\pi}(\mathcal{S}_n \otimes \mathcal{S}_n, \mathcal{C}_n) & \le & \sqrt{2}, & & \boldsymbol{\pi}(\mathcal{S}_n \otimes \mathcal{S}_n^\circ, \mathcal{C}_n) & \le & \sqrt{2}, \\
\boldsymbol{\pi}(\mathcal{C}_n \otimes \mathcal{S}_n, \mathcal{S}_n) & \le & \sqrt{2}, & & \boldsymbol{\pi}(\mathcal{C}_n \otimes \mathcal{S}_n^\circ, \mathcal{S}_n) & \le & \sqrt{2}, \\
\boldsymbol{\pi}(\mathcal{S}_n \otimes \mathcal{C}_n, \mathcal{S}_n) & \le & \sqrt{2}, & & \boldsymbol{\pi}(\mathcal{S}_n \otimes \mathcal{C}_n^\circ, \mathcal{S}_n) & \le & \sqrt{2}.
\end{array}$$

and

$$\begin{array}{rclcrcl}
\boldsymbol{\pi}(\mathcal{E}_n, \mathcal{E}_n \otimes \mathcal{E}_n) & \le & 1, & \quad & \boldsymbol{\pi}(\mathcal{E}_n, \mathcal{E}_n \otimes \mathcal{E}_n^\circ) & \le & 1, \\
\boldsymbol{\pi}(\mathcal{S}_n, \mathcal{S}_n \otimes \mathcal{C}_n) & \le & \sqrt{2}, & & \boldsymbol{\pi}(\mathcal{S}_n, \mathcal{S}_n \otimes \mathcal{C}_n^\circ) & \le & 2\sqrt{2}, \\
\boldsymbol{\pi}(\mathcal{C}_n, \mathcal{C}_n \otimes \mathcal{C}_n) & \le & 2\sqrt{2}, & & \boldsymbol{\pi}(\mathcal{C}_n, \mathcal{C}_n \otimes \mathcal{C}_n^\circ) & \le & 2\sqrt{2}.
\end{array}$$

Unfortunately, we do not know whether

$$\begin{array}{rclcrcl}
\boldsymbol{\pi}(\mathcal{C}_n, \mathcal{S}_n \otimes \mathcal{S}_n) & \le & c, & \quad & \boldsymbol{\pi}(\mathcal{C}_n, \mathcal{S}_n \otimes \mathcal{S}_n^\circ) & \le & c, \\
 & & & & \boldsymbol{\pi}(\mathcal{S}_n, \mathcal{C}_n \otimes \mathcal{S}_n^\circ) & \le & c
\end{array}$$

for some constant $c \ge 1$.

6 Main results

The significance of the following theorems is due to the fact that they compare ideal norms with quite different properties.

For example, $\boldsymbol{\pi}(\mathcal{E}_n^\circ, \mathcal{U}_n^\circ)$ is symmetric and $\boldsymbol{\pi}(\mathcal{E}_n, \mathcal{U}_n)$ is not. On the other hand, the sequence of the latter ideal norms increases as $n \to \infty$, but we have no idea about the monotonicity of the first one. Moreover, the sequence of the ideal norms $\boldsymbol{\pi}(\mathcal{E}_n^\circ, \mathcal{U}_n^\circ)$ is submultiplicative which means that

$$\boldsymbol{\pi}(X|\mathcal{E}_{mn}^\circ, \mathcal{U}_{mn}^\circ) \le \boldsymbol{\pi}(X|\mathcal{E}_m^\circ, \mathcal{U}_m^\circ)\, \boldsymbol{\pi}(X|\mathcal{E}_n^\circ, \mathcal{U}_n^\circ)$$

for all Banach space X and $m, n \in \mathbb{N}$.

Similarly, we were not able to prove the monotonicity of $\boldsymbol{\delta}(\mathcal{E}_n, \mathcal{E}_n)$, while that of $\boldsymbol{\pi}(\mathcal{C}_n, \mathcal{S}_n)$ is trivial.

Last but not least, we mention that the Dirichlet ideal norms $\boldsymbol{\delta}(\mathcal{E}_n, \mathcal{U}_n)$ and $\boldsymbol{\delta}(\mathcal{E}_n, \mathcal{E}_n)$ behave well with respect to complex interpolation. This important property is quite useful to compute or estimate the values of that ideal norms for concrete Banach spaces; see section 8.

We now state our main results, which can be checked by sometimes obvious and sometimes tricky manipulations of trigonometric formulas; see the next section.

Theorem 1 *The sequences of the following ideal norms are uniformly equivalent:*

$$\begin{array}{llllll}
\boldsymbol{\pi}(\mathcal{E}_n, \mathcal{U}_n) & = & \boldsymbol{\delta}(\mathcal{E}_n, \mathcal{U}_n), & & \boldsymbol{\pi}(\mathcal{U}_n, \mathcal{E}_n), & & \boldsymbol{\delta}(\mathcal{U}_n, \mathcal{E}_n), \\
\boldsymbol{\pi}(\mathcal{E}_n^\circ, \mathcal{U}_n^\circ) & = & \boldsymbol{\delta}(\mathcal{E}_n^\circ, \mathcal{U}_n^\circ) & = & \boldsymbol{\pi}(\mathcal{U}_n^\circ, \mathcal{E}_n^\circ) & = & \boldsymbol{\delta}(\mathcal{U}_n^\circ, \mathcal{E}_n^\circ), \\
\boldsymbol{\pi}(\mathcal{C}_n, \mathcal{U}_n) & = & \boldsymbol{\delta}(\mathcal{C}_n, \mathcal{U}_n), & & \boldsymbol{\pi}(\mathcal{U}_n, \mathcal{C}_n), & & \boldsymbol{\delta}(\mathcal{U}_n, \mathcal{C}_n), \\
\boldsymbol{\pi}(\mathcal{C}_n^\circ, \mathcal{U}_n^\circ) & = & \boldsymbol{\delta}(\mathcal{C}_n^\circ, \mathcal{U}_n^\circ), & & \boldsymbol{\pi}(\mathcal{U}_n^\circ, \mathcal{C}_n^\circ), & & \boldsymbol{\delta}(\mathcal{U}_n^\circ, \mathcal{C}_n^\circ), \\
\boldsymbol{\pi}(\mathcal{S}_n, \mathcal{U}_n) & = & \boldsymbol{\delta}(\mathcal{S}_n, \mathcal{U}_n), & & \boldsymbol{\pi}(\mathcal{U}_n, \mathcal{S}_n), & & \boldsymbol{\delta}(\mathcal{U}_n, \mathcal{S}_n), \\
\boldsymbol{\pi}(\mathcal{S}_n^\circ, \mathcal{U}_n^\circ) & = & \boldsymbol{\delta}(\mathcal{S}_n^\circ, \mathcal{U}_n^\circ) & = & \boldsymbol{\pi}(\mathcal{U}_n^\circ, \mathcal{S}_n^\circ) & = & \boldsymbol{\delta}(\mathcal{U}_n^\circ, \mathcal{S}_n^\circ).
\end{array}$$

Theorem 2 *The sequences of the following ideal norms are uniformly equivalent:*

$$\begin{array}{llllll}
\boldsymbol{\delta}(\mathcal{E}_n, \mathcal{E}_n), & \boldsymbol{\delta}(\mathcal{C}_n, \mathcal{C}_n), & \boldsymbol{\delta}(\mathcal{S}_n, \mathcal{S}_n), & ----, & ----, & ----, \\
\boldsymbol{\delta}(\mathcal{E}_n^\circ, \mathcal{E}_n), & \boldsymbol{\delta}(\mathcal{C}_n^\circ, \mathcal{C}_n), & \boldsymbol{\delta}(\mathcal{S}_n^\circ, \mathcal{S}_n), & ----, & ----, & ----, \\
\boldsymbol{\delta}(\mathcal{E}_n, \mathcal{E}_n^\circ), & \boldsymbol{\delta}(\mathcal{C}_n, \mathcal{C}_n^\circ), & \boldsymbol{\delta}(\mathcal{S}_n, \mathcal{S}_n^\circ), & \boldsymbol{\pi}(\mathcal{E}_n, \mathcal{E}_n^\circ), & \boldsymbol{\pi}(\mathcal{C}_n, \mathcal{C}_n^\circ), & \boldsymbol{\pi}(\mathcal{S}_n, \mathcal{S}_n^\circ), \\
\boldsymbol{\delta}(\mathcal{C}_n, \mathcal{S}_n), & \boldsymbol{\delta}(\mathcal{S}_n, \mathcal{C}_n), & & \boldsymbol{\pi}(\mathcal{C}_n, \mathcal{S}_n), & \boldsymbol{\pi}(\mathcal{S}_n, \mathcal{C}_n), & \\
\boldsymbol{\delta}(\mathcal{C}_n^\circ, \mathcal{S}_n), & \boldsymbol{\delta}(\mathcal{S}_n^\circ, \mathcal{C}_n), & & \boldsymbol{\pi}(\mathcal{C}_n^\circ, \mathcal{S}_n), & \boldsymbol{\pi}(\mathcal{S}_n^\circ, \mathcal{C}_n), & \\
\boldsymbol{\delta}(\mathcal{C}_n, \mathcal{S}_n^\circ), & \boldsymbol{\delta}(\mathcal{S}_n, \mathcal{C}_n^\circ), & & \boldsymbol{\pi}(\mathcal{C}_n, \mathcal{S}_n^\circ), & \boldsymbol{\pi}(\mathcal{S}_n, \mathcal{C}_n^\circ), & \\
\boldsymbol{\delta}(\mathcal{C}_n^\circ, \mathcal{S}_n^\circ), & \boldsymbol{\delta}(\mathcal{S}_n^\circ, \mathcal{C}_n^\circ), & & \boldsymbol{\pi}(\mathcal{C}_n^\circ, \mathcal{S}_n^\circ), & \boldsymbol{\pi}(\mathcal{S}_n^\circ, \mathcal{C}_n^\circ). &
\end{array}$$

The remaining Parseval ideal norms

$$\begin{array}{lll}
\boldsymbol{\pi}(\mathcal{E}_n, \mathcal{E}_n), & \boldsymbol{\pi}(\mathcal{C}_n, \mathcal{C}_n), & \boldsymbol{\pi}(\mathcal{S}_n, \mathcal{S}_n), \\
\boldsymbol{\pi}(\mathcal{E}_n^\circ, \mathcal{E}_n), & \boldsymbol{\pi}(\mathcal{C}_n^\circ, \mathcal{C}_n), & \boldsymbol{\pi}(\mathcal{S}_n^\circ, \mathcal{S}_n),
\end{array}$$

are uniformly equivalent to the operator norm.

7 Some typical proofs

In order to indicate the flavour of our methods, we present proofs of the following results.

(1) $\boldsymbol{\pi}(\mathcal{S}_n, \mathcal{S}_n \otimes \mathcal{C}_n) \leq \sqrt{2}$,

(2) $\boldsymbol{\pi}(\mathcal{S}_n \otimes \mathcal{S}_n, \mathcal{C}_n) \leq \sqrt{2}$,

(3) $\boldsymbol{\pi}(\mathcal{S}_n, \mathcal{C}_n) \leq 2\boldsymbol{\pi}(\mathcal{C}_n, \mathcal{S}_n)$.

Proof of (1): Using the translation invariance of the Lebesgue measure, it follows for any $s \in \mathbb{R}$ that

$$\begin{aligned} \|(x_k)|\mathcal{S}_n\|^2 &= \frac{1}{\pi}\int_{-\pi}^{+\pi}\left\|\sum_{k=1}^{n} x_k \sin k(s+t)\right\|^2 dt \\ &\leq \frac{2}{\pi}\int_{-\pi}^{+\pi}\left\|\sum_{k=1}^{n} x_k \sin ks \cos kt\right\|^2 dt + \frac{2}{\pi}\int_{-\pi}^{+\pi}\left\|\sum_{k=1}^{n} x_k \cos ks \sin kt\right\|^2 dt. \end{aligned}$$

Integration over s yields

$$\|(x_k)|\mathcal{S}_n\|^2 \leq \|(x_k)|\mathcal{S}_n \otimes \mathcal{C}_n\|^2 + \|(x_k)|\mathcal{C}_n \otimes \mathcal{S}_n\|^2.$$

Proof of (2): Using the translation invariance of the Lebesgue measure, it follows from

$$2\sin ks \sin kt = \cos k(s-t) - \cos k(s+t)$$

that

$$\begin{aligned} \|(x_k \sin ks)|\mathcal{S}_n\| &= \left(\frac{1}{\pi}\int_{-\pi}^{+\pi}\left\|\sum_{k=1}^{n} x_k \sin ks \sin kt\right\|^2 dt\right)^{1/2} \\ &\leq \frac{1}{2}\left\{\begin{array}{c} \left(\frac{1}{\pi}\int_{-\pi}^{+\pi}\|\sum_{k=1}^{n} x_k \cos k(s-t)\|^2 dt\right)^{1/2} \\ + \\ \left(\frac{1}{\pi}\int_{-\pi}^{+\pi}\|\sum_{k=1}^{n} x_k \cos k(s+t)\|^2 dt\right)^{1/2} \end{array}\right\} = \|(x_k)|\mathcal{C}_n\|. \end{aligned}$$

Squaring and integrating over s yields the desired result.

Proof of (3): Obviously, we have

$$\boldsymbol{\pi}(\mathcal{S}_n, \mathcal{C}_n) \leq \boldsymbol{\pi}(\mathcal{S}_n, \mathcal{S}_n \otimes \mathcal{C}_n)\boldsymbol{\pi}(\mathcal{S}_n \otimes \mathcal{C}_n, \mathcal{S}_n \otimes \mathcal{S}_n)\boldsymbol{\pi}(\mathcal{S}_n \otimes \mathcal{S}_n, \mathcal{C}_n).$$

The assertion now follows from (1), (2) and

$$\boldsymbol{\pi}(\mathcal{S}_n \otimes \mathcal{C}_n, \mathcal{S}_n \otimes \mathcal{S}_n) \leq \boldsymbol{\pi}(\mathcal{C}_n, \mathcal{S}_n).$$

8 Examples

First of all, we stress the fact that

$$\pi(H|\mathcal{B}_n, \mathcal{A}_n) = \delta(H|\mathcal{B}_n, \mathcal{A}_n) = 1$$

for any Hilbert space $H \neq \{o\}$ and arbitrary orthonormal systems $\mathcal{A}_n$ and $\mathcal{B}_n$.

We now consider the classical function spaces $L_p = L_p(M, \mu)$, which are assumed to be infinite dimensional.

It turns out that

$$\begin{aligned} \pi(L_p|\mathcal{E}_n, \mathcal{U}_n) &\asymp n^{|1/p-1/2|} && \text{for } 1 \leq p \leq \infty, \\ \delta(L_p|\mathcal{E}_n, \mathcal{E}_n) &\asymp 1 && \text{for } 1 < p < \infty, \\ \delta(L_p|\mathcal{E}_n, \mathcal{E}_n) &\asymp 1 + \log n && \text{for } p = 1 \text{ and } p = \infty. \end{aligned}$$

The same results hold for all ideal norms which are uniformly equivalent to $\pi(\mathcal{E}_n, \mathcal{U}_n)$ and $\delta(\mathcal{E}_n, \mathcal{E}_n)$, by theorems 1 and 2, respectively.

9 Some classes of Banach spaces

First of all, we fix the following notation.

- L: class of all Banach spaces,
- H: class of *Hilbertian Banach spaces*,
- B: class of *B–convex Banach spaces*,
- S: class of *superreflexive Banach spaces.*

The concepts of B–convexity and superreflexivity are defined, for example, in [BEA85, p. 225].

The classical *Fourier transform*

$$\mathcal{F} : f(t) \longrightarrow \hat{f}(s) := \frac{1}{\sqrt{2\pi}} \int_{\mathbb{R}} f(t) \exp(ist)\, dt$$

yields an operator from $L_1(\mathbb{R})$ into $C(\mathbb{R})$. Plancherel observed that

$$\|\hat{f}|L_2\| = \|f|L_2\| \quad \text{for } f \in L_1(\mathbb{R}) \cap L_2(\mathbb{R}).$$

Hence there exists an extension from $L_2(\mathbb{R})$ onto $L_2(\mathbb{R})$. Finally, it follows by complex interpolation that

$$\mathcal{F} : L_r(\mathbb{R}) \longrightarrow L_{r'}(\mathbb{R})$$

whenever $1 \leq r \leq 2$.

A Banach space X is said to have the *Hausdorff–Young property of order* r if

$$\mathcal{F} \otimes X : L_r(\mathbb{R}) \otimes X \longrightarrow L_{r'}(\mathbb{R}) \otimes X,$$

defined by

$$\mathcal{F} \otimes X : \sum_{k=1}^{n} f_k \otimes x_k \longrightarrow \sum_{k=1}^{n} \hat{f}_k \otimes x_k$$

for $f_1, \ldots, f_n \in L_r(\mathbb{R})$ and $x_1, \ldots, x_n \in X$, extends continuously to an operator

$$[\mathcal{F}, X] : [L_r(\mathbb{R}), X] \longrightarrow [L_{r'}(\mathbb{R}), X].$$

The class of these Banach spaces, which is denoted by HY_r, was introduced by J. Peetre in 1969; see [Pee69], [Mil84] and [RdF–T87].

Replacing $L_r(\mathbb{R})$ and $L_{r'}(\mathbb{R})$ by the Lorentz spaces $L_{r,1}(\mathbb{R})$ and $L_{r',\infty}(\mathbb{R})$, respectively, yields the *weak Hausdorff–Young property of order* r. In other terms, we assume that

$$[\mathcal{F}, X] : [L_{r,1}(\mathbb{R}), X] \longrightarrow [L_{r',\infty}(\mathbb{R}), X].$$

The class of these Banach spaces is denoted by $\mathsf{HY}_r^{\text{weak}}$.

Obviously

$$\mathsf{HY}_2 \subseteq \mathsf{HY}_2^{\text{weak}} \subset \mathsf{HY}_r \subseteq \mathsf{HY}_r^{\text{weak}} \subset \mathsf{HY}_s \subset \mathsf{HY}_1 = \mathsf{HY}_1^{\text{weak}}$$

for $1 < s < r < 2$. Note that $L_s \in \mathsf{HY}_s \setminus \mathsf{HY}_r$. We conjecture that HY_r is a proper subclass of $\mathsf{HY}_r^{\text{weak}}$ whenever $1 < r < 2$, while equality could hold in the limiting case $r = 2$.

For $0 \leq \varrho \leq 1/2$, we define

$$\mathsf{F}_\varrho := \{X : (n^{-\varrho} \boldsymbol{\pi}(X|\mathcal{E}_n, \mathfrak{U}_n)) \quad \text{is bounded}\}.$$

Moreover, let

$$\mathsf{F} := \{X : (n^{-1/2} \boldsymbol{\pi}(X|\mathcal{E}_n, \mathfrak{U}_n)) \quad \text{converges to } 0\}.$$

Note that $\boldsymbol{\pi}(X|\mathcal{E}_n, \mathfrak{U}_n)$ can be replaced by any ideal norm listed in theorem 1. Using the submultiplicativity of the sequence $\boldsymbol{\pi}(\mathcal{E}_n^\circ, \mathfrak{U}_n^\circ)$, we are able to show that

$$\mathsf{F} = \bigcup_{0 \leq \varrho < 1/2} \mathsf{F}_\varrho.$$

Obviously,

$$\mathsf{F}_0 \subset \mathsf{F}_\sigma \subset \mathsf{F}_\varrho \subset \mathsf{F} \subset \mathsf{F}_{1/2} = \mathsf{L}$$

for $0 < \sigma < \varrho < 1/2$. It follows from $L_r \in \mathsf{F}_\varrho \setminus \mathsf{F}_\sigma$ with $\varrho = |1/r - 1/2|$ that the scale $\{\mathsf{F}_\varrho : 0 \leq \varrho \leq 1/2\}$ is strictly increasing.

The following relations are known from [Kwa72], [Bou82] and [Pis79]:

$$\mathsf{H} = \mathsf{F}_0 = \mathsf{HY}_2 \quad \text{and} \quad \mathsf{B} = \mathsf{F}.$$

Furthermore, it follows easily that

$$\mathsf{HY}_r^{\text{weak}} \subseteq \mathsf{F}_\varrho \quad \text{for } 1 < r < 2 \text{ and } \varrho = 1/r - 1/2.$$

We strongly conjecture that there even holds equality. However, until this moment, we only were able to show that $\mathsf{F}_\varrho \subset \mathsf{HY}_{r-\varepsilon}$ for $0 < \varepsilon < r-1$. This can be done by employing a technique from [Bou88].

The classical *Hilbert transform*

$$\mathcal{H} : f(t) \longrightarrow \tilde{f}(s) := \frac{1}{2\pi} \,\text{p.v.} \int_{-\pi}^{+\pi} \cot \tfrac{1}{2}(s-t) f(t)\, dt$$

yields an operator on $L_2(-\pi, +\pi)$.

A Banach space X is said to *respect the Hilbert transform* if

$$\mathcal{H} \otimes X : L_2(-\pi, +\pi) \otimes X \longrightarrow L_2(-\pi, +\pi) \otimes X,$$

defined by

$$\mathcal{H} \otimes X : \sum_{k=1}^{n} f_k \otimes x_k \longrightarrow \sum_{k=1}^{n} \tilde{f}_k \otimes x_k$$

for $f_1, \ldots, f_n \in L_2(-\pi, +\pi)$ and $x_1, \ldots, x_n \in X$, extends continuously to an operator

$$[\mathcal{H}, X] : [L_2(-\pi, +\pi), X] \longrightarrow [L_2(-\pi, +\pi), X].$$

The class of these Banach spaces is denoted by HT. We stress the fact that HT coincides with UMD, the class of all Banach spaces with the *unconditionality property for martingale differences*; see [Bou83], [Bur83] and [Bur86].

For $0 \le \varrho \le 1$, we define

$$\mathsf{R}_\varrho := \{X \in \mathsf{L} : ((1 + \log n)^{-\varrho} \boldsymbol{\delta}(X|\mathcal{E}_n, \mathcal{E}_n)) \quad \text{is bounded}\}$$

and

$$\mathsf{R} := \{X \in \mathsf{L} : ((1 + \log n)^{-1} \boldsymbol{\delta}(X|\mathcal{E}_n, \mathcal{E}_n)) \quad \text{converges to } 0\}.$$

Note that $\boldsymbol{\delta}(X|\mathcal{E}_n, \mathcal{E}_n)$ can be replaced by any ideal norm listed in theorem 2.

Obviously,

$$\mathsf{R}_0 \subset \mathsf{R}_\sigma \subset \mathsf{R}_\varrho \subset \mathsf{R} \subset \mathsf{R}_1 = \mathsf{L}$$

for $0 < \sigma < \varrho < 1$. Since

$$L_p \in \mathsf{R}_0 \quad \text{for } 1 < p < \infty \text{ and } \quad L_1, L_\infty \in \mathsf{R} \setminus \mathsf{R}_\varrho \quad \text{for } 0 \le \varrho < 1,$$

the classical Banach spaces cannot be used to show that the scale $\{\mathsf{R}_\varrho : 0 \le \varrho \le 1\}$ is strictly increasing. Nevertheless, there exist sophisticated examples which yield that monotonicity.

The following relations are known from folklore and [P–W9?]:

$$\mathsf{HT} = \mathsf{R}_0 \quad \text{and} \quad \mathsf{R} \subseteq \mathsf{S}.$$

The problem whether R is a proper subclass of S remains open.

Finally, we stress the fact that the classes

$$\mathsf{H},\ \mathsf{B},\ \mathsf{S},\ \mathsf{HY}_r,\ \mathsf{HY}_r^{\text{weak}},\ \mathsf{HT},\ \mathsf{F}_\varrho,\ \mathsf{F},\ \mathsf{R}_\varrho \quad \text{and} \quad \mathsf{R}$$

are stable under passing to subspaces, quotients, duals and finite direct sums. For those classes which are defined by the asymptotic behaviour of ideal norms, these permanence properties follow directly from the injectivity, surjectivity and symmetry of $\boldsymbol{\pi}(\mathcal{E}_n^\circ, \mathcal{U}_n^\circ)$ and $\boldsymbol{\delta}(\mathcal{E}_n, \mathcal{E}_n)$, respectively.

References

[AMI86] Amir, D.: *Isometric Characterization of Inner Product Spaces.* Birkhäuser: Basel, 1986.

[BEA85] Beauzamy, B.: *Introduction to Banach Spaces and their Geometry.* North–Holland: Amsterdam, 1985.

[B–L76] Bergh, J.; Löfström, J.: *Interpolation Spaces.* Springer: Berlin, Heidelberg, 1976.

[Boc33] Bochner, S.: Abstrakte Funktionen und die Besselsche Ungleichung. *Göttinger Nachrichten, Fachgruppe I,* Nr. 40 (1933), pages 178–180.

[Bou81] Bourgain, J.: On trigonometric series in super reflexive spaces. *J. London Math. Soc. (2)* **24** (1981), pages 165–174.

[Bou82] Bourgain, J.: A Hausdorff–Young inequality for B–convex Banach spaces. *Pacific J. Math.* **101** (1982), pages 255–262.

[Bou83] Bourgain, J.: Some remarks on Banach spaces in which martingale difference sequences are unconditional. *Ark. Mat.* **22** (1983), pages 163–168.

[Bou88] Bourgain, J.: Vector–valued Hausdorff–Young inequalities and applications. In *Geometric Aspects of Functional Analysis*, Lecture Notes in Math., no. **1317**, pages 239–249. Springer, Berlin, Heidelberg, 1988.

[Bur83] Burkholder, D. L.: A geometrical condition that implies the existence of certain singular integrals of Banach space valued functions. In: *Proc. Conf. Harmonic Analysis in Honor of Antony Zygmund*, pages 270–286, Chicago, 1983.

[Bur86] Burkholder, D. L.: Martingales and Fourier analysis in Banach spaces. In *Probability and Analysis, Varenna (Italy) 1985*, Lecture Notes in Math., no. **1206**, pages 61–108. Springer, Berlin, Heidelberg, 1986.

[Kwa72] Kwapień, S.: Isomorphic characterizations of inner product spaces by orthogonal series with vector valued coefficients. *Studia Math.* **44** (1972), pages 583–595.

[M–P76] Maurey, B.; Pisier, G.: Séries de variables aléatoires vectorielles indépendantes et propriétés géométriques des espaces de Banach. *Studia Math.* **58** (1976), pages 45–90.

[Mil84] Milman, M.: Complex interpolation and geometry of Banach spaces. *Ann. Mat. Pura Appl. (4)* **136** (1984), pages 317–328.

[Pee69] Peetre, J.: Sur la transformation de Fourier des fonctions à valeurs vectorielles. *Rend. Sem. Mat. Univ. Padova* **42** (1969), pages 15–26.

[PIE78] Pietsch, A.: *Operator Ideals.* D. Verlag der Wissenschaften, Berlin, 1978.

[P–W9?] Pietsch, A.; Wenzel, J.: *Orthonormal Systems and Banach Space Geometry.* in preparation.

[Pis79] Pisier, G. Sur les espaces de Banach de dimension finie à distance extrémal d'un espace euclidien, d'après V. D. Milman et H. Wolfson. Séminaire d'analyse fonctionelle 1978/79, Exposé XVI, Ecole Polytechnique, Palaiseau, 1979.

[RdF–T87] Rubio de Francia, J. L.; Torrea, J. L.: Some Banach techniques in vector valued Fourier series. *Collect. Math.* **54** (1987), pages 273–284.

[Wen93] Wenzel, J.: *Idealnormen und trigonometrische Orthonormalsysteme.* PhD thesis, Friedrich–Schiller–Universität Jena, 1993.

Spaces of Lipschitz Functions on Banach Spaces

CHARLES STEGALL Institut für Mathematik, Johannes Kepler Universität, A-4040 Linz, Austria

A remarkable theorem of R. C. James is the following: suppose that X is a Banach space and $C \subseteq X$ is a norm bounded, closed and convex set such that every linear functional $x^* \in X^*$ attains its supremum on C; then C is a weakly compact set. Actually, this result is significantly stronger than this statement; indeed, the proof can be used to obtain other surprising results. For example, suppose that X is a separable Banach space and S is a norm separable subset of the unit ball of X^* such that for each $x \in X$ there exists $x^* \in S$ such that $x^*(x) = \|x\|$ then X^* is itself norm separable (this is in **[R]**). If we call S a support set, in this case, with respect to the entire space X, one can ask questions about the size and structure of a support set, a support set not only with respect to X itself but perhaps with respect to some other subset of X. We analyze one particular case of this as well as give some applications. We begin by restating an old result, some consequences of which are, perhaps, not well known.

Main Theorem. *Let X be a Banach space and K a weak* compact subset of the dual X^*. The following are equivalent:*

(i) if Y is any separable Banach space and $U : Y \to X$ is any operator, then $U^*(K)$ is a norm separable subset of Y^*;

(ii) if $R : X \to C(K)$ is the canonical operator then R maps bounded subsets of X into equimeasurable subsets of $C(K)$;

(iii) the operator R factors through an Asplund space (a Banach space whose dual has the Radon-Nikodym property);

(iv) for any regular Borel (with respect to the weak topology) measure μ on K and any $\epsilon > 0$ there exists a norm compact subset K_ϵ of K such that $|\mu|(K_\epsilon) > |\mu|(K) - \epsilon$;*

(v) given any subset S of K and any $\epsilon > 0$ there exists an $x \in X$ and a real r so that $\{x^* \in S : x^*(x) > r\}$ is not empty and has norm diameter less than ϵ.

In **[S5]** it is proved that if K satisfies one of the above then so does the weak* closed convex hull of K. An immediate consequence of the equivalence of (i) and (iv) above is

Theorem. *Let $U : X \to Y$ be an operator and $K \subseteq Y^*$ a weak* compact set. Suppose that T^* is both a weak* and a norm homeomorphism on K. Then K satisfies any one of the above if and only if $T^*(K)$ does.*

Actually, we may formulate the result above as only requiring that T^* is one to one on K and $(T^*)^{-1}$ is (norm to norm) in the first Baire class (or, for that matter, in any Baire class). If (i) is false all of the others are false **[S5]**. The other directions, except, perhaps, those involving (v), are essentially results of Smulian, Dunford, Pettis, Phillips and Grothendieck; see **[G]**, **[S5]** and **[S6]** for more detail. Condition (v) is implicit in the results of Smulian. What is somewhat newer is that these conditions are equivalent to:

(vi) there exists a (norm to norm) first Baire class function $\lambda : X \to X^*$ so that $\lambda(x) \in K$ and $\lambda(x)(x) = \sup_{y^* \in K} y^*(x)$ for all $x \in X$.

That (vi) follows from (v) can be obtained from **[JR]** and see **[S3]** for a transparent proof. The first use of selection theorems in this connection is in **[F]**. That (vi) implies (i) is almost explicitly stated in **[R]** and is based on results of R. C. James (see **[S1]** for a discussion of James' results). We give a proof that a somewhat different version of (vi) implies (i).

Let X be a Banach space and K a weak* compact subset of the unit ball of X^*. Suppose that S is a subset of X^* that is Lindelöf in the weak* topology (in particular, S is norm separable) and $K \setminus S \neq \emptyset$. Fix $y^* \in K \setminus S$. Observe that

$$S \subseteq \bigcup_{\|x\|=1} \{x^* : |(y^* - x^*)x| > 0\}.$$

We may choose a sequence $\{x_n\}$ with $\|x_n\| = 1$ so that

$$S \subseteq \bigcup_n \{x^* : |(y^* - x^*)x_n| > 0\}.$$

Define

$$\phi(x^*) = 1 - \sum_n 2^{-n} |(y^* - x^*)x_n|.$$

Observe that $-1 \leq \phi(x^*) \leq 1$ and

$$\sup_{z^* \in K} \phi(z^*) \geq 1 = \phi(y^*) > \phi(x^*)$$

if $x^* \in S$. Thus, ϕ does not attain its supremum on K at any point of S. Moreover,

$$\begin{aligned} |\phi(x^*) - \phi(z^*)| &= \left| \sum_n 2^{-n} |(y^* - x^*)x_n| - \sum_n 2^{-n} |(y^* - z^*)x_n| \right| \\ &\leq \sum_n 2^{-n} |(z^* - x^*)x_n| \leq \|z^* - x^*\| \end{aligned}$$

which proves that the Lipschitz norm of ϕ is no more than one. Also, ϕ is concave and is obtained by applying lattice operations to the sequence $\{x_n\}$ and the constant functions. Suppose, in addition, that X is a separable Banach space. If, given any ϕ that is weak* continuous on K and Lipschitz in the norm, there exists $x^* \in S \cap K$ so that $\phi(x^*) = \sup_K \phi$ then $S \cap K = K$. This follows from the fact that the unit ball of X^* is a separable metric space in the weak* topology. If we fix $z^* \in K$ and let $S = K \setminus \{z^*\}$ then there is an ϕ that attains its supremum only at z^*. In fact, it suffices to take only those ϕ which can be generated as above from X and the constants.

Let X^* be the dual of the Banach space X. Let $L(X)$ denote the smallest lattice of functions that are both weak* continuous on the unit ball B of X^* and Lipschitz with respect to the norm on X^*; the norm on $L(X)$ is given by

$$L(f) = \max\{ \sup_{\|x^*\| \leq 1} |f(x^*)|, \sup_{\substack{x^* \neq y^* \\ \|x^*\| \leq 1 \\ \|y^*\| \leq 1}} \|x^* - y^*\|^{-1} |f(x^*) - f(y^*)| \}. \tag{1}$$

Observe that $L(X)$ is generated by X and the constants. Consider the canonical operators

$$X \xrightarrow{I} L(X) \xrightarrow{R} C(B).$$

This means that each x^* in the unit ball of X^* can be considered as three different objects: δ_{x^*}, denoting the point mass in $C(B)$, as an element of $L(X)^*$ and $x^* \in X^*$. Also, I is an isometry on X and I^* is an isometry on B. Of course, B, considered as a subset of $L(X)^*$, does not have the same convexity structure as it does as a subset of X^*. Suppose that $K \subseteq B$ is weak* compact and satisfies the hypothesis of the Main Theorem. The canonical operator $R : L(X) \to C(K)$ (the restriction operator) has norm no more than one and transforms the unit ball of $L(X)$ into an equimeasurable subset of $C(K)$; this is nothing more difficult than the results of **[S6]** and the Arzela-Ascoli theorem. There exist (see **[JR]**, **[F]** and **[S3]**) a sequence of norm to norm continuous functions $\lambda_n : L(X) \to L(X)^*$ such that $\|\lambda_n(f)\| \leq 1$ for all n and $f \in L(X)$, $\lim_n \lambda_n(f) = \lambda_0(f)$ (limit in norm) exists for all f, $\lambda_0(f) \subset K$ and $f(\lambda_0(f)) = \sup_K f$. Since B is a retract of X^* and I^* is an isometry on B, considered as a subspace of $L(X)^*$, it follows that B is a retract of $L(X)^*$. We may assume that each λ_n takes its values in B. Define

$$S = \overline{\bigcup_{n=0}^{\infty} \lambda_n(L(X))}^{\text{norm}} .$$

What is critical here is that $K \subseteq S$. Since K and S are both subsets of B it does not matter whether we write x^* or δ_{x^*}. Suppose that there exist $z^* \in K$ and and $\eta > 0$ so that $\|z^* - x^*\| > \eta > 0$ for all $x^* \in S$. Fix $\delta > 0$ and a norm compact subset D of S. Choose a finite subset F of the unit ball of X such that

$$\eta < \|x^* - z^*\| < \sup_{x \in F} |(x^* - z^*)x| + \delta$$

for all $x^* \in D$. Define

$$\phi(u^*) = \sup_{x\in F} |(z^* - u^*)x|.$$

Now, the elementary computation:

$$\begin{aligned}\phi(u^*) &= \sup_{x\in F} |(z^* - u^*)x| \leq \sup_{x\in F}\{|(v^* - u^*)x| + |(z^* - v^*)x|\} \\ &\leq \sup_{x\in F}\{\|v^* - u^*\| + |(z^* - v^*)x|\} \leq \|v^* - u^*\| + \phi(v^*).\end{aligned}$$

Since $\sup_{x^*\in K} \phi(x^*) \leq 2$, this proves that $L(1-\phi) \leq 1$. Also, we have

$$\begin{aligned}&1 - \phi(u^*) \leq 1 - \eta + \delta \qquad \text{for all} \qquad u^* \in D \qquad \text{and} \\ &1 - \phi(z^*) = 1.\end{aligned}$$

Let $\delta = \eta/3$. We have produced a filter (over the norm compact subsets D of S)

$$\mathcal{F} = \{\psi_D : \psi_D(D) \leq \frac{2}{3}\eta,\ \psi_D(z^*) = 1, \text{ and } L(\psi_D) \leq 1\}.$$

Moreover, each ψ_D is concave and generated by the lattice operations on a finite subset of X and the constants. Let h be any cluster point of $\mathcal{F}$ in the topology of pointwise convergence on K. If $F \subseteq X$ then we denote by $L(F)$ the smallest norm closed, linear sublattice of $L(X)$ containing F and the constants; if F is norm separable then so is $L(F)$. Choose $\psi_1 \in \mathcal{F}$ so that $|h(z^*) - \psi_1(z^*)| < \frac{\delta}{2}$. Choose F_1 a finite subset of X such that $\psi_1 \in L(F_1)$ and let

$$L(F_1) = \overline{\bigcup_j G_{1,j}}$$

where each $G_{1,j}$ is norm compact. In general, we construct $\{\psi_k\} \subseteq \mathcal{F}$, $\{F_k\}$ finite subsets of X, compact subsets $\{G_{k,j}\}$ of $L(\cup_{i\leq k} F_i)$ so that $\psi_k(z^*) = 1$ and

$$L(\cup_{i\leq k} F_i)) = \overline{\bigcup_j G_{k,j}} \quad \text{and}$$

$$\psi_{k+1} \leq \frac{2}{3}\eta \quad \text{on} \quad \bigcup_{p,q,r\leq k} \lambda_p(G_{q,r}).$$

Since $\{\psi_k\}$ is bounded in the Lipschitz norm it follows that the subset of S where $\{\psi_k\}$ converges to null is norm closed. Everything reduces to the case of the separable Banach space Y generated by $\cup_k F_k$. The functions $\{\psi_k\}$ may be considered as weak* continuous and Lipschitz on the unit ball of Y^* and converge to null on every $\lambda_0(f)|Y$ but this set contains all of $K|Y$ and we have that $\limsup \psi_n(x^*) \leq \frac{2}{3}\eta$ for all $x^* \in K$ which contradicts $\liminf \psi_n(z^*) = 1$. Denote by $\mathcal{C} \subseteq L(X)$ the set of concave functions such that $L(f) \leq 1$.

Theorem. *If $I : X \to L(X)$ is as above and K is a Radon-Nikodym set in the unit ball of X^* then $K = \overline{\lambda_0(\mathcal{C})}^{\text{norm}}$.*

Proof. Suppose not. Suppose that $z^* \in K$, $\epsilon > 0$ and $\|\lambda_0(f) - z^*\| > \epsilon$ for all $f \in \mathcal{C}$. Suppose that $\|z^*\| < 1$. Choose $0 < \delta < \epsilon$ such that $\|z^*\| + \delta < 1$. Then λ_0 is a function of the first Baire class with values in $C = B \setminus B(z^*, \delta)$. The relevant geometric property of C is that it is arcwise connected and locally arcwise connected; see **[Ve]** and its references or it is easy to rearrange the proof given in **[S3]** to obtain continuous functions $\xi_n : \mathcal{C} \to C$ such that $\lim \xi_n(f) = \lambda_0(f)$ for all $f \in \mathcal{C}$. We have already proved that

$$K \subseteq \overline{\bigcup_n \xi_n(\mathcal{C})}^{\text{norm}}$$

but this contradicts $\|\xi_n(f) - z^*\| \geq \delta$ for all n and all $f \in \mathcal{C}$. Suppose that $\|z^*\| = 1$. Define $\rho : B \to B$ by $\rho(x^*) = x^*$ for $\|x^* - z^*\| \geq \frac{\epsilon}{2}$ and

$$\rho(x^*) = (1 - \frac{\epsilon}{2} + \|x^* - z^*\|)x^* + (\frac{\epsilon}{2} - \|x^* - z^*\|)(-z^*)$$

for $\|x^* - z^*\| \leq \frac{\epsilon}{2}$. Obviously, if $\|x^* - z^*\| \geq \frac{\epsilon}{2}$ then $\|\rho(x^*) - z^*\| \geq \frac{\epsilon}{2}$; suppose that

$$\delta = \frac{\epsilon}{2} - \|x^* - z^*\| < \frac{\epsilon}{2}.$$

If $\delta \leq \frac{\epsilon}{8}$ we have that $\|x^* - z^*\| \geq \frac{\epsilon}{2} - \frac{\epsilon}{8}$ and

$$\|\rho(x^*) - z^*\| = \|(1 - \delta)x^* - (1 + \delta)z^*\| \geq \|x^* - z^*\| - \delta\|x^* + z^*\| \geq \frac{\epsilon}{2} - \frac{\epsilon}{8} - \frac{2\epsilon}{8} = \frac{\epsilon}{8}.$$

If $\delta \geq \frac{\epsilon}{8}$ then

$$\|\rho(x^*) - z^*\| = \|(1 - \delta)x^* - (1 + \delta)z^*\| \geq (1 + \delta) - (1 - \delta) = 2\delta \geq \frac{2\epsilon}{8}.$$

In either case, $\|\rho(x^*) - z^*\| \geq \frac{\epsilon}{8}$. Observe that $\lim_n \rho \circ \lambda_n(f) = \lambda_0(f)$. Again, we know that

$$K \subseteq \overline{\bigcup_n \rho \circ \lambda_n(\mathcal{C})}^{\text{norm}}$$

but this is a contradiction. In the formula for ρ we have considered B as a subset of X^*; of course, when B is considered as a subspace of $L(X)$ then we have to consider the canonical inverse of the isometry composed with ρ.

With a non trivial expansion of these techniques, the following can be shown: with the hypothesis as above, if $x^* \in K$ and $\epsilon > 0$ then there exits $\phi \in \mathcal{C}$ such that ϕ attains its supremum only on $K \cap B(x^*, \epsilon)$, which means that $\lambda_0(\phi) \in K \cap B(x^*, \epsilon)$ (for any λ_0). The case considered above where $\|z^*\| < 1$ could be dealt with differently, with considerable overkill. In any infinite dimensional Banach space the sphere S of a closed ball B is a retract of B; this is an old result for a Hilbert space and it follows for any Banach space by the Anderson-Kadec-Torunczyk theorem (see **[BP]** and **[vM]**). Thus we could retract B onto $B \setminus B(z^*, \delta)$ and replace each λ_n with its composition with such a retraction. The following is almost proved in **[F]**; what is missing can be found in **[R]**. More details can be found in **[FG]** and short proofs in **[S9]** and **[S10]**. The proof below does not use the results of R. C. James.

Theorem. *Let X be a Banach space such that X^* has the Radon-Nikodym property (the unit ball of X^* satisfies the Main Theorem). Then there exists an interval of ordinals $[1, \eta]$ such that for each $1 \le \alpha \le \eta$ we may construct subspaces X_α of X and subspaces $X_\alpha^\sharp$ of X^* such that*

(i) $X_\alpha^\sharp | X_\alpha$ is an isometry onto X_α^*;

(ii) for each limit ordinal β it follows that

$$X_\beta = \overline{\bigcup X_\alpha}^{\text{norm}} \qquad \text{and} \qquad X_\beta^\sharp = \overline{\bigcup X_\alpha^\sharp}^{\text{norm}};$$

(iii) X_1 is norm separable;

(iv) the norm density of X_α (which equals the norm density of $X_\alpha^\sharp$) is countable or no more than that of α and

(v) $X_\eta = X$.

Proof. Here, K is the unit ball of X^*. The proof is by induction. Assume that we have constructed $Y_1 = X_\alpha$. Let L_i be the smallest lattice in $L(X)$ containing Y_i. Choose $Y_1 \subseteq Y_2$ such that $Y_2 \neq Y_1$,

$$X_\alpha^\sharp \subseteq [\bigcup_j \lambda_j(L_2)] \subseteq X^*$$

and both Y_2 and Y_1 have the same norm density. In general, choose $Y_n \subseteq Y_{n+1}$ so that Y_{n+1} norms $[\cup_j \lambda_j(L_n)]$. Remember that the mapping from $L(X)^*$ to X^* is an isometry on the unit ball of X^*. Define $X_{\alpha+1} = [\cup_n Y_n]$ and $X_{\alpha+1}^\sharp = [\cup_n \cup_j \lambda_j(L_n)]$. It follows that $X_{\alpha+1}^\sharp | X_{\alpha+1}$ is an isometry onto.

Before returning to Banach spaces, we discuss an abstraction of the above. Suppose that K is a compact Hausdorff space that has a finer topology given by a metric d. We denote by $L(K, d)$ the Banach space of functions that are both continuous on K and Lipschitz with respect to d and the norm on $L(K, d)$ is given by

$$L(f) = \max\{\sup_k |f(k)|, \sup_{k \neq k'} d(k, k')^{-1} |f(k) - f(k'|\}. \tag{2}$$

We make two assumptions:

(i) $L(K, d)$ separates the points of K and

(ii) for each regular Borel measure μ on K and each $\epsilon > 0$ there exists a subset M of K that is compact in the metric and

$$|\mu|(M) > \|\mu\| - \epsilon.$$

If we replace d by the metric

$$\rho(k, k') = \sup_{L(f) \le 1} |f(k) - f(k')| \tag{3}$$

it is quite easy to see that $L(K, \rho)$ also satisfies (i) and (ii). It may happen of course that $L(K, \rho)$ is larger than $L(K, d)$ and this, in general, is desirable. Thus, we shall assume that $\rho = d$. If we assume that (K, d) is complete and (iia) for each subset S of K and each $\epsilon > 0$ there exists an open subset V (in the topology that makes K compact) so that $V \cap S \neq \emptyset$ and diameter $V \cap S < \epsilon$

then (iia) implies (ii) ([**S7**]) and (i) and (iia) together give a useful topological characterization of compact spaces that are homeomorphic to a weak* compact subspace of a dual Banach space satisfying the Main Theorem (see [**S7**] and [**Na**]). For our purposes, and probably all purposes, (i) and (ii) are more useful than (i) and (iia). Condition (iia) does not imply condition (i). Indeed, the examples in [**T**] and [**SL**] are compact spaces satisfying (iia) (this is a special case of [**Gu**], [**V**], [**Na**] and for lots of details, [**S7**]) and this example is not an Eberlein compact but is a Corson compact (same references). As we shall see below, a Corson compact space that is the continuous image of a space satisfying (i) and (ii) must be an Eberlein compact. Conditions (i) and (ii) together are much stronger than (iia) alone. Observe that the canonical operator $U : L(K, \rho) \to C(K)$ transforms the unit ball of $L(K, \rho)$ into an equimeasurable set (see [**S6**]) and it follows that K considered as a subset of $L(K, \rho)^*$ satisfies the Main Theorem. Suppose that there exists a subset S of $L(K, d)$ such that S also separates the points of K and S is point countable on K, which means that

$$\{f \in L : |f(k)| > 0\}$$

is countable for each $k \in K$ (this is the definition that K is a Corson compact). We may also assume that

$$\rho(k, k') = \sup_{\substack{L(f) \leq 1 \\ f \in S}} |f(k) - f(k')| \tag{4}$$

We shall show that if a compact Hausdorff space K is both a Corson compact space and homeomorphic (in the weak* topology) to a weak* compact subset of some X^* satisfying the conditions of the Main Theorem then K is an Eberlein compact.

Suppose that X is a Banach space and $K \subseteq B \subseteq X^*$ are as above and K satisfies the Main Theorem. Suppose that K is also a Corson compact. Define $L(X)$ as above and let

$$X \xrightarrow{I} L(X) \xrightarrow{R} C(K)$$

be the canonical operators; R is the restriction operator which is also a lattice homomorphism. Suppose that $S \subseteq C(K)$ is point countable and point separating. It is easy to see that the smallest algebra over the rational numbers is also point countable and point separating. Thus we may assume that S is uniformly dense in $C(K)$; also $R(L(X))$ is uniformly dense in $C(K)$ (see [**N**] concerning the Stone-Weierstraß theorem). Suppose $S = \{f_\alpha\}$ and for each α choose an $n \in \mathcal{N}$ and $g_{n,\alpha}$ in the unit ball of $L(X)$ such that

$$\|2^n R(g_{n,\alpha}) - f_\alpha\| < \frac{1}{4}$$

and define

$$h_{n,\alpha} = (g_{n,\alpha} \vee 2^{-n-1}) - 2^{-n-1}.$$

If, for some $k \in K$, $h_{n,\alpha}(k) > 0$, then $R(2^n g_{n,\alpha})(k) > \frac{1}{2}$ and it follows that $f_\alpha(k) \neq 0$; thus, $\{h_{n,\alpha}\}$ is point countable. Suppose, k and k' are distinct points of K. Choose $f \in C(K)$ such that $\|f\| = 1$ and $f(k) = 1$ and $f(k') = -1$. Choose α such that $\|f_\alpha - f\| < \frac{1}{4}$. For the appropriate $2^n g_{n,\alpha}$ we have that

$$\|2^n R(g_{n,\alpha}) - f\| < \frac{1}{2}$$

and it follows that $2^n R(g_{n,\alpha})(k) > \frac{1}{2}$ and $2^n R(g_{n,\alpha})(k') < -\frac{1}{2}$. Thus, $h_{n,\alpha}(k) > 0$ and $h_{n,\alpha}(k') = 0$. We have that $\{h_{n,\alpha}\}$ is point countable and point separating. We may assume the following: we have a Banach space Y (in this case, $Y = L(X)$) and a Radon-Nikodym subset K of Y^* and a subset T ($T = \{h_{n,\alpha}\}$) of Y that is point countable and point separating on K. We shall apply the entire procedure to $L(Y)$; this roughly corresponds to the double interpolation in **[S10]**. At this point, however, we shall simplify the notation.

Suppose that X is a Banach space, $K \subseteq X^*$ a Radon-Nikodym set, and $T \subseteq X$ is point countable and point separating on K. We may also assume that the linear span $[T]$ of T is X; in fact we may assume that T is a norm dense linear (over the rationals) subspace of X. The weak* closed convex hull of K is also a Radon-Nikodym set (see **[S6]** but not the review thereof in *Mathematical Reviews*). We check that the norm closed convex hull of K is the same as the weak* closed convex hull. Let μ be a Radon probability measure on K; the resolvent $r(\mu) \in X^*$ is the linear functional given by

$$r(\mu)(x) = \int x \, d\mu.$$

The set of all resolvents is exactly the weak* closed convex hull of K. For any $\epsilon > 0$ there exists a norm compact set $F \subseteq K$ such that $\mu(F) > 1 - \epsilon$. The norm closed convex hull of F is norm compact and contains $r(\nu)$ where $\nu = (\mu(F))^{-1}\chi_F \mu$. For any $x \in X$ we have that

$$|r(\mu)(x) - r(\nu)(x)| = |\int x \, d\mu - \int x\nu|$$
$$\leq \left(\int_{K \setminus F} |x| \, d\mu\right) + |1 - \mu(F)^{-1}| \left(\int_F |x| \, d\mu\right) \leq (\epsilon + |1 - \frac{1}{1-\epsilon}|)\|x\|.$$

It remains to verify that T is point countable on the weak* closed convex hull of K. Suppose that y^* is in this hull; then there exists a countable set $C \subseteq K$ such that y^* is in the norm convex hull of C. If some $x \in T$ is nonzero on y^* then x is nonzero on some point in C; since C is countable this proves that

$$\{x \in T : |y^*(x)| > 0\}$$

is countable. Finally, we may assume that K is convex. Define $I : X \to L(X)$ and $\lambda_0, \lambda_1, \lambda_2, \ldots$ as before. Since K is convex and norm closed it is a (norm) retract of X^*

(see **[BP]**); also, I^* is an isometry on K (considered as a subset of $L(X)$) and it is easy to see that K is a retract $L(X)^*$. We may assume that each λ_n takes values in K. An elementary property of Corson compacta is that given an infinite set $T' \subseteq T$ then there exist $S \subseteq T$, $T' \subseteq S$ and a continuous (in this case weak* continuous) retraction r on K such that $f \circ r = f$ for all $f \in S$ and $f \circ r = 0$ for all $f \in T \setminus S$; such sets S are said to be good subsets (see **[Ne]**). The increasing union of good sets is also good. The crucial facts about any weak* compact subset F of K and any family S of continuous functions that separate the points of F are the following:

$$\text{norm density } F \leq \text{ weight } F \leq \text{ cardinality } S$$

(the weight is with respect to the weak* topology) because K satisfies the hypothesis of the Main Theorem. There exists an interval of ordinals $[0, \eta]$ for which we can construct good increasing subsets S_α of T with the following properties:

(i) $S_0 = \emptyset$;

(ii) S_1 is countable;

(iii) S_α is countable or the cardinality of S_α is the cardinality of α;

(iv) if β is a limit ordinal then $S_\beta = \cup_{\alpha<\beta} S_\alpha$;

(v) $S_\eta = T$;

(vi) there exist continuous retractions $\{r_\alpha\}$ on K such that $fr_\alpha = f$ if $f \in S_\alpha$ and if $f \in T \setminus S_\alpha$ then $fr_\alpha = 0$ and

(vii) if L_α denotes the smallest lattice containing S_α and the constants then, for any limit ordinal β,

$$\overline{\bigcup_n \lambda_n(L_\beta)}^{\text{norm}} = \overline{\bigcup_n \bigcup_{\alpha<\beta} \lambda_n(L_\alpha)}^{\text{norm}} = r_\beta(K).$$

The construction is by induction and follows from the following discussion. Fix an infinite subset S of T. Construct an increasing subsequence of good subsets $\{T_j\}$ of T of the same cardinality as S, with associated retractions $\{r_j\}$, the norm closure of T_j is a linear subspace X_j of X, X_j generates the lattice L_j and

$$r_j(K) \subseteq \overline{\bigcup_n \lambda_n(L_j)}^{\text{norm}} \subseteq r_{j+1}(K).$$

It follows that $\cup_j T_j = T_0$ is also good with retraction r. Let $L = \overline{\cup_j L_j}$; it follows that L is the lattice generated by $Y = \overline{\cup_j X_j}^{\text{norm}}$. Observe that $f \circ r = f$ for all $f \in L$. Let $J : L \to L(X)$ be the containment operator. From the above we know that

$$J^*(K) = J^*r(K) \subseteq \overline{\bigcup_n J^*\lambda_n(L)}^{\text{norm}}$$

$$\subseteq \overline{\bigcup_j \bigcup_n J^*\lambda_n(L_j)}^{\text{norm}} \subseteq \overline{\bigcup_j J^*r_{j+1}(K)}^{\text{norm}} \subseteq J^*r(K).$$

For a given $\epsilon > 0$ and $k \in K$ choose n, j and $\phi \in L_j$ such that

$$\|J^*r(k) - J^*(\lambda_n(\phi))\| < \epsilon;$$

because $f \circ r \in L$ and $r(\lambda_n(\phi)) = \lambda_n(\phi)$ it follows that

$$\begin{aligned}\|r(k) - \lambda_n(\phi)\| &= \sup_{\substack{f \in L(X) \\ L(f) \leq 1}} |f(r(k)) - f(\lambda_n(\phi))| \\ &= \sup_{\substack{f \in L \\ L(f) \leq 1}} |f(r(k)) - f(\lambda_n(\phi))| = \|J^*r(k) - J^*(\lambda_n(\phi))\| < \epsilon.\end{aligned}$$

The relevant formula is

$$r(K) = \overline{\bigcup_j r_j(K)}^{\text{norm}}.$$

If we have constructed S_α then $S_{\alpha+1}$ is constructed by this process.
If we define

$$\rho_\alpha(k, k') = \sup_{\substack{L(f) \leq 1 \\ f \in S_\alpha}} |f(k) - f(k')| \tag{5}$$

then we have that

$$\rho_\alpha(k, k') = \|r_\alpha(k) - r_\alpha(k')\|.$$

Define the sets $T_\beta = S_\beta \setminus \cup_{\alpha < \beta} S_\alpha$ and the functions

$$\phi_\beta(k) = \sup_{\substack{L(f) \leq 1 \\ f \in T_\beta}} |f(k)|.$$

Observe that each ϕ_β is lower semi-continuous, $\phi_\beta(k) = 0$ if β is a limit ordinal and

$$\phi_{\beta+1}(k) = \rho_{\beta+1}(r_{\beta+1}(k), r_\beta(k)) = \|r_{\beta+1}(k) - r_\beta(k)\|.$$

Suppose that each α_i is not a limit ordinal and

$$0 < \alpha_1 < \alpha_2 < \cdots < \beta = \sup_n \alpha_n < \eta.$$

Fix $k \in K$ so that $r_\beta(k) = k$. Choose $k_n \in r_{\alpha_n}(K)$ so that $\|k - k_n\| \to 0$. Then

$$\begin{aligned}&\|k_n - r_{\alpha_{n+1}-1}(k)\| = \|r_{\alpha_{n+1}-1}(k_n) - r_{\alpha_{n+1}-1}(k)\| \\ &= \rho_{\alpha_{n+1}-1}(k_n, k) \leq \|k_n - k\| \xrightarrow{n} 0.\end{aligned}$$

Therefore, $\|k - r_{\alpha_n - 1}(k)\| \xrightarrow{n} 0$ and

$$\phi_{\alpha_n}(k) \leq \|r_{\alpha_n-1}(k) - r_{\alpha_n}(k)\| = \rho_{\alpha_n-1}(k, r_{\alpha_n}(k)) \xrightarrow{n} 0.$$

This proves that

$$\{\alpha : |\phi_\alpha(k)| > \epsilon\}$$

is finite for each k and each $\epsilon > 0$. Assume that for any subset T of S such that the cardinality of T is less than that of S we have that the image of K in $\prod_T[-1,1]$ defined by $k \to (f(k))_{f\in T}$ is an Eberlein compact. Let

$$\{E_\beta \subseteq \prod_{T_\beta}[-1,1]\}$$

be the images under these maps. From this, it follows that K itself is an Eberlein compact because the image of K in $\prod E_\beta$ defined by $k \to ((f(k))_{f\in T_\beta})_\beta$ is a homeomorphism whose image is in the c_0 sum of the E_β's; this follows because $\{\alpha : |\phi_\alpha(k)| > \epsilon\}$ is finite for each k and each $\epsilon > 0$. The following is in **[S10]** but the proof here is even easier.

Theorem. *Suppose that K and T are compact Hausdorff spaces and $q : K \to T$ is continuous. Suppose that at least one of K and T is a Corson compact and at least one is a Radon-Nikodym compact. Then q continuously factors through a space S that is an Eberlein compact.*

Proof. We have shown that if K, or T, has both properties then K, or T, is an Eberlein compact. It is known that if K is Corson compact then $q(K)$ is Corson compact; see **[Ne]**, but a nearly trivial proof is given in **[S10]**, which bears repeating. We shall assume that T is Corson compact and K is a Radon-Nikodym compact. We assume that q is onto and $A \subseteq C(K)$ is the subalgebra such that $f \in A$ if and only if there exists $h \in C(T)$ such that $gq = f$. We have already used a variant of the following. Suppose that C is a subset of the unit ball of $C(K)$ that is equimeasurable and separates the points of K. We may assume that C is a convex and symmetrical subset of the unit ball with $1_K \in C$. It is easy to check that $C \cdot C$ is also equimeasurable and, by induction, C^n is equimeasurable; it is also routine to check that $(C \vee C) - C$ is equimeasurable. Let E be an equimeasurable set that is closed, convex, symmetrical and contains

$$\sum_n 2^{-n}C^n.$$

Again, it follows from the Stone-Weierstraß theorems (see **[N]**) that $\cup_n 2^n E$ is uniformly dense in $C(K)$. Let $\mathbf{F}$ be any subset of A that is point countable and separates the (positive and multiplicative) states of A. We shall show that there exists a subset $\mathbf{G}$ of $C(K)$ that is both equimeasurable and point countable and the algebra A_1 generated by $\mathbf{G}$ contains $\mathbf{F}$. Observe that the family of all polynomials in $\mathbf{F}$ with rational coefficients is also point countable. Thus, we may assume that $\mathbf{F}$ is a dense subset of the unit ball of A (yes, again, the Stone-Weierstraß theorem). Partition $\mathbf{F}$ by

$$\mathbf{F} = \bigcup_n \mathbf{F}_n$$

so that

$$\mathbf{F}_n \subseteq (2^n E) + B(0, \frac{1}{4}).$$

For each $f_{n,\alpha} \in \mathbf{F}_n$ choose $h_{n,\alpha} \in E$ so that $\|2^n h_{n,\alpha} - f_{n,\alpha}\| < \frac{1}{4}$. Define

$$u_{n,\alpha} = (2^n h_{n,\alpha} \vee \frac{1}{2}) - \frac{1}{2} \geq 0$$

and observe that

$$2^{-n} u_{n,\alpha} = (h_{n,\alpha} \vee \frac{1}{2^{n+1}}) - \frac{1}{2^{n+1}} \in (E \vee \frac{1}{2^{n+1}}) - \frac{1}{2^{n+1}} \subseteq (E \vee E) - E$$

because E is convex and contains the origin and 1_K. Thus, $\{2^{-n}u_{n,\alpha} : n, \alpha\}$ is equimeasurable. Fix $k \in K$; if $2^{-n}u_{n,\alpha}(k) > 0$ then $2^n h_{n,\alpha}(k) > \frac{1}{2}$ which implies that $|f_{n,\alpha}(k)| > \frac{1}{4}$. Thus, $\{2^{-n}u_{n,\alpha} : n, \alpha\}$ is also point countable. Fix two points k_0 and k_1 in K so that $k_0|A = a_0$ and $k_1|A = a_1$ are distinct (positive, multiplicative) states of A and choose a function $h \in A$ so that

$$-1 = h(a_0) < h(a_1) = 1 = \|h\|.$$

There exist n and $f_{n,\alpha} \in \mathbf{F}_n$ so that $\|f_{n,\alpha} - h\| < \frac{1}{4}$. Therefore, $\|2^n h_{n,\alpha} - h\| < \frac{1}{2}$ and it follows that

$$0 = u_{n,\alpha}(a_0) < u_{n,\alpha}(a_1).$$

Let $\mathbf{G} = \{2^{-n}u_{n,\alpha} : n, \alpha\}$. Let A_2 be the algebra generated by $\mathbf{F} \cup \mathbf{G}$ and let A_1 be the algebra generated by $\mathbf{G}$. The arguments above show that the functions in A_2 and A_1 separate exactly the same points of K. If some $h \in A$ separates two points in K then those two points are also separated by some $u_{n,\alpha}$. This means that $A \subseteq A_1 = A_2$. The algebra A_1 is generated by the point countable and equimeasurable set $\{2^{-n}u_{n,\alpha} : n, \alpha\}$. We have proved that the state space of A_1 is an Eberlein compact. From **[Gul]** or **[BRW]** (see also **[MR]** for another approach) it follows that the state space of A is also an Eberlein compact and this, of course, is $q(K) = T$.

BIBLIOGRAPHY

[AL] D. Amir and J. Lindenstrauss, The structure of weakly compact sets in Banach spaces, Ann. of Math. 88 (1968), 35–46

[BRW] Y. Benjamini, M. E. Rudin and M. Wage, Continuous images of weakly compact subsets of Banach spaces, Pacific J. Math. 70 (1977), 309–324

[BP] C. Bessaga and A. Pelczynski, *Selected Topics in infinite-dimensional Topology*, PWN Warszawa, 1975

[DFJP] W. Davis, T. Figiel, W. Johnson and A. Pełczynski, Factoring weakly compact operators, J. Funct. Analysis 17 (1974), 311–327

[F] M. Fabian, Each weakly countably determined Asplund space admits a Fréchet differentiable norm, Bull. Austral. Math. Soc. 36 (1987), 367–384

[**FG**] M. Fabian and G. Godefroy, The dual of every Asplund space admits a projectional resolution of identity, to appear in Studia Math.

[**G**] A. Grothendieck, *Produits tensoriels topologiques et espaces nucléaires*, Memoire AMS 16, 1955

[**Gu**] S. P. Gul'ko, On the structure of spaces of continuous functions and their complete paracompactness, Russian Math. Surveys 34 (1979), 36–44

[**Gu1**] S.P. Gul'ko, On properties of subsets of σ-products, Dokl. Akad. Nauk. USSR 237 (1977), 505–508

[**Gr**] G. Gruenhage, A note on Gul'ko compact spaces, Proc. AMS 100, (1987) 371–376

[**MR**] E. Michael and M. E. Rudin, A note on Eberlein compacts, Pacific J. Math. 72 (1977), 487–495

[**vM**] J. van Mill, *Infinite-Dimensional Topology*, North-Holland, Amsterdam, 1989

[**N**] M. A. Naimark, *Normed Rings*, Noordhoff, Groningen, 1964

[**Na**] I. Namioka, Radon-Nikodym compact spaces and fragmentability, Mathematika 34 (1987), 258–281

[**Ne**] Negrepontis, Banach Spaces and Topology, in *Handbook of Set Theoretic Topology*, North Holland, Amsterdam, 1984

[**R**] G. Rodé, Superkonvexität und schwache Kompaktheit, Arch. Math. 36 (1981), 62–72

[**SL**] G. A. Sokolov and Leiderman, Adequate families of sets and Corson compacts, Commentationes Mathematicae Universitatis Carolinae 25 (1984), 233–246

[**S1**] C. Stegall, *Applications of Descriptive Topology in Functional Analysis*, Universität Linz, 1985

[**S2**] C. Stegall, The duality between Asplund spaces and spaces with the Radon-Nikodym property, Israel J. Math. 29 (1978), 408–412

[**S3**] C. Stegall, Functions of the first Baire class with values in Banach spaces, Proc. AMS 111 (1991), 981–991

[**S4**] C. Stegall, A Proof of the Theorem of Amir and Lindenstrauss, Israel J. Math. 68 (1989), 185–192

[**S5**] C. Stegall, The Radon-Nikodym property in conjugate Banach spaces, Trans. AMS 206 (1975), 213-223

[**S6**] C. Stegall, The Radon-Nikodym property in conjugate Banach spaces II, Trans. AMS 264 (1981), 507-519

[**S7**] C. Stegall, The Topology of Certain Spaces of Measures, Topology and its Applications 41 (1991), 73–112

[**S8**] C. Stegall, Dunford-Pettis Sets, in *General Topology and its Relations to Modern Analysis and Algebra VI*, Z. Frolík, editor, Heldermann, Berlin, 1988

[**S9**] C. Stegall, More Facts about conjugate Banach Spaces with the Radon-Nikodym Property, Acta Universitatis Carolinae–Math. et Phys. 31 (1990), 107–117

[**S10**] C. Stegall, More Facts about conjugate Banach Spaces with the Radon-Nikodym Property II, Acta Universitatis Carolinae–Math. et Phys. 32 (1991), 47–54

[T] M. Talagrand, Espaces de Banach faiblement k-analytique, Ann. of Math. 110 (1979), 407–438

[V] L. Vasak, On one generalization of weakly compactly generated Banach spaces, Studia Math. 70 (1981), 11–19

[Ve] Libor Veselý, Characterization of Baire-one functions between topological spaces, to appear, Acta Universitatis Carolinae–Math. et Phys.

Some Lifting Theorems for Bounded Linear Operators

DIRK WERNER

I. Mathematisches Institut, Freie Universität Berlin, Arnimallee 3, D-1000 Berlin 33, Germany. e-mail: werner@math.fu-berlin.de

1 INTRODUCTION

Our aim in the present paper is to survey some results in connection with linear liftings of operators. That is, we are interested in the following problem.

> PROBLEM. Let Y and X be Banach spaces, $J \subset X$ a closed subspace and $T : Y \to X/J$ a bounded linear operator. Does there exist a bounded linear operator $L : Y \to X$ such that (q denoting the quotient map from X onto X/J) $qL = T$, in other words, such that the diagram
>
> 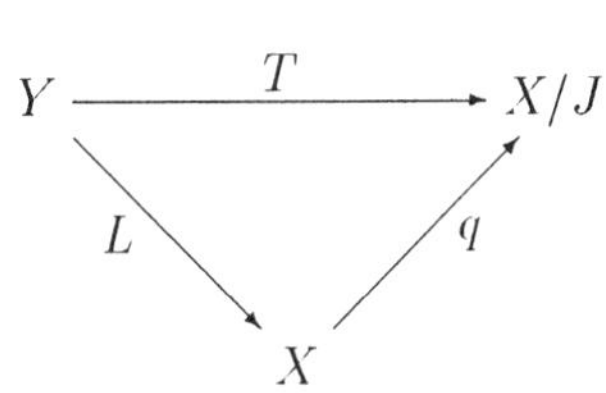
>
>
> commutes?

This problem is most often posed in the special case that $Y = X/J$ and $T = Id$. Under this assumption a lifting L is a right inverse for the quotient map.

It is clear that – with no additional assumptions made – this is not to be expected. For if L is a continuous linear right inverse for the quotient map, then the operator

$P = Lq$ is a continuous linear projection on X with ker $P = J$. However, for example in the case that $X = \ell^\infty$ and $J = c_0$ no such projection exists [42].

In this paper we shall discuss the Ando-Choi-Effros theorem (see Section 2) giving sufficient conditions which ensure that a continuous linear lifting exists. Then we will apply it in various instances, and we reprove several classical theorems on linear extension operators (Section 3). In Section 4 we deal with Sobczyk's theorem on projections onto c_0 and obtain isomorphic representations of some spaces of analytic functions. This paper is – for the most part – of expository nature, most of its results are already known.

In order to state the Ando-Choi-Effros theorem, we need the notion of an M-ideal in a Banach space.

Definition 1.1 *A closed subspace J of a Banach space X is called an M-ideal if there is a linear projection P from X^* onto $J^\perp := \{x^* \in X^* \mid x^*(x) = 0 \ \forall x \in J\}$ satisfying*

$$\|x^*\| = \|Px^*\| + \|x^* - Px^*\| \qquad \forall x^* \in X^*.$$

This definition is due to Alfsen and Effros [2]. Here are some examples of M-ideals.

Example 1.2(a) Let S be a locally compact Hausdorff space. Then $J \subset C_0(S)$ is an M-ideal if and only if there is a closed subset D of S such that

$$J = J_D := \{x \in C_0(S) \mid x(s) = 0 \quad \text{for all } s \in D\}.$$

Clearly, $\mu \mapsto \chi_D\mu$ is the required projection onto $J_D^\perp$. For a detailed proof of the converse see [7, p. 40] or [20, Ex. I.1.4(a)]. Note that in particular c_0 is an M-ideal in ℓ^∞.

Example 1.2(b) Let A be the disk algebra, that is the complex Banach space of continuous functions on the closed unit disk which are analytic in the open unit disk. It will be convenient to consider A (via boundary values) as a subspace of $C(\mathbb{T})$, where $\mathbb{T}$ is the unit circle. Then J is a nontrivial M-ideal in A if and only if there is a closed subset $D \neq \emptyset$ of $\mathbb{T}$ with linear Lebesgue measure 0 such that (J_D as above)

$$J = J_D \cap A = \{x \in A \mid x(t) = 0 \quad \text{for all } t \in D\}.$$

This follows from Example 1.2(a) and Propositions 3.3 and 3.4 below which are applicable by the F. and M. Riesz theorem. More generally, the M-ideals of a function algebra coincide with the annihilators of p-sets, see [21] or [20, Th. V.4.2].

Example 1.2(c) Let K be a compact convex set in a Hausdorff locally convex topological vector space. As usual, $A(K)$ denotes the space of real-valued affine continuous functions on K. Let us recall the definition of a split face of K [1, p. 133]. A face F of K is called a split face if there is another face F' such that every $k \in K\backslash(F \cup F')$ has a *unique* representation

$$k = \lambda k_1 + (1 - \lambda)k_2 \quad \text{with } k_1 \in F,\ k_2 \in F',\ 0 < \lambda < 1.$$

It is known that every closed face of a simplex is a split face [1, p. 144].

Then J is an M-ideal in $A(K)$ if and only if there exists a closed split face F of K such that

$$J = J_F \cap A(K) = \{x \in A(K) \mid x(k) = 0 \quad \text{for all } k \in F\}.$$

The proof of this fact can be given along the lines of (b), the crucial step being the measure theoretic characterisation of closed split faces (see [1, Th. II.6.12]) which replaces the use of the F. and M. Riesz theorem.

Example 1.2(d) In a C^*-algebra the M-ideals coincide with the closed two-sided ideals.

This was first proved in [2] and [38], see also [20, Th. V.4.4].

For future reference we state an equivalent characterisation of M-ideals by means of an intersection property of balls, the so-called 3-ball property.

Theorem 1.3 *A closed subspace J of a Banach space X is an M-ideal if and only if for all $x \in B_X$ (the closed unit ball of X), all $y_1, y_2, y_3 \in B_J$ and all $\varepsilon > 0$ there is some $y \in J$ such that*

$$\|x + y_i - y\| \leq 1 + \varepsilon \qquad (i = 1, 2, 3).$$

PROOF: [25, Th. 6.17] or [20, Th. I.2.2]. □

A detailed exposition of M-ideal theory will appear in [20].

2 AROUND THE ANDO-CHOI-EFFROS THEOREM

As already mentioned, the Ando-Choi-Effros theorem presents additional sufficient conditions for a continuous linear lifting to exist. As usual, AP stands for 'approximation property', BAP for 'bounded approximation property', and MAP for 'metric approximation property'.

Theorem 2.1 (Ando-Choi-Effros)
Suppose J is an M ideal in the Banach space X, Y is a separable Banach space, and $T \in L(Y, X/J)$ with $\|T\| = 1$. Assume further

(a) *Y has BAP*

or

(b) *J is an L^1-predual.*

Then there is a continuous linear lifting L for T. More precisely, we obtain a lifting with

$$\|L\| \leq \lambda \qquad \text{if } Y \text{ has } \lambda\text{-AP},$$

resp.

$$\|L\| = 1 \qquad \text{under assumption (b)}.$$

PROOF: See the original papers [5] and [12]. Ando's proof is elaborated in [20, II.2.1]. □

In the next sections some consequences of this theorem will be presented. But first of all let us discuss to what extent the additional assumptions in Theorem 2.1 are really needed. As for the separability one just has to activate the example of the noncomplemented M-ideal c_0 in ℓ^∞. (In view of the examples in [13] or [26, Prop. 3.5]

Theorem 2.1 does not even extend to weakly compactly generated spaces Y.) Instead of the separability of Y and (a) or (b) one may of course use the weaker assumptions

(a′) T factors through a separable space with BAP

resp.

(b′) J is an L^1-predual, and T has separable range.

As for the approximation property we first note a proposition.

Proposition 2.2 *For every separable Banach space Y there are a separable Banach space X enjoying MAP and an M-ideal J in X such that $Y \cong X/J$.*

Applying Proposition 2.2 to a space Y without BAP we infer that there can be no continuous linear lifting for the quotient map, because otherwise Y would be isomorphic to a complemented subspace of X and hence would have BAP.

PROOF OF PROPOSITION 2.2: We let (E_n) be an increasing sequence of finite dimensional subspaces of Y such that $\bigcup E_n$ is dense and define

$$\begin{aligned} X &= \{(x_n) \mid x_n \in E_n, \ \ \lim x_n \text{ exists}\}, \\ J &= \{(x_n) \mid x_n \in E_n, \ \ \lim x_n = 0\}. \end{aligned}$$

(These spaces are, of course, equipped with the sup norm.)

It is quickly verified that Y is isometric to X/J. Moreover, J is an M-ideal. To prove this we use the 3-ball property (Theorem 1.3). In fact, given normed vectors $\xi = (x_n) \in X$, $\eta_i = (y_n^i) \in J$ $(i = 1, 2, 3)$ and $\varepsilon > 0$, choose N such that

$$\|y_n^i\| \le \varepsilon \qquad \text{for } n > N \text{ and } i = 1, 2, 3.$$

If $\eta = (y_n)$ with $y_n = x_n$ $(n \le N)$, $y_n = 0$ $(n > N)$, then

$$\|\xi + \eta_i - \eta\| \le 1 + \varepsilon.$$

It remains to notice that X has MAP since the contractive finite rank projections $P_m :$ $(x_n) \mapsto (x_1, \dots, x_m, x_m, x_m, \dots)$ converge strongly to the identity. □

Also, it is not enough to assume that $J^\perp$ is norm one complemented by *some* projection as is shown by the example of a subspace J of ℓ^1 such that $\ell^1/J = L^1[0, 1]$. However, in this setting we can prove the following.

Proposition 2.3 *Let $J \subset X$ be a closed subspace, and suppose that there is a contractive linear projection from X^* onto $J^\perp$. Then every operator $T \in L(Y, X/J)$ which is a limit of finite rank operators can be lifted to an operator $L \in L(Y, X)$, which is a limit of finite rank operators, too. Also, given $\varepsilon > 0$, one may require that $\|L\| \le (1 + \varepsilon)\|T\|$.*

PROOF: We let $\overline{F}(.,.)$ denote the norm closure of the finite rank operators. We claim that

$$\overline{F}(Y, X/J) \cong \overline{F}(Y, X)/\overline{F}(Y, J) \tag{$*$}$$

which immediately gives the assertion of Proposition 2.3.

Now $\overline{F}(Y, X) \cong Y^* \widehat{\otimes}_\varepsilon X$, the completed injective tensor product of Y^* and X; thus $(*)$ is equivalent to the claim that $Id \otimes q$ is a quotient map on $Y^* \widehat{\otimes}_\varepsilon X$, where of course

$q : X \to X/J$ denotes the canonical mapping. This in turn is equivalent to $(Id \otimes q)^*$ being an isometric embedding of $(Y^* \widehat{\otimes}_\varepsilon X/J)^*$ into $(Y^* \widehat{\otimes}_\varepsilon X)^*$. But these duals can be represented by spaces of integral operators:

$$
\begin{aligned}
(Y^* \widehat{\otimes}_\varepsilon X/J)^* &\cong I(Y^*, (X/J)^*), \\
(Y^* \widehat{\otimes}_\varepsilon X)^* &\cong I(Y^*, X^*).
\end{aligned}
$$

Thus, yet another reformulation of $(*)$ reads: The canonical map of $I(Y^*, (X/J)^*)$ into $I(Y^*, X^*)$ is an isometry. But this follows from the supposed existence of a contractive projection P from X^* onto $J^\perp \cong (X/J)^*$. Indeed, if $S \in I(Y^*, (X/J)^*)$,

$$
\begin{aligned}
\|S \mid I(Y^*, X^*)\| &\leq \|S \mid I(Y^*, J^\perp)\| \\
&= \|PS \mid I(Y^*, J^\perp)\| \\
&\leq \|P\| \cdot \|S \mid I(Y^*, X^*)\| \\
&= \|S \mid I(Y^*, X^*)\|.
\end{aligned}
$$

□

Let us observe that Proposition 2.3 does not hold for merely compact operators. Suppose it did. Let E be a Banach space without the approximation property, and let Y be a Banach space such that there is a compact operator $T : Y \to E$ which is not approximable by finite rank operators (cf. [30, p. 32]). Let X and J be as in Proposition 2.2 such that $X/J \cong E$. If T were liftable to a compact operator $L : Y \to X$, then L would be approximable since X has the MAP, and so would be $T = qL$. This is a contradiction.

It is noteworthy that there is always a nonlinear continuous projection onto an M-ideal and thus a nonlinear continuous lifting for the quotient map; this follows from Michael's selection theorem and the fact that the metric projection onto an M-ideal is continuous with respect to the Hausdorff metric. (For details, cf. [23], [17], [43] or [20, II.1.9].)

3 LINEAR EXTENSION OPERATORS

We now turn to the consequences of Theorem 2.1. It contains several well-known results on the existence of linear extension operators as a special case.

Corollary 3.1 (Borsuk-Dugundji)
Let K be a compact Hausdorff space and let $D \subset K$ be a closed metrizable subset. Then there is a linear extension operator $T : C(D) \to C(K)$ with $\|T\| = 1$, i.e. $(Tx)(t) = x(t)$ for $x \in C(D)$ and $t \in D$.

PROOF: Consider the M-ideal $J_D = \{x \in C(K) \mid x|_D = 0\}$ (Example 1.2(a)). Then $C(D) \cong C(K)/J_D$ meets the requirements of Theorem 2.1. (As a matter of fact, both (a) and (b) are fulfilled: $C(D)$ has MAP, and J_D is an L^1-predual.) □

Corollary 3.2 (Pelczyński)
Let A be the disk algebra, and suppose D is a subset of the unit circle with Lebesgue measure 0. Then there is a contractive linear extension operator from $C(D)$ to A.

PROOF: Consider the M-ideal $J = \{x \in A \mid x|_D = 0\}$ (Example 1.2(b)). By the Rudin-Carleson theorem [18, p. 58] we have $C(D) \cong A/J$, hence the result. □

More general corollaries can be formulated along the same lines on the basis of the Glicksberg peak interpolation theorem [18, p. 58, Th. 12.5 and Th. 12.7] and Proposition 4.8 below, along with the discussion preceding it.

Next we show how to obtain the Michael-Pelczyński-Ryll-Nardzewski theorem as a consequence of Theorem 2.1. For this we need to discuss M-ideals in subspaces of $C(K)$.

We recall from the fundamental Example 1.2(a) that the M-ideals in $C(K)$ coincide with the ideals J_D. The following proposition says that the J_D are in fact the ancestors of all M-ideals, since every Banach space is a subspace of some $C(K)$.

Proposition 3.3 *Let X be a closed subspace of $C(K)$, and let J be an M-ideal in X. Then there is a closed subset D of K such that $J = J_D \cap X$.*

PROOF: Let $D = \{k \in K \mid \delta_k|_X \in J^\perp\}$. Then D is a closed set, and $J \subset J_D \cap X$ by construction. If the inclusion were proper, we could separate a certain $x_0 \in J_D \cap X$ from J by a functional $p \in J^\perp$. We may even assume $p \in \text{ex } B_{J^\perp}$ by the Krein-Milman theorem and thus $p \in \text{ex } B_{X^*}$. (It is easily verified that for an M-ideal $\text{ex } B_{J^\perp} \subset \text{ex } B_{X^*}$.) Such a p is of the form $p(x) = \lambda \cdot x(k)$ for some $k \in K$, $|\lambda| = 1$. Since $p \in J^\perp$ we must have $k \in D$ and hence $x_0(k) = 0$. On the other hand $p(x_0) \neq 0$ since p is a separating functional: a contradiction. □

We turn to the converse of this proposition, which, of course, will generally not hold. The following is obvious.

Proposition 3.4 *Let X be a closed subspace of $C(K)$. Then $J_D \cap X$ is an M-ideal in X if the L-projection $\mu \mapsto \chi_D\mu$ from $C(K)^*$ onto $J_D^\perp$ leaves $X^\perp$, the annihilator of X in $C(K)^*$, invariant.*

Let us formulate an important example where M-ideals are induced in a subspace. Suppose X is a closed subspace of $C(K)$ and suppose $D \subset K$ is closed. We let $X|_D$ be the space of all restrictions $\{x|_D \mid x \in X\}$. Following [31] one says that $(X|_D, X)$ has the *bounded extension property* if there exists a constant C such that, given $\xi \in X|_D$, $\varepsilon > 0$ and an open set $U \supset D$, there is some $x \in X$ such that

$$\begin{aligned} x|_D &= \xi, \\ \|x\| &\leq C \cdot \|\xi\|, \\ \|x(k)\| &\leq \varepsilon \qquad \text{for } k \notin U. \end{aligned}$$

(Note that $X|_D$ is closed under this assumption.) For example, if $X \subset C(K)$ is a subalgebra and $D \subset K$ is a subset of the form $f^{-1}(\{1\})$ for some $f \in B_X$ (a 'peak set'), then the pair $(X|_D, X)$ has the bounded extension property: Let $\xi \in X|_D$ and $g \in X$ such that $g|_D = \xi$. First of all we remark that replacing f by $(1+f)/2$ permits us to assume that in addition $|f(k)| < 1$ if and only if $k \notin D$. Then $g \cdot f^n$ meets the requirements of the above definition if n is large enough. (The argument easily extends to intersections of peak sets, the so-called p-sets.)

We now have:

Proposition 3.5 *If $(X_{|D}, X)$ has the bounded extension property, then $J_D \cap X$ is an M-ideal in X.*

PROOF: To begin with we observe that the constant C appearing in the definition of the bounded extension property may be chosen as close to 1 as we wish. To see this let $U \supset D$ be an open set and let $\varepsilon > 0$. Applying the bounded extension property with $U_1 = U$ and ε yields an extension x_1 of a given $\xi \in X_{|D}$ (w.l.o.g. $\|\xi\| = 1$) such that $\|x_1\| \leq C$ and $|x_1| \leq \varepsilon$ off U_1. Then we repeat this procedure with $U_2 = \{k \mid |x_1(k)| < 1+\varepsilon/2\} \cap U_1$ and obtain an extension x_2 such that $\|x_2\| \leq C$ and $|x_2| \leq \varepsilon$ off U_2. In the third step one applies the bounded extension property with $U_3 = \{k \mid |x_2(k)| < 1 + \varepsilon/2\} \cap U_2$ etc. This yields a sequence of extensions (x_n) with $\|x_n\| \leq C$ and $|x_n| \leq \varepsilon$ off U_n. Let $x = \frac{1}{N}\sum_{n=1}^{N} x_n$. Obviously we have $x_{|D} = \xi$ and $|x| \leq \varepsilon$ off U. Finally, if $k \in U$, then by construction $|x_n(k)|$ is big (but $\leq C$) for at most one n, and $|x_n(k)| \leq 1+\varepsilon/2$ otherwise. It follows

$$|x(k)| \leq \frac{1}{N}\Big(C + (N-1)(1+\varepsilon/2)\Big) \leq 1+\varepsilon$$

for sufficiently large N. (As a consequence of this one may remark that $X/(J_D \cap X) \cong X_{|D}$.)

Now we can prove that $J_D \cap X$ is an M-ideal employing the 3-ball property of Theorem 1.3. Thus, let y_1, y_2, y_3 in the unit ball of $J_D \cap X$, $x \in B_X$ and $\varepsilon > 0$. With the help of the $(1+\varepsilon)$-bounded extension property we may find $\widehat{x} \in X$ such that

$$\widehat{x}_{|D} = x_{|D},$$

$$\|\widehat{x}\| \leq (1+\varepsilon)\|x_{|D}\| \leq 1+\varepsilon,$$

$$|\widehat{x}(k)| \leq \varepsilon \quad \text{if } \max_i |y_i(k)| \geq \varepsilon.$$

Let $y = x - \widehat{x}$ so that $y \in J_D \cap X$. Distinguishing whether or not $\max_i |y_i(k)| \geq \varepsilon$ one can immediately verify that

$$|(x + y_i - y)(k)| = |y_i(k) + \widehat{x}(k)| \leq 1 + 2\varepsilon$$

for all $k \in K$, i.e.

$$\|x + y_i - y\| \leq 1 + 2\varepsilon \qquad (i = 1, 2, 3).$$

□

Corollary 3.6 (Michael and Pełczyński, Ryll-Nardzewski)
Suppose $X \subset C(K)$ and $D \subset K$ is closed such that the pair $(X_{|D}, X)$ has the bounded extension property. If $X_{|D}$ is separable and has the MAP, then there is a contractive linear extension operator from $X_{|D}$ into X.

PROOF: This follows from Proposition 3.5 and Theorem 2.1. □

Next we present an application of Theorem 2.1 to potential theory. Let $U \subset \mathbb{R}^n$ be open and bounded. We put

$$H(U) = \{f \in C(\overline{U}) \mid f \text{ is harmonic in } U\}.$$

The classical Dirichlet problem requires to find, given $\varphi \in C(\partial U)$, some $f \in H(U)$ such that $f_{|\partial U} = \varphi$. This is generally impossible. However, there is always a generalised solution, called the Perron-Wiener-Brelot solution. Those points $x_0 \in \partial U$ such that for all $\varphi \in C(\partial U)$ and corresponding Perron-Wiener-Brelot solutions f the relation $\lim_{x \to x_0} f(x) = \varphi(x_0)$ is valid are called regular boundary points. The set of all regular boundary points is denoted by $\partial_r U$, and it is classical that $\partial_r U = \partial U$ if the boundary of U is sufficiently smooth or if U is simply connected and $n = 2$.

Concerning the weak solvability of the Dirichlet problem we now have:

Proposition 3.7 *Let $U \subset \mathbb{R}^n$ be open and bounded, and let $E \subset \partial_r U$ be compact. Then there is a contractive linear operator $L : C(E) \to H(U)$ such that $(L\varphi)_{|E} = \varphi$ for all $\varphi \in C(E)$.*

PROOF: This result will turn out to be almost obvious after we have reformulated it in terms of convexity theory. The space $H(U)$ is an order unit space and can hence be represented as a space of affine continuous functions $A(K)$. The crux of the matter is that here K is a Choquet simplex; see [11] or [15]. Moreover, it is known from [6] that $\partial_r U$ can be identified with the Choquet boundary of $H(U)$, that is ex K. If we regard E as a compact subset of ex K, then a corollary to Edwards' separation theorem states that every continuous function on E has a norm preserving extension to an affine continuous function on K [1, p. 91]. Let us denote $F = \overline{\text{co}}\, E$ so that F is a split face since K is a simplex. Consequently, by Example 1.2(c),

$$\{f \in A(K) \mid f_{|E} = 0\} = J_F \cap A(K) =: J$$

is an M-ideal in $A(K)$, and the quotient space $A(K)/J$ is isometric with $C(E)$ by the above. By Theorem 2.1 there is a linear contractive lifting L for the quotient map, and reidentifying $A(K)$ with $H(U)$ yields the desired solution operator from $C(E)$ to $H(U)$. □

One can also recover results of Andersen [3] using Example 1.2(d). For the original proofs of the above linear extension theorems see [14], [32], [33], [31] and [34]; for Proposition 3.7 there is a direct potential theoretic approach in [10]. A particularly nice proof of Corollary 3.1 in the case of metrizable K can be found in [22, p. 103].

4 SOBCZYK'S THEOREM AS A THEOREM ON M-IDEALS

A slightly different type of corollary is Sobczyk's theorem. In order to prove this classical theorem by M-ideal methods, we first provide a renorming result.

Proposition 4.1 *Let X be a Banach space and $Y \subset X$ a subspace isometric to c_0. Then there is an equivalent norm on X which agrees with the original norm on Y so that Y becomes an M-ideal.*

PROOF: $Y^{\perp\perp}$, which is canonically isometric to Y^{**}, is isometric to ℓ^∞, and Y (more precisely $i_X(Y)$) is an M-ideal in $Y^{\perp\perp}$, since c_0 is an M-ideal in ℓ^∞. Since $Y^{\perp\perp}$ is isometric to ℓ^∞ there is a contractive projection P from X^{**} onto $Y^{\perp\perp}$. Hence $Y^\perp$ is the kernel of a contractive projection on X^*, viz. $Q = i_X^* P^* i_{X^*}$. (To see this check $\operatorname{ran}(Id - Q) \subset Y^\perp \subset \ker Q$ which shows $Q(Id - Q) = 0$ and $\ker Q = Y^\perp$.) We now renorm X^* so that Q becomes an L-projection:

$$|x^*| := \|Qx^*\| + \|x^* - Qx^*\|$$

i.e.

$$(X^*, |\,.\,|) = \operatorname{ran}(Q) \oplus_1 Y^\perp.$$

Unfortunately $|\,.\,|$ need not be a dual norm, therefore we cannot conclude directly that Y is an M-ideal in some renorming of X. However, we have with respect to the dual norm

$$(X^{**}, |\,.\,|) = \ker(Q^*) \oplus_\infty Y^{\perp\perp}$$

so that $Y^{\perp\perp}$ is an M-summand in $(X^{**}, |\,.\,|)$. Note that $|\,.\,|$ and $\|\,.\,\|$ coincide on $Y^{\perp\perp}$; it follows that Y is an M-ideal in $(X^{**}, |\,.\,|)$, a fortiori Y is an M-ideal in the intermediate space $(X, |\,.\,|)$. □

Corollary 4.2 (Sobczyk)
If X is a separable Banach space and $Y \subset X$ is a closed subspace isometric to c_0, then there is a continuous linear projection π from X onto Y with $\|\pi\| \le 2$.

PROOF: Let $(X, |\,.\,|)$ be the renorming devised by Proposition 4.1. By Theorem 2.1 there is a contractive (with respect to $|\,.\,|$) linear lifting L of the quotient map $q : X \to X/Y$ (Y is an L^1-predual!), consequently $\pi = Id - Lq$ is a linear projection onto Y. It remains to estimate the norm:

$$\|\pi(x)\| = |\pi(x)| \le |\pi| \cdot |x| \le 2 \cdot \|x\|$$

since $\|x^*\| \le |x^*|$ for all $x^* \in X^*$ whence $|x^{**}| \le \|x^{**}\|$ for all $x^{**} \in X^{**}$. □

We hasten to add that the above argument is probably the most complicated proof of Sobczyk's theorem that has appeared in the literature; a very simple one can be found in [30, Th. 2.f.5]. The most important feature of our proof is that one can read it from the bottom to the top. This yields the following result on Banach spaces which are M-ideals in their biduals; we shall call such a Banach space M-embedded. (For the definition and basic properties of $\mathcal{L}^p$-spaces we refer to [27], [28], and [29].)

Theorem 4.3 *Let X be a separable $\mathcal{L}^\infty$-space which is an M-ideal in its bidual. Then X is isomorphic to c_0.*

PROOF: We will show that X is a complemented subspace of any separable superspace Y containing X. Zippin's famous characterisation of separably injective spaces [44] then gives the claim.

As every separable Banach space embeds into $C[0,1]$ it is sufficient to prove that every subspace of $C[0,1]$ isometric to X is the range of a continuous linear projection. Since X is an $\mathcal{L}^\infty$-space, its bidual is injective. Note $X^{**} \cong X^{\perp\perp}$ and X is an M-ideal in $X^{\perp\perp}$. Now $X^{\perp\perp}$ is injective and is therefore complemented in $C[0,1]^{**}$. Let P denote a projection from $C[0,1]^{**}$ onto $X^{\perp\perp}$; hence

$$C[0,1]^{**} \simeq X^{\perp\perp} \oplus \ker P.$$

We now renorm $C[0,1]^{**}$ (and thus its subspace $C[0,1]$) to the effect that

$$(C[0,1]^{**}, |\,.\,|) \cong X^{\perp\perp} \oplus_\infty \ker P.$$

Observe that $\|\,.\,\|$ and $|\,.\,|$ coincide on $X^{\perp\perp}$. We conclude that, since $X^{\perp\perp}$ is an M-summand in $(C[0,1]^{**}, |\,.\,|)$, X (being an M-ideal in $X^{\perp\perp}$) is an M-ideal in $(C[0,1]^{**}, |\,.\,|)$; a fortiori X is an M-ideal in $(C[0,1], |\,.\,|)$.

We remark that $(C[0,1]/X)^{**}$ is isomorphic to $\ker P$, which has BAP (it is a complemented subspace of the space $C[0,1]^{**}$ which has MAP in its original norm). Therefore $C[0,1]/X$ is a separable space with BAP, hence, by Theorem 2.1, X is a complemented subspace of $C[0,1]$. □

We do not know if the above theorem holds also in the nonseparable case. Its proof essentially used results which are limited to separable spaces; and it cannot be transferred to nonseparable spaces on a formal level since there are spaces not isomorphic to $c_0(\Gamma)$ such that every separable subspace embeds into c_0 isomorphically [24]. However, the example given in the latter paper is not weakly compactly generated whereas M-embedded spaces are [16].

Theorem 4.3 was also proved by G. Godefroy using a similar, yet somewhat more complicated argument [19]; the present proof has appeared in [39].

We wish to apply Theorem 4.3 in the following setting. We denote by $\mathbb{D}$ the open unit disk and let $\phi : [0,1] \to \mathbb{R}$ be a positive continuous decreasing function such that $\phi(0) = 1$ and $\phi(1) = 0$. We define

$$\begin{aligned} A_\infty(\phi) &= \{f : \mathbb{D} \to \mathbb{C} \text{ analytic} \mid \|f\|_\phi = \sup_{z\in\mathbb{D}} |f(z)| \cdot \phi(|z|) < \infty\}, \\ A_0(\phi) &= \{f \in A_\infty(\phi) \mid \lim_{|z|\to 1} |f(z)| \cdot \phi(|z|) = 0\}. \end{aligned}$$

These are Banach spaces under the norm $\|\,.\,\|_\phi$, and $A_\infty(\phi)$ is canonically isometric with $A_0(\phi)^{**}$ ([35], [9]). It is shown in [41] that $A_0(\phi)$ is an M-ideal in its bidual. On the other hand, for a number of weight functions ϕ it is known that $A_0(\phi)$ is an $\mathcal{L}^\infty$-space [36, Theorem 1]. Hence in this case $A_0(\phi)$ is even isomorphic to c_0; see also [37] for this result. The weight function $\phi(r) = 1 - r^2$ constitutes a case of special importance, since the corresponding space $A_0(\phi)$ is isometric to the well-known little Bloch space [4].

Actually, more general weighted spaces of analytic functions of one or several variables can be shown to be M-embedded, see [40, Section 3]. Therefore, we may add one more equivalence to [8, Prop. 3.4], where the authors prove that a certain exact sequence of Banach spaces of analytic functions is a so-called $\otimes$-sequence if and only if $A_0(\phi)$ is an $\mathcal{L}^\infty$-space. The previous discussion shows that '$A_0(\phi)$ is isomorphic to c_0' is another equivalent condition.

REFERENCES

[1] E. M. ALFSEN. *Compact Convex Sets and Boundary Integrals.* Springer, Berlin-Heidelberg-New York, 1971.

[2] E. M. ALFSEN AND E. G. EFFROS. *Structure in real Banach spaces. Part I and II.* Ann. of Math. **96** (1972) 98–173.

[3] T. B. ANDERSEN. *Linear extensions, projections, and split faces.* J. Funct. Anal. **17** (1974) 161–173.

[4] J. M. ANDERSON, J. CLUNIE, AND CH. POMMERENKE. *On Bloch functions and normal functions.* J. Reine Angew. Math. **270** (1974) 12–37.

[5] T. ANDO. *A theorem on non-empty intersection of convex sets and its applications.* J. Approx. Th. **13** (1975) 158–166.

[6] H. BAUER. *Šilovscher Rand und Dirichletsches Problem.* Ann. Inst. Fourier **11** (1961) 89–136.

[7] E. BEHRENDS. *M-Structure and the Banach-Stone Theorem.* Lecture Notes in Math. 736. Springer, Berlin-Heidelberg-New York, 1979.

[8] K. D. BIERSTEDT, J. BONET, AND A. GALBIS. *Weighted spaces of holomorphic functions on balanced domains.* Preprint, 1991.

[9] K. D. BIERSTEDT AND W. H. SUMMERS. *Biduals of weighted Banach spaces of analytic functions.* J. Austral. Math. Soc. (1992) (to appear).

[10] J. BLIEDTNER AND W. HANSEN. *The weak Dirichlet problem.* J. Reine Angew. Math. **348** (1984) 34–39.

[11] N. BOBOC AND A. CORNEA. *Convex cones of lower semicontinuous functions.* Rev. Roum. Math. Pures Appl. **12** (1967) 471–525.

[12] M.-D. CHOI AND E. G. EFFROS. *Lifting problems and the cohomology of C^*-algebras.* Canadian J. Math. **29** (1977) 1092–1111.

[13] H. H. CORSON AND J. LINDENSTRAUSS. *On simultaneous extension of continuous functions.* Bull. Amer. Math. Soc. **71** (1965) 542–545.

[14] J. DUGUNDJI. *An extension of Tietze's theorem.* Pacific J. Math. **1** (1951) 353–367.

[15] E. G. EFFROS AND J. L. KAZDAN. *Applications of Choquet simplexes to elliptic and parabolic boundary value problems.* J. Diff. Equations **8** (1970) 95–134.

[16] M. FABIAN AND G. GODEFROY. *The dual of every Asplund space admits a projectional resolution of the identity.* Studia Math. **91** (1988) 141–151.

[17] H. FAKHOURY. *Existence d'une projection continue de meilleure approximation dans certains espaces de Banach.* J. Math. pures et appl. **53** (1974) 1–16.

[18] T. W. GAMELIN. *Uniform Algebras.* Prentice-Hall, Englewood Cliffs, 1969.

[19] G. GODEFROY AND D. LI. *Some natural families of M-ideals.* Math. Scand. **66** (1990) 249–263.

[20] P. HARMAND, D. WERNER, AND W. WERNER. *M-ideals in Banach Spaces and Banach Algebras.* Monograph in preparation.

[21] B. HIRSBERG. *M-ideals in complex function spaces and algebras.* Israel J. Math. **12** (1972) 133–146.

[22] R. B. HOLMES. *Geometric Functional Analysis and its Applications.* Springer, Berlin-Heidelberg-New York, 1975.

[23] R. B. HOLMES, B. SCRANTON, AND J. WARD. *Approximation from the space of compact operators and other M-ideals.* Duke Math. J. **42** (1975) 259–269.

[24] W. B. JOHNSON AND J. LINDENSTRAUSS. *Some remarks on weakly compactly generated Banach spaces.* Israel J. Math. **17** (1974) 219–230. *Corrigendum.* Ibid. **32** (1979) 382–383.

[25] Å. LIMA. *Intersection properties of balls and subspaces in Banach spaces.* Trans. Amer. Math. Soc. **227** (1977) 1–62.

[26] J. LINDENSTRAUSS. *Weakly compact sets — their topological properties and the Banach spaces they generate.* In: R. D. Anderson, editor, *Symposium on infinite dimensional topology*, pages 235–273. Ann. of Math. Studies **69**, 1972.

[27] J. LINDENSTRAUSS AND A. PEŁCZYŃSKI. *Absolutely summing operators in $\mathcal{L}_p$-spaces and their applications.* Studia Math. **29** (1968) 275–326.

[28] J. LINDENSTRAUSS AND H. P. ROSENTHAL. *The $\mathcal{L}_p$ spaces.* Israel J. Math. **7** (1969) 325–349.

[29] J. LINDENSTRAUSS AND L. TZAFRIRI. *Classical Banach Spaces.* Lecture Notes in Math. 338. Springer, Berlin-Heidelberg-New York, 1973.

[30] J. LINDENSTRAUSS AND L. TZAFRIRI. *Classical Banach Spaces I.* Springer, Berlin-Heidelberg-New York, 1977.

[31] E. MICHAEL AND A. PEŁCZYŃSKI. *A linear extension theorem.* Illinois J. Math. **11** (1967) 563–579.

[32] A. PEŁCZYŃSKI. *On simultaneous extension of continuous functions.* Studia Math. **24** (1964) 285–304.

[33] A. PEŁCZYŃSKI. *Supplement to my paper 'On simultaneous extension of continuous functions'.* Studia Math. **25** (1964) 157–161.

[34] A. PEŁCZYŃSKI AND P. WOJTASZCZYK. *Banach spaces with finite dimensional expansions of identity and universal bases of finite dimensional subspaces.* Studia Math. **40** (1971) 91–108.

[35] L. A. RUBEL AND A. L. SHIELDS. *The second duals of certain spaces of analytic functions.* J. Austral. Math. Soc. **11** (1970) 276–280.

[36] A. L. SHIELDS AND D. L. WILLIAMS. *Bounded projections, duality, and multipliers in spaces of analytic functions.* Trans. Amer. Math. Soc. **162** (1971) 287–302.

[37] A. L. SHIELDS AND D. L. WILLIAMS. *Bounded projections, duality, and multipliers in spaces of harmonic functions.* J. Reine Angew. Math. **299/300** (1978) 256–279.

[38] R. R. SMITH AND J. D. WARD. *M-ideal structure in Banach algebras.* J. Funct. Anal. **27** (1978) 337–349.

[39] D. WERNER. *De nouveau: M-idéaux des espaces d'opérateurs compacts.* Séminaire d'Initiation à l'Analyse, année 1988/89. Université Paris VI **28** (1989) Exposé 17.

[40] D. WERNER. *Contributions to the Theory of M-Ideals in Banach Spaces.* Habilitationsschrift. FU Berlin, 1991.

[41] D. WERNER. *New classes of Banach spaces which are M-ideals in their biduals.* Math. Proc. Cambridge Phil. Soc. **111** (1992) 337–354.

[42] R. WHITLEY. *Projecting m onto c_0.* Amer. Math. Monthly **73** (1966) 285–286.

[43] D. YOST. *Best approximation and intersections of balls in Banach spaces.* Bull. Austral. Math. Soc. **20** (1979) 285–300.

[44] M. ZIPPIN. *The separable extension problem.* Israel J. Math. **26** (1977) 372–387.

Invariant Subspaces for Certain Representations of $H^\infty(G)$

ERNST ALBRECHT Fachbereich Mathematik, Universität des Saarlandes, D-6600 Saarbrücken, Germany

BERNARD CHEVREAU U.E.R. de Mathématiques et Informatique, Université de Bordeaux-I, 351, cours de la Libération, 33405 Talence, France

0 INTRODUCTION

In this note we give some sufficient conditions for the existence of non-trivial (closed) invariant subspaces for certain representations of the algebra $H^\infty(G)$ of all bounded analytic functions on a sufficiently "nice" bounded domain in $\mathbf{C}^N$ (or of weak* closed subalgebras of $H^\infty(G)$). In the case N=1 our result is a generalization to arbitrary Banach spaces of an invariant subspace theorem due to the second author, C. Pearcy and A. Shields [10]. In the case of several variables completely new phenomena occur. In this situation it is not even known if singular Fredholm systems have non-trivial invariant subspaces. The problem is caused by the possible existence of non-zero interior cohomology groups (even if the exterior groups may be trivial) in the Koszul complex of the system (see the next section for the definition). Therefore, in most of the research concerning invariant subspaces for systems of several commuting operators [3,11, 15,...] (and also in our third section), the authors consider only situations in which these interior cohomology groups are trivial. However, in section 2, we obtain a positive result in the Hilbert space case for certain C_{00}-representations in a situation, where non-trivial interior cohomology groups are allowed.

Recall that $H^\infty(G)$ is a weak* closed subalgebra of $L^\infty(G)=L^1(G)^*$. Therefore, every weak* closed subalgebra $\mathcal{B}(G)$ of $H^\infty(G)$ has a predual $\mathcal{B}_*(G)$ which can be canonically identified with $L^1(G)/{}^\perp\mathcal{B}(G)$. In particular, $\{\mathcal{E}_\lambda;\lambda\in G\}\subset\mathcal{B}_*(G)$, where $\mathcal{E}_\lambda:\mathcal{B}(G)\to\mathbf{C}$ are the point evaluations $f\mapsto\langle f,\mathcal{E}_\lambda\rangle:=f(\lambda)$, $\lambda\in G$.

If $\Phi:\mathcal{B}(G) \to L(X)$ is a representation of $\mathcal{B}(G)$ with values in the Banach algebra $L(X)$ of all bounded linear operators on a Banach space X, the range $\operatorname{ran}\Phi(1)$ and the kernel $\ker\Phi(1)$ of $\Phi(1)$ are obviously invariant subspaces for $\operatorname{ran}\Phi$. Hence, in the sequel, we will always assume that Φ is a unital homomorphism. We write $\operatorname{Lat}\Phi$ for the lattice of all (closed) invariant subspaces for $\operatorname{ran}\Phi$. The well known principle of the Scott Brown technique [2 - 5, 7 - 12,...] is to find, for some $\lambda \in G$, a solution $x \in X$, $x^* \in X^*$ of the equation

$$\mathcal{E}_\lambda = x \square x^*, \tag{0.1}$$

where $x\square x^*:\mathcal{B}(G) \to \mathbf{C}$ is the linear functional given by $\langle f, x\square x^*\rangle := \langle \Phi(f)x, x^*\rangle$ for $f \in \mathcal{B}(G)$. Indeed, if $x \in X$, $x^* \in X^*$ is a solution of (0.1), we define

$$Y := \bigvee\{\Phi(f)x; f \in \ker\mathcal{E}_\lambda\} \in \operatorname{Lat}\Phi,$$

where $\bigvee$ denotes the closed linear span. Because of $Y \subset \ker x^*$ and $\langle x, x^*\rangle = 1$, the subspace Y is different from X. Hence, Y is a non-trivial invariant subspace, or we are in the trivial situation that Φ is given by $\Phi(f)=f(\lambda)1$ for all $f\in \mathcal{B}(G)$. If we can even find solutions $(x_j)_{j\in\mathbf{N}}$ in X, $(x_k^*)_{k\in\mathbf{N}}$ in X^* of

$$\delta_{j,k}\mathcal{E}_\lambda = x_j \square x_k^* \qquad (j,k\in\mathbf{N}), \tag{0.2}$$

where $\delta_{j,k}$ denotes the Kronecker symbol, then, with

$$Y_0 := \bigvee\{\Phi(f)x_j; f \in \ker\mathcal{E}_\lambda, j \in \mathbf{N}\} \text{ and } Y_1 := \bigvee\{\Phi(f)x_k^*; f \in \mathcal{B}(G), k\in\mathbf{N}\},$$

we have $\dim Y_1/Y_0 = \infty$ and $Y\in\operatorname{Lat}\Phi$ for closed subspaces Y of X with $Y_0 \subseteq Y \subseteq Y_1$. In that case we shall say that $\operatorname{Lat}\Phi$ is rich. Solutions of (0.1) and (0.2) are often obtained by means of approximation methods. In this connection the following sets $\mathfrak{X}_\theta(\Phi)$, $0<\theta<1$, are of importance. $\mathfrak{X}_\theta(\Phi)$ is the set of all those $L\in\mathcal{B}_*(G)$ such that for given finitely many $a_1,\dots,a_m \in X$, $b_1^*,\dots,b_m^* \in X^*$ and $\varepsilon > 0$, there exist $x\in X$, $x^* \in X^*$ with the following properties,

$$\|x\| \le 1, \|x^*\| \le 1, \tag{0.3}$$

$$\|L - x\square x^*\| < \theta + \varepsilon, \tag{0.4}$$

$$\|x\square b_k^*\| < \varepsilon, \|a_k \square x^*\| < \varepsilon \quad \text{for } 1 \le k \le m. \tag{0.5}$$

It is easy to verify that $\mathfrak{X}_\theta(\Phi)$ is circled and closed. If $X = \mathcal{H}$ is a Hilbert space, then $\mathfrak{X}_\theta(\Phi)$ is actually an absolutely convex closed subset of $\mathcal{B}_*(G)$. Using the idea of the proof of Proposition 5.4 in [2], this can also be shown for the Banach spaces $l^p, 1 < p < \infty$. The following fact, which is an immediate consequence of Theorem 8.3 and the proof of Theorem 7.2 in H. Bercovici, C. Foias, C. Pearcy [4], shows the importance of this convexity property.

0.1. PROPOSITION. *Let $\Phi:\mathcal{B}(G) \to L(X)$ be a continuous unital homomorphism such that*

$$\forall x \in X, x^* \in X^*:\ x \square x^* \in \mathcal{B}_*(G). \tag{0.6}$$

If there exist $0 \le \theta < 1$ *and* $\gamma \in \mathbf{R}$ *with* $\theta < \gamma$ *such that*

$$\mathcal{B}_*(G)_\gamma := \{L \in \mathcal{B}_*(G); \|L\| \le \gamma\} \subseteq \mathbf{X}_\theta(\Phi), \tag{0.7}$$

then, for each family $(L_{j,k})_{j,k \in \mathbf{N}}$ *in* $\mathcal{B}_*(G)$, *there are sequences* $(x_j)_{j\in\mathbf{N}}$ *in* X, $(x_k^*)_{k\in\mathbf{N}}$ *in* X^* *satisfying*

$$L_{j,k} = x_j \square x_k^* \quad \textit{for all } j,k \in \mathbf{N}. \tag{0.8}$$

If for every infinite matrix $(L_{j,k})_{j,k\in\mathbf{N}}$ with entries in $\mathcal{B}_*(G)$, there are solutions $(x_j)_{j\in\mathbf{N}}$ in X, $(x_k^*)_{k\in\mathbf{N}}$ in X* of (0.8), we say that Φ *has the property* $(\mathbb{A}_{\aleph_0})$.

0.2. REMARKS. **(a)** Condition (0.6) is equivalent to the continuity of $\Phi: H^\infty(G) \to L(X)$ with respect to the weak* topology on $\mathcal{B}(G)$ and the weak operator topology on $L(X)$.

(b) A subset E of G is said to be *dominating for* $\mathcal{B}(G)$ if

$$\forall f \in \mathcal{B}(G):\quad \|f\|_\infty := \sup_{z\in G}|f(z)| = \sup_{z\in E}|f(z)|.$$

Therefore, if $\mathbf{X}_\theta(\Phi)$ is convex (and hence absolutely convex and closed), condition (0.7) will be satisfied if for all $f \in \mathcal{B}(G)$ we have,

$$\gamma\{\mathcal{E}_\lambda; \lambda \in E\} \subset \mathbf{X}_\theta(\Phi). \tag{0.9}$$

Thus, in the case of Hilbert spaces or l^p spaces, $1 < p < \infty$, it suffices to verify (0.9) in order to obtain solutions to (0.2). In the case of general Banach spaces, condition (0.7) will follow, if we can show that

$$\gamma \cdot \operatorname{conv}\{\alpha \cdot \mathcal{E}_\lambda; \lambda \in E, |\alpha| \le 1\} \subset \overline{\mathbf{X}}_\theta(\Phi)$$

for some dominating subset E for $\mathcal{B}(G)$.

(c) In analogy with the theory of contractions (see B. Sz.-Nagy, C. Foias [18]) a continuous unital homomorphism $\Phi: \mathcal{B}(G) \to L(X)$ will be called a $C_{0.}$*-representation of* $\mathcal{B}(G)$ if, for every sequence $(f_n)_{n\in\mathbf{N}}$ in $\mathcal{B}(G)$ converging to 0 in the weak* topology, we have, for all $x \in X$, $\Phi(f_n)x \to 0$ in the norm topology of X. Φ is called a $C_{.0}$*-representation of* $\mathcal{B}(G)$ if the representation $\Phi^*: \mathcal{B}(G) \to L(X^*)$ given by $\Phi^*(f) := \Phi(f)^*$ for $f \in \mathcal{B}(G)$ is a $C_{0.}$-representation of $\mathcal{B}(G)$. See Vasilescu [21] for examples occurring in several variable spectral theory. If Φ satisfies both conditions, then we say that Φ is a C_{00}*-representation of* $\mathcal{B}(G)$. Notice, that for $C_{0.}$- and $C_{.0}$-representations of $\mathcal{B}(G)$, by (a), the condition (0.6) is fulfilled.

We write $\pi: \mathbf{C}^N \to \mathbf{C}^N$ for the identity on $\mathbf{C}^N$. Then $\pi = (\pi_1, \ldots, \pi_N)$, where $\pi_j: z = (z_1, \ldots, z_N) \mapsto z_j$ are the coordinate functions. Our criteria for the existence of non-trivial invariant subspaces for Φ will depend on the richness of the Taylor spectrum of $T = (T_1, \ldots, T_N) := (\Phi(\pi_1), \ldots, \Phi(\pi_N))$ with respect to $\mathcal{B}(G)$. Therefore, we will recall in the following section some facts from several variable spectral theory [19, 20, 21]. In section 2 we then show that LatΦ will be rich if Φ: $\mathcal{B}(G) \to L(\mathcal{H})$ is a C_{00}-representation of $\mathcal{B}(G)$ of norm 1 on a Hilbert space and $\sigma_e(T) \cap G$ is dominating for $\mathcal{B}(G)$, where $\sigma_e(T)$ denotes the essential Taylor spectrum of T. In the case N=1 or imposing stronger conditions on $\sigma_e(T)$, we obtain (in the last part of this note) the property ($\mathbb{A}_{\aleph_0}$) even for $C_{0\cdot}$- and $C_{\cdot 0}$-representations of $\mathcal{B}(G)$ on arbitrary Banach spaces without imposing any conditions on the norm of Φ.

1 PRELIMINARY RESULTS

Let $T=(T_1,\ldots,T_N)$ be a finite system of commuting operators in $L(X)$, X a Banach space. Recall [19,20] that the Koszul (cochain) complex K(T,X) of T with respect to X is given by

$$0 \to \Lambda^0(X) \xrightarrow{\delta^0(T)} \Lambda^1(X) \xrightarrow{\delta^1(T)} \Lambda^2(X) \xrightarrow{\delta^2(T)} \cdots \xrightarrow{\delta^{N-1}(T)} \Lambda^N(X) \to 0.$$

Here $\Lambda^p(X)$ is the set of all p-forms with coefficients in X and $\delta^p(T):\Lambda^p(X) \to \Lambda^{p+1}(X)$ is the cochain mapping defined by

$$\delta^p(T) \sum_{|I|=p}{}' x_I s_I := \sum_{j=1}^{N} \sum_{|I|=p}{}' T_j x_I s_j \wedge s_I$$

where $s_1,\ldots,s_N$ is a fixed basis of $\Lambda^1(\mathbf{C})$ and $\sum_{|I|=p}{}'$ means that the sum has to be taken over all $I=(i_1,\ldots,i_p) \in \mathbf{N}^p$ with $1 \le i_1 < \ldots < i_p \le N$, $s_I := s_{i_1} \wedge \ldots \wedge s_{i_p}$. Notice, that $\Lambda^p(X)$ is canonically isomorphic to a direct sum of $\binom{N}{p}$ copies of X and hence can be endowed with a natural norm. If $X = \mathcal{H}$ is a Hilbert space, then $\Lambda^p(X)$ is a Hilbert space by means of the scalar product given by

$$\left\langle \sum_{|I|=p}{}' x_I s_I, \sum_{|I|=p}{}' y_I s_I \right\rangle := \sum_{|I|=p}{}' \langle x_I, y_I \rangle.$$

The Taylor spectrum $\sigma(T)$ of T with respect to X is by definition [19,20] the set of all $\lambda \in \mathbf{C}^N$ such that the complex $K(\lambda - T,X)$ is not exact. The essential Taylor spectrum $\sigma_e(T)$ of T with respect to X is the set of all $\lambda \in \mathbf{C}^N$ such that at least one of the cohomology groups $H^p(\lambda-T):=\ker \delta^p(\lambda-T)/\operatorname{ran} \delta^{p-1}(T)$ has infinite codimension. It follows that

$$\sigma_e(T) := \bigcup_{p=0}^{N} \sigma_e^p(T),$$

where $\sigma_e^p(T)$ is the set of all $\lambda \in \mathbf{C}^N$ such that the mapping

$$\hat{\delta}^p(\lambda - T) : \ker \delta^p(\lambda - T) / \overline{\operatorname{ran} \delta^{p-1}(\lambda - T)} \to \Lambda^{p+1}(X)$$

has non-closed range or infinite dimensional kernel. For Hilbert space $X=\mathcal{H}$ this is the case, if and only if there exists an orthonormal sequence $(\omega_n)_{n=1}^{\infty}$ in

$\Lambda^p(\mathcal{H})\ominus \operatorname{ran}\delta^{p-1}(\lambda-T)$ such that

$$\delta^p(\lambda\text{-}T)\omega_n \to 0 \text{ for } n \to \infty. \tag{1.1}$$

If $j\in\{1,\dots,N\}$ is fixed, then each $\omega\in\Lambda^p(X)$ has a unique decomposition of the form $\omega = \omega_j' + s_j \wedge \omega_j''$, where s_j does not occur in $\omega_j'\in\Lambda^p(X)$, $\omega_j''\in\Lambda^{p-1}(X)$. The chain mapping $\delta_p(T):\Lambda^p(X) \to\Lambda^{p-1}(X)$ is then defined by

$$\delta_p(T)\omega = \sum_{j=1}^{N} T_j\omega''_j .$$

By $(T)'$ we denote the commutant algebra of T, i.e. $(T)'=\{A\in L(X);\ T_jA=AT_j \text{ for } j=1,\dots,N\}$. We shall need the following well known algebraic fact (see e.g. [21], Lemma III.6.5).

1.1. LEMMA. *Let* $T=(T_1,\dots,T_N)$ *be a commutative system in* $L(X)$ *and suppose that* $S=(S_1,\dots,S_N)$ *is a commutative system in* $(T)'$. *Then, for* $p=0,1,\dots,N$, *we have*

$$\delta^{p-1}(T)\delta_p(S) + \delta_{p+1}(S)\delta^p(T) = \sum_{j=1}^{N} S_jT_j$$

on $\Lambda^p(X)$.

1.2. COROLLARY. *Let* $T=(T_1,\dots,T_N)$ *be a commutative system in* $L(\mathcal{H})$, $\mathcal{H}$ *a Hilbert space and let* $\lambda\in\mathbf{C}^N$.

(a) *For each* $\omega \in\ker\delta^p(\lambda - T)\ominus\operatorname{ran}\delta^{p-1}(\lambda -T)$ *and each polynomial* $p\in\mathbf{C}[Z_1,\dots,Z_N]$ *we have* $\langle p(T)\omega,\omega\rangle = p(\lambda)\|\omega\|^2$.

(b) *If* $(\omega_n)_{n=1}^{\infty}$ *is an orthonormal sequence in* $\Lambda^p(\mathcal{H})\ominus\operatorname{ran}\delta^{p-1}(\lambda -T)$ *satisfying* (1.1) *then, for each* $p\in\mathbf{C}[Z_1,\dots,Z_N]$, $\langle p(T)\omega_n,\omega_n\rangle \to p(\lambda)$ for $n \to \infty$.

PROOF. Indeed, we have for $z\in\mathbf{C}^N$,

$$p(z) = p(\lambda) + \sum_{j=1}^{N}(\lambda_j - z_j)q_j(z)$$

with polynomials $q_1,\dots,q_N\in\mathbf{C}[Z_1,\dots,Z_N]$. Applying Lemma 1.1 to the systems $\lambda -T=(\lambda_1 -T_1,\dots,\lambda_N -T_N)$ and $S=(q_1(T),\dots,q_N(T))$ we easily obtain the statements of the corollary.

Let now G be a bounded domain in $\mathbf{C}^N$, $\mathcal{B}(G)$ be a weak*-closed subalgebra of $H^\infty(G)$ containing the constants and the coordinate functions $\pi_1,\dots,\pi_N$. For $\lambda\in G$ consider the ideals $\ker \mathcal{E}_\lambda=\{f\in\mathcal{B}(G);\ f(\lambda)=0\}$ (where $\mathcal{E}_\lambda$ is the point evaluation at λ) and

$$\mathcal{J}(\lambda,\mathcal{B}(G)):=\Big\{\sum_{j=1}^{N}(\lambda_j - \pi_j)g_j;g_1,\dots,g_N \in\mathcal{B}(G)\Big\}.$$

Notice that $\ker \mathcal{E}_\lambda$ is weak*-closed and contains the finitely generated ideal $\mathcal{J}(\lambda,\mathcal{B}(G))$. We shall say that $\mathcal{B}(G)$ has the *Gleason property* if $\mathcal{J}(\lambda,\mathcal{B}(G))=\ker \mathcal{E}_\lambda$. The Gleason property holds in many cases. For example, if $G=G_1\times\dots\times G_N$ is a polydomain or if G is pseudoconvex with sufficiently smooth boundary, then $H^\infty(G)$ has the Gleason property [1,

13, 14, 16]. On the other hand, for non pseudoconvex domains it may happen that, for some $\lambda \in G$, the ideal $\mathcal{J}(\lambda, \mathcal{B}(G))$ is not even dense in ker $\mathcal{E}_\lambda$.

1.3. REMARK. If $\mathcal{B}(G)$ has the Gleason property then, by the open mapping theorem, for every $\lambda \in G$, there exists some constant $C(\lambda)>0$ such that, for all $f \in \ker \mathcal{E}_\lambda$, there exist $g_1,\dots,g_N \in \mathcal{B}(G)$ with $\|g_j\|_\infty \le C(\lambda)\cdot\|f\|_\infty$ for $j=1,\dots,N$ and

$$\sum_{j=1}^{N} (\lambda_j - \pi_j) g_j = f.$$

1.4. LEMMA. *Let* $\mathcal{B}(G)$ *have the Gleason property and let* $\Phi: \mathcal{B}(G) \to L(\mathcal{H})$ *be a unital continuous homorphism,* $T:=(T_1,\dots,T_N):=(\Phi(\pi_1),\dots,\Phi(\pi_N))$. *If* $\lambda \in \sigma_e(T) \cap G$ *then, for some* $p \in \{0,1,\dots,N\}$, *there exists an orthonormal sequence* $(\omega_n)_{n=1}^{\infty}$ *in* $\Lambda^p(\mathcal{H}) \ominus \operatorname{ran}\delta^{p-1}(\lambda - T)$ *satisfying* (1.1). *Furthermore, for such a sequence,* $\|\mathcal{E}_\lambda - \omega_n \Box \omega_n\| \to 0$ *as* $n \to \infty$.

PROOF. For each $n \in \mathbf{N}$, there exists some $h_n \in \mathcal{B}(G)$ with $\|h_n\| = 1$ such that

$$\|\mathcal{E}_\lambda - \omega_n \Box \omega_n\| - 1/n \le |\langle h_n, \mathcal{E}_\lambda - \omega_n \Box \omega_n \rangle| = |\langle \Phi(h_n - h_n(\lambda))\omega_n, \omega_n \rangle|.$$

Since $\mathcal{B}(G)$ has the Gleason property, there exists some $C(\lambda) > 0$ and $g_{n,1},\dots,g_{n,N} \in \mathcal{B}(G)$ with $\|g_{n,i}\| \le C(\lambda)\cdot\|h_n - h_n(\lambda)\|_\infty \le 2C(\lambda)$ such that

$$h_n - h_n(\lambda) = \sum_{i=1}^{N} g_{n,i}(\lambda_i - \pi_i).$$

Hence, by Lemma 1.1,

$$\begin{aligned}\|\mathcal{E}_\lambda - \omega_n \Box \omega_n\| - 1/n &\le \left|\left\langle \sum_{i=1}^{N} \Phi(g_{n,i})(\lambda_i - T_i)\omega_n, \omega_n \right\rangle\right| \le \\ &\le \left|\left\langle \delta_{p-1}(\Phi(g_{n,1}),\dots,\Phi(g_{n,N}))\delta^p(\lambda - T)\omega_n, \omega_n \right\rangle\right| \le \\ &\le 2N\cdot C(\lambda)\left\|\delta^p(\lambda - T)\omega_n\right\| \to 0 \quad \text{for } n \to \infty .\end{aligned}$$

2 C_{00}-representations with dominating essential spectrum

2.1. THEOREM. *Let* $\mathcal{B}(G)$ *be a weak*-closed subalgebra of* $H^\infty(G)$ *with the Gleason property,* G *a bounded domain in* $\mathbf{C}^N$, *and suppose that* $\Phi: \mathcal{B}(G) \to L(\mathcal{H})$ *is a continuous unital Hilbert space representation of* $\mathcal{B}(G)$ *with* $\|\Phi\|=1$. *If* $\sigma_e(T) \cap G$ *is dominating for* $\mathcal{B}(G)$, *where* $T:=(T_1,\dots,T_N):=(\Phi(\pi_1),\dots,\Phi(\pi_N))$ *then* Φ *has the property* $(\mathbb{A}_{\aleph_0})$ *and therefore a rich invariant subspace lattice.*

PROOF. The Hilbert space $\mathcal{K} := \Lambda(\mathcal{H}) := \bigoplus_{p=0}^{N} \Lambda^p(\mathcal{H})$ is canonically isometrically isomorphic to $\mathcal{H}^{2^N}$. If $\lambda \in \sigma_e(T) \cap G$, then by Lemma 1.4, for some $p \in \{0,1,\ldots,N\}$, there exists an orthonormal sequence $(\omega_n)_{n=1}^{\infty}$ in $\Lambda^p(\mathcal{H}) \ominus \operatorname{ran}\delta^{p-1}(\lambda - T) \subset \mathcal{K}$ such that $\omega_n \square \omega_n \to \mathcal{C}_\lambda$ for $n \to \infty$ in the Banach space $\mathcal{B}_*(G)$. Passing (if necessary) to some subsequence we may assume, that for one of the 2^N components (say the k-th $x_n^{(k)}$) of ω_n we have $\left\|x_n^{(k)}\right\|^2 \geq 2^{-N}$ for all $n \in \mathbf{N}$. It follows that, for a given $\varepsilon > 0$, there exists $n_0 \in \mathbf{N}$ such that for $n \geq n_0$,

$$\left\|\mathcal{C}_\lambda - x_n^{(k)} \square x_n^{(k)}\right\| \leq \left\|\mathcal{C}_\lambda - \omega_n^{(k)} \square \omega_n^{(k)}\right\| \leq \sum_{j=1}^{2^N} \left\|x_n^{(j)} \square x_n^{(j)}\right\|^2 \leq \varepsilon + \sum_{\substack{j=1 \\ j \neq k}}^{N} \left\|x_n^{(j)} \square x_n^{(j)}\right\|^2 \cdot \|\Phi\| \leq \varepsilon + \frac{2^N - 1}{2^N}$$

Moreover, $x_n^{(k)} \to 0$ weakly, since $(\omega_n)_{n=1}^{\infty}$ is an orthonormal sequence. Let now $a_1, \ldots, a_m \in \mathcal{H}$ be given vectors. Because of $x_n^{(k)} \square a_j \in \mathcal{B}_*(G)$ there exists a function $f_{n,j}$ in $\mathcal{B}(G)$ with $\|f_{n,j}\|_\infty = 1$ such that $\left\|x_n^{(k)} \square a_j\right\| = \left\langle f_{n,j}, x_n^{(k)} \square a_j \right\rangle = \left\langle \Phi(f_{n,j}) x_n^{(k)}, a_j \right\rangle$. Dropping to some subsequence we may assume that $f_{n,j} \to f_j$ in the weak* topology for $n \to \infty$ where f_j is some function in the unit ball of $\mathcal{B}(G)$. It follows that

$$\left\|x_n^{(k)} \square a_j\right\| = \left\langle \Phi(f_{n,j} - f_j) x_n^{(k)}, a_j \right\rangle + \left\langle x_n^{(k)}, \Phi(f_j)^* a_j \right\rangle \to 0 \quad \text{for } n \to \infty,\ j=1,\ldots,N,$$

since $x_n^{(k)} \to 0$ weakly and because of the $C_{\cdot 0}$-property of Φ. In a dual way (using the $C_{0\cdot}$-property of Φ) we can see that also $\|a_j \square x_n^{(k)}\| \to 0$ for $n \to \infty$, $j=1,\ldots,N$. Hence, we have proved that

$$\{\mathcal{C}_\lambda ; \lambda \in \sigma_e(T) \cap G\} \subset \mathbf{X}_\theta(\Phi) \text{ with } \theta = \frac{2^N - 1}{2^N}.$$

Since we are in the Hilbert space case, the set $\mathbf{X}_\theta(\Phi)$ is absolutely convex and closed. By Remark 0.2, the assumptions of Theorem 0.1 are satisfied and the conclusion now follows from Theorem 0.1.

3 $C_{0\cdot}$- and $C_{\cdot 0}$-representation with dominating exterior essential spectrum

The reason why - in the preceding section - we had to require $X = \mathcal{H}$ to be a Hilbert space and $\|\Phi\| = 1$, was the presence of $\sigma_e^p(T)$-points with $1 \leq j \leq N-2$ and the fact that $\dim \ker \delta^p(\lambda - T) / \overline{\operatorname{ran}\delta^{p-1}(\lambda - T)}$ might be infinite dimensional. Notice, that $\sigma_{le}(T) = \sigma_e^0(T)$

is by definition the set of all $\lambda \in \mathbf{C}^N$ such that for every closed subspace Y of finite codimension in X there exists a sequence $(x_n)_{n\in\mathbf{N}}$ in Y with $\|x_n\|=1$ for all $n\in\mathbf{N}$, satisfying

$$\|(\lambda_j - T_j)x_n\| \to 0 \quad \text{for } n \to \infty,\ j=1,\ldots,N, \tag{3.1}$$

or equivalently, $\lambda\in\sigma_{le}(T)$ if and only if the map $x\mapsto ((\lambda_j - T_j)x)_{j=1}^N x\mapsto ((\lambda_j - T_j)x)_{j=1}^N$ from X to X^N has non-closed range or $\ker(\lambda_1 - T_1)\cap\ldots\cap\ker(\lambda_N - T_N)$ has infinite dimension. Dually, we may define the right essential spectrum $\sigma_{re}(T)$ to be the set of all $\lambda \in \mathbf{C}^N$ such that $\sum_{j=1}^N \operatorname{ran}(\lambda_j - T_j)$ has infinite codimension in X. It follows that $\sigma_{re}(T) = \sigma_{le}(T^*)$, where T^* is the transposed system $T^* = (T_1^*,\ldots,T_N^*)$ in X^*. Thus $\lambda \in \mathbf{C}^N$ is in $\sigma_{re}(T)$ if and only if for every closed subspace Y^* of finite codimension in X^* there exists sequence $(x_n^*)_{n=1}^\infty$ in the unit sphere of Y^* with the property that

$$\|(\lambda_j - T_j^*)x_n^*\| \to 0 \quad \text{for } n \to \infty,\ j=1,\ldots,N. \tag{3.2}$$

In general, for N>1, the union $\sigma_{re}(T) \cup \sigma_{le}(T)$ will be a proper subset of $\sigma_e(T)$. If instead of the dominance of $\sigma_e(T) \cap G$ we make the stronger requirement that $(\sigma_{re}(T) \cup \sigma_{le}(T)) \cap G$ or even $\sigma_{re}(T) \cap G$ resp. $\sigma_{le}(T) \cap G$ is dominating for G, then the Hilbert space requirement for X and the norm condition for Φ can be dropped, as we want to show now. Throughout this section X will always denote an arbitrary infinite dimensional Banach space.

Let G be a bounded domain in $\mathbf{C}^N$, $\mathcal{B}(G)$ be a weak* closed subalgebra of $H^\infty(G)$ and Φ: $\mathcal{B}(G) \to L(X)$ be a unital homomorphism. For s,t>0 we denote by $\mathcal{Y}_{s,t}(\Phi)$ the set of all $L\in\mathcal{B}_*(G)$ such that, for any finitely many $a_1,\ldots,a_p\in X$, $a_1^*,\ldots,a_p^*\in X^*$, and $\varepsilon>0$, there exist $x\in X$, $x^*\in X^*$ satisfying the following conditions (3.3) - (3.6).

$$\|x\| < s + \varepsilon, \|x^*\| < t + \varepsilon, \tag{3.3}$$

$$\|L - x\square x^*\| < \varepsilon, \tag{3.4}$$

$$\|x\square a_j^*\| < \varepsilon \quad \text{for } j=1,\ldots,q, \tag{3.5}$$

$$\|a_k\square x^*\| < \varepsilon \quad \text{for } k=1,\ldots,p. \tag{3.6}$$

In this situation we have,

3.1. LEMMA. (a) *$\mathcal{Y}_{s,t}(\Phi)$ is circled.*
(b) $\mathcal{Y}_{s,t}(\Phi) + \mathcal{Y}_{\sigma,\tau}(\Phi) \subseteq \mathcal{Y}_{s+\sigma,t+\tau}(\Phi)$ *for all* $s,t,\sigma,\tau > 0$.

PROOF. (a) is obvious. To prove (b), fix $L_1\in \mathcal{Y}_{s,t}(\Phi)$, $L_2 \in \mathcal{Y}_{\sigma,\tau}(\Phi)$, and let $a_1^*,\ldots,a_p^*\in X^*, a_1,\ldots,a_p \in X$, $a_1^*,\ldots,a_p^*\in X^*$ be finite families in X resp. X^*, $\varepsilon > 0$. If $\varepsilon' > 0$ then, because of $L_1\in \mathcal{Y}_{s,t}(\Phi)$, there are $x_1\in X$, $x_1^*\in X^*$, such that (3.3)-(3.4) is fulfilled for x_1, x_1^*, L_1, ε'. Since $L_2 \in \mathcal{Y}_{\sigma,\tau}(\Phi)$, we find $x_2 \in X$, $x_2^* \in X^*$ with $\|x_2\| < \sigma + \varepsilon'$, $\|x_2^*\| < \tau + \varepsilon'$ such that (3.4)-(3.6) are fulfilled for x_2, x_2^*, L_2, ε' and such that in addition $\|x_1 \square x_2^*\| < \varepsilon'$, $\|x_2 \square x_1^*\| < \varepsilon'$. With $x := x_1 + x_2$, $x^* := x_1^* + x_2^*$, $L := L_1 + L_2$ we see that (3.4)-(3.6) are satisfied if we choose $\varepsilon' < \varepsilon/4$. Since $\|x\| < s + \sigma + 2\varepsilon' < s + \sigma + \varepsilon$ and also $\|x_2^*\| < t + \tau + \varepsilon$, we conclude that $L := L_1 + L_2 \in \mathcal{Y}_{s+\sigma,t+\tau}(\Phi)$.

In the proof of the following proposition we use ideas of the proof of Proposition 2.8 of Eschmeier and Prunaru [12].

3.2. PROPOSITION. *Let* G *be a bounded domain in* $\mathbf{C}^N$ ($N \in \mathbf{N}$), $\mathcal{B}(G)$ *be a weak* closed subalgebra of* $H^\infty(G)$ *having the Gleason property, and let* Φ: $\mathcal{B}(G) \to L(X)$ *be a continuous unital homomorphism,* $T = (T_1,\ldots,T_N) = (\Phi(\pi_1),\ldots,\Phi(\pi_N))$.

(a) *If* Φ *has the* $C_{0.}$*-property then,* abs conv$\{\mathcal{E}_\lambda\,;\, \lambda \in \sigma_{le}(T) \cap G\} \subset \mathcal{Y}_{4,8}(\Phi)$.

(b) *If* Φ *has the* $C_{.0}$*-property then,* abs conv$\{\mathcal{E}_\lambda\,;\, \lambda \in \sigma_{re}(T) \cap G\} \subset \mathcal{Y}_{8,4}(\Phi)$.

PROOF. Since the proofs are completely dual to each other we only prove (b).

1.) In a first step we show conv$\{\mathcal{E}_\lambda\,;\, \lambda \in \sigma_{re}(T) \cap G\} \subset \mathcal{Y}_{2,1}(\Phi)$. Hence, assume that $L = \alpha_1 \mathcal{E}_{\lambda_1} + \ldots + \alpha_m \mathcal{E}_{\lambda_m}$ with $\lambda_1,\ldots,\lambda_m \in \sigma_{re}(T) \cap G$, $\alpha_1,\ldots,\alpha_m > 0$, $\alpha_1 + \ldots + \alpha_m = 1$. Let also $a_1,\ldots,a_p \subset X$, $a_1^*,\ldots,a_p^* \subset X^*$ and $\varepsilon > 0$ be given. By the $C_{.0}$-property of Φ and the weak* compactness of the unit ball in $\mathcal{B}(G)$ the set

$$K^* := \bigcup_{j=1}^{q} \{\Phi(f)^*\, a_j^*\,; f \in \mathcal{B}(G),\ \|f\|_\infty \le 1\}$$

is compact in X^*. Hence, if $0 < \varepsilon' < 1$, there are finitely many $u_1^*,\ldots,u_r^* \in X^*$ such that, for all $v^* \in K^*$, we have $\|v^* - u_k^*\| < \varepsilon'$ for some $k \in \{1,\ldots,r\}$. Write Y_0^* for the span $\bigvee\{u_1^*,\ldots,u_r^*\}$ of $u_1^*,\ldots,u_r^*$ and $X_0^* := \bigvee\{a_1,\ldots,a_p\}^\perp$. By Singer [17], Lemma III.1.1, there exists a closed subspace X_1^* of finite codimension in X^* with $\delta(X_0^*, X_1^*) > 1 - \varepsilon'$, where $\delta(X_0^*, X_1^*)$ denotes the gap between X_0^* and X_1^*. Since $X_0^* \cap X_1^*$ is of finite codimension in X^* and $\lambda_1 \in \sigma_{re}(T) \cap G$, there exists some $x_1^* \in X_0^* \cap X_1^*$ with $\|x_1^*\| = 1$ and

$$\|(\lambda_{1,j} - T_j^*)x_1^*\| < \varepsilon' \quad \text{for } j=1,\ldots,N.$$

Again by [17], Lemma III.1.1, there is a closed subspace X_2^* of finite codimension in X^* with

$$\delta\left(\bigvee\{u_1^*,\ldots,u_r^*,x_1^*\}, X_2^*\right) > 1 - \varepsilon'.$$

The subspace $X_0^* \cap X_1^* \cap X_2^*$ is still of finite codimension, so that, as above, we find $x_2^* \in X_0^* \cap X_1^* \cap X_2^*$ such that $\|x_2^*\| = 1$ and $\|(\lambda_{2,j} - T_j^*)x_2^*\| < \varepsilon'$ for $j=1,\ldots,N$. In this way we obtain successively closed subspaces $X_0^*,\ldots,X_m^*$ of finite codimension in X^* and vectors $x_k^* \in X_0^* \cap X_1^* \cap \ldots \cap X_k^*$ with $\|x_k^*\| = 1$ for $k=1,\ldots,m$ such that

$$\delta\left(\bigvee\{u_1^*,\ldots,u_r^*,x_1^*,\ldots,\ x_k^*\}, X_{k+1}^*\right) > 1 - \varepsilon' \quad \text{for } 1 \le k < m, \text{ and}$$

$$\left\|(\lambda_{k,j}-T_j^*)x_k^*\right\| < \varepsilon' \quad \text{for } j=1,\ldots,N,\ k=1,\ldots,m.$$

By a result of Zenger (see [6]), there exists some linear functional F on $\bigvee\{x_1^*,\ldots,x_m^*\}$ with $\|F\|=1$ and complex numbers $\mu_1,\ldots,\mu_m$ such that, with $x^*:=\mu_1x_1^*+\ldots+\mu_mx_m^*$, we have $\|x^*\|=1$ and $F(\mu_kx_k^*)=\alpha_k$ for k=1,...,m. Because of $\delta(Y_0{}^*,\bigvee\{x_1^*,\ldots,x_m^*\})>1-\varepsilon'$, the canonical projection P of $Y^*:=\bigvee\{u_1^*,\ldots,u_r^*,x_1^*,\ldots,x_m^*\}$ onto $\bigvee\{x_1^*,\ldots,x_m^*\}$ along Y_0^* has norm $<2/(1-\varepsilon')$. Since Y* is a finite dimensional subspace of X*, there exists some $x\in X$ with $\|x\|<2/(1-\varepsilon')$ and $\langle x,u^*\rangle=F(Pu^*)$ for all $u^*\in Y^*$. Obviously, x and x* satisfy condition (3.3) with s=2, t=1 if ε' is sufficiently small. Using

$$\delta\left(\bigvee\{u_1^*,\ldots,u_r^*,x_1^*,\ldots,x_k^*\},\{x_{k+1}^*,\ldots,x_m^*\}\right)>1-\varepsilon' \text{ for } k=1,\ldots,m-1,$$

one can show by means of direct computation, that $|\mu_k|<2/(1-\varepsilon')$. To prove that (3.4) is satisfied for x, x*, L, ε, fix $h\in\mathcal{B}(G)$ with $\|h\|_\infty=1$ and $\|L-x\square x^*\|=\langle h, L-x\square x^*\rangle$. It follows that

$$\begin{aligned}\|L-x\square x^*\| &= \left|\sum_{k=1}^m \alpha_k h(\lambda_k)-\langle\Phi(h)x,x^*\rangle\right|\\ &= \left|\sum_{k=1}^m \alpha_k h(\lambda_k)-\sum_{k=1}^m \mu_k\langle x,\Phi(h-h(\lambda_k))^*x_k^*\rangle+\sum_{k=1}^m h(\lambda_k)\langle x,\mu_kx_k{}^*\rangle\right|\\ &\le \sum_{k=1}^m|\mu_k|\sum_{j=1}^N\left|\left\langle x,\Phi(g_{k,j})^*(\lambda_{k,j}-T_j^*)x_k^*\right\rangle\right|,\end{aligned}$$

where, according to Remark 1.3, the functions $g_{k,1},\ldots,g_{k,N}\in\mathcal{B}(G)$ can be choosen with $\|g_{k,j}\|_\infty\le C(\lambda_k)\|h-h(\lambda_k)\|_\infty\le 2C(\lambda_k)$. Hence,

$$\left\|L-x\square x^*\right\|\le\sum_{k=1}^m\frac{2}{1-\varepsilon'}\cdot 2C(\lambda_k)\cdot\|x\|\cdot\|\Phi\|\cdot N\varepsilon'\le\frac{8\|\Phi\|\cdot N\varepsilon'}{(1-\varepsilon')^2}\cdot\sum_{k=1}^m C(\lambda_k)<\varepsilon$$

for ε' sufficiently small.

Let now $1\le j\le q$. Then there exists a function $h\in\mathcal{B}(G)$ with $\|h\|_\infty=1$ and $\|x\square a_j^*\|=\langle h,x\square a_j^*\rangle=\langle x,\Phi(h)^*a_j^*\rangle$. Because of $\Phi(h)^*a_j^*\in K^*$, there is some $k\in\{1,\ldots,r\}$ such that $\|\Phi(h)^*a_j^*-u_k^*\|<\varepsilon'$. Hence,

$$\left\|x\square a_j^*\right\|\le\|x\|\cdot\varepsilon'+\left|\langle x,u_k^*\rangle\right|\le\frac{2\varepsilon'}{1-\varepsilon'}<\varepsilon$$

for ε' small enough. (Notice that $\langle x,u_k^*\rangle=0$ by the construction of x). This shows that (3.5) is fulfilled for L, x, x*.

To verify (3.6), fix $j\in\{1,\dots,p\}$ and $h\in\mathcal{B}(G)$ with $\|h\|_\infty = 1$ and $\|a_j\square x^*\| = \langle h, a_j\square x^*\rangle$. From the definition of x* we obtain , as in the verification of (3.4),

$$\|a_j\square x^*\| \le \sum_{k=1}^{m}|\mu_k|\cdot\left|\left\langle \Phi(h)a_j, x_k^*\right\rangle\right| = \sum_{k=1}^{m}|\mu_k|\cdot\left|\left\langle \Phi(h-h(\lambda_k))a_j, x_k^*\right\rangle\right| =$$

$$= \sum_{k=1}^{m}|\mu_k|\cdot\left|\left\langle a_j, \Phi(h-h(\lambda_k))^* x_k^*\right\rangle\right| \le \frac{4\|a_j\|\cdot\|\Phi\|\varepsilon'}{1-\varepsilon'}\cdot\sum_{k=1}^{m}C(\lambda_k) < \varepsilon$$

for ε' sufficiently small. Thus we proved that $L\in\mathcal{Y}_{2,1}(\Phi)$.

2.) Fix now $\lambda_1,\dots,\lambda_m\in\sigma_{re}(T)\cap G$ and consider $L := \sum_{k=1}^{m} c_k\mathcal{E}_{\lambda_k}$ with $c_1,\dots,c_m\in\mathbf{C}$ and $|c_1|+\dots+|c_m|\le 1$. It follows, with $i=\sqrt{-1}$, that L has a representation of the form

$$L = L_1 + iL_2 + i^2L_3 + i^3L_4$$

with $L_\nu = \sum_{k=1}^{m}\alpha_k^{(\nu)}\mathcal{E}_{\lambda_k}$, where now $\alpha_k^{(\nu)}>0$ and $\sum_{k=1}^{m}\alpha_k^{(\nu)}\le 1$ for $\nu = 1, 2, 3, 4$. From the first part of the proof and Lemma 3.1, we now conclude that $L\in\mathcal{Y}_{8,4}(\Phi)$.

3.3. COROLLARY. *Let* G *and* $\mathcal{B}(G)$ *be as in Proposition* 3.2 *and suppose that* $\Phi:\mathcal{B}(G)\to L(X)$ *is a continuous unital* C_{00}*-representation of* $\mathcal{B}(G)$ *such that* $(\sigma_{le}(T)\cup\sigma_{re}(T))\cap G$ *is dominating for* $\mathcal{B}(G)$, *where* $T = (T_1,\dots,T_N) = (\Phi(\pi_1),\dots,\Phi(\pi_N))$. *Then* Φ *is of class* $\mathbb{A}_{\aleph_0}$ *and hence has a rich invariant subspace lattice.*

PROOF. Fix $L\in\operatorname{absconv}\{\mathcal{E}_\lambda;\lambda\in(\sigma_{re}(T)\cup\sigma_{le}(T))\cap G\}$. Then $L = \alpha L_l + \beta L_r$, where $\alpha,\beta\in\mathbf{C}$ with $|\alpha|+|\beta|\le 1$ and

$$L_l\in\operatorname{absconv}\{\mathcal{E}_\lambda;\lambda\in\sigma_{le}(T)\cap G\},\quad L_r\in\operatorname{absconv}\{\mathcal{E}_\lambda;\lambda\in\sigma_{re}(T)\cap G\}.$$

By Proposition 3.2 we have $L_l\in\mathcal{Y}_{4,8}(\Phi)$ and $L_r\in\mathcal{Y}_{8,4}(\Phi)$ which, because of Lemma 3.1, implies $L\in\mathcal{Y}_{12,12}(\Phi)$. From this we see that

$$\frac{1}{144}\operatorname{absconv}\{\mathcal{E}_\lambda;\lambda\in(\sigma_{re}(T)\cup\sigma_{le}(T))\cap G\}\subset\mathfrak{X}_0(\Phi).$$

Hence, by Remark 0.2, the assumptions of Proposition 0.1 are fulfilled and we conclude from Proposition 0.1 that Φ has the property $(\mathbb{A}_{\aleph_0})$.

3.4. COROLLARY. *Let* G *and* $\mathcal{B}(G)$ *be as in Proposition* 3.2 *and suppose that* $\Phi:\mathcal{B}(G)\to L(X)$ *is a continuous unital homomorphism. Write again* $T = (T_1,\dots,T_N) = (\Phi(\pi_1),\dots,\Phi(\pi_N))$. *If*

(a) $\sigma_{le}(T)\cap G$ *is dominating for* $\mathcal{B}(G)$ *and* Φ *has property* $C_{0.}$ *or if*

(b) $\sigma_{re}(T)\cap G$ *is dominating for* $\mathcal{B}(G)$ *and* Φ *has property* $C_{.0}$, *then Φ has the property ($\mathbb{A}_{\aleph_0}$) and hence has a rich invariant subspace lattice.*

PROOF. If $L=L_l \in \mathrm{absconv}\{\mathcal{E}_\lambda; \lambda \in \sigma_{le}(T) \cap G\}$ resp. $L=L_r \in \mathrm{absconv}\{\mathcal{E}_\lambda; \lambda \in \sigma_{re}(T) \cap G\}$, then by Proposition 3.2, $L \in \mathcal{Y}_{8,4}(\Phi)$ resp. $L \in \mathcal{Y}_{4,8}(\Phi)$. Thus, in both cases,

$$\frac{1}{32}\mathrm{absconv}\{\mathcal{E}_\lambda ; \lambda \in E\} \subset \mathbf{X}_0(\Phi)$$

for some subset E of G which is dominating for $\mathcal{B}(G)$. As in the proof of Corollary 3.3 we conclude by means of 0.2 and 0.1 that Φ is of class $\mathbb{A}_{\aleph_0}$.

Specializing to the case N=1, we obtain a generalization of a result in [10] to the case of general Banach spaces.

3.5. COROLLARY. *Let G be a bounded domain in* **C** *and let* $\Phi: H^\infty(G) \to L(X)$ *be a continuous unital homomorphism with the* $C_{.0}$*- or the* $C_{0.}$*-property. Write* $T:=\Phi(\pi)$.
(a) *If* $\sigma(T)$ *is dominating for* $H^\infty(G)$, *then* Lat(T) *is non-trivial.*
(b) *If* $\sigma_e(T)$ *is dominating for* $H^\infty(G)$, *then* T *has a rich invariant subspace lattice.*

PROOF. If $\sigma(T) \neq \sigma_e(T)$, then T has non-trivial hyperinvariant subspaces. Hence, it suffices to prove (b). A standard argument shows, that in the cases $\sigma_e(T) \neq \sigma_{le}(T)$ and $\sigma_e(T) \neq \sigma_{re}(T)$ the invariant subspace lattice of Φ is rich. Therefore, we may assume $\sigma_e(T) = \sigma_{le}(T) = \sigma_{re}(T)$. The statement now follows from Corollary 3.4 (notice that for N=1, the algebra $H^\infty(G)$ always has the Gleason property).

3.6. REMARKS. (a) Corollary 3.5 remains of course valid if we replace $H^\infty(G)$ by any weak* closed full subalgebra of $H^\infty(G)$ containing the polynomials.

(b) Already the first reduction step in the proof of the preceding corollary causes problems in the case of several variables. The authors do not know, if every singular Fredholm system has non-trivial invariant (or even hyperinvariant) subspaces.

REFERENCES

1. P. Ahern and R. Schneider, Holomorphic Lipschitz functions in pseudo-convex domains, *Amer. J. Math.* **101** (1979), 543-564.
2. E. Albrecht and B. Chevreau, Invariant subspaces for l^p – operators having Bishop's property (β) on a large part of their spectrum. *J. Operator Theory* **18** (1987), 339-372.
3. C. Apostol, A functional calculus and invariant subspaces, *J. Operator Theory* **4** (1980), 159-190
4. H. Bercovici, C. Foias, and C. Pearcy, *Dual algebras with applications to invariant subspaces and dilation theory,* C.B.M.S. Regional Conference Series in Math., No. **56**, Amer. Math. Soc. 1985.
5. H. Bercovici, C. Foias, and C. Pearcy, Two Banach space methods and dual operator algebras, *J. Functional Anal.* **78** (1988), 306-345.

6. F.F. Bonsall and J. Duncan, *Numerical ranges* II, Cambridge University Press, 1973.
7. S. Brown, Some invariant subspaces for subnormal operators, *Integral Equations Operator Theory* **1** (1978), 310-333.
8. S. Brown, Hyponormal operators with thick spectra have invariant subspaces, *Ann. of Math.* **125** (1987), 93-103.
9. S. Brown, B. Chevreau, and C. Pearcy, Contractions with rich spectrum have invariant subspaces, *J. Operator Theory* **1** (1979) , 123-136.
10. B. Chevreau, C. Pearcy, and A. Shields, Finitely connected domains G, representations of $H^\infty(G)$, and invariant subspaces. *J. Operator Theory* **6** (1981), 375-405.
11. J. Eschmeier, Bishop's condition (β) and joint invariant subspaces, *J. reine angewandte Math.* **426** (1992), 1-22.
12. J. Eschmeier and B. Prunaru, Invariant subspaces for operators with Bishop's property (β) and thick spectrum, *J. Functional Anal.* **94** (1990), 196-222.
13. M. Grangé, Diviseurs de Leibenson et problème de Gleason pour $H^\infty(\Omega)$ dans le cas convexe, *Bull. Soc. math. France* **114** (1986), 225-245.
14. N. Kerzman and A. Nagel, Finitely generated ideals in certain function algebras, *J. Functional Anal.* **7** (1971), 212-215.
15. M. Kosiek and M. Ptak, Reflexivity of N-tuples of contractions with rich joint left essential spectrum, Preprint
16. M. R. Range, *Holomorphic functions and integral representations in several complex variables,* Springer-Verlag, New York-Berlin-Heidelberg-Tokyo, 1986.
17. I. Singer, *Bases in Banach spaces* II, Springer-Verlag, Berlin-Heidelberg-New York, 1981.
18. B. Sz.-Nagy and C. Foias, *Harmonic analysis of operators on Hilbert spaces,* North-Holland, Amsterdam, 1970.
19. J. L. Taylor, A joint spectrum of several commuting operators, *J. Functional Anal.* **6** (1970), 172-191.
20. J. L. Taylor, The analytic functional calculus for several commuting operators, *Acta Math.* **125** (1970), 1-38.
21. F.-H. Vasilescu, *Analytic functional calculus and spectral decompositions,* D. Reidel Publ. Corp., Dordrecht, and Editura Academie, Bucuresti, 1982.
22. F.-H. Vasilescu, An operator-valued Poisson kernel, *J. Functional Anal.* **110** (1992), 47-72.

On Unbounded Calkin Representations

KLAUS-DETLEF KÜRSTEN Fachbereich Mathematik-Informatik, Universität Leipzig Augustusplatz, D-O-7010 Leipzig, Germany

ABSTRACT

Representations of maximal Op^*-algebras $L^+(D)$ which generalize representations of $\mathfrak{B}(H)$ constructed by J. W. Calkin and H. Behncke are considered. The domain of such a representation is shown to be complete in some cases. For a large class of domains D, several irreducibility conditions on these representations, such as algebraical or topological cyclicity of each non-zero vector, triviality of the weak or the strong commutant, and non-decomposibility into a direct orthogonal sum of representations are shown to be equivalent.

1 INTRODUCTION

Unbounded operator algebras are usually considered as sets of operators acting on a certain non-normable locally convex spaces D which may be understood as a generalization of a Hilbert space. Based on certain properties of this locally convex domain of the algebra, it is possible to approximate elements of the algebra by bounded Hilbert space operators and to use this approximation to prove general properties of unbounded operator algebras. Here we will present one application of this method, the investigation of unbounded Calkin representations. J. W. Calkin ([2]) constructed a representation of the algebra $\mathfrak{B}(H)$ of all bounded linear operators on a (separable) Hilbert space H which has the ideal $\mathfrak{K}(H)$ of the compact operators as its kernel. G. A. Reid ([17]) investigated the problem of irreducibility of such a representation. H. Behncke ([1]) considered these questions for non-separable Hilbert spaces and for representations which have larger ideals than $\mathfrak{K}(H)$ as their kernels. These results were generalized by the author and M. Milde in [13] to unbounded Calkin representations. It should be noted, however, that there exist several non-equivalent properties of general *-representations, which are equivalent to irreducibility in the case of closed *-representations of C^*-algebras. The main part of this paper is concerned with the proof of the somewhat surprising fact that similar properties are equivalent also for unbounded Calkin representations. Moreover, we consider the problem of completeness of the domain of the representation or, equivalentely, the question whether or not the representation is closed.

Note that unbounded Calkin representations were studied also by F. Löffler and W. Timmermann in [15, 16] and by K. Schmüdgen in [18].

The pattern of the paper is as follows. In Section 2, we give the necessary definitions concerning unbounded *-representations and Calkin representations. In Section 3, we introduce some topological properties of operator algebras and of their domains as well as some further notations. Also the question of completeness of the domain of an unbounded Calkin representation will be considered in this section. In particular it will be shown that the domain of the representation is again a Fréchet space or a direct orthogonal sum of Fréchet spaces if D has this property. In Section 4, it will be shown (under a certain general topological assumption) that topological cyclicity of each non-zero vector implies algebraical cyclicity of each non-zero vector. In Section 5, we show that several commutants coincide and that cyclicity of each non-zero vector is equivalent to non-decomposibility into a direct orthogonal sum of non-zero representations for unbounded Calkin representations.

Throughout the paper, we will need some set-theoretical notations which will be introduced now. The set theory within which we work is the Zermelo-Fraenkel set theory together with the axiom of choice. For definitions and basic facts concerning ordinals, cardinals, and ultrafilters we refer to [3]. In particular, an ordinal is the set of all smaller ordinals and a cardinal is an ordinal that is not in a one-to-one correspondence with a smaller ordinal. Ordinals will be denoted by $\alpha, \beta, \gamma, \delta, \xi, \eta, \zeta$ and infinite cardinals will be denoted by λ, μ, ν. Also ω_0 denotes the least infinite cardinal and ω_β is the least cardinal that is greater then ω_η for each $\eta < \beta$. The cardinality of a set M is denoted by card M. As usual we write $\aleph_\beta$ for the cardinality of ω_β. The axiom of choice implies that any infinite set M satisfies card $M = \aleph_\beta$ for some ordinal β. In particular, each infinite cardinal is equal to some ω_β. The *cofinality* of an infinite cardinal μ is the least cardinal λ for which there exists a function $s : \lambda \to \mu$ such that $s(\eta) \leq s(\zeta)$ for $\eta < \zeta < \lambda$ and $\sup\{s(\eta) : \eta < \lambda\} = \mu$. It is denoted by cof (μ). An infinite cardinal μ with $\mu = \text{cof}(\mu)$ is said to be *regular*. The cardinals $\omega_{\alpha+1}$ are regular (cf. [5], section 7).

We will also use the following notations concerning Hilbert spaces and operators on them. The scalar product, the norm, and the unit ball of a (complex) Hilbert space H are denoted by $\langle\, . \,, \,.\, \rangle$, $\|\,.\,\|$, and U_H, resp. For a closable operator a on H, $D(a)$ denotes the domain, a^* the adjoint, and $\bar{a} = a^{**}$ the closure. If $D \subset D(a)$, $a \restriction D$ denotes the restriction of a to D.

2 UNBOUNDED CALKIN REPRESENTATIONS

A *-representation π of a *-algebra $\mathfrak{A}$ on a complex pre Hilbert space $D = D_\pi$ is a homomorphism of the algebra $\mathfrak{A}$ into the algebra $\mathrm{Lin}(D)$ of linear operators on D such that

$$\langle \pi(a)\varphi, \psi\rangle = \langle \varphi, \pi(a^*)\psi\rangle \quad (a \in \mathfrak{A},\ \varphi, \psi \in D).$$

We always assume that $\mathfrak{A}$ has a unit e and that $\pi(e) = id_D$ (the identical map of D). The range of a *-representation π is a *-algebra w. r. t. the involution $a \to a^+ := a^* \upharpoonright D$, where a^* denotes the Hilbert space adjoint of a in the Hilbert space completion H_π (or H) of D. Involutive algebras obtained in this way are termed Op^*-algebras on the domain D. On each domain D there exists a maximal Op^*-algebra which will be denoted by $L^+(D)$. Its elements are the linear operators a acting on D which satisfy $D \subset D(a^*)$ and $a^*(D) \subset D$.

The domain D of an Op^*-algebra $\mathcal{A}$ will be endowed with the topology generated by the seminorms $\varphi \to \|a\varphi\|$ $(a \in \mathcal{A})$. This topology will be denoted by $t_\mathcal{A}$ or by t_π when $\mathcal{A} = \pi(\mathfrak{A})$ and by t when $\mathcal{A} = L^+(D)$. Moreover, the algebra $\mathcal{A}$ will be endowed with the uniform topology $\tau_{b,\mathcal{A}}$ defined by the seminorms

$$a \to \sup_{\varphi,\psi \in M} |\langle a\varphi, \psi\rangle|,$$

where $M \subset D$ is $t_\mathcal{A}$-bounded. This topology will also be denoted by $\tau_{b,\pi}$ when $\mathcal{A} = \pi(\mathfrak{A})$ and by τ_b when $\mathcal{A} = L^+(D)$.

A *-representation π of $\mathfrak{A}$ is said to be closed, if it coincides with its closure $\bar{\pi}$ defined by

$$D_{\bar{\pi}} = \bigcap_{a \in \mathfrak{A}} D(\overline{\pi(a)}), \quad \bar{\pi}(a) = \overline{\pi(a)} \upharpoonright D_{\bar{\pi}}.$$

It is well-known and easy to see that the closure $\bar{\pi}$ of a *-representation π is always a closed *-representation and that $D_{\bar{\pi}}[t_{\bar{\pi}}]$ is the completion of $D_\pi[t_\pi]$.

In order to define Calkin representations, we consider a domain D, an index set μ, a singular state θ on the W^*-algebra $l_\infty(\mu)$ of all bounded complex valued μ-sequences, and an ordinal β.

Then we define the following linear spaces:

$$\begin{aligned} l_\infty(\mu, D) &= \Big\{(\varphi_\xi) \in \prod_{\xi \in \mu} D : \{\varphi_\xi\}_{\xi \in \mu} \text{ is } t\text{-bounded}\Big\}, \\ X_\beta &= \{(\varphi_\xi) \in l_\infty(\mu, D) : \mathrm{card}\{\xi : |\langle \varphi_\xi, \varphi\rangle| > \varepsilon\} < \aleph_\beta \text{ for each } \varepsilon > 0 \text{ and } \varphi \in D\}, \\ N_{\beta,\theta} &= \{(\varphi_\xi) \in X_\beta : \theta((\|\varphi_\xi\|^2)_{\xi \in \mu}) = 0\}, \\ D_{\beta,\theta} &= X_\beta / N_{\beta,\theta}. \end{aligned}$$

The equivalence class $(\varphi_\xi) + N_{\beta,\theta} \in D_{\beta,\theta}$ of $(\varphi_\xi) \in X_\beta$ will be denoted by $(\varphi_\xi)_\theta$. Now it is easy to see that the formulas

$$\begin{aligned} \langle (\varphi_\xi)_\theta, (\psi_\xi)_\theta\rangle &= \theta((\langle \varphi_\xi, \psi_\xi\rangle)_{\xi \in \mu}), \\ \pi_{\beta,\theta}(a)(\varphi_\xi)_\theta &= (a\varphi_\xi)_\theta \end{aligned}$$

define a scalar product and a *-representation $\pi_{\beta,\theta}$ of $L^+(D)$ on $D_{\beta,\theta}$ (cf. [13] for more details). The representations $\pi_{\beta,\theta}$ will be termed (unbounded) Calkin representations. Note that $D_{\beta,\theta}$ may degenerate to $\{0\}$. Thus we can not exclude zero Calkin representations (on the space $D_\pi = 0$) from our considerations. However, the domain D is always assumed to be different from $\{0\}$.

Let us note the following easy result.

PROPOSITION 2.1 *Suppose that there exist $(\varphi_\xi) \in X_\beta$ and $\mu_1 \subset \mu$ such that*

$$\theta((\chi_{\mu_1}(\xi) \cdot \|\varphi_\xi\|^2)_{\xi\in\mu}) \neq 0 \text{ and } \theta((\chi_{\mu\setminus\mu_1}(\xi) \cdot \|\varphi_\xi\|^2)_{\xi\in\mu}) \neq 0.$$

(χ_M denotes the characteristic function of the set M.)

Then $\pi_{\beta,\theta}$ decomposes into a direct orthogonal sum of two non-zero representations.

Proof: The spaces

$$D_1 = \{(\chi_{\mu_1}(\xi) \cdot \varphi_\xi)_\theta : (\varphi_\xi)_{\xi\in\mu} \in X_\beta\} \text{ and } D_2 = \{(\chi_{\mu\setminus\mu_1}(\xi) \cdot \varphi_\xi)_\theta : (\varphi_\xi)_{\xi\in\mu} \in X_\beta\}$$

are non-zero invariant subspaces for $\pi_{\beta,\theta}$ which satisfy $D_1 \oplus D_2 = D_{\pi_{\beta,\theta}}$. This proves the assertion.

Let us consider now a non-zero representation $\pi_{\beta,\theta}$ which has the property that the assumptions of the previous proposition are never satisfied. Then this representation is unitarily equivalent to a representation $\pi_{\beta,\theta_{\mathfrak{U}}}$ in a natural way, where $\theta_{\mathfrak{U}}$ is given as an ultrafilter limit w.r.t. a non-trivial ultrafilter $\mathfrak{U}$ of subsets of μ

$$\theta_{\mathfrak{U}}((x_\xi)_{\xi\in\mu}) = \lim_{\mathfrak{U}} x_\xi,$$

as shown in [13], Proposition 4.1. For Calkin representations of this type, we will use the abbreviations $\pi_{\beta,\theta_{\mathfrak{U}}} = \pi_{\beta,\mathfrak{U}}$, $D_{\beta,\theta_{\mathfrak{U}}} = D_{\beta,\mathfrak{U}}$, $(\varphi_\xi)_{\theta_{\mathfrak{U}}} = (\varphi_\xi)_{\mathfrak{U}}$ etc. It will be explained in Section 3 that completeness of the space $D_{\beta,\theta} = D_{\pi_{\beta,\theta}}$ is also related to the property that θ is given as an ultrafilter limit.

Since the sets $\{\varphi_\xi\}_{\xi\in\mu}$ $((\varphi_\xi) \in X_\beta)$ are assumed to be bounded, the linear functionals

$$L^+(D) \ni a \to \langle \pi_{\beta,\theta}(a)\,(\varphi_\xi)_\theta, (\psi_\xi)_\theta \rangle = \theta((\langle a\,\varphi_\xi, \psi_\xi \rangle)_{\xi\in\mu}) \quad ((\varphi_\xi), (\psi_\xi) \in X_\beta)$$

are continuous w.r.t. τ_b. Since the kernel of $\pi_{\beta,\theta}$ is the intersection of the kernels of all these functionals, we obtain the following result.

PROPOSITION 2.2 *The kernel* Ker $\pi_{\beta,\theta}$ *is a closed (two-sided, *-invariant) ideal of* $L^+(D)[\tau_b]$.

Note that the trivial ideals $\{0\}$ and $L^+(D)$ may also appear as Ker $\pi_{\beta,\theta}$. In the main part of this paper we will be concerned, however, with representations that have a non-trivial ideal of $L^+(D)$ as their kernels.

REMARKS : 1. J. W. Calkin ([2]) defined *-representations of the *-algebra $\mathfrak{B}(H)$ of all bounded linear operators on a separable Hilbert space H which have the unique closed ideal $\mathfrak{K}(H)$ as kernel. These representations depend on an arbitrary singular state on the algebra $l_\infty(\mathbb{N})$.

2. G. A. Reid ([17]) investigated irreducibility of the representations $\pi_{\beta,\mathfrak{U}}$ in the case $D = H$, $\beta = 0$, $\mu = \mathbb{N}$. He also observed that these representations are closed.

H. Behncke considered similar questions for non-separable Hilbert spaces in [1]. In fact he investigated irreducibility of the representations $\pi_{0,\mathfrak{U}}$ and $\pi_{\beta+1,\mathfrak{U}}$ in case $D = H$. Here the use of uncountable index sets is necessary in order to get also representations which have larger ideals than $\mathfrak{K}(H)$ as their kernels. (Such ideals exist for non-separable H.) Unbounded Calkin representations (with $\mu = \mathbb{N}$ and with more special classes of domains D) were studied first by F. Löffler and W. Timmermann in [15, 16] and by K. Schmüdgen in [18].

3. It seems to be natural to consider also modifications of the definition of Calkin representations which are obtained by replacing the space X_β by other invariant subspaces of $l_\infty(\mu, D)$. Examples of such spaces are given by

$$\begin{aligned} Z_\beta &= \{(\varphi_\xi) \in l_\infty(\mu, D) : \operatorname{card}\{\xi : |\langle \varphi_\xi, \varphi \rangle| \neq 0\} < \aleph_\beta \text{ for each } \varphi \in D\}, \\ X_\theta &= \{(\varphi_\xi) \in l_\infty(\mu, D) : \theta((\langle \varphi_\xi, \varphi \rangle)_{\xi \in \mu}) = 0 \text{ for each } \varphi \in D\}. \end{aligned}$$

A certain topological condition described in the next section (the condition that D is dual hilbertisable) implies that each space Z_β ($\beta > 0$) coincides with X_β or with a countable union of spaces X_{β_n} and that the representation defined by using Z_0 is the zero representation (cf. [13], Proposition 3.4). So the investigation of representations obtained by replacing the space X_β in the definition of $\pi_{\beta,\theta}$ by the space Z_β reduces almost completely to the study of the very representations $\pi_{\beta,\theta}$. Our knowledge about representations defined by using X_θ instead of X_β is less complete.

3 TOPOLOGICAL PROPERTIES

Usually the locally convex spaces $D[t_\pi]$ or $D[t]$ are assumed to satisfy additional topological conditions such as semi-reflexivity, completeness or metrizibility. We will say that a unitary space D is a semi-reflexive domain, a complete domain, a Fréchet domain etc., if the locally convex space $D[t]$ has the corresponding property. In a first part of this section we will introduce further topological properties of domains of *-representations, which are already known to be useful in the theory of unbounded operator algebras. Then we review some results about kernels of Calkin representations, which will be needed later on. In the final part of this section we will be concerned with topological properties of domains $D_\pi[t_\pi]$ of Calkin representations $\pi = \pi_{\beta,\theta}$. In particular we obtain results for Fréchet domains and for direct orthogonal sums of Fréchet domains. In some cases, completeness of $D_\pi[t_\pi]$ may be characterized in terms of properties of the state θ. However, in the general case the situation is more complicated, as an example shows.

Before giving the definitions of further topological properties, we discuss the relation between Hilbert discs and Hilbert space operators.

DEFINITION 3.1 *A bounded subset B of a locally convex space E is said to be a Hilbert disc if there exists a structure of a Hilbert space on the linear span of B such that B becomes the unit ball. This Hilbert space will be denoted by E_B.*

A subset $B \subset E$ is a Hilbert disc if and only if there exists a Hilbert space H_1 and a continuous linear operator $c : H_1 \to E$ such that $B = c(U_{H_1})$. Indeed, given such an operator c, the scalar product on E_B may be defined by setting $\langle c\varphi, c\psi \rangle_{E_B} = \langle \varphi, \psi \rangle_{H_1}$, where φ and ψ are in the orthogonal complement of the kernel of c. Conversely, given a Hilbert disc

$B \subset E$, the operator c may be defined to be the identical imbedding $E_B \to E$. Let us now consider the case when $E = H$ is itself a Hilbert space. In this case we can use the polar decomposition $c = \sqrt{c\,c^*}\,u$, where $c^* : H \to H_1$ is the adjoint and $u : H_1 \to H$ is a partial isometry with $\sqrt{c\,c^*}\,u\,u^* = \sqrt{c\,c^*}$. It follows that the operator $b = \sqrt{c\,c^*}$ satisfies $b(U_H) = B$ in this case. Finally, let us consider the domain $D_\pi[t_\pi]$ of a *-representation π. Note that, given $a \in \mathfrak{A}$ and $b \in \mathfrak{B}(H_\pi)$ with $b(H_\pi) \subset D_\pi$, the operator $\pi(a)\,b$ is norm-bounded (because its domain is H_π and the adjoint is densely defined). This implies that a Hilbert disc B in H_π with $B \subset D_\pi$ is automatically bounded w.r.t. t_π. These (essentially known, cf. [6, 13]) observations my be summarized as follows.

PROPOSITION 3.2 *Consider a *-representation π. A subset $B \subset D_\pi$ is a Hilbert disc in $D_\pi[t_\pi]$ if and only if there exists a (positive self-adjoint) operator $b \in \mathfrak{B}(H)$ such that $B = b(U_H)$.*

Let us introduce now some topological conditions.

DEFINITION 3.3

a) *A locally convex space E is said to be a (QF)-space (or a quasi-Fréchet space) if it has the property that the closed linear span of any bounded subset is a Fréchet space w.r.t. the topology induced by E.*

b) *A locally convex space E is said to be a (DH)-space (or a dual hilbertisable space) if each bounded subset is contained in a Hilbert disc.*

The class of (QF)-domains (introduced by the author in [8]) is general enough to contain most examples of domains of Op^*-algebras considered in the literature. We refer to [11] for more details, examples and further references on these domains. Here we note only that (QF)-domains and complete (DF)-domains are also (DH)-domains. For complete (DF)-domains this was shown by H. Junek in [6]. In the context of the theory of locally convex spaces, hilbertisable (QF)-spaces were considered by the author in [12], too.

The notation "dual hilbertisable space" was chosen in [13], because E is a (DH)-space if and only if it is semi-reflexive and the topology of the strong dual can be given by semi-norms which are defined by semi-scalar products in a usual way.

The following concepts will be useful.

DEFINITION 3.4 *Given an Op^*-algebra $\mathcal{A}$ on D, we define the subsets*

$$\begin{aligned} \mathcal{B}_\mathcal{A} &= \{b \in \mathcal{A} : b \text{ is bounded w.r.t. the norm and satisfies } \bar{b}(H) \subset D \text{ and } b^*(H) \subset D\}, \\ \mathcal{P}_\mathcal{A} &= \{p \in \mathcal{B}_\mathcal{A} : p^2 = p = p^+\}. \end{aligned}$$

We will also use the notations $\mathcal{B}_\pi = \mathcal{B}_\mathcal{A}$ and $\mathcal{P}_\pi = \mathcal{P}_\mathcal{A}$ if $\mathcal{A} = \pi(\mathfrak{A})$ and $\mathcal{B}_D = \mathcal{B}_\mathcal{A}$ and $\mathcal{P}_D = \mathcal{P}_\mathcal{A}$ if $\mathcal{A} = L^+(D)$. The algebra $\mathcal{A}$ is said to have the (uniform) density property, if there exists a net $(p_\iota)_{\iota \in J}$ in $\mathcal{P}_\mathcal{A}$ such that the net $(p_\iota\, a\, p_\iota)_{\iota \in J}$ converges to a w.r.t. $\tau_{b,\mathcal{A}}$ for each $a \in \mathcal{A}$ (and uniformly on each bounded subset of $\mathcal{A}[\tau_{b,\mathcal{A}}]$).

It is known that $L^+(D)$ has the uniform density property if D is a (DH)-domain (cf. [13]).

In order to describe the closed ideals of $L^+(D)[\tau_b]$ and the kernel of $\pi_{\beta,\theta}$, we need the following notations.

Given a pre Hilbert space D, $\gamma = \gamma(D)$ denotes the least ordinal such that

$$\dim p(D) < \aleph_\gamma \quad \text{for all } p \in \mathcal{P}_D.$$

Here $\dim p(D)$ is the Hilbert space dimension, i.e. the cardinality of a maximal orthonormal system of the Hilbert space $p(D)$.

Given an ordinal η, we define $\mathcal{I}_\eta$ to be the τ_b-closed, two-sided ideal of $L^+(D)$ generated by

$$\{p \in \mathcal{P}_D : \dim p(D) < \aleph_\eta\}.$$

Given a singular state θ on $l_\infty(\mu)$, we let $\alpha = \alpha(\theta)$ denote the minimal ordinal such that there exists a subset $M \subset \mu$ with card $M = \aleph_\alpha$ and $\theta(\chi_M) \neq 0$. Also we define an ordinal $\delta = \delta(\theta)$ by the condition $\operatorname{cof}(\omega_\alpha) = \omega_\delta$. If $\theta = \theta_{\mathfrak{U}}$ is given as an ultrafilter limit, we will use also the notations $\alpha(\mathfrak{U}) = \alpha(\theta_{\mathfrak{U}})$ and $\delta(\mathfrak{U}) = \delta(\theta_{\mathfrak{U}})$.

The following result was obtained by the author in [10] for Fréchet domains. The proof given there applies, however, also in the general case.

PROPOSITION 3.5 *Suppose that $L^+(D)$ has the density property. Then*

$$\{\mathcal{I}_\eta : \eta < \gamma(D)\}$$

is the set of all non-trivial, two-sided, τ_b-closed ideals of $L^+(D)$. Moreover, $\mathcal{I}_\eta$ is strictly contained in $\mathcal{I}_\zeta$ if $\eta < \zeta \leq \gamma(D)$.

The following description of kernels of Calkin representations was obtained in [13].

PROPOSITION 3.6 *Suppose that $L^+(D)$ has the density property and that $\pi_{\beta,\theta}$ is the Calkin representation of $L^+(D)$. Then the following assertions are satisfied:*

a) $\operatorname{Ker} \pi_{\beta,\theta} = \{0\}$ *if* $\beta > \alpha(\theta)$.

b) $\operatorname{Ker} \pi_{\beta,\theta} = \mathcal{I}_{\delta(\theta)}$ *if* $\beta = \alpha(\theta)$.

c) $\operatorname{Ker} \pi_{\beta,\theta} = \mathcal{I}_{\alpha(\theta)}$ *if* $\beta < \alpha(\theta)$.

In particular, $\operatorname{Ker}\pi_{\beta,\theta} = L^+(D)$, *if* $\beta = \alpha(\theta)$ *and* $\delta(\theta) \geq \gamma(D)$ *or* $\beta < \alpha(\theta)$ *and* $\alpha(\theta) \geq \gamma(D)$.

Now we are going to prove that the representation space $D_\pi[t_\pi]$ is complete for some unbounded Calkin representations π. In case $D = H$, the space $D_\pi[t_\pi]$ is known to be complete if and only if θ is a (finite) convex combination of ultrafilter limits (cf. [9]). However, the situation is more complicated in the unbounded case.

PROPOSITION 3.7 *Suppose that D is a Fréchet domain and that $\pi = \pi_{\beta,\mathfrak{U}}$ is a non-zero Calkin representation of $L^+(D)$. Then $D_\pi[t_\pi]$ is a Fréchet space.*

Proof: There exists a sequence $(a_n)_{n\in\mathbb{N}}$ in $L^+(D)$ such that the sets $S_n = \{\varphi \in D : \|a_n\varphi\| \leq 1\}$ form a 0-neighbourhood base in $D[t]$ and such that $S_{n+1} \subset S_n$. Thus, given $b \in L^+(D)$, there exists $n \in \mathbb{N}$ such that $\|b\,\varphi\| \leq \|a_n\,\varphi\|$, or equivalentely,

$$\langle b^+\, b\,\varphi, \varphi\rangle \leq \langle (a_n)^+\, a_n\,\varphi, \varphi\rangle$$

for all $\varphi \in D$. This implies

$$\|\pi(b)\,\varphi\|^2 = \langle \pi(b^+\,b)\,\varphi, \varphi\rangle \leq \langle \pi((a_n)^+\,a_n)\,\varphi, \varphi\rangle = \|\pi(a_n)\,\varphi\|^2$$

for all $\varphi \in D_\pi$. Hence the sets $S'_n = \{\varphi \in D_\pi : \|\pi(a_n)\varphi\| \leq 1\}$ form a 0-neighbourhood base in $D_\pi[t_\pi]$. Thus $D_\pi[t_\pi]$ is metrizable. In order to prove that this space is also complete, it suffices to show that a series $\sum_{n=1}^{\infty} \varphi_n$ converges in $D_\pi[t_\pi]$ whenever $\|\pi(a_n)\,\varphi_n\| < 2^{-n}$. In order to do so, we find $(\varphi_{n,\xi}) \in X_\beta$ such that $(\varphi_{n,\xi})_{\mathfrak{U}} = \varphi_n$. Then we define $U_n \in \mathfrak{U}$ such that $\|a_n\,\varphi_{n,\xi}\| < 2^{-n}$ for $\xi \in U_n$. Defining $\psi_{n,\xi} = \chi_{U_n}(\xi) \cdot \varphi_{n,\xi}$, we have $(\psi_{n,\xi})_{\mathfrak{U}} = \varphi_n$ and

$$\sup_{\xi} \sum_{n=k}^{\infty} \|a_k\,\psi_{n,\xi}\| \leq \sum_{n=k}^{\infty} 2^{-n}.$$

This implies easily that each series $\sum_{n=1}^{\infty} \psi_{n,\xi}$ converges in $D[t]$ to some element ψ_ξ, that $(\psi_\xi)_{\xi\in\mu} \in X_\beta$ and that $\sum_{n=1}^{\infty} \varphi_n = (\psi_\xi)_{\mathfrak{U}}$ in $D_\pi[t_\pi]$. This completes the proof.

Using the idea of Proposition 2.1, we obtain the following generalization.

COROLLARY 3.8 *Suppose that D is a Fréchet domain and that the state θ on $l_\infty(\mu)$ is given as a (finite) convex combination of ultrafilter limits. Suppose further that $\pi = \pi_{\beta,\theta}$ is a non-zero Calkin representation of $L^+(D)$. Then $D_\pi[t_\pi]$ is a Fréchet space.*

Similar results will be seen to be true for direct orthogonal sums of Fréchet domains. Given a family $(D_\iota)_{\iota\in J}$ of pre Hilbert spaces, the direct orthogonal sum

$$D = \sum_{\iota\in J}^{\perp} D_\iota$$

is defined to be the direct sum endowed with the scalar product induced from the complete orthogonal sum

$$H = \sum_{\iota\in J}^{\oplus} H_\iota$$

of the Hilbert space completions H_ι of D_ι.

As shown previously by the author in [11], Lemma 1.3, the topology t of an arbitary direct orthogonal sum of Fréchet domains is generated by the semi-norms

$$\sum_{\iota\in J}^{\perp} D_\iota \ni (\varphi_\iota)_{\iota\in J} \to \|(a_\iota\,\varphi_\iota)_{\iota\in J}\| = \Big(\sum_{\iota\in J} \|a_\iota\,\varphi_\iota\|^2\Big)^{1/2} \quad (a_\iota \in L^+(D_\iota)). \tag{1}$$

It was recently observed by the author and W.-D. Heinrichs (unpublished up to now) that this result does not even generalize to (QF)-domains. The corresponding example is based on a construction given by the author in [7].

In the following considerations we will use the well-known fact that the topology t_π coincides with t whenever it is a Fréchet topology.

THEOREM 3.9 *Suppose that D is given as the direct orthogonal sum of a family $(D_\iota)_{\iota\in J}$ of Fréchet domains and that the state θ on $l_\infty(\mu)$ is given as a (finite) convex combination of ultrafilter limits. Suppose further that $\pi = \pi_{\beta,\theta}$ is a non-zero Calkin representation of $L^+(D)$. Let π_ι denote the Calkin representation $\pi_{\beta,\theta}$ of $L^+(D_\iota)$. Set $J' = \{\iota \in J : D_{\pi_\iota} \neq \{0\}\}$. Then D_π is unitarily equivalent to the direct orthogonal sum*

$$D_1 = \sum_{\iota\in J'}^{\perp} D_{\pi_\iota}.$$

Moreover, the topology t_π coincides with the topology t of D_π.

Proof: By Corollary 3.8, the non-zero spaces $D_{\pi_\iota}[t_{\pi_\iota}]$ are Fréchet domains. The (natural) unitary equivalence of D_π and D_1 follows from the fact that each bounded subset of D is already contained in some finite direct orthogonal sum of spaces D_ι. The coincidence of the topologies follows from the fact that the t_π-continuous semi-norms

$$\sum_{\iota\in J'}^{\perp} D_{\pi_\iota} \ni (\varphi_\iota)_{\iota\in J'} \to \|(\pi_\iota(a_\iota)\,\varphi_\iota)_{\iota\in J'}\| \quad (a_\iota \in L^+(D_\iota)).$$

define already the topology t of $L^+(D_1)$ as mentioned above. This completes the proof.

Note that the space $D_\pi[t_\pi]$ in the previous theorem is complete because the direct orthogonal sum of complete domains is easily seen to be a complete domain. Note also that it is a Fréchet space if and only if the set J' defined in the theorem is finite.

Next, we want to show that certain spaces $D_{\pi_{\beta,\theta}}[t_{\pi_{\beta,\theta}}]$ are not complete.

PROPOSITION 3.10 *Let a Calkin representation $\pi = \pi_{\beta,\theta}$ be given. Suppose that there exists $M_0 \subset \mu$ and $(\varphi_\xi) \in X_\beta$ with $\|\varphi_\xi\| \geq 1$ for all $\xi \in M_0$ such that the set of values of the finitely additive measure*

$$M_0 \supset M \to \theta(\chi_M)$$

is infinite. Then the space $D_\pi[t_\pi]$ is not complete.

Proof: It follows from the assumptions that there exists a sequence $(M_n)_{n\in\mathbb{N}}$ of pairwise disjoint subsets of M_0 such that $\theta(\chi_{M_n}) > 0$. Additionally we may ensure that $\theta(\chi_{M_n}) < 2^{-n}$. Let us now consider the elements $\psi_n = ((n\cdot\chi_{M_n}(\xi)\cdot\varphi_\xi)_{\xi\in\mu})_\theta$ of D_π. Since

$$\|\pi(a)\,\psi_n\|^2 = \theta((n^2\cdot\chi_{M_n}(\xi)\cdot\|a\,\varphi_\xi\|^2)_{\xi\in\mu}) \leq n^2\cdot 2^{-n}\cdot\sup_{\xi\in\mu}\|a\,\varphi_\xi\|^2,$$

the series $\sum_{n=1}^{\infty}\psi_n$ converges absolutely in $D_\pi[t_\pi]$. Thus it suffices to show that it does not converge. Suppose, on the contrary, that it converges to $\psi = (\psi_\xi)_\theta$, where $(\psi_\xi) \in X_\beta$. Set $x = \sup_{\xi\in\mu}\|\psi_\xi\|$ and choose $m, k \in \mathbb{N}$ with $m > k > x$. Then it follows that

$$\Big\|\Big(\sum_{n=1}^{m}\psi_n\Big) - \psi\Big\| \geq \theta((\chi_{M_k}(\xi)\cdot\|k\,\varphi_\xi - \psi_\xi\|^2)_{\xi\in\mu}) \geq (k-x)^2\,\theta(\chi_{M_k}).$$

This means that the series $\sum_{n=1}^{\infty}\psi_n$ does not converge to $(\psi_\xi)_\theta$ and the proof is complete.

Note that a singular state θ on $l_\infty(\mu)$ has a canonical representation $\theta = \theta_0 + \theta_1$, where θ_0 is a positive linear functional on $l_\infty(\mu)$ such that $\lim_{U\in\mathfrak{U}}\theta_0(\chi_U) = 0$ for all ultrafilters $\mathfrak{U}$ on μ and $\theta_1 = 0$ or it can be written as a finite or infinite sum $\sum x_n\theta_{\mathfrak{U}_n}$ $(x_n > 0)$ of weighted, pairwise distinct ultrafilter limits. The previous proposition implies that the domain $D_\pi[t_\pi]$ of the representation $\pi = \pi_{\beta,\theta}$ is not complete in each of the following cases:

a) *There exists $(\varphi_\xi) \in X_\beta$ such that $\theta_0((\|\varphi_\xi\|^2)_{\xi\in\mu}) \neq 0$.*

b) *$\gamma(D) > \alpha(\theta_0)$ or $\beta > \alpha(\theta_0)$.*

c) $\theta_1 = \sum_{n=1}^{\infty} x_n \, \theta_{\mathfrak{U}_n}$, *where* $x_n > 0$ *and the ultrafilter limits* $\theta_{\mathfrak{U}_n}$ *are pairwise distinct, and there exists* $(\varphi_\xi) \in X_\beta$ *such that the set*

$$\{n \in \mathbb{N} : \lim_{\mathfrak{U}_n} \|\varphi_\xi\|^2 \geq 1\}$$

is infinite.

d) $\theta_1 = \sum_{n=1}^{\infty} x_n \, \theta_{\mathfrak{U}_n}$, *where* $x_n > 0$ *and the ultrafilter limits* $\theta_{\mathfrak{U}_n}$ *are pairwise distinct, and* $\gamma(D) > \sup\{\alpha(\mathfrak{U}_n) : n \in \mathbb{N}\}$ *or* $\beta > \sup\{\alpha(\mathfrak{U}_n) : n \in \mathbb{N}\}$.

We conclude this discussion with an example which shows that completeness of the representation space D_π of the Calkin representation $\pi = \pi_{\beta,\theta}$ may also occur for states θ which are not (finite) convex combinations of ultrafilter limits.

EXAMPLE: Let a sequence $(\mathfrak{U}_n)_{n\in\mathbb{N}}$ of non-trivial ultrafilters on μ and a sequence $(H_n)_{n\in\mathbb{N}}$ of Hilbert spaces be given such that $\alpha(\mathfrak{U}_n) < \gamma(H_n) \leq \alpha(\mathfrak{U}_{n+1})$ for all $n \in \mathbb{N}$. Define a domain D and a state θ on $l_\infty(\mu)$ by

$$D = \sum_{n\in\mathbb{N}}^{\perp} H_n \qquad \theta((x_\xi)_{\xi\in\mu}) = \sum_{n=1}^{\infty} 2^{-n} \lim_{\mathfrak{U}_n} x_\xi .$$

Let $H_{k,n}$ $(k \leq n)$ be the domain $D_{\pi_{k,n}}$ obtained for the representation $\pi_{k,n} = \pi_{0,\mathfrak{U}_k}$ of the algebra $L^+(H_n) = \mathfrak{B}(H_n)$. In view of Propositions 3.6 and 3.7, the spaces $H_{k,n}$ are non-zero Hilbert spaces. Finally, let π denote the Calkin representation $\pi_{0,\theta}$ of $L^+(D)$.

Then D_π is unitarily equivalent to the direct orthogonal sum

$$D_1 = \sum_{n\in\mathbb{N}}^{\perp} \sum_{k=1}^{n}{}^{\oplus} H_{k,n}$$

and t_π is the topology of the locally convex direct sum. Consequently, $D_\pi[t_\pi] = D_\pi[t]$ is complete.

Proof: Fix pairwise disjoint sets U_n with $U_n \in \mathfrak{U}_n$ and card $U_n = \aleph_{\alpha(\mathfrak{U}_n)}$. Let $(\varphi_\xi)_{\xi\in\mu}$ be a D-valued bounded μ-sequence which belongs to X_0. Let φ_ξ be given as $\varphi_\xi = (\varphi_{\xi,n})_{n\in\mathbb{N}}$, where $\varphi_{\xi,n} \in H_n$. Because $\{\varphi_\xi\}_{\xi\in\mu}$ is bounded in D, there exists $m \in \mathbb{N}$ such that $\varphi_{\xi,n} = 0$ whenever $n > m$. Define $\varphi_{k,n,\xi} = \chi_{U_k}(\xi) \cdot \varphi_{\xi,n}$. Since

$$\|(\varphi_\xi)_\theta\|^2 = \sum_{n=1}^{m}\sum_{k=1}^{n} 2^{-k} \lim_{\mathfrak{U}_k} \|\varphi_{k,n,\xi}\|^2,$$

the unitary equivalence between D_π and D_1 may be defined by the formula

$$u\,(\varphi_\xi)_\theta = \sum_{n=1}^{m}\sum_{k=1}^{n} 2^{-k/2}\,(\varphi_{k,n,\xi})_{\mathfrak{U}_k}.$$

For an arbitrary scalar valued sequence $(x_n)_{n\in\mathbb{N}}$, we define the operator

$$a : \; D = \sum_{n\in\mathbb{N}}^{\perp} H_n \;\ni\; (\varphi_n)_{n\in\mathbb{N}} \to (x_n\,\varphi_n)_{n\in\mathbb{N}},$$

which belongs to $L^+(D)$, of course. Finally, given $\varphi = ((\varphi_{k,n})_{k=1}^n)_{n\in\mathbb{N}} \in D_1$, where $\varphi_{k,n} \in H_{k,n}$, we get

$$u\,\pi(a)\,u^{-1}\,\varphi = ((x_n \cdot \varphi_{k,n})_{k=1}^n)_{n\in\mathbb{N}},$$
$$\|u\,\pi(a)\,u^{-1}\,\varphi\| = \Big(\sum_{n=1}^{\infty} |x_n|^2 \cdot \|(\varphi_{k,n})_{k=1}^n\|^2\Big)^{1/2}.$$

Since these semi-norms define the topology of the locally convex direct sum of Hilbert spaces, the proof is complete.

At the end of this section, we prove two further properties of Calkin representations defined with the help of ultrafilter limits.

LEMMA 3.11 *Consider a non-zero Calkin representation $\pi = \pi_{\beta,\mathfrak{U}}$ of $L^+(D)$ and an operator $b \in \mathcal{B}_D$. Then*

$$(\pi(b))^{**}(H_\pi) \subset D_\pi.$$

In particular, for each $p \in \mathcal{P}_D$ the space $(\pi(p))^{**}(H_\pi)$ is a Hilbert space contained in D_π, that coincides with $\pi(p)(D_\pi)$.

Proof: Let us consider the Calkin representation $\tilde{\pi} = \pi_{\beta,\mathfrak{U}}$ of $\mathfrak{B}(H)$, where H is the Hilbert space completion of D. As a consequence of Proposition 3.7, $D_{\tilde{\pi}}$ is a Hilbert space. Given a t-bounded μ-sequence $(\varphi_\xi)_{\xi\in\mu}$ in X_β, the element $(\varphi_\xi)_\mathfrak{U}$ of D_π may be identified with the element $(\varphi_\xi)_\mathfrak{U}$ of $D_{\tilde{\pi}}$. Thus D_π becomes a (not necessarily dense) linear subspace of $D_{\tilde{\pi}}$. Now it is easy to see that $\tilde{\pi}(\bar{b})$ maps $D_{\tilde{\pi}}$ into D_π and that the restriction $\tilde{\pi}(\bar{b}) \restriction D_\pi$ coincides with $\pi(b)$. Consequently, the closure $(\pi(b))^{**}$ of $\pi(b)$ in H_π coincides with the restriction of $\tilde{\pi}(\bar{b})$ to the norm-closure in $D_{\tilde{\pi}}$ of D_π. This implies that $(\pi(b))^{**}(H_\pi) \subset D_\pi$ and the proof is complete.

LEMMA 3.12 *Suppose that $D[t]$ is dual hilbertisable. Then, given a t-bounded μ-sequence $(\varphi_\xi) \in X_\beta$ (with values in D), there exist a t-bounded μ-sequence $(\psi_\xi) \in X_\beta$ and a bounded, positive self-adjoint operator b on the Hilbert space completion H of D with $b(H) \subset D$ such that $b\,\psi_\xi = \varphi_\xi$ for all $\xi \in \mu$.*

Proof: According to Proposition 3.2, we can find a bounded, positive self-adjoint operator c on H with $c(H) \subset D$ such that $\{\varphi_\xi\}_{\xi\in\mu} \subset c(U_H)$. By K. Schmüdgen ([19], Proposition 3.1.4), the operator $b = c^{1/2}$ also satisfies $b(H) \subset D$. We represent b as orthogonal sum $b_1 \oplus \mathbf{0}$, where $\mathbf{0}$ is the zero operator on the kernel of b. Furthermore,we define the operator $u = (b_1)^{-1} \oplus \mathbf{0}$. Then we set $\psi_\xi = u\,\varphi_\xi$.

The set $\{\psi_\xi\}_{\xi\in\mu} = \{b\,u^2\,\varphi_\xi\}_{\xi\in\mu}$ is a bounded subset of $D[t]$ because it is contained in $b(U_H)$.

In order to show that $(\psi_\xi) \in X_\beta$, we note first that $D_1 = b(D) + \operatorname{Ker} b$ is a dense linear subspace of H and that $\langle \psi_\xi, \varphi\rangle = \langle \varphi_\xi, u\,\varphi\rangle$ for $\varphi \in D_1$. Consequently,

$$\operatorname{card}\{\xi : |\langle \varphi_\xi, \varphi\rangle| > \varepsilon\} < \aleph_\beta.$$

for each $\varepsilon > 0$ and $\varphi \in D_1$. But then the same statement is even true for each $\varepsilon > 0$ and all $\varphi \in H$. This shows that $(\psi_\xi) \in X_\beta$. Since obviously $b\,\psi_\xi = \varphi_\xi$, the proof is complete.

4 ALGEBRAICAL IRREDUCIBILITY

Irreducibility of closed *-representations of C^*-algebras may be characterized by each of the following properties:

a) *Each non-zero vector* $\varphi \in D_\pi = H_\pi$ *is cyclic, i.e.* $\pi(\mathfrak{A})(\varphi) = H_\pi$.

b) *Each non-zero vector* $\varphi \in D_\pi = H_\pi$ *is topologically cyclic, i.e.* $\pi(\mathfrak{A})(\varphi)$ *is dense in* H_π.

c) *The commutant of* $\pi(\mathfrak{A})$ *is trivial*

(cf. e.g. [4] 2.3.1 and 2.8.4). But a corresponding result for general (unbounded) *-representations does not hold. Moreover, there exist several non-equivalent definitions of commutants of unbounded operator algebras. We will consider two of them, the strong and the weak commutants defined by

$$\begin{aligned} \mathcal{A}'_s &= \{b \in \mathfrak{B}(H) : ba\varphi = ab\varphi \text{ for all } a \in \mathcal{A} \text{ and } \varphi \in D\} \\ \mathcal{A}'_w &= \{b \in \mathfrak{B}(H) : \langle ba\varphi, \psi\rangle = \langle b\varphi, a^+\psi\rangle \text{ for all } a \in \mathcal{A} \text{ and } \varphi, \psi \in D\}. \end{aligned}$$

For more details, examples, and further references on commutants we refer to [19]. Here we only note the well-known fact that the usual representation of the Heisenberg algebra defined on the Schwartz space $D = \mathfrak{S}(\mathbb{R}^n)$ of tempered test functions by the formulas

$$\begin{aligned} (\pi(p_j)\varphi)(x) &= -i\frac{\partial\varphi}{\partial x_j}(x), \\ (\pi(q_j)\varphi)(x) &= x_j \cdot \varphi(x). \end{aligned}$$

gives an example of a closed *-representation with trivial commutant which has non-trivial closed invariant subspaces. Clearly, an invariant subspace of a *-representation π is understood to be a linear subspace of D_π which is invariant for each operator $\pi(a)$ $(a \in \mathfrak{A})$.

It will be shown in this section that, in contrast to the case of general *-representations, algebraical irreducibility is equivalent to topological irreducibility for certain unbounded Calkin representations. Before stating this result we want to recall a result of the author and M. Milde. To do so, we need the concept of an $\aleph_\beta$-R-K-minimal ultrafilter, which is related to the Rudin-Keisler order defined for classes of ultrafilters.

DEFINITION 4.1 *An ultrafilter* $\mathfrak{U}$ *of subsets of* μ *is said to be* $\aleph_\beta$-R-K*-minimal if and only if it has the following property: Given a map* $s : \mu \to \mu$ *such that*

$$\operatorname{card} s^{-1}(\xi) < \aleph_\beta$$

for all $\xi \in \mu$*, there exists* $U \in \mathfrak{U}$ *such that the restriction* $s \restriction U$ *is one to one.*

Note that the $\aleph_\beta$-R-K-minimality of $\mathfrak{U}$ is equivalent to the following property: Given a disjoint decomposition

$$\mu = \bigcup_{\eta \in \nu} M_\eta$$

such that $\operatorname{card} M_\eta < \aleph_\beta$ for all $\eta \in \nu$, there exists $U \in \mathfrak{U}$ such that $\operatorname{card} M_\eta \cap U \leq 1$.

Now we can formulate a result on irreducibility, obtained in [13].

THEOREM 4.2 *Suppose that $L^+(D)$ has the density property. Let $\mathfrak{U}$ be an ultrafilter of subsets of μ with $\alpha(\mathfrak{U}) < \gamma(D)$ and let β be an ordinal with $\beta \leq \alpha(\mathfrak{U})$. Let π denote the Calkin representation $\pi_{\beta,\mathfrak{U}}$. Then the following three conditions are equivalent:*

a) *$\mathfrak{U}$ is $\aleph_\beta$-R-K-minimal.*

b) *The only t_π-closed invariant subspaces of D_π are $\{0\}$ and D_π.*

c) *The weak commutant of $\pi(L^+(D))$ is trivial (i.e. it contains only scalar multiples of the identity).*

Moreover, each of these equivalent conditions implies that $\pi = \pi_{0,\mathfrak{U}}$.

Note that the conditions $\beta \leq \alpha(\mathfrak{U}) < \gamma(D)$ ensure that $\pi_{\beta,\mathfrak{U}}$ has a non-trivial ideal of $L^+(D)$ as its kernel and that there exists a t-bounded orthonormal family $(\varphi_\xi)_{\xi\in U}$ for some $U \in \mathfrak{U}$. Other non-zero Calkin representations will be considered in some remarks at the end of Section 5.

Now we state the main result of this section.

THEOREM 4.3 *Suppose that D is dual hilbertisable. Let $\mathfrak{U}$ be an ultrafilter of subsets of μ with $\alpha(\mathfrak{U}) < \gamma(D)$ and let β be an ordinal with $\beta \leq \alpha(\mathfrak{U})$. Let $\pi = \pi_{\beta,\mathfrak{U}}$ be the corresponding Calkin representation of $L^+(D)$. Then the $\aleph_\beta$-R-K-minimality of $\mathfrak{U}$ is equivalent to the following condition:*

The representation π is algebraically irreducible, i.e. $\pi(\mathfrak{A})\varphi = D_\pi$ for each $\varphi \in D_\pi \setminus \{0\}$.

In order to prove algebraical irreducibility in the case of $\aleph_\beta$-R-K-minimal ultrafilters, we will combine methods developed by G. A. Reid in [17] and H.Behncke in [1] with methods of the theory of unbounded operator algebras and with perturbation arguments for Riesz basis. The proof will be divided into several steps.

LEMMA 4.4 *Consider a sequence $(\varepsilon_{1,n})$ of positive real numbers and real constants c_1, c_2 with $0 < c_1 \leq c_2$. Then there exists a sequence $(\varepsilon_{2,n})$ of positive real numbers such that the following assertion is satisfied:*

For each sequence (φ_n) of elements of a Hilbert space H, which satisfies

$$c_1 \leq \|\varphi_n\| \leq c_2 \text{ and } |\langle\varphi_n, \varphi_m\rangle| < \varepsilon_{2,m} \text{ for } n < m, \tag{2}$$

the Gram-Schmidt orthogonalization process may be applied to define an orthogonal sequence (ψ_n). Moreover, the coefficients $x_{k,j}$ in the equations

$$\psi_{n+1} = \varphi_{n+1} - \sum_{j=1}^{n} x_{n,j}\varphi_j \tag{3}$$

are uniquely defined and satisfy $|x_{n,j}| < \varepsilon_{1,n}$.

Proof: The assertion follows easily from Cramér's rule applied to the linear system

$$\sum_{j=1}^{n} \langle\varphi_j, \varphi_k\rangle x_{n,j} = \langle\varphi_{n+1}, \varphi_k\rangle \qquad (k \in \{1, \ldots, n\}). \tag{4}$$

Indeed, since $\langle\varphi_k, \varphi_k\rangle \geq (c_1)^2$, one can define by induction positive numbers $\varepsilon_{2,n}$ such that condition (2) implies that the absolute value of the determinant of the system (4) is larger than $\frac{1}{2}(c_1)^{2n}$ and that the solutions of this system satisfy $|x_{n,j}| \leq \varepsilon_{1,n}$.

COROLLARY 4.5 *Given positive real numbers c_1 and c_2, there exists a sequence (ε_n) of positive real numbers such that the following assertion is satisfied:*

Whenever sequences (φ_n), (ψ_n) with values in a Hilbert space H satisfy $c_1 \leq \|\varphi_n\| \leq c_2$, $\|\psi_n - \varphi_n\| \leq \varepsilon_n$, and $|\langle\varphi_n, \varphi_m\rangle| \leq \varepsilon_m$ for $n < m$, the sequence (ψ_n) satisfies also

$$\left(\frac{1}{4}c_1\right)^2 \cdot \sum_{j=1}^{\infty} |x_j|^2 \leq \left\|\sum_{j=1}^{\infty} x_j\,\psi_j\right\|^2 \leq (4\,c_2)^2 \cdot \sum_{j=1}^{\infty} |x_j|^2. \tag{5}$$

Proof: By a perturbation argument for general Schauder bases (cf. [14], Proposition 1.a.9), we find some $\varepsilon > 0$ such that (5) is a consequence of the following property of the sequence (ψ_n): There exists an orthogonal sequence $(\psi_{1,n})$ with $\frac{1}{2}\,c_1 \leq \|\psi_{1,n}\| \leq 2\,c_2$ such that $\|\psi_n - \psi_{1,n}\| \leq 2\,\varepsilon^n$.

By the previous lemma there exists a sequence $(\varepsilon_{2,n})$ of positive real numbers such that the conditions $c_1 \leq \|\varphi_n\| \leq c_2$ and $|\langle\varphi_n, \varphi_m\rangle| \leq \varepsilon_{2,m}$ for $n < m$ imply that $\|\varphi_n - \psi_{1,n}\| \leq \varepsilon^n$ and $\frac{1}{2}c_1 \leq \|\psi_{1,n}\| \leq 2\,c_2$, where $(\psi_{1,n})$ is obtained by applying the Gram-Schmidt orthogonalization process to the sequence (φ_n). Finally we conclude that the sequence (ε_n) defined by $\varepsilon_n = \min\{\varepsilon^n, \varepsilon_{2,n}\}$ satisfies the assertion of the corollary.

The following proposition contains the main step of the proof of the fact that $\aleph_\beta$-R-K-minimality of the ultrafilter $\mathfrak{U}$ implies algebraical irreducibility of the Calkin representation $\pi_{\beta,\mathfrak{U}}$.

PROPOSITION 4.6 *Suppose that D is dual hilbertisable. Let $\mathfrak{U}$ be an ultrafilter of subsets of μ with $\alpha(\mathfrak{U}) < \gamma(D)$ and let $\beta \leq \alpha(\mathfrak{U})$ be an ordinal. Suppose that $\mathfrak{U}$ is $\aleph_\beta$-R-K-minimal. Let π denote the Calkin representation $\pi_{\beta,\mathfrak{U}}$.*

Then, given $(\varphi_\xi)_{\xi\in\mu} \in X_\beta$ with $(\varphi_\xi)_\mathfrak{U} \neq 0$, there exists $(\psi_\xi)_{\xi\in\mu} \in X_0$ such that the following conditions are satisfied:

a) $(\varphi_\xi)_\mathfrak{U} = (\psi_\xi)_\mathfrak{U}$.

b) The elements ψ_ξ are pairwise orthogonal and satisfy $\|\psi_\xi\| = \|(\varphi_\xi)_\mathfrak{U}\| \cdot \chi_U(\xi)$ for some $U \in \mathfrak{U}$.

c) The closed linear span of $\{\psi_\xi\}_{\xi\in\mu}$ taken in the Hilbert space H is a Hilbert space contained in D.

Proof: Note that $\mathfrak{U}$ is $\aleph_{\beta+1}$-R-K-minimal if ω_β is not regular as shown in [13], Lemma 4.2. Consequently, we can (and do) assume that ω_β is regular. We can also assume that $\|(\varphi_\xi)_\mathfrak{U}\| = 1$ and that $\|\varphi_\xi\| = 1$ for $\xi \in U_1$, where $U_1 \in \mathfrak{U}$. According to Proposition 3.2, we can find a positive bounded self-adjoint operator c on H with $c(H) \subset D$ such that $\{\varphi_\xi\}_{\xi\in\mu} \subset c(U_H)$. By Schmüdgen ([19], Proposition 3.1.4), the operator $b = c^{1/3}$ satisfies also $b(H) \subset D$. We represent b as orthogonal sum $b_1 \oplus \mathbf{0}$, where $\mathbf{0}$ is the zero operator on the kernel of b. Furthermore,we define the operator $u = (b_1)^{-1} \oplus \mathbf{0}$. Note that

$$u^3\varphi_\xi \in U_H \text{ and } \varphi_\xi = b^3\,u^3\,\varphi_\xi = b^2\,u^2\,\varphi_\xi = b\,u\,\varphi_\xi,$$

which also implies

$$\begin{aligned} \|u^2\,\varphi_\xi\| &= \|b\,u^3\,\varphi_\xi\| \leq \|b\|, \\ \|u\,\varphi_\xi\| &= \|b^2\,u^3\,\varphi_\xi\| \leq \|b\|^2, & (6) \\ 1 = \|\varphi_\xi\| &= \|b\,u\,\varphi_\xi\| \leq \|b\|\;\|u\,\varphi_\xi\| \text{ for } \xi \in U_1, \\ 1/\|b\| &\leq \|u\,\varphi_\xi\| \leq \|b\|^2 \text{ for } \xi \in U_1. & (7) \end{aligned}$$

We choose a sequence (ε_n) with $0 < \varepsilon_n < 1/2$ such that the assertion of Corollary 4.5 is satisfied for the constants $c_1 = 1/\|b\|$, $c_2 = \|b\|^2$ and that $\lim_{n\to\infty} \varepsilon_n = 0$. Then we choose a monotonically decreasing sequence $(\varepsilon_{2,n})$ such that the assertion of Lemma 4.4 is satisfied for $\varepsilon_{1,n} = n^{-1} \cdot (1 + \|b\|^2)^{-1} \cdot \varepsilon_n$ and $c_1 = c_2 = 1$ and such that additionally $\varepsilon_{2,n} < \varepsilon_n$.

By transfinite induction on $\eta \in \mu$ and induction on $n \in \mathbb{N}_0$ we define the sets

$$\begin{aligned}
M_{\eta,0} &= \Big\{ \min\Big(\{\xi \in \mu : \varphi_\xi \neq 0\} \setminus \bigcup_{\eta' < \eta} M_{\eta'} \Big) \Big\}, \\
(M_{\eta,0} &= \emptyset \text{ if the argument of the minimum is empty}), \\
M_{\eta,n+1} &= \Big\{ \xi \in \mu \setminus \bigcup_{k \le n} M_{\eta,k} : \text{there exists } \xi' \in \bigcup_{k\le n} M_{\eta,k} \text{ such that} \\
&\qquad |\langle \varphi_\xi, \varphi_{\xi'} \rangle| \ge \varepsilon_{2,n+2} \text{ or } |\langle \varphi_\xi, u^2\, \varphi_{\xi'} \rangle| \ge \varepsilon_{2,n+2} \Big\}, \\
M_\eta &= \bigcup_{n \in \mathbb{N}_0} M_{\eta,n}, \\
L_n &= \bigcup_{\eta \in \mu} M_{\eta,n}.
\end{aligned}$$

Note that φ_ξ is orthogonal to $\varphi_{\xi'}$ if $\xi \in M_\eta$, $\xi' \in M_{\eta'}$, and $\eta \neq \eta'$.

Let us consider the case $\beta > 0$ first. In this case regularity of ω_β implies card $M_\eta < \aleph_\beta$ for all $\eta \in \mu$. By the $\aleph_\beta$-R-K-minimality of $\mathfrak{U}$, there exists $U \in \mathfrak{U}$ with $U \subset U_1$ such that

$$\text{card } (U \cap M_\eta) \le 1 \tag{8}$$

for all $\eta \in \mu$. Now we set $\psi_\xi = \chi_U(\xi) \cdot \varphi_\xi$. Then assertions a) and b) of the proposition are satisfied by construction. Moreover, $\langle u\, \psi_\xi, u\, \psi_{\xi'} \rangle = \langle \psi_\xi, u^2\, \psi_{\xi'} \rangle = 0$ for $\xi \neq \xi'$. This implies further that

$$\sum_{\xi \subset \mu} x_\xi\, \psi_\xi = b\Big(\sum_{\xi \subset \mu} x_\xi\, u\, \psi_\xi \Big) \in D$$

for all μ-sequences $(x_\xi) \in l_2(\mu)$ (the space of all complex valued absolutely square-summable μ-sequences). Thus the closed linear span of $\{\psi_\xi\}_{\xi \in \mu}$ (taken in H) is contained in D. This completes the proof in case $\beta > 0$.

If $\beta = 0$ and if $M_\eta \cap U_2$ is finite for some $U_2 \in \mathfrak{U}$ and all $\eta \in \mu$, one can find $U \in \mathfrak{U}$ with $U \subset (U_1 \cap U_2)$ such that (8) is satisfied. Thus the proof may be completed in this case in the same way as in the previous case.

Let us now consider the remaining (more complicated) case when $\beta = 0$ and when for each $U \in \mathfrak{U}$ there exists $\eta \in \mu$ such that the set $M_\eta \cap U$ is infinite. In this case all the sets $M_{\eta,n}$ are finite (because $(\varphi_\xi)_{\xi \in \mu} \in X_0$). By the $\aleph_0$-R-K-minimality of $\mathfrak{U}$, there exists $U_2 \in \mathfrak{U}$ such that card $(U_2 \cap M_{\eta,n}) \le 1$ for all $\eta \in \mu$ and $n \in \mathbb{N}_0$. Furthermore, there exists $U_3 \in \mathfrak{U}$ with $U_3 \subset (U_1 \cap U_2)$ such that

$$\text{either } U_3 \subset \bigcup_{n \in \mathbb{N}_0} L_{2n} \text{ or } U_3 \subset \bigcup_{n \in \mathbb{N}_0} L_{2n+1}.$$

This ensures that $|\langle \varphi_\xi, \varphi_{\xi'} \rangle| < \varepsilon_{2,n}$ and $|\langle \varphi_\xi, u^2\, \varphi_{\xi'} \rangle| < \varepsilon_{2,n}$ if $\xi \in M_{\eta,k}$, $\xi' \in M_{\eta,n}$, and $k < n$. Clearly, these scalar products are equal to zero if $\xi \in M_{\eta,k}$, $\xi' \in M_{\eta',n}$, and $\eta \neq \eta'$.

Next we write the elements of infinite sets $M_\eta \cap U_3$ as members of sequences $(\xi_{\eta,n})_{n \in \mathbb{N}}$ such that $n < n'$ whenever $\xi_{\eta,n} \in M_{\eta,m}$, $\xi_{\eta,n'} \in M_{\eta,m'}$, and $m < m'$. Let $(\psi_{\eta,n})_{n \in \mathbb{N}}$ denote the sequence obtained by applying the Gram-Schmidt orthogonalization process to the sequence $(\varphi_{\xi_{\eta,n}})_{n\in\mathbb{N}}$. Then we set $\psi_{1,\xi_{\eta,n}} = \psi_{\eta,n}$ and $\psi_{1,\xi} = 0$ if ξ does not belong to any infinite set $M_\eta \cap U_3$. Using Lemma 4.4, the definition of $\varepsilon_{1,n}$, and (6) we can conclude that

$$\| \psi_{1,\xi_{\eta,n}} - \varphi_{\xi_{\eta,n}}\| \leq (1+\|b\|^2)^{-1}\cdot \varepsilon_n < \tfrac{1}{2}, \tag{9}$$

$$\|u\,\psi_{1,\xi_{\eta,n}} - u\,\varphi_{\xi_{\eta,n}}\| \leq \varepsilon_n, \tag{10}$$

$$\tfrac{1}{2} < \| \psi_{1,\xi_{\eta,n}}\| < \tfrac{3}{2}. \tag{11}$$

Using (7), (10), and the fact that $|\langle u\,\varphi_{\xi_{\eta,k}}, u\,\varphi_{\xi_{\eta,n}}\rangle| < \varepsilon_{2,n} < \varepsilon_n$ for $k < n$, we can apply Corollary 4.5 with $(u\,\varphi_{\xi_{\eta,n}})$ and $(u\,\psi_{1,\xi_{\eta,n}})$ in place of (φ_n) and (ψ_n). Thus we obtain

$$\Big(\frac{1}{4\,\|b\|}\Big)^2 \cdot \sum_{n=1}^{\infty} |x_n|^2 \leq \Big\| \sum_{n=1}^{\infty} x_n\, u\,\psi_{1,\xi_{\eta,n}} \Big\|^2 \leq (4\,\|b\|^2)^2 \cdot \sum_{n=1}^{\infty} |x_n|^2. \tag{12}$$

Note that the sets

$$U = U_3 \setminus \Big(\bigcup_{\eta\in\mu,\, M_\eta\cap U_3 \text{ is finite}} M_\eta \Big)$$

$$V_n = U \setminus \Big(\bigcup_{\eta\in\mu,\, M_\eta\cap U_3 \text{ is infinit}} \{\xi_{\eta,1},\dots,\xi_{\eta,n}\} \Big)$$

are elements of the ultrafilter $\mathfrak{U}$. Since (9) implies $\| \psi_{1,\xi} - \varphi_\xi\| < \varepsilon_n$ for all $\xi \in V_n$, $\lim_{\mathfrak{U}} \|\psi_{1,\xi} - \varphi_\xi\| = 0$.

Since the elements $\psi_{1,\xi_{\eta,n}}$ and $\psi_{1,\xi_{\eta',n}}$ with $\eta \neq \eta'$ are orthogonal to each other, we can conclude from (12) that

$$\Big(\frac{1}{4\,\|b\|}\Big)^2 \cdot \sum_{\xi\in U} |x_\xi|^2 \leq \Big\| \sum_{\xi\in U} x_\xi\, u\,\psi_{1,\xi} \Big\|^2 \leq (4\,\|b\|^2)^2 \cdot \sum_{\xi\in U} |x_\xi|^2. \tag{13}$$

As a consequence of these estimations and of (11) each element of the norm-closed linear span of $\{\psi_{1,\xi}\}_{\xi\in U}$ (taken in H) has a representation

$$\sum_{\xi\in U} x_\xi\, \psi_{1,\xi} = b\Big(\sum_{\xi\in U} x_\xi\, u\,\psi_{1,\xi}\Big),$$

where $(x_\xi)_{\xi\in U} \in l_2(U)$. Consequently this closed linear span is contained in D and the unit ball of this closed linear span is a Hilbert disc in $D[t]$. Finally, we set $\psi_\xi = \|\psi_{1,\xi}\|^{-1}\,\psi_{1,\xi}$ for $\xi \in U$ and $\psi_\xi = 0$ for $\xi \in \mu \setminus U$. Then $(\psi_\xi)_{\xi\in\mu}$ satisfies all assertions of the proposition, which completes the proof.

Proof of Theorem 4.3: Algebraical irreducibility of $\pi = \pi_{\beta,\mathfrak{U}}$ implies the $\aleph_\beta$-R-K-minimality of $\mathfrak{U}$ by Theorem 4.2.

Conversely, suppose that $\mathfrak{U}$ is $\aleph_\beta$-R-K-minimal. Let $(\varphi_{1,\xi})$ and $(\varphi_{2,\xi})$ in X_β with $(\varphi_{1,\xi})_{\mathfrak{U}} \neq 0$ and $(\varphi_{2,\xi})_{\mathfrak{U}} \neq 0$ be given. It suffices to find $v \in L^+(D)$ such that $\pi_{\beta,\mathfrak{U}}(v)\,(\varphi_{1,\xi})_{\mathfrak{U}} = (\varphi_{2,\xi})_{\mathfrak{U}}$. First we find $(\psi_{1,\xi}) \in X_0$, $(\psi_{2,\xi}) \in X_0$, and $U_1, U_2 \in \mathfrak{U}$ such that assertions *a*), *b*), and *c*) of Proposition 4.6 are satisfied with $(\varphi_{k,\xi})$, $(\psi_{k,\xi})$, and U_k $(k \in \{1,2\})$ in place of (φ_ξ), (ψ_ξ), and U. Then we set $U = U_1 \cap U_2$ and define the operator

$$u:\ H \ni \varphi \to \frac{1}{\|(\varphi_{1,\xi})_{\mathfrak{U}}\|} \sum_{\xi\in U} \langle \varphi, \psi_{1,\xi}\rangle \psi_{2,\xi}.$$

Since $u(H) \subset D$ and $u^*(H) \subset D$, the restriction $v = u \restriction D$ belongs to $L^+(D)$. Since it also satisfies

$$\pi(v)\,(\varphi_{1,\xi})_{\mathfrak{U}} = (v\,\psi_{1,\xi})_{\mathfrak{U}} = (\psi_{2,\xi})_{\mathfrak{U}} = (\varphi_{2,\xi})_{\mathfrak{U}},$$

the proof is complete.

5 COMMUTANTS

In this section, we will obtain further equivalent characterizations of the irreducibility conditions in Theorems 4.2 and 4.3. These results will be stated in the first two theorems of this section. They will be obtained as a consequence of more general results about the coincidence of several commutants for unbounded Calkin representations. Finally, we will make some remarks on non-zero Calkin representations $\pi_{\beta,\mathfrak{U}}$ which do not satisfy the assumption $\alpha(\mathfrak{U}) < \gamma(D)$ of Theorems 4.2 and 4.3.

THEOREM 5.1 *Suppose that $L^+(D)$ has the density property and that $\pi = \pi_{\beta,\mathfrak{U}}$ is a non-zero Calkin representation of $L^+(D)$. Then the following three conditions are equivalent:*

a) *The weak commutant of $\pi(L^+(D))$ is trivial (i.e. it contains only scalar multiples of the identity).*

b) *The strong commutant $(\bar{\pi}(L^+(D)))'_s$ of the range of the closure $\bar{\pi}$ of π is trivial.*

c) *The closure $\bar{\pi}$ of π does not decompose into an orthogonal sum of non-zero representations.*

THEOREM 5.2 *Suppose that D is dual hilbertisable and that $\pi = \pi_{\beta,\mathfrak{U}}$ is a non-zero Calkin representation of $L^+(D)$. Then the following three conditions are equivalent:*

a) *The weak commutant of $\pi(L^+(D))$ is trivial.*

b) *The strong commutant of $\pi(L^+(D))$ is trivial.*

c) *The representation π does not decompose into an orthogonal sum of non-zero representations.*

Before proving these theorems, we establish some more general results about commutants of Calkin representations.

THEOREM 5.3 *Suppose that $L^+(D)$ has the density property and that $\pi = \pi_{\beta,\theta}$ is a non-zero Calkin representation of $L^+(D)$. Then the commutants*

$$(\pi(L^+(D)))'_w, \quad (\bar{\pi}(L^+(D)))'_w, \quad \text{and} \quad (\bar{\pi}(L^+(D)))'_s$$

coincide and form a von Neumann algebra. Moreover, the set of restricted operators

$$\{a \restriction D_{\bar{\pi}} \;:\; a \in (\bar{\pi}(L^+(D)))'_s\}$$

forms an Op^*-algebra on $D_{\bar{\pi}}$.

Proof: Note that the following assertions are satisfied for arbitrary *-representations:

$(\pi(\mathfrak{A}))'_s \subset (\pi(\mathfrak{A}))'_w$, $(\bar{\pi}(\mathfrak{A}))'_w = (\pi(\mathfrak{A}))'_w$,

$c(D_\pi) \subset D_\pi$ for all $c \in (\pi(\mathfrak{A}))'_s$,

$(\pi(\mathfrak{A}))'_s = (\pi(\mathfrak{A}))'_w$ implies that $(\pi(\mathfrak{A}))'_s$ is a von Neumann algebra

(cf. [19], Proposition 7.2.10). Thus it suffices to prove that $(\bar{\pi}(L^+(D)))'_w \subset (\bar{\pi}(L^+(D)))'_s$.

Let $(p_\iota)_{\iota\in J}$ be a net in $\mathcal{P}_D$ such that

$$\tau_b\text{-}\lim_J p_\iota\, a\, p_\iota = a$$

for all $a \in L^+(D)$. Then the net $(a\, p_\iota\, \varphi)_{\iota\in J}$ converges (w.r.t. the norm topology) to $a\,\varphi$ for each $\varphi \in D$ and $a \in L^+(D)$ and this convergence is uniform on each bounded subset of $D[t]$ (cf. [13], Lemma 4.3). It follows that, w.r.t. the norm topology,

$$\lim_J \pi(a)\, \pi(p_\iota)\, \varphi = \pi(a)\, \varphi \tag{14}$$

for all $\varphi \in D_\pi$ and $a \in L^+(D)$ and

$$\lim_J \pi(p_\iota)\, \varphi = \varphi \tag{15}$$

for all $\varphi \in H_\pi$.

Note that the range of the orthogonal projection $(\pi(p_\iota))^{**}$ is contained in $D_{\bar{\pi}}$ because all the operators $\pi(a)\pi(p_\iota)$, where $a \in L^+(D)$, are bounded. Consequently, given $a \in L^+(D)$, $b \in (\bar{\pi}(L^+(D)))'_w$, and $\varphi, \psi \in D_\pi$, we obtain

$$\begin{aligned} \langle b\, \pi(a)\, \pi(p_\iota)\, \varphi, \psi\rangle &= \langle b\, \varphi, \pi(p_\iota)\pi(a^+)\psi\rangle = \langle \bar{\pi}(a)\, (\pi(p_\iota))^{**}\, b\, \varphi, \psi\rangle, \\ b\, \pi(a)\, \pi(p_\iota)\, \varphi &= \bar{\pi}(a)\, (\pi(p_\iota))^{**}\, b\, \varphi. \end{aligned}$$

Using (14) and (15), we can conclude that $b\varphi \in D(\overline{\pi(a)})$. Since this is true for each $a \in L^+(D)$, $b\,\varphi \in D_{\bar{\pi}}$. This means that

$$\bar{\pi}(a)\, b\, \varphi = b\, \bar{\pi}(a)\, \varphi$$

for all $\varphi \in D_\pi$. Since D_π is dense in $D_{\bar{\pi}}[t_{\bar{\pi}}]$, the last equation is even true for all $\varphi \in D_{\bar{\pi}}$, which proves that $b \in (\bar{\pi}(L^+(D)))'_s$. This completes the proof.

THEOREM 5.4 *Suppose that D is dual hilbertisable and that $\pi = \pi_{\beta,\mathfrak{U}}$ is a non-zero Calkin representation of $L^+(D)$.*

Then the strong and the weak commutants of $\pi(L^+(D))$ coincide. Moreover, the set of restricted operators

$$\{a \restriction D_\pi \;:\; a \in (\pi(L^+(D)))'_s\}$$

forms an Op-algebra on D_π.*

Proof: First we note that, given an arbitrary *-representation π, an operator $c \in (\pi(\mathfrak{A}))'_w$ belongs to $(\pi(\mathfrak{A}))'_s$ if an only if it satisfies $c(D_\pi) \subset D_\pi$ (cf. [19], Proposition 7.2.10). We also note that $(\pi(L^+(D)))'_w$ is a *-algebra by Theorem 5.3. Thus we need to show only that $c\,\varphi \in D_\pi$ for all $\varphi \in D_\pi$ and $c \in (\pi(L^+(D)))'_w$. To do so, fix $\varphi \in D_\pi$ and $c \in (\pi(L^+(D)))'_w$.

According to Lemma 3.12, we can find $\psi \in D_\pi$ and $b \in \mathcal{B}_D$ such that $\pi(b)\psi = \varphi$. It follows that

$$\begin{aligned} \langle c\,\varphi, \varphi' \rangle &= \langle c\,\pi(b)\,\psi, \varphi' \rangle = \langle c\,\psi, \pi(b^+)\,\varphi' \rangle = \langle (\pi(b))^{**}\,c\,\psi, \varphi' \rangle \qquad (\varphi' \in D_\pi), \\ c\,\varphi &= (\pi(b))^{**}\,c\,\psi. \end{aligned}$$

Since the range of $(\pi(b))^{**}$ is contained in D_π by Lemma 3.11, the proof is complete.

Proof of Theorems 5.1 and 5.2: Let us note the well-known fact that an arbitrary *-representation π decomposes into an orthogonal sum of non-zero *-representations if and only if the strong commutant of $\pi(\mathfrak{A})$ contains non-trivial orthogonal projections ([19], Lemma 8.3.3). Let us also note that each non-trivial von Neumann algebra contains non-trivial orthogonal projections. Thus, the assumptions of Theorem 5.1 imply that the von Neumann algebra $(\pi(L^+(D)))'_w$ is trivial if and only if it does not contain non-trivial orthogonal projections. In view of the above mentioned fact concerning projections in strong commutants, Theorems 5.1 and 5.2 are now immediate consequences of Theorems 5.3 and 5.4.

Let us finally make some remarks concerning non-zero Calkin representations $\pi_{\beta,\mathfrak{U}}$ which do not satisfy the assumption $\alpha(\mathfrak{U}) < \gamma(D)$ of Theorems 4.2 and 4.3.

REMARKS : 1. Suppose that $L^+(D)$ has the density property and consider the Calkin representation $\pi_{\beta,\mathfrak{U}}$ of $L^+(D)$. Suppose that $\alpha(\mathfrak{U}) \geq \gamma(D)$. Then $\pi_{\beta,\mathfrak{U}}$ is a non-zero representation if and only if either $\beta = \alpha(\mathfrak{U}) \geq \gamma(D) > \delta(\mathfrak{U})$ or $\beta > \alpha(\mathfrak{U})$.

2. Suppose that D is a dual hilbertisable domain and consider a representation $\pi_{\beta,\mathfrak{U}}$ of $L^+(D)$. Suppose further that $\beta = \alpha(\mathfrak{U}) \geq \gamma(D) > \delta(\mathfrak{U})$. Use the notations $\alpha = \alpha(\mathfrak{U})$, $\delta = \delta(\mathfrak{U})$, and $\gamma = \gamma(D)$. Assume (for convenience) that $\mu = \omega_\alpha$. Consider a mapping

$$s : \mu = \omega_\alpha \to \omega_\delta \subset \mu$$

such that card $s^{-1}(\xi) < \aleph_\beta$ for each $\xi \in \omega_\delta$. Let $\mathfrak{V}$ be the ultrafilter of subsets of μ with base $\{s(U) : U \in \mathfrak{U}\}$. Define the space

$$E_s - \{((\psi_{s(\xi)})_{\xi\in\mu})_{\mathfrak{U}} \cdot (\psi_\xi)_{\xi\in\mu} \in X_0\}.$$

Then E_s is an invariant subspace of $D_{\beta,\pi}$ and the restricted *-representation

$$L^+(D) \ni a \to \pi_{\beta,\mathfrak{U}}(a) \restriction E_s$$

is unitarily equivalent to the representation $\pi_{0,\mathfrak{V}}$. Since $\alpha(\mathfrak{V}) = \delta < \gamma$, Theorems 4.2 and 4.3 may be applied to this representation. Furthermore, the orthogonal projection p onto the norm-closure of E_s is an element of the weak commutant of $\pi_{\beta,\mathfrak{U}}(L^+(D))$. Using Theorem 5.4 (and the arguments of the proof of Theorems 5.1 and 5.2), we can conclude that $D_{\beta,\mathfrak{U}} = E_s \dot{+} (E_s)^\perp$, where $(E_s)^\perp = (id - p)\,D_{\beta,\mathfrak{U}}$ is the orthogonal complement of E_s, taken in $D_{\beta,\mathfrak{U}}$. Thus, if $\pi_{\beta,\mathfrak{U}}$ were irreducible in some sense, it would be unitarily equivalent to $\pi_{0,\mathfrak{V}}$ and $\mathfrak{V}$ would be $\aleph_0$-R-K-minimal. However, the author does not know whether $(E_s)^\perp = \{0\}$ is possible in the case considered here.

3. Let us consider now a representation $\pi = \pi_{\beta,\mathfrak{U}}$ of $L^+(D)$ such that $\beta > \alpha(\mathfrak{U})$. Then observations undertaken by H. Behncke ([1]) in the case $D = H$ may be applied also in the unbounded case. Fix a set $U \in \mathfrak{U}$ with card $U = \aleph_{\alpha(\mathfrak{U})}$. D_π containes the invariant subspace

$$E = \{((\chi_U(\xi) \cdot \varphi)_{\xi\in\mu})_{\mathfrak{U}} : \varphi \in D\}.$$

Morover, E has the invariant w.r.t. π orthogonal complement

$$\{(\varphi_\xi)_{\mathfrak{U}} : (\varphi_\xi) \in X_\beta \cap X_{\mathfrak{U}}\},$$

where $X_{\mathfrak{U}}$ is the space X_θ with $\theta = \theta_{\mathfrak{U}}$, defined in the last remark in Section 2.

REFERENCES

[1] H. Behncke. On the Calkin representations of $B(H)$. *Proc. Amer. Math. Soc.*, 101:101–107, 1987.

[2] J. W. Calkin. Two-sided ideals and congruences in the ring of bounded operators in Hilbert space. *Ann. of Math.*, 42:839–873, 1941.

[3] W. W. Comfort and S. Negrepontis. *The Theory of Ultrafilters.* Springer-Verlag, Berlin, Heidelberg, New York, 1974.

[4] J. Dixmier. *Les C^* - algèbres et leurs représentations.* Gauthier-Villars, Paris, 1969. deuxième édition.

[5] T. Jech. *Lectures in Set Theory.* Springer-Verlag, Berlin, Heidelberg, New York, 1971.

[6] H. Junek. Maximal Op^*-algebras on (DF)-domains. *Zeitschr. f. Anal. u. ihre Anwendungen*, 9:403–414, 1990.

[7] K.-D. Kürsten. Ein Gegenbeispiel zum Reflexivitätsproblem für gemeinsame Definitionsbereiche von Operatorenalgebren im separablen Hilbert-Raum. *Wissenschaftliche Zeitschrift der Karl-Marx-Universität Leipzig, Mathematisch-Naturwissenschaftliche Reihe*, 31(1):49–54, 1982.

[8] K.-D. Kürsten. Lokalkonvexe *-Algebren und andere lokalkonvexe Räume von auf einem unitären Raum definierten linearen Operatoren. Universität Leipzig, 1985. Dissertation B.

[9] K.-D. Kürsten. On completeness of the representation space of J. W. Calkin. In Z. Frolík, V. Souček, and J. Vinárek, editors, *PROCEEDINGS OF THE 13th WINTER SCHOOL ON ABSTRACT ANALYSIS Srni, 20-27 January, 1985, section of analysis*, pages 79–81, 1985. *Supplemento ai Rendiconti del Circolo Matematico di Palermo, Serie II, numero 10, 1985.*

[10] K.-D. Kürsten. Two-sided closed ideals of certain algebras of unbounded operators. *Mathematische Nachrichten*, 129:157–166, 1986.

[11] K.-D. Kürsten. On topological linear spaces of operators with a unitary domain of definition. *Wissenschaftliche Zeitschrift der Karl-Marx-Universität Leipzig, Mathematisch-Naturwissenschaftliche Reihe*, 39(6):623–655, 1990.

[12] K.-D. Kürsten. Bounded sets in projective tensor products of hilbertizable locally convex spaces. *Studia Mathematica*, 99(3):185–198, 1991.

[13] K.-D. Kürsten and M. Milde. Calkin representations of unbounded operator algebras acting on non-separable domains. *Math.Nachr.*, 154:285–300, 1991.

[14] J. Lindenstrauss and L. Tzafriri. *Classical Banach Spaces I, Sequence Spaces.* Springer-Verlag, Berlin, Heidelberg, New York, 1977.

[15] F. Löffler and W. Timmermann. The Calkin representation for a certain class of algebras of unbounded operators. *Rev. Roumaine math. Pures Appl.*, 31:891–903, 1986.

[16] F. Löffler and W. Timmermann. On the irreducibility of generalized Calkin representations. *Rev. Roumaine math. Pures Appl.*, 33:413–421, 1988.

[17] G.A. Reid. On the Calkin representations. *Proc.London Math.Soc.*, 23:547–564, 1971.

[18] K. Schmüdgen. Topological realizations of Calkin algebras on Fréchet domains of unbounded operator algebras. *Zeitschr. f. Anal. u. ihre Anwendungen*, 5(6):481–490, 1986.

[19] K. Schmüdgen. *Unbounded Operator Algebras and Representation Theory.* Akademie-Verlag, Berlin, 1990.

Differentiability of the Norm in C^*-Algebras

KEITH F. TAYLOR[1] Department of Mathematics, University of Saskatchewan, Saskatoon, Saskatchewan, S7N 0W0, Canada

WEND WERNER Universität-Gesamthochschule-Paderborn, Fachbereich 17, Postfach 1621, 4790 Paderborn, Germany

1 INTRODUCTION

Let E be a Banach space with norm $\|\cdot\|$ and unit ball B_E. Let E^* denote the dual space with unit ball B_{E^*}. For $x \in E$, x is called a smooth point in E if there is only one $\varphi \in B_{E^*}$ such that $\varphi(x) = \|x\|$. It is well known that x is a smooth point in E if and only if the norm is Gâteaux differentiable at x; i.e. the directional derivatives

$$(*) \qquad \lim_{h \to 0} \frac{\|x + hy\| - \|x\|}{h}$$

exist for all directions y in E. In this paper, we develop several characterizations of

[1]Partially supported by NSERC Canada Operating Grant.

smooth points in C*-algebras.

The smooth points in $C_0(X)$, for a locally compact Hausdorff space X, essentially have been known at least since the appearance of [3]. They are easily characterized as those $f \in C_0(X)$ for which there exists a *unique* $x_0 \in X$ such that $|f(x_0)| = \|f\|$. In this paper, we generalize this statement to noncommutative C*-algebras in a number of ways. Because of the great diversity of C*-algebras, different generalizations of the commutative result are useful for different classes of C*-algebras.

The most basic noncommutative C*-algebras are the algebras, $\mathcal{K}(H)$, of all compact operators on each Hilbert space H. In [7], the smooth points in $\mathcal{K}(H)$ are characterized as those $T \in \mathcal{K}(H)$ for which there exists a $\xi_0 \in H$(unique up to multiplication by a scalar of absolute value 1) such that $\|T\xi_0\| = \|T\|$. We prefer to phrase this as: $T \in \mathcal{K}(H)$ is a smooth point if and only if there is a minimal projection P in $\mathcal{B}(H)$, the space of all bounded operators on H, such that $\|TP\| = \|T\|$ and $\|TQ\| < \|T\|$, for all minimal projections Q in $\mathcal{B}(H)$ distinct from P. Theorem 3.1 below is a direct generalization of this characterization to arbitrary C*-algebras $\mathfrak{A}$, with the minimal projection now living in the bidual of $\mathfrak{A}$. Of course, $\mathcal{B}(H)$ is the bidual of $\mathcal{K}(H)$.

For many C*-algebras, such as $C_0(X)$, the bidual is not an easily accessible object. A characterization of smooth points which is intrinsic to the C*-algebra itself can be more useful. Such a characterization, involving the ideals of the C*-algebra, is formulated in Theorem 3.2 and is easily seen to be equivalent to the usual result for $C_0(X)$.

One may view X as the space of irreducible representations of $C_0(X)$ and formulate possible representation theoretic generalizations of the commutative characterization of smooth points. Interestingly, the most obvious statement does not hold in general. Its failure is connected with the natural topology on $\hat{\mathfrak{A}}$, the space of (equivalence classes of) irreducible representations of a C*-algebra $\mathfrak{A}$. This topology need not be T_1 and when points in $\hat{\mathfrak{A}}$ are not closed, it is possible to have a smooth point $a \in \mathfrak{A}$ such that $\|\pi(a)\| = \|\sigma(a)\| = \|a\|$, for two distinct $\pi, \sigma \in \hat{\mathfrak{A}}$. However, if $\mathfrak{A}$ is a liminal C*-algebra (which implies $\hat{\mathfrak{A}}$ is T_1), then we prove in Theorem 3.4 that $a \in \mathfrak{A}$ is a smooth point if and only if there exists a $\pi_0 \in \hat{\mathfrak{A}}$ such that $\|\pi_0(a)\| = \|a\|$, $\|\sigma(a)\| < \|a\|$, for all $\sigma \in \hat{\mathfrak{A}}$, $\sigma \neq \pi$ and $\pi(a)$ is a smooth point in $\mathcal{K}(H_\pi)$. Note that $\mathfrak{A}$ liminal means that $\rho(\mathfrak{A}) = \mathcal{K}(H_\rho)$, for each $\rho \in \hat{\mathfrak{A}}$. Thus, there is a very satisfactory representation theoretic characterization of smooth points for liminal C*-algebras.

These results will be established in section 3 after we have presented in section 2 all the necessary definitions and preliminary results. For a nice presentation of the geometrical background see [10].

In contrast to mere smoothness, points where the norm is Fréchet differentiable follow more closely the pattern of their commutative counterparts. (The reader should recall that the norm is Fréchet differentiable at the point $x \in E$ whenever the limit in $(*)$ is uniform in y. Furthermore, this stronger differentiability property is reflected in a reinforcement of the uniqueness of the norming functional φ: By a classical result of Šmulyan φ must be weak*-strongly exposed, i.e. whenever $\varphi_n(x) \to \|x\|$ with the φ_n in B_{E^*} then $\|\varphi_n - \varphi\| \to 0$.) In particular, (the analogue of) Theorem 3.4 now is valid for all C*-algebras: $a \in \mathfrak{A}$ is a point of Fréchet differentiability if and only if there exists a $\pi_0 \in \widehat{\mathfrak{A}}$ and a positive number ε_0 such that $\|\pi_0(a)\| = \|a\|$, $\|\pi(a)\| < \|a\| - \varepsilon_0$, for all $\pi \in \widehat{\mathfrak{A}}$ different from π_0 and $\pi(a)$ is a point of Fréchet differentiability in $\mathcal{B}(H_\pi)$.

Our results on Fréchet differentiability are contained in section 4. In the final section we briefly have a look at the derivatives that arise through differentiation of the norm and present one more rather weak sufficient condition for Fréchet differentiability of the norm.

2 C*-ALGEBRAS

As basic references on C*-algebras, we use the books of Dixmier [4], Kadison and Ringrose [8], Takesaki [11] and Pedersen [6]. In this section, we gather together definitions and several special results that are needed in the main sections.

2.1 A C*-algebra is a Banach algebra $\mathfrak{A}$ with involution, $x \to x^*$, such that $\|x^*x\| = \|x\|^2$, for all $x \in \mathfrak{A}$. Let $\mathfrak{A}^*$ denote the dual space of $\mathfrak{A}$ and $\mathfrak{A}^{**}$, the bidual. There is a natural von Neumann algebra structure on $\mathfrak{A}^{**}$ (see 3.7.8 of [6], for example) and the canonical embedding of $\mathfrak{A}$ into $\mathfrak{A}^{**}$ is an isomorphism of $\mathfrak{A}$ with a C*-subalgebra of $\mathfrak{A}^{**}$. Whenever it is convenient, we will consider $\mathfrak{A}$ as identified with a C*-subalgebra of $\mathfrak{A}^{**}$.

The elements of $\mathfrak{A}^*$ are, at the same time, normal linear functionals on $\mathfrak{A}^{**}$. There are natural left and right actions of $\mathfrak{A}^{**}$ (or $\mathfrak{A}$) on $\mathfrak{A}^*$ which make $\mathfrak{A}^*$ into a 2-sided Banach $\mathfrak{A}^{**}$-module. For $x \in \mathfrak{A}^{**}$ and $\varphi \in \mathfrak{A}^*$, let $(x\varphi)(y) = \varphi(yx)$ and $(\varphi x)(y) = \varphi(xy)$, for all $y \in \mathfrak{A}$. Then $x\varphi$ and φx are elements of $\mathfrak{A}^*$ and $\|x\varphi\| \le \|x\| \, \|\varphi\|$, $\|\varphi x\| \le \|x\| \, \|\varphi\|$.

2.2 For $\psi \in \mathfrak{A}^*$ define the adjoint functional ψ^* by $\psi^*(x) = \overline{\psi(x^*)}$, for all $x \in \mathfrak{A}$. If $\psi^* = \psi$, the ψ is called hermitian.

2.3 An element φ in $\mathfrak{A}^*$ is called positive if $\varphi(x^*x) \ge 0$, for all $x \in \mathfrak{A}$ and this is denoted by $\varphi \ge 0$. For φ and ψ hermitian, we write $\varphi \le \psi$ if $\psi - \varphi \ge 0$. If $\varphi \in \mathfrak{A}^*$ satisfies $\varphi \ge 0$ and $\|\varphi\| = 1$, then φ is called a state of $\mathfrak{A}$. Let $\mathcal{S}(\mathfrak{A})$ denote the set of all states of $\mathfrak{A}$. A state φ of $\mathfrak{A}$ is called pure if $0 \le \psi \le \varphi$ implies $\psi = \alpha\varphi$ for some $0 \le \alpha \le 1$. Equivalently, ψ is pure if and only if it is an extreme point of the set $\mathcal{S}(\mathfrak{A})$. Let $\mathcal{P}(\mathfrak{A})$ denote the set of pure states of $\mathfrak{A}$.

2.4 If $\psi \in \mathcal{S}(\mathfrak{A})$, then the support projection of ψ is a projection $\mathrm{s}(\psi)$ in $\mathfrak{A}^{**}$ such that $\psi\mathrm{s}(\psi) = \mathrm{s}(\psi)\psi = \psi$ and, if p is a projection in $\mathfrak{A}^{**}$ satisfying $\psi(p) = 0$, then $\mathrm{s}(\psi)p = 0$ (section III 4 of [11]). Furthermore, for $\psi \in \mathcal{S}(\mathfrak{A})$, ψ is a pure state if and only if $\mathrm{s}(\psi)$ is a minimal projection in $\mathfrak{A}^{**}$ (3.13.6 of [6]).

2.5 The following lemma is well known (a proof may be found e.g. as part of (e)$\Rightarrow$(f) in the proof of [12]'s central result) :

LEMMA *Suppose that $a \ge 0$ is of unit norm. If p is a minimal projection such that $\|ap\| = \|a\| = 1$, then $ap = pa = p$.*

2.6 The following lemma was established in [12], 2.7.

LEMMA *Let (ψ_n) be a sequence in $\mathfrak{A}^*$, with $\|\psi_n\| \le 1$, for all n. Suppose p and q are projections in $\mathfrak{A}^{**}$ such that $\|p\psi_n q\| \to 1$ as $n \to \infty$. Then $\|p\psi_n q - \psi_n\| \to 0$, as $n \to \infty$.*

2.7 For $x \in \mathfrak{A}$, $|x| = (x^*x)^{\frac{1}{2}} \in \mathfrak{A}$ and there exists a unique partial isometry $u \in \mathfrak{A}^{**}$ such that $x = u|x|$ and u^*u is the smallest projection p in $\mathfrak{A}^{**}$ which satisfies $p|x| = |x|$ (6.1.2 and 6.1.3 of [8]). This is the polar decomposition of x. There is also a polar decomposition for elements of $\mathfrak{A}^*$.

2.8 If $\varphi \in \mathfrak{A}^*$, there exists a unique (v, ψ) with $\psi \in \mathfrak{A}^*$, $\psi \geq 0$ and v a partial isometry in $\mathfrak{A}^{**}$ such that $\varphi = v\psi$ and $v^*v = s(\psi)$ (see III 4.2 of [11]). Usually this ψ is denoted $|\varphi|$ and called the absolute value of φ.

By checking the proof of III 4.2 in [11], one can see that the following lemma holds.

2.9 LEMMA *If $\varphi \in \mathfrak{A}^*$ and $x \in \mathfrak{A}$ is such that $\|x\| = 1$ and $\varphi(x) = \|\varphi\|$, then the polar decompositions of $\varphi^* = v|\varphi^*|$ and $x = u|x|$ are related by $v = us(|\varphi^*|)$ and $|\varphi^*| = \varphi u$.*

2.10 A representation of $\mathfrak{A}$ is a $*$–homomorphism π of $\mathfrak{A}$ into $\mathcal{B}(H_\pi)$, the bounded operators on some Hilbert space H_π. If $\pi(\mathfrak{A})H_\pi$ is dense in H_π, then π is called non-degenerate. A representation π of $\mathfrak{A}$ is called irreducible if H_π has no nontrivial closed subspaces which are $\pi(\mathfrak{A})$ invariant. Two representations π and ρ of $\mathfrak{A}$ are called equivalent if there exists a unitary map U from H_π to H_ρ such that $\rho(x)U = U\pi(x)$, for all $x \in \mathfrak{A}$. Let $\widehat{\mathfrak{A}}$ denote the set of equivalence classes of irreducible representations of $\mathfrak{A}$. The definition and basic properties of the topology on $\widehat{\mathfrak{A}}$ can be found in Chapter 3 of Dixmier [4].

2.11 A C*-algebra $\mathfrak{A}$ is called liminal if $\pi(\mathfrak{A}) = \mathcal{K}(H_\pi)$, for each $\pi \in \widehat{\mathfrak{A}}$, where $\mathcal{K}(H_\pi)$ denotes the elementary C*-algebra of all compact operators on H_π. See Chapter 4 of [4] for the basic properties of liminal C*-algebras.

2.12 If $\mathfrak{A}$ has an identity 1, then 1 is also the identity of $\mathfrak{A}^{**}$. If $\mathfrak{A}$ does not have an identity, then let $\widetilde{\mathfrak{A}}$ denote the C*-subalgebra of $\mathfrak{A}^{**}$ generated by $\mathfrak{A}$ and 1, the identity of $\mathfrak{A}^{**}$.

2.13 PROPOSITION *Let $\mathfrak{A}$ be a C*-algebra with no identity and let $a \in \mathfrak{A}$, $a \geq 0$ and $\|a\| = 1$. Then a is a smooth point in $\mathfrak{A}$ if and only if a is a smooth point in $\widetilde{\mathfrak{A}}$.*

PROOF: By 2.1.7 of [4], the map $\psi \to \psi|_{\mathfrak{A}}$ is a bijection between $\{\psi \in \mathcal{S}(\widetilde{\mathfrak{A}}) : \|\psi|_{\mathfrak{A}}\| = 1\}$ and $\mathcal{S}(\mathfrak{A})$. If a is a smooth point in $\mathfrak{A}$ and $\varphi \in \mathfrak{A}^*$, $\|\varphi\| = 1$ with $\varphi(a) = 1$, then, since $a \geq 0$, φ must be positive (by lemma III 3.2 of [11]); that is, $\varphi \in \mathcal{S}(\mathfrak{A})$. If $\psi \in \mathcal{S}(\widetilde{\mathfrak{A}})$ and $\psi(a) = 1$, then $\|\psi|_{\mathfrak{A}}\| = 1$ so $\psi|_{\mathfrak{A}} = \varphi$. Thus ψ is unique and a must be a smooth point in $\widetilde{\mathfrak{A}}$. The converse is trivial. □

We will show in theorem 3.1 that an element in a C*-algebra is a smooth point if and only if its absolute value is a smooth point. Thus the assumption that $a \geq 0$ in the above proposition is not essential.

2.14 For any locally compact space X, there is a one to one correspondence between the closed ideals in $C_0(X)$ and the closed subsets of X. For a closed subset A of X, the corresponding ideal is $\mathfrak{J}_A = \{f \in C_0(X) | f_{|A} \equiv 0\}$. If $f \in C_0(X)$, let $Z_f = \{x \in X | f(x) = 0\}$. Then the closed ideal generated by f is determined by Z_f. That is $\overline{C_0(X)f} = \mathfrak{J}_{Z_f}$. For noncommutative C*-algebras, we will need the following lemma on the closed left ideal generated by a positive element.

2.15 LEMMA *Let $\mathfrak{A}$ be a C*-algebra and $x \in \mathfrak{A}$, $x \geq 0$. For any real number $\alpha > 0$,*

$$\overline{\mathfrak{A}x^\alpha} = \overline{\mathfrak{A}x} .$$

PROOF: Suppose, first, that $\alpha > 1$. Then

$$\mathfrak{A}x^\alpha = \mathfrak{A}(x^{\alpha-1})x \subseteq \mathfrak{A}x .$$

Thus $\overline{\mathfrak{A}x^\alpha} \subseteq \overline{\mathfrak{A}x}$. For the reverse containment, let C*(x) denote the (commutative) C*-subalgebra of $\mathfrak{A}$ generated by x. Now $x^\alpha \in C^*(x)$ and, from 2.14, $\overline{C^*(x)x^\alpha} = \overline{C^*(x)x}$. By using an approximate identity in $C^*(x)$, we see that $x \in \overline{C^*(x)x^\alpha}$. Thus $\overline{\mathfrak{A}x} \subseteq \overline{\mathfrak{A}x^\alpha}$.

□

2.16 If φ is any state of $\mathfrak{A}$, let $\mathfrak{J}_\varphi$ denote the left ideal defined by φ. We have

$$\mathfrak{J}_\varphi = \{y \in \mathfrak{A} \mid \varphi(y^*y) = 0\}$$

2.17 LEMMA *Let $x \in \mathfrak{A}$, $x \geq 0$ and suppose $\varphi \in \mathcal{S}(\mathfrak{A})$ is such that $\varphi(x) = 0$. Then $\overline{\mathfrak{A}x} \subseteq \mathfrak{J}_\varphi$.*

PROOF: For any $y \in \mathfrak{A}$, by the Cauchy-Schwarz inequality,

$$|\varphi(yx^{\frac{1}{2}})| \leq \varphi(y^*y)\varphi(x) = 0 .$$

Thus $\mathfrak{A}x^{\frac{1}{2}} \subseteq N(\varphi)$, where $N(\varphi)$ denotes the null space of φ. By 2.15, $\overline{\mathfrak{A}x} \subseteq N(\varphi)$. For any $y \in \overline{\mathfrak{A}x}$, $y^*y \in \overline{\mathfrak{A}x}$ also. Thus, $\varphi(y^*y) = 0$. Therefore $\overline{\mathfrak{A}x} \subseteq \mathfrak{J}_\varphi$. □

3 SMOOTHNESS IN C*-ALGEBRAS

3.1 THEOREM *Let $\mathfrak{A}$ be a C^*-algebra and let $a \in \mathfrak{A}$. The following assertions are equivalent:*

(a) *a is a smooth point of $\mathfrak{A}$.*

(b) *$|a|$ is a smooth point of $\mathfrak{A}$.*

(c) *There is a minimal projection p in $\mathfrak{A}^{**}$ with the property that $\|ap\| = \|a\|$ and $\|aq\| < \|a\|$ for all minimal projections $q \in \mathfrak{A}^{**}$ different from p.*

PROOF: Let us suppose that $\|a\| = 1$ throughout the proof.

(a)⇒(b) Suppose that $a \in \mathfrak{A}$ is smooth and has polar decomposition $a = u|a|$. If $\varphi_1, \varphi_2 \in B_{\mathfrak{A}}$ are (necessarily) positive functionals both norming $|a|$, then evaluation at a yields $\varphi_1 u^* = \varphi_2 u^*$ and hence, $\varphi_1 u^* u = \varphi_2 u^* u$. Since $\varphi_i u^* u(|a|) = 1$ for $i = 1, 2$, we have $\|\varphi_i u^* u\| = 1$. But u^*u is a projection and the functionals φ_i are positive. Therefore, by Lemma 2.6 they must be equal and $|a|$ is smooth.

(b)⇒(c) Let φ be the pure state that norms $|a|$ and put $p = \mathrm{s}(\varphi)$. By 2.4, p is minimal and

$$\|ap\| \geq \varphi u^*(ap) = \varphi(|a|) = 1,$$

hence $\|ap\| = 1$. To see that there is only one such projection, fix any minimal projection q as well as a pure normal state ψ norming q. By 2.6 $q\psi = \psi$ which implies $q = \mathrm{s}(\psi)$. Suppose $\|aq\| = 1$. Then, a fortiori, $\||a|q\| = 1$. Since $|a|$ is positive and q is minimal, Lemma 2.5 yields that actually $|a|q = q$. But $\psi(|a|) = q\psi(|a|) = \psi(q) = 1$ and therefore $\psi = \varphi$. It follows that $p = \mathrm{s}(\varphi) = \mathrm{s}(\psi) = q$.

(c)⇒(a) Let $a = u|a|$ be the polar decomposition of a and fix $\psi \in \mathrm{ex}\, B_{\mathfrak{A}^*}$ with $\psi(a) = 1$. Then $|\psi^*|$ is a pure state and $\mathrm{s}(|\psi^*|)$ is a minimal projection. As a consequence of the dependence of polar decompositions for norming functionals (2.9),

$$\|a\mathrm{s}(|\psi^*|)\| \geq \psi(a\mathrm{s}(|\psi^*|)) = |\psi^*|(|a|\mathrm{s}(|\psi^*|)) = \psi(a) = 1,$$

and it follows that $s(|\psi^*|) = p$. Observe next that there is only one state φ_p norming p. (In fact, for a minimal projection p one has $pap = \varphi_p(a)p$ for all $a \in \mathfrak{A}$ and some functional φ_p. If $\psi \in B_{\mathfrak{A}^*}$ norms p then, by Lemma 2.6, $\psi = p\psi p = \varphi_p$.) Hence, since $|\psi^*|(p) = |\psi^*|(1) = 1$, we have $|\psi^*| = \varphi_p$. It finally follows from 2.9 that ψ is of the form

$$\psi = |\psi^*| s(|\psi^*|)u^* = \varphi_p u^*,$$

and thus depends on p and a alone. □

3.2 THEOREM *Let $\mathfrak{A}$ be a C^*-algebra and let $a \in \mathfrak{A}$ with $\|a\| = 1$. Then a is a smooth point of $\mathfrak{A}$ if and only if $\overline{\mathfrak{A}(1-|a|)}$ is a maximal left ideal in $\mathfrak{A}$.*

Note. Even if $1 \notin \mathfrak{A}$, $\overline{\mathfrak{A}(1-|a|)}$ is a left ideal in $\mathfrak{A}$.

PROOF: By 3.1, we may assume that $0 \le a \le 1$. Then any norming functional for a is necessarily a state and the extreme points of the w*-compact convex set of all norming functionals of a are necessarily pure states. Thus a is a smooth point of $\mathfrak{A}$ if and only if there is a unique $\varphi \in \mathcal{P}(\mathfrak{A})$ such that $\varphi(a) = 1$. But if $\varphi \in \mathcal{P}(\mathfrak{A})$ is such that $\varphi(a) = 1$, then, by Lemma 2.17, $\overline{\mathfrak{A}(1-a)} \subseteq \mathfrak{J}_\varphi$. By 3.13.5 and 3.13.6 of [6], $\overline{\mathfrak{A}(1-a)}$ is a maximal left ideal of $\mathfrak{A}$ if and only if there is a unique $\varphi \in \mathcal{P}(\mathfrak{A})$ with $\overline{\mathfrak{A}(1-a)} \subseteq \mathfrak{J}_\varphi$. This establishes 3.2. □

3.3 REMARK If a is a smooth point in $\mathfrak{A}$ and φ is the unique pure state such that $\varphi(|a|) = 1$, then the minimal projection p of **3.1**(c) satisfies $p = s(\varphi)$ (see the proof of 3.1) and $\overline{\mathfrak{A}(1-|a|)} = \mathfrak{A} \cap \mathfrak{A}^{**}p$ (see the proof of 3.13.6 in [6]).

3.4 THEOREM *Let $\mathfrak{A}$ be a liminal C^*-algebra. Let $a \in \mathfrak{A}$. Then a is a smooth point in $\mathfrak{A}$ if and only if there is a $\pi \in \widehat{\mathfrak{A}}$ such that $\|\pi(a)\| = \|a\|$, $\pi(a)$ is a smooth point in $\mathcal{K}(H_\pi)$ and $\|\rho(a)\| < \|a\|$, for all $\rho \in \widehat{\mathfrak{A}}$, $\rho \ne \pi$.*

PROOF: Note that $\mathfrak{A}$ liminal implies that $\rho(\mathfrak{A}) = \mathcal{K}(H_\rho)$ and $\{\rho\}$ is a closed set in $\widehat{\mathfrak{A}}$, for all $\rho \in \widehat{\mathfrak{A}}$, [4, 4.2.7, 4.7.15].

Assume $a \in \mathfrak{A}$ is a smooth point. Then $|a|$ is a smooth point and there exists a unique $\varphi \in \mathcal{P}(\mathfrak{A})$ such that $\varphi(|a|) = \|a\|$. Let π_φ be the corresponding irreducible representation and ξ_φ the unit vector in H_{π_φ} such that $\varphi(x) = \langle \pi_\varphi(x)\xi_\varphi, \xi_\varphi \rangle$, for all $x \in \mathfrak{A}$. Then $\|\pi_\varphi(|a|)\| = \|a\|$ and $\pi_\varphi(|a|)$ is a smooth point in $\mathcal{K}(H_{\pi_\varphi})$ (otherwise, there would exist at least two states on $\mathcal{K}\left(H_{\pi_\varphi}\right)$ which norm $\pi_\varphi(|a|)$ and composing with π_φ, two distinct states on $\mathfrak{A}$ which norm $|a|$.) But $\pi_\varphi(|a|) = |\pi_\varphi(a)|$. So $\|\pi_\varphi(a)\| = 1$ and $\pi_\varphi(a)$ is a smooth point in $\mathcal{K}(H_{\pi_\varphi})$.

For $\sigma \in \widehat{\mathfrak{A}}$, $\sigma \neq \pi$, if $\|\sigma(a)\| = \|a\|$, then $\|\sigma(|a|)\| = \|a\|$ and since $\sigma(|a|) \in \mathcal{K}(H_\sigma)$, there exists $\eta \in H_\sigma$ such that $\sigma(|a|)\eta = \|a\|\eta$. Thus $\rho(x) = \langle\sigma(x)\eta, \eta\rangle$, for $x \in \mathfrak{A}$, defines a state ρ on $\mathfrak{A}$ which norms $|a|$. By smoothness of $|a|$, $\rho = \varphi$. Then by [4, 2.4.1] the representations σ and π_φ are unitarily equivalent; that is, equal as elements of $\widehat{\mathfrak{A}}$.

The converse direction is easy. □

This theorem is the natural generalization of the well known unique peak point criterion for abelian C*-algebras. The assumption that $\mathfrak{A}$ be liminal cannot be weakened. Even for post-liminal C*-algebras smoothness of the point a no longer implies that there exists a unique $\pi \in \widehat{\mathfrak{A}}$ such that $\|\pi(a)\| = \|a\|$. This is illustrated with the following example.

3.4 EXAMPLE Let H be an infinite dimensional separable Hilbert space and let $\mathfrak{A} = \{\alpha I + T : \alpha \in \mathbb{C} \text{ and } T \in \mathcal{K}(H)\}$. Then $\widehat{\mathfrak{A}}$ consists of two points $\{[\pi_1], [\pi_2]\}$, where $H_{\pi_1} = H$ and $H_{\pi_2} = \mathbb{C}$ and

$$\pi_1(\alpha I + T) = \alpha I + T$$

$$\pi_2(\alpha I + T) = \alpha \ ,$$

for all $\alpha I + T \in \mathfrak{A}$. In the natural topology on $\widehat{\mathfrak{A}}$, $[\pi_2]$ is a closed point and $\{[\pi_1]\}$ is a dense open set. Let $\{e_n : n \in \mathbb{N}\}$ be an orthonormal basis of H and define $A \in \mathcal{B}(H)$ by $Ae_n = (1 - \frac{1}{n})e_n$, for all $n \in \mathbb{N}$. Then $A = I - T$, where $T = \sum_{n\in\mathbb{N}} \left(\frac{1}{n}\right)(e_n \otimes e_n) \in \mathcal{K}(H)$. Thus $A \in \mathfrak{A}$. Since $T = I - A$, $\mathfrak{A}(I - A) \subseteq \mathcal{K}(H)$ and $\mathfrak{A}(I - A)$ contains all finite rank operators. Thus

$$\mathfrak{J}_{(I-A)} = \overline{\mathfrak{A}(I - A)} = \mathcal{K}(H) \ .$$

Thus $\mathfrak{J}_{(I-A)}$ is a maximal left ideal in $\mathfrak{A}$ and, by Theorem 3.2, A is a smooth point in $\mathfrak{A}$. However, $\|\pi_1(A)\| = \|\pi_2(A)\| = \|A\| = 1$. Note that there are no vector states associated with π_2 which norm A; however, if $\varphi_n(S) = \langle\pi_1(S)e_n, e_n\rangle$, for all $S \in \mathfrak{A}$ and $n \in \mathbb{N}$, then $\varphi_n(A) = 1 - \frac{1}{n} \to 1$. Note that the φ_n converge in the w*-sense, to the one vector state associated with π_2 (which does norm A). This actually implies that $[\pi_2] \in \overline{\{[\pi_1]\}}$ in $\widehat{\mathfrak{A}}$.

3.5 REMARK A criterion for smoothness in terms of representations and vector states can be formulated for any C*-algebra $\mathfrak{A}$. For $a \in \mathfrak{A}$, $a \geq 0$, $\|a\| = 1$, a is a smooth point in $\mathfrak{A}$ if and only if there exits a unique pair (π, ξ), $\pi \in \widehat{\mathfrak{A}}$, $\xi \in H_\pi$, $\|\xi\| = 1$, such that $\langle \pi(a)\xi, \xi \rangle = 1$. But this is just using the fact that all pure states of $\mathfrak{A}$ are vector states associated with irreducible representations and lacks the "peaking" appeal of the statement in Theorem 3.4.

4 FRECHET SMOOTHNESS

4.1 THEOREM *Let $\mathfrak{A}$ be a C^*-algebra and let $a \in \mathfrak{A}$. The following assertions are equivalent:*

(a) *The norm in $\mathfrak{A}$ is Fréchet differentiable at a.*

(b) *The norm in $\mathfrak{A}$ is Fréchet differentiable at $|a|$.*

(c) *There exists a projection p, minimal in $\mathfrak{A}^{**}$ and contained in $\mathfrak{A}$, such that*

$$\|ap\| = \|a\| \quad \text{and} \quad \|a(1-p)\| < \|a\|.$$

(d) *$\|a\|$ is an isolated point in the spectrum of $|a|$ and the corresponding spectral projection is a projection minimal in the enveloping von Neumann algebra $\mathfrak{A}^{**}$ contained in $\mathfrak{A}$.*

PROOF: We again assume that a is of unit norm and has polar decomposition $a = u|a|$.

(a)⇒(b) By Theorem 3.1 we know that there is exactly one pure state φ that norms $|a|$. Let us show that φ is a weak*-strongly exposed point of the unit ball of $\mathfrak{A}^*$. Fix a sequence (φ_n) in the unit ball of $\mathfrak{A}^*$ such that $\varphi_n(|a|) \to 1$. Then $\varphi_n u^*(a) \to 1$, and by the Fréchet differentiability of the norm at a we must have $\|\varphi_n u^* u - \varphi u^* u\| \to 0$. Lemma 2.6 shows $\|\varphi_n - \varphi_n u^* u\| \to 0$ as well as $\varphi u^* u = \varphi$. These remarks prove $\varphi_n \to \varphi$ in norm and so $|a|$ is a point of Fréchet differentiability.

(b)⇒(c) $|a|$ clearly is a point of Fréchet differentiability in $C^*(|a|)$, the C*-algebra generated by $|a|$. This means that there is a projection p in $C^*(|a|)$ such that $|a|p = p|a| = p$,

$\||a|p\| = 1$ and $\||a|(1-p)\| < 1$, which implies that also $\|ap\| = 1$ and $\|a(1-p)\| < 1$. Suppose that $p = p_1 + p_2$, where p_1 and p_2 are projections, different from zero and mutually orthogonal. Represent $\mathfrak{A}^{**}$ faithfully on a Hilbert space and fix vectors in the respective range of these projections. If ψ_1 and ψ_2 are the normal vector states that go with these vectors then $\psi_i(p_j) = \delta_{ij}$. It follows that $\psi_i(|a|) = \psi_i(p) - \psi_i(p - |a|) \geq \psi_i(p) = 1$ and hence $|a|$ can be smooth only when p cannot be decomposed in such a way, i.e. p is minimal.

The impications (c)$\Rightarrow$(d)$\Rightarrow$(a) are either obvious or follow from the main result of [12]. □

4.2 THEOREM *Let $\mathfrak{A}$ be a C^*-algebra and let $a \in \mathfrak{A}$. The norm in $\mathfrak{A}$ is Fréchet differentiable at a if and only if there exists $\pi \in \widehat{\mathfrak{A}}$ and $\varepsilon > 0$ such that $\|\pi(a)\| = \|a\|$, the norm in $\mathcal{B}(H_\pi)$ is Fréchet differentiable at $\pi(a)$ and $\|\sigma(a)\| < \|a\| - \varepsilon$, for all $\sigma \in \widehat{\mathfrak{A}} \backslash \{\pi\}$. In this case, π is necessarily an open point in $\widehat{\mathfrak{A}}$.*

PROOF: We may assume $a \geq 0$ and $\|a\| = 1$. Suppose the norm is Fréchet differentiable at a.

Let us first show that, up to equivalence, there cannot be two different representations that norm a: In fact, should $\|\pi_i(a)\| = 1$ for $i = 1, 2$, then, for the minimal projection p in part (c) of Theorem 3.6, $\|\pi_i(p)\| = 1$, since p belongs to the C*-algebra generated by a. Recall that there is a unique normal extension π_i^{**} of π_i to a representation of $\mathfrak{A}^{**}$ onto $\pi_i(\mathfrak{A}^{**}) = \mathcal{B}(H_{\pi_i})$. Furthermore, there exists a central projection $c(\pi_i)$ such that π_i^{**} is an isomorphism of $\mathfrak{A}^{**}c(\pi_i)$ onto $\mathcal{B}(H_{\pi_i})$. Since p is minimal in $\mathfrak{A}^{**}$ it therefore follows that $\pi_i(p)$ is minimal in $\mathcal{B}(H_{\pi_i})$ and hence, there is ξ_i such that the functionals $\varphi_i : x \mapsto \langle \pi(x)\xi_i, \xi_i \rangle$ both norm p. But p itself is smooth so $\varphi_1 = \varphi_2$ and π_1 is equivalent to π_2.

On the other hand, it is a well known consequence of the GNS-construction that there is always at least one element $\pi \in \widehat{\mathfrak{A}}$ that norms a. Also, if p is as above, it is easy to check that $\|\pi(a)\pi(p)\| = 1$ and $\|\pi(a)(1 - \pi(p))\| < 1$. Since as before, $\pi(p)$ is minimal in $\mathcal{B}(H_\pi)$, the point $\pi(a)$ must be Fréchet-smooth in $\mathcal{B}(H_\pi)$ by the central result in [12].

Finally, suppose there is a sequence (π_n) of representations contained in $\widehat{\mathfrak{A}} \setminus \{\pi\}$ such that $\|\pi_n(a)\| \to 1$. Fix vectors ξ_n of unit norm in H_{π_n} such that $\langle \pi_n(a)\xi_n, \xi_n \rangle \to 1$

and denote the corresponding vector states on $\mathfrak{A}$ by φ_n. Since a is a point of Fréchet differentiability of the norm in $\mathfrak{A}$, it follows that $\|\varphi_n - \varphi\| \to 0$ as well as $\|\varphi_n p - \varphi\| \to 0$, where φ is the state norming p. In particular, $\pi_n(p) \neq 0$ whenever n is large enough. But for all these n,

$$\|\pi_n(a)\| \geq \|\pi_n(p)\pi_n(a)\| = \|\pi_n(p)\| = 1,$$

and by uniqueness of the representation norming a, we arrive at a contradiction.

Conversely, suppose the condition holds. Let π^{**} and $c(\pi)$ be as above. Then $ac(\pi)$ is a point of Fréchet differentiability of the norm in $\mathfrak{A}^{**}c(\pi)$. Now $\mathfrak{A}^{**}$ is the ℓ^∞-direct sum of $\mathfrak{A}^{**}c(\pi)$ and $\mathfrak{A}^{**}(1-c(\pi))$. The condition that $\|\sigma(a)\| < 1-\varepsilon$, for all $\sigma \in \widehat{\mathfrak{A}}\backslash\{\pi\}$ is easily seen to imply that $\|a(1-c(\pi))\| \leq 1-\varepsilon$. Using the theorem in [12], the norm in $\mathfrak{A}^{**}$ is Fréchet differentiable at a; hence, likewise for the norm in $\mathfrak{A}$. □

5 NOTES AND REMARKS

5.1 The above can be used to prove a rather weak sufficient condition for Fréchet differentiability of the norm at a point $a \in \mathfrak{A}$. (That this condition also is necessary is by now quite obvious.)

PROPOSITION *A norm one element $a \in \mathfrak{A}$ is a point of Fréchet differentiability for the norm of $\mathfrak{A}$, if there are a complex scalar λ and a partial isometry $v \in \mathfrak{A}^{**}$ such that the projection v^*v is minimal and $\|a - \lambda v\| < 1$.*

PROOF: Suppose that $\mathfrak{A} \subseteq \mathfrak{A}^{**} \subseteq \mathfrak{A}^{(4)}$ in the canonical way, denote by $c_2(p)$ and $c_4(p)$ the central cover of the projection p in $\mathfrak{A}^{**}$ and $\mathfrak{A}^{(4)}$, respectively, and write $a = u|a|$ for the polar decomposition of a. Observe first that under the present circumstances

$$(*) \qquad \||a|\left[1 - c_2\left(v^*v\right)\right]\| < 1.$$

In fact, there would otherwise exist a minimal projection $p \in \mathfrak{A}^{(4)}$ with

$$|a|\left[1 - c_2\left(v^*v\right)\right]p = p.$$

By minimality of p one either has $pc_2\left(v^*v\right) = p$, which is impossible, or else $pc_2\left(v^*v\right) = 0$.

In the latter case, $|a|p = p$ and hence

$$1 > \|a - \lambda v\| \geq \|p - \lambda u^* v c_4 (v^*v)\, c_4(p)p\|.$$

But then $c_4 (v^*v)\, c_4(p) \neq 0$ and [8, 6.18] yields $p \sim v^*v$. (Note that v^*v remains minimal as an element of $\mathfrak{A}^{(4)}$.) Hence $c_2 (v^*v) \geq c_4 (v^*v) = c_4(p)$, a contradiction.

We next use the fact that $c_2 (v^*v)\, \mathfrak{A}^{**} = \mathcal{B}(H)$ for some Hilbert space H. In light of $(*)$, the decomposition $\mathfrak{A}^{**} = c_2 (v^*v)\, \mathfrak{A}^{**} \oplus_\infty (1 - c_2 (v^*v))\, \mathfrak{A}^{**}$ and the fact that for von Neumann algebras smoothness of the norm at $|a|$ implies its Fréchet differentiability at a [12], it will suffice to show that $c_2 (v^*v)\, |a|$ is a smooth point in $\mathcal{B}(H)$. Indeed, an immediate consequence of $(*)$ is that for all $q \in \mathfrak{A}^{**}$ with $q|a| = q$ the estimate $\|q - qc_2 (v^*v)\, \| < 1$ holds, which is only possible when $c_2(q) = c_2 (v^*v)$. It follows that there is at least one minimal projection q contained in $c_2 (v^*v)\, \mathfrak{A}^{**}$ such that $q|a| = q$. On the other hand,

$$\|c_2(v^*v)q - \lambda c_2(v^*v)u^*vq\| = \|u^*aq - \lambda u^*vq\| < 1,$$

and since $\lambda c_2 (v^*v)\, u^*vq$ corresponds to a rank one operator in $\mathcal{B}(H)$, it follows that $q \sim v^*v$. Therefore, there can only be one minimal projection $q_0 \in c_2(v^*v)\mathfrak{A}^{**}$ with $q_0|a| = q_0$, or, equivalently, there exists exactly one vector state ψ on $\mathcal{B}(H)$ norming $c_2 (v^*v)\, |a|$. Because the essential norm of $c_2 (v^*v)\, |a|$ in $\mathcal{B}(H)$ is less than one by assumption on $c_2 (v^*v)\, \lambda v = \lambda v$, there can be no other state norming $|a|$, and we are done. $\square$

The above should be compared to the main result in [9] where it has been shown that an operator T is a smooth point in $\mathcal{B}(H)$ if and only if the essential norm of T is strictly less than its operator norm, and there is, up to scalars of modulus one, a unique point $x_0 \in H$ where $\|T\| = \|Tx_0\|$. In fact, in the end of the above proof we made use of the easier direction of this result, which follows rapidly from the fact that for the extreme points of the dual unit ball of $\mathcal{B}(H)$ there is a disjoint decomposition

$$\text{ex}\, B_{\mathcal{B}(H)} = \text{ex}\, B_{\mathcal{K}(H)^\circ} \cup \text{ex}\, B_{\mathcal{K}(H)^*}.$$

5.2 The previous results admit a dual formulation which we state in the following result. Recall that a norm-one element x_0 of a Banach space X is called an exposed point of

B_X iff there is a functional $x_0^* \in B_{X^*}$ with $1 = x_0^*(x_0) > x_0^*(x)$ for all $x \in B_X$ different from x_0. The point x_0 is strongly exposed, iff it is exposed and if $\lim_{n\to\infty} x_0^*(x_n) = 1$ then $\lim_{n\to\infty} x_n = x_0$ in norm.

THEOREM *Let $\mathfrak{M}$ denote a von Neumann algebra and write $\mathfrak{M}_*$ for its predual.*

(a) *A functional $\varphi \in B_{\mathfrak{M}_*}$ is a strongly exposed point of $B_{\mathfrak{M}_*}$ if and only if it is an extreme point of the unit ball of $\mathfrak{M}_*$.*

Suppose in addition that $\mathfrak{M}_* = \mathfrak{A}^*$ for some C^*-algebra $\mathfrak{A}$.

(b) *The functional $\varphi \in B_{\mathfrak{A}^*}$ is weak*-strongly exposed in $B_{\mathfrak{A}^*}$ if and only if* $\mathrm{s}(|\varphi|)$ *is minimal and lies in $\mathfrak{A}$.*

(c) *$\varphi \in B_{\mathfrak{A}^*}$ is weak* exposed in $B_{\mathfrak{A}^*}$ if and only if* $\mathrm{s}(|\varphi|)$ *is a G_δ-projection, which is minimal among all G_δ-projections.*

In the above, we call a projection $p \in \mathfrak{A}^{**}$ a G_δ-projection if it is the monotone, w* limit of a decreasing sequence (x_n) contained in $\mathfrak{A}$. Equivalently, using Akemann's sandwich-theorem [1, Lemma 3.1], it follows that, at least for compact projections p, the G_δ-property is equivalent to the existence of a sequence (q_n) of open projections in $\mathfrak{A}^{**}$ with $p = \bigwedge_{n=1}^{\infty} q_n$.

We omit the proof.

5.3 In [2] (see also [5]) it was shown that the (norm-closed) faces of the unit ball of a C^*-algebra $\mathfrak{A}$ are all of the form

$$F_v = v + (1 - vv^*)B_{\mathfrak{A}}(1 - v^*v) = \{a \in B_{\mathfrak{M}} \mid vv^* = va\},$$

where v runs through the partial isometries of $\mathfrak{A}^{**}$. (Of course, F_v might be empty for some of these v.)

If $a = u|a|$ is smooth in $\mathfrak{A}$ and p denotes the minimal projection of Theorem 3.1, then $va = vv^*$, where $v = pu^*$. Using e.g. Lemma 2.5 it is not difficult to see that this observation yields that a is a smooth point if and only if there is exactly one minimal partial isometry v (in the sense of 5.1) such that $a \in F_v$. Furthermore, a must be a point of Fréchet differentiability iff this unique partial isometry belongs to $\mathfrak{A}$.

A major part of this research was carried out while the second named author visited the University of Saskatchewan in Saskatoon. He would like to express his gratitude to Keith Taylor for his hospitality.

References

[1] C.A. Akemann. A Gelfand representation theory for C*-algebras. *Pacific J. Math.*, 39:1–11, 1971.

[2] C. A. Akemann and G.K. Pedersen. Facial structure in operator algebra theory. Preprint, 1990.

[3] S. Banach. *Théorie des opérations linéaires.* Monografie matematyczne, Warszawa, 1932.

[4] J. Dixmier. *Les C*-algèbres et leurs représentations.* Gauthier-Villars, Paris, second edition, 1969.

[5] C. M. Edwards and G.T. Rüttimann. On the facial structure of the unit balls in a JBW^*-triple and its dual. *J. London Math. Soc.*, 38:317–332, 1988.

[6] G.K.Pedersen. *C*-Algebras and Their Automorphism Groups*, volume 14 of *London Math. Soc. Monographs.* Academic Press, London, 1979.

[7] S. Heinrich. The differentiability of the norm in spaces of operators. *Functional Analysis Appl.*, 9:360–362, 1975.

[8] R. V. Kadison and J. R. Ringrose. *Fundamentals of the Theory of Operator Algebras I,II.* Academic Press, New York-London, 1983/86.

[9] F. Kittaneh and R. Younis. Smooth points of certain operator spaces. *Int. Equ. and Oper. Th.*, 13:849–855, 1990.

[10] R. R. Phelps. *Convex Functions, Monotone Operators and Differentiability.* Lecture Notes in Math. 1364. Springer, Berlin-Heidelberg-New York, 1989.

[11] M. Takesaki. *Theory of Operator Algebras I.* Springer, Berlin-Heidelberg-New York, 1979.

[12] K. F. Taylor and W. Werner. Smooth points in von Neumann algebras. *Proc. Amer. Math. Soc.*, 1992. to appear.

Properties of Function Spaces Depending on the Dimension of Their Domains and Related Topics

ALEKSANDER PEŁCZYŃSKI Institute of Mathematics, Polish Academy of Sciences, Śniadeckich 8, 00-950 Warsaw, Poland; e-mail: olek@impan.impan.gov.pl

INTRODUCTION

One of the questions arising in Functional Analysis is how to measure the "size" of an (infinite dimensional) object. The objects in question are most often Banach and Fréchet spaces consisting of functions defined on various domains. The functional analytic nature of these function spaces depends on topological invariants like topological dimension (the number of independent parameters), the number of connected components, etc. The present article is a survey of a number of results and problems of that sort. Originally it was presented to an audience of non-specialists. Therefore, in many instances, we oversimplify the subject trying to avoid more scholarly discussions. The literature at the end of the paper is far from being complete and up to date.

1 THE BANACH–MAZUR COMPACTA

Let X and Y be Banach spaces. Recall that X is *isometrically isomorphic* to Y, in symbols $X \equiv Y$, if and only if there is a linear map $T: X \to Y$ such that $T(X) = Y$ and $\|Tx - Tz\| = \|x - z\|$ for $x, z \in X$.

Write $X = Y$, provided that X is *isomorphic* to Y, i.e., there exists a linear surjection $T: X \to Y$ (called an *isomorphism*) such that

$$(1) \qquad \|x\| \leq \|Tx\| \leq C\|x\| \quad \forall x \in X$$

for some $C > 0$.

For a fixed isomorphism $T: X \to Y$ we put

$$C_T = \inf\{C \ : \ C \text{ satisfies } (1)\}.$$

The *Banach–Mazur distance* between X and Y is

$$d(X, Y) = \inf\{C_T \ : \ T \text{ is an isomorphism from } X \text{ to } Y\}.$$

All n-dimensional Banach spaces are isomorphic ($n = 1, 2, \ldots$).

The *Banach–Mazur compactum* BM_n is the metric space of all n-dimensional Banach spaces with the metric $\log d(X, Y)$, where isometrically isomorphic Banach spaces are identified.

All 1-dimensional Banach spaces are isometrically isomorphic, that is, BM_1 is a singleton.

1.1 Mazur's problem on finite dimensional universal spaces

In 1935 Mazur asked:

Does there exist a 3-dimensional Banach space U such that every 2-dimensional Banach space is isometrically isomorphic to a subspace of U ? More generally, given an integer $m \geq 2$, does there exist an integer $n > m$ and an n-dimensional Banach space U such that every m-dimensional Banach space is isometrically isomorphic to a subspace of U.([Sc], Problem 41).

Formally the problem depends on the field of scalars. In fact this is inessential.

Recall that every real n-dimensional Banach space corresponds to the space $\mathbb{R}^n$ with a distinguished bounded, symmetric with respect to the origin, convex body. Two n-dimensional Banach spaces are isomorphic iff the corresponding convex bodies are affinely equivalent. Thus Mazur's problem is equivalent to the question:

Does there exist a symmetric bounded convex body B_U in $\mathbb{R}^3$ such that every 2-dimensional symmetric bounded convex body is affinely equivalent to the intersection of B_U with a plane passing through the centre of symmetry of B_U ?

In the late fifties Bessaga [B] and independenly Grünbaum [G] solved the problem in the negative.

THEOREM (cf. [B]). Given an integer $m \geq 2$, there is no finite-dimensional Banach space U with the property that every m-dimensional Banach space is isometrically isomorphic to a subspace of U. ■

For related results see Klee's comment [Sc], p.118.
A routine compactness argument yields

COROLLARY. Let $2 \leq m < n$.There exists a natural number $K(m, n)$ which is the largest k such that, for every collection $\mathcal{B}$ of cardinality k of symmetric bounded convex bodies in $\mathbb{R}^m$ there exists an n-dimensional symmetric bounded convex body $B_{\mathcal{B}}$ such that every member of $\mathcal{B}$ is the intersection of $B_{\mathcal{B}}$ with an m-dimensional hyperplane passing through the center of symmetry of $B_{\mathcal{B}}$. ■

Problem. Find the exact value of $K(2,3)$. Find the exact value or a reasonable approximation of $K(m, n)$.

CONJECTURE. $K(2,3) = 3$ or 4; more generally $K(m, n) = \binom{n}{m}$, or $\binom{n}{m} + 1$.

It is known that $3 \leq K(2,3) \leq (1992)^6$.

Comment. Roughly speaking, Bessaga's Theorem says that, for every n-dimensional symmetric bounded convex body B, the size of BM_m is greater than the size of the space $\mathrm{Sect}_m B$ of all sections of B by m-dimensional planes passing through the center of symmetry of B $(2 \leq m < n)$.

1.2 Topological properties of BM_n

What is the topological size of BM_n ? More precisely:

Problem. Do all the compacta BM_n $(n = 2, 3, \ldots)$ represent the same topological type ?

CONJECTURE. Each BM_n $(n = 2, 3, \ldots)$ is homeomorphic to the Hilbert cube $\mathbb{I}^\infty$. In particular each BM_n is an absolute retract.

1.3 A question of J. J. Schäffer

Let E be an infinite-dimensional Banach space . Denote by BM_E the space consisting of all Banach spaces isomorphic to E and with the pseudo-metric $\log d(X, Y)$. The following curious question due to J. J. Schäffer is still open:

Problem. Is BM_E unbounded for every infinite-dimensional E ? Equivalently, given a Banach space E and a constant $C > 0$, does there exist a Banach space X isomorphic to E such that $d(E, X) > C$?

2 AROUND MILJUTIN'S THEOREM

Recall the classical

BANACH–MAZUR UNIVERSALITY THEOREM. Every uncountable compact metric space K (in particular the unit interval $\mathbb{I}$) has the property that every separable Banach space is isometrically isomorphic to a subspace of the space $C(K)$ of all continuous scalar-valued functions defined on K. ■

In particular, $C(\mathbb{I})$ and $C(\mathbb{I}^2)$ have the same "maximal" size. Banach asked whether $C(\mathbb{I})$ and $C(\mathbb{I}^2)$ are isomorphic. He proved only that they are not isometrically isomorphic.

Banach's problem has been solved by Miljutin([Mi1] (1952), cf. [Mi2] and [P3]).

THEOREM (Miljutin). Let $\mathcal{K}$ denote the class of all uncountable compact metric spaces. Then $C(K) = C(\mathbb{I})$ for every $K \in \mathcal{K}$. ■

An immediate consequence of Miljutin's theorem is:

COROLLARY. There exist $a, b > 0$ such that

$$(i) \qquad d(C(K), C(\mathbb{I})) \leq a \quad \forall K \in \mathcal{K},$$

$$(ii) \qquad d(C(K), C(K')) \leq b \quad \forall K, K' \in \mathcal{K}.$$

■

Let $\alpha = \inf\{a : a \text{ satisfies } (i)\}$, $\beta = \inf\{b : b \text{ satisfies } (ii)\}$.

Problem. Find exact values of α and β.

It is known that $3 \leq \alpha \leq 12$; $3 \leq \beta \leq 144$.

3 HILBERT'S 13th PROBLEM AND ITS RELATIVES

Miljutin's theorem seems to be related to the 13th Hilbert problem and its relatives. Our presentation of this topic follows the excellent survey of Y. Sternfeld [St].

In 1900 Hilbert asked:

"Prove that the equation of the seventh degree

$$(*) \qquad x^7 + ax^3 + bx^2 + cx + 1 = 0$$

is not solvable by means of any continuous functions of only two variables !"

The precise statement:

Hilbert's 13th Problem. Prove that any function $x(a,b,c)$ which is implicitly given by (*) cannot be represented as the composition of functions of a fewer number of variables !

In fact the problem has a negative solution. It was obtained in the late fifties by Kolmogorov [K1] and Arnold [A]. The final result is (cf. [K2]):

KOLMOGOROV'S SUPERPOSITION THEOREM. For each $n \geq 2$ there exists an $n \times [2n+1]$ matrix of functions $(\varphi_{kj})_{1\leq k\leq n;1\leq j\leq 2n+1}$; $\varphi_{kj} \in C(\mathbb{I})$ such that, if

$$\Phi_j(t_1, t_2, ..., t_n) = \sum_{k=1}^{n} \varphi_{kj}(t_k),$$

then every $f \in C(\mathbb{I}^n)$ is of the form

$$\binom{*}{*} \qquad f = \sum_{j=1}^{2n+1} g_j(\Phi_j) \quad \text{for some } g_j \in C(\mathbb{I}) \ \ (j = 1, 2, \ldots, 2n+1).$$

■

COROLLARY. Define the linear operator

$$\Psi: [C(\mathbb{I})]^{2n+1} \to C(\mathbb{I}^n)$$

by

$$\Psi(g_1, g_2, \ldots, g_{2n+1}) = \sum_{j=1}^{2n+1} g_j(\Phi_j) \quad \text{for} \quad (g_1, g_2, \ldots, g_{2n+1}) \in [C(\mathbb{I})]^{2n+1}.$$

Then Ψ is onto. ■

Problem. Does, for an appropriate choice of the matrix (φ_{ik}), the operator Ψ admit a right inverse, i.e. does there exist a linear operator $R: C(\mathbb{I}^n) \to [C(\mathbb{I})]^{2n+1}$ such that $\Psi R = \mathrm{id}_{C(\mathbb{I}^n)}$?

A positive answer to this problem might give a new approach to Miljutin's theorem.

Remarks. 1° Fridman ([F], 1967) proved that the functions φ_{ik} can be constructed so that all of them belong to Lip_1; for a simple proof see (Kahane [Ka], 1975). But φ_{ik} cannot be taken smooth (Vitushkin–Henkin [HVi], 1967); for a simple proof see (Kaufman [Kf], 1974).

2° Hilbert's 13th problem has a positive solution if "continuous" is replaced by "r-times continuously differentiable" $(r = 1, 2, \ldots)$.

THEOREM (Vitushkin ([Vi], 1954)). In every space $C^r(\mathbb{I}^n)$ $(r = 1, 2, \ldots$ and $n = 2, 3, \ldots)$ there exist functions which cannot be represented as superpositions of r-times continuously differentiable functions of fewer than n variables. ■

4 ISOMORPHIC CLASSIFICATION OF SPACES OF DIFFERENTIABLE FUNCTIONS

Vitushkin's theorem stated above suggests that, in contrast to Miljutin's theorem, the next problem may have positive solution, i.e., the number of variables is, perhaps, a linear topological invariant of Banach spaces of differentiable functions.

Problem. Given integers $n, n' \geq 2$ and $r, r' \geq 1$. Is it true that

$$C^r(\mathbb{I}^n) = C^{r'}(\mathbb{I}^{n'}) \;\Rightarrow\; r = r' \;\text{ and }\; n = n' \,?$$

Very little is known which could help to solve the problem.

1° For $n = n' = 1$ the situation is different:

THEOREM (Borsuk [Bo], 1933). $C^r(\mathbb{I}) = C(\mathbb{I})$ for $r = 1, 2, \ldots$ ■

2° On the other hand, we have

THEOREM (Grothendieck [Gr], 1956; Henkin [H1], 1967). If $n \geq 2$ then $C^r(\mathbb{I}^n) \neq C(\mathbb{I})$ for $r = 1, 2, \ldots$ ■

3° Something can be saved from Miljutin's theorem:

THEOREM (Mitiagin [M2], 1970). For every n-dimensional compact differential manifold Ω,

$$C^r(\Omega) = C(\mathbb{I}^n).$$ ■

The first two theorems generalize to anisotropic Sobolev spaces.

Let $\mathcal{S}$ be a *smoothness*, i.e., a finite nonempty set of partial derivatives satisfying some saturation conditions. Then

THEOREM 1 ([PS], [KiS]). For every smoothness $\mathcal{S}$ the following are equivalent:

(*i*) $$\mathrm{L}^1_{\mathcal{S}} = \mathrm{L}^1(\mathbb{I}),$$

(*ii*) $$C^r_{\mathcal{S}}(\mathbb{I}^n) = C(\mathbb{I}),$$

(*iii*) $\mathcal{S}$ is a parallelepiped.

■

For a fixed multiindex $a = (a(j))$ the *parallelepiped* P_a is the set of partial derivatives

$$P_a = \{D^\alpha : \alpha(j) \leq a(j) \;\text{ for }\; j = 1, 2, \ldots, n\}.$$

Here, for $\alpha = (\alpha(j))$,

$$D^\alpha = \frac{\partial^{\alpha(1)+\cdots+\alpha(n)}}{\partial x_1^{\alpha(1)} \cdots \partial x_n^{\alpha(n)}}.$$

We also have (cf. [PS]).

THEOREM 2. If $1 < p < \infty$ then, for every smoothness $\mathcal{S}$, the Sobolev space $L^p_{\mathcal{S}}(\mathbb{I}^r)$ is isomorphic to $L^p(\mathbb{I})$. ∎

Concluding this section note that, for Fréchet spaces of infinitely differentiable functions we do have an analogue of Miljutin's theorem. The pioneering result due to Mitiagin ([M1], 1960) states:

THEOREM. 1. If Ω is a compact differential manifold with a nice boundary then $C^\infty(\Omega) = C^\infty(\mathbb{I})$;
2. If the manifold Ω is not compact then $C^\infty(\Omega) = C^\infty(\mathbb{R})$ and $C^\infty(\mathbb{R}) \neq C^\infty(\mathbb{I})$. ∎

However Tidten ([T], 1979) proved that there exists a continuum of mutually nonisomorphic spaces $C^\infty(\Xi)$ for pathological domains Ξ in the plane.

At present there is a huge literature concerning the isomorphic classification of Fréchet spaces of differentiable functions. We mention only an old paper by Zerner [Ze] and Vogt's survey [V2], concerning more general problems. In fact the isomorphic classification of infinitely differentiable functions is closely related to the existence of bases in these spaces, as well as, various generalizations of Whitney's extension theorem, the existence of fundamental solutions for linear differential operators with constant coefficients, etc. In recent years there appeared several papers by R. Meise, B. A. Taylor and D. Vogt concerning this subject.

5 ISOMORPHIC CLASSIFICATION OF SPACES OF ANALYTIC FUNCTIONS

5.1 Banach spaces

Let $D = \{z \in \mathbb{C} : |z| \leq 1\}$, the unit disc. For n variables ($n = 1, 2, \ldots$) there are two analogues of the unit disc. The n-dimensional *polydisc*:

$$D^n = \{\zeta = (z_j) \in \mathbb{C}^n : \max_{1 \leq j \leq n} |z_j| \leq 1\}$$

and the *unit ball* in $\mathbb{C}^n$:

$$B_n = \{\zeta = (z_j) \in \mathbb{C}^n : \sum_{1 \leq j \leq n} |z_j|^2 \leq 1\}.$$

Obviously $D^1 = B_1 = D$.

Let $\Omega \subset \mathbb{C}^n$ be a closed domain. The symbol $A(\Omega)$ denotes the Banach space of complex-valued uniformly continuous functions on Ω which are holomorphic in the interior of Ω.

The isomorphic classification of Banach spaces of analytic functions is more advanced than that of spaces of differentiable functions, although still not satisfactory. For simplicity we restrict ourselves to the spaces $A(D^n)$ and $A(B_n)$ for $n = 1, 2, \ldots$

Various analytic techniques have been used to obtain the results presented below.

THEOREM (Pełczyński [P2], 1964). The space $A(D)$ is not isomorphic to any complemented subspace of $C(\mathbb{I})$. ■

COROLLARY. $A(D^n) \neq C(\mathbb{I})$; $A(B_n) \neq C(\mathbb{I})$ $n = 1, 2, \ldots$ ■

THEOREM (Henkin [H2], 1967). $A(D^n) \neq A(B_m)$ for $n = 2, 3, \ldots$; $m = 1, 2, \ldots$ ■

Analytic tools: The concept of analytic measures and new integral representations.

THEOREM (Mitiagin and Pełczyński [MP], 1975) $A(D) \neq A(B_m)$ for $m = 2, 3, \ldots$ ■

Tools: absolutely summing operators.

THEOREM (Bourgain [Br], 1984). If $m \neq m'$ then $A(D^m) \neq A(D^{m'})$ for $m, m' = 1, 2, \ldots$ ■

Tools: theory of complex martingales.

Therefore in the case of polydiscs D^n the number of variables n is a linear topological invariant of the spaces $A(D^n)$. We do not know what is the case for balls.

Problem. Is it true that $m \neq m'$ implies $A(B_m) \neq A(B_{m'})$ for $m, m' = 1, 2, \ldots$?

Note:

THEOREM ([P4], 1992). $A(D^n) \neq C^r(\mathbb{I}^m)$ for $n, m, r = 1, 2, \ldots$ ■

Problem. Is it true that $A(B_n) \neq C^r(\mathbb{I}^m)$ for $m, r = 1, 2, \ldots$; $n = 2, 3, \ldots$?

5.2 Fréchet spaces of holomorphic functions

Let M^n denote a complex n-manifold, in particular $M^n = \mathbb{C}^n$, and let G be an open set in M^n. The symbol $H(G)$ stands for the Fréchet space of complex-valued holomorphic functions on G.

We have rather satisfactory information concerning the isomorphic classification of Fréchet spaces $H(G)$ for typical domains G.

In contrast to Mitiagin's theorem on Fréchet spaces of infinitely differentiable functions the topological dimension of the domain G is a linear topological invariant of H(G). That means:

THEOREM 1. If G and G' are open subsets of M^n and M^k, respectively, where $n \neq k$ and $n, k = 1, 2, \ldots$ then $H(G) \neq H(G')$. ■

For a fixed positive integer n and for sufficiently regular bounded open domains G in $\mathbb{C}^n$, the space $H(G)$ represents the same linear topological type as the space of holomorphic functions on the open n-disc. In particular,

THEOREM 2. If G is a convex and bounded open domain in $\mathbb{C}^n$, then $C(G) = C(\operatorname{int} D^n)$ for $n = 1, 2, \ldots$ ■

On the other hand the spaces of entire functions are not isomorphic to spaces of analytic functions on polydiscs. More generally:

THEOREM 3. $H(\mathbb{C}^k \times \operatorname{int} D^n) \neq H(\mathbb{C}^p \times \operatorname{int} D^m)$, where $(k,n) \neq (p,m)$ and $k,n,p,m = 1,2,\ldots$ ∎

The above results accomplish the research program which goes back to the article of Gelfand ([Ge], 1956) and to Kolmogorov's note [K3], 1958. The quoted theorems and more sophisticated ones are due to several mathematicians. We mention here only Dragilev [D], Mitiagin [M1], Rolewicz [R1], Tillmann [Ti], Zaharyuta [Z1], [Z2]. An extensive, however not complete, literature of the subject can be found in Rolewicz's monograph [R2].

The results on non-isomorphisms are based on invariants of the type of *approximative dimension* (defined in late fifties by Kolmogorov [K3] and Pełczyński [P1]) and on invariants expressed by means of inequalities between the values of seminorms on individual vectors or functionals, cf. Dragilev [D] and Vogt [V1], [V2], Vogt and Wagner [VW]. Establishing isomorphisms between Fréchet spaces of infinitely differentiable functions and analytic functions is related to the existence and "uniqueness" of bases in these spaces and to representations of them as Köthe echelon spaces.

There is a huge, and still growing, literature concerning various classes of nuclear Köthe spaces, the uniqueness of bases in these spaces (a phenomenon discovered by Dragilev), description of their quotient spaces and (complemented) subspaces, and on analytic objects related to these spaces. In this respect we mention the names of Vogt, Meise and their collaborators, Palamodov, Kondakov (cf. [Ko], [Me], [V1], [V2], [VW], [W]). For earlier results we refer to the book [Du] and References in there.

Aknowledgement. The author would like to express his gratitude to professor Czesław Bessaga for his substantial help in preparation of this paper.

REFERENCES

[A] V. I. Arnold, *Representation of continuous functions of three variables by the superpositions of continuous functions of two variables*, Math. Sb. (N.S.) 48(90) (1959), 3–74 (= AMS Transl. 2(28) (1963), 61–149).

[B] C. Bessaga, *A note on universal Banach spaces of a finite dimension*, Bull. Acad. Polon. Sci., sér. math., astr. et phys. 6 (1958), 97–101.

[Bo] K. Borsuk, *Über Isomorphie der Funktionalräume*, Bull. Int. Acad. Polon. Sci. (1933), 1-10.

[Br] J. Bourgain, *The dimension conjecture for polydisc algebra*, Israel J. Math. 48(4) (1984), 289–304.

[D] M. M. Dragilev, *Standard form of a basis for the space of analytic functions* (Russian), Uspehi Mat. Nauk 15, 2(92) (1960), 181-188.

[Du] E. Dubinsky, *The structure of nuclear Fréchet spaces*, Lecture Notes in Mathematics 720, Springer-Verlag 1979.

[F] B. L. Fridman, *Improvement in the smoothness of functions in Kolmogorov's superposition theorem*, Dokl. Akad. Nauk SSSR 117 (1967), 1019–1022 (= Soviet Math. Dokl. 8 (1967), 1550–1553).

[Ge] I. M. Gelfand, *On certain problems of functional analysis* (Russian), Uspehi Mat. Nauk. 11, 6 (1956), 3-12.

[Gr] A. Grothendieck, *Erratum Memoir: Produits tensoriels topologiques et espaces nucléaires*, Ann. Inst. Fourier 6 (1956), 117–120.

[G] B. Grünbaum, *On a problem of S. Mazur*, Bull. Res. Council Israel (Ser. F), 7 (1958), 133-135.

[H1] G. M. Henkin, *Proof of the nonisomorphism of smooth function spaces on a segment and on a square*, Dokl. Akad. Nauk SSSR 172, No 1 (1967), 48–51 (= Soviet Math. Dokl. 8, No 1 (1967), 46–50).

[H2] G. M. Henkin, *Banach spaces of analytic functions on the ball and on the bicylinder are nonisomorphic*(Russian), Funkc. Analiz 2, vol. 4 (1968), 82–91.

[HVi] A. G. Vitushkin and G. M. Henkin, *Linear superpositions of functions*, Uspehi Mat. Nauk 22, 1(133) (1967), 8–124 (= Russian Math. Surveys 22(1) (1967), 77-126).

[Ka] J. P. Kahane, *Sur le théoreme de Kolmogorov*, J. Approx. Th. 13 (1975), 229-234.

[Kf] R. Kaufman, *Linear superpositions of smooth functions*, Proceedings AMS 46 (1974), 360-362.

[KiS] S. V. Kisliakov and N. G. Sidorenko, *The non-existence of local unconditional structure in anisotropic spaces of smooth functions* (Russian), Sibir. Math. J. 29, No. 3 (1988), 64–77.

[K1] A. N. Kolmogorov, *On the representation of continuous functions of several variables by superpositions of continuous functions of fewer variables*, Dokl. Akad. Nauk. SSSR 108 (1956), 179–182 (= AMS Transl (2) 17 (1961), 369-373).

[K2] A. N. Kolmogorov, *On the representation of continuous funcions of many variables by superpositions of continuous functions of one variable and addition*, Dokl. Akad. Nauk SSSR 114 (1957), 953–956. (= AMS Transl. 2(28) (1963), 55–61).

[K3] A. N. Kolmogorov, *On the linear dimension of topological vector spaces* (Russian), Dokl. Akad. Nauk. SSSR 120, 2 (1958), 239–241.

[Ko] V. P. Kondakov, *On the structure of unconditional bases in some Fréchet spaces* (Russian), Studia Math. 76 (1983), 137–151.

[Me] R. Meise, *Structure of closed linear translation invariant subspaces of* $A(\mathbb{C})$ *and kernels of convolution operators*, p. 331–347 in: "Functional Analysis: Surveys and Recent Results III", K. D. Bierstedt and B. Fuchssteiner (eds.), North Holland Math. Studies 90 (1984).

[Mi1] A. A. Miljutin, *On spaces of continuous functions* (Russian), Dissertation, Moscow State University, 1952.

[Mi2] A. A. Miljutin, *On isomorphism of spaces of continuous functions on compacta of cardinality continuum* (Russian), Teor. Funkc., Funkc. Anal. Priloz., Kharkov, 2 (1966), 150-156.

[M1] B. S. Mitiagin, *Approximative dimension and bases in nuclear spaces* (Russian), Uspehi Mat. Nauk 16, No. 4 (1961), 63-132.

[M2] B. S. Mitiagin, *The homotopy structure of the linear group of a Banach space* (Russian), Uspehi Mat. Nauk 25 (1970), 63–106 .

[MP] B. Mitiagin and A. Pełczyński, *On the non-existence of linear isomorphism between Banach spaces of analytic functions of one and several complex variables*, Studia Math. 56 (1975), 85–96.

[P1] A. Pełczyński, *On the approximation of F-spaces by finite-dimensional spaces*, Bull. Acad. Polon. Sci., sér math., astr. et phys. 5 (1957), 879–881.

[P2] A. Pełczyński, *On simultaneous extension of continuous functions*, Studia Math. 24 (1964), 285–304.

[P3] A. Pełczyński, *Linear extensions, linear averagings, and their applications to linear topological classification of spaces of continuous functions*, Dissertationes Math. 58 (1968).

[P4] A. Pełczyński, *Non isomorphism of the disc algebra with spaces of differentiable functions*, Proc. Centre for Mathematics and its Applications. Australian National University 29 (1992), 183–195.

[PS] A. Pełczyński and K. Senator, *On isomorphisms of anisotropic Sobolev spaces with "classical" Banach spaces and a Sobolev type embedding theorem*, Studia Math. 84 (1986), 169–215.

[R1] S. Rolewicz, *On spaces of holomorphic functions*, Studia Math. 21 (1962), 135–160.

[R2] S. Rolewicz, *Metric linear spaces*, PWN Warszawa & D. Reidel Publishing Company Dordrecht/ Boston/ Lancaster 1984.

[Sc] *The Scottish Book*, edited by R. Daniel Mauldin, Birkhäuser Boston, 1981.

[St] Y. Sternfeld, *Hilbert's 13th problem and dimension*, in: Geometric aspects of functional analysis, Lecture Notes in Mathematics 1376, Springer–Verlag 1989.

[T] M. Tidten, *Fortsetzungen von* C^∞*-Funktionen, welche auf einer abgeschlossenen Menge in* $\mathbb{R}^n$ *definiert sind*, Manuscripta Math. 27 (1978), 291-312.

[Ti] H. G. Tillmann, *Randverteilungen analytischer Funktionen und Distributionen*, Math. Zeitschr. 59 (1963), 61–83.

[Vi] Vitushkin, *On the 13th problem of Hilbert* (Russian), Dokl. Akad Nauk SSSR 95 (1954), 701–704.

[V1] D. Vogt, *Charakterisierung der Unterräume eines nuklearen stabilen Potenzreihenraumes von endlichem Typ*, Studia Math. 71 (1982), 251–270.

[V2] D. Vogt, *Sequence space representations of spaces of test functions and distributions*, pp. 405–443 in "Functional analysis, holomorphy and approximation theory", G. Zapata (ed.), Lecture Notes in Pure and Appl. Math. 83, M. Dekker, New York 1983.

[VW] D. Vogt and D. J. Wagner, *Charakterisierung der Unterräume und Quotientenräume der nuklearen stabilen Potenzreihenräume von unendlichem Typ*, Studia Math. 70 (1981), 63-80.

[W] M. J. Wagner, *Quotientenräume von stabilen Potenzreihenräumen endlichen Typs*, Manusripta Math. 31 (1980), 97-109.

[Z1] V. P. Zaharyuta, *Spaces of functions of one variable, analytic on open sets and on compacta* (Russian), Matem. Sb. 82, (124) (1970), 84–89.

[Z2] V. P. Zaharyuta, *On bases and isomorphism of spaces of analytic functions on convex domains of several variables* (Russian), Teor. Funkc., Funkc. Anal. Priloz., Kharkov, 5 (1967), 5-12.

[Ze] M. Zerner, *Développement en séries de polynomes orthonormaux de fonctions différentiables, C. R. Acad. Sci. Paris 268 (1969), 218–220.*

Continuous Linear Right Inverses for Partial Differential Operators with Constant Coefficients and Phragmén-Lindelöf Conditions

R. MEISE Mathematisches Institut, Heinrich-Heine-Universität Düsseldorf, Germany

B. A. TAYLOR Department of Mathematics, University of Michigan, West Engineering, Ann Arbor, MI 48109

D. VOGT Fachbereich Mathematik, Bergische Universität/GH, Wuppertal 1, Germany

In the early fifties L. Schwartz posed the problem of determining when a linear differential operator $P(D)$ admits a (continuous linear) right inverse, that is, when does there exist a continuous linear map R such that

$$P(D)Rf = f \text{ for all } f \in \mathcal{E}(\Omega) \text{ or all } f \in \mathcal{D}'(\Omega).$$

For example, when Ω is $\mathbb{R}^n$ and $P(D)$ is hyperbolic in some direction, such an operator R exists; one can take $R(f)$ to be the unique solution of the Cauchy problem $P(D)u = f$ with zero initial data. Negative results for important special classes were given by several authors. For $n \geq 2$ Grothendieck has shown (see e.g. Treves [32]) that no elliptic operator can have a continuous linear right inverse on $\mathcal{E}(\Omega)$. The same holds for hypoelliptic operators, as Vogt [33], [34] has proved. For parabolic and other operators this had been shown before by Cohoon [10], [11].

A complete solution of Schwartz's problem for distributions and C^∞-functions on arbitrary open subsets Ω of $\mathbb{R}^n$ was obtained by the present authors in [22]. Earlier, certain parts of this solution had been announced in [20] and [21]. The related problem of the splitting of differential complexes with constant coefficients was solved by Palamodov [30]. To indicate the form of the solution, we recall from [22]:

Theorem. *Let P be a non-constant complex polynomial in n variables and let $\Omega \neq \emptyset$ be an open subset of $\mathbb{R}^n$. Then the following assertions are equivalent:*

(1) $P(D) : \mathcal{D}'(\Omega) \to \mathcal{D}'(\Omega)$ admits a continuous linear right inverse.

(2) For each compact subset K of Ω there exists an open set Q with $K \subset Q \subset\subset \Omega$ such that for each $f \in \mathcal{D}'(\Omega)$ with $f|_Q = 0$ there exists $g \in \mathcal{D}'(\Omega)$ with $P(D)g = f$ and $g|_K = 0$.

(3) $P(D) : \mathcal{E}(\Omega) \to \mathcal{E}(\Omega)$ admits a continuous linear right inverse.

If, in addition, Ω is convex then (1) - (3) are equivalent to

(4) The variety $V(P) := \{z \in \mathbb{C}^n : P(-z) = 0\}$ satisfies the Phragmén-Lindelöf condition $\mathrm{PL}(\Omega, (\omega))$ for $\omega(t) = \log(2 + t)$ (see Definition 2.4(a) below).

Condition (4) is obtained by an application of Fourier analysis and is a variation of a Phragmén-Lindelöf condition that was introduced by Hörmander [15] to characterize the solvability of the equation $P(D)f = g$ for real-analytic functions.

In fact, there are quite a number of other, rather technical conditions that are equivalent to those given above and that are needed in the proof of the Theorem. Since the properties (1) and (3) can be regarded as stronger versions of strong P–convexity, respectively P–convexity, it is remarkable that they are equivalent.

Recently, the solution of Schwartz's problem in the framework of ω–ultradistributions and ω–ultradifferentiable functions was given in [24]. It is the aim of the present article to give a clear picture of the actual state of our research and to point out several open problems related to it. Roughly speaking, we have the following situation: The conditions in the general characterization are rather technical and difficult to check in concrete cases and further evaluation is needed. Concerning their evaluation, there are essentially two cases that are treated by different methods. In the case of convex open sets we evaluate the Phragmén–Lindelöf condition $\mathrm{PL}(\Omega, (\omega))$. A comprehensive study of this condition was presented in [25] and by the results of this article we are able to decide in many cases whether $P(D)$ admits a right inverse on $\mathcal{E}_\omega(\mathbb{R}^n)$.

In the case of bounded open sets with C^1-boundary and of non-characteristic half spaces and their intersections, it is Holmgren's uniqueness theorem which shows that the polynomial P has to be very hyperbolic if $P(D)$ has a right inverse on such open sets. In the case of ω-ultradistributions the appropriate notion of ω-hyperbolicity was investigated in [28].

The paper is organized as follows. In section 1 we introduce the spaces of ω-ultradifferentiable functions and ω-ultradistributions that we are going to use. The general solution of Schwartz's problem in this context is stated in section 2, where we also outline its proof, give several corollaries and comments on the subsequent sections of the article. In section 3 we give a complete proof of the fact that the existence of a right inverse for $P(D)$ on an open convex set Ω is equivalent to the Phragmén-Lindelöf condition and to the existence of a right inverse for $P(D)$ on $\mathcal{D}'(\Omega)$. The main results in [25] on the Phragmén-Lindelöf conditions are outlined in section 4. In section 5 we indicate why hyperbolicity and ω-hyperbolicity are important in connection with the existence of right inverses.

1 Ultradifferentiable functions and ultradistributions

In this section we introduce weight functions, ultradifferentiable functions and ultradistributions.

1.1 Definition. A continuous increasing function $\omega : [0, \infty[\rightarrow [0, \infty[$ is called a *weight function* if it satisfies the following conditions:

(α) there exists $K \geq 1$ with $\omega(2t) \leq K(1 + \omega(t))$ for all $t \geq 0$,

(β) $\displaystyle\int_1^\infty \frac{\omega(t)}{t^2} dt < \infty$,

(γ) $\log(t) = O(\omega(t))$,

(δ) $\varphi : t \mapsto \omega(e^t)$ is convex.

For a weight function ω we define $\tilde{\omega} : \mathbb{C}^n \rightarrow [0, \infty[$ by $\tilde{\omega}(z) = \omega(|z|)$ and again call this function ω, by abuse of notation.
The *Young conjugate* $\varphi^* = \varphi^*_\omega : [0, \infty[\rightarrow]-\infty, \infty]$ of $\varphi = \varphi_\omega$ is defined by

$$\varphi^*(y) := \sup\{xy - \varphi(x) : x \geq 0\}.$$

1.2 Remark. (a) Each weight function ω satisfies $\lim_{t\rightarrow\infty} \omega(t)/t = 0$ by the remark following 1.3 of [19].
(b) For each weight function ω there exists a weight function σ satisfying $\sigma(t) = \omega(t)$ for all large $t > 0$ and $\sigma \mid [0,1] \equiv 0$. This implies $\varphi_\sigma(y) = \varphi_\omega(y)$ for all large $y, \varphi^*_\sigma([0, \infty[) \subset [0, \infty[$ and $\varphi^{**}_\sigma = \varphi_\sigma$. From this it follows that all subsequent definitions do not change if ω is replaced by σ. On the other hand they also do not change if ω is replaced by $\omega + c$, c some positive number. Therefore we can and will assume that $\omega(0) \geq 1$.

In order to introduce ultradifferentiable functions and ultradistributions we define the Fourier transform $\hat{f}$ of a function $f \in L^1(\mathbb{R}^n)$ by

$$\hat{f}(\xi) := \int_{\mathbb{R}^n} f(x) e^{-i<x,\xi>} dx \ , \quad \xi \in \mathbb{R}^n,$$

where $< x, \xi >$ denotes the Euclidean scalar product on $\mathbb{R}^n$. The norm which is induced by this scalar product will be denoted by $|\cdot|$.

1.3 Definition. Let ω be a weight function.
(a) For a compact set K in $\mathbb{R}^n$ we let

$$\mathcal{D}_{(\omega)}(K) := \{f \in C(\mathbb{R}^n) \ : \quad \text{Supp}(f) \subset K \text{ and for each } \lambda > 0 : \\ \|f\|_\lambda := \int_{\mathbb{R}^n} |\hat{f}(\xi)| \exp(\lambda \omega(\xi)) d\xi < \infty\}$$

and we endow $\mathcal{D}_{(\omega)}(K)$ with its natural Fréchet space topology.
(b) If ω satisfies $\log = o(\omega)$ then for any compact set K in $\mathbb{R}^n$ we let

$$\mathcal{D}_{\{\omega\}}(K) := \{f \in C(\mathbb{R}^n) : \text{Supp}(f) \subset K \text{ and there is } \lambda > 0 \text{ with } \|f\|_\lambda < \infty\}$$

and we endow $\mathcal{D}_{\{\omega\}}(K)$ with its natural inductive limit topology.
(c) For any open set Ω in $\mathbb{R}^n$ and a fundamental sequence $(K_j)_{j \in \mathbb{N}}$ of compact subsets of Ω we let

$$\mathcal{D}_{(\omega)}(\Omega) := \underset{j \to}{\text{ind}}\, \mathcal{D}_{(\omega)}(K_j) \ , \ \mathcal{D}_{\{\omega\}}(\Omega) := \underset{j \to}{\text{ind}}\, \mathcal{D}_{\{\omega\}}(K_j).$$

The elements of $\mathcal{D}_{(\omega)}(\Omega)$ (resp. $\mathcal{D}_{\{\omega\}}(\Omega)$) are called *$\omega$-ultradifferentiable test functions of Beurling-type (resp. Roumieu-type).* The corresponding strong duals $\mathcal{D}'_{(\omega)}(\Omega)$ and $\mathcal{D}'_{\{\omega\}}(\Omega)$ are called the spaces of *ω-ultradistributions on Ω of Beurling-type or Roumieu-type*, respectively.
In the sequel we write $\mathcal{D}_*(\Omega)$ or $\mathcal{D}'_*(\Omega)$, whenever a statement holds if at all places we let $* = (\omega)$ or $* = \{\omega\}$.
(d) For an open set Ω in $\mathbb{R}^n$ we let

$$\mathcal{E}_*(\Omega) := \{f \in C(\Omega) : fg \in \mathcal{D}_*(\Omega) \text{ for each } g \in \mathcal{D}_*(\Omega)\}.$$

By Braun, Meise and Taylor [6], 4.6, we have

$$\mathcal{E}_{(\omega)}(\Omega) = \{f \in C^\infty(\Omega) : \text{for each compact set } K \subset \Omega \text{ and each } m \in \mathbb{N} : \\ \|f\|_{K,m} := \sup_{x \in K} \sup_{\alpha \in \mathbb{N}_0^n} |f^{(\alpha)}(x)| \exp(-m\varphi^*(\tfrac{|\alpha|}{m})) < \infty\},$$

$$\mathcal{E}_{\{\omega\}}(\Omega) = \{f \in C^\infty(\Omega) : \text{for each compact set } K \subset \Omega \text{ there is } m \in \mathbb{N} : \\ \|f\|_{K,\frac{1}{m}} := \sup_{x \in K} \sup_{\alpha \in \mathbb{N}_0^n} |f^{(\alpha)}(x)| \exp(-\tfrac{1}{m}\varphi^*(|\alpha| m)) < \infty\}.$$

We endow $\mathcal{E}_{(\omega)}(\Omega)$ with its natural Fréchet space topology. The natural topology of $\mathcal{E}_{\{\omega\}}(\Omega)$ is that of a projective limit of (DF)–spaces. By [24], 3.2, this topology can be described as follows: We let

$$S_\omega := \{\sigma : \sigma \text{ is a weight function with } \log = o(\sigma) \text{ and } \sigma = o(\omega)\}.$$

Then the system $(\|\cdot\|_{K,\sigma})_{K \subset\subset \Omega, \sigma \in S_\omega}$ of semi-norms,

$$\|f\|_{K,\sigma} := \sup_{x \in K} \sup_{\alpha \in \mathbb{N}_0^n} |f^{(\alpha)}(x)| \exp(-\varphi_\sigma^*(|\alpha|)) \ , \ f \in \mathcal{E}_{\{\omega\}}(\Omega)$$

induces the topology of $\mathcal{E}_{\{\omega\}}(\Omega)$.

1.4 Remark. (a) The definitions in 1.1 and 1.3 are variations of the classical ones of Beurling [2] (see also Petzsche and Vogt [31]) and were used by Braun, Meise and Taylor [6] to show that Beurling's ideas work equally well in this setup, which is better adapted to the application of Fourier analysis. For further properties of the spaces $\mathcal{E}_*(\Omega), \mathcal{D}_*(\Omega)$ and $\mathcal{D}'_*(\Omega)$, we refer to [6]. In particular it is shown there that all these spaces are not trivial.

(b) If K is a compact subset of the open set Ω then $\mathcal{D}_*(K)$ is a linear subspace of $\mathcal{E}_*(\Omega)$. The induced topology coincides with the one defined in 1.3(a), (b).

1.5 Examples. (1) If $\omega(t) = \log(2+t)$ then $\mathcal{E}_{(\omega)}(\Omega)$ is the space $\mathcal{E}(\Omega)$ of all C^∞-functions on Ω, while $\mathcal{D}_{(\omega)}(\Omega)$ is the space $\mathcal{D}(\Omega)$ of test functions of L. Schwartz. Note that $\mathcal{E}_{\{\omega\}}(\Omega)$ is not defined for this weight function ω.

(2) If $\omega(t) = t^\beta$ for $0 < \beta < 1$ then we get the *Gevrey classes* $\Gamma^{(1/\beta)}$ and $\Gamma^{\{1/\beta\}}$ that are usually defined for $1 < d < \infty$ by

$$\Gamma^{(d)}(\Omega) := \{f \in C^\infty(\Omega) \; : \; \text{For each compact set } K \subset \Omega \text{ and each } h > 0 : \sup_{x\in K} \sup_{\alpha\in\mathbb{N}_0^n} |f^{(\alpha)}(x)|(h(p!)^d)^{-|\alpha|} < \infty\},$$

$$\Gamma^{\{d\}}(\Omega) := \{f \in C^\infty(\Omega) \; : \; \text{For each compact set } K \subset \Omega \text{ there is } h > 0 : \sup_{x\in K} \sup_{\alpha\in\mathbb{N}_0^n} |f^{(\alpha)}(x)|(h(p!)^d)^{-|\alpha|} < \infty\}.$$

(3) Further examples of weight functions are:

$$\omega(t) := (\log(2+t))^\beta \; , \; \beta > 1 \; , \; \omega(t) := t(\log(e+t))^{-\beta} \; , \; \beta > 1.$$

1.6 Polynomials and partial differential operators. By $\mathbb{C}[z_1, ..., z_n]$ we denote the ring of all complex polynomials in n variables, which will also be regarded as functions on $\mathbb{C}^n$. For $P \in \mathbb{C}[z_1, ..., z_n], P(z) = \sum_{|\alpha|\le m} a_\alpha z^\alpha$, with $\sum_{|\alpha|=m} |a_\alpha| \neq 0$ we call

$$P_m : z \mapsto \sum_{|\alpha|=m} a_\alpha z^\alpha$$

the *principal part* of P. Note that P_m is a homogeneous polynomial of degree m.

For P as above and an open set Ω in $\mathbb{R}^n$ we define the linear partial differential operator

$$P(D) : \mathcal{D}'_*(\Omega) \to \mathcal{D}'_*(\Omega) \quad , \quad P(D)f := \sum_{|\alpha|\le m} a_\alpha i^{-|\alpha|} f^{(\alpha)}.$$

Then $P(D)$ is a continuous endomorphism of $\mathcal{D}'_*(\Omega)$ and its restriction to $\mathcal{E}_*(\Omega)$ is a continuous endomorphism of $\mathcal{E}_*(\Omega)$.

An element E of $\mathcal{D}'_*(\mathbb{R}^n)$ is called *a fundamental solution* for $P(D)$ if $P(D)E = \delta$, where δ denotes the point evaluation at zero.

2 The general characterization

In this section we will present the solution of a problem posed by L. Schwartz, namely to characterize the linear partial differential operators that admit a continuous linear right inverse on $\mathcal{E}(\Omega)$ or on $\mathcal{D}'(\Omega)$, as it was obtained by the present authors in [22] and then extended to such operators acting on $\mathcal{E}_*(\Omega)$ or on $\mathcal{D}'_*(\Omega)$ in [24]. In order to formulate this result in a convenient way we first introduce some notation.

2.1 Definition. Let Ω be an open subset of $\mathbb{R}^n$ and $P \in \mathbb{C}[z_1, ..., z_n]$. Then we let

$$\mathcal{N}_*(\Omega) := \{f \in \mathcal{D}'_*(\Omega) : P(D)f = 0\}.$$

Furthermore, for $\varepsilon > 0$ we define

$$\Omega_\varepsilon := \{x \in \Omega : |x| < \frac{1}{\varepsilon} \text{ and } \operatorname{dist}(x, \partial\Omega) > \varepsilon\}.$$

For a space $X(\Omega)$ of functions or distributions on Ω and for any subset U of Ω we let

$$X(\Omega, U) := \{f \in X(U) : f\,|_U \equiv 0\}.$$

2.2 Definition. For a compact convex set K in $\mathbb{R}^n$ we define the *support functional* h_K of K by

$$h_K(x) := \sup_{y \in K} \langle x, y \rangle.$$

2.3 Definition. Let V be an analytic variety in $\mathbb{C}^n$. A function $u : V \to [-\infty, \infty[$ is called *plurisubharmonic* (psh) on V if it is locally bounded from above and psh at the regular points $V_{\text{reg}} \subset V$. The values of u at the singular points $V_{\text{sing}} \subset V$ are not important in our considerations. However, it is convenient in the sequel to assume that

$$u(z) = \limsup_{\zeta \in V_{\text{reg}} \zeta \to z} u(\zeta) \text{ for all } z \in V_{\text{sing}}.$$

By PSH(V) we denote the set of all psh functions on V which satisfy this condition.

2.4 Definition. Let Ω be a convex open subset of $\mathbb{R}^n$, let $P \in \mathbb{C}[z_1, ..., z_n]$ be non-constant and let ω be a weight function. Then we let

$$V(P) := \{z \in \mathbb{C} : P(-z) = 0\}$$

and we denote by $\mathcal{K}(\Omega) = \{K \subset \Omega : K \text{ is compact and convex}, \overset{\circ}{K} \neq \emptyset\}$.

(a) We say that P or $V(P)$ satisfies the condition PL$(\Omega, (\omega))$ if the following holds: For each $K \in \mathcal{K}(\Omega)$ there exists $K' \in \mathcal{K}(\Omega)$ so that for each $K'' \in \mathcal{K}(\Omega)$ there exists $B > 0$ so that each $u \in$ PSH$(V(P))$ satisfying (α) and (β) also satisfies (γ), where

(α) $u(z) \le h_K(\operatorname{Im} z) + O(\omega(z))$, $z \in V$,

(β) $u(z) \le h_{K''}(\operatorname{Im} z)$, $z \in V$,

(γ) $u(z) \le h_{K'}(\operatorname{Im} z) + B\omega(z)$, $z \in V$.

If the above holds only for all $u = \log|f|, f$ an entire function on $\mathbb{C}^n$, then we say that P or $V(P)$ satisfies the condition $\mathrm{APL}(\Omega, (\omega))$.

(b) We say that P or $V(P)$ satisfies the condition $\mathrm{PL}(\Omega, \{\omega\})$ if $\log = o(\omega)$ and if the following holds:

For each $K \in \mathcal{K}(\Omega)$ there exists $K' \in \mathcal{K}(\Omega)$ so that for each $K'' \in \mathcal{K}(\Omega)$ there exists $\sigma \in S_\omega$ so that each $u \in \mathrm{PSH}(V(P))$ satisfying (α) and (β) also satisfies (γ), where

(α) $u(z) \le h_K(\operatorname{Im} z) + o(\omega(z))$, $z \in V$,

(β) $u(z) \le h_{K''}(\operatorname{Im} z)$, $z \in V$,

(γ) $u(z) \le h_{K'}(\operatorname{Im} z) + \sigma(z)$, $z \in V$.

If the above holds only for all $u = \log|f|, f$ an entire function on $\mathbb{C}^n$, then we say that P or $V(P)$ satisfies the condition $\mathrm{APL}(\Omega, \{\omega\})$.

(c) As in 1.3 we write $\mathrm{PL}(\Omega, *)$ or $\mathrm{APL}(\Omega, *)$ if a statement holds when at all places $*$ can be equal to (ω) or $\{\omega\}$.

Now we can formulate the result that was announced at the beginning of this section.

2.5 Theorem. *Let $P \in \mathbb{C}[z_1, \ldots, z_n]$ be non-constant and let $\Omega \neq \emptyset$ be an open subset of $\mathbb{R}^n$. Then the following assertions are equivalent:*

(1) $P(D) : \mathcal{D}'_*(\Omega) \to \mathcal{D}'_*(\Omega)$ admits a continuous linear right inverse.

(2) For each $\varepsilon > 0$ there exists $0 < \delta < \varepsilon$ so that for each $f \in \mathcal{D}'_*(\Omega, \Omega_\delta)$ there exists $g \in \mathcal{D}'_*(\Omega, \Omega_\varepsilon)$ with $P(D)g = f$.

(3) For each $\varepsilon > 0$ there exists $0 < \delta < \varepsilon$ so that for each $\mu \in \mathcal{N}_*(\Omega_\delta)$ there exists $\nu \in \mathcal{N}_*(\Omega)$ with $\nu|_{\Omega_\varepsilon} = \mu|_{\Omega_\varepsilon}$.

(4) For each $\varepsilon > 0$ there exists $0 < \delta_0 < \varepsilon$ so that for all $0 < \zeta < \sigma < \eta < \delta < \delta_0$ and each $\xi \in \overline{\Omega}_\eta \setminus \Omega_\delta$ there exists $E_\xi \in \mathcal{D}'_*(\mathbb{R}^n)$ so that

(i) $\operatorname{Supp} E_\xi \subset (\mathbb{R}^n \setminus \Omega_\varepsilon) - \xi$,

(ii) $P(D)E_\xi = \delta + T_\xi$, *where* $\operatorname{Supp} T_\xi \subset (\Omega_\zeta \setminus \Omega_\sigma) - \xi$.

(5) $P(D) : \mathcal{E}_*(\Omega) \to \mathcal{E}_*(\Omega)$ admits a continuous linear right inverse.

If, in addition, Ω is convex then (1) – (5) are also equivalent to:

(6) P satisfies the condition $\mathrm{APL}(\Omega, *)$.

(7) P satisfies the condition $\mathrm{PL}(\Omega, *)$.

In order to formulate another significant condition that is equivalent to 2.5(1) – (5), we introduce the following notation.

Notation. Let ω be a weight function, $\Omega \subset \mathbb{R}^n$ open, $\varepsilon > 0$, $m \in \mathbb{N}$ and $\sigma \in S_\omega$. Then we let

$$B_{\varepsilon,m} := \{\mu \in \mathcal{E}'_{(\omega)}(\Omega) : \operatorname{Supp}\mu \subset \Omega_\varepsilon, |\mu(f)| \leq \|f\|_{\Omega_\varepsilon,m}\}$$

$$B_{\varepsilon,\sigma} := \{\mu \in \mathcal{E}'_{\{\omega\}}(\Omega) : \operatorname{Supp}\mu \subset \Omega_\varepsilon, |\mu(f)| \leq \|f\|_{\Omega_\varepsilon,\sigma}\}.$$

Note that $(B_{\varepsilon,m})_{\varepsilon>0,m\in\mathbb{N}}$ is a fundamental system of equicontinuous subsets of $\mathcal{E}'_{(\omega)}(\Omega)$, while $(B_{\varepsilon,\sigma})_{\varepsilon>0,\sigma\in S_\omega}$ has the same property for $\mathcal{E}'_{\{\omega\}}(\Omega)$.

2.6 Proposition. *Let $P \in \mathbb{C}[z_1, ..., z_n]$ be non-constant, let $\Omega \neq \emptyset$ be an open subset of $\mathbb{R}^n$ and let ω be a weight function.*

(a) For $* = (\omega)$ the conditions 2.5(1) - (5) are equivalent to:
For each $\varepsilon > 0$ there exists $0 < \delta < \varepsilon$ such that for each $0 < \eta < \delta$ and each $m \in \mathbb{N}$ (there is $m \in \mathbb{N}$), there exist $k \in \mathbb{N}$, $C > 0$ such that for each $\mu \in \mathcal{E}'_{(\omega)}(\Omega_\varepsilon)$ with $(\mu + \operatorname{im} P(D)^t) \cap B_{\eta,m} \neq \emptyset$ there exists $\lambda \in \mathcal{E}'_{(\omega)}(\Omega_\delta)$ satisfying $\mu + P(D)^t\lambda \in CB_{\delta,k}$.

(b) For $* = \{\omega\}$ the conditions 2.5(1) - (5) are equivalent to:
For each $\varepsilon > 0$ there exists $0 < \delta < \varepsilon$ such that for each $0 < \eta < \delta$ and each $\sigma \in S_\omega$ (there is $\sigma \in S_\omega$) there exist $\kappa \in S_\omega$ and $C > 0$ such that for each $\mu \in \mathcal{E}'_{\{\omega\}}(\Omega_\varepsilon)$ with $(\mu + \operatorname{im} P(D)^t) \cap B_{\eta,\sigma} \neq \emptyset$ there exists $\lambda \in \mathcal{E}'_{\{\omega\}}(\Omega_\delta)$ satisfying $\mu + P(D)^t\lambda \in CB_{\delta,\kappa}$.

The conditions stated in Proposition 2.6 (a) / (b) imply by [24], 2.9 and 3.5, that Ω is P-convex in the sense of Hörmander [14] Def. 3.5.1. In fact, they can be considered as P-convexity with bounds which are related with the weight function ω. Therefore we define:

2.7 Definition. Let $\Omega \neq \emptyset$ be an open subset of $\mathbb{R}^n$ and let $P \in \mathbb{C}[z_1, ..., z_n]$ be non-constant. Then Ω is said to be *P-convex with $*$-bounds* if one of the equivalent conditions 2.5(1) – (5) is satisfied. We will drop $*$ if $\omega(t) = \log(2+t)$.

The proof of Theorem 2.5 can be outlined as follows: Rather elementary arguments show that the implications (1) $\Rightarrow$ (2) $\Rightarrow$ (3) $\Rightarrow$ (4) hold. We refer to [22], 2.1 and [24], 3.3 for details. Since linearity does not occur any more in the conditions (2), (3) and (4), one has to work in order to regain it from (4). Essentially one uses the existence of sufficiently many cut-off functions, the properties of the convolution and a linear Mittag–Leffler argument to show (4) $\Rightarrow$ (1). In fact, the same arguments also prove (4) $\Rightarrow$ (5). For details we refer to [22], 2.3 and [24], 2.4 and 3.7. Also, the idea behind this will become clear when we prove this in the special case $\Omega = \mathbb{R}^n$ in 2.10 below.

The most difficult part in the proof of Theorem 2.5 is to show (5) $\Rightarrow$ (4) which gives (5) $\Rightarrow$ (1) by the above. In [22] and also in [24] this is achieved via the conditions stated in Proposition 2.6, using the theorem of Hahn–Banach. In the case of ultradifferentiable functions, one also uses a recent result of Braun [4] which extends a previous one of Komatsu [17] on the local structure of ultradistributions. We refer to [22], 2.4 – 2.6 and to [24], 2.5 – 2.7 and 3.5 – 3.6.

In the case of a convex open set Ω, one uses an essential part of the proof of the so-called "fundamental principle" to show that the conditions in Proposition 2.6, and hence any of 2.5(1) – (5), are equivalent to (6). For the details we refer to [22], sect. 4 and [24], sect. 5.

The equivalence of (6) and (7) might appear as obvious, since, roughly speaking, plurisubharmonic functions are upper envelopes of functions of the form $\log|f|, f$ analytic. However, because of the singularities that might occur in the variety $V(P)$, things are more subtle. In fact, the equivalence of (6) and (7) follows from [23], 6.2, if $* = (\omega)$ and by the arguments that are used in [23], sect. 6, if $* = \{\omega\}$.

Before we give some corollaries of Theorem 2.5 we want to add a few comments on it and on the organization of the rest of the article.

The outline of the proof of Theorem 2.5 does not reflect how its conditions came up. In fact, our first observation was that for $\Omega = \mathbb{R}^n$ and $\omega = \log(2+t)$, condition 2.5(5) implies APL$(\Omega, (\omega))$. Knowing this, we found out how to treat the general case, using the conditions in Proposition 2.6. Therefore, the original argument was not mentioned in [22]. Since this argument has a different flavour and gives an interpretation of the condition PL$(\Omega, *)$ that has not been noted so far, we present it in section 3. This provides an independent proof of the equivalence of the conditions 2.5(5) and 2.5(1) in the case of convex open sets Ω.

Evidently, some of the conditions in Theorem 2.5 are rather involved or difficult to verify. It seems to be almost impossible to use them in order to decide whether a given open set is P-convex with $*$-bounds for a given polynomial P. Therefore these conditions have to be evaluated further. As the investigations in [22] and [24] show, there are essentially two cases that are handled by different methods.

The first case is that of a convex open set Ω, in particular $\Omega = \mathbb{R}^n$. In this case the condition PL$(\Omega, *)$ is the main tool for further evaluation. In fact, the entire article [25] is devoted to this and quite a number of interesting questions remain open. We will outline this in section 4.

The second case is that of an open set Ω with non-empty C^1-boundary, in particular Ω equal to a half space. In this case condition 2.5(2) together with Holmgren's uniqueness theorem implies that P has to satisfy certain hyperbolicity conditions with respect to non-characteristic normal vectors to $\partial\Omega$. We will outline these results in section 5.

We note the following corollaries of Theorem 2.5.

2.8 Corollary. *Let $P \in \mathbb{C}[z_1, ..., z_n]$ be non-constant and let Ω be an open subset of $\mathbb{R}^n$. Then the following holds:*

(a) *If Ω is P-convex with $\{\omega\}$-bounds then Ω is P-convex with (σ)-bounds whenever $\omega = O(\sigma)$.*

(b) *If Ω is P-convex with (ω)-bounds then Ω is P-convex with $\{\sigma\}$-bounds whenever $\omega = o(\sigma)$.*

PROOF. By Braun, Meise and Taylor [6], 3.9, the inclusion $\mathcal{D}_{(\sigma)}(\mathbb{R}^n) \hookrightarrow \mathcal{D}_{\{\omega\}}(\mathbb{R}^n)$ is continuous and has dense range whenever $\omega = O(\sigma)$. Therefore, its adjoint is continuous and injective, so that we can regard $\mathcal{D}'_{\{\omega\}}(\mathbb{R}^n)$ as a linear subspace of $\mathcal{D}'_{(\sigma)}(\mathbb{R}^n)$. Hence (a) follows from 2.5(4). Similarly, $\omega = o(\sigma)$ implies that we can consider $\mathcal{D}'_{(\omega)}(\mathbb{R}^n)$ as a linear subspace of $\mathcal{D}'_{\{\sigma\}}(\mathbb{R}^n)$. From this and 2.5(4) we get (b).

2.9 Corollary. *Let $P \in \mathbb{C}[z_1, ..., z_n]$ be non-constant, Ω an open set in $\mathbb{R}^n$ and ω a weight function with* $\log = o(\omega)$. *Then Ω is P-convex with $\{\omega\}$-bounds if and only if there exists a weight function σ with $\sigma = o(\omega)$ so that Ω is P-convex with (σ)-bounds.*

PROOF. One implication holds by Corollary 2.8. To prove the other one, assume that Ω is P-convex with $\{\omega\}$-bounds. Then condition 2.5(4) holds. In proving that 2.5(4) implies 2.5(1), we only use countably many distributions $(E_{\xi_j})_{j\in\mathbb{N}}$, as the proof of [23], 2.4, shows. By Braun, Meise and Taylor [6], 7.6, there exist weight functions σ_j with $\sigma_j = o(\omega)$ and $E_{\xi_j} \in \mathcal{D}'_{(\sigma_j)}(\Omega)$ for each $j \in \mathbb{N}$. Now note that by [6], 1.9, there exists a weight function σ with $\sigma = o(\omega)$ and $\sigma_j = o(\sigma)$ for all $j \in \mathbb{N}$. Hence the proof of [23], 2.4, shows that $P(D) : \mathcal{D}'_{(\sigma)}(\Omega) \to \mathcal{D}'_{(\sigma)}(\Omega)$ admits a continuous linear right inverse, i.e. Ω is P-convex with (σ)-bounds.

In the case $\Omega = \mathbb{R}^n$, condition 2.5(4) simplifies considerably.

2.10 Corollary. *Let $P \in \mathbb{C}[z_1, ..., z_n]$ be non-constant. Then $\mathbb{R}^n$ is P-convex with $*$-bounds if and only if the following condition holds: For each $r > 0$ there is $R > 0$ so that for each $\xi \in \mathbb{R}^n$ with $|\xi| > R$ there exists a fundamental solution $E_\xi \in \mathcal{D}'_*(\mathbb{R}^n)$ of $P(D)$ with*

$$\operatorname{Supp} E_\xi \subset \{x \in \mathbb{R}^n : |x - \xi| \geq r\}.$$

PROOF. If $\mathbb{R}^n$ is P–convex with $*$–bounds then condition 2.5(3) holds. To show that this implies the present condition, we fix a fundamental solution $E \in \mathcal{D}'(\mathbb{R}^n)$ for $P(D)$. If $r > 0$ is given we let $\varepsilon := 1/r$ and we choose $0 < \delta < \varepsilon$ according to condition 2.5(3). Next we let $R := 1/\delta$ and fix $\xi \in \mathbb{R}^n$ with $|\xi| > R$. If τ_ξ denotes the translation by ξ then $\tau_\xi E$ (defined by $(\tau_\xi E)(\varphi) = \langle E_x, \varphi(x + \xi)\rangle$) is in $\mathcal{N}_*(\Omega_\delta)$. Hence 2.5(3) gives the existence of $\nu \in \mathcal{N}_*(\mathbb{R}^n)$ with $(\tau_\xi E)\,|_{\Omega_\varepsilon} = \nu\,|_{\Omega_\varepsilon}$, which implies

$$\operatorname{Supp}(\tau_\xi E - \nu) \subset \{x \in \mathbb{R}^n : |x| \geq r\}.$$

Therefore, $E_\xi := E - \tau_{-\xi}\nu$ is in $\mathcal{D}'_*(\mathbb{R}^n)$ and satisfies

$$\operatorname{Supp} E_\xi \subset \{x \in \mathbb{R}^n : |x - \xi| \geq r\}$$

and $P(D)E_\xi = \delta$. Hence the present condition holds.

To prove the converse implication we choose a continuous, strictly increasing function $R :]0, \infty[\to]0, \infty[$ such that the condition of the corollary holds for $R = R(r)$. Then we define $r(\xi) := k$ if $R(k) < |\xi| < R(k + 2)$ and we choose a locally finite partition of unity $(\varphi_j)_{j\in\mathbb{N}_0}$ of functions in $\mathcal{E}_*(\mathbb{R}^n)$, subordinate to the following open cover of $\mathbb{R}^n$:

$$\{x \in \mathbb{R}^n : |x - \xi| < r(\xi)/2\}_{|\xi|>R(1)}, \ \{x \in \mathbb{R}^n : |x| < R(1) + 1\}.$$

For each $j \in \mathbb{N}$ we choose $\xi_j \in \mathbb{R}^n, |\xi_j| > R(1)$ with

$$\operatorname{Supp} \varphi_j \subset \{x \in \mathbb{R}^n : |x - \xi_j| < r(\xi_j)/2\}.$$

Next we fix a fundamental solution $E_{\xi_0} \in \mathcal{D}'(\mathbb{R}^n)$ for $P(D)$, and according to the hypothesis of the corollary, we choose fundamental solutions $E_{\xi_j} \in \mathcal{D}'_*(\mathbb{R}^n)$ of $P(D)$ satisfying

$$\operatorname{Supp} E_{\xi_j} \subset \{x \in \mathbb{R}^n : |x - \xi_j| \geq r(\xi_j)\}.$$

Then we note that for each $\mu \in \mathcal{D}'_*(\mathbb{R}^n)$ we have $\check{\varphi}_j \mu \in \mathcal{E}'_*(\mathbb{R}^n)$, where $\check{\varphi}_j(x) = \varphi_j(-x)$. Moreover, we have

$$\operatorname{Supp}((\check{\varphi}_j \mu) * E_{\xi_j}) \subset \{x \in \mathbb{R}^n : |x| \geq r(\xi_j) - r(\xi_j)/2\}$$

for each $j \in \mathbb{N}$. Hence the following series

$$R(\mu) := \sum_{j=0}^{\infty} (\check{\varphi}_j \mu) * E_{\xi_j}$$

converges in $\mathcal{D}'_*(\mathbb{R}^n)$ since the family of supports of its members is locally finite. Moreover, the map $R : \mathcal{D}'_*(\mathbb{R}^n) \to \mathcal{D}'_*(\mathbb{R}^n)$ is continuous and linear and satisfies

$$P(D)R(\mu) = \sum_{j=0}^{\infty} (\check{\varphi}_j \mu) * P(D)E_{\xi_j} = (\sum_{j=0}^{\infty} \check{\varphi}_j)\mu = \mu.$$

Hence R is a continuous linear right inverse for $P(D)$. Note that by the same arguments R also defines a continuous linear right inverse for $P(D)$ on $\mathcal{E}_*(\mathbb{R}^n)$

2.11 Corollary. *Assume that for $P \in \mathbb{C}[z_1, ..., z_n]$ there exist fundamental solutions $E_1, ..., E_k \in \mathcal{D}'_*(\mathbb{R}^n)$ so that $\operatorname{Supp} E_j \subset \mathbb{R}^n \setminus \Gamma_j$, where Γ_j is an open convex cone with vertex at zero for $1 \leq j \leq k$. If $\bigcup_{j=1}^k \Gamma_j$ covers the unit sphere of $\mathbb{R}^n$ then $\mathbb{R}^n$ is P-convex with $*$-bounds.*

2.12 Remark. Each polynomial $P \in \mathbb{C}[z_1, ..., z_n]$ which is hyperbolic with respect to some direction satisfies the hypothesis of Corollary 2.11 by Hörmander [14], Thm. 5.6.1. Hence $\mathbb{R}^n$ is P-convex with bounds for each hyperbolic polynomial. For further details we refer to section 5.

2.13 Problem. Let $P \in \mathbb{C}[z_1, ..., z_n]$ be a homogeneous polynomial for which $\mathbb{R}^n$ is P-convex with bounds. Does this already imply that there exist fundamental solutions in $\mathcal{D}'(\mathbb{R}^n)$ that satisfy the hypothesis of Corollary 2.11?

Remark. Let $P(z_1, z_2, z_3, z_4) := z_1^2 + z_2^2 - z_3^2 - z_4^2$ be the ultrahyperbolic polynomial. In [22], 4.9, it was shown that P satisfies $\mathrm{PL}(\Omega, \log(2+t))$. Hence $\mathbb{R}^4$ is P-convex with bounds (see also 4.9(c) below). However, it seemed not to be known whether $P(D)$ admits fundamental solutions that vanish in cone-shaped regions. Recently, it was shown in [26] that P actually satisfies the hypothesis of Corollary 2.11.

3 An application of Fourier analysis

In this section we will use Fourier analysis to prove that for any convex open set Ω in $\mathbb{R}^n$ and each weight function ω the conditions (5), (6), (7) and (1) of Theorem 2.5 are equivalent. Note that it is sufficient to show this for all polynomials that are of the following form

$$(*) \qquad P = \prod_{j=1}^{l} P_j \ , \ P_1, \dots, P_l \text{ irreducible and pairwise not proportional,}$$

because each product of polynomials satisfies any of these conditions if and only if each factor does. To state the main result of this section, we introduce some auxiliary spaces.

3.1 Definition. Let Q be a convex compact subset of $\mathbb{R}^n$ with $\mathring{Q} \neq \emptyset$, let ω be a weight function and $\varphi = \varphi_\omega$ (see 1.1). Then we define

$$\begin{aligned} \mathcal{E}_{(\omega)}(Q) \ &:= \ \{f \in C(Q) \cap C^\infty(\mathring{Q}) : \text{ for all } m \in \mathbb{N} : \\ &\quad \|f\|_m := \sup_{x \in Q, \alpha \in \mathbb{N}_0^n} |f^{(\alpha)}(x)| \exp\big(-m\varphi^*(\frac{|\alpha|}{m})\big) < \infty\} \end{aligned}$$

and we endow $\mathcal{E}_{(\omega)}(Q)$ with its natural Fréchet space topology.
It is easy to check that for each $P \in \mathbb{C}[z_1, \dots, z_n]$ the operator $P(D)$ acts continuously on $\mathcal{E}_{(\omega)}(Q)$ and on $\mathcal{D}'_{(\omega)}(Q)$ (see [6], sect.6). We let

$$N_{(\omega)}(Q) := \{f \in \mathcal{E}_{(\omega)}(Q) : P(D)f = 0\} \ , \ \mathcal{N}_{(\omega)}(Q) := \{f \in \mathcal{D}'_{(\omega)}(Q) : P(D)f = 0\}.$$

Furthermore, for any open subset Ω of $\mathbb{R}^n$ we let

$$N_{(\omega)}(\Omega) := \{f \in \mathcal{E}_{(\omega)}(\Omega) : P(D)f = 0\}.$$

3.2 Definition. Let V be an analytic subvariety of $\mathbb{C}^n$ and Q a compact convex subset of $\mathbb{R}^n$. Then we denote by $A(V)$ the Fréchet space of all holomorphic functions on V (which are locally extendable to a neighborhood) and we let

$$\begin{aligned} A_Q^-(V) \ &:= \ \{f \in A(V) : \|f\|_m := \sup\nolimits_{z \in V} |f(z)| \exp(-h_Q(\operatorname{Im} z) - m\omega(z)) < \infty \\ &\qquad \text{for some } m \in \mathbb{N}\}, \\ A_Q^+(V) \ &:= \ \{f \in A(V) : \|f\|_m := \sup\nolimits_{z \in V} |f(z)| \exp(-h_Q(\operatorname{Im} z) + m\omega(z)) < \infty \\ &\qquad \text{for all } m \in \mathbb{N}\}. \end{aligned}$$

We endow $A_Q^-(V)$ with its natural inductive limit topology and $A_Q^+(V)$ with its natural Fréchet space topology.

3.3 Theorem. *Let Ω be an open convex subset of $\mathbb{R}^n$, let ω be a weight function and assume that $P \in \mathbb{C}[z_1, \dots, z_n]$ satisfies (*). Then the following assertions are equivalent:*

(1) $P(D) : \mathcal{E}_{(\omega)}(\Omega) \longrightarrow \mathcal{E}_{(\omega)}(\Omega)$ admits a continuous linear right inverse.

(2) For each $K \in \mathcal{K}(\Omega)$ there exists $K' \in \mathcal{K}(\Omega)$ with $K' \supset K$ such that for each $f \in N_{(\omega)}(K')$ there exists $g \in N_{(\omega)}(\Omega) : f|_K = g|_K$.

(3) For each $K \in \mathcal{K}(\Omega)$ there exists $K' \in \mathcal{K}(\Omega)$ with $K' \supset K$ such that for each $K'' \in \mathcal{K}(\Omega)$ with $K'' \supset K'$ the spaces $A^-_{K''}(V(P))$ and $A^-_{K'}(V(P))$ induce the same topology on the common subspace $A^-_K(V(P))$.

(4) P satisfies $\mathrm{APL}(\Omega, (\omega))$.

(5) P satisfies $\mathrm{PL}(\Omega, (\omega))$.

(6) For each $K \in \mathcal{K}(\Omega)$ there exists $K' \in \mathcal{K}(\Omega)$ with $K' \supset K$ such that for each $K'' \in \mathcal{K}(\Omega)$ with $K'' \supset K'$ the spaces $A^+_{K''}(V(P))$ and $A^+_{K'}(V(P))$ induce the same topology on the common subspace $A^+_K(V(P))$.

(7) For each $\varepsilon > 0$ there exists $0 < \delta < \varepsilon$ such that for each $\mu \in \mathcal{N}_{(\omega)}(\Omega_\delta)$ there exists $\nu \in \mathcal{N}_{(\omega)}(\Omega) : \mu|_{\Omega_\varepsilon} = \nu|_{\Omega_\varepsilon}$.

(8) $P(D) : \mathcal{D}'_{(\omega)}(\Omega) \longrightarrow \mathcal{D}'_{(\omega)}(\Omega)$ admits a continuous linear right inverse.

The proof of Theorem 3.3 will be based on the following six lemmas.

3.4 Lemma. *Let Q be a convex compact subset of $\mathbb{R}^n$ with $\mathring{Q} \neq \emptyset$, let ω be a weight function and assume that $P \in \mathbb{C}[z_1, \dots, z_n]$ satisfies the condition (*). Then the Fourier-Laplace transform $\mathcal{F}$, defined by $\mathcal{F}(\mu) : z \mapsto \langle \mu_x, \exp(-i\langle x, z\rangle)\rangle$, is a linear topological isomorphism between the following spaces:*

(1) $\mathcal{F} : \mathcal{D}_{(\omega)}(Q) \longrightarrow A^+_Q(\mathbb{C}^n)$

(2) $\mathcal{F} : \mathcal{E}_{(\omega)}(Q)' \longrightarrow A^-_Q(\mathbb{C}^n)$

(3) $\mathcal{F} : N_\omega(Q)' \longrightarrow A^-_Q(V(P))$.

PROOF. (1) This holds by Braun, Meise and Taylor [6], 3.9(1).
(2) This holds by Meise and Taylor [18], 3.6.
(3) By (2), the restriction map

$$\rho : A^-_Q(\mathbb{C}^n) \longrightarrow A^-_Q(V(P)) \, , \, \rho(f) := f|_{V(P)}$$

is linear and continuous. Next note that for each $m \in \mathbb{N}$ the function $\varphi_m(z) := h_Q(\operatorname{Im} z) + m\omega(z)$ is plurisubharmonic on $\mathbb{C}^n$ and there exists $C_m > 0$ such that

$$|\varphi_m(z_1) - \varphi_m(z_2)| \le C_m \text{ for all } z_1, z_2 \in \mathbb{C}^n \text{ with } |z_1 - z_2| \le 1 \, .$$

Since P satisfies (*), an application of an extension theorem based on Hörmander's solution of the $\overline{\partial}$–problem (see Hansen [13], Thm. 2.3) shows that σ is surjective.
To determine $\ker\rho$, we define the polynomial $\check{P}$ by $\check{P}(z) = P(-z)$. Then obviously $\check{P}A_Q^-(\mathbb{C}^n) \subset \ker\rho$. To prove the converse inclusion we fix $f \in \ker\rho$. Since P satisfies (*) there exists $g \in A(\mathbb{C}^n)$ with $f = \check{P}g$ and the Ehrenpreis–Malgrange inequality implies $g \in A_Q^-(V(P))$ (see Hansen [13], Lemma 2.2 and its proof). Hence we have $\ker\rho = \check{P}A_Q^-(\mathbb{C}^n)$. In particular, $\check{P}A_Q^-(\mathbb{C}^n)$ is a closed linear subspace of $A_Q^-(\mathbb{C}^n)$.
Next note that $\mathcal{F} : \mathcal{E}_{(\omega)}(Q)' \longrightarrow A_Q^-(\mathbb{C}^n)$ is an isomorphism by (2) and that $\mathcal{F}(\mathrm{im}P(D)^t) = \check{P}A_Q^-(\mathbb{C}^n)$. Thus, $\mathrm{im}P(D)^t$ is closed. Therefore, classical duality theory together with the above implies that we have the following isomorphisms

$$\begin{aligned} N_{(\omega)}(Q)' &\cong \mathcal{E}_{(\omega)}(Q)'/(\ker P(D))^\perp \cong \mathcal{E}_{(\omega)}(Q)'/\mathrm{im}P(D)^t \\ &\cong A_Q^-(\mathbb{C}^n)/\check{P}A_Q^-(\mathbb{C}^n) \cong A_Q^-(\mathbb{C}^n)/\ker\rho \cong A_Q^-(V(P)). \end{aligned}$$

An easy evaluation of these maps shows that in fact $\mathcal{F}$ is the isomorphism that we have obtained.

Since the functions $\psi_m(z) := h_Q(\mathrm{Im}\, z) - m\omega(z)$ need not be plurisubharmonic on $\mathbb{C}^n$ for $m \in \mathbb{N}$, the Fourier–Laplace transform seems to establish an isomorphism between $\mathcal{N}_{(\omega)}(Q)$ and $A_Q^+(V(P))$ only if ω and Ω satisfy additional conditions. However, the following weaker version, which is proved by some of the arguments that we used in the proof of 3.4(3), is sufficient for our purposes.

3.5 Lemma. *Let Q, ω and P be as in Lemma 3.4. Then the restriction map $\rho : A_Q^+(\mathbb{C}^n) \longrightarrow A_Q^+(V(P))$, $\rho(f) := f|_{V(P)}$ is linear and continuous and $\ker\rho = \check{P}A_Q^+(\mathbb{C}^n)$, where $\check{P}(z) = P(-z)$.*

The next lemma shows how one can overcome the difficulty that the functions ψ_m need not be plurisubharmonic. However, one has to pay for this by enlarging the set Q to $Q + B_\varepsilon$, where $B_\varepsilon := \{x \in \mathbb{R}^n : |x| \le \varepsilon\}$.

3.6 Lemma. *Let Q be a convex compact subset of $\mathbb{R}^n$ and σ a weight function satisfying $\log t = o(\sigma(t))$. If $P \in \mathbb{C}[z_1, \dots, z_n]$ satisfies (*) then for each $\varepsilon > 0$ there exist $D \ge 1$ and $0 < d \le \frac{1}{2}$ such that for each $f \in A(V(P))$ satisfying*

$$(1) \qquad |f(z)| \le \exp\left(h_Q(\mathrm{Im}\, z) - \sigma(z)\right) \text{ for all } z \in V(P)$$

there exists $g \in A(\mathbb{C}^n)$ satisfying $g|_{V(P)} = f$ and

$$(2) \qquad |g(z)| \le D\exp\left(h_{Q+B_\varepsilon}(\mathrm{Im}\, z) - d\sigma(z)\right) \text{ for all } z \in \mathbb{C}^n .$$

PROOF. Fix $\varepsilon > 0$ and note that by Braun and Meise [5], Prop.5, there exist $u \in \mathrm{PSH}(\mathbb{C}^n)$ and $C_\varepsilon \ge 1$ such that

$$(3) \qquad -C_\varepsilon\sigma(z) \le u(z) \le \varepsilon|\mathrm{Im}\, z| - \sigma(z) \text{ for all } z \in \mathbb{C}^n .$$

From (3) it follows that each $f \in A(V(P))$ satisfying (1) also satisfies

$$|f(z)| \le \exp\left(h_Q(\mathrm{Im}) + C_\varepsilon^{-1}u(z)\right) \text{ for all } z \in V(P) .$$

Since σ is a weight function we obtain from Hansen [13], Thm. 2.3, the existence of $M \geq 1$ depending only on P such that there exists $g \in A(\mathbb{C}^n)$ satisfying $g|_{V(P)} = f$ and

$$(4) \qquad |g(z)| \leq \exp\left(h_Q(\operatorname{Im} z) + C_\varepsilon^{-1} u(z) + M \log(2 + |z|^2)\right) .$$

Next we let $d := (2C_\varepsilon)^{-1}$. Since $\log t = o(\sigma(t))$, we can find $D \geq 1$ so that

$$-C_\varepsilon^{-1}\sigma(t) + M\log(2 + |t|^2) \leq -d\sigma(t) + \log D \text{ for all } t \geq 0.$$

Using this and (3) it is easy to see that (4) implies (2), since

$$h_Q(\operatorname{Im} z) + \varepsilon|\operatorname{Im} z| = h_{Q+B_\varepsilon}(\operatorname{Im} z) .$$

An obvious modification of the proof of Bonet, Braun, Meise and Taylor [3], 4.4, shows that the following lemma holds.

3.7 Lemma. *Let ω be a weight function and assume that $h : [0, \infty[\longrightarrow [0, \infty[$ satisfies $\omega = o(h)$. Then there exists a weight function σ satisfying $\omega = o(\sigma)$ and $\sigma = o(h)$.*

3.8 Lemma. *Let Q be a convex compact subset of $\mathbb{R}^n$ and let ω be a weight function. If $P \in \mathbb{C}[z_1, \ldots, z_n]$ satisfies (*) then for each bounded set B in $A_Q^+(V(P))$ and each $\varepsilon > 0$ there exists a bounded set M in $A_{Q+B_\varepsilon}^+(\mathbb{C}^n)$ such that for each $f \in B$ there exists $g \in M$ with $g|_{V(P)} = f$.*

PROOF. If B is a given bounded subset of $A_Q^+(V(P))$ then for each $m \in \mathbb{N}$ there exists $C_m > 0$ such that for all $f \in B$ we have

$$|f(z)| \leq C_m \exp\left(h_Q(\operatorname{Im} z) - m\omega(z)\right) \text{ for all } z \in V(P) .$$

For each $m \in \mathbb{N}$ and $t > 0$ this implies

$$-h(t) := \sup\{\log|f(z)| - h_Q(\operatorname{Im} z) : f \in B \,,\ |z| = t \,,\ z \in V(P)\} \leq -m\omega(z) + \log C_m$$

and hence $\omega = o(h)$. By Lemma 3.7 we can choose a weight function σ with $\omega = o(\sigma)$ and $\sigma = o(h)$. In particular, there exists $L \geq 1$ so that $\sigma(t) \leq h(t) + L$ for all $t \geq 0$. By the definition of h this implies that each $f \in B$ satisfies

$$|f(z)| \leq \exp\left(h_Q(\operatorname{Im} z) - \sigma(z) + L\right) \text{ for all } z \in V(P).$$

Now let $\varepsilon > 0$ be chosen arbitrarily. Then Lemma 3.6 implies that for each $f \in B$ there exists $g \in A(\mathbb{C}^n)$ satisfying $g|_{V(P)} = f$ and

$$|g(z)| \leq D \exp\left(h_{Q+B_\varepsilon}(\operatorname{Im} z) - d\sigma(z) + L\right) \text{ for all } z \in \mathbb{C}^n .$$

Since $\omega = o(\sigma)$, the set M of all $g \in A(\mathbb{C}^n)$ satisfying this estimate is a bounded subset of $A_{Q+B_\varepsilon}^+(\mathbb{C}^n)$ and has all the required properties.

3.9 Lemma. *Let Q be a convex compact subset of $\mathbb{R}^n$, ω a weight function and P a polynomial on $\mathbb{C}^n$ which satisfies (*). Then for each $\varepsilon > 0$ and each $\mu \in \mathcal{N}_{(\omega)}(Q + B_\varepsilon)$ there exists $\lambda \in A_Q^+(V)'$ such that for each $\psi \in \mathcal{D}_{(\omega)}(Q + B_\varepsilon)$ with $\mathcal{F}(\psi)|_{V(P)} \in A_Q^+(V)$ we have $\lambda(\mathcal{F}(\psi)|_{V(P)}) = \mu(\psi)$.*

PROOF. We fix $\varepsilon > 0$ and $\mu \in \mathcal{N}_{(\omega)}(Q + B_\varepsilon)$. To define λ, first note that for each $f \in A_Q^+(V(P))$, Lemma 3.8 implies the existence of $g \in A_{Q+B_\varepsilon}^+(\mathbb{C}^n)$ satisfying $g|_{V(P)} = f$. By Lemma 3.4(1), there exists a unique $\psi \in \mathcal{D}_{(\omega)}(Q + B_\varepsilon)$ with $g = \mathcal{F}(\psi)$. If $\psi_1, \psi_2 \in \mathcal{D}_{(\omega)}(Q + B_\varepsilon)$ satisfy $\mathcal{F}(\psi_1)|_{V(P)} = \mathcal{F}(\psi_2)|_{V(P)}$, then Lemma 3.5 together with Lemma 3.4(1) imply the existence of $\chi \in \mathcal{D}_{(\omega)}(Q + B_\varepsilon)$ such that

$$\mathcal{F}(\psi_1) - \mathcal{F}(\psi_2) = \mathcal{F}(\psi_1 - \psi_2) = \check{P}\mathcal{F}(\chi) = \mathcal{F}(\check{P}(D)\chi) .$$

Since μ is in $\mathcal{N}_{(\omega)}(Q + B_\varepsilon)$, this implies

$$\mu(\psi_1 - \psi_2) = \mu(P(-D)\chi) = \mu(P(D)^t\chi) = (P(D)\mu)(\chi) = 0 .$$

The above consideration shows that

$$\lambda(f) := \mu(\psi) \text{ , whenever } \psi \in \mathcal{D}(Q + B_\varepsilon) \text{ and } \mathcal{F}(\psi)|_{V(P)} = f ,$$

is a well–defined linear form on $A_Q^+(V)$. Since $A_Q^+(V)$ is a Fréchet space and hence bornological, λ is continuous if λ is bounded on each bounded subset B of $A_Q^+(V)$. This, however, is an immediate consequence of Lemma 3.8, the isomorphism in Lemma 3.4(1) and the continuity of μ.

PROOF OF THEOREM 3.3: (1) $\Rightarrow$ (2): Let R denote a continuous linear right inverse for $P(D)$ on $\mathcal{E}_{(\omega)}(\Omega)$. The continuity of R implies:

(9) For each $K \in \mathcal{K}(\Omega)$ there exist $Q \in \mathcal{K}(\Omega)$ and $m \in \mathbb{N}$ such that $\|R(F)\|_{K,1} \leq \|f\|_{Q,m}$ for all $f \in \mathcal{E}_{(\omega)}(\Omega)$.

To derive (2) from this, we fix $K \in \mathcal{K}(\Omega)$ and choose $K' \in \mathcal{K}(\Omega)$ so that $(K')^\circ \supset K+Q$. Then we choose $\psi \in \mathcal{D}_{(\omega)}((K')^\circ)$ with $\psi|_{K+Q} \equiv 1$. Next we fix $f \in N_{(\omega)}(K')$. Then ψf is in $\mathcal{E}_{(\omega)}(\Omega)$, if we define it to be zero outside of the support of ψ. Since $P(D)(\psi f)|_Q \equiv 0$, we get from (9) that $h := R(P(D)(\psi f))$ vanishes on K'. Then $g := \psi f - h$ satisfies

$$P(D)g = P(D)(\psi f) - P(D)R(P(D)(\psi f)) = 0 ,$$

and hence $g \in N_{(\omega)}(\Omega)$. We also have

$$g|_K = (\psi f)|_K - h|_K = f|_K .$$

(2) $\Rightarrow$ (3): If $K \in \mathcal{K}(\Omega)$ is given we choose K' according to (2) and we fix $K'' \in \mathcal{K}(\Omega)$ with $K'' \supset K'$. Then we define the maps $\rho : N_{(\omega)}(K'') \to N_{(\omega)}(K)$, $\sigma : N_{(\omega)}(K') \to N_{(\omega)}(K)$ and $\tau : N_{(\omega)}(K'') \to N_{(\omega)}(K')$ to be the obvious restrictions, which are continuous and linear. By a standard factorization argument we get the following commutative diagram with continuous linear maps:

$$\begin{array}{ccc} N_{(\omega)}(K'')/\ker\rho & \xrightarrow{\;\overline{\tau}\;} & N_{(\omega)}(K')/\ker\sigma \\ & {\scriptstyle\overline{\rho}}\searrow \qquad \swarrow{\scriptstyle\overline{\sigma}} & \\ & N_{(\omega)}(K) & \end{array}$$

Now note that $\overline{\tau}$ is injective by its definition. From (2) it follows that $\overline{\tau}$ is also surjective. Hence $\overline{\tau}$ is an isomorphism by the open mapping theorem for Fréchet spaces. Next we note that $\mathcal{E}_{(\omega)}(Q)$ and hence $N_{(\omega)}(Q)$ is a Fréchet–Schwartz space for each $Q \in \mathcal{K}(\mathbb{R}^n)$, which can be derived easily from 3.4(2). Therefore, we can identify the strong dual of $N_{(\omega)}/F$ with $F^{\perp} \subset N_{(\omega)}(Q)'$ for each closed subspace F of $N_{(\omega)}(Q)$. Hence we get by duality the following commutative diagram

$$\begin{array}{ccc} N_{(\omega)}(K')' \supset (\ker\sigma)^{\perp} & \xrightarrow{\;\overline{\tau}\,'\;} & (\ker\rho)^{\perp} \subset N_{(\omega)}(K'')' \\ & {\scriptstyle\overline{\rho}\,'}\nwarrow \qquad \nearrow{\scriptstyle\overline{\sigma}\,'} & \\ & N_{(\omega)}(K)' & \end{array}$$

where $\overline{\tau}'$ is a linear topological isomorphism. Applying the Fourier–Laplace transform to this diagram and using Lemma 3.4(3), we obtain (3).

(3) $\Rightarrow$ (4): First we note that $A_{\overline{Q}}(V(P))$ is a (DFS)–space for each $Q \in \mathcal{K}(\mathbb{R}^n)$. Hence a subset B of $A_{\overline{Q}}(V(P))$ is bounded if and only if there exist $m \in \mathbb{N}$ and $C > 0$ so that each $f \in B$ satisfies

$$|f(z)| \leq C \exp\left(h_Q(\operatorname{Im} z) + m\omega(z)\right) \text{ for all } z \in V(P)\,.$$

To prove (4), let $K \in \mathcal{K}(\Omega)$ be given. Then we choose $K' \in \mathcal{K}(\Omega)$ according to (3) and fix $K'' \in \mathcal{K}(\Omega)$ with $K'' \supset K'$. Next we note that the set

$$\begin{aligned} M := \{f \in A(V(P)) : {} & \log|f(z)| \leq h_{K''}(\operatorname{Im} z) \text{ and} \\ & \log|f(z)| \leq h_K(\operatorname{Im} z) + O(\omega(z))\,,\ z \in V(P)\} \end{aligned}$$

is contained in $A_{\overline{K}}(V(P))$ and bounded in $A_{\overline{K''}}(V(P))$. Consequently it is bounded in $A_{\overline{K'}}(V(P))$ by (3). Hence there exists $B > 0$ such that

$$M \subset \{f \in A(V(P)) : \log|f(z)| \leq h_{K'}(\operatorname{Im} z) + B\omega(z)\}\,.$$

This proves APL$(\Omega, (\omega))$, since each $f \in A(V(P))$ which satisfies (α) and (β) belongs to M and consequently satisfies (γ) of APL$(\Omega, (\omega))$.

(4) $\Rightarrow$ (5): This holds by [23], Thm. 6.2.

(5) $\Rightarrow$ (6): For a convex compact subset Q of $\mathbb{R}^n$ and $k \in \mathbb{N}$ we let

$$U_k^Q := \{f \in A_Q^+(V(P)) : \|f\|_k \leq 1\} .$$

If $K \in \mathcal{K}(\Omega)$ is given, we choose $K' \in \mathcal{K}(\Omega)$, $K' \supset K$, according to PL$(\Omega, (\omega))$ and fix $K'' \in \mathcal{K}(\Omega)$ with $K'' \supset K'$. Since the inclusion $A_{K'}^+(V(P)) \hookrightarrow A_{K''}^+(V(P))$ is continuous, it suffices to show that for each $p \in \mathbb{N}$ there exists $m \in \mathbb{N}$ such that

$$U_m^{K''} \cap A_K^+(V(P)) \subset U_p^{K'} .$$

To prove this, let $p \in \mathbb{N}$ be given. Then we choose $B \in \mathbb{N}$ according to PL$(\Omega, (\omega))$ and let $m := B + p$. Next we fix $f \in U_m^{K''} \cap A_K^+(V(P))$ and define

$$v(z) := \log |f(z)| + m\omega(z) , \; z \in V(P) .$$

Obviously, v is plurisubharmonic on $V(P)$. Since f is in $A_K^+(V(P))$ the function v satisfies

$$(\alpha) \qquad v(z) \leq h_K(\operatorname{Im} z) + O(\omega(z)) , \; z \in V(P) .$$

Since f is in $U_m^{K''}$, v satisfies

$$(\beta) \qquad v(z) \leq h_{K''}(\operatorname{Im} z) - m\omega(z) + m(z) = h_{K''}(\operatorname{Im} z) , \; z \in V(P) .$$

Therefore, PL$(\Omega, (\omega))$ implies

$$(\gamma) \qquad \log |f(z)| + m\omega(z) = v(z) \leq h_{K'}(\operatorname{Im} z) + B\omega(z) , \; z \in V(P)$$

and hence

$$\log |f(z)| \leq h_{K'}(\operatorname{Im} z) - p\omega(z) , \; z \in V(P) .$$

(6) $\Rightarrow$ (7): This will be shown in two steps. At first we prove that (6) implies

(10) For each $K \in \mathcal{K}(\Omega)$ there exists $L \in \mathcal{K}(\Omega)$ with $L \supset K$ such that for each $K'' \in \mathcal{K}(\Omega)$ with $K'' \supset L$ and each $\mu \in \mathcal{N}_{(\omega)}(L)$ there exists $\nu \in \mathcal{N}_{(\omega)}(K'')$ satisfying $\mu|_K = \nu|_K$.

To prove this, let $K \in \mathcal{K}(\Omega)$ be given and choose $K' \in \mathcal{K}(\Omega)$ with $K' \supset K$ according to (6). Then choose $\varepsilon > 0$ so small that $L := K' + B_\varepsilon \in \mathcal{K}(\Omega)$ and fix $K'' \supset L$. Next let $\mu \in \mathcal{N}_{(\omega)}(L) = \mathcal{N}_{(\omega)}(K' + B_\varepsilon)$. By Lemma 5.9 there exists $\lambda \in A_{K'}^+(V(P))'$ such that by 3.4(1) we have

$$(11) \qquad \lambda(\mathcal{F}(\psi)|_{V(P)}) = \mu(\psi) \text{ for all } \psi \in \mathcal{D}_{(\omega)}(K') .$$

Because of (6), the restriction of λ to $A_K^+(V(P))$ is continuous for the topology induced by $A_{K''}^+(V(P))$. Therefore, the theorem of Hahn–Banach implies the existence of $\kappa \in A_{K''}^+(V(P))'$ satisfying

$$(12) \qquad \lambda|_{A_K^+(V(P))} = \kappa|_{A_K^+(V(P))} .$$

Now note that by Lemma 3.4(1), the map

$$\nu(\chi) := \kappa(\mathcal{F}(\chi)|_{V(P)}) , \; \chi \in \mathcal{D}_{(\omega)}(K'')$$

is in $\mathcal{D}'_{(\omega)}(K'')$. In fact, ν is in $\mathcal{N}_{(\omega)}(K'')$ since it satisfies

$$P(D)\nu[\varphi] = \nu[P(D)^t\varphi] = \kappa(\mathcal{F}(P(-D)\varphi)|_{V(P)}) = \kappa(\check{P}\mathcal{F}(\varphi)|_{V(P)}) = 0$$

for all $\varphi \in \mathcal{D}_{(\omega)}(K'')$. Next note that $\mathcal{F}(\varphi)|_{V(P)}$ is in $A_K^+(V(P))$ for each $\varphi \in \mathcal{D}_{(\omega)}(K)$, by Lemma 3.4(1). Therefore, (11) and (12) imply

$$\nu(\varphi) = \kappa(\mathcal{F}(\varphi)|_{V(P)}) = \lambda(\mathcal{F}(\varphi)|_{V(P)}) = \mu(\varphi) \text{ for all } \varphi \in \mathcal{D}_{(\omega)}(K)\,,$$

which proves (10).
To derive (7) from (10), let $\varepsilon > 0$ be given so that $\Omega_\varepsilon \neq \emptyset$. Then let $K := \overline{\Omega}_\varepsilon$ and choose $L_1 \in \mathcal{K}(\Omega)$, $L_1 \supset K$, according to (10). Next find $\delta > 0$ with $\Omega_\delta \supset L_1$ and fix $\mu \in \mathcal{N}_{(\omega)}(\Omega_\delta)$. Then apply (10) with $K = L_1$ to find $L_2 := L(L_1) \supset L_1$. By (10) there exist $\nu_1 \in \mathcal{N}_{(\omega)}(L_2)$ with $\nu_1|_{\overline{\Omega}_\varepsilon} = \mu|_{\overline{\Omega}_\varepsilon}$. Applying the same argument inductively, we find an increasing sequence $(L_m)_{m\in\mathbb{N}}$ in $\mathcal{K}(\Omega)$ with $\bigcup_{m\in\mathbb{N}} \mathring{L}_m = \Omega$ and a sequence $(\nu_m)_{m\in\mathbb{N}}$ with $\nu_m \in \mathcal{N}_{(\omega)}(L_{m+1})$ satisfying $\mu_{m+1}|_{L_m} = \nu_m|_{L_m}$ for all $m \in \mathbb{N}$. Obviously the sequence $(\nu_m)_{m\in\mathbb{N}}$ convergences in $\mathcal{D}'_{(\omega)}(\Omega)$. Its limit ν belongs to $\mathcal{N}_{(\omega)}(\Omega)$ and satisfies

$$\nu|_{\Omega_\varepsilon} = \nu_1|_{\Omega_\varepsilon} = \mu|_{\Omega_\varepsilon}\,,$$

which proves (7).

(7) $\Rightarrow$ (8) $\Rightarrow$ (1): This holds by Theorem 2.5.

Using the ideas of the proof of Theorem 3.3 one can show that in the Roumieu case the completely analogous result to Theorem 3.3 holds, too. However, by Corollary 2.9 this can be obtained more easily by a reduction to the Beurling case. More precisely, we have:

3.10 Theorem. *Let Ω be a convex open subset of $\mathbb{R}^n$, ω a weight function satisfying $\log = o(\omega)$ and $P \in \mathbb{C}[z_1, \ldots, z_n]$, non-constant. Then the following assertions are equivalent:*

(1) $P(D) : \mathcal{E}_{\{\omega\}}(\Omega) \to \mathcal{E}_{\{\omega\}}(\Omega)$ admits a continuous linear right inverse.

(2) There exists a weight function σ with $\sigma = o(\omega)$ such that P satisfies $\mathrm{PL}(\Omega, (\sigma))$.

(3) P satisfies $\mathrm{PL}(\Omega, \{\omega\})$.

(4) P satisfies the following condition $\widetilde{\mathrm{PL}}(\Omega, \{\omega\})$:

For each $K \in \mathcal{K}(\Omega)$ there exists $K' \in \mathcal{K}(\Omega)$ such that for each $K'' \in \mathcal{K}(\Omega)$ and each $l \in \mathbb{N}$ there exist $m \in \mathbb{N}$ and $C > 0$ such that each $u \in$ $\mathrm{PSH}(V(P))$ *satisfying (α) and (β) also satisfies (γ), where*

(α) $u(z) \le h_K(\operatorname{Im} z) + o(\omega)\,,\ z \in V(P)$

(β) $u(z) \le h_{K''}(\operatorname{Im} z) + \frac{1}{m}\omega(z)\,,\ z \in V(P)$

(γ) $u(z) \le h_{K'}(\operatorname{Im} z) + \frac{1}{l}\omega(z) + C\,,\ z \in V(P)\,.$

For the proof of Theorem 3.10 we refer to [25], sect.6.

Remark. If a variety $V(P)$ satisfies the condition $\widetilde{\mathrm{PL}}(\mathbb{R}^n,\{\omega\})$ then it also satisfies the Phragmén-Lindelöf condition $(\mathrm{PL})_\omega$ of Braun, Meise and Vogt [8], 3.7, which by [8] characterizes the surjectivity of $P(D)$ on $\mathcal{E}_{\{\omega\}}(\mathbb{R}^n)$. For a systematic study of the condition $(\mathrm{PL})_\omega$ we refer to Braun, Meise and Vogt [9].

In the next section we will evaluate the Phragmén–Lindelöf conditions $\mathrm{PL}(\Omega,(\omega))$ and $\mathrm{PL}(\Omega,\{\omega\})$ further.

4 Phragmén–Lindelöf conditions

In this section we outline the main results of [25] and we rephrase them in terms of P-convexity with $*$-bounds. Before we go into this, recall the classical Phragmén–Lindelöf theorem and its local version.

4.1 Theorem. *If $u:\mathbb{C}\longrightarrow[-\infty,\infty[$ is subharmonic and satisfies the conditions (α) and (β) then it also satisfies (γ), where*

(α) $u(z)\le|z|+o(|z|)\ ,\ z\in\mathbb{C}$

(β) $u(z)\le 0\ ,\ z\in\mathbb{R}$

(γ) $u(z)\le|\operatorname{Im} z|\ ,\ z\in\mathbb{C}.$

PROOF. Assume that the subharmonic function u satisfies (α) and (β). Then for each $\varepsilon>0$ there exists $D_\varepsilon>0$ such that

$$u(z)\le(1+\varepsilon)|z|+D_\varepsilon \text{ for all } z\in\mathbb{C}.$$

Now fix $\varepsilon>0$ and note that the function

$$v_\varepsilon(z):=u(z)-(1+\varepsilon)\operatorname{Im} z,\ z\in\mathbb{C}$$

is subharmonic and satisfies

$$v_\varepsilon(z)\le 0 \text{ for } x\in\mathbb{R} \text{ and } v_\varepsilon(iy)\le D_\varepsilon \text{ for } y\ge 0.$$

To apply the arguments of the proof of Hörmander [16], 16.3.11, to this function, we fix $\eta>0$ and let

$$v_{\varepsilon,\eta}(z):=v_\varepsilon(z)-\eta\operatorname{Re}(ze^{-i\pi/4})^{3/2},\ \operatorname{Im} z\ge 0.$$

Then $v_{\varepsilon,\eta}$ is subharmonic in the open upper half plane. Applying the maximum principle to $v_{\varepsilon,\eta}$ on the set

$$G(R):=\{z\in\mathbb{C}:\operatorname{Re} z>0,\ \operatorname{Im} z>0,\ |z|<R\}$$

for sufficiently large $R>0$, we get for each $\eta>0$:

$$v_{\varepsilon,\eta}(z)\le D_\varepsilon \text{ for } z\in\mathbb{C} \text{ with } \operatorname{Im} z\ge 0 \text{ and } \operatorname{Re} z\ge 0.$$

Since we can argue similarly for the other quadrant, this implies

$$v_\varepsilon(z) \le D_\varepsilon \text{ for } \operatorname{Im} z \ge 0.$$

Next fix $\delta > 0$ and note that

$$v_\varepsilon(z) - \delta \log|z+i|$$

is subharmonic in the upper half plane. Moreover, this function is bounded above by zero and tends to $-\infty$ as $|z|$ tends to ∞. Hence the maximum principle implies as before

$$u(z) - (1+\varepsilon)\operatorname{Im} z = v_\varepsilon(z) \le 0 \text{ for } \operatorname{Im} z \ge 0.$$

Since this holds for each $\varepsilon > 0$, we conclude

$$u(z) \le |\operatorname{Im} z| \text{ for } \operatorname{Im} z \ge 0.$$

Arguing similarly for $\operatorname{Im} z \le 0$ we obtain (γ).

The following local version of the Pragmén–Lindelöf theorem is a consequence of the estimates for the harmonic measure of the half disk, given in the proof of Ahlfors [1], Thm.3.4.

4.2 Theorem. *Let $D_+ := \{z \in \mathbb{C} : \operatorname{Im} z \ge 0, |z| \le 1\}$ and let u be subharmonic in a neighborhood of D_+. If u satisfies (α) and (β) then it also satisfies (γ), where*

(α) $u(z) \le 1$, $z \in D_+$

(β) $u(z) \le 0$, $z \in \mathbb{R} \cap D_+$

(γ) $u(iy) \le \frac{4}{\pi} y$, $0 \le y \le 1$.

In 1973 Hörmander [15] observed that properties of algebraic varieties similar to those of the classical Phragmén–Lindelöf theorem are important in studying certain properties of linear partial differential operators with constant coefficients. More precisely, he proved the following theorem.

4.3 Theorem. *For a polynomial $P \in \mathbb{C}[z_1, \ldots, z_n]$ let P_m denote its principal part. Let Ω be a convex open set and let $\mathcal{A}(\Omega)$ denote the real–analytic functions on Ω having complex values. Then $P(D) : \mathcal{A}(\Omega) \longrightarrow \mathcal{A}(\Omega)$ is surjective if and only if $V(P_m)$ satisfies the following condition* HPL(Ω)*:*
For each $K \in \mathcal{K}(\Omega)$ there exist $K' \in \mathcal{K}(\Omega)$ and $\delta > 0$ so that each $u \in \mathrm{PSH}(V(P_m))$ satisfying (α) and (β) also satisfies (γ), where

(α) $u(z) \le h_K(\operatorname{Im} z) + \delta|z|$, $z \in V(P_m)$

(β) $u(z) \le 0$, $z \in V(P_m) \cap \mathbb{R}^n$

(γ) $u(z) \le h_{K'}(\operatorname{Im} z)$, $z \in V(P_m)$.

Hörmander [15] called this condition a Phragmén–Lindelöf principle for the homogeneous variety $V(P_m)$. He also proved several equivalent conditions and introduced a local version of the PL–condition (see 4.4(b) below). In fact, Hörmander's result in connection with the results of Meise and Vogt [29] and Braun, Meise and Vogt [7] on convolution operators led us to the guess that the existence of a continuous linear right inverse for $P(D)$ might be equivalent to a certain Phragmén–Lindelöf condition for the variety $V(P)$.

In order to outline how the condition PL$(\Omega, *)$ is related to the condition HPL(Ω) we concentrate our attention first to homogeneous varieties. For such varieties PL$(\Omega, *)$ is fairly well understood by the work in [25]. To state the main results of [25], we introduce the following definitions.

4.4 Definition. Let $P \in \mathbb{C}[z_1, \ldots, z_n]$ be homogeneous and let Ω be an open convex subset of $\mathbb{R}^n$.
(a) We say that P or $V(P)$ satisfies the condition PL(Ω) if the following holds:
For each $K \in \mathcal{K}(\Omega)$ there exists $K' \in \mathcal{K}(\Omega)$ so that each $u \in \mathrm{PSH}(V)$ satisfying (α) and (β) also satisfies (γ), where

(α) $u(z) \le h_K(\operatorname{Im} z) + o(|z|)$, $z \in V$

(β) $u(z) \le 0$, $z \in V \cap \mathbb{R}^n$

(γ) $u(z) \le h_{K'}(\operatorname{Im} z)$, $z \in V$.

(b) We say that $V(P)$ satisfies the condition HPL(Ω, loc) at $\xi \in V(P) \cap \mathbb{R}^n$ if the following holds:
There exist open sets $U_1 \subset\subset U_2 \subset\subset U_3 \subset \mathbb{C}^n$ with $\xi \in U_1$ so that for each $K \in \mathcal{K}(\Omega)$ there exist $K' \in \mathcal{K}(\Omega)$ and $\delta > 0$ so that each $u \in \mathrm{PSH}(U_3 \cap V)$ satisfying (α) and (β) also satisfies (γ), where

(α) $u(z) \le h_K(\operatorname{Im} z) + \delta$, $z \in U_3 \cap V$

(β) $u(z) \le 0$, $z \in U_2 \cap V \cap \mathbb{R}^n$

(γ) $u(z) \le h_{K'}(\operatorname{Im} z)$, $z \in U_1 \cap V$.

4.5 Theorem. *For each non-constant homogeneous* $P \in \mathbb{C}[z_1, \ldots, z_n]$ *and each convex open subset* Ω *of* $\mathbb{R}^n$ *the following conditions are equivalent:*

(1) P satisfies PL(Ω).

(2) P satisfies PL$(\Omega, (\omega))$ *for some/all weight functions* ω.

(3) P satisfies PL$(\Omega, \{\omega\})$ *for some/all weight functions* ω *satisfying* $\log = o(\omega)$.

(4) $V(P)$ *satisfies* HPL(Ω, loc) *at each* $\xi \in V \cap \mathbb{R}^n$.

(5) $V(P)$ *satisfies* HPL(Ω, loc) *at zero.*

For the proof of Theorem 4.5 we refer to [25], 3.3, where it is shown that the conditions (1), (2), (4) and (5) are equivalent. From this, Corollary 2.9 and Theorem 2.5 we get that also (3) is equivalent to any one of these conditions.

The condition PL($\mathbb{R}^n$) is very similar to Theorem 4.1, since a homogeneous polynomial $P \in \mathbb{C}[z_1, \dots, z_n]$ satisfies PL($\mathbb{R}^n$) if and only if there exists $A \geq 1$, so that each $u \in \mathrm{PSH}(V(P))$ satisfying (α) and (β) also satisfies (γ), where

(α) $u(z) \leq |\operatorname{Im} z| + o(|z|)$, $z \in V(P)$

(β) $u(z) \leq 0$, $z \in V(P) \cap \mathbb{R}^n$

(γ) $u(z) \leq A|\operatorname{Im} z|$, $z \in V$.

In the case $\Omega = \mathbb{R}^n$ the conditions (1)-(5) in Theorem 4.5 are equivalent to the one in the following theorem, which is a reformulation of [25], Thm.3.13.

4.6 Theorem. *Let $P \in \mathbb{C}[z_1, \dots, z_n]$ be homogeneous and non-constant. Then $\mathbb{R}^n$ is P-convex with $*$-bounds for each choice of $*$ if and only if the following two conditions are fulfilled:*

(i) $\dim_{\mathbb{R}} V(Q) \cap \mathbb{R}^n = n - 1$ *for each irreducible factor Q of P.*

(ii) $V(P)$ satisfies the condition HPL(Ω, *loc*) *at each $\xi \in V \cap \mathbb{R}^n$ with $|\xi| = 1$.*

As a corollary of Theorem 4.6 we obtain the following comparison between PL($\mathbb{R}^n$) and HPL($\mathbb{R}^n$) (see [25], 3.14):

4.7 Corollary. *For a homogeneous polynomial $P \in \mathbb{C}[z_1, \dots, z_n]$ the following assertions are equivalent:*

(1) $\mathbb{R}^n$ is P-convex with bounds.

(2) $V(P)$ satisfies the condition PL($\mathbb{R}^n$).

(3) $V(P)$ satisfies HPL($\mathbb{R}^n$) *and none of its irreducible components W is elliptic (i.e. $W \cap \mathbb{R}^n = \{0\}$).*

Using Theorem 4.2 one derives from Theorem 4.6 the following corollary (see [22], 4.8).

4.8 Corollary. *Let $P \in \mathbb{C}[z_1, \dots, z_n]$ be irreducible and homogeneous. If P has real coefficients and satisfies*

$$\operatorname{grad} P(\xi) \neq 0 \text{ for each } \xi \in V(P) \cap \mathbb{R}^n \quad , \quad |\xi| = 1$$

then P is either elliptic or $\mathbb{R}^n$ is P-convex with bounds.

From Corollary 4.8 one derives the following examples (see [22] 4.9):

4.9 Examples. For $n, m \in \mathbb{N}$ with $n \geq 2$ consider the following polynomials in $\mathbb{C}[z_1, \ldots, z_n]$ that are homogeneous of degree m:

$$P(z_1, \ldots, z_n) := \sum_{k=1}^{n} a_k z_k^m \ , \ a_k \in \mathbb{R} \setminus \{0\} \text{ for } 1 \leq k \leq n \, .$$

(a) $\mathbb{R}^2$ is P–convex with bounds for such a polynomial if and only if $m = 1$, or $m = 2$ and sign $a_1 \neq$ sign a_2 .
(b) $\mathbb{R}^n$ for $n \geq 3$ is P–convex with bounds for such a polynomial if and only if m is odd or there exist j, l with sign $a_j \neq$ sign a_l .
(c) From (b) one obtains examples of polynomials P on $\mathbb{C}^n$ that are not hyperbolic with respect to any direction and for which $\mathbb{R}^n$ is P–convex with bounds, namely for $n = 3 : z_1^3 + z_2^3 + z_3^3$ and for $n = 4 : z_1^4 + z_2^4 - z_3^4 - z_4^4$, the ultrahyperbolic polynomial.

The case of the nonhomogeneous polynomials is not as well understood. In the general situation, only necessary conditions are known and a satisfactory characterization is missing. From [25], 4.1 and Corollary 2.12 we get the following necessary condition:

4.10 Theorem. *Let $P \in \mathbb{C}[z_1, \ldots, z_n]$ be non-constant, let P_m denote its principal part and let Ω be a convex open subset of $\mathbb{R}^n$. If Ω is P–convex with $*$–bounds then Ω is P_m–convex with $*$–bounds.*

For further necessary conditions we refer to [25], sect.4.

4.11 Corollary. *Let $P \in \mathbb{C}[z_1, \ldots, z_n]$ be non-constant and let Ω be an open convex subset of $\mathbb{R}^n$. If Ω is P–convex with $*$–bounds then $P(D) : \mathcal{A}(\Omega) \longrightarrow \mathcal{A}(\Omega)$ is surjective.*

PROOF. Let P_m denote the principal part of P. If Ω is P–convex with $*$–bounds, then Ω is P_m–convex with $*$–bounds by Theorem 4.10. Since P_m is a homogeneous polynomial, Theorem 4.5 implies that P_m satisfies $\text{PL}(\Omega, \log(2+t))$. Therefore P_m satisfies $\text{HPL}(\Omega)$ by [22], 4.12. Since the condition $\text{HPL}(\Omega)$ depends only on the principal part of a polynomial, P satisfies $\text{HPL}(\Omega)$. By Hörmander's result (see Theorem 4.3) this shows that $P(D) : \mathcal{A}(\Omega) \to \mathcal{A}(\Omega)$ is surjective.

Theorem 4.10 suggests that one should treat an arbitrary polynomial P as a perturbation of its principal part P_m , i.e. one would like to solve the following problem.

4.12 Problem. Does there exist a condition $c(P, P_m, \Omega, *)$ so that Ω is P–convex with $*$–bound if and only if Ω is P_m–convex with $*$–bounds and condition $c(P, P_m, \Omega, *)$ holds?

The authors have not been able to find such a condition in the general case. In fact, some reasonable guesses failed because of the following example.

4.13 Example. For $\lambda \in \mathbb{C}$ and $|\lambda| = 1$ let

$$P_\lambda(z_1, z_2, z_3) = \lambda z_1 + z_2 z_3 \, .$$

Then all the polynomials P_λ have the same principal part Q, $Q(z_1, z_2, z_3) = z_2 z_3$, which is the product of two linear functions with real coefficients and hence hyperbolic. Note that $V(Q)$ is the union of two hyperplanes in $\mathbb{C}^3$ which intersect in a complex line. Hence $V(Q)$ has a very mild singularity. From [25], 4.9, and 4.10, it follows that $\mathbb{R}^3$ is P_λ–convex for $\lambda = \pm 1$ and not P_λ–convex for $\operatorname{Im} \lambda \neq 0$.

It is in fact the singularities of the variety $V(P_m)$ of the principal part of P that make it difficult to solve Problem 4.12, as the following partial solution of this problem indicates. For its proof we refer to [25], 5.8.

4.14 Theorem. *Let $P_m \in \mathbb{C}[z_1, ..., z_n]$ be homogeneous of degree m. If* grad $P_m(z) \neq 0$ *for all $z \in \mathbb{C}^n \setminus \{0\}$ then the following assertions are equivalent:*

(1) $\mathbb{R}^n$ is P_m–convex with bounds.

(2) $\mathbb{R}^n$ is $(P_m + Q)$–convex with bounds for each $Q \in \mathbb{C}[z_1, ..., z_n]$ with deg $Q < m$.

(3) $\mathbb{R}^n$ is $(P_m + Q)$–convex with bounds for some $Q \in \mathbb{C}[z_1, ..., z_n]$ with deg $Q < m$.

Using this result, one obtains the following examples.

4.15 Example. For $n \geq 3$ and $m \geq 2$ let P_m be a polynomial of the form

$$P_m(z_1, \ldots, z_n) = \sum_{k=1}^{n} a_k z_k^m \quad , \quad a_k \in \mathbb{R} \setminus \{0\} \text{ for } 1 \leq k \leq n\,.$$

Then $\mathbb{R}^n$ is $(P_m + Q)$–convex with bounds for some/each $Q \in \mathbb{C}[z_1, \ldots, z_n]$ with deg$Q < m$ if and only if either m is odd or m is even and there are j, k with sign $a_j \neq$ sign a_k.

5 P–convexity with $*$–bounds and hyperbolicity conditions

In this section we use Holmgren's uniqueness theorem to evaluate the conditions of Theorem 2.5 for open subsets Ω of $\mathbb{R}^n$ with a non–empty C^1–boundary. We illustrate the basic idea in the important special case of an open half space. To do this we recall the following notation.

Notation. (a) For $N \in \mathbb{R}^n \setminus \{0\}$ we define the *positive (negative) open half space determined by* N as

$$H_{\underset{(-)}{+}}(N) := \{x \in \mathbb{R}^n : \langle x, N \rangle \underset{(<)}{>} 0\}.$$

(b) If $P \in \mathbb{C}[z_1, ..., z_n]$ is non–constant and if P_m denotes the principal part of P then $N \in \mathbb{R}^n$ is said to be *non–characteristic* for P if $P_m(N) \neq 0$.

5.1 Lemma. *Let $P \in \mathbb{C}[z_1, ..., z_n]$ and let $N \in \mathbb{R}^n$ be non-characteristic for P. If $H_-(N)$ is P-convex with $*$-bounds then there exist a closed convex cone Γ with $\Gamma \setminus \{0\} \subset H_+(N)$ and a fundamental solution $E \in \mathcal{D}'_*(\mathbb{R}^n)$ for $P(D)$ which satisfies* $\operatorname{Supp} E \subset \Gamma$.

PROOF. Without restriction we can assume $N = (0, ..., 0, 1)$ so that $H_-(N) = \{x \in \mathbb{R}^n : x_n < 0\}$. Then the sets

$$Q_\varepsilon := \{x \in \mathbb{R}^n : -\frac{1}{\varepsilon} < x_n < -\varepsilon \text{ and } |x_j| < \frac{1}{\varepsilon} \text{ for } 1 \le j < n\}, \ 0 < \varepsilon < 1$$

form a fundamental system of the relatively compact subsets of $H_-(N)$. If $H_-(N)$ is P–convex with $*$–bounds, we get from this and 2.5(2) that the following condition is satisfied:

(1) For each $0 < s < 1$ there exists $0 < t < s$ so that for each $f \in \mathcal{D}'_*(H_-(N), Q_t)$ there exists $g \in \mathcal{D}'_*(H_-(N), Q_s)$ with $P(D)g = f$.

To evaluate this condition we note that N is non–characteristic for P by hypothesis. Hence there exists $0 < \alpha < \pi/4$ such that each non–zero vector in the closed cone

$$\Gamma_\alpha(N) := \{L \in \mathbb{R}^n : |L| \cos \alpha \le \langle L, N \rangle\}$$

is non–characteristic for P. Next we choose $s > 0$ so small that $4s < \tan \alpha$ and we find $0 < t < s$ according to (1). If we let $\xi := (0, ..., 0, -t/2)$ then $\delta_\xi \in \mathcal{D}'_*(H_-(N), Q_t)$. Hence (1) implies the existence of $g \in \mathcal{D}'_*(H_-(N), Q_s)$ with $P(D)g = \delta_\xi$. Now we choose $\rho \in \mathcal{D}_*(\mathbb{R}^n)$ with $\operatorname{Supp} \rho \subset \{x \in \mathbb{R}^n : |x| < 1\}$ and $\int \rho d\lambda = 1$, and we let $\rho_\varepsilon(x) := \varepsilon^{-n}\rho(x/\varepsilon)$. Then $g_\varepsilon := g * \rho_\varepsilon$ is a C^∞–function on the set $\{x \in \mathbb{R}^n : x_n < -\varepsilon\}$ and satisfies $P(D)g_\varepsilon = 0$ there. Next we let

$$\Gamma_- := \{y \in \mathbb{R}^n : |y| \sin \alpha < -y_n = -\langle y, N \rangle\},$$

fix $x_0 = (x', -t/4)$ with $x' \in \mathbb{R}^{n-1}$ and $|x'| = \frac{t}{4} \cot \alpha$ and define

$$\Omega_2 := x_0 + \Gamma_- \text{ and } \Omega_1 := \{y \in x_0 + \Gamma_- : y_n < -s\}.$$

For each $0 < \varepsilon < t/4$ our choice of s implies that we can define $\tilde{g}_\varepsilon \in C^\infty(\Omega_2)$ by $\tilde{g}_\varepsilon$ on $\Omega_2 \cap \{x \in \mathbb{R}^n : x_n > -s - t/4\}$ and by zero on the complement of this set. Then we note that by our choices it is easy to check that each hyperplane which is characteristic for P and satisfies $H \cap \Omega_2 \neq \emptyset$ also satisfies $H \cap \Omega_1 \neq \emptyset$. Hence, Hörmander [14], Thm. 5.3.3, — a consequence of Holmgren's uniqueness theorem — implies that $\tilde{g}_\varepsilon$ is identically zero on $\Omega_2 = x + \Gamma_-$. Since this holds for all $0 < \varepsilon < t/4$, it follows that g vanishes on $(x_0 + \Gamma_-) \cap \{x \in \mathbb{R}^n : x_n > -s - t/4\}$ for all x_0 described above. Therefore we can choose a function ψ in $\mathcal{D}_*(H_-(N))$ that is identically 1 in a suitable neighborhood of ξ so that $T_0 := \psi g \in \mathcal{E}'_*(\mathbb{R}^n)$ satisfies

$$P(D)T_0 = \delta_\xi - S_0,$$

for some $S_0 \in \mathcal{E}'_*(\mathbb{R}^n)$, where $\operatorname{Supp} T_0 \subset \xi + \Gamma$ for

$$\Gamma := \{x \in \mathbb{R}^n : |x| \sin \alpha \le \langle x, N \rangle\}$$

and where

$$\operatorname{Supp} S_0 \subset (\xi + \Gamma) \cap \{x \in \mathbb{R}^n : x_n > -3t/8\}.$$

Translating ξ into zero, we obtain $T_1, S_1 \in \mathcal{E}'_*(\mathbb{R}^n)$ satisfying

$$P(D)T_1 = \delta - S_1, \ \operatorname{Supp} T_1 \subset \Gamma,$$

$$\operatorname{Supp} S_1 \subset \{x \in \Gamma : x_n > t/8\}.$$

Now we define $S_1^0 := \delta$ and S_1^j as the j-fold convolution product of S_1 with itself. Then we note that the series $\sum_{j=0}^{\infty} S_1^j$ converges in $\mathcal{D}'_*(\mathbb{R}^n)$ since the family of supports of its members is locally finite. Hence

$$E := T_1 * (\sum_{j=0}^{\infty} S_1^j)$$

is in $\mathcal{D}'_*(\mathbb{R}^n)$ and $\operatorname{Supp} E \subset \Gamma + \Gamma \subset \Gamma$. From

$$P(D)E = (P(D)T_1) * (\sum_{j=0}^{\infty} S_1^j) = (\delta - S_1) * (\sum_{j=0}^{\infty} S_1^j) = \delta$$

we conclude that E is a fundamental solution for $P(D)$.

The necessary condition that we have obtained in Lemma 5.1 is in fact well–known. To see this, we recall the following definition.

5.2 Definition. Let $P \in \mathbb{C}[z_1, ..., z_n]$ be a polynomial of degree $m > 0$, let P_m denote its principal part and let $N \in \mathbb{R}^n$ be non–characteristic for P. $P(D)$ is called *$*$–hyperbolic with respect to N* if $P(D)$ admits a fundamental solution $E \in \mathcal{D}'_*(\mathbb{R}^n)$ with $\operatorname{Supp} E \subset \overline{H_+(N)}$. $P(D)$ is called *$*$–hyperbolic* if it is $*$–hyperbolic with respect to some non–characteristic direction.

5.3 Remark. (a) $P(D)$ is hyperbolic with respect to $N \in \mathbb{R}^n$ in the sense of Hörmander [14], Def. 5.4.1, if and only if $P(D)$ is $(\log(1+t))$–hyperbolic with respect to N. This follows from Hörmander [14], Thm. 5.6.1 and Thm. 5.6.2.
(b) For a systematic study of $*$–hyperbolicity of linear partial differential operators with constant coefficients we refer to [28]. In particular it is shown there that $P(D)$ is $*$–hyperbolic with respect to N if and only if $P(D)$ is $*$–hyperbolic with respect to $-N$. It is also indicated that $P(D)$ is $*$–hyperbolic with respect to N if and only if there exist a closed convex cone Γ satisfying $\Gamma \setminus \{0\} \subset H_+(N)$ and a fundamental solution $E \in \mathcal{D}'_*(\mathbb{R}^n)$ of $P(D)$ with $\operatorname{Supp} E \subset \Gamma$.

From Lemma 5.1 and the results in [28] we derive:

5.4 Proposition. *For $P \in \mathbb{C}[z_1, ..., z_n]$ and a vector $N \in \mathbb{R}^n$ which is not characteristic for P, the following assertions are equivalent:*

(1) $H_+(N)$ and/or $H_-(N)$ are P–convex with $*$–bounds.

(2) $P(D)$ is $*$–hyperbolic with respect to N.

PROOF. (1) $\Rightarrow$ (2): If $H_-(N)$ is P–convex with $*$–bounds then Lemma 5.1 implies that $P(D)$ is $*$–hyperbolic with respect to N. If $H_+(N) = H_-(-N)$ is P–convex with $*$–bounds then $P(D)$ is $*$–hyperbolic with respect to $-N$ and by 5.3 (b) also with respect to N.
(2) $\Rightarrow$ (1): By 5.3 (b) there exist closed convex cones Γ_+ and Γ_- with $\Gamma_+ \setminus \{0\} \subset H_+(N)$ and $\Gamma_- \setminus \{0\} \subset H_-(N)$ and fundamental solutions E_+ and E_- for $P(D)$ in $\mathcal{D}'_*(\mathbb{R}^n)$ with $\operatorname{Supp} E_+ \subset \Gamma_+$ and $\operatorname{Supp} E_- \subset \Gamma_-$. We use these fundamental solutions to find a continuous linear right inverse for $P(D)$ acting on $\mathcal{E}_*(H_+(N))$. The case $H_-(N)$ is treated in the same way.
First we choose $\psi \in \mathcal{E}_*(H_+(N))$ with

$$\operatorname{Supp} \psi \subset \{x \in \mathbb{R}^n : \langle x, H\rangle < 2\}$$

and $\psi(x) = 1$ for all $x \in \mathbb{R}^n$ with $0 < \langle x, N\rangle \leq 1$. Next we fix $f \in \mathcal{E}_*(\mathbb{R}^n)$. Then $(1-\psi)f$ and ψf are in $\mathcal{E}_*(H_+(N))$ by Braun, Meise and Taylor [6], 4.4. Moreover, $E_+ * (1-\psi)f$ is in $\mathcal{E}_*(\mathbb{R}^n)$ with $\operatorname{Supp}(E_+ * (1-\psi)f) \subset \{x \in \mathbb{R}^n : \langle x, N\rangle \geq 1\}$. Furthermore, $E_- * (\psi f)$, defined by

$$E_- * (\psi f)(x) := \langle E_{-,y}\,, (\psi f)(x-y)\rangle \text{ for } \langle x, H\rangle > 0$$

is in $\mathcal{E}_*(H_+(N))$. Therefore, we can define

$$R : \mathcal{E}_*(H_+(N)) \to \mathcal{E}_*(H_+(N))\ ,\ R(f) := E_+ * (1-\psi)f + E_- * (\psi f).$$

It is easy to check that R is a continuous linear right inverse for $P(D)$. By Theorem 2.5, this is equivalent to (1).

Remark. If N is a characteristic vector for P then $H_+(N)$ may or may not be P–convex with bounds, as [22], 3.6 and 3.7 show. The sufficient condition in [22], 3.6, was recently extended to a more general one in [26].

The argument that we used at the beginning of the proof of Lemma 5.1 is in fact a local one. Therefore it applies not only to half spaces but also to open subsets that have a C^1–boundary. Hence we have the following lemma which can be proved in the same way as [22], 3.1.

5.5 Lemma. *Let Ω be an open subset with C^1-boundary $\partial\Omega \neq \emptyset$. If Ω is P-convex with $*$-bounds then $P(D)$ is $*$-hyperbolic with respect to each vector $N \in \mathbb{R}^n$ that is non-characteristic for P and normal to $\partial\Omega$ at some point.*

From Lemma 5.5 and some further arguments for which we refer to [24], sect. 4, it follows that an operator $P(D)$ has to be very hyperbolic if the open Euclidean unit ball is P-convex with bounds. To formulate the main result in this direction precisely, we recall that for an open set Ω in $\mathbb{R}^n$ with C^1-boundary the *Gauss-map* $G : \partial\Omega \rightarrow S^{n-1}$ is defined by $G(x) = N_x$, where N_x is the outer unit normal to $\partial\Omega$ at x.

5.6 Theorem. *For each non-constant $P \in \mathbb{C}[z_1, ..., z_n]$ the following assertions are equivalent:*

(1) There exists a bounded open subset $\Omega \neq \emptyset$ of $\mathbb{R}^n$ with C^1-boundary which is P-convex with $*$-bounds.

(2) There exists an open subset $\Omega \neq \emptyset$ of $\mathbb{R}^n$ with C^1-boundary and surjective Gauss-map which is P-convex with $*$-bounds.

(3) $P(D)$ is $*$-hyperbolic with respect to each $N \in \mathbb{R}^n$ which is non-characteristic for P.

(4) Each open convex subset of $\mathbb{R}^n$ is P-convex with $*$-bounds.

The polynomials P for which $P(D)$ is hyperbolic with respect to each non-characteristic direction have been characterized by de Christoforis [12]. From his result we get the following corollary (see [22], 3.8).

5.7 Corollary. *For each non-constant $P \in \mathbb{C}[z_1, ..., z_n]$ the following assertions are equivalent:*

(1) There exists an open bounded subset $\Omega \neq \emptyset$ of $\mathbb{R}^n$ with C^1-boundary which is P-convex with bounds.

(2) P and P_m are equally strong in the sense of Hörmander [14], Def. 3.2.1, and P_m is proportional to a product of m linear functions with real coefficients.

(3) Each open convex subset of $\mathbb{R}^n$ is P-convex with bounds.

From Corollary 5.7 we derive a further partial solution of the perturbation problem 4.12. In fact, we have the following proposition.

5.8 Proposition. *Let $P \in \mathbb{C}[z_1, ..., z_n]$ be non-constant, let P_m denote its principal part and let $N \in \mathbb{R}^n$ be non-characteristic for P. Assume that $\Omega \subset \mathbb{R}^n$ is either $H_+(N)$, $H_-(N)$ or a convex bounded open set with C^1-boundary. Then the following assertions are equivalent:*

(1) Ω is P-convex with bounds.

(2) Ω is P_m-convex with bounds and P and P_m are equally strong.

PROOF. Assume first that $\Omega = H_-(N)$; the case $\Omega = H_+(N)$ can be treated in the same way.
(1) $\Rightarrow$ (2): By Theorem 4.10 we know that $\Omega = H_-(N)$ is P_m-convex with bounds. By Lemma 5.1, this and (1) imply that P_m and P are hyperbolic with respect to N. Now Hörmander [16], 12.14.6, shows that P and P_m are equally strong.
(2) $\Rightarrow$ (1): By Lemma 5.1, P_m is hyperbolic with respect to N. Since P and P_m are equally strong, this and Hörmander [16], 12.14.6, implies that P is hyperbolic with respect to N. Hence $\Omega = H_-(N)$ is P-convex with bounds by Lemma 5.4.

If Ω is a convex bounded open set with C^1-boundary then the equivalence of (1) and (2) follows easily from Corollary 5.7 and Theorem 4.10.

The following example shows that Proposition 5.8 does not hold for characteristic half spaces.

5.9 Example. Let $P \in \mathbb{C}[z_1, z_2, z_3]$ be defined as

$$P(z_1, z_2, z_3) := z_1 + z_2 z_3.$$

Then $N := (1, 0, 0)$ is non-characteristic for P. In [26] it is shown that $H_+(N)$ is P-convex with bounds. However, it is easy to check that P and its principal part Q, $Q(z_1, z_2, z_3) = z_2 z_3$, are not equally strong.

By Example 4.9(b) there exist polynomials P on $\mathbb{C}^n$ for $n \geq 3$ that are not hyperbolic with respect to any direction and for which $\mathbb{R}^n$ is P-convex with bounds. For $n = 2$ the situation is different, as the following result shows (see [22], 4.11 and [24], sect. 4).

5.10 Theorem. *For each non-constant $P \in \mathbb{C}[z_1, z_2]$ the following assertions are equivalent:*

(1) $P(D)$ is $*$-hyperbolic with respect to some direction.

(2) $\mathbb{R}^2$ is P-convex with $*$-bounds.

(3) $P(D)$ is $*$-hyperbolic with respect to each non-characteristic direction.

(4) Each open convex set Ω in $\mathbb{R}^2$ is P-convex with $*$-bounds.

References

[1] L.V. Ahlfors: 'Conformal Invariants; Topics in Geometric Function Theory', McGraw-Hill, 1973.

[2] A. Beurling: 'Quasi-analyticity and general distributions', Lectures 4. and 5, AMS Summer Institute Stanford (1961).

[3] J. Bonet, R.W. Braun, R. Meise, B.A. Taylor: 'Whitney's extension theorem for nonquasianalytic classes of ultradifferentiable functions', Studia Math. **99** (1991), 155–184.

[4] R.W. Braun: 'An extension of Komatsu's second structure theorem for ultradistributions', preprint.

[5] R.W. Braun, R. Meise: 'Generalized Fourier expansions for zero-solutions of surjective convolution operators on $\mathcal{D}'_{\{w\}}(\mathbb{R})$, Arch. Math. **55** (1990), 55–63.

[6] R.W. Braun, R. Meise, B.A. Taylor: 'Ultradifferentiable functions and Fourier analysis', Res. Math. **17** (1990), 207–237.

[7] R.W. Braun, R. Meise, D. Vogt: 'Existence of fundamental solutions and surjectivity of convolution operators on classes of ultradifferentiable functions', Proc. London Math. Soc. **61** (1990), 344–370.

[8] R.W. Braun, R. Meise, D. Vogt: 'Applications of the projective limit functor to convolution and partial differential equations', pp. 29–46 in 'Advances in the Theory of Fréchet Spaces', T. Terzioğlu (Ed.), NATO ASI Series C Vol. 287 (Kluwer) 1989.

[9] R.W. Braun, R. Meise, D. Vogt: 'Characterization of the linear partial differential operators with constant coefficients which are surjective on non-quasianalytic classes of Roumieu type, preprint.

[10] D.K. Cohoon: 'Nonexistence of a continuous right inverse for parabolic differential operators', J. Diff. Equ. **6** (1969), 503–511.

[11] D.K. Cohoon: 'Nonexistence of a continuous linear right inverse for linear partial differential operators with constant coefficients', Math. Scand. **29** (1971), 337–342.

[12] M.L. de Cristoforis: 'Soluzioni con lacune di certi operatori differenziali lineari', Rendiconti Accademia Nazionale delle Scienze detta dei XL, Memorie di Matematica **102**, vol. VIII (1984), 137–142.

[13] S. Hansen: 'On the fundamental principle of L. Ehrenpreis', Banach Center Publ. **10** (1983), 185–201.

[14] L. Hörmander: 'Linear Partial Differential Operators', Springer 1963.

[15] L. Hörmander: 'On the existence of real analytic solutions of partial differential equations with constant coefficients', Invent. Math. **21** (1973), 151–182.

[16] L. Hörmander: 'The Analysis of Linear Partial Differential Operators II', Springer 1983.

[17] H. Komatsu: 'Ultradistributions I. Structure theorems and a characterization', J. Fac. Sci. Tokyo Sec. IA, **20** (1973), 25–105.

[18] R. Meise, B.A. Taylor: 'Whitney's extension theorem for ultradifferentiable functions of Beurling type', Ark. Mat. **26** (1988), 265–287.

[19] R. Meise, B.A. Taylor, D. Vogt: 'Equivalence of slowly decreasing conditions and generalized Fourier expansions', Indiana Univ. J. Math. **36** (1987), 729-750.

[20] R. Meise, B.A. Taylor, D. Vogt: 'Charactérisation des opérateurs linéaires aux dérivées partielles avec coefficients constants sur $\mathcal{E}(\mathbb{R}^n)$ admettant un inverse à droite qui est linéaire et continu', C. R. Acad. Paris **307** (1988), 239-242.

[21] R. Meise, B.A. Taylor, D. Vogt: 'Partial differential operators with continuous linear right inverse', pp. 47–62 in 'Advances in the Theory of Fréchet Spaces', T. Terzioğlu (Ed.), NATO ASI Series C Vol. 287 (Kluwer) 1989.

[22] R. Meise, B.A. Taylor, D. Vogt: 'Characterization of the linear partial differential operators with constant coefficients that admit a continuous linear right inverse', Ann. Inst. Fourier (Grenoble) **40** (1990), 619-655.

[23] R. Meise, B.A. Taylor, D. Vogt: 'Equivalence of analytic and plurisubharmonic Phragmén–Lindelöf principles on algebraic varieties', Proceedings of Symposia in Pure Mathematics **52** (1991), 287-308.

[24] R. Meise, B.A. Taylor, D. Vogt: 'Continuous linear right inverses for partial differential operators on non–quasianalytic classes and on ultradistributions', preprint.

[25] R. Meise, B.A. Taylor, D. Vogt: 'Phragmén-Lindelöf principles for algebraic varieties', preprint.

[26] R. Meise, B.A. Taylor, D. Vogt: 'Examples in continuous linear right inverses for partial differential operators', manuscript.

[27] R. Meise, B.A. Taylor, D. Vogt: 'Extremal plurisubharmonic functions of linear growth on algebraic varieties', preprint.

[28] R. Meise, B.A. Taylor, D. Vogt: 'ω-Hyperbolicity of linear partial differential operators with constant coefficients', manuscript.

[29] R. Meise, D. Vogt: 'Characterization of convolution operators on spaces of C^∞-functions admitting a continuous linear right inverse', Math. Ann. **279** (1987), 141–155.

[30] V.P. Palamodov: 'A criterion for splitness of differential complexes with constant coefficients', Geometrical and Algebraical Aspects in Several Complex Variables, C.A. Berenstein and D.C. Struppa (Eds.), EditEl (1991), pp. 265–291.

[31] H.J. Petzsche, D. Vogt: 'Almost analytic extension of ultra-differentiable functions and the boundary values of holomorphic functions', Math. Ann. **267** (1984), 17–35.

[32] F. Treves: 'Locally Convex Spaces and Linear Partial Differential Equations', Springer (1967).

[33] D. Vogt: 'On the solvability of $P(D)f = g$ for vector valued functions, RIMS Kokyoroku **508** (1983), 168 – 182.

[34] D. Vogt: 'Some results on continuous linear maps between Fréchet spaces', p. 349–381 in 'Functional Analysis: Surveys and Recent Results II'; K.D. Bierstedt and B. Fuchssteiner (Eds), North – Holland Mathematics Studies **90** (1984).

Convergent Sequences in Duals of Fréchet Spaces

JOSÉ BONET Departamento de Matemática Aplicada, E.T.S. Arquitectura, Universidad Politécnica de Valencia, E-46071 Valencia, Spain

MIKAEL LINDSTRÖM Department of Mathematics, Åbo Akademi, SF-20500 Åbo, Finland

INTRODUCTION

The aim of this paper is to report on recent progress concerning extensions to Fréchet spaces of the theorem of Josefson and Nissenzweig. Furthermore we characterize some properties of Fréchet spaces in terms of the behaviour of sequences of the topological dual. This article contains some new results as well as new proofs of known results.

The theorem of Josefson and Nissenzweig was proved independently by these two authors in 1975 and can be stated as follows: *A Banach space X is finite dimensional if and only if every sequence in its topological dual X', which is $\sigma(X',X)$-convergent to zero, is also convergent to zero for the norm topology on X'.*

Josefson was interested in a certain question arising in infinite dimensional holomorphy. We refer to Dineen [13] for an excellent description of this topic. To be more precise let E be a complex locally convex space. A subset B of E is called *bounding* if every $\mathbb{C}$-valued holomorphic function is bounded on B, and *limited* if every equicontinuous sequence in E', which tends to zero in $\sigma(E',E)$, also converges uniformly to zero on B. Every bounding subset of a complex locally convex space is limited. The converse does not hold as it was recently shown independently by Josefson [27] and Schlumprecht [39]. It was Dineen who asked if there was an infinite dimensional Banach space X such that its unit ball was bounding. A direct application of the Josefson-Nissenzweig theorem shows that every bounding subset of an infinite dimensional Banach space X has empty interior, and further, that on every infinite dimensional Banach space X there is a holomorphic function which is not bounded on bounded sets (i.e. it is not of bounded type).

In order to obtain extensions of the Josefson-Nissenzweig theorem to Fréchet spaces one has to observe first that the unit ball of a Banach space is simultaneously the typical neighbourhood of zero and also the typical bounded set. Accordingly we are led to two possible extensions of the Josefson-Nissenzweig theorem to Fréchet spaces. In 1980 Jarchow [25] conjectured the following results:

(a) A Fréchet space E is Montel if and only if every $\sigma(E', E)$-null sequence in E' is strongly convergent to zero.
(b) A Fréchet space E is Schwartz if and only if every $\sigma(E', E)$-null sequence in E' converges uniformly to zero on a zero-neighbourhood.

Under the assumption that E, in addition, is separable (a) was proved by Dieudonne and Gomes [12] and the Schwartz case was obtained by Floret [18]. These two conjectures were proved without any further assumption on E in [3, 5, 31].

In section 2 we will present a new proof of these extensions of the theorem of Josefson and Nissenzweig. We also obtain related results which characterize reflexive Fréchet spaces and Fréchet spaces not containing l_1 in terms of the behaviour of sequences in the duals. Corresponding results for DF-spaces are also included. In the last section we mention some applications to infinite holomorphy and spaces of vector valued continuous functions.

NOTATION AND TERMINOLOGY

For basic definitions and notations for locally convex spaces (lcs) we refer to [25], for Banach spaces to [11, 29] and for infinite holomorphy to [13]. For a lcs E over the field $\mathbb{K}$ of complex or real numbers, $\mathcal{U}_E$ denotes a fundamental system of absolutely convex neighbourhoods of zero in E and by $\mathcal{B}_E$ we denote the family of all closed absolutely convex bounded subsets of E. For $U \in \mathcal{U}_E$, we write E_U for the normed space canonically associated with U and ϕ_U for the corresponding canonical mapping $E \to E_U$. We also consider ϕ_U as a map from E into $\hat{E}_U$ the completion of E_U. When $B \in \mathcal{B}_E$ let E_B denote the associated normed space. By E' we mean the topological dual of E. For each $U \in \mathcal{U}_E$ the Banach spaces E'_{U° and $(E_U)'$ are isometrically isomorphic. If E is a Fréchet space we denote by $(\|\cdot\|_k)$ an increasing sequence of seminorms such that $U_k := \{x \in E : \|x\|_k \leq 1\}$, $k \in \mathbb{N}$, form a basis of neighbourhoods of zero in E. We write $E_k := E_{U_k}$, $k \in \mathbb{N}$, and for the Banach space $E'_{U_k^\circ}$ we set E'_k. If E is a lcs the inductive dual E'_i of E is defined by $E'_i := \text{ind}_{U \in \mathcal{U}_E} E'_{U^\circ}$. In particular, for a Fréchet space E, we have that E'_i is the (LB)-space $\text{ind}_k E'_k$, and it coincides with the bornological space associated to the strong dual $E'_b = (E', \beta(E', E))$ of E. By a result of Grothendieck, $E'_i = (E', \beta(E', E''))$ whenever E is a Fréchet space.

For a set Γ, Σ is the σ-algebra of Borel sets of $\{0,1\}^\Gamma$, and μ_Γ is the unique probability Haar measure on the topological group $\{0,1\}^\Gamma$. As usual, we denote the spaces $L_1(\mu_\Gamma)$ and $L_\infty(\mu_\Gamma)$ by $L_1(\{0,1\}^\Gamma)$ and $L_\infty(\{0,1\}^\Gamma)$ respectively. By $\|\cdot\|_1$ we denote the usual norm in $L_1(\{0,1\}^\Gamma)$.

1 A CHARACTERIZATION OF QUASINORMABLE FRÉCHET SPACES

If one compare Jarchow's conjectures (a) and (b) in the introduction, then it becomes clear that it would be desirable to know for which class of Fréchet spaces E every strongly null sequence in the dual E' converges uniformly to zero on some zero-neighbourhood. These Fréchet spaces behave more "like Banach spaces".

Around 1954 Grothendieck [22] introduced the class of quasinormable lcs as a large class with good stability properties containing the usual function and distribution spaces. A lcs E is *quasinormable* if for every $U \in \mathcal{U}_E$ there is $V \in \mathcal{U}_E$, $V \subset U$, such that for each $\varepsilon > 0$ there is $B \in \mathcal{B}_E$ with $V \subset B + \varepsilon U$. We recall that a Fréchet space is Schwartz if and only if it is Montel and quasinormable. By the very definition it follows that every strongly null sequence in E' converges uniformly to zero on some zero-neighbourhood whenever the Fréchet space E is quasinormable. Valdivia [40] proved in 1981 that the converse was true for separable Köthe echelon spaces of order one and he asked if this could be true in general. This question was solved positively by the first author in [3], and it is the content of the next result.

THEOREM 1 A Fréchet space E is quasinormable if and only if every strongly-null sequence in E' converges uniformly to zero on some zero-neighbourhood in E.

Proof: We show that the condition is sufficient for quasinormability. It is easy to see that the assumption is equivalent to the fact that every null sequence in the inductive limit $E'_i = \text{ind}_k E'_k$ is contained in some step E'_k and converges to zero in E'_k (i.e. E'_i is a sequentially retractive inductive limit). By a theorem of Neus (see e.g. [35, 8.5.48]) this implies that for each k there is $l > k$ such that E'_i and E'_l induce the same topology on U_k°. Accordingly we have

$$\forall k \quad \exists l > k \quad \forall m > l \quad \forall \alpha > 0 \quad \exists \beta > 0 : \beta U_m^\circ \cap U_k^\circ \subset \alpha U_l^\circ .$$

Taking polars in E we get

$$\forall k \quad \exists l > k \quad \forall m > l \quad \forall \varepsilon > 0 \quad \exists \lambda > 0 : U_l \subset \lambda U_m + \varepsilon U_k .$$

This is a condition introduced by Vogt and Wagner and which was proved to be equivalent to quasinormability by Meise and Vogt [32]. In [3] it is shown that quasinormability of E follows directly from the condition above by a Mittag-Leffler argument. □

Theorem 1 was used to obtain a proof of the conjecture (b) in [3], but it is also relevant in connection with other questions.

(1) *If the bidual E'' of a Fréchet space E is quasinormable, then E is also quasinormable.*

(2) *A Fréchet space E is quasinormable if every separable subspace is contained in a quasinormable separable subspace of E* [4].

(3) *A Fréchet space E is quasinormable if and only if for one (every) compact subset $K \neq \emptyset$ of E the space of holomorphic germs $H(K)$ on K is a sequentially retractive inductive limit.* This solves an open problem from 1979 of Bierstedt and Meise [2].

(4) *If a complex Fréchet space E satisfies that every $\mathbb{C}$-holomorphic function on E which is bounded on bounded set must be even bounded on a neighbourhood and its multiples, then the space E is quasinormable.* We will come back to this question in section 4.

The *density condition* was introduced by Heinrich in his study of ultrapowers of lcs [23]. This condition was thoroughly investigated for Fréchet and Köthe spaces by Bierstedt and Bonet [1]. It is proved there that a Fréchet space E has the density condition if and only if the bounded subsets of the strong dual E'_b are metrizable. Every quasinormable and every Fréchet-Montel space has the density condition. Peris has observed the following characterization of the density condition in terms of convergent sequences of the dual: *A Fréchet space E has the density condition if and only if there is a bounded subset $B \in \mathcal{B}_E$ such that whenever a bounded sequence in E'_b converges uniformly to zero on B it also converges to zero in E'_b.*

2 ON WEAK*-CONVERGENT CONVEX BLOCKS FOR FRÉCHET SPACES

In this section we consider for simplicity that all lcs are defined over the real field $\mathbf{R}$. The main results (Theorems 5, 9, 10, 12, Propositions 4, 11) remain valid for complex lcs.

In [37] Schlumprecht studied the relation between properties of the weak*-topology of the dual X' of a Banach space and the property of X' containing $L_1(\{0,1\}^\Gamma)$ for a set Γ. His main result is that a dual Banach space which contains bounded sequences without weak*-convergent convex blocks contains a copy of $L_1(\{0,1\}^{\omega_1})$, where ω_1 is the first uncountable ordinal. We show that a Fréchet space E which contains an equicontinuous sequence in E' without a weak*-convergent convex block contains a copy of l_1. For Banach spaces this result is due to Bourgain [6] and Rosenthal [36] respectively. Valdivia [41] proved the theorem for Fréchet spaces with a long proof. Here we obtain it as a consequence of the above mentioned theorem of Schlumprecht [37]. We also include new proofs of characterizations of Fréchet-Montel and Fréchet-Schwartz spaces by sequences in the duals. They constitute new proofs of Jarchow's conjectures (a) and (b). The proofs are different from the original ones in [3, 5, 31]. Characterizations of Fréchet spaces which do not contain l_1 (due to Valdivia [42] and Domanski, Drewnowski [14]) and of reflexive Fréchet spaces (cf. [41]) are also included.

DEFINITION Let E be a lcs and $(u_n) \subset E'$ an equicontinuous sequence. A sequence $(\Sigma_{i=k_n}^{k_{n+1}-1} \alpha_i u_i)$ is called a *convex block* of (u_n), if (k_n) is increasing in $\mathbf{N}$, (α_n) are non-negative real numbers and $\Sigma_{i=k_n}^{k_{n+1}-1} \alpha_i = 1$ for each $n \in \mathbf{N}$. A lcs E is said to have the *weak*-convex block property* (abbreviated by (w^*cb) from now on), if every equicontinuous sequence in E' has a weak*-convergent convex block.

Every lcs E for which equicontinuous subsets of E' are w^*-sequentially compact has the (w^*cb)-property. In particular, every separable lcs or every reflexive Fréchet or DF-space has the (w^*cb)-property. Moreover a direct application of the Hahn-Banach theorem shows that the (w^*cb)-property is inherited by subspaces.

LEMMA 2 A lcs E has property $(w^\star cb)$ if and only if $\hat{E}_U$ has property $(w^\star cb)$ for every $U \in \mathcal{U}_E$.

Proof: First take an equicontinuous sequence (u_n) in E'. Then $u_n \in U^\circ$ for some $U \in \mathcal{U}_E$. Hence, by assumption, there is a convex block (v_n) of (u_n) which converges to some $v \in E'_{U^\circ}$

in $\sigma(E'_{U^\circ}, \hat{E}_U)$ and $v \in U^\circ$. Since $\sigma(E'_{U^\circ}, \hat{E}_U)$ is finer than $\sigma(E', E)$, we have $v_n \to v$ in $\sigma(E', E)$. Conversely, assume that (u_n) is bounded in $E'_{U^\circ} = (\hat{E}_U)'$ for some $U \in \mathcal{U}_E$. We may suppose that $u_n \in U^\circ$ for each n. By assumption there is a convex block (v_n) of (u_n) which converges to some $v \in U^\circ$ for the topology $\sigma(E', E)$. By the Alaoglu-Bourbaki theorem, $(U^\circ, \sigma(E'_{U^\circ}, \hat{E}_U))$ is compact and hence $\sigma(E'_{U^\circ}, \hat{E}_U)$ and $\sigma(E', E)$ coincide on U°. Thus $v_n \to v$ in $(U^\circ, \sigma(E'_{U^\circ}, \hat{E}_U))$, and the proof is complete. □

LEMMA 3 Let E be a lcs without property $(w^\star cb)$. Then there exist a continuous linear mapping $f : E \to L_\infty(\{0,1\}^{\omega_1})$ and a family $\{x_\alpha : \alpha < \omega_1\}$ in E such that $\|f(x_\alpha) - f(x_\beta)\|_1 \geq \theta > 0$ whenever $\alpha, \beta \in [0, \omega_1[, \alpha \neq \beta$.

Proof: According to Lemma 2 there is a $U \in \mathcal{U}_E$ such that $\hat{E}_U$ does not have property $(w^\star cb)$. By Schlumprecht's result in [37], E'_{U° contains a copy of $L_1(\{0,1\}^{\omega_1})$. Now proceeding by transfinite induction as in [38, p. 43], one can show that there exist a continuous linear mapping $S : \hat{E}_U \to L_\infty(\{0,1\}^{\omega_1})$, a $\delta > 0$, and a bounded family $\{y_\alpha : \alpha < \omega_1\}$ in $\hat{E}_U$ such that $\|S(y_\alpha) - S(y_\beta)\|_1 \geq \delta$ whenever $\alpha, \beta \in [0, \omega_1[$ and $\alpha \neq \beta$. Since E_U is dense in $\hat{E}_U$ we can find a bounded family $\{\phi_U(x_\alpha) : \alpha < \omega_1\}$ in E_U with $\|S \circ \phi_U(x_\alpha) - S \circ \phi_U(x_\beta)\|_1 \geq \theta > 0$ for every $\alpha, \beta \in [0, \omega_1[$ with $\alpha \neq \beta$. Now put $f := S \circ \phi_U : E \to L_\infty(\{0,1\}^{\omega_1})$ and the proof is complete. □

The following generalization of the Rosenthal l_1 theorem will be needed later. It was obtained by Diaz in [10]. Recall that a subset B of a lcs E is called *conditionally weakly compact*, if every sequence in B has a weak Cauchy subsequence.

PROPOSITION 4 Let E be a Fréchet space. Every bounded sequence in E has a subsequence that is either weakly Cauchy or equivalent to the unit vector basis of l_1. In particular, a Fréchet space E does not contain a copy of l_1 if and only if every bounded subset of E is conditionally weakly compact.

THEOREM 5 Let E be a Fréchet space not containing a copy of l_1. Then E has property $(w^\star cb)$.

Proof: Assume that E does not have property $(w^\star cb)$. Take a continuous linear mapping $f : E \to L_\infty(\{0,1\}^{\omega_1})$ and a family $\{x_\alpha : \alpha < \omega_1\}$ in E such as in Lemma 3. By Lemma 2 in [31] there is an infinite set I such that (x_{α_n}), $n \in I$, is a bounded sequence in E. Next we claim that the sequence (x_{α_n}) in E has a subsequence $(x_{\alpha_{n_k}})$ which is equivalent to the standard basis of l_1. If not, by Proposition 4, a subsequence $(x_{\alpha_{n_k}})$ is weakly Cauchy in E. Hence $(f(x_{\alpha_{n_k}}))$ is weakly Cauchy in $L_\infty(\{0,1\}^{\omega_1})$. The inclusion mapping of $L_\infty(\{0,1\}^{\omega_1})$ into $L_1(\{0,1\}^{\omega_1})$ factors through a Hilbert space and is therefore weakly compact. Hence, since $L_\infty(\{0,1\}^{\omega_1})$ has the Dunford-Pettis property [11], $(f(x_{\alpha_{n_k}}))$ is norm convergent in $L_1(\{0,1\}^{\omega_1})$, which is a contradiction. This means that E contains a copy of l_1, and the theorem is proved. □

COROLLARY 6 If E is a Fréchet space not containing a copy of l_1, then E is isomorphic to a closed subspace of a product of a sequence of Banach spaces with property $(w^\star cb)$.

The counterexample of Köthe of a Fréchet-Montel space having l_1 as a quotient space shows that the property of having no copy of l_1 is not preserved under Hausdorff quotient. In view of Corollary 6 it is therefore of interest to ask if the property that no Hausdorff quotient of a Fréchet space E contains a copy of l_1 implies that E is isomorphic to a closed subspace of a product of a sequence of Banach spaces not containing a copy of l_1. We have not been able to prove or disprove this statement. The corresponding result characterizing totally reflexive Fréchet spaces as closed subspaces of a product of a sequence of reflexive Banach spaces has been proved by Valdivia in [42]. See also [34] and our Proposition 15 below for related results.

Now we use the $(w^\star cb)$-property to give new proofs of the extensions to Fréchet spaces [3, 5, 31] of the Josefson-Nissenzweig theorem (see conjectures (a) and (b) in the introduction).

LEMMA 7 (a) If E is a Fréchet space, then every limited set in E is conditionally weakly compact.

(b) Let E be a complete lcs with property $(w^\star cb)$. Then every limited subset of E is relatively weakly compact.

Proof: (a) This result follows by a diagonal procedure from the corresponding result for Banach spaces proved in [7].

(b) Let $B \subset E$ be limited. Then $\phi_U(B) \subset \hat{E}_U$ is limited for each $U \in \mathcal{U}_E$. By Lemma 2 we can assume that every $\hat{E}_U$ has property $(w^\star cb)$. Therefore, by proceeding as in the proof of [7, proposition 7], the weak closure of $\phi_U(B)$ is compact in $\sigma(\hat{E}_U, E'_{U^\circ})$ for every $U \in \mathcal{U}_E$. But, since $(E, \sigma(E, E'))$ is closed in $\prod_{U \in \mathcal{U}_E}(\hat{E}_U, \sigma(\hat{E}_U, E'_{U^\circ}))$, the set B is relatively weakly compact in E by Tychonoff's theorem. □

As a direct consequence of Theorem 5 and Lemma 7 (b) we obtain

COROLLARY 8 Let E be Fréchet space not containing l_1. Then every limited subset of E is relatively weakly compact.

THEOREM 9 A Fréchet space E is Montel (respectively Schwartz) if and only if every $\sigma(E', E)$-null sequence in E' is also strongly convergent to zero (respectively converges uniformly to zero on some zero-neighbourhood in E).

Proof: We only need to consider the Montel case. The characterization of Fréchet-Schwartz spaces then follows either from Theorem 1 or [25, 11.6.3], since every Fréchet-Montel space is separable. We assume that every $\sigma(E', E)$-null sequence in E' is strongly convergent to zero. This implies that every bounded subset of E is limited, hence conditionally weakly compact by Lemma 7 (a). Therefore we can apply Proposition 4 to conclude that E does not contain a copy of l_1. According to Corollary 8 it follows that every bounded subset of E is relatively weakly compact. Thus E is reflexive. To conclude that E is Montel it is enough to show that for every $U \in \mathcal{U}_E$, U° is $\beta(E', E)$-sequentially compact (cf. [25, 11.5.2] and a result of Pfister which ensures that the precompact subsets of a DF-space are metrizable to be found, e.g., in [35]). Let (u_n) be a sequence in U°. Since E is reflexive, $\sigma(E', E'') = \sigma(E', E)$. By a result of Cascales and Orihuela [8], E'_b is weakly angelic, and therefore there is a subsequence (v_n) of (u_n) which is weakly convergent to $v \in U^\circ$. By our assumption $v_n \to v$ in $\beta(E', E)$, and the proof is complete. □

For a proof of Theorem 9 based on results of Valdivia about projectional resolutions of identity we refer to [5].

Now we include a characterization of Fréchet spaces not containing l_1 in terms of sequences in the dual space. This result was obtained independently by Valdivia [42] and Domanski,

Drewnowski [14]. We indicate the elegant proof of Domanski and Drewnowski. Also compare with [17].

THEOREM 10 A Fréchet space E does not contain l_1 if and only if every $\mu(E',E)$-null sequence in E' is strongly convergent to zero.

Proof: First assume that l_1 is contained in E. The identity map $i : l_1 \to c_0$ factors through l_∞ [11, p. 223]. Since l_∞ is injective there is a continuous linear map $T : E \to c_0$ whose restriction to l_1 coincide with i and which factors through l_∞. Since l_∞ has the Dunford-Pettis property and every operator from l_∞ to c_0 is weakly compact (see [29, 2.f.4]), we conclude that T maps weakly compact sets to strongly compact sets. From this, using a well-known description of compact sets in c_0, it follows that if (e'_n) are the coefficient functionals on c_0, then the functionals $u_n := e'_n \circ T \to 0$ in $\mu(E',E)$, but not in $\beta(E',E)$.

Now we suppose that E does not contain l_1 and that (u_n) is a sequence in E' which is $\mu(E',E)$-null but not $\beta(E',E)$-null. We select a bounded sequence (x_n) in E such that $|u_n(x_n)| \geq \varepsilon > 0$. By Proposition 4 we may assume w.l.o.g. that (x_n) is a $\sigma(E,E')$-Cauchy sequence. Proceeding by induction we select an increasing sequence (n_k) in $\mathbb{N}$ such that $|u_{n_{k+1}}(x_{n_k})| < \varepsilon/3$ for every $k \in \mathbb{N}$. The closed absolutely convex hull A of $\{x_{n_{k+1}} - x_{n_k} : k \in \mathbb{N}\}$ is $\sigma(E,E')$-compact, hence there is $k_0 \in \mathbb{N}$ with $|u_{n_{k+1}}(z)| < \varepsilon/3$ for every $k \geq k_0$ and $z \in A$. For $k \geq k_0$ we get

$$\varepsilon \leq |u_{n_{k+1}}(x_{n_{k+1}})| \leq |u_{n_{k+1}}(x_{n_{k+1}} - x_{n_k})| + |u_{n_{k+1}}(x_{n_k})| < 2\varepsilon/3,$$

a contradiction. □

In view of the characterization given in the theorem above, one of the extensions of the Josefson-Nissenzweig theorem to Fréchet spaces given in Theorem 9 can now be stated as follows: A Fréchet space E is Montel if and only if it does not contain a copy of l_1 and it satisfies condition $(*)$ below,

$(*)$ every $\sigma(E',E)$-null sequence in E' is also $\mu(E',E)$-null.

Accordingly the following remarks on Fréchet spaces that satisfy condition $(*)$ are in order. A lcs E is called *Schur*, if every $\sigma(E,E')$-convergent sequence also converges in E, or equivalently for an angelic space E, if every $\sigma(E,E')$-compact set is compact in E. A lcs E is called a *Grothendieck space* if weak*-convergent sequences in E' are weakly convergent. A Fréchet space E is said to have the *Dunford-Pettis property* (DP-property) if every continuous linear map from E into a Banach space F which transforms bounded sets into relatively weakly compact sets, transforms absolutely convex $\sigma(E,E')$-compact sets into relatively compact subsets of F (see [16, p. 633]). Proceeding as in [16, p. 635–636] and using that every DF-space is weakly angelic (cf. [8]), it follows that a Fréchet space has the DP-property if and only if for every $(x_n) \subset E$ with $x_n \to 0$ in $\sigma(E,E')$ and every $(u_n) \subset E'$ with $u_n \to 0$ in $\sigma(E',E'')$, we have $u_n(x_n) \to 0$.

PROPOSITION 11 Let E be a Fréchet space.

(a) If E is a Schur space, then E satisfies condition $(*)$. The converse holds if $U°$ is $\sigma(E',E)$-sequentially compact for every $U \in \mathcal{U}_E$, in particular, if E is separable.

(b) If E satisfies $(*)$, then E has the DP-property. The converse holds if E is a Grothendieck space. In particular, l_∞ satisfies condition $(*)$ but it is not a Schur space.

Proof: (a) If E is a Schur space, then the topology $\mu(E',E)$ coincides with the topology of uniform convergence on the compact subsets of E. Accordingly for every $U \in \mathcal{U}_E$ we have $\sigma(E',E)_{|U^\circ} = \mu(E',E)_{|U^\circ}$ and condition $(*)$ holds.

Now assume that U° is $\sigma(E',E)$-sequentially compact for all $U \in \mathcal{U}_E$ and E is not a Schur space. Then we can find $(x_n) \subset E$, $x_n \to 0$ in $\sigma(E,E')$, $(u_n) \subset E'$, $u_n \to u_0 \in E'$ in $\sigma(E',E)$ such that $|u_n(x_n)| \geq \varepsilon > 0$ for every $n \in \mathbb{N}$. By condition $(*)$ $u_n \to u_0$ in $\mu(E',E)$ and the closed absolutely convex hull of (x_n) is $\sigma(E,E')$-compact. This is a contradiction.

(b) By our remarks above about the DP-property it is very easy to see that it is necessary for condition $(*)$. Assume now that E is Grothendieck and has the DP-property. If condition $(*)$ does not hold, there is $(u_n) \subset E'$ which is $\sigma(E',E)$-null, hence $\sigma(E',E'')$-null, but not $\mu(E',E)$-null. By the Eberlein-Smulian theorem, and passing to a subsequence if necessary, we find $(x_n) \subset E$ with $x_n \to x_0 \in E$ in $\sigma(E,E')$ such that $|u_n(x_n)| \geq \varepsilon > 0$ for every $n \in \mathbb{N}$. Since $u_n(x_0) \to 0$ as $n \to \infty$, it follows that $(u_n(x_n - x_0))$ does not tend to zero, and this contradicts the DP-property. □

We recall that for a Fréchet space E the following conditions are equivalent: (i) E is a Schur space, (ii) every $\sigma(E',E)$-relatively compact sequence in E' is $\mu(E',E)$-relatively compact.

The following result is due to Valdivia [41], but we present a direct proof. In [19, p. 41] it is proved that a complete separable lcs is semi-reflexive if and only if it is a Grothendieck space.

THEOREM 12 A Fréchet space E is reflexive if and only if the following two conditions are satisfied

(i) every $\mu(E',E)$-null sequence in E' converges strongly to zero,
(ii) E is a Grothendieck space.

Proof: Clearly if E is reflexive conditions (i) and (ii) are satisfied. Assume that E is not reflexive. By a characterization of weakly compact subsets due to James [24], it follows that there are a bounded sequence (x_n) in E, an equicontinuous sequence (u_n) in E' and $\theta > 0$ such that $u_n(x_k) > \theta$ if $n \leq k$ and $u_n(x_k) = 0$ if $n > k$. By Theorem 10, condition (i) is equivalent to the fact that E does not contain a copy of l_1. We can apply Proposition 4 to select a subsequence (x_{n_i}) which is $\sigma(E,E')$-Cauchy. There is $z \in E''$ such that $x_{n_i} \to z$ as $i \to \infty$ in $\sigma(E'',E')$. Since l_1 is not contained in E, we can conclude that E has property (w^*cb) by Theorem 5. Therefore given $(u_{n_i}) \subset E'$ we can find an increasing sequence (k_i) in $\mathbb{N}$, $\alpha_k \geq 0$, and $v \in E'$ such that $\sum_{k=k_i}^{k_{i+1}-1} \alpha_k = 1$ and $v_i := \sum_{k=k_i}^{k_{i+1}-1} \alpha_k u_{n_k}$, $i \in \mathbb{N}$, satisfies $v_i \to v$ for the topology $\sigma(E',E)$. By (ii), $v_i \to v$ also for the topology $\sigma(E',E'')$, hence, in particular, $v_i(z) \to v(z)$ as $i \to \infty$. Fix $j \in \mathbb{N}$. If $i \in \mathbb{N}$ satisfies $n_{k_i} > n_j$ we have

$$v_i(x_{n_j}) = \sum_{k=k_i}^{k_{i+1}-1} \alpha_k u_{n_k}(x_{n_j}) = 0.$$

This yields $v(x_{n_j}) = 0$ for every $j \in \mathbb{N}$ and, since $x_{n_j} \to z$ for $\sigma(E'',E')$, we get $v(z) = 0$. Fix now $i \in \mathbb{N}$ and select $j \in \mathbb{N}$ with $j > k_{i+1}$. We have

$$v_i(x_{n_j}) = \sum_{k=k_i}^{k_{i+1}-1} \alpha_k u_{n_k}(x_{n_j}) > \theta.$$

Since $x_{n_j} \to z$ for $\sigma(E'', E')$ it follows that $v_i(z) \geq \theta$ for every $i \in \mathbb{N}$. Consequently, we have $v(z) = 0$, $v_i(z) \geq \theta > 0$ for every $i \in \mathbb{N}$ and $z \in E''$. This is a contradiction since $v_i \to v$ in $\sigma(E', E'')$. □

3 ON CONVERGENT SEQUENCES IN DUALS OF QUASINORMABLE SPACES

In this section we obtain an extension to quasinormable spaces of the Josefson-Nissenzweig theorem. Our main aim is to treat DF-spaces. Since every DF-space is quasinormable, our present frame covers the most important examples.

Bourgain and Diestel proved in [7], using basically a Banach space version of Theorem 5, that every limited set in a Banach space not containing l_1 is relatively weakly compact. With this result in hand the second author showed in [30] that a lcs E is Schwartz if and only if for every $U \in \mathcal{U}_E$ there is a $V \in \mathcal{U}_E$ with $V \subset U$ such that the inclusion mapping $\phi'_{UV} : E'_{U^\circ} \to E'_{V^\circ}$ takes $\sigma(E'_{U^\circ}, \hat{E}_U)$-null sequences to E'_{V°-null sequences. Since $\sigma(E', E)$ and $\sigma(E'_{U^\circ}, \hat{E}_U)$ coincide on U° for every $U \in \mathcal{U}_E$, one can easily derive from this characterization of Schwartz spaces the following result:

THEOREM 13 Let E be a lcs and τ' a locally convex topology on E' such that for every $U \in \mathcal{U}_E$ there is a $V \in \mathcal{U}_E$, $V \subset U$, such that the topologies induced on U° by E'_{V° and $E'_{\tau'}$ coincide. If every equicontinuous, $\sigma(E', E)$-null sequence in E' is also $E'_{\tau'}$-null convergent, then E is Schwartz.

COROLLARY 14 (a) A lcs E is Schwartz if and only if E is quasinormable and every equicontinuous, $\sigma(E', E)$-null sequence in E' is also E'_b-null.

(b) Let E be a complete DF-space. Then E is Montel if and only if every $\sigma(E', E)$-null sequence in E' is also strongly convergent to zero.

Proof: (a) Apply Theorem 13 to the strong topology $\beta(E', E)$ on E', and observe that E is quasinormable exactly when for every $U \in \mathcal{U}_E$ there exists a $V \subset U$ in $\mathcal{U}_E$ such that the topologies induced on U° by E'_{V° and E'_b coincide.

(b) This is a direct consequence of (a), since every DF-space is quasinormable and every DF-Schwartz space is quasi-barrelled [25, 12.5.5]. □

Theorem 13 can also be used to obtain the Schwartz part of Theorem 9. In fact, apply Theorem 13 to the inductive dual E'_i, and it is enough to use that the Mackey convergence condition for E'_i implies the strict Mackey convergence condition for E'_i by a result of Neus [35, 8.5.49]. P. Domanski has shown the authors a direct proof of the Josefson-Nissenzweig theorem for quasi-barrelled DF-spaces.

It is natural to ask if Theorem 5 is also valid for quasinormable spaces. Our next proposition gives an answer to this question.

PROPOSITION 15 Let E be a quasinormable lcs such that every bounded set in E is conditionally weakly compact. Then E has property (w^*cb).

Proof: We first show that E has a zero-basis $\mathcal{V}_E$ of closed, absolutely convex sets in E such that $\hat{E}_U$ does not contain a copy of l_1 for any $U \in \mathcal{V}_E$ (cf. Theorem 5 in [34]).

Indeed, since E is quasi-normable and every bounded set is conditionally weakly compact, it follows from Theorem 7.3.6 in [28] that for every $W \in \mathcal{U}_E$ there is a $V \in \mathcal{U}_E$, $V \subset W$, such that $\phi_{WV} : \hat{E}_V \to \hat{E}_W$ takes the unit ball of $\hat{E}_V$ into a conditionally weakly compact set. Since every such mapping factorizes through a Banach space not containing a copy of l_1, the statement follows. Now, the proposition is a direct consequence of Theorem 5 and Lemma 2. □

COROLLARY 16 Let E be a DF-space. If E is a Grothendieck space such that every bounded set in E is conditionally weakly compact, then E'_b is reflexive.

Proof: Assume that the Fréchet space E'_b is non-reflexive and that E is a Grothendieck space. Since E'_b is weakly angelic, there exists a bounded sequence (u_n) in E'_b without $\sigma(E', E'')$-convergent subsequences. Since E is a DF-space the sequence (u_n) in E' is equicontinuous. From the Alaouglu-Bourbaki theorem and the Grothendieck property of E we conclude that (u_n) has no $\sigma(E', E'')$-Cauchy subsequences. Hence, by Proposition 4, we can assume that (u_n) has a subsequence (v_n) which is equivalent to the unit vector basis of l_1. By the Schur property of l_1, no convex block of (v_n) is $\sigma(E', E'')$-convergent. Since E is a Grothendieck space we get a contradiction by Proposition 15. □

4 SOME APPLICATIONS

4.1 Infinite Holomorphy

For a complex lcs E let $H(E)$ denote the space of all holomorphic mappings $f : E \to \mathbb{C}$. A holomorphic mapping $f \in H(E)$ is said to be of *bounded type*, and we write $f \in H_b(E)$, if f is bounded on the bounded subsets of E. A holomorphic mapping $f \in H(E)$ is said to be of *uniformly bounded type*, and we write $f \in H_{ub}(E)$, if f is bounded on all multiples of some neighbourhood of zero in E. Clearly if E is Montel, then $H(E) = H_b(E)$. In [9] Colombeau and Mujica proved that $H(E) = H_{ub}(E)$ for every DFM-space. If $E = H(\mathbb{C})$, then $H(E) \neq H_{ub}(E)$. Meise, Vogt [33] investigated the identity $H(E) = H_{ub}(E)$ for Fréchet nuclear spaces. More recently Galindo, Garcia, Maestre [21] have studied the identity $H_b(E) = H_{ub}(E)$ for arbitrary Fréchet spaces, and they proved that it holds if the bidual E'' of E is a quotient of a countable product of Banach spaces. Our first lemma is well-known [13, 30].

LEMMA 17 Let (u_n) be an equicontinuous sequence in the dual of a complex lcs E. Define $f(x) := \sum_n u_n(x)^n$, $x \in E$. We have

(*a*) $f \in H(E)$ if and only if $u_n \to 0$ in $\sigma(E', E)$,

(*b*) $f \in H_b(E)$ if and only if $u_n \to 0$ in $\beta(E', E)$,

(*c*) $f \in H_{ub}(E)$ if and only if $u_n \to 0$ uniformly on a neighbourhood of zero in E.

PROPOSITION 18 Let E be a Fréchet space.

(*a*) If $H(E) = H_{ub}(E)$, then E is a Schwartz space.

(*b*) $H(E) = H_b(E)$ if and only if E is a Montel space.

(*c*) If $H_b(E) = H_{ub}(E)$, then E is a quasinormable space.

Proof: (*a*) and (*b*) follow from Lemma 17 and Theorem 9. (*c*) is a direct consequence of Lemma 17 and Theorem 1. □

By the examples mentioned above the reverse implications in (a) and (c) do not hold in general.

Our next result can be found in [30]. Its proof depends on Corollary 14.

PROPOSITION 19 (a) If E is a quasinormable lcs and $H(E) = H_{ub}(E)$, then E is a Schwartz space.

(b) Let E be a complete DF-space. Then E is a DFM-space if and only if $H(E) = H_{ub}(E)$.

At this point it is worth mentioning that in [20] it is shown that, $H_b(E) = H_{ub}(E)$ for every DF-space E.

4.2 Spaces of vector valued continuous functions

Let X be a Hausdorff completely regular topological space and let E be a Fréchet space. $C(X, E)$ and $C(X, E, w)$ respectively are the space of all functions $f : X \to E$ that are continuous with respect to the strong and weak topology in E respectively. They are endowed with the topology of uniform convergence on compact sets in X, relative to the strong topology on E. In a recent paper Domanski and Drewnowski [14] have investigated when these spaces are injective or, more generally, complemented in a dual Fréchet space (i.e., in a Fréchet space which is the strong dual of a bornological DF-space). Motivated by the fact that c_0 is never complemented in a dual Fréchet space they proved in [14] the following result.

THEOREM 20 Let X be a Hausdorff completely regular topological space containing an infinite compact set and let E be a Fréchet space.

(a) If E is not Montel, then $C(X, E)$ contains a complemented copy of c_0.

(b) If E contains a copy of l_1, then $C(X, E, w)$ contains a complemented copy of c_0.

The proof of (a) is based on Theorem 9 and the proof of (b) depends on Theorem 10. As a consequence they could prove [14, thms 4 and 6]:

THEOREM 21 Let X be a Hausdorff locally compact space with a fundamental sequence of compact sets which is not discrete and let E be an infinite-dimensional Fréchet space.

(a) $C(X, E, w)$ is complemented in a dual Fréchet space if and only if $C(X)$ is injective and E is reflexive.

(b) $C(X, E)$ is injective (or equivalently $C(X, E, w)$ is injective) if and only if $E = \mathbb{K}^{\mathbb{N}}$ and $C(X)$ is injective.

4.3 Spaces of compact operators

The theorem of Josefson-Nissenzweig is also useful in connection with the following problem: Let E and F be Fréchet or complete DF-spaces. Is it true that the space $K(E, F)$ of compact operators either equals the space $L_b(E, F)$ of all continuous linear operators or is uncomplemented in $L_b(E, F)$? Several authors have studied this question for Banach spaces E and F. We refer to a recent paper of K. John [26] for details and further references. Some results for Fréchet and DF-spaces will appear in a forthcoming paper by Domanski and Drewnowski [15]. We only include here some observations in this direction.

If E is a lcs, then $L(E, c_0) = K(E, c_0)$ if and only if every equicontinuous $\sigma(E', E)$-null sequence converges uniformly to zero on a neighbourhood of zero in E. In particular, if E

is a Fréchet space, then $L(E,c_0) = K(E,c_0)$ if and only if E is Schwartz by Theorem 9. For every non-Schwartz Fréchet-Montel space E we have that $L(E,c_0) \neq K(E,c_0)$. It seems to be unknown if there is a non-Schwartz Fréchet-Montel space E (necessarily without the approximation property) such that $K(E,c_0)$ is complemented, or even closed, in $L_b(E,c_0)$ (recall that if E is a Fréchet-Montel space with the approximation property then $K(E,c_0)$ is dense in $L_b(E,c_0)$).

If E and F are complete lcs, we denote by $M(E,F)$ the set of all continuous linear maps from E into F which takes bounded sets into relatively compact sets. Clearly $M(E,F)$ is a closed subspace of $L_b(E,F)$. If E is quasinormable and F is a Banach space, then $M(E,F) = K(E,F)$. Indeed, given $f \in M(E,F)$ select $U \in \mathcal{U}_E$ with $f(U)$ contained in the unit ball B_F of F. Since E is quasinormable there is $V \in \mathcal{U}_E$, $V \subset U$, such that for all $\varepsilon > 0$ there is $C_\varepsilon \in \mathcal{B}_E$ with $V \subset C_\varepsilon + \varepsilon U$. This implies that the bounded subset $f(V)$ of F satisfies that for all $\varepsilon > 0$ there is a relatively compact subset $K_\varepsilon := f(C_\varepsilon)$ in F with $f(V) \subset K_\varepsilon + \varepsilon B_F$. Thus $f(V)$ is itself relatively compact (see [11,p. 5]).

PROPOSITION 22 (*a*) Let E be a quasinormable lcs. Then $K(E,c_0)$ is complemented in $L_b(E,c_0)$ if and only if E is Schwartz.

(*b*) If E is a Fréchet space such that $K(E,c_0)$ is complemented in $L_b(E,c_0)$, then E is Montel.

(*c*) Let E be a Fréchet space. Then $K(E,c_0) = M(E,c_0)$ if and only if E is quasinormable.

Proof: The proof of (*a*) depends on Corollary 14 and the proof of (*b*) on Theorem 9. These results were given in [30]. The proof of (*c*) follows from Theorem 1 and our remarks above.

Acknowledgement. A major part of this paper was written when the first named author visited Åbo Akademi University in late June and early July 1992, supported by Stiftelsens för Åbo Akademi Forskningsinstitut. The research of J. Bonet was also partially supported by the project of DGICYT PS92-0538. Several friends read the manuscript and contributed with suggestions and ideas. Thanks are given to J.C. Diaz, P. Domanski, A. Galbis and C. Fernández.

REFERENCES

1. Bierstedt, K. D. and Bonet, J., Stefan Heinrich's density condition for Fréchet spaces and the characterization of distinguished Köthe echelon spaces. *Math. Nachr.* 135, 149–180 (1988).
2. Bierstedt, K. D. and Meise, R., Aspects of inductive limits in spaces of germs of holomorphic functions on locally convex spaces and applications to a study of $(H(U), \tau_\omega)$. pp. 111–178 in *Advances in Holomorphy.* North-Holland Math. Studies 34 Amsterdam 1979.
3. Bonet, J., A question of Valdivia on quasi-normable Fréchet spaces. *Canad. Math. Bull.* 34, 301–304 (1991).
4. Bonet, J. and Diaz, J. C., The density condition in subspaces and quotients of Fréchet spaces. Preprint 1992.

5. Bonet, J., Lindström, M. and Valdivia, M., Two theorems of Josefson-Nissenzweig type for Fréchet spaces. To appear in *Proc. Amer. Math. Soc.*
6. Bourgain, J., *La propiété de Radon-Nikodym.* Publications Mathématiques de l'Université Pierre et Marie Currie 36, 1979.
7. Bourgain, J. and Diestel, J., Limited operators and strict cosingularity. *Math. Nachr.* 119, 55–58 (1984).
8. Cascales, B. and Orihuela, J., On compactness in locally convex spaces. *Math. Z.* 195, 365–381 (1987).
9. Colombeau, J. F. and Mujica, J., Holomorphic and differentiable mappings of uniform bounded type. pp. 179–200 in *Functional Analysis, Holomorphy and Approximation Theory.* North-Holland Math. Studies 71 Amsterdam 1982.
10. Diaz, J. C., Montel subspaces in the countable projective limits of $L_p(\mu)$-spaces. *Canad. Math. Bull.* 32, 169–176 (1989).
11. Diestel, J., *Sequences and Series in Banach Spaces.* Springer New York 1984.
12. Dieudonné, J. and Gomes, A.P., Sur certains espaces vectoriels topologiques. *C.R. Acad. Sci. Paris* 230, 1129–1130 (1950).
13. Dineen, S., *Complex Analysis in Locally Convex Spaces.* North-Holland Math. Studies 57 Amsterdam 1981.
14. Domanski, P. and Drewnowski, L., Fréchet spaces of continuous vector-valued functions: Complementability in dual Fréchet spaces and injectivity. *Studia Math.* 102, 257–267 (1992).
15. Domanski, P. and Drewnowski, L., Injectivity of spaces of operators. Preprint 1992.
16. Edwards, R. E., *Functional Analysis.* Holt, Rinehart and Winston New York 1965.
17. Emmanuele, G., A dual characterization of Banach spaces not containing l_1. *Bull. Acad. Sci. Pol. Math.* 34, 155–159 (1986).
18. Floret, K., Lokalkonvexe Sequenzen mit kompakten Abbildungen. *J. Reine Angew. Math.* 247, 155–195 (1971).
19. Floret, K., *Weakly Compact Sets.* Lecture Notes in Math. 801 Springer Berlin 1980.
20. Galindo, P., Garcia, D. and Maestre, M., Holomorphic mappings of bounded type on (DF)-spaces. pp. 135–148 in *Progress in Functional Analysis.* North-Holland Math. Studies 170 Amsterdam 1992.
21. Galindo, P., Garcia, D. and Maestre, M., Entire functions of bounded type on Fréchet spaces. To appear in *Math. Nachr.*
22. Grothendieck, A., Sur les espaces (F) et (DF). *Summa Brasil Math.* 3, 57–123 (1954).
23. Heinrich, S., Ultrapowers of locally convex spaces and applications I. *Math. Nachr.* 118, 211–229 (1984).
24. James, R. C., Weak compactness and reflexivity. *Israel J. Math.* 2, 101–119 (1964).
25. Jarchow, H., *Locally convex spaces.* Teubner Stuttgart 1981.
26. John, K., On the uncomplemented subspace $K(X, Y)$. *Czech. Math. J.* 42, 167–173 (1992).
27. Josefson, B., A Banach space containing non-trivial limited sets but no non-trivial bounding sets. *Israel J. Math.* 71, 321–327 (1990).
28. Junek, H., *Locally Convex Spaces and Operator Ideals.* Teubner Leipzig 1983.
29. Lindenstrauss, J. and Tzafriri, L., *Classical Banach Spaces I.* Springer Berlin 1977.
30. Lindström, M., A characterization of Schwartz spaces. *Math. Z.* 198, 423–430 (1988).

31. Lindström, M. and Schlumprecht, Th., A Josefson-Nissenzweig theorem for Fréchet spaces. To appear in *Bull. London Math. Soc.*
32. Meise, R. and Vogt, D., A characterization of quasinormable Fréchet spaces. *Math. Nachr.* 122, 141–150 (1985).
33. Meise, R. and Vogt, D., Holomorphic functions of uniformly bounded type on nuclear Fréchet spaces. *Studia Math.* 83, 147–166 (1986).
34. Miñarro, M. A., Two results on quasi-normable Fréchet spaces. Preprint 1992.
35. Perez Carreras, P. and Bonet, J., *Barrelled Locally Convex Spaces.* North-Holland Math. Studies 131 Amsterdam 1987.
36. Rosenthal, H.P., On the structure of weak*-compact subsets of a dual Banach space. Preprint.
37. Schlumprecht, Th., On dual spaces with bounded sequences without weak* convergent convex blocks. *Proc. Amer. Math. Soc.* 107, 395–408 (1989).
38. Schlumprecht, Th., Limited sets in Banach spaces. Dissertation, München, 1987.
39. Schlumprecht, Th., A limited set that is not bounding. *Proc. R. Ir. Acad.* 90, 125–129 (1990).
40. Valdivia, M., On quasinormable echelon spaces. *Proc. Edinburgh Math. Soc.* 24, 73–80 (1981).
41. Valdivia, M., Fréchet spaces without subspaces isomorphic to l_1. To appear in *Math. Japonica.*
42. Valdivia, M., On totally reflexive Fréchet spaces. pp. 39–55 in *Actti del Convegno Internazionale in Analisi Matematico e sue Applicazioni, Dedicato Prof. G. Aquaro.* Bari 1991.
43. Valdivia, M., A characterization of totally reflexive Fréchet spaces. *Math. Z.* 200, 327–346 (1989).

Complementability and Injectivity of Fréchet Spaces of Continuous Vector-Valued Functions

PAWEŁ DOMAŃSKI and LECH DREWNOWSKI Institute of Mathematics, A. Mickiewicz University, ul. Matejki 48/49, 60–769 Poznań, POLAND

0 INTRODUCTION

This paper is related to the authors' recent articles [5] and [6]. In [5], it is shown that the Banach space of vector-valued continuous functions is usually uncomplemented in the larger spaces of functions continuous with respect to some weaker topologies in the range Banach space. On the other hand, it is proved in [6] that various Fréchet spaces of vector-valued functions are usually non-injective. The aim of the present paper is to give a unified, general and elementary approach which yields those earlier results as well as many new ones. This approach, or some variants of it, can be applied for similar purposes in other settings, too. Thus, for instance, the authors have used it in [7] for some spaces of operators, mostly components of closed operator ideals. For other results connected with the subject see [1].

Before we start the presentation, we fix our basic notation and recall some definitions and facts. Let G be a lcs. Then G' always denotes its strong dual, and $G'^{\times}$ denotes the strong dual of the bornological space associated with the equicontinuous bornology on G'. It is known that G is a complemented subspace of the dual of a bornological lcs iff G is complemented in $G'^{\times}$ [4, Prop. 1.3]. We call G a *dual Fréchet space* if G is isomorphic to

The paper was written while the first named author held the Alexander von Humboldt Fellowship at University of Wuppertal. The research was partially supported by the KBN grant no. 2 1083 91 01.

F', where F is a bornological DF-space. The space G is said to have the *Schur property* if every weakly convergent sequence in G is convergent in the original topology, and is called a *Grothendieck space* if every $\sigma(G',G)$-convergent sequence in G' is $\sigma(G',G'')$-convergent. It is easily seen that if weakly compact subsets of G are weakly sequentially compact, then G has the Schur property iff every weakly compact set in the space is compact. In particular, it is so when G is a Fréchet space or, more generally, admits a weaker metrizable locally convex topology [11, 9.3.8]. If E is a Fréchet space, then $c_0(E)$ and $l_\infty(E)$ denote the Fréchet spaces of E-valued sequences which tend to zero or are bounded, respectively, both equipped with the uniform topology. Finally, a Fréchet space is called *injective* if it is complemented in every Fréchet space containing it, and operator always means a continuous linear map. For other, unexplained notions and notation, see [11].

1 VECTOR-VALUED SEQUENCE SPACES

The paper is based on some results concerning vector-valued sequence spaces proved in [7]. We present some of them for the sake of completeness. If E is a Fréchet space, then by *a Fréchet space of bounded E-valued sequences* we mean a closed topological linear subspace $k(E)$ of $l_\infty(E)$, containing $c_0(E)$ and closed with respect to taking subsequences. The latter means that if $(x_n)_{n\in\mathbb{N}} \in k(E)$ and $M = \{m_n : n \in \mathbb{N}\}$ is an infinite subset of $\mathbb{N}$, then also $(x_n)_{n\in M} = (x_{m_n})_{n\in N} \in k(E)$. Such a space $k(E)$ is called *l_∞-determined* if a sequence $(x_n)_{n\in\mathbb{N}} \in l_\infty(E)$ belongs to $k(E)$ whenever for every operator $S: E \to l_\infty$ and every subsequence $(x_n)_{n\in M}$ there is a further subsequence $(x_n)_{n\in L}$ such that $\hat{S}((x_n)_{n\in L}) \in \hat{S}(k(E))$, where $\hat{S}: l_\infty(E) \to l_\infty(l_\infty)$ is the operator defined by $\hat{S}((y_n)_{n\in\mathbb{N}}) := (Sy_n)_{n\in\mathbb{N}}$. Also, we say that $k(E)$ is *l_∞-stable* if for any $(x_n)_{n\in\mathbb{N}} \in k(E)$ and $(a_n)_{n\in\mathbb{N}} \in l_\infty$ the multiplied sequence $(a_n x_n)_{n\in\mathbb{N}}$ belongs to $k(E)$.

Now, let us formulate the main result on vector-valued sequence spaces which will be used in the paper. We denote by e_n the n-th unit vector of l_∞.

THEOREM 1.1. [7] *Let E be a Fréchet space and let $k(E)$ be an l_∞-determined Fréchet space of bounded E-valued sequences. If $(x_n)_{n\in\mathbb{N}}$ is a sequence in E and $T: l_\infty \to k(E)$ is an operator such that*

$$T(e_n) = (0,0,\dots,0,\underbrace{x_n}_{n\text{-th place}},0,\dots) \qquad \text{for every} \quad n \in \mathbb{N},$$

then $(x_n)_{n\in\mathbb{N}} \in k(E)$.

Proof: Let us assume that $(x_n)_{n\in\mathbb{N}} \notin k(E)$. Then there is an operator $S: E \to l_\infty$ and an infinite set $M \subseteq \mathbb{N}$ such that for each infinite set $L \subseteq M$,

$$\hat{S}((x_n)_{n\in L}) \notin \hat{S}(k(E)).$$

For simplicity, we may assume that $M = \mathbb{N}$. Then $\hat{S}\circ T: l_\infty \to l_\infty(l_\infty) \simeq l_\infty$ is a map such that $\hat{S}\circ T(e_n) = (0,\dots,0,Sx_n,0,\dots)$, where Sx_n stands in the n-th slot. By a result of Phillips [13, 31.2.2] (see also [12, Prop. 4]), there is an infinite set $L \subseteq M = \mathbb{N}$ such that

$$\hat{S}\circ T(1_L) = (y_n)_{n\in\mathbb{N}}, \qquad \text{where } y_n = Sx_n \text{ for } n \in L \quad \text{and} \quad y_n = 0 \text{ for } n \notin L.$$

Since $T(1_L) \in k(E)$ and since $k(E)$ is closed with respect to taking subsequences, we get $\hat{S}((x_n)_{n\in L}) \in \hat{S}(k(E))$; a contradiction.

COROLLARY 1.2. [7] *Let E be a Fréchet space and let $k_1(E)$, $k_2(E)$ be two Fréchet spaces of bounded E-valued sequences such that $k_1(E)$ is l_∞-determined and $k_2(E)$ is l_∞-stable. If there exists an operator $T\colon k_2(E) \to k_1(E)$ such that $T|_{c_0(E)} = \mathrm{id}_{c_0(E)}$, then $k_2(E) \subseteq k_1(E)$. In particular, if $k_1(E)$ is a complemented subspace of $k_2(E)$, then $k_1(E) = k_2(E)$.*

Proof: Given $(x_n)_{n\in\mathbb{N}} \in k_2(E)$, we define an operator

$$S\colon l_\infty \to k_2(E), \qquad S((a_n)_{n\in\mathbb{N}}) = (a_n x_n)_{n\in\mathbb{N}}.$$

Then we apply Theorem 1.1 to the map $T \circ S\colon l_\infty \to k_1(E)$.

In order to make the above results applicable we need some examples. Let τ be a linear topology on E weaker than the original one; then we denote

$$c_0(E,\tau) := \{(x_n)_{n\in\mathbb{N}} \in l_\infty(E) : x_n \xrightarrow{\tau} 0\}.$$
$$\kappa(E,\tau) := \{(x_n)_{n\in\mathbb{N}} \in l_\infty(E) : \{x_n : n \in \mathbb{N}\} \text{ is relatively } \tau\text{-compact}\},$$

and equip both spaces with the uniform topology. We are interested in four particular cases:

(1) τ is the original topology (in which case we omit τ in the notation);
(2) $\tau = \sigma(E, E')$;
(3) $E = F'$ and $\tau = \mu(F', F)$;
(4) $E = F'$ and $\tau = \sigma(F', F)$.

THEOREM 1.3. *Let E be a Fréchet space and F a bornological DF-space. Then the following classes of vector-valued sequences:*

(1) $\kappa(E)$;
(2) $\kappa(E, \sigma(E, E'))$;
(3) $\kappa(F', \mu(F', F))$;

are l_∞-stable, l_∞-determined Fréchet spaces of bounded sequences.

Remark. Parts (1) and (2) were implicitly proved in [5, Prop. 2.] and [6, Lemma 3]. A longer list of examples of l_∞-determined sequence spaces is given in [7].

Proof: It suffices to prove l_∞-determinedness.

(1) and (2): Let $k(E)$ denote either $\kappa(E)$ or $\kappa(E, \sigma(E, E'))$. Let $(x_n)_{n\in\mathbb{N}} \in l_\infty(E)$ be such that for every operator $S\colon E \to l_\infty$, and every subsequence $(x_n)_{n\in M}$ there is a further subsequence $(x_n)_{n\in L}$ with $\hat{S}((x_n)_{n\in L}) \in \hat{S}(k(E))$. It is enough to show that $(x_n)_{n\in\mathbb{N}} \in k(Y)$, where $Y = \overline{\mathrm{lin}}\{x_n : n \in \mathbb{N}\}$. Since Y is separable, we can embed Y into $\prod_{l\in\mathbb{N}} E_l$, $E_l \simeq l_\infty$. Of course, it suffices to show that for each l the image of $(x_n)_{n\in\mathbb{N}}$ under the natural projection P_l belongs to $k(E_l)$. To see that this does indeed hold, extend (by the injectivity of l_∞) $P_l|_Y$ to an operator $S_l\colon E \to l_\infty$ and next observe that every subsequence of $(S_l(x_n))_{n\in\mathbb{N}}$ contains a subsequence which is an element of $k(E_l)$. Hence, using in the case (2) the Eberlein-Šmuljan Theorem [11, 9.8.3], $(S_l(x_n))_{n\in\mathbb{N}} \in k(E_l)$.

(3): Let $(x_n)_{n\in\mathbb{N}} \in l_\infty(F')$. Then the $\sigma(F',F)$-closed absolutely convex hull of $(x_n)_{n\in\mathbb{N}}$ is $\sigma(F',F)$-compact and thus $\mu(F',F)$-complete. Hence, $(x_n)_{n\in\mathbb{N}} \in \kappa(F',\mu(F',F))$ iff $(x_n)_{n\in\mathbb{N}}$ is $\mu(F',F)$-precompact.

Assume that $(x_n)_{n\in\mathbb{N}}$ is not $\mu(F',F)$-precompact. We may assume (and we will) that there is an absolutely convex $\sigma(F,F')$-compact set $B \subseteq F$ such that

$$\sup_{x\in B} |\langle x_n - x_m, x\rangle| > 1 \qquad \text{for } m \neq n.$$

Then we can find a double sequence $(y_{nm})_{n,m\in\mathbb{N}} \subseteq B$ such that

$$|\langle x_n - x_m, y_{nm}\rangle| \geq 1 \qquad \text{for } m \neq n.$$

Let $G := \overline{\mathrm{lin}}\{y_{nm} : n, m \in \mathbb{N}\}$. Then there is an operator $T\colon l_1 \to G$ with a dense range, and $T'\colon G' \to l_\infty$ is a continuous injection.

Define an operator $S\colon F' \to l_\infty$ by $S(x) = T'(x|_G)$. Then for any infinite set $L \subseteq \mathbb{N}$,

$$\hat{S}((x_n)_{n\in L}) \notin \hat{S}(\kappa(F',\mu(F',F))).$$

Indeed, if $(z_n)_{n\in\mathbb{N}} \in \kappa(F',\mu(F',F))$, then

$$(z_n|_G)_{n\in\mathbb{N}} \in \kappa(G',\mu(G',G)).$$

On the other hand, $(x_n|_G)_{n\in L} \notin \kappa(G',\mu(G',G))$ and $\ker T' = 0$.

2 VECTOR-VALUED FUNCTION SPACES

Throughout this section,

— E is a Fréchet space;

— X is a locally compact σ-compact non-discrete Hausdorff topological space.

We will consider various Fréchet spaces consisting of functions from X to E. First of all, we define

$$l_c(X,E) := \{f : X \to E :\ f \text{ is bounded on compact sets}\},$$
$$l_\infty(X,E) := \{f : X \to E :\ f \text{ is bounded}\}.$$

Next, if τ is a locally convex topology on E weaker than the original one, then we define:

$$C_c(X,E,\tau) := \{f \in l_c(X,E) : f \text{ is } \tau\text{-continuous}\},$$
$$C_\infty(X,E,\tau) := l_\infty(X,E) \cap C_c(X,E,\tau),$$
$$C_0(X,E,\tau) := \{f \in C_\infty(X,E,\tau) : f \text{ is } \tau\text{-vanishing at infinity}\}.$$

We equip the spaces $l_c(X,E)$ and $C_c(X,E,\tau)$ with the compact-open topology, and the spaces $l_\infty(X,E)$, $C_\infty(X,E,\tau)$ and $C_0(X,E,\tau)$ with the uniform topology (both topologies defined with respect to the strong topology on E!). As usual, if τ is the original topology we omit τ in the notation. More generally, by a *Fréchet function space* (consisting of functions from X to E) we shall understand a Fréchet space $F(X,E)$ such that

$$C_0(X,E) \subseteq F(X,E) \subseteq l_c(X,E),$$

where both inclusions are continuous.

We fix an arbitrary relatively compact sequence $(t_n) \subseteq X$ and a sequence of compactly supported scalar-valued functions with pairwise disjoint supports:

$$\varphi_n \colon X \to [0,1], \qquad \varphi_n(t_n) = 1,$$

such that $\bigcup_{n\in\mathbb{N}} \operatorname{supp}\varphi_n$ is relatively compact. In this setting, we define two operators:

$$S \colon l_\infty(E) \to l_c(X,E); \qquad Q \colon l_c(X,E) \to l_\infty(E)$$

as follows:

$$S((x_n)_{n\in\mathbb{N}}) = \sum_{n\in\mathbb{N}} x_n\varphi_n(\cdot) \quad \text{(pointwise convergence)}, \qquad Q(f) = (f(t_n))_{n\in\mathbb{N}}\,.$$

THEOREM 2.1. *Let E be a Fréchet space, and let $F_1(X,E)$ and $F_2(X,E)$ be two Fréchet function spaces such that $F_1(X,E) \subseteq F_2(X,E)$ and $F_1(X,E)$ is complemented in $F_2(X,E)$. Also, let $k_1(E)$ and $k_2(E)$ be two Fréchet spaces of bounded E-valued sequences such that $k_1(E)$ is l_∞-determined and $k_2(E)$ is l_∞-stable. If*

$$S(k_2(E)) \subseteq F_2(X,E) \qquad \text{and} \qquad Q(F_1(X,E)) \subseteq k_1(E),$$

then

$$k_2(E) \subseteq k_1(E).$$

Proof: By the Closed Graph Theorem, the operators $S|_{k_2(E)} \colon k_2(E) \to F_2(X,E)$ and $Q|_{F_1(X,E)} \colon F_1(X,E) \to k_1(E)$ are continuous. Now, if $P \colon F_2(X,E) \to F_1(X,E)$ is a projection, then the operator $Q \circ P \circ S|_{k_2(E)} \colon k_2(E) \to k_1(E)$ satisfies the assumptions of Corollary 1.2.

COROLLARY 2.2. *The following assertions are equivalent for every Fréchet space E:*

(1) *$C_c(X,E)$ is complemented in $C_c(X,E,\sigma(E,E'))$;*
(2) *$C_c(X,E) = C_c(X,E,\sigma(E,E'))$;*
(3) *$C_\infty(X,E)$ is complemented in $C_\infty(X,E,\sigma(E,E'))$;*
(4) *$C_\infty(X,E) = C_\infty(X,E,\sigma(E,E'))$;*
(5) *$C_0(X,E)$ is complemented in $C_0(X,E,\sigma(E,E'))$;*
(6) *$C_0(X,E) = C_0(X,E,\sigma(E,E'))$;*
(7) *E has the Schur property.*

Proof: Obviously (7) $\Longrightarrow$ (2) $\Longrightarrow$ (1), (7) $\Longrightarrow$ (4) $\Longrightarrow$ (3) and (7) $\Longrightarrow$ (6) $\Longrightarrow$ (5). In order to prove that (1), (3) or (5) implies (7) we apply Theorem 2.1 to $k_2(E) = c_0(E,\sigma(E,E'))$ and $k_1(E) = \kappa(E)$. The statement $c_0(E,\sigma(E,E')) \subseteq \kappa(E)$ means that $c_0(E,\sigma(E,E')) \subseteq c_0(E)$.

COROLLARY 2.3. *The following assertions are equivalent for every Banach space F:*

(1) $C_c(X, F')$ *is complemented in* $C_c(X, F', \mu(F', F))$;
(2) $C_c(X, F') = C_c(X, F', \mu(F', F))$;
(3) $C_\infty(X, F')$ *is complemented in* $C_\infty(X, F', \mu(F', F))$;
(4) $C_\infty(X, F') = C_\infty(X, F', \mu(F', F))$;
(5) $C_0(X, F')$ *is complemented in* $C_0(X, F', \mu(F', F))$;
(6) $C_0(X, F') = C_0(X, F', \mu(F', F))$;
(7) F *contains no copy of* l_1.

Proof: In order to show that (1), (3) or (5) implies (7), we apply Theorem 2.1 to $k_2(F') = c_0(F', \mu(F', F))$ and $k_1(F') = \kappa(F')$. The statement $c_0(F', \mu(F', F)) \subseteq \kappa(F')$ means that $c_0(F', \mu(F', F)) = c_0(F')$. Thus, by [6, Lemma 1 and Remark (c) on p. 260] (see also [1, Th. 10]), F contains no copy of l_1.

By [9], if F contains no copy of l_1, then $\mu(F', F)$ and strong compact sets coincide, which implies (2), (4) and (6).

COROLLARY 2.4. *Let X be a metrizable space, then the following assertions are equivalent for every bornological DF-space F:*

(1) $C_c(X, F', \sigma(F', F''))$ *is complemented in* $C_c(X, F', \sigma(F', F))$;
(2) $C_c(X, F', \sigma(F', F'')) = C_c(X, F', \sigma(F', F))$;
(3) $C_\infty(X, F', \sigma(F', F''))$ *is complemented in* $C_\infty(X, F', \sigma(F', F))$;
(4) $C_\infty(X, F', \sigma(F', F'')) = C_\infty(X, F', \sigma(F', F))$;
(5) $C_0(X, F', \sigma(F', F''))$ *is complemented in* $C_0(X, F', \sigma(F', F))$;
(6) $C_0(X, F', \sigma(F', F'')) = C_0(X, F', \sigma(F', F))$;
(7) F *is a Grothendieck space.*

Proof: We apply Theorem 2.1 to $k_2(F') = c_0(F', \sigma(F', F))$ and $k_1(F') = \kappa(F', \sigma(F', F''))$. The rest of the proof is similar as for the previous corollaries.

COROLLARY 2.5. *Let F be a complete LB-space, then the following assertions are equivalent:*

(1) $C_c(X, F')$ *is complemented in* $C_c(X, F', \sigma(F', F))$;
(2) $C_c(X, F') = C_c(X, F', \sigma(F', F))$;
(3) $C_\infty(X, F')$ *is complemented in* $C_\infty(X, F', \sigma(F', F))$;
(4) $C_\infty(X, F') = C_\infty(X, F', \sigma(F', F))$;
(5) $C_0(X, F')$ *is complemented in* $C_0(X, F', \sigma(F', F))$;
(6) $C_0(X, F') = C_0(X, F', \sigma(F', F))$;
(7) F *is a Schwartz space.*

Proof: We apply Theorem 2.1 to $k_2(F') = c_0(F', \sigma(F', F))$ and $k_1(F') = \kappa(F')$. Then we use [14] (see also [1, Cor. 14 (a)]) and the known fact that F is Schwartz iff F' is Montel.

Remark. Corollaries 2.2 and 2.5 refine [5, Cor. 1 and 2].

A similar method can be used to show that the considered spaces are usually neither injective nor complemented in a dual Fréchet space.

THEOREM 2.6. *Let E be a Fréchet space, $F(X,E)$ a Fréchet function space complemented in a dual space, and $k(E)$ an l_∞-determined Fréchet space of bounded E-valued sequences. If $Q(F(X,E)) \subseteq k(E)$, where Q is the operator defined before Theorem 2.1, then $l_\infty(E) = k(E)$.*

Proof: Let (φ_n) and (t_n) be chosen as in the definition of Q. For every bounded sequence $(x_n)_{n\in\mathbb{N}}$ in E we can define a map

$$R\colon c_0 \to C_0(X,E), \qquad R((a_n)_{n\in\mathbb{N}}) = \sum_{n\in\mathbb{N}} a_n x_n \varphi_n(\cdot) \qquad \text{(pointwise convergence)}.$$

Let $P\colon F(X,E)'^{\times} \to F(X,E)$ be a projection, and let $R'^{\times}: l_\infty = c_0'^{\times} \to C_0(X,E)'^{\times}$ be the second adjoint of R. Then $Q{\circ}P{\circ}R'^{\times}: l_\infty \to k(E)$ is a map satisfying assumptions of Theorem 1.1. Therefore, $(x_n)_{n\in\mathbb{N}} \in k(E)$ and thus $l_\infty(E) = k(E)$.

It is worth mentioning that $C_c(X)$ and $C_\infty(X)$ are always complemented, respectively, in C_c-type and C_∞-type vector-valued spaces. Thus if any of the latter spaces is complemented in a dual Fréchet space (or is injective), then also the corresponding former space must be complemented in a dual Fréchet space, i.e. injective (see [4, Cor. 5.8 and 8.8]).

COROLLARY 2.7. *Assume that $C_c(X)$ is injective and let E be a Fréchet space and F a bornological DF-space.*

(1) [6, Th. 5] *$C_c(X,E)$ is complemented in a dual Fréchet space iff*

$$C_c(X,E) = C_c(X,E'',\sigma(E'',E'))$$

iff E is a Montel space.

(2) [6, Th. 4] *$C_c(X,E,\sigma(E,E'))$ is complemented in a dual Fréchet space iff*

$$C_c(X,E,\sigma(E,E')) = C_c(X,E'',\sigma(E'',E'))$$

iff E is reflexive.

(3) *$C_c(X,F',\mu(F',F))$ is complemented in a dual Fréchet space iff*

$$C_c(X,F',\mu(F',F)) = C_c(X,F',\sigma(F',F))$$

iff $\beta(F',F)$-bounded (or equicontinuous) sets in F' are relatively $\mu(F',F)$-compact.

Remarks. (a) The same holds for C_∞-type spaces instead of C_c-type spaces. C_0-type spaces always contain a complemented copy of c_0 and, therefore, they are never complemented in a dual Fréchet space (and obviously never injective).

(b) Since F is quasi-barrelled, we have

$$C_c(X,F',\sigma(F',F)) \simeq L_\beta(F,C_c(X))$$

and

$$C_\infty(X,F',\sigma(F',F)) \simeq C_c(\beta X,F',\sigma(F',F)) \simeq L_\beta(F,C_c(\beta X)) \simeq L_\beta(F,C_\infty(X)),$$

where βX denotes the Čech-Stone compactification of X. The space $L_\beta(F,C_c(X))$ is complemented in a dual Fréchet space whenever $C_c(X)$ is injective (see [6, Th. 3]).

Proof: It is obvious that, in each part, the third condition implies the second and, by Remark (b) above, the second condition implies the first. (Observe that in (1) and (2) the second condition implies that E is reflexive and, hence, E' is quasi-barrelled.)

In order to prove that the first condition implies the third, we apply Theorem 2.6 to

$$k(E) = \kappa(E) \text{ (for (1))}, \quad = \kappa(E, \sigma(E, E')) \text{ (for (2)) and } \quad = \kappa(F', \mu(F', F)) \text{ (for (3))}.$$

Additionally note that $l_\infty(F') = \kappa(F', \mu(F', F))$ holds iff every $\beta(F', F)$-bounded set is relatively $\mu(F', F)$-compact.

Remark. In connection with part (3) of Corollary 2.7, let us make the following observation (cf. [18, Prop. 3.1]).

Let F be a lcs in which weakly compact convex sets are sequentially weakly compact. Then the following are equivalent.

(1) Weakly compact convex subsets of F are compact.

(2) Equicontinuous $\sigma(F', F)$-compact subsets of F' are $\mu(F', F)$-compact.

Thus, in particular, if F is a complete lcs admitting a weaker metrizable locally convex topology, then the above conditions are equivalent with the Schur property of F.

Proof: To see that (1) $\Longrightarrow$ (2) it is enough to remember that on equicontinuous subsets of F' the topology $\sigma(F', F)$ coincides with the topology of uniform convergence on compact sets.

(2) $\Longrightarrow$ (1): Let $A \subset F$ be convex and weakly compact. Then A is complete (in the original topology) and it is enough to verify that it is precompact. Let $(x_n)_{n\in\mathbb{N}}$ be any sequence in A. Without loss of generality we may assume that $(x_n)_{n\in\mathbb{N}} \in c_0(F, \sigma(F, F'))$, and we have to show that $(x_n)_{n\in\mathbb{N}} \in c_0(F)$. If $(x_n)_{n\in\mathbb{N}} \notin c_0(F)$, then there is an equicontinuous sequence $(y_n)_{n\in\mathbb{N}} \in F'$ such that

$$|\langle x_n, y_n\rangle| \geq 1 \qquad \text{for } n \in \mathbb{N}.$$

We can chose a subsequence $(x_n)_{n\in M}$ such that

$$|\langle x_m, y_n\rangle| \leq 1/4 \qquad \text{for } m > n,\ n \in \mathbb{N},\ m \in M.$$

Finally,

$$|\langle x_m, y_n - y_m\rangle| \geq 3/4 \qquad \text{for } m > n,\ m,\ n \in M$$

and, therefore, $(y_n)_{n\in\mathbb{N}}$ is not $\mu(F', F)$-precompact; a contradiction with (2).

COROLLARY 2.8. *Let $C_c(X)$ be injective, and E, F be as in Corollary 2.7.*

(1) [6, Th. 6] *$C_c(X, E)$ is injective iff $E \simeq \omega$.*

(2) [6, Th. 6] *$C_c(X, E, \sigma(E, E'))$ is injective iff $E \simeq \omega$.*

(3) *$C_c(X, F', \mu(F', F))$ is injective iff $C_c(X, F', \mu(F', F)) = C_c(X, F', \sigma(F', F))$ and F' is injective.*

Remark. An analogous result holds also for spaces of the type C_∞. A weaker version of part (1) was proved in [2] and [10]. Comp. also [16] and [17].

Proof: Necessity: Apply Corollary 2.7 and the known fact that ω is the only reflexive (Montel) injective infinite dimensional Fréchet space (see [15] or [8]).

Sufficiency: If F' is injective then $C_c(X, F', \sigma(F', F)) \simeq L_\beta(F, C_c(X))$ is injective as shown in [6, Th. 7].

REFERENCES

1. J. Bonet, M. Lindström, *Convergent sequences in duals of Fréchet spaces*, these proceedings.
2. P. Cembranos, *$C(K, E)$ contains a complemented copy of c_0*, Proc. Amer. Math. Soc. **91** (1984), 556–558.
3. J. Diestel, *Sequences and Series in Banach Spaces*, Springer, New York, 1984.
4. P. Domański, *$\mathcal{L}_p$-spaces and injective locally convex spaces*, Dissertationes Math. **298** (1990), 1–76.
5. P. Domański and L. Drewnowski, *Uncomplementability of the spaces of norm continuous functions in some spaces of "weakly" continuous functions*, Studia Math. **97** (1991), 245–251.
6. ———, *Fréchet spaces of continuous vector-valued functions: Complementability in dual Fréchet spaces and injectivity*, Studia Math. **102** (1992), 257–267.
7. ———, *Injectivity of spaces of operators* (to appear).
8. P. Domański and A. Ortyński, *Complemented subspaces of products of Banach spaces*, Trans. Amer. Math. Soc. **316** (1989), 215–231.
9. G. Emmanuele, *A dual characterization of Banach spaces not containing l_1*, Bull. Acad. Sci. Pol. Math. **34** (1986), 155–159.
10. F. J. Freniche, *Barrelledness of the space of vector–valued and simple functions*, Math. Ann. **267** (1984), 479–486.
11. H. Jarchow, *Locally Convex Spaces*, B. G. Teubner, Stuttgart, 1981.
12. N. J. Kalton, *Spaces of compact operators*, Math. Ann. **208** (1974), 267–278.
13. G. Köthe, *Topological Vector Spaces, vol. I*, Springer, Berlin, 1969.
14. M. Lindström, *A characterization of Schwartz spaces*, Math. Z. **198** (1988), 423–430.
15. G. Metafune and V. B. Moscatelli, *Complemented subspaces of sums and products of Banach spaces*, Ann. Mat. Pura Appl. **159** (1988), 175–190.
16. R. Ryan, *Complemented copies of c_0 in injective tensor products and spaces of compact operators*, Revista Mat. (Complutense) **2** no. suppl., Proc. Congreso de Análysis Funcional, El Escorial (1988), 13–18.
17. E. Saab and P. Saab, *On complemented copies of c_0 in injective tensor products*, Geometry of Normed Linear Spaces (R. G. Bartle, N. T. Peck, A. L. Peressini, J. J. Uhl, eds.), Contemporary Math., vol. 52, AMS, Providence, 1986, pp. 131–135.
18. G. Schlüchtermann and R. F. Wheeler, *The Mackey dual of a Banach space* (to appear).

Differentiable Functions and the $\overline{\partial}$-Complex

MICHAEL LANGENBRUCH Fachbereich Mathematik
Bergische Universität-GH Wuppertal, Gaußstr. 20, D-W-Wuppertal 1, Germany

The first part of this paper is concerned with the solution of the $\bar{\partial}$-equation with bounds. If the bounds are given by a plurisubharmonic (psh.) weight function, then Hörmander's $\bar{\partial}$-methods ([9]) give a solution with precise estimates. However problems arise for bounds, which are not psh. Such bounds are used for instance in connection with Fourier transforms of test functions (see e.g. Hörmander [10], sections 15.2 and 15.4, and Braun, Meise [5]) and the application of function theoretic methods to partial differential operators and convolution operators in spaces of (ultra)distributions. Often equivalent psh. weight systems are constructed (Hörmander [10], Braun, Meise [5], Bonet et al. [4], Remark 2.9). On the other hand, Palamodov [24] has given a solution of the $\bar{\partial}$-problem for special non psh. weight systems. So the question arises, to characterize the weight systems $\mathfrak{W} = \left\{ W_n \;\middle|\; n \geq 1 \right\}$ on a pseudoconvex open set $\Omega \subset \mathbb{C}^N$ such that the associated $\bar{\partial}$-complex of (weighted) square integrable functions

$$0 \to H(\mathfrak{W},\Omega) \to \tilde{L}_0(\mathfrak{W},\Omega) \xrightarrow{\bar{\partial}} \tilde{L}_1(\mathfrak{W},\Omega) \xrightarrow{\bar{\partial}} \cdots \xrightarrow{\bar{\partial}} \tilde{L}_N(\mathfrak{W},\Omega) \to 0 \qquad (*)$$

is exact (see section 1 for the definitions).

This question is solved in the first section. We consider inductive and projective weight systems $\mathfrak{W}$ (i.e. $\mathfrak{W}$ is an IWS or DWS, respectively. For the mild technical assumptions see Definition 1.1). Then the methods of Hörmander [9] and Langenbruch [17] can be used to characterize the exactness of (*) in the following way (see Lemma 1.2):

LEMMA. Let $\mathfrak{W}$ be an IWS. The following are equivalent:

i) The complex (*) is exact.

ii) For any $n \geq 1$ there are $k \geq 1$ and C such that for any $t \in \Omega$ there are psh. functions ψ_t on Ω such that

$$\psi_t(z) - \psi_t(t) \leq W_k(z) - W_n(t) + C \text{ on } \Omega.$$

iii) ii) holds with $\psi_t = \ln|g_t|$ for some $g_t \in H(\Omega)$.

iv) $\mathfrak{W}$ is equivalent to some IWS $\tilde{\mathfrak{W}}$ consisting of psh. functions.

So the solution of Hörmander [9] is optimal for an IWS $\mathfrak{W}$. The conditions ii) and iii) are very similar to the splitting condition obtained in Langenbruch [17, 18].

For a DWS $\mathfrak{W}$ a similar statement holds, in fact we only have to add some density condition (see Lemma 1.3).

In section 2 the exactness of (*) is characterized for weights coming from Fourier transforms of ultradifferentiable test functions. The corresponding weights have the form $W_n(z) = |\text{Im } z| - \tilde{W}_n(|z|)$ with some positive function $\tilde{W}_n$. So they are (a priori) not psh. For any of the classes of ultradifferentiable functions the exactness of (*) is equivalent to the condition (M3) of Komatsu [12] and also to the existence of optimal cut off functions, which are often needed in connection with extension problems or structural results for ultradifferentiable functions (see e.g. Bonet et al. [4], Proposition 2.2, Franken [8], Langenbruch [14], Petzsche [25], proof of Theorem 2.2).

In section 3 the splitting condition is evaluated for these weight systems, and interesting differences come up. While for the (F)-spaces of testfunctions, splitting and exactness are the same, for the (DF)-spaces splitting is much stronger (and e.g. not satisfied in the case of Gevrey-classes).

In section 4 we show that the $\bar{\partial}$-complexes considered in section 2 and 3 are a basic tool for the theory of ultradifferentiable functions, that is, they are important for the extension problems of Borel and Whitney, the existence of continuous linear extension operators and the linear topological structure of these classes of ultradifferentiable functions. Especially, condition iii) in the lemma above provides a very simple solution of the problems of Borel and Whitney (see Bonet et al. [3, 4], Langenbruch [19], Meise, Taylor [21-23], Petzsche [25] for different solutions and the rich literature on this problem). Finally we prove the existence of continuous linear extension operators from intervals, especially also for the class $\mathcal{E}_{\{\omega\}}$ for $\omega(x) = \left(\ln(1+|x|)\right)^{\delta}$, $\delta > 1$. The existence of extension operators for this type of functions apparently has not been known before.

1 EXACTNESS OF THE $\bar{\partial}$-COMPLEX

In this section Ω is always an open pseudoconvex set in $\mathbb{C}^N$. The mild technical assumptions on the weight systems needed in this paper are the same as in Langenbruch [17]. They are collected in the following definition:

1.1. DEFINITION. An increasing system $\mathfrak{W} := \left\{ W_n \;\middle|\; n \geq 1 \right\}$ of measurable functions on Ω is called an increasing weight system (IWS), if there is $r\colon \Omega \to \left(0, \min\{1, \text{dist}(z, \partial\Omega)\}\right)$ such that for any $n \geq 1$ there are $k \geq 1$ and A_n such that for any $z \in \Omega$

$$\sup_{|\xi|\leq r(z)} W_n(z + \xi) \leq \inf_{|\xi|\leq r(z)} W_k(z + \xi) + A_n \text{ for any } z \in \Omega \tag{1.1}$$

$$\ln(1/r(z)) + \ln(1+|z|) + W_n(z) \leq W_k(z) + A_n \text{ for any } z \in \Omega. \tag{1.2}$$

A system $\mathfrak{W}$ is called a decreasing weight system (DWS), if $-\mathfrak{W}$ is an IWS.

In this paper $\mathfrak{W}$ always denotes an IWS or a DWS. By (1.1) we can (and will) assume from now on, that the functions of an IWS or DWS are locally bounded on Ω.

Let PSH(Ω) be the plurisubharmonic functions on Ω. For an IWS $\mathfrak{W}$ (respectively a DWS $\mathfrak{W}$)

let PSH($\mathfrak{W}$) denote the functions $f \in PSH(\Omega)$ such that

$$f(z) \le W_n(z) + C_n \text{ on } \Omega \text{ for some (respectively for any) } W_n \in \mathfrak{W}.$$

An IWS $\mathfrak{W}$ defines a weighted space of locally square integrable functions by

$$L(\mathfrak{W},\Omega) := \left\{ f \in L^2_{loc}(\Omega) \;\middle|\; \|f\|_n^2 := \int |f(z)|^2 e^{-2W_n(z)} dz < \infty \text{ for some } n \ge 1 \right\}.$$

Let $H(\mathfrak{W},\Omega)$ be the holomorphic functions in $L(\mathfrak{W},\Omega)$. $L(\mathfrak{W},\Omega)$ and $H(\mathfrak{W},\Omega)$ are defined similarly for a DWS $\mathfrak{W}$. Let $L_k(\mathfrak{W},\Omega)$ be the k-forms in $d\bar{z}$ with coefficients in $L(\mathfrak{W},\Omega)$ (with the topology induced from $L(\mathfrak{W},\Omega)$) and let

$$\tilde{L}_k(\mathfrak{W},\Omega) := \left\{ f \in L_k(\mathfrak{W},\Omega) \;\middle|\; \bar{\partial} f \in L_{k+1}(\mathfrak{W},\Omega) \right\}$$

be the natural domain of definition for the $\bar{\partial}$-operators endowed with the topology given by $\left\{ \|f\|_n + \|\bar{\partial} f\|_n \;\middle|\; n \ge 1 \right\}$. We will often skip the set Ω in the notation for the above spaces, if the set Ω is not important.

We now have the following complex of continuous linear mappings:

$$0 \to H(\mathfrak{W},\Omega) \to \tilde{L}_0(\mathfrak{W},\Omega) \xrightarrow{\bar{\partial}} \tilde{L}_1(\mathfrak{W},\Omega) \xrightarrow{\bar{\partial}} \dots \xrightarrow{\bar{\partial}} \tilde{L}_N(\mathfrak{W},\Omega) \to 0. \tag{1.3}$$

The exactness of (1.3) for an IWS $\mathfrak{W}$ can now be characterized in a similar way as we characterized the splitting of (1.3) in Langenbruch [17]. The only difference lies in the order of quantifiers needed.

Two IWS $\mathfrak{W}$ and $\tilde{\mathfrak{W}}$ are called equivalent, if for any $n \ge 1$ there are $k \ge 1$ and C such that

$$W_n(z) \le \tilde{W}_k(z) + C \text{ and } \tilde{W}_n(z) \le W_k(z) + C \text{ on } \Omega.$$

Two DWS $\mathfrak{W}$ and $\tilde{\mathfrak{W}}$ are called equivalent, if $-\mathfrak{W}$ and $-\tilde{\mathfrak{W}}$ are equivalent.

1.2. LEMMA. Let $\mathfrak{W}$ be an IWS. The following are equivalent:

i) (1.3) is exact for $\mathfrak{W}$.

ii) For any $n \ge 1$ there are $k \ge 1$ and A_n such that for any $t \in \Omega$ there is $\psi_t = \psi_{t,n} \in PSH(\Omega)$ such that

$$\psi_t(t) \ge 0 \text{ and } \psi_t(z) \le W_k(z) - W_n(t) + A_n \text{ for any } z \in \Omega.$$

iii) ii) holds for $\psi_t = \ln|g_t|$ with $g_t \in H(\Omega)$.

iv) $\mathfrak{W}$ is equivalent to an IWS $\tilde{\mathfrak{W}}$ contained in $PSH(\Omega)$.

Proof. "iv) $\Rightarrow$ i)" by L. Hörmander's solution of the weighted $\bar{\partial}$-problem ([9]).

"i) $\Rightarrow$ iii)" Let $K_q(\mathfrak{W}) := \ker \bar{\partial} \cap \tilde{L}_q(\mathfrak{W}) = \ker \bar{\partial} \cap L_q(\mathfrak{W})$ endowed with its natural inductive limit topology. Since $\bar{\partial}: \tilde{L}_q(\mathfrak{W}) \to K_{q+1}(\mathfrak{W})$ is continuous and surjective by assumption, $\bar{\partial}$ is open by the open mapping theorem for (LF)-spaces and any bounded set $B := \left\{ f \in K_{q+1}(\mathfrak{W}) \;\middle|\; \|f\|_n \le 1 \right\}$ is in the image of a bounded set $\tilde{B} = \left\{ g \in \tilde{L}_q(\mathfrak{W}) \;\middle|\; \|g\|_k \le C \right\}$ by Grothendieck's lemma (Köthe [11], 19.5(4)). So we can use the double complex methods as in the proof of "iii) $\Rightarrow$ ii)" of Theorem 1.7 in Langenbruch [17]. In fact, we only have to substitute any application of a right inverse for $\bar{\partial}$ (also in Lemma 1.6, loc. cit.) by an appropriate solution of the $\bar{\partial}$-equation with the above estimates.

"iii) $\Rightarrow$ ii)" is trivial

"ii) $\Rightarrow$ iv)" (see Remark 1.2 in Langenbruch [17]) Set $\tilde{W}_n(z) := \left(\sup_{t\in\Omega} \psi_t(z) + W_n(t)\right)^*$, where ψ_t is chosen for n according to ii) and $f^*(z) := \limsup_{\eta\to z} f(\eta)$ is the upper regularization of f. $\tilde{W}_n$ exists by ii) and (1.1) and is psh. (Lelong, Gruman [20], Theorem 1.26). By (1.1) and ii) there are $j \geq 1$ and C_n such that

$$W_n(z) \leq \tilde{W}_n(z) \leq W_j(z) + C_n \text{ on } \Omega.$$

For a DWS $\mathfrak{W}$ the corresponding result is slightly different in formulation, since we have to use a Mittag-Leffler procedure to produce solutions in $\tilde{L}_k(\mathfrak{W})$ rather than in $\tilde{L}_k(W_n)$. Let again $K_q(\mathfrak{W}) := \ker \bar{\partial} \cap \tilde{L}_q(\mathfrak{W})$.

1.3. LEMMA. Let $\mathfrak{W}$ be a DWS. The following are equivalent:

i) (1.3) is exact for $\mathfrak{W}$.

ii) For any $n \geq 1$ there are $k \geq 1$ and A_n such that for any $t \in \Omega$ there is $\psi_t \in PSH(\mathfrak{W})$ such that for any $z \in \Omega$:

$$\psi_t(\xi) \geq 0 \text{ near } t \text{ and } \psi_t(z) \leq W_n(z) - W_k(t) + A_n.$$

iii) For any $n \geq 1$ there are $k \geq 1$ and A_n such that for any $t \in \Omega$ there are $\psi_t \in PSH(\Omega)$ such that the following holds:

$$\psi_t(t) \geq 0 \text{ and } \psi_t(z) \leq W_n(z) - W_k(t) + A_n \text{ on } \Omega. \qquad (*)$$

$$H(\mathfrak{W}) \text{ is dense in } H(W_k) \text{ for the toplogy of } H(W_n). \qquad (1.4)$$

iv) (*) holds with $\psi_t := \ln|g_t|$ and $g_t \in H(\mathfrak{W})$.

v) $\mathfrak{W}$ is equivalent to a DWS $\tilde{\mathfrak{W}}$ contained in PSH(Ω) and (1.4) holds (for $\mathfrak{W}$ or $\tilde{\mathfrak{W}}$).

Proof. "v) $\Rightarrow$ i)" Since $\mathfrak{W}$ and $\tilde{\mathfrak{W}}$ are equivalent, (1.4) holds for $\tilde{\mathfrak{W}}$ in any case. Hörmander's $\bar{\partial}$-methods ([9]) and (1.2) show, that for any $n \geq 1$ there is $k \geq 1$ such that

$$\text{for any } f \in K_{q+1}(W_k) \text{ there is a solution } g \in L_q(W_n) \text{ of } \bar{\partial}g = f. \qquad (1.5)$$

Now (1.4) implies that for any $n \geq 1$ there is $k \geq 1$ such that for any q

$$K_q(\mathfrak{W}) \text{ is dense in } K_q(W_k) \text{ for the topology of } K_q(W_n). \qquad (1.4')$$

In fact, for $q = 0$ this is (1.4). Let $S_j := \left\{ x \in \Omega \;\middle|\; |x| + 1/\text{dist}(x, \partial\Omega) \leq e^j \right\}$. Then

$$\text{dist}(S_j, \complement S_{4j}) \geq e^{-j}/2 \text{ for } j \geq 1.$$

For $q \geq 1$ we now take $\Phi_j \in D(S_{4j})$ such that for some C and any $j \geq 1$

$$0 \leq \Phi_j \leq 1 \text{ on } \Omega,\ \Phi_j(x) = 1 \text{ for } x \in S_j \text{ and } \|\Phi_j\|_\infty + \|\nabla\Phi_j\|_\infty \leq C\, e^j.$$

For $n \geq 1$ we can choose $n' \geq 1$ by (1.2) such that

$$W_{n'}(z) + 2\big(\ln|z| + \ln(1/\text{dist}(z, \partial\Omega))\big) \leq W_n(z) + C_1. \qquad (1.6)$$

Then we take f and $k \geq 1$ (and g) as in (1.5) for n' and define $f_j := \bar{\partial}(\Phi_j g)$. Since

$$\bar{\partial}(hg) = (\bar{\partial}g)h + g \wedge \bar{\partial}h = fh + g \wedge \bar{\partial}h \text{ for } h \in C^\infty(\Omega),$$

we get:

$$\|f_j - f\|_n = \|\bar{\partial}\big((1 - \Phi_j)g\big)\|_n \leq \|f(1 - \Phi_j)\|_k + \|g \wedge \bar{\partial}\Phi_j\|_n$$

The first term tends to zero, since $f \in K_{q+1}(W_k)$. For the second term we get:

$$\|g \wedge \bar{\partial}\Phi_j\|_n \leq C_2\, e^j\, \|g\|_{n'} \sup\left\{ e^{W_{n'}(z) - W_n(z)} \;\middle|\; z \in S_{4j}\setminus S_j \right\} \leq C_3\, e^{-j}\, \|g\|_n.$$

Thus $f_j \to f$ in $K_{q+1}(W_n)$ and $f_j \in K_{q+1}(\mathfrak{W})$, since each W_d is locally bounded on Ω by (1.1). By (1.4') and (1.5) the Mittag-Leffler procedure can be applied to show the exactness of (1.3) for $\tilde{\mathfrak{W}}$, hence for $\mathfrak{W}$.

"i) $\Rightarrow$ iv)" follows by the same arguments as "i) $\Rightarrow$ iii)" in Lemma 1.2, using the fact, that the $\bar\partial$-operators are open.

"iv) $\Rightarrow$ ii)" is trivial, since $\ln|g_t|$ is continuous near t.

"ii) $\Rightarrow$ iii)" We have to show (1.4): For $n \geq 1$ choose $n_1 \geq 1$ such that

$$W_{n_1}(z) + \ln(1+|z|^2) \leq C_1 + W_n(z) \tag{1.7}$$

and then $k \geq 1$ and ψ_t according to ii). By ii) and (1.1) there is a finite set $S \subset S_{4j}$ such that

$$\Psi_j(z) := \sup_{t \in S} \psi_t(z) + W_k(t) \geq W_{k_1}(z) \text{ on } S_{4j}.$$

Since S is finite, $\Psi_j \in PSH(\mathfrak{W})$ and

$$\Psi_j(z) \leq W_{n_1}(z) + C_2 \text{ on } \Omega . \tag{1.7'}$$

Choose $k_2 \geq 1$ for k_1 as in (1.6). For $f \in H(W_{k_2})$ we can find a solution f_j of

$$\bar\partial f_j = (\bar\partial\Phi_j)f$$

such that

$$\int |f_j(z)|^2 e^{-2\Psi_j(z)}(1+|z|^2)^{-2}dz \leq \int |\bar\partial\Phi_j(z)f(z)|^2 e^{-2\Psi_j(z)}dz$$
$$\leq C_3 e^{2j} \int_{S_{4j}\setminus S_j} |f(z)|^2 \exp\left(-2W_{k_1}(z)\right) dz \leq C_4 e^{-2j} \|f\|_{k_2}^2$$

as above. By (1.7) and (1.7') we conclude, that $\|f_j\|_n \to 0$ for $j \to \infty$. Also, $\|(1-\Phi_j)f\|_n \to 0$ for $j \to \infty$. Let $F_j := \Phi_j f - f_j$. Then $F_j \in H(\mathfrak{W})$, since $\Psi_j \in PSH(\mathfrak{W})$ and since each weight W_d is bounded on the compact set $\operatorname{supp}(\Phi_j f)$ by (1.1). Since we have shown, that $\|F_j - f\|_n \to 0$ for $j \to \infty$, the claim is proved.

"iii) $\Rightarrow$ v)" follows as "ii) $\Rightarrow$ iv)" in Lemma 1.2.

The equivalent conditions obtained in Lemmata 1.2 and 1.3 are very similar to the following conditions, which are equivalent to the splitting of the complex (1.3) for an IWS (respectively, a DWS, see Langenbruch [17]):

For any $t \in \Omega$ there are $\varphi_t \in PSH(\Omega)$ and for any $n \geq 1$ there are $A_n \geq 0$ and $k \geq 1$ such that for any $n \geq 1$ and any $z, t \in \Omega$:

$$\varphi_t(t) \geq 0 \text{ and } \varphi_t(z) \leq W_k(z) - W_n(t) + A_n \tag{1.8}$$

$$\text{(respectively, } \varphi_t(t) \geq 0 \text{ and } \varphi_t(z) \leq W_n(z) - W_k(t) + A_n). \tag{1.8'}$$

Again, we can equivalently choose $\varphi_t := \ln|g_t|$ for some $g_t \in H(\Omega)$. So the only difference between the splitting condition and the surjectivity condition lies in the order of quantifiers, i.e. the functions φ_t are independent of $n \geq 1$ in the splitting condition.

One interpretation of Lemmata 1.2 and 1.3 is, that the solution of the weighted $\bar\partial$-problem by L. Hörmander is optimal for IWS or DWS.

An instructive example considers Hörmander algebras:

Let $\mathfrak{W} := \left\{ nW(z) \;\middle|\; n \geq 1 \right\}$ be an IWS on Ω. Then (1.3) is exact for $\mathfrak{W}$ if and only if W is equivalent to some $\tilde{W} \in PSH(\Omega)$, i.e.

$$W(z) \leq \tilde{W}(z) + C \leq C\,W(z) + C^2 \text{ on } \Omega \text{ for some } C.$$

For radial weights (i.e. $W(z) = W(|z|_2)$) this means, that W is equivalent to a logarith-

mically convex function $\tilde{W}$.
Palamodov [24] shows the exactness of the $\bar{\partial}$-equation in certain spaces of C^∞-functions defined by a weight system "of type I", a notion, which is adapted to the spaces of holomorphic functions coming from Fourier transforms of (generalized) functions. Systems of type I are IWS (except for the fact, that the weights need not be finite). The main condition (Palamodov [24], p. 93) guarantees the existence of $e_n \in H(\Omega)$ such that $\psi_t(z) := \ln|e_n(z-t)|$ satisfies Lemma 1.2ii).

2 FOURIER TRANSFORMS OF ULTRADIFFERENTIABLE FUNCTIONS: EXACTNESS.

The evaluation of Lemmata 1.2 and 1.3 for IWS or DWS coming from Fourier transforms of ultradifferentiable functions with compact support is the main content of this section. There are several types of ultradifferentiable functions used in the literature (see e.g. Komatsu [12], Roumieu [27] or Beurling [1], Björck [2], Braun et al. [6]). Our notation is mainly taken from H. Komatsu [12]. It is as follows:
Let $(M_a)_{a\geq 0}$ be a sequence such that for $a \in \mathbb{N}$

$$m_a := M_a/M_{a-1} \text{ is increasing and unbounded.} \tag{2.1}$$

The first part of (2.1) is the condition (M1) (logarithmic convexity) of Komatsu [12]. By the second part of (2.1) we can define

$$m(z) := \left|\left\{ a \geq 1 \;\middle|\; m_a \leq |z|_\infty \right\}\right| \text{ and } M(z) := \int_{-\infty}^{\ln|z|_\infty} m(e^\lambda)d\lambda, \tag{2.2}$$

where $|z|_\infty := \max_{i\leq N} |z_i|$ for $z \in \mathbb{C}^N$. Then

$$M(z) = \sup\left\{ \ln\left(M_0|z|_\infty^a /M_a\right) \;\middle|\; a \in \mathbb{N}_0 \right\} \tag{2.3}$$

(see Komatsu [12], (3.1) and (3.11)). M(z) is called the associated function for (M_a).
We will also use the assumption

$$\ln(x) = O(m(x)), \tag{2.4}$$

i.e. the condition (M2') (continuity of differential operators; see Komatsu [12], Proposition 3.4). We will also need the condition of non-quasianalyticity ((M3') of Komatsu):

$$\int_1^\infty m(x)\; x^{-2}dx < \infty. \tag{2.5}$$

For a convex function f(y) on $\mathbb{R}_+$ with $\lim_{y\to\infty} f(y)/y = \infty$ let

$$f^*(x) := \sup\left\{ xy - f(y) \;\middle|\; y \geq 0 \right\}$$

denote the Young conjugate of f.
The set of all compact convex sets $K \subset \mathbb{R}^N$ with nonvoid interior is denoted by $\mathfrak{K} = \mathfrak{K}(\mathbb{R}^N)$.
The definition of ultradifferentiable functions with compact support is now as follows:

2.1. DEFINITION. Let $K \in \mathfrak{K}$ and let $|a| := \sum a_j$ for $a \in \mathbb{N}_0^N$.
a) (2.1) and (2.5) be satisfied.

$$\mathfrak{D}_{(M_a)}(K) := \left\{ f \in \mathfrak{D}(K) \;\middle|\; \forall\; n\geq 1\text{: } \|f\|_n := \sup\left(\|f^{(a)}\|_\infty n^{|a|}/M_{|a|} \;\middle|\; a \in \mathbb{N}_0^N \right) < \infty \right\}.$$

$$\mathfrak{D}_{\{M_a\}}(K) := \left\{ f \in \mathfrak{D}(K) \;\middle|\; \|f\|_{1/n} < \infty \text{ for some } n \geq 1 \right\}.$$

b) Let $\omega(x)$ be logarithmically convex and satisfy (2.5). Set $\psi := \left(\omega \circ \exp\right)^*$ and

$$\mathfrak{D}_{(\omega)}(K) := \left\{ f \in \mathfrak{D}(K) \;\middle|\; \forall\ n\geq 1: |f|_n := \sup\left(\|f^{(a)}\|_\infty e^{-n\psi(|a|/n)} \;\middle|\; a \in \mathbb{N}_0^N\right) < \infty \right\}.$$

If $\omega(x)$ satifies

$$\ln|x| = o(\omega(x)), \tag{2.4'}$$

we set

$$\mathfrak{D}_{\{\omega\}}(K) := \left\{ f \in \mathfrak{D}(K) \;\middle|\; |f|_{1/n} < \infty \text{ for some } n \geq 1 \right\}.$$

For the remaining part of the section we will always assume, that (M_a), $m(x)$ and $M(z)$ are connected as in (2.2) and (2.3) and satisfy (2.1), (2.4) and (2.5) (resp. that ω is logarithmically convex and satifies (2.4') and (2.5)).
For $K \in \mathfrak{K}$ let $H_K(z) := \sup\left\{ \langle \operatorname{Im} z, x\rangle \;\middle|\; x \in K \right\}$ be the support functional of K.

2.1'. REMARK. It is easily seen, that the Fourier transformation gives the following topological isomorphisms for $K \in \mathfrak{K}$:

$$\mathfrak{D}_{(M_a)}(K) \simeq H(\mathfrak{W}_{(M_a)}) \text{ for } \mathfrak{W}_{(M_a)} := \left\{ H_K(z) - M(nz) \;\middle|\; n \geq 1 \right\}.$$
$$\mathfrak{D}_{\{M_a\}}(K) \simeq H(\mathfrak{W}_{\{M_a\}}) \text{ for } \mathfrak{W}_{\{M_a\}} := \left\{ H_K(z) - M(z/n) \;\middle|\; n \geq 1 \right\}.$$
$$\mathfrak{D}_{(\omega)}(K) \simeq H(\mathfrak{W}_{(\omega)}) \text{ for } \mathfrak{W}_{(\omega)} := \left\{ H_K(z) - n\omega(z) \;\middle|\; n \geq 1 \right\}.$$
$$\mathfrak{D}_{\{\omega\}}(K) \simeq H(\mathfrak{W}_{\{\omega\}}) \text{ for } \mathfrak{W}_{\{\omega\}} := \left\{ H_K(z) - \omega(z)/n \;\middle|\; n \geq 1 \right\}.$$

The $\bar\partial$-complex (1.3) for these four weight systems will be considered now, and we will show, that in each case the exactness of this complex is equivalent to condition (M3) of Komatsu [12] (i.e. to (2.6) below, see Langenbruch [14], Meise, Taylor [23], Petzsche [25]) and to the existence of "optimal cut off functions" used in connection with Whitney's theorem and structure theory for these classes of functions.

2.2. THEOREM. The following are equivalent:
i) The complex (1.3) is exact for $\mathfrak{W}_{(M_a)} = \left\{ H_K(z) - M(nz) \;\middle|\; n \geq 1 \right\}$.
ii) For any $n \geq 1$ there are $k \geq 1$ and for any $d > 0$ and any $t \in \mathbb{C}^N$ there are functions $\varphi_{t,d} \in PSH(\mathbb{C}^N)$ such that for any $z, t \in \mathbb{C}^N$ and any $n, d \geq 1$:

$$\varphi_{t,d}(t) \geq 0 \text{ near } t \text{ and } \varphi_{t,d}(z) \leq d\left(H_K(\operatorname{Im} z) - H_K(\operatorname{Im} t)\right) + M(kt) - M(nz) + A_{n,d}$$

iii) There is $C \geq 1$ such that for any $y \geq 1$ there are $h_y \in \mathfrak{D}([-Cm(y)/y, Cm(y)/y]^N)$ and for any $n \geq 1$ there are $k \geq 1$ and C_1 such that for any y and n

$$h_y(0) = 1 \text{ and } \|h_y\|_n \leq C_1 e^{M(ky)}.$$

iv) (M_a) satisfies (M3), that is, there are $C_1 > C > 1$ such that

$$m(C_1 x) \leq C\, m(x) \text{ for large } x. \tag{2.6}$$

Since we will have similar statements for the other classes of ultradifferentiable functions and different situations (see Theorem 2.5 and section 3), we first give some steps in the proof of Theorem 2.2 as separate lemmata in a general formulation.
To prove the necessity of (2.6), we use the following easy reduction:
Let $K \in \mathfrak{K}$ and let $\mathfrak{W} = \left\{ H_K(\operatorname{Im} z) - W_n(|z|_\infty) \;\middle|\; n \geq 1 \right\}$ be an IWS (or a DWS, respec-

tively). Let (1.3) be exact for $\mathfrak{B}$. With $\tilde{g}_\xi$ chosen from Lemmata 1.2iii) (and 1.3iv), respectively) and suitable $\varepsilon \in \mathbb{R}$ we set $g_t(z) := \tilde{g}_{(t,0,..,0)}(z,0,..,0)\exp\big(i\varepsilon(z-t)\big)$ for $z, t \in \mathbb{C}$ and obtain for some $d > 0$:

For any $n \geq 1$ there are $k \geq 1$ and C such that for any t there is $g_t \in H(\mathbb{C})$ such that

$$g_t(t) = 1 \text{ and } \ln|g_t(z)| \leq C + d\big(|\text{Im } z| - |\text{Im } t|\big) + W_n(|t|) - W_k(|z|) \tag{2.7}$$

$$\big(\text{respectively, } \ln|g_t(z)| \leq C + d\big(|\text{Im } z| - |\text{Im } t|\big) + W_k(|t|) - W_n(|z|)\big). \tag{2.7'}$$

In the following lemma we assume, that (N_a), $N(x)$ and $n(x)$ are connected as (M_a), $M(x)$ and $m(x)$ in (2.1) – (2.3) and that $W(t) = W(|t|)$.

2.3. LEMMA. Let $\lim_{t\to\infty} W(t)/t = 0$ and suppose that for any $t \in \mathbb{C}$ there is $g_t \in H(\mathbb{C})$ such that $g_t(t) = 1$ and

$$\ln|g_t(z)| \leq C + d\big(|\text{Im } z| - |\text{Im } t|\big) - \ln(1+|z|^2) + W(t) - N(z) \text{ on } \mathbb{C}. \tag{2.8}$$

Then for large y there are $h_y \in C^\infty(\mathbb{R})$ such that $h_y(\xi_y) = 1$, $h_y(x) = 0$ for $x \leq 0$ and

$$\sup\big(\|h_y^{(a)}\|_\infty/N_a \bigm| a \in \mathbb{N}_0\big) \leq 2\, e^{C + W(y)}/N_0$$

for some $0 \leq \xi_y \leq c_y := 2\big(W(y) - N(y/2) + C + \ln(2d)\big)/y$.

Proof. The Paley-Wiener theorem implies the existence of $h_y \in \mathfrak{D}[0, 2d]$, such that $\hat{h}_y(z) = g_{iy}(-z+iy)e^{-izd}$ (Notice, that $\ln|x| = o(N(x))$, since $n(x)$ is unbounded). The Fourier inversion formula and Cauchy's integral formula imply for $x \in \mathbb{R}$:

$$2\pi|h_y(x)| = \Big|\int g_{iy}(-\xi+iy/2)e^{i(x-d)(\xi+iy/2)}d\xi\Big| \leq \pi\, e^C\, e^{-xy/2+W(y)-N(y/2)} \leq \pi/(2d)$$

for $x \geq 2\,\big(W(y)-N(y/2) + C + \ln(2d)\big)/y = c_y$.

Now $1 = g_{iy}(iy) = \hat{h}_y(0) = \int h_y(x)dx$. Since $\operatorname{supp} h_y \subset [0, 2d]$ and $c_y \to 0$ for $y \to \infty$, we get for large y:

$$1 = \Big(\int_{c_y}^{2d} + \int_0^{c_y}\Big) h_y(x)dx \leq 1/2 + 1/2 \sup\big\{\, |h_y(x)| \bigm| 0 \leq x \leq c_y \,\big\}.$$

Hence there is $\xi_y \in [0, c_y]$ such that $|h_y(\xi_y)| = 1$, and the lemma is proved, since

$$2\pi|h_y{}^{(a)}(x)|/N_a = \Big|\int \xi^a \hat{h}_y(\xi)e^{ix\xi}d\xi\Big|/N_a \leq \pi\, e^{C + W(y)}/N_0.$$

2.4. LEMMA. Suppose there are $\xi_y > 0$ and $h_y \in C^\infty(\mathbb{R})$ such that $h_y(x) = 0$ for $x \leq 0$, $h_y(\xi_y) = 1$ and

$$\sup\big(\|h_y^{(a)}\|_\infty/N_a \bigm| a \in \mathbb{N}_0\big) \leq C_1 e^{W(y)}.$$

Then we have for any $\tau \geq y$:

$$\Big(\, Q_n(\tau) := \tau \int_\tau^\infty n(\lambda)/\lambda^2\, d\lambda = \int_1^\infty n(\tau\lambda)/\lambda^2\, d\lambda > 8\tau\xi_y \,\Big) \Rightarrow \big(N(\tau) \leq W(y) + \ln(C_1 N_0)\big)$$

Proof. Set $a_j := 1/(8\tau)$ for $1 \leq j \leq n(\tau)$ and $a_j := 1/(8n_j)$ for $j > n(\tau)$. Then a_j is decreasing. Since $\varepsilon_\nu := \frac{1}{8}\sum_{p>\nu} 1/n_p \to 0$ by (2.5), we get from Komatsu [12], p. 56, l.(-2):

$$\sum_{1\leq p\leq\nu} a_p = \frac{1}{8}\Big(n(\tau)/\tau + \sum_{p>n(\tau)} 1/n_p\Big) - \varepsilon_\nu \geq \frac{1}{8}\int_\tau^\infty n(\lambda)/\lambda^2\, d\lambda - \varepsilon_\nu \geq \xi_y \text{ for large } \nu$$

by assumption. Now Lemma 1.3.6 of Hörmander [10] implies:

$$1 = |h_y(\xi_y)| \leq C_1 \sum_{j \geq n(\tau)} 2^{-j}\, \tau^{-n(\tau)}\, N_{n(\tau)}\, e^{W(y)} \leq C_1 e^{W(y) - N(\tau)}/N_0,$$

since $N(\tau) + \ln(N_{n(\tau)}/N_0) = (\ln \tau)\, n(\tau)$ for any $\tau \geq 1$.

We can now give the proof of Theorem 2.2:

$\mathfrak{B}_{(M_a)}$ is a DWS by (2.4).
"ii) $\Rightarrow$ i)" by Lemma 1.3ii), since $\varphi_{t,d} \in PSH(\mathfrak{B})$ by ii).
"i) $\Rightarrow$ iv)" We first use (2.7'). Since $M(y)/y \to 0$ for $y \to \infty$ by (2.5) (Komatsu [12], Lemma 4.1), the assumptions of Lemma 2.3 are satisfied for $N(z) = M(z)$ and $W(t) = M(kt)$ for some k by (2.4). Then

$$c_y = 2\big(M(ky) - M(y/2) + C_2\big)/y$$

and we use Lemma 2.4 with $\tau = 2ky$ to show that

$$Q_m(2ky) \leq 16ky\, c_y \leq C_1 m(ky) \text{ for large } y \tag{2.9}$$

(use also (2.2)). Indeed, we would otherwise get the contradiction

$$M(2ky) \leq M(ky) + \ln C_3 \text{ for large } y.$$

(2.9) implies (2.6) (see Meise, Taylor [23], Proposition 1.3).
"iii) $\Rightarrow$ iv)" We can assume, that $N = 1$ in iii), and insert $\tilde{h}_y(x) := h_y(x - Cm(y)/y)$ in Lemma 2.4 as in "i) $\Rightarrow$ iv)". This gives iv) as above.
"iv) $\Rightarrow$ iii)" By the proof of Lemma 4.10 in Langenbruch [18], (2.6) implies, that there is C_1 such that for any $n \geq 1$ there is $k \geq 1$ and for any (large) y there is $H_y \in H(\mathbb{C}^N)$ (set $H_y(z) := \hat{g}_y(z_1) \cdot .. \cdot \hat{g}_y(z_N)$ in the notation of [18]) such that

$$H_y(0) = 1 \text{ and } \ln|H_y(z)| \leq C_1|\operatorname{Im} z|\, m(y)/y + M(ky) - M(nz) + A_n.$$

By the Paley-Wiener theorem we have $H_y = \mathfrak{F}(h_y)$ for some $h_y \in \mathfrak{D}([-C_1 m(y)/y, C_1 m(y)/y]^N)$ satisfying ii).
"iv) $\Rightarrow$ ii)" We can assume, that $0 \in \operatorname{int}(K)$. With $H_{|t|} \in H(\mathbb{C}^N)$ as in "iv) $\Rightarrow$ iii)" we also get: For any $n \geq 1$ there are $k \geq 1$ and A_n such that

$$\ln |H_{|t|}(z-t)| \leq M(kt) - M(nz) + C_1\, |\operatorname{Im} z - \operatorname{Im} t|\, m(t)/|t| + A_n \leq$$
$$M(kt) - M(nz) + C_2 H_K(\operatorname{Im} z)\, m(t)/|t| + C_1 m(t) + A_n.$$

For $d > 0$ we set

$$\varphi_{t,d} := \ln |H_t(z-t)| + (d - C_2 m(t)/|t|)\big(H_K(\operatorname{Im} z) - H_K(\operatorname{Im} t)\big).$$

Then $\varphi_{t,d} \in PSH(\mathbb{C}^N)$ for large $|t|$, since $m(t)/|t| \to 0$ for $|t| \to \infty$ by (2.5) (Komatsu [12], Lemma 4.1). ii) follows, since $\varphi_{t,d}(t) = 0$, $\varphi_{t,d}$ is continuous near t, and

$$\varphi_{t,d}(z) \leq d\big(H_K(\operatorname{Im} z) - H_K(\operatorname{Im} t)\big) + M(kt) - M(nz) + C_3 m(t) + A_n \leq$$
$$d\big(H_K(\operatorname{Im} z) - H_K(\operatorname{Im} t)\big) + M(e^{C_3} kt) - M(nz) + A_n.$$

A discussion of the different conditions in Theorem 2.2 is postponed after the following theorem, which essentially follows from Theorem 2.2 and the first section and gives the corresponding statements for the other types of ultradifferentiable functions.

2.5. THEOREM. a) The following are equivalent:

i) The complex (1.3) is exact for $\mathfrak{W}_{\{M_a\}} = \left\{ H_K(z) - M(z/n) \mid n \geq 1 \right\}$.

ii) There is $C \geq 1$ such that for any $n \geq 1$ there is $k \geq 1$ such that for any $y \geq 1$ there are $h_y \in \mathfrak{D}([-Cm(y/n)/y, Cm(y/n)/y]^N)$ such that

$$h_y(0) = 1 \text{ and } \|h_y\|_{1/k} \leq C_n e^{M(y/n)}$$

iii) (M_a) satisfies (2.6)

b) The following are equivalent to (2.6) for $\omega(x)$:

iv) The complex (1.3) is exact for $\mathfrak{W}_{(\omega)} = \left\{ H_K(z) - n\omega(z) \mid n \geq 1 \right\}$

v) There is $C \geq 1$ such that for any $y \geq 1$ there are $h_y \in \mathfrak{D}([-C\omega(y)/y, C\omega(y)/y]^N)$ and for any $n \geq 1$ there are $k \geq 1$ and C_n such that for any y and n:

$$h_y(0) = 1 \text{ and } |h_y|_n \leq C_n e^{k\omega(y)}$$

vi) The complex (1.3) is exact for $\mathfrak{W}_{\{\omega\}} = \left\{ H_K(z) - \omega(z)/n \mid n \geq 1 \right\}$

vii) There is $C \geq 1$ such that for any $n \geq 1$ there is $k \geq 1$ such that for any $y \geq 1$ there are $h_y \in \mathfrak{D}([-C\omega(y)/(yn), C\omega(y)/(yn)]^N)$ such that

$$h_y(0) = 1 \text{ and } |h_y|_{1/k} \leq C_n e^{\omega(y)/n}$$

Proof. $\mathfrak{W}_{\{M_a\}}$ and $\mathfrak{W}_{\{\omega\}}$ are IWS by (2.4) (and (2.4'), respectively). $\mathfrak{W}_{(\omega)}$ is a DWS by (2.4).

I) *Necessity of (2.6):*

"i) $\Rightarrow$ (2.6)" By Lemma 1.2iii), (2.7) and (2.4) we have (2.8) for $N(z) = M(z/k)$ and $W(t) = M(t)$. Then $\tilde{g}_t(z) := g_{kt}(kz)$ satisfies the estimates used in "2.2i) $\Rightarrow$ 2.2iv)" and (2.6) follows as above.

"iv) $\Rightarrow$ (2.6)" By Lemma 1.3iv), (2.7') and (2.4') we can apply Lemma 2.3 with $W(t) = k\omega(t)$ and $N(z) = \omega(z)$. Now Lemma 2.4 (with $\tau = 2ky$ and $\xi_y \leq c_y \leq 2k\omega(y)/y$) implies the following alternative for some C_1 and large y:

$$\omega(2ky) \leq k\omega(y) \tag{2.10}$$

or

$$Q_n(2ky) \leq C_1\omega(y). \tag{2.11}$$

The equation (Komatsu [12], before Proposition 4.5)

$$Q_n(y) = -N(y) + Q_N(y). \tag{2.12}$$

and (2.11) imply that

$$Q_\omega(y) \leq (C_1 + 1)\omega(y). \tag{2.13}$$

Fix any $y \geq 1$ and set $d_j := (2k)^j y$. If (2.13) is not true, we can choose $J \leq \infty$ such that

$$\omega(d_{\nu+1}) \leq k\, \omega(d_\nu) \text{ for } 0 \leq j \leq J \text{ and } Q_\omega(d_{J+1}) \leq (C_1+1)\, \omega(d_{J+1})$$

(With obvious interpretation for $J = \infty$). Then we have for $J < \infty$:

$$\begin{aligned} Q_\omega(y) &= y \sum_{0\leq\nu\leq J} \int_{d_\nu}^{d_{\nu+1}} \omega(t)/t^2\, dt + Q_\omega(d_{J+1})y/d_{J+1} \\ &\leq y\, \omega(y) \sum_{0\leq\nu\leq J} k^{\nu+1} \int_{d_\nu}^{d_{\nu+1}} t^{-2}\, dt + C_2\omega(y) \leq C_3\omega(y). \end{aligned}$$

For $J = \infty$ we just skip the second terms. This again implies (2.6).

"vi) $\Rightarrow$ (2.6)" By Lemma 1.2iii), (2.7) and (2.4) we get $g_t \in H(\mathbb{C})$ such that

$$g_t(t) = 1 \text{ and } \ln|g_t(z)| \leq C + d\big(|\mathrm{Im}\, z| - |\mathrm{Im}\, t|\big) - 2\ln(1+|z|^2) + \omega(t) - \omega(z)/k.$$

g_t^k now satisfies the estimates used in "iv) $\Rightarrow$ (2.6)" and (2.6) follows from that proof. "ii) or v) or vii) $\Rightarrow$ (2.6)" This can easily be reduced to the proofs above.

II) *Sufficiency of (2.6):*

"(2.6) $\Rightarrow$ i)" With $k = k(1)$ and $\varphi_{t,kn}$ from Theorem 2.2ii) we set $\varphi_t(z) := \varphi_{t/(kn),kn}(z/(kn))$ and obtain Lemma 1.2iii), hence i).

"(2.6) $\Rightarrow$ iv)" Take $\varphi_t(z) := \varphi_{t,1}(z) + M(z) - M(t)$ with $\varphi_{t,1}$ from Theorem 2.2ii) and $m(z) := \omega(z)$. By (2.6) we know (Langenbruch [13]), that for any $n \ge 1$ there is $k \ge 1$ such that

$$M(nx) \le M(x) + k\ m(x) \text{ and } M(x) + n\ m(x) \le M(kx) \text{ for large } x. \tag{2.14}$$

Thus φ_t satisfies (1.8') and hence also Lemma 1.3ii). So iv) follows.

"(2.6) $\Rightarrow$ vi)" With $k = k(1)$ and φ_t from "(2.6) $\Rightarrow$ iv)" we set $\tilde{\varphi}_t(z) := \varphi_{tkn}(zkn)/(nk)$ and obtain:

$$\tilde{\varphi}_t(t) \ge 0 \text{ and } \tilde{\varphi}_t(z) \le H_K(z) - H_K(t) + \omega(nkt)/n - \omega(nkz)/(kn) + C. \tag{2.15}$$

Now (2.6) implies the existence of $\varepsilon < 1$ and $C \ge 1$ such that for any $\tau \ge 1$:

$$\omega(\tau x) \le C\tau^{\varepsilon}\omega(x) \text{ for large } x.$$

So (2.15) implies Lemma 1.2ii), hence vi).

III) *Existence of cut off functions:*

"(2.6) $\Rightarrow$ ii)" Fix $n \ge 1$ and take h_y and k from Theorem 2.2iii). Then $H_y(x) := h_{y/(kn)}(ykn)$ satisfies ii).

"vi) $\Rightarrow$ vii)" For $N = 1$ in vii) we just take $H_y(x) := h_y(x+\xi_y)h_y(-x+\xi_y)$ with h_y from Lemma 2.3 (since

$$|fg|_{1/(Lk)} \le C\ |f|_{1/k}|g|_{1/k} \text{ for some } L \text{ and any } k \ge 1,$$

see Braun et al. [6], Proposition 4.4). The general case then follows by defining $H_y(x) := H_y(x_1) \cdot \ldots \cdot H_y(x_N)$.

"(2.6) $\Rightarrow$ v)" We already remarked, that (1.8') is satisfied, hence (1.8') holds for $\varphi_t := \ln|g_t|$ for some $g_t \in H(\mathbb{C})$. We now argue as in the preceding paragraph.

The theorem is proved.

The functions in Theorems 2.2iii), 2.5ii), v) and vii) are called optimal cut off functions. It is sometimes more convenient to normalize these functions in such a way, that $\operatorname{supp} h_\tau \subset [-C/\tau, C/\tau]$ with large τ. This renormalization is easily done as follows for Theorem 2.2: If (2.6) is satisfied, we can assume, that $M_a^* := M_a/a!$ satisfies (2.1) (this is condition (M1*) of Komatsu, see Petzsche [25]). Let $m^*(x)$ be connected to (M_a^*) by (2.1). We then set $y := \tau m^*(\tau)$ and obtain from the equation $m(\tau m^*(\tau)) = m^*(\tau)$ (see Schroer [28], Lemma 3.1), that $1/\tau = m(y)/y$. Satz 3.2 of Schroer [28] and (M1*) imply

$$M(ky) \le M(k\tau m^*(k\tau)) \le m^*(k\tau) + M^*(k\tau) \le M^*(ek\tau).$$

For Theorem 2.5ii) we proceed in the same way with $M_a n^a$ instead of M_a.

For Theorem 2.5v) and vii) we set $\tilde{\omega}(y) := \sup\left\{ \omega(x) - xy \mid x \ge 0 \right\}$ and get for $y = \tau\omega^*(\tau)$:

$$\omega(y)/2 = \omega(\tau\omega^*(\tau)) - (\tau\omega^*(\tau))/(2\tau) \le \tilde{\omega}(1/(2\tau)).$$

Optimal cut off functions were used in connection with the structure theory for ultradifferentiable functions (Langenbruch [14], Lemma 1.2), with the restriction problems of Borel and Whitney (Bonet et al. [4], Proposition 2.2, Petzsche [25], proof of Theorem

2.2) and the existence of extension operators (Franken [8]). We have proved here, that the existence of optimal cut off functions is in fact equivalent to these problems. Also, their existence for any of the classes of ultradifferentiable ultradifferentiable functions follows from the construction in Theorem 2.2iii).
The surjectivity condition in Lemma 1.2iv) for $\mathfrak{B}_{\{\omega\}}$ has been verified by Bonet et al. [4], Remark 2.9 (For a weaker result than Lemma 1.3v) for $\mathfrak{B}_{(\omega)}$ see Braun, Meise [5], Proposition 5).

3 FOURIERTRANSFORMS OF ULTRADIFFERENTIABLE FUNCTIONS: SPLITTING

The splitting of the complex (1.3) for the spaces considered in the preceding section has important consequences for the structure of the corresponding spaces of ultradifferentiable functions and the existence of extension operators (see section 4). The similarity between the surjectivity conditions in Lemmata 1.2 and 1.3, respectively, and the splitting conditions (1.8) and (1.8') might suggest, that surjectivity and splitting of the complex (1.3) for the Fourier transforms of ultradifferentiable functions is almost the same problem. This however is only partly true, as we will see now.
We will generally assume in this section, that (M_a) satisfies (2.1), (2.4) and (2.5), and that $\omega(x)$ is logarithmically convex and satisfies (2.4') and (2.5).
The following result has essentially been proved already:

3.1. PROPOSITION. a) The complex (1.3) splits for $\mathfrak{B}_{(M_a)}$ if and only if (M_a) satisfies (2.6).
b) The complex (1.3) splits for $\mathfrak{B}_{(\omega)}$ if and only if $\omega(x)$ satisfies (2.6).

Proof. The necessity of (2.6) follows from Theorem 2.2 and 2.5. The sufficiency follows for (M_a) from Theorem 2.2ii) with $d = 1$ and (1.8'). For $\omega(x)$ we observed already in "(2.6) $\Rightarrow$ vi)" of the proof of Theorem 2.5, that (1.8') is satisfied.

The situation is quite different for the two other types of ultradifferentiable functions, as will be shown in the subsequent theorems.

3.2. THEOREM. The complex (1.3) splits for $\mathfrak{B}_{\{M_a\}}$ if and only if for any $n \geq 1$ there is $k \geq 1$ such that

$$Q_m(x) \leq M(x/n) - M(x/k) \text{ for large } x \tag{3.1}$$

if and only if (M_a) satisfies (2.6) and for any $n \geq 1$ there is $k \geq 1$ such that

$$n\, m(x) \leq M(x) - M(x/k) \text{ for large } x. \tag{3.2}$$

Proof. "(3.1) $\Longleftrightarrow$ (3.2)" By (2.6) (see e.g. Meise, Taylor [23]) and (3.2) we have:

$$Q_m(nx) \leq C_1 m(nx) \leq C_1 C_n m(x) \leq M(x) - M(x/k) \text{ for large } x.$$

(3.1) implies:

$$nQ_m(x) \leq \left(M(x) - M(x/k_1)\right) + \left(M(x/k_1) - M(x/k_2)\right) + .. + \left(M(x/k_{n-1}) - M(x/k_n)\right) = \\ = M(x) - M(x/k_n) \text{ for large } x. \tag{3.3}$$

This implies (3.2), since $m(x) \le Q_m(x)$. (2.6) obviously also follows from (3.1) and Meise, Taylor [23], since

$$M(x) - M(x/k) \le \ln k \; m(x).$$

"Necessity of (3.1)" As before we apply Lemma 2.3 and get $h_y \in C^\infty(\mathbb{R})$ such that for any $n \ge 1$ there are $\xi_y \ge 0$ and $j \ge 1$ such that $h_y(\xi_y) = 1$ and

$$\xi_y \le c_y = 2\big(M(y/n) - M(y/(2j)) + C_n\big)/y \text{ for large } y.$$

We now apply Lemma 2.4 for $N_a = M_a k^a$, $k = k(1)$ from (2.7), $\tau = 2ky$ and $W(y) = M(y)$ and get with C_3 independent of n:

$$Q_m(y) \le C_2 c_y \, y \le C_3\big(M(y/n) - M(y/(2j))\big) \text{ for large } y.$$

As in (3.3) this implies (3.1).

"Sufficiency of (3.2)" The construction is similar but simpler than in the proof of Lemma 4.10 of Langenbruch [18]: For $p \le m(y)$ we set $a_p := 1/y$ and for $p > m(y)$ we set $a_p := 1/m_p$. Then Theorem 1.3.5 of Hörmander [10] shows the existence of $g_y \in \mathcal{D}([0, 2d_y])$ such that

$$\int g_y(x)\, dx = 1 \text{ and } \|g_y^{(a)}\| \le M_{y,a},$$

where $M_{y,p} = \prod_{0 \le k \le p} (1/a_k)$ and

$$\sum a_p = \big(m(y) + 1\big)/y + \sum_{p > m(y)} 1/m_p = \big(1 + Q_m(y)\big)/y =: d_y .$$

Then $d_y \to 0$ for $y \to 0$ by (2.6), since $m(y)/y \to 0$ by (2.5) (Komatsu [12]).

Let $M_y(x)$ be the associated function for $(M_{y,a})$ and $f_t(z) := \sum \ln|\hat{g}_{|t|}(z_j)|$ for $t \in \mathbb{C}^N$. Then $M_y(x) = 0$ for $x \le y$ and $M_y(x) = M(x) - M(y)$ for $x \ge y$. For $|z| \ge 2|t|$ we get for any $k \ge 1$ (by (2.2)):

$$M_{|t|}(z-t) \ge M_{|t|}(z/2) = M(z/2) - M(t) \ge M(z/(2k)) - M(t/k).$$

For $|z| \le 2|t|$ we get for any $k \ge 1$:

$$M_{|t|}(z-t) \ge 0 \ge M(z/(2k)) - M(t/k).$$

Let $\varphi_t(z) = f_t(z-t) + (1 - 2Bd_{|t|})\big(H_K(z) - H_K(t)\big)$, where $|\text{Im } z|_1 \le B\, H_K(z)$. Such B exists, since we can assume, that $0 \in \text{int}(K)$. Then φ_t is psh. for large $|t|$, $\varphi_t(t) = 0$ and

$$\begin{aligned}\varphi_t(z) &\le -M_{|t|}(z-t) + 2d_{|t|}|\text{Im } z - \text{Im } t|_1 + (1 - 2B\, d_{|t|})\big(H_K(z) - H_K(t)\big) + C \\ &\le C' - M(z/(2k)) + M(t/k) + H_K(z) - H_K(t) + B_1 d_{|t|}|t|\end{aligned}$$

for some B_1. Since (3.1) also holds for $B_1 Q_m$ instead of Q_m, we can choose $k \ge 1$ for $n \ge 1$ such that

$$\varphi_t(z) \le H_K(z) - H_K(t) - M(z/(2k)) + M(t/n) + C''.$$

This proves the theorem by (1.8).

3.3. THEOREM. The complex (1.3) is split for $\mathfrak{B}_{\{\omega\}}$ if and only if

$$\omega(Cy) = (1 + o(1))\omega(y) \text{ for any } C \ge 1 \tag{3.4}$$

if and only if

$$Q_m(y) = o(\omega(y)), \tag{3.5}$$

where $m(y)$ is the logarithmic derivative of $\omega(y)$.

Proof. "(3.4) $\Longleftrightarrow$ (3.5)" (3.5) implies, that $Q_m(Cy) = O(Q_m(y)) = o(\omega(y))$ for any C and therefore $m(Cy) = o(\omega(y))$ for any C. This implies (3.4), since

$$\omega(Cy) - \omega(y) \le \ln(C)\ m(Cy) = o(\omega(y)).$$

If (3.4) is applied with C = 2 we get for $1 < q < 2$ and large y:

$$Q_\omega(y) = \sum_{k\ge 0} \int_{2^k}^{2^{k+1}} \omega(yt)/t^{-2}\,dt \le \sum_{k\ge 0} (q/2)^{k+1}\,\omega(y) = q\,\omega(y)/(2-q),$$

hence $Q_\omega(y) = (1 + o(1))\omega(y)$ and $Q_m(y) = o(\omega(y))$ by (2.12).

"Sufficiency of (3.5)" We define f_t and φ_t exactly as in the sufficiency part of the proof of Theorem 3.2 (M(x) is $\omega(x)$). This gives (1.8), since we have

$$\omega(z/2) - \omega(t) \ge \omega(z)/(kC) - \omega(t)/k \text{ for any } k \ge 1 \text{ and } |z| \ge 2|t|$$

$$0 \ge \omega(z)/(kC) - \omega(t)/k \text{ for any } k \ge 1 \text{ and } |z| \le 2|t|,$$

if C is chosen such that

$$\omega(2x) \le C\omega(x).$$

"Necessity of (3.4)" In the application of Lemma 2.3 we have $c_y = o(\omega(y)/y)$. Hence we get $k = k(1)$ from Lemma 2.4, such that for any $j \ge 1$, $\mu \ge 1$ and large y:

$$Q_m(\mu k y) \le \omega(y)/j \text{ or } \omega(\mu k y) \le k\omega(y), \tag{3.6}$$

hence by (2.12)

$$Q_\omega(y) \le (1 + 1/j)\ \omega(y) \text{ or } \omega(\mu k y) \le k\omega(y). \tag{3.7}$$

(3.6) implies with $\mu = e^{(k-1)j}$:

$$m(y) \le \omega(y)/j \text{ or } (k-1)\omega(y) \ge \omega(\mu k y) - \omega(y) \ge \ln(\mu k)\ m(y) \ge (k-1)j\ m(y).$$

Thus $m(y) = o(\omega(y))$. By (3.7) we have $\omega(Cy) = O(\omega(y))$ for any C. So $m(Cy) = o(\omega(y))$ for any C, which implies (3.4) as above.

Notice that the above proof implies, that

$$\text{we can choose } k(n) = Cn \text{ for some C in (1.8) for } \mathfrak{W}_{\{\omega\}}. \tag{3.8}$$

4 APPLICATIONS

In this section we will apply the results of sections 2 and 3 to the problems of Borel and Whitney for ultradifferentiable functions. In fact we give here a suprisingly simple proof, using the surjectivity conditions obtained from Lemmata 1.2 and 1.3. For different proofs and the corresponding literature see Bonet et al. [3, 4], Langenbruch [19], Meise, Taylor [22] and Petzsche [25].

We will always assume in this section, that (M_a) satisfies (2.1), (2.4) and (2.5), and that $\omega(x)$ is logarithmically convex and satisfies (2.4') and (2.5).

For $K \in \mathfrak{K}$ let $\mathfrak{E}_{(M_a)}(K)$, $\mathfrak{E}_{\{M_a\}}(K)$, $\mathfrak{E}_{(\omega)}(K)$ and $\mathfrak{E}_{\{\omega\}}(K)$ denote the set of all $f \in C^\infty(K)$ corresponding to $\mathfrak{D}_{(M_a)}(K)$, $\mathfrak{D}_{\{M_a\}}(K)$, $\mathfrak{D}_{(\omega)}(K)$ and $\mathfrak{D}_{\{\omega\}}(K)$, respectively. We will use $\mathfrak{E}_*(K)$ as a common substitute for any of these spaces. For $K = \{0\}$ we define $\mathfrak{E}_*(K)$ as the set of all sequences $(c_a) = (f^{(a)}(0))_{a\in\mathbb{N}^N}$ satisfying the corresponding estimates on $K = \{0\}$.

Let $\mathfrak{K}_0 := \mathfrak{K} \cup \{0\}$. For $K \in \mathfrak{K}_0$ let $K_1 \in \mathfrak{K}$ be chosen such that $\operatorname{int}(K_1) \supset K$ and let

$$\rho: \mathfrak{E}_*(K_1) \to \mathfrak{E}_*(K) \text{ be the restriction mapping,}$$

given for $K = \{0\}$ by $\rho(f) := (f^{(a)}(0))_{a \in \mathbb{N}_0^N}$.
The extension problems of Borel and Whitney consist in proving the surjectivity of ρ. To solve this problem, we will use the following *Paley-Wiener type theorem* as in Ehrenpreis [7] (see e.g. Langenbruch [19], Corollary 1.4 and the remarks before Theorem 2.6): The Fourier transformation

$$\mathfrak{F}(T)(z) := \hat{T}(z) := \langle_x T, e^{-i\langle x,z\rangle}\rangle$$

is a topological isomorphism from $\mathfrak{E}_*(K)'_b$ onto $H(\mathfrak{W}(*,K))$ where $\mathfrak{W}(*,K)$ is obtained from $\mathfrak{W}_*$ in Remark 2.1' by changing the "-" sign into a "+" sign.
Then we have the following theorem:

4.1. THEOREM. If (2.6) is satisfied for $m(x)$ (or $\omega(x)$, respectively), then

$$\rho: \mathfrak{E}_*(K_1) \to \mathfrak{E}_*(K) \text{ is surjective.} \tag{4.1}$$

Proof. a) We first consider the case (M_a): By the closed range theorem and the Paley-Wiener theorem above we only have to show, that

$$\mathrm{id}: H(\mathfrak{W}((M_a),K)) \to H(\mathfrak{W}((M_a),K_1)) \text{ is a topological isomorphism.} \tag{4.1'}$$

By Baernstein's lemma we have to show the following:
For any $n \geq 1$ there are $k \geq 1$ and C such that for any $f \in H(\mathfrak{W}((M_a),K))$ we have:

$$|f(z)| \leq \exp\Big(H_{K_1}(z) + M(nz)\Big) \text{ on } \mathbb{C}^N \tag{4.2}$$

$$\Longrightarrow |f(z)| \leq \exp\Big(H_K(z) + M(kz) + C\Big) \text{ on } \mathbb{C}^N \tag{4.2'}$$

(4.2') is easily verified as follows:
Let $B := \Big\{ f\, \varphi_t\, e^{-M(kt) + |\mathrm{Im}\, t|} \;\Big|\; t \in \mathbb{C}^N,\ f \in H(\mathfrak{W}((M_a),K)) \text{ satifies (4.2)} \Big\}$, where

$$M(nz) + (2N+1)\ln(1+|z|) \leq M(n'z) + C_1$$

and k and φ_t are chosen for n' and $K = B_1(0) = \Big\{ x \in \mathbb{R}^N \;\Big|\; |x|_2 \leq 1 \Big\}$ by Lemma 1.2iii).
Then $g \in B$ satisfies:

$$|g(z)| \leq \exp\Big(|\mathrm{Im}\, z|_2 + H_{K_1}(z) - (2N+1)\ln(1+|z|) + C_1\Big) \text{ on } \mathbb{C}^N$$

and

$$|g(z)| \leq \exp\Big(|\mathrm{Im}\, z|_2 + H_K(z) + M(c_f z) + C_f\Big) \text{ on } \mathbb{C}^N.$$

Therefore $\mathfrak{F}^{-1}(B)$ is a bounded set in $C(K_1 + B_1(0))$, consisting of (M_a)-ultradistributions with support in $\Big(K + B_1(0)\Big)$ (by the Paley-Wiener theorem above). Hence $\mathfrak{F}^{-1}(B)$ is a bounded set in $C(K + B_1(0))$, which means that there is C_2 such that for any $z,t \in \mathbb{C}^N$:

$$|f(z)\varphi_t(z)| \leq C_2\, e^{|\mathrm{Im}\, z| - |\mathrm{Im}\, t| + M(kt) + H_K(z)}.$$

(4.2) follows by evaluation at $t = z$, since $\varphi_z(z) = 1$.
b) The case of (ω) is treated exactly as above, while for $\{M_a\}$ and $\{\omega\}$ Baernstein's lemma is not used.

It is well known, that the converse of Theorem 4.1 is also true (see Bonet et al. [3, 4], Langenbruch [19], Meise et al. [22], Petzsche [25]).
To prove the existence of continuous and linear extension operators, we first need a result on the linear topological structure of $\mathfrak{D}_*(K)$ following from section 3 (see Proposition 4.2). We will use the theory of power series spaces as developed by Vogt et al. [26, 29 - 31] and partly also work with l-tame instead of continuous mappings. These notions are defined as follows: For an increasing and unbounded sequence α_n let

$$\Lambda_r(\alpha_n) := \left\{ (c_j) \in \mathbb{C}^{\mathbb{N}} \mid \sum |c_j|^2 e^{2\rho\,\alpha_j} < \infty \text{ for any } \rho < r \right\}.$$

$\Lambda_r(\alpha_n)$ is called power series space of finite type (respectively, of infinite type), if $r < \infty$ (if $r = \infty$, respectively).
Let $(E, \|\ \|_n)$ and $(F, \|\ \|_n)$ be (FS)-spaces with fixed increasing fundamental systems of seminorms (also called grading). A linear mapping $T: E \longrightarrow F$ is called linear tame (l-tame) if there is A such that for any $n \geq 1$ there is C such that

$$\|T(x)\|_n \leq C\,\|x\|_n \text{ for any } x \in E.$$

T is called an t-isomorphism, if T is bijective and T and T^{-1} are l-tame.
Subspaces, quotient spaces and dual spaces are endowed with their canonical gradings (see e.g. Langenbruch [16] for these and connected notations).

4.2. PROPOSITION. a) If (2.6) holds for m(x) and $\omega(x)$, then

$$\mathfrak{D}_{(M_a)}(K) \simeq \Lambda_\infty(m(n^{1/N})) \text{ and } \mathfrak{D}_{(\omega)}(K) \simeq \Lambda_\infty(\omega(n^{1/N})).$$

b) (3.1) implies that

$$\mathfrak{D}_{\{M_a\}}(K) \simeq \Lambda_\infty(m(n^{1/N}))'_b$$

c) (3.4) implies that $\mathfrak{D}_{\{\omega\}}(K)$ is l-tamely isomorphic to $\Lambda_0(\omega(n^{1/N}))'_b$.

Proof. a) By (2.14) $\tilde{L}(\mathfrak{W}_{(M_a)})$ has the properties (DN) and (Ω) introduced by Vogt and Wagner (see Vogt [29]). Since the complex (1.3) is split for $\mathfrak{W}_{(M_a)}$ by Proposition 3.1, also the complemented subspace $H(\mathfrak{W}_{(M_a)})$ $\mathfrak{D}_{(M_a)}(K)$ has (DN) and (Ω). The remaining part of the proof is by now standard: Let $\mathfrak{E}^c_{(M_a)}$ be the c-periodical functions of class (M_a). Then $\mathfrak{D}_{(M_a)}(K)$ is contained in $\mathfrak{E}^c_{(M_a)} \simeq \Lambda_\infty(m(n^{1/N}))$ (by Langenbruch [13]), hence $\mathfrak{D}_{(M_a)}(K)$ is $\Lambda_{\mathbb{N}}(m(n^{1/N}))$-nuclear. Since $m(n^{1/N})$ is stable, $\mathfrak{D}_{(M_a)}(K)$ is isomorphic to a complemented subspace of $\Lambda_\infty(m(n^{1/N}))$ by Satz 3.5 of Vogt, Wagner [31]. It is well known, that $\mathfrak{D}_{(M_a)}(K)$ contains $\mathfrak{E}^d_{(M_a)} \simeq \Lambda_\infty(m(n^{1/N}))$ as a complemented subspace for some d (Vogt [30]). The result for $\mathfrak{D}_{(M_a)}(K)$ now follows from Pelczynski's trick.
The statement for $\mathfrak{D}_{(\omega)}(K)$ is proved similarly, since $\tilde{L}(\mathfrak{W}_{(\omega)})$ always has (DN) and (Ω).
b) (3.1) implies (3.2). Therefore (see Langenbruch [13]) for any $n \geq 1$ there is $k \geq 1$ such that

$$M(x/k) \leq M(x) - n\,m(x) \text{ and } M(x) - k\,m(x) \leq M(x/n) \text{ for large } x.$$

So $\tilde{L}(\mathfrak{W}_{\{M_a\}})'_b$ has the properties (DN) and (Ω) and the conclusion follows as above.
c) (1.3) is split for $\mathfrak{W}_{\{\omega\}}$ by Theorem 3.3 and the $\bar{\partial}$-operator has a l-tame right inverse by (3.8) and Langenbruch [17], Corollary 1.9. Hence $H(\mathfrak{W}_{\{\omega\}})'_b \simeq \mathfrak{D}_{\{\omega\}}(K)'_b$ has the properties $(\underline{DN})_t$ and $(\bar{\Omega})_t$ (see Langenbruch [16]), since these properties clearly hold for $\tilde{L}(\mathfrak{W}_{\{\omega\}})'_b$, and $\mathfrak{D}_{\{\omega\}}(K)'_b$ is t-isomorphic to some $\Lambda_0(\alpha_n)$ (Langenbruch [16], Theorem 1.5). It is seen as above, that α_n can be chosen as $\omega(n^{1/N})$.

The converse statements in Proposition 4.2 rely on the problem, whether the isomorphisms in 4.2 imply (2.6). This seems not to be known.
Sequence space representations for $\mathfrak{D}_*(K)$ have been proved by different methods by several authors (Langenbruch [14], Meise, Taylor [22], Petzsche [25]) except for the l-tame isomorphism which is new. A topological isomorphism in c) is already implied by (2.6) (Meise, Taylor [22], Theorem 4.8).
The existence of continuous linear extension operators

$$E_{*,c}: \mathfrak{E}_*([-c, c]) \to \mathfrak{E}_*([-d, d]),\ 0 \le c < d \tag{4.3}$$

is now a well known consequence of Theorem 4.1, Proposition 4.2 and the splitting theorems of Vogt [29] and Poppenberg, Vogt [26].

4.3. THEOREM. a) (2.6) implies, that $E_{(M_a),c}$ exists for $c \ge 0$ and $E_{(\omega),c}$ exists for $c > 0$.
b) (3.1) implies, that $E_{\{M_a\},c}$ exists for $c \ge 0$.

Proof. The complex

$$0 \to \mathfrak{D}_*([-d, -c]) \times \mathfrak{D}_*([c, d]) \to \mathfrak{D}_*([-d, d]) \xrightarrow{\rho} \mathfrak{E}_*([-c, c]) \to 0 \tag{4.4}$$

is exact by Theorem 4.1.
a) For $* = (M_a)$ and $* = (\omega)$ the first space in (4.4) has the property (Ω) by Proposition 4.2. Let $I = [-c, c]$. It is easily seen, that $\mathfrak{E}_{(M_a)}(I)$ has (DN) (Vogt [29], Proposition 8.1). $\mathfrak{E}_{(\omega)}(I)$ has (DN) for $c > 0$ by Meise, Taylor [21], Corollary 5.6. This can be seen also by considering suitable $\bar\partial$-complexes: (1.3) is split for $\mathfrak{W} := \left\{ M(nz) \;\middle|\; n \ge 1 \right\}$ by Langenbruch [18], Theorem 3.7, since $m(2t) \le C\, m(t)$ for some C and large t. With φ_t from (1.8) for $\mathfrak{W}$ and ψ_t from Lemma 1.3ii) with $n = 1$ for $\mathfrak{W}_{(M_a)}$ we set $\tilde\varphi_t := \varphi_t + \psi_t$ and obtain (1.8') for $\mathfrak{W}((\omega),I))$ by (2.14). Since $L(\mathfrak{W}((\omega),I))$ has (DN), also the (complemented) subspace $\mathfrak{E}_{(\omega)}(I) \simeq H(\mathfrak{W}((\omega),I))'_b$ has (DN). Hence the sequence (4.4) splits by the splitting theorem of Vogt [29].
b) For $* = \{M_a\}$ the dual of the first space in (4.4) has (DN) by Proposition 4.2, if (3.1) holds. $\mathfrak{E}_{\{M_a\}}(I)'_b$ has (Ω) by a similar argument as in Vogt [29], Proposition 8.1. So the splitting theorem of Vogt [29] again applies.

For other proofs of results like Theorem 4.3 (partly for more general sets) see Franken [8], Langenbruch [14], Meise, Taylor [22] and Petzsche [25]. To have extension operators for $\mathfrak{E}_{(\omega)}(\{0\})$, one needs the extra condition (Meise, Taylor [22]) that for any $C > 1$ there is $\delta > 0$ such that

$$\omega^{-1}(Cx)\ \omega^{-1}(\delta x) \le \left(\omega^{-1}(x)\right)^2 \text{ for large } x. \tag{4.5}$$

The existence of extension operators $E_{\{\omega\},c}$ (appearently not known sofar) is proved in the following theorem:

4.4. THEOREM. Let $\omega(x)$ satisfies (3.4) and let $m(x)$ be the logarithmical derivative of $\omega(x)$. If there is $C > 1$ such that

$$2\ m(\omega^{-1}(x)) \le m(\omega^{-1}(Cx)) \text{ for large } x, \tag{4.6}$$

then $E_{\{\omega\},c}$ exists for $c > 0$.

Proof. We have to use l-tame mappings in the proof, since power series spaces of finite type are involved. Let $I = [-c, c]$.

a) $\mathfrak{E}_{\{\omega\}}(I)'_b$ is l-tamely isomorphic to a power series space of finite type.

Proof. $\mathfrak{E}_{\{\omega\}}(I)'_b$ is l-tamely isomorphic to $H(\mathfrak{B}(\{\omega\},I))$ by the precise Paley-Wiener theorem 1.3 in Langenbruch [19]. We now only have to show, that

$$\text{the } \bar{\partial}\text{-operator has a l-tame right inverse for } \mathfrak{B}(\{\omega\},I), \tag{4.7}$$

since then $H(\mathfrak{B}(\{\omega\},I))$ has the properties $(\underline{DN})_t$ and $(\bar{\Omega})_t$ (which are trivial for $L(\mathfrak{B}(\{\omega\},I))$, hence $H(\mathfrak{B}(\{\omega\},I))$ is l-tamely isomorphic to some power series space of finite type by Langenbruch [16], Theorem 1.3. To show (4.7), we first notice that (4.6) implies the existence of $\gamma > 1$ and $\psi_t \in SH(\mathbb{C})$, such that for any $n \geq 1$ and any $z, t \in \mathbb{C}$

$$\psi_t(t) = 0 \text{ and } \psi_t(z) \leq (1+1/n)\omega(z) - (1+1/(\gamma n))\omega(t) + A_n$$

(This is (1.8') for $\mathfrak{B} = \left\{ (1+1/n)\omega(z) \;\middle|\; n \geq 1 \right\}$ and $k = \gamma n$). Indeed, this follows similarly as Theorem 3.6 in Langenbruch [18], where the case $\left\{ (1-1/n)\omega(z) \;\middle|\; n \geq 1 \right\}$ is treated. Let $f_t(z) := \ln|g_t(z)|$ with g_t defined in the proof of Theorem 3.2 (for $M(z) := \omega(z)$) and set $\varphi_t(z) := f_t(z-t) + (1-2d_{|t|})\big(|\mathrm{Im}\, z| - |\mathrm{Im}\, t|\big)$. Then we get for large $|t|$:

$$\begin{aligned}\psi_t(z) + \varphi_t(z) &\leq |\mathrm{Im}\, z| - |\mathrm{Im}\, t| + 4\, Q_m(t) - \omega(z/2) + (1+1/n)\omega(z) - \omega(t)/(\gamma n) + B_n \\ &\leq |\mathrm{Im}\, z| - |\mathrm{Im}\, t| + 2\, \omega(z)/n - \omega(t)/(2\gamma n) + C_n\end{aligned}$$

by (3.4) and (3.5). This is the desired l-tame version of (1.8') for $\mathfrak{B}(\{\omega\},I)$, and (4.7) follows from Langenbruch [17].

b) $\mathfrak{D}_{\{\omega\}}([-d, -c]) \times \mathfrak{D}_{\{\omega\}}([c, d])$ is l-tamely isomorphic to $\Lambda_0(\omega(n))'_b \times \Lambda_0(\omega(n))'_b$ by Proposition 4.2, the latter space being l-tamely isomorphic to $\Lambda_0(\omega(n))'_b$ by (3.4). For $* = \{\omega\}$ the complex (4.4) is l-tamely exact, since ρ is l-tame and the restriction operator

$$\tilde{\rho}: \mathfrak{E}_{\{\omega\}}([-d, d]) \to \mathfrak{E}_{\{\omega\}}([-c, c]) \text{ is l-tamely surjective.} \tag{4.8}$$

To see (4.8) we again use the fact, that the Fourier transform is a l-tame isomorphism between $\mathfrak{E}_{\{\omega\}}([-a, a])'_b$ and $H(\mathfrak{B}(\{\omega\},[-a, a]))$ for any $a > 0$ and notice, that the proof of Theorem 4.1 shows that

$${}^t\tilde{\rho} = \mathrm{id}: H(\mathfrak{B}(\{\omega\},[-c, c])) \to \mathrm{Range}(\rho) \subset H(\mathfrak{B}(\{\omega\},[-d, d])) \text{ is a l-tame isomorhism,}$$

if (3.4) is satisfied (φ_t from Lemma 1.3 is chosen with l-tame estimates as in (3.8)).

c) Summing up we have shown, that (4.4) is a l-tame exact sequence of spaces, which are l-tamely isomorphic to (nuclear) power series spaces of finite type, hence the sequence is split by Poppenberg, Vogt [26].

The assumptions of Theorem 4.4 are satisfied, if $\omega(x) = \big(\ln |x|\big)^{\delta}\big(\ln \ln |x|\big)^{\gamma}$ with $\delta > 1$ and $\gamma \geq 0$ for large $|x|$.

REFERENCES

1. A. Beurling, Quasianalyticity and general distributions, lectures 4 and 5, AMS Summer Institute, Stanford 1961.
2. G. Björck, Linear partial differential operators and generalized distributions, *Ark. Mat. 6*: 351-407 (1966)
3. J. Bonet, R. Meise, B.A. Taylor, Whitney's extension theorem for ultradifferentiable functions of Roumieu type, *Proc. R. Ir. Acad. 89A*: 53-66 (1989).
4. J. Bonet, R.W. Braun, R. Meise, B.A. Taylor, Whitney's extension theorem for non-quasianalytic classes of ultradifferentiable functions, preprint.
5. R.W. Braun, R. Meise, Generalized Fourier expansion for zero-solutions of surjective convolution operators on $D_{\{\omega\}}(\mathbb{R})'$, to appear in *Arch. Math.*
6. R.W. Braun, R. Meise, B.A. Taylor, Ultradifferentiable functions and Fourier analysis, *Res. Math. 17*: 207-237 (1990).
7. L. Ehrenpreis, Fourier analysis in several complex variables, Van Nostrand 1967
8. U. Franken, Stetige lineare Ausdehnungsoperatoren auf Räumen ultradifferenzierbarer Funktionen vom Beurling Typ, Dissertation Düsseldorf 1992.
9. L. Hörmander, An introduction to complex analysis in several variables, North Holland/American Elsevier, New York 1973.
10. L. Hörmander, The analysis of linear partial differential operators I+II, Grundlehren 256+257, Springer, Berlin/New York 1983.
11. G. Köthe, Topological vector spaces I, Springer, Berlin/New York 1969.
12. H. Komatsu, Ultradistributions I. Structure theorems and a characterization, *J. Fac. Sci. Tokyo Ser. IA 20*: 25-105 (1973).
13. M. Langenbruch, Ultradifferentiable functions on compact intervals, *Math. Nachr. 37* : 21-45 (1989).
14. M. Langenbruch, Bases in spaces of ultradifferentiable functions, *Math. Ann. 281* : 31-42 (1988).
15. M. Langenbruch, Extension of ultradifferentiable functions of Roumieu type, *Arch. Math. 51* : 353-361 (1988).
16. M. Langenbruch, Solution operators for partial differential equations in weighted Gevrey spaces, *Mich. Math. J. 37*: 3-24 (1990).
17. M. Langenbruch, Splitting of the $\bar{\partial}$-complex in weighted spaces of square integrable functions , *Rev. Mat. Univ. Complutense Madrid 5* (1992).
18. M. Langenbruch, The splitting condition for the weighted $\bar{\partial}$-complex, *Result. Math. 22* : 560-597 (1992).
19. M. Langenbruch, Extension of ultradifferentiable functions, preprint.
20. P. Lelong, L. Gruman, Entire functions in several complex variables, Grundlehren 286, Springer, Berlin/New York 1986
21. R. Meise, B.A. Taylor, A decomposition lemma for entire functions and its application to spaces of ultradifferentiable functions, *Math. Nachr. 142*: 45-72 (1989).
22. R. Meise, B.A. Taylor, Linear extension operators for ultradifferentiable functions of Beurling type on compact sets, *Amer. J. Math. 111* : 309-337 (1989).
23. R. Meise, B.A. Taylor, Whitney's extension theorem for ultradifferentiable functions of Beurling type, *Ark. Mat. 26*: 265-287 (1988).

24. V.P. Palamodov, Linear differential operators with constant coefficients, Grundlehren 168, Springer, Berlin/New York 1970.

25. H.J. Petzsche, On E. Borel's theorem, *Math. Ann. 282*: 299-313 (1988).

26. M. Poppenberg, D. Vogt, A tame splitting theorem for exact sequences of Fréchet spaces, preprint.

27. C. Roumieu, Sur quelques extensions de la notion de distributions, *Ann. Sci. Ecole Norm. Sup. Paris, 3 Ser. 77* : 41-121 (1960).

28. G. Schroer, Ultradistributionen und ultradistributionelle Randwerte holomorpher Funktionen, Diplomarbeit Münster 1969.

29. D. Vogt, Subspaces and quotient spaces of (s), in "Functional analysis: Surveys and recent results" (K.-D. Bierstedt, B. Fuchssteiner (eds.), *North Holland Math. Studies 27* : 167-187 (1977).

30. D. Vogt, Eine Charakterisierung der Potenzreihenräume von endlichem Typ und ihre Folgerungen, *Manuscr. Math. 37*: 269-301 (1982).

31. D. Vogt, M.J. Wagner, Charakterisierung der Unterräume und Quotientenräume der nuklearen stabilen Potenzreihenräume von unendlichem Typ, *Studia Math. 67*: 225-240 (1980).

Division Problems in Spaces of Entire Functions of Finite Order

SIEGFRIED MOMM Mathematisches Institut, Heinrich-Heine-Universität Düsseldorf, Germany

INTRODUCTION

The question whether a convolution operator between locally convex spaces of C^∞-functions is surjective – considering the dual operator and using Fourier transform – is often equivalent to the following problem of division for its symbol F: Let A and B be weighted spaces of entire functions on $\mathbb{C}^N$ and let F be entire with $F \cdot A \subset B$. When does $g \in B$, with g/F entire, imply that $g/F \in A$?

In his fundamental paper Ehrenpreis [3] (Thm. I) characterized the surjective convolution operators on the space $C^\infty(\mathbb{R}^N)$ by means of a certain "slow decrease" of its symbol F on $\mathbb{R}^N$. Hörmander [6] extended this result to convolution operators $C^\infty(\Omega_1) \to C^\infty(\Omega_2)$ for open convex sets Ω_i in $\mathbb{R}^N$, $i = 1, 2$.

Epifanov [4], [5] and Tkačenko [17], [18] applied constructive methods due to Azarin [1] to characterize the surjective convolution operators $A(G_1) \to A(G_2)$ between the spaces of analytic functions on the open convex sets G_i in $\mathbb{C}$, $i = 1, 2$, by means of the "completely regular growth" of its symbol F on a certain cone which is related to G_2. Lelong and Gruman [10], Morzhakov [14] and recently Sigurdsson [16] and Krivosheev [9] applied Hörmander's $\overline{\partial}$-technique to extend these results to several variables.

There are several authors who investigated weighted spaces of both types presented above. For contributions to this theme we refer to [13] and [16] and to the literature cited there.

In the present paper we give a systematic approach to division problems. It applies to all situations mentioned above and gives new proofs and extensions of the corresponding results. In particular it shows that the cone which occurs in the "slowly decreasing" condition or in the condition of "completely regular growth" both have the same origin. This cone is the support of the Monge-Ampère measure $(dd^cH)^N$ for the positively homogeneous plurisubharmonic function H on $\mathbb{C}^N$ which occurs in the description of the dual space.

To state a special case of our result, we fix $\rho > 0$. Let $H, L: \mathbb{C}^N \to \mathbb{R}$ be positively homogeneous functions of order ρ and which satisfy a Lipschitz condition on the unit sphere. In a more general context we consider the following spaces of entire functions of at most order ρ and finite type:

$$A_H^0 = \left\{ f \in A(\mathbb{C}^N) \,\middle|\, \|f\|_k = \sup_{z\in\mathbb{C}^N} |f(z)| \exp(-H(z) - |z|^\rho/k) < \infty \text{ for each } k \in \mathbb{N} \right\},$$

$$A_H = \left\{ f \in A(\mathbb{C}^N) \,\middle|\, |f|_k = \sup_{z\in\mathbb{C}^N} |f(z)| \exp(-H(z) + |z|^\rho/k) < \infty \text{ for some } k \in \mathbb{N} \right\}.$$

In the latter case we assume that $A_H \neq \{0\}$. Then for each $F \in A_L^0$ we have $F \cdot A_H^0 \subset A_{L+H}^0$ and $F \cdot A_H \subset A_{L+H}$. Thus we can consider the corresponding division problems. If $L = 0$, it is well known that for each $F \in A_L^0 \backslash \{0\}$ this division problem has an answer in the affirmative (see Lelong and Gruman [10], Theorem 9.11). If $L \neq 0$, in general $F \in A_L^0 \backslash \{0\}$ has to satisfy an extra condition depending on H and L in order to solve the division problem. In order to specify this extra condition, let H be in addition plurisubharmonic. By Γ_H we denote the support of the Monge-Ampère measure $(dd^cH)^N$, that is $\mathbb{C}^N \backslash \Gamma_H$ is the largest open cone in $\mathbb{C}^N$ with vertex at the origin on which H is a maximal plurisubharmonic function. We assume that the following Phragmén-Lindelöf theorem holds in A_H^0 (and a corresponding one in A_H): Each entire function f of at most order ρ and finite type which satisfies

$$\sup_{z\in\Gamma_H} |f(z)| \exp(-H(z) - |z|^\rho/k) < \infty \qquad \text{for all } k \in \mathbb{N},$$

is already contained in A_H^0.

THEOREM. *Assume that H and L satisfy the assumptions above. Then for $F \in A_L^0 \backslash \{0\}$ the following assertions are equivalent:*

(i) Whenever $g \in A_{L+H}^0$ and $f := g/F$ is entire, then $f \in A_H^0$.

(ii) FA_H^0 is a closed subspace of A_{L+H}^0.

(E) For each $k \in \mathbb{N}$ there is $R > 0$ such that for each $z \in \Gamma_H$ with $|z| \geq R$ there exists $w \in \mathbb{C}^N$ with $|w - z| \leq |z|/k$ and $|F(w)| \geq \exp(L(w) - |w|^\rho/k)$.

They are also equivalent to the assertions (i) and (ii) where "A_H^0" is replaced by "A_H".

If $\rho = 1$ and H is the support functions of a bounded convex domain of $\mathbb{C}^N$, then by Krivosheev [9] and by [12] an appropriate Phragmén-Lindelöf theorem is available. Thus our theorem extends a result of [9]. If $\Gamma_H = \mathbb{C}^N$, then the desired Phragmén-Lindelöf theorem holds trivially. Hence our theorem extends a result of Lelong and Gruman [10] and Morzhakov [14].

In our approach we roughly proceed as Ehrenpreis did in [3]. The sufficiency of (E) is essentially Ehrenpreis' proof. To prove the necessity of (E), as in [3] we apply the open mapping theorem to establish norm estimates. To show that these norm estimates imply condition (E), we argue by contraposition. We construct certain plurisubharmonic functions as the solution of a Dirichlet problem for the complex Monge-Ampère equation and apply Hörmander's $\overline{\partial}$-technique to construct entire functions which violate the norm estimates.

I like to thank R. W. Braun for interesting conversation and helpful comments.

1 PRELIMINARIES

In this section we introduce the function spaces involved and we fix the notation which will be used throughout this paper.

For the sequel let $\rho > 0$ and $N \in \mathbb{N}$ be fixed. We will use the abbreviation $\mathbb{R}_+ := \{x \in \mathbb{R} \mid x \geq 0\}$. For $z \in \mathbb{C}^N$ we put $|z| := (\sum_{j=1}^N |z_j|^2)^{1/2}$. For $a \in \mathbb{C}^N$ and $R > 0$ we put $U(a,R) := \{z \in \mathbb{C}^N \mid |z-a| < R\}$, $B(a,R) := \overline{U(a,R)}$, $S(a,R) := \partial U(a,R)$, $S := S(0,1)$.

1.1. DEFINITION. Let $\rho > 0$. a) Let $\Omega = (\omega_k)_{k\in\mathbb{N}}$ be a sequence of continuous nonnegative increasing functions on $\mathbb{R}_+$. We assume that for each $n \subset \mathbb{N}$ there are $k \in \mathbb{N}$ and $R \geq 0$ such that for all $r \geq R$

$$(1)\quad \omega_{n+1}(r) \leq \omega_n(r) \qquad (2)\quad \log(1+r) = o(\omega_n(r)) \qquad (3)\quad \omega_n(r) = O(r^\rho)$$

$$(4)\quad 2\omega_k(r) \leq \omega_n(r) \qquad (5)\quad \omega_k(2r) \leq \omega_n(r).$$

Then Ω will be called a *weight system.* Two weight systems $\Omega = (\omega_k)_{k\in\mathbb{N}}$ and $\tilde{\Omega} = (\tilde{\omega}_k)_{k\in\mathbb{N}}$ will be called equivalent if for each $n \in \mathbb{N}$ there are $k \in \mathbb{N}$ and $R \geq 0$ such that $\omega_k(r) \leq \tilde{\omega}_n(r)$ and $\tilde{\omega}_k(r) \leq \omega_n(r)$ for all $r \geq R$.

b) Throughout this paper, for $z \in \mathbb{C}^N$ we put $r(z) := (1+|z|)^{1-\rho}$ if $\rho > 1$, $r(z) := 1$ if $0 < \rho \leq 1$, and $\omega_k(z) := \omega_k(|z|)$ $(k \in \mathbb{N})$.

c) By $\mathcal{H}_\rho$ we denote the set of all functions $H: \mathbb{C}^N \to \mathbb{R}$, which are positively homogeneous of order ρ, i.e. $H(tz) = t^\rho H(z)$ for all $z \in \mathbb{C}^N$ and $t \geq 0$, and which are Lipschitz continuous on S, i.e. there is $C > 0$ with $|H(z) - H(w)| \leq C|z-w|$ for all $z, w \in S$.

1.2 EXAMPLES. Let $\rho > 0$. The following $\Omega = (\omega_k)_{k\in\mathbb{N}}$ are weight systems:

(1) $\omega_k(r) = \frac{1}{k} r^d$, $r \in \mathbb{R}_+$, where $0 < d \leq \rho$.

(2) $\omega_k(r) = \frac{1}{k}\log(1+r)^d$, $r \in \mathbb{R}_+$, where $d > 1$.

(3) $\omega_k(r) = r^{d_k}$, $r \in \mathbb{R}_+$, where $0 < d_k < \rho$ strictly decreases with $k \in \mathbb{N}$.

1.3 EXAMPLE. Let $\rho = 1$. If $G \subset \mathbb{C}^N$ is a bounded convex set then its support function

$$H_G(z) := \sup_{w\in G} \mathrm{Re}\langle w, z\rangle = \sup_{w\in G} \mathrm{Re} \sum_{j=1}^{N} w_j \bar{z}_j, \qquad z \in \mathbb{C}^N,$$

is plurisubharmonic and belongs to $\mathcal{H}_\rho$.

1.4 LEMMA. *Let $\rho > 0$ and let $H: \mathbb{C}^N \to \mathbb{R}$ be positively homogeneous of order ρ. Then $H \in \mathcal{H}_\rho$ if and only if there is $C > 0$ such that*

$$|H(z) - H(w)| \leq C \max\{|z|^{\rho-1}, |w|^{\rho-1}\}|z - w|, \qquad z, w \in \mathbb{C}^N\backslash\{0\}.$$

In particular, if $H \in \mathcal{H}_\rho$ then for each $k \geq 1$ there is $L > 0$ such that ($r(z)$ as in 1.1)

$$|H(z) - H(w)| \leq L \quad \text{if } z \in \mathbb{C}^N,\ w \in B(z, k\, r(z)). \tag{$*$}$$

Proof. If $H \in \mathcal{H}_\rho$, let $C > 0$ as in Definition 1.1. Let $z, w \in \mathbb{C}^N\backslash\{0\}$. Then

$$|H(z/|z|) - H(w/|w|)| \leq C|z/|z| - w/|w||$$

and, since $|w - |z|w/|w|| = ||w| - |z|| \leq |z - w|$,

$$|H(z) - H(|z|w/|w|)| \leq C|z|^{\rho-1}|z - |z|w/|w|| \leq 2C|z|^{\rho-1}|z - w|.$$

On the other hand, since $||z|^\rho - |w|^\rho| \leq \rho \max\{|z|^{\rho-1}, |w|^{\rho-1}\}||z| - |w||$, we have

$$|H(|z|w/|w|) - H(w)| = |H(w/|w|)|\,||z|^\rho - |w|^\rho| \leq \tilde{C}\rho \max\{|z|^{\rho-1}, |w|^{\rho-1}\}|z - w|$$

with some $\tilde{C} > 0$ only depending on H. Both parts together give the desired estimate.

1.5 DEFINITION. Let $\rho > 0$. Let $\Omega = (\omega_k)_{k\in\mathbb{N}}$ be a weight system and let $H \in \mathcal{H}_\rho$. By $A(\mathbb{C}^N)$ we denote the space of all entire functions on $\mathbb{C}^N$. We define

$$A^0_{H+\Omega} := \left\{ f \in A(\mathbb{C}^N) \;\middle|\; \|f\|_k = \sup_{z\in\mathbb{C}^N} |f(z)|e^{-H(z)-\omega_k(z)} < \infty \text{ for each } k \in \mathbb{N} \right\},$$

$$A_{H-\Omega} := \left\{ f \in A(\mathbb{C}^N) \;\middle|\; |f|_k = \sup_{z\in\mathbb{C}^N} |f(z)|e^{-H(z)+\omega_k(z)} < \infty \text{ for some } k \in \mathbb{N} \right\}.$$

Endowed with the topology induced by $(\|\ \|_k)_{k\in\mathbb{N}}$ and $(|\ \ |_k)_{k\in\mathbb{N}}$, $A^0_{H+\Omega}$ is a nuclear Fréchet space and $A_{H-\Omega}$ is a (DFN)–space, respectively (however, we will not apply this information in the sequel). Obviously for each $L \in \mathcal{H}_\rho$ we have $A^0_{L+\Omega} \cdot A^0_{H+\Omega} \subset A^0_{H+L+\Omega}$ and $A^0_{L+\Omega} \cdot A_{H-\Omega} \subset A_{H+L-\Omega}$. If $L = 0$ we write $A^0_\Omega = A^0_{L+\Omega}$.

The crucial part of our procedure is a comparison of the weight function $H \pm \omega_k$ and the solution of the Dirichlet problem for the complex Monge–Ampère equation on the ball $B(a, R)$ with boundary values $H \pm \omega_n$. We need the following notations.

1.6 NOTATIONS AND REMARKS. Let $H\colon \mathbb{C}^N \to \mathbb{R}$ be positively homogeneous of order $\rho > 0$ and plurisubharmonic (psh).

a) For $a \in \mathbb{C}^N$ and $R \geq 0$ we consider the function $h = h(H; a, R) : \mathbb{C}^N \to \mathbb{R}$, which equals H on $\mathbb{C}^N \backslash U(a,R)$ and with

$$h(z) := \sup\left\{u(z) \,\middle|\, u \text{ psh on } U(a,R),\ u^*(\zeta) := \limsup_{w\to\zeta} u(w) \leq H(\zeta) \text{ for } \zeta \in S(a,R)\right\}$$

for $z \in U(a,R)$. By [12], Lemma 2, h is plurisubharmonic, and h is continuous if H is continuous. The function $d(H; a, R)\colon z \mapsto h(H; a, R)(z) - H(z)$, $z \in \mathbb{C}^N$, is nonnegative and vanishes outside $U(a,R)$. We define

$$\Delta(H; a, R) := \sup_{z\in U(a,R)} d(H; a, R)$$

and note that $\Delta(H; ta, tR) = t^\rho \Delta(H; a, R)$ for all $t \geq 0$, $a \in \mathbb{C}^N$ and $R \geq 0$. Since $h(H; a, R)$ is plurisubharmonic, we have $\Delta(H; a, R) \leq \Delta(H; \tilde{a}, \tilde{R})$ if $U(a,R) \subset U(\tilde{a}, \tilde{R})$.

b) Let Γ_H denote the set of all $a \in \mathbb{C}^N$ such that $\Delta(H; a, R) > 0$ for all $R > 0$. Because of the homogeneity of Δ, Γ_H is a cone, i.e., $\mathbb{R}_+ \cdot \Gamma_H \subset \Gamma_H$. Since $a \in \mathbb{C}^N \backslash \Gamma_H$ if and only if $\Delta(H; a, R) = 0$ for some $R > 0$, and by a), the set $\mathbb{C}^N \backslash \Gamma_H$ is open. Thus Γ_H is closed. By Bedford and Taylor [2], §9, the closed cone Γ_H is the support of the Monge-Ampère measure $(dd^c H)^N$. We will sometimes use the abreviation $S_H := \Gamma_H \cap S$.

c) If $\tilde{H}\colon \mathbb{C}^N \to \mathbb{R}$ is another plurisubharmonic function which is positively homogeneous of order ρ, and if $a, \tilde{a} \in \mathbb{C}^N$ and $R, \tilde{R} > 0$, then

$$\begin{aligned}
\Delta(H; a, R) &= \sup_{z\in B(0,1)} \left(h(H(a + R\zeta); 0, 1)(z) - H(a + Rz)\right) \\
&\leq \sup_{z\in B(0,1)} \left(h(\tilde{H}(\tilde{a} + \tilde{R}\zeta); 0, 1)(z) - \tilde{H}(\tilde{a} + \tilde{R}z)\right) + \\
&\qquad\qquad 2 \sup_{z\in B(0,1)} |H(a + Rz) - \tilde{H}(\tilde{a} + \tilde{R}z)| \\
&= \Delta(H; \tilde{a}, R) + 2 \sup_{z\in B(0,1)} |H(a + Rz) - \tilde{H}(\tilde{a} + \tilde{R}z)|
\end{aligned}$$

and thus

$$|\Delta(H; a, R) - \Delta(\tilde{H}; \tilde{a}, \tilde{R})| \leq 2 \sup_{z\in B(0,1)} |H(a + Rz) - \tilde{H}(\tilde{a} + \tilde{R}z)|.$$

In particular, if H is continuous then $\Delta(H; a, R)$ is a continuous function of $(a, R) \in \mathbb{C}^N \times \mathbb{R}_+$.

d) Let $a \in \Gamma_H$. The function $R \mapsto \Delta(H; a, R)$, $R \geq 0$, is strictly increasing and unbounded. To see this, we prove that $\Delta(a,R)/(|a| + R)^\rho$ is a nondecreasing function of $R > 0$: If $R < \tilde{R}$ and $\lambda := (|a| + \tilde{R})/(|a| + R)$ we have $U(\lambda a, \lambda R) \subset U(a, \tilde{R})$ and thus by a), that $\lambda^\rho \Delta(a, R) \leq \Delta(a, \tilde{R})$. If H is continuous, thus by c), the function $\Delta(H; a, \cdot)$ maps $\mathbb{R}_+$ onto $\mathbb{R}_+$. By $\Delta^{-1}(H; a, \cdot) : \mathbb{R}_+ \to \mathbb{R}_+$ we denote its inverse function. If $a \in \Gamma_H$, $x \geq 0$ and $t \geq 0$, then $\Delta^{-1}(H; ta, t^\rho x) = t\Delta^{-1}(H; a, x)$.

e) If $H \in \mathcal{H}_\rho$ and is plurisubharmonic (i.e., if H is in addition Lipschitz continuous on S), then by 1.4(∗) and by c), there is $L > 0$ with

$$\Delta(H; a, R + r(a)) \leq \Delta(H; a, R) + L, \qquad a \in \mathbb{C}^N,\ R \in \mathbb{R}_+,$$

with

$$\Delta^{-1}(H;a,x)+r(a)\le\Delta^{-1}(H;a,x+L),\qquad a\in\Gamma_H,\ x\in\mathbb{R}_+,$$

and

$$\Delta(H;a,R)\le\sup_{z\in B(a,R)}\Big(h(H(a)+LR;a,R)(z)-H(a)+LR\Big)=2LR,$$

for all $a\in S$ and $0\le R\le 1/2$.

In the sequel we will omit the index H if no confusion is possible.

1.7 LEMMA. *If $\rho=1$ and $H\in\mathcal{H}_\rho$ is the support function of a bounded convex domain in $\mathbb{C}^N$, then for each weight system $\Omega=(\omega_k)_{k\in\mathbb{N}}$ which consists of concave functions ω_k, $k\in\mathbb{N}$, there is a sequence $(u_k)_{k\in\mathbb{N}}$ of plurisubharmonic functions which is equivalent to $(H-\omega_k)_{k\in\mathbb{N}}$ in the following sense: For each $k\in\mathbb{N}$ there are $n\in\mathbb{N}$ and $C>0$ such that*

$$H-\omega_k\le u_n+C\quad\text{and}\quad u_k\le H-\omega_n+C\qquad\text{on }\mathbb{C}^N.$$

For $\rho>0$ the same is true in the following two cases:

(a) $\Omega=(\frac{1}{k}|z|^\rho)_{k\in\mathbb{N}}$, $H\in\mathcal{H}_\rho$ is plurisubharmonic and $A_{H-\Omega}\neq\{0\}$.

(b) $\Omega=(\omega_k)_{k\in\mathbb{N}}$ has concave functions $t\mapsto\omega_k(t^{1/\rho})$, $t\ge0$ $(k\in\mathbb{N})$, and $H\in\mathcal{H}_\rho$ is twice continuously differentiable on $\mathbb{C}^N\backslash\{0\}$ and strictly plurisubharmonic (hence $H(z)-\varepsilon|z|^\rho$ is plurisubharmonic for some $\varepsilon>0$).

In particular (by 1.1 and 1.4(*)), in these three cases the following holds: For each $k\in\mathbb{N}$ there are $n\in\mathbb{N}$, $R,C>0$ such that for each $x\in\Gamma_H$, $|x|\ge R$ there is a plurisubharmonic function u with

$$H(x)-\omega_k(x)\le\inf_{z\in B(x,r(x))}u(z)\quad\text{and}\quad u\le H-\omega_n+C.\tag{$*$}$$

Proof. To prove the first assertion, we fix $k\in\mathbb{N}$ and put $h_k(t):=\sup_{x>0}(\omega_k(x)-xe^{-t})\in]-\infty,\infty]$, $t\in\mathbb{R}$. Since ω_k is concave, we have

$$\omega_k(r)=\inf_{y>0}(rx+h_k(-\log y)),\qquad r>0.$$

The proof of Lemma 6 of Napalkov [15] then shows that there is some $K>0$ only depending on G and some $C_k>0$ with

$$H(z)-\omega_k(Kz)-C_k\le u_k(z)\le H(z)-\omega_k(z)+C_k,\qquad z\in\mathbb{C}^N,$$

where

$$u_k(z):=\sup_{\zeta\in G}(\mathrm{Re}\langle z,\zeta\rangle-h_k(-\log d(\zeta))),\qquad z\in\mathbb{C}^N,$$

and where $d(\zeta)$ is the distance of ζ to the boundary of G. Since u_k is plurisubharmonic, we have proved the first assertion.

(a) Let $f\in A_{H-\Omega}\backslash\{0\}$. The regularized radial indicator

$$v(z):=\left(\limsup_{r\to\infty}\frac{\log|f(rz)|}{r^\rho}\right)^*,\qquad z\in\mathbb{C}^N,$$

is plurisubharmonic and positively homogeneous of order ρ with

$$-\infty < v(z) \leq H(z) - \varepsilon|z|^\rho, \qquad z \in \mathbb{C}^N,$$

for some $\varepsilon > 0$. By Lelong and Gruman [10], Thm. 7.6, we may assume that v is C^∞ on $\mathbb{C}^N \backslash \{0\}$, hence $v \in \mathcal{H}_\rho$. We put $u_k := (1 - 1/k)H + 1/kv$, $k \in \mathbb{N}$. Then $u_k(z) \leq H(z) - \varepsilon/k|z|^\rho$, $z \in \mathbb{C}^N$, $k \in \mathbb{N}$. For $k \in \mathbb{N}$ choose $n \geq k \max_{z\in S}(H(z) - v(z))$. This gives for $z \in S$

$$H(z) - 1/k|z|^\rho = (1 - 1/n)H(z) + 1/n(H(z) - v(z)) + 1/nv(z) - 1/k|z|^\rho \leq u_n(z).$$

Since all the functions are positively homogeneous of order ρ, the same estimate holds for all $z \in \mathbb{C}^N$.

(b) Since $t \mapsto \omega_k(t^{1/\rho})$ is concave, for each sufficiently large k we can modify ω_k in a neighborhood of zero to get $\tilde{\omega}_k$ such that the unbounded convex function $t \mapsto \varepsilon t - \tilde{\omega}_k(t^{1/\rho})$ is nondecreasing on $\mathbb{R}_+$. Since $z \mapsto |z|^\rho$ is plurisubharmonic on $\mathbb{C}^N$, thus by Hörmander [7], Thm. 1.6.7, for these k the following function is plurisubharmonic

$$u_k(z) := H(z) - \tilde{\omega}_k(z) = H(z) - \varepsilon|z|^\rho + \varepsilon|z|^\rho - \tilde{\omega}_k(z), \qquad z \in \mathbb{C}^N.$$

Checking the proof of Theorem 4.4.4 of Hörmander [7], by standard arguments, one gets:

1.8 THEOREM. *Let $\varphi : \mathbb{C}^N \to \mathbb{R}$ be plurisubharmonic. Let $r : \mathbb{C}^N \to]0, \infty[$ (for example the function from 1.1b)), and let $x \in \mathbb{C}^N$. Then there is $f \in A(\mathbb{C}^N)$ with*

$$f(x) = \exp\Big(\inf_{w\in B(x,r(x))} \varphi(w) - 2N\log(1/r(x)) - N\log(1 + |x|^2)\Big),$$

$$|f(z)| \leq C\exp\Big(\sup_{w\in B(z,r(z))} \varphi(w) + N\log(1/r(z)) + 3N/2\log(1 + |z|^2)\Big), \qquad z \in \mathbb{C}^N,$$

where C is a positive constant which only depends on N.

2 CHARACTERIZATION OF THE FUNCTIONS FOR WHICH DIVISION IS POSSIBLE

The crucial part of this section are Propositions 2.3 and 2.4, where we prove that the Ehrenpreis condition is necessary for the division problem to hold. The Lemmas 2.1 and 2.2 are needed for sufficiency part; the first one has been proved in [13], Lemma 2.3, the second one in Hörmander [6], Lemma 3.2.

2.1 LEMMA. *For $0 < \lambda < 1$ and $z \in \mathbb{C}^N \backslash \{0\}$ let $0 < \tau(z) < \sigma(z) < |z|$ with $\sigma(z) = \tau(z)^\lambda |z|^{1-\lambda}$. Let F be analytic in $B(z, |z|)$. Let $\omega, \tilde{\omega}$: $B(z, |z|) \to \mathbb{R}$ with*

$$\frac{1}{\lambda} \sup_{w\in B(z,\sigma(z))} \omega(w) + \frac{1-\lambda}{\lambda} \log \sup_{w\in B(z,|z|)} |F(w)| \leq \inf_{w\in B(z,\tau(z))} \tilde{\omega}(w). \qquad (1)$$

If there is $w \in B(z,\sigma(z))$ with $|F(w)| \geq \exp(-\omega(w))$ then there is $w \in B(z,\tau(z))$ with $|F(w)| \geq \exp(-\tilde{\omega}(w))$.

2.2 LEMMA. Let g, F and g/F be entire. Then we have for every $r > 0$

$$|g(z)/F(z)| \leq \sup_{|z-\zeta|<4r} |g(\zeta)| \sup_{|z-\zeta|<4r} |F(\zeta)|/(\sup_{|z-\zeta|<r} |F(\zeta)|)^2.$$

2.3 PROPOSITION. *Let $\rho > 0$ and $H, L \in \mathcal{H}_\rho$, where H is plurisubharmonic, and let $\Omega = (\omega_k)_{k\in\mathbb{N}}$ be a weight system. For $F \in A^0_{L+\Omega}\backslash\{0\}$ we have "(i)⇒(ii)⇒(iii)⇒(iv)":*

(i) $g \in A^0_{H+L+\Omega}$ and $f := g/F \in A(\mathbb{C}^N)$ imply $f \in A^0_{H+\Omega}$.

(ii) The subspace $FA^0_{H+\Omega}$ is closed in $A^0_{H+L+\Omega}$.

(iii) For each $n \in \mathbb{N}$ there are $k \in \mathbb{N}$ and $C > 0$ such that for all $f \in A^0_{H+\Omega}$ we have $\|f\|_n \leq C\|Ff\|_k$ (the first norm is taken in $A^0_{H+\Omega}$, the second one in $A^0_{H+L+\Omega}$).

(iv) For each $k \in \mathbb{N}$ there is $R > 0$ such that for each $z \in \Gamma_H$ with $|z| \geq R$ there exists $w \in B(z, \Delta^{-1}(H; z, \omega_k(z)))$ such that $|F(w)| \geq \exp(L(w) - \omega_k(w))$.

Proof. (i)⇒(ii): $A^0_{H+L+\Omega}$ is continuously embedded in $A(\mathbb{C}^N)$.

(ii)⇒(iii): Since $FA^0_{H+\Omega}$ is closed, by the open mapping theorem, the multiplication operator M_F: $A^0_{H+\Omega} \to A^0_{H+L+\Omega}$, $M_F(f) = Ff$, is a linear topological isomorphism onto $FA^0_{H+\Omega}$.

(iii)⇒(iv): For $k \in \mathbb{N}$ we put

$$\sigma_k(z) := \Delta^{-1}(z, \omega_k(z)) = |z|\Delta^{-1}(z/|z|, \omega_k(z)/|z|^\rho), \qquad z \in \Gamma_H\backslash\{0\}.$$

Since for each $\varepsilon > 0$ the continuous function $\Delta(\cdot, \varepsilon)$ is positive on the compact set S_H, we have $\lim_{x\to 0} \Delta^{-1}(a, x) = 0$ uniformly for $a \in S_H$. Hence by Definition 1.1, we may assume that $\sigma_k(z) \leq |z|/4$, $z \in \mathbb{C}^N$, $k \in \mathbb{N}$. Assume that (iv) does not hold. Then there are $\tilde{k} \in \mathbb{N}$ and a sequence $(z^j)_{j\in\mathbb{N}}$ in Γ_H with $\lim_{j\to\infty} |z^j| = \infty$ such that

$$|F(w)| < \exp(L(w) - \omega_{\tilde{k}}(w)) \text{ for all } w \in B(z^j, \sigma_{\tilde{k}}(z^j)) \text{ and } j \in \mathbb{N}. \tag{2}$$

By 1.1 and 1.4(∗) (see 1.6) we can choose $m \geq \tilde{k}$ and $t_0 > 0$ such that for $|z| \geq t_0$ we have $\omega_m(2z) \leq \omega_{\tilde{k}}(z)$ and

$$\sigma_m(z) + 2r(z/2) \leq \Delta^{-1}(z, \omega_m(z) + L') \leq \sigma_{\tilde{k}}(z) \text{ if } z \in \Gamma_H, |z| \geq t_0. \tag{3}$$

We are going to derive a contradiction to (iii). Fix $j \in \mathbb{N}$ with $|z^j| \geq 2t_0$. Put $h_j := h(H; z^j, \sigma_m(z^j))$. Choose $x^j \in B(z^j, \sigma_j(z^j))$ with

$$h_j(x^j) = H(x^j) + \Delta(z^j, \sigma_m(z^j)) = H(x^j) + \omega_m(z^j). \tag{4}$$

Note that $|z^j|/2 \leq |z| \leq 3/2|z^j|$ if $z \in B(z^j, \sigma_j(z^j) + 2r(z^j/2))$. In particular $r(x^j) \leq \min\{r(z^j/2), r(z/3)\}$ for all $z \in S(z^j, \sigma_m(z^j) + r(z^j/2))$. Hence by Lemma 1.4(∗), there is $L_1 > 0$ only depending on H with

$$\sup_{w\in B(z,r(x^j))} h_j(w) = \sup_{w\in B(z,r(x^j))} H(w) \leq H(z) + L_1, \qquad z \in S(z^j, \sigma_m(z^j) + r(z^j/2)).$$

Since the left-hand side is a plurisubharmonic function of $z \in \mathbb{C}^N$, we get

$$\sup_{w\in B(z,r(x^j))} h_j(w) \leq h\big(H + L_1; z^j, \sigma_m(z^j) + r(z^j/2)\big)(z) =: \varphi_j(z)$$

for all $z \in B(z^j, \sigma_m(z^j) + r(z^j/2))$, and

$$\inf_{z\in B(x^j,r(x^j))} \varphi_j(z) \geq \inf_{z\in B(x^j,r(x^j))} \sup_{w\in B(z,r(x^j))} h_j(w) \geq h_j(x^j).$$

Applying Theorem 1.8 to the plurisubharmonic function φ_j, we obtain a function $f_j \in A(\mathbb{C}^N)$ with

$$|f_j(x^j)| \geq \exp\Big(\inf_{w\in B(x^j,r(x^j))} \varphi_j(w) - C(N,\rho)\log(1 + |x^j|^2)\Big) \tag{5}$$

and

$$|f_j(z)| \leq C(N,\rho)\exp\Big(\sup_{w\in B(z,r(z))} \varphi_j(w) + C(N,\rho)\log(1 + |z|^2)\Big), \quad z \in \mathbb{C}^N, \tag{6}$$

where $C(N,\rho) > 0$ only depends on N and ρ. By 1.4(*) (see 1.6), there is $L_2 \geq L_1$ only depending on H with

$$\begin{aligned} \varphi_j(z) &\leq H(z) + L_1 + \Delta(z^j, \sigma_m(z^j) + r(z^j/2)) \\ &\leq H(z) + \Delta(z^j, \sigma_m(z^j)) + L_2 \\ &= H(z) + \omega_m(z^j) + L_2, \qquad z \in \mathbb{C}^N. \end{aligned}$$

If $z \in B(z^j, \sigma_m(z^j) + 2r(z^j/2))$ we have $r(z) \leq r(z^j/2)$. Since we may assume that the function $t \mapsto t + r(t)$ is nondecreasing for $t \geq 0$, $B(z, r(z))$ does not intersect $B(z^j, \sigma_m(z^j) + r(z^j/2))$ for each $z \notin B(z^j, \sigma_m(z^j) + 2r(z^j/2))$. Hence by Lemma 1.4(*), there is $L_3 > L_2$ only depending on H with

$$\sup_{w\in B(z,r(z))} \varphi_j(w) \leq \begin{cases} H(z) + L_3 & \text{for } z \notin B(z^j, \sigma_m(z^j) + 2r(z^j/2)) \\ H(z) + L_3 + \omega_m(z^j) & \text{for } z \in B(z^j, \sigma_m(z^j) + 2r(z^j/2)). \end{cases} \tag{7}$$

In particular, $f_j \in A^0_{H+\Omega}$.

We now estimate Ff_j. If $z \in B(z^j, \sigma_m(z^j) + 2r(z^j/2))$, then $|z| \geq |z^j|/2 \geq t_0$, and by (2), (3), (6), (7), we obtain

$$\begin{aligned} |F(z)f_j(z)| &\leq C(N,\rho)e^{L_3}\exp(H(z) + L(z) + C(N,\rho)\log(1 + |z|^2) + \omega_m(2z) - \omega_{\tilde{k}}(z)) \\ &\leq C(N,\rho)e^{L_3}\exp(H(z) + L(z) + C(N,\rho)\log(1 + |z|^2)). \end{aligned}$$

Hence by (6), (7), and 1.1, for each $k \in \mathbb{N}$

$$\sup_{j\in\mathbb{N}} \sup_{z\in\mathbb{C}^N} |F(z)f_j(z)|\exp(-H(z) - L(z) - \omega_k(z)) < \infty. \tag{8}$$

Since $|z^j| \geq 2/3|x^j|$, on the other hand, we obtain from (5) and (4)

$$|f_j(x^j)| \geq \exp\Big(H(x^j) - C(N,\rho)\log(1 + |x^j|^2) + \omega_m((2/3)x^j)\Big).$$

Since $|x^j| \geq |z^j|/2$, we have $\lim_{j\to\infty} |x^j| = \infty$. Hence by 1.1, we conclude that for some $n \in \mathbb{N}$

$$\lim_{j\to\infty} |f_j(x^j)| \exp\left(-H(x^j) - \omega_n(x^j)\right) = \infty.$$

Together with (8) this is a contradiction to (iii).

2.4 PROPOSITION. *Let $\rho > 0$ and $H, L \in \mathcal{H}_\rho$, where H is plurisubharmonic, and let $\Omega = (\omega_k)_{k\in\mathbb{N}}$ be a weight system such that 1.7(*) holds. For $F \in A^0_{L+\Omega}\backslash\{0\}$ we have "(i)⇒(ii)⇒(iii)⇒(iv)":*

- *(i) $g \in A_{H+L-\Omega}$ and $f := g/F \in A(\mathbb{C}^N)$ imply $f \in A_{H-\Omega}$.*
- *(ii) The subspace $FA_{H-\Omega}$ is closed in $A_{H+L-\Omega}$.*
- *(iii) For each $k \in \mathbb{N}$ there are $n \in \mathbb{N}$ and $C > 0$ such that for all $f \in A_{H-\Omega}$ we have $|f|_n \leq C|Ff|_k$ (the first norm is taken in $A_{H-\Omega}$, the second one in $A_{H+L-\Omega}$).*
- *(iv) For each $k \in \mathbb{N}$ there is $R > 0$ such that for each $z \in \Gamma_H$ with $|z| \geq R$ there exists $w \in B(z, \Delta^{-1}(H; z, \omega_k(z)))$ such that $|F(w)| \geq \exp(L(w) - \omega_k(w))$.*

Proof. (i)⇒(ii): As Proposition 2.3.

(ii)⇒(iii): Since $FA_{H-\Omega}$ is closed, $\operatorname{ind}_{k\to\infty}(A_{H+L-\omega_k} \cap FA_{H-\Omega})$ is an (LF)–space, where $A_{H+L-\omega_k} := \{f \in A_{H+L-\Omega} \,|\, |f|_k < \infty\}$. Hence we can apply the open mapping theorem for (LF)–spaces to the continuous and bijective map $M_F\colon A_{H-\Omega} \to \operatorname{ind}_{k\to\infty}(A_{H+L-\omega_k} \cap FA_{H-\Omega})$, $M_F(f) = Ff$, which shows that M_F is a linear topological isomorphism onto $FA_{H-\Omega}$ endowed with the topology of $\operatorname{ind}_{k\to\infty}(A_{H+L-\omega_k} \cap FA_{H-\Omega})$. Thus (iii) follows from Grothendieck's factorization theorem (see e.g., Köthe [8], §19,5.(4)).

(iii)⇒(iv): We use the notations and agreements of the proof of 2.3 and fix $j \in \mathbb{N}$. According to 1.7(*) and Definition 1.1, we choose a plurisubharmonic function u_j with

$$H(x^j) - \omega_m(z^j) \leq \inf_{x\in B(x^j, r(x^j))} u_j(x) \quad \text{and} \quad u_j \leq H - \omega_{\tilde{n}} + C,$$

for some $\tilde{n} \in \mathbb{N}$ and $C > 0$. We define the plurisubharmonic function $\tilde{\varphi}_j := 1/2\varphi_j + 1/2u_j$. Then

$$\inf_{x\in B(x^j, r(x^j))} \tilde{\varphi}_j(x) \geq 1/2h_j(x^j) + 1/2 \inf_{x\in B(x^j, r(x^j))} u_j(x) \geq H(x^j).$$

Applying Theorem 1.8, we get $f_j \in A(\mathbb{C}^N)$ such that (5) and (6) hold with φ_j replaced by $\tilde{\varphi}_j$. By 1.4(*) and (7), there is $L_3' > 0$ not depending on j with

$$\sup_{w\in B(z, r(z))} \tilde{\varphi}_j(w) \leq \begin{cases} H(z) - \frac{1}{2}\omega_{\tilde{n}}(|z| - r(z)) + L_3' & \text{for } z \notin B(z^j, r_j) \\ H(z) - \frac{1}{2}\omega_{\tilde{n}}(|z| - r(z)) + L_3' + \omega_m(z^j) & \text{for } z \in B(z^j, r_j), \end{cases}$$

where $r_j := \sigma_m(z^j) + 2r(z^j/2)$. In particular $f_j \in A_{H-\Omega}$. As in 2.3 we get the following contradiction to (iii): For some $k \in \mathbb{N}$ and for each $n \in \mathbb{N}$ we have

$$\sup_{j\in\mathbb{N}} \sup_{z\in\mathbb{C}^N} |F(z)f_j(z)| \exp(-H(z) - L(z) + \omega_k(z)) < \infty,$$

$$\lim_{j\to\infty} |f_j(x^j)| \exp(-H(x^j) + \omega_n(x^j)) = \infty.$$

2.5 LEMMA. *Let $\rho > 0$ and $H, L \in \mathcal{H}_\rho$, where H is plurisubharmonic, and let $\Omega = (\omega_k)_{k\in\mathbb{N}}$ be a weight system. Let $F \in A^0_{L+\Omega}$. Then "(ii)⇒(i)". If $L = 0$ and there is $\lambda > 0$ such that for each $n \in \mathbb{N}$ there are $k \in \mathbb{N}$ and $r_0 > 0$ with*

$$\omega_k(z)/|z|^\rho \le \Delta(H; z/|z|, (\omega_n(z)/|z|^\rho)^\lambda), \quad z \in \Gamma_H, |z| \ge r_0,$$

then also "(i)⇒(ii)".

(i) *For each $k \in \mathbb{N}$ there is $R > 0$ such that for each $z \in \Gamma_H$ with $|z| \ge R$ there exists $w \in B(z, \Delta^{-1}(H; z, \omega_k(z)))$ such that $|F(w)| \ge \exp(L(w) - \omega_k(w))$.*

(ii) *(i) holds with $\Delta^{-1}(H; z, \omega_k(z)) = |z|\Delta^{-1}(H; z/|z|, \omega_k(z)/|z|^\rho)$ replaced by the radius $|z|^{1-\rho}\omega_k(z)$.*

Proof. (ii)⇒(i): By 1.6, there is $K > 0$ such that for large $k \in \mathbb{N}$ and large $|z|$ with $z \in \Gamma_H$

$$\Delta^{-1}(z, \omega_k(z)) = |z|\Delta^{-1}(z/|z|, \omega_k(z)/|z|^\rho) \ge 1/K|z|\omega_k(z)/|z|^\rho = 1/K|z|^{1-\rho}\omega_k(z).$$

By 1.1 this gives the assertion.

(i)⇒(ii): Let $L = 0$ and let the extra hypothesis be fulfilled. We may assume that $0 < \lambda < 1$. Let $\tilde{k} \in \mathbb{N}$. By the properties of Δ (see 1.6), and by the extra hypothesis on H and Ω, there are $k \in \mathbb{N}$ and $R > 0$ with $\Delta^{-1}(z, \omega_k(z)) \le (|z|^{1-\rho}\omega_{\tilde{k}}(z))^\lambda |z|^{1-\lambda}$, $z \in \Gamma_H$, $|z| \ge R$. We may assume that R is so large that (i) holds. We put $\tau(z) := |z|^{1-\rho}\omega_{\tilde{k}}(z)$. By 1.1, we may furthermore assume that k and R are chosen so large that for all $|z| > R$ we have $\sigma(z) := \tau(z)^\lambda |z|^{1-\lambda} < |z|/2$ and that

$$\frac{1}{\lambda}\omega_k(2|z|) + \frac{1-\lambda}{\lambda} \log \sup_{w\in B(z,|z|)} |F(w)| \le \omega_{\tilde{k}}(|z|/2).$$

Applying 2.1 with $\omega = \omega_k$ and $\tilde{\omega} = \omega_{\tilde{k}}$, we get (ii) from (i).

2.6 DEFINITION. Let $\rho > 0$, $H \in \mathcal{H}_\rho$ and let $\Omega = (\omega_k)_{k\in\mathbb{N}}$ be a weight system. We will say that $A^0_{H+\Omega}$ has the Phragmén-Lindelöf-property if each entire function $f \in A(\mathbb{C}^N)$ with

$$\sup_{z\in\mathbb{C}^N} |f(z)| \exp(-C|z|^\rho) < \infty \text{ for some } C > 0$$

and

$$\sup_{z\in\Gamma_H} |f(z)| e^{-H(z)-\omega_k(z)} < \infty$$

for each $k \in \mathbb{N}$, already belongs to $A^0_{H+\Omega}$. Similary we say that $A_{H-\Omega}$ has the Phragmén-Lindelöf-property if each entire function $f \in A(\mathbb{C}^N)$ with

$$\sup_{z\in\mathbb{C}^N} |f(z)| \exp(-C|z|^\rho) < \infty \text{ for some } C > 0$$

and

$$\sup_{z\in\Gamma_H} |f(z)| e^{-H(z)+\omega_k(z)} < \infty$$

for some $k \in \mathbb{N}$, already belongs to $A_{H-\Omega}$.

The following four theorems, collect our results on the solution of the division problem. In section 3 they will be applied to more concrete situations, in which the two hypotheses on Ω and H, the Phragmén-Lindelöf property and a certain lower bound for the Monge-Ampère measure of H, can be fulfilled. 2.7 and 2.8 concern the special case that $L = 0$. Theorem 2.8 shows that our procedure is optimal at least in the case of one complex variable. 2.9 and 2.10 concern the case of arbitrary $L \in \mathcal{H}_\rho$. That is why we assume a stronger lower bound for the Monge-Ampère measure of H in this case.

2.7 THEOREM. *Let $\rho > 0$, $L = 0$, let $H \in \mathcal{H}_\rho$ be plurisubharmonic, and let $\Omega = (\omega_k)_{k\in\mathbb{N}}$ be a weight system. We assume that $A^0_{H+\Omega}$ has the Phragmén-Lindelöf-property and that there is some $\lambda > 0$ such that for each $n \in \mathbb{N}$ there are $k \in \mathbb{N}$ and $r_0 > 0$ with*

$$\omega_k(z)/|z|^\rho \le \Delta(H; z/|z|, (\omega_n(z)/|z|^\rho)^\lambda), \quad z \in \Gamma_H, |z| \ge r_0.$$

Then for each $F \in A^0_\Omega \backslash \{0\}$ the following are equivalent:

(i) Whenever $g \in A^0_{H+\Omega}$ and $f := g/F \in A(\mathbb{C}^N)$, then $f \in A^0_{H+\Omega}$.

(ii) $FA^0_{H+\Omega}$ is a closed subspace of $A^0_{H+\Omega}$.

(iii) For each $n \in \mathbb{N}$ there are $k \in \mathbb{N}$ and $C > 0$ such that for all $f \in A^0_{H+\Omega}$ we have $\|f\|_n \le C\|Ff\|_k$.

(iv) For each $k \in \mathbb{N}$ there is $R > 0$ such that for each $z \in \Gamma_H$ with $|z| \ge R$ there exists $w \in B(z, \Delta^{-1}(H; z, \omega_k(z)))$ such that $|F(w)| \ge \exp(-\omega_k(w))$.

(v) For each $k \in \mathbb{N}$ there is $R > 0$ such that for each $z \in \Gamma_H$ with $|z| \ge R$ there exists $w \in B(z, |z|^{1-\rho}\omega_k(z))$ such that $|F(w)| \ge \exp(-\omega_k(w))$.

The same assertion is true if "$A^0_{H+\Omega}$" is replaced by "$A_{H-\Omega}$" and (iii) is replaced by 2.4(iii), if we assume moreover that 1.7(*) holds.

Proof. From 2.3, 2.4 and 2.5 we obtain "(i)⇒(ii)⇒(iii)⇒(iv)⇒(v)". As in [13], Thm. 2.5, the proof of "(v)⇒(i)" follows by 2.2 and 1.4 from the Phragmén-Lindelöf-property. For sake of completeness, we will repeat the proof in the case of the space $A_{H-\Omega}$. The proof for the space $A^0_{H+\Omega}$ is analogous.

Let $g \in A_{H-\Omega}$ with $f := g/F \in A(\mathbb{C}^N)$. Let $n \in \mathbb{N}$ with $|g|_n < \infty$. For $k \ge n$ (which will determined later) we choose $R > 0$ according to (v). We may assume $|F(\zeta)| \le \exp \omega_k(\zeta)$, $\zeta \in \mathbb{C}^N$, and that $0 < |z|^{1-\rho}\omega_k(z) \le |z|/8$, $|z| \ge R$. Fix $z \in \Gamma_H$ with $|z| \ge R$. We apply Lemma 2.2 with $r = |z|^{1-\rho}\omega_k(z)$. Since H is Lipschitz continuous on S, by 1.4(*) there is a constant $L > 0$ depending on H such that

$$|f(z)| \le |g|_n \exp(H(z) + 4L\omega_k(z) - \omega_n(z/2) + \omega_k((3/2)z) + 2\omega_k((3/2)z)).$$

If k is chosen sufficiently large, by 1.1 there is $C > 0$ such that

$$\log|f(z)| \le C + \log|g|_n + H(z) - \omega_k(z), \quad z \in \Gamma_H.$$

Since it is well known that f is of at most order ρ and finite type, we thus get from the Phragmén-Lindelöf property of $A_{H-\Omega}$ that $f \in A_{H-\Omega}$.

Note that we do not assume a lower bound for $\Delta(H;\cdot,\cdot)$ in the following theorem.

2.8 THEOREM. *Let $N = 1$ and $\rho > 1$. Let $L = 0$, let $H \in \mathcal{H}_\rho$ be subharmonic, and let $\Omega = (\omega_k)_{k\in\mathbb{N}}$ be a weight system. We assume that $A^0_{H+\Omega}$ has the Phragmén-Lindelöf-property. Then for each $F \in A^0_\Omega\backslash\{0\}$ the conditions (i)-(iv) of Theorem 2.7 are equivalent. The same assertion is true if "$A^0_{H+\Omega}$" is replaced by "$A_{H-\Omega}$" and (iii) is replaced by 2.4(iii), if we assume moreover that 1.7(*) holds.*

Proof. In view of Theorem 2.7, we have to prove "(iv)$\Rightarrow$(i)". For sake of completeness, we copy the proof from [13], Corollary 2.6, for the space $A_{H-\Omega}$. The proof for the space $A^0_{H+\Omega}$ is analogous.

Let $g \in A_{H-\Omega}$ with $f := g/F \in A(\mathbb{C})$. Let $n \in \mathbb{N}$ with $|g|_n < \infty$. For $k \geq n$ (which will determined later) we choose $R > 0$ according to (iv). By 1.1, we may assume $|F(\zeta)| \leq \exp\omega_k(\zeta)$, $\zeta \in \mathbb{C}$, and furthermore by 1.1, 1.6(d) and by the continuity of Δ that $\Delta^{-1}(z,\omega_k(z)) \leq |z|/(2(4e+1))$, $z \in \Gamma_H$, $|z| \geq R$. Fix $z \in \Gamma_H$ with $|z| \geq R$. Choose $w \in \mathbb{C}$ according to (iv). Note that for $|\zeta - w| \leq (4e/(2(4e+1)))|z|$ we have

$$|z| \leq |z-w| + |w-\zeta| + |\zeta| \leq |z|/(2(4e+1)) + (4e/(2(4e+1)))|z| + |\zeta| \leq |z|/2 + \zeta,$$

hence $|z| \leq 2|\zeta|$; on the other hand $|\zeta| \leq |z| + |z-w| + |w-\zeta| \leq (3/2)|z|$.

We may assume that $w \neq z$. We apply the minimum modulus theorem from Levin [11], Thm. 11, with $\eta = ((5/4)-1)/((5/4)16)$ and $R = (5/4)|w-z|$ and get $|w-z| < r_1 < |w-z|5/4$ and absolute constants $a, A > 0$ such that

$$|F(\zeta)| \geq |F(w)|^{a+1}\Big(\max_{|x-w|\leq 2eR}|F(x)|\Big)^{-a} \geq \exp(-A\omega_k(A\zeta)), \qquad |\zeta - w| = r_1.$$

We choose $\zeta_0 \in \mathbb{C}$ with $|\zeta_0 - w| = r_1$ and $|\zeta_0 - z| < |w-z|/4$. We apply [11], Thm. 11, again, with $\eta = 1/32$ and $R = 2|w-z|/4$ and get $|w-z|/4 < \tilde{r}_1 < |w-z|/2$ and an absolute constant $\tilde{A} > 0$ such that

$$|F(\zeta)| \geq \exp(-\tilde{A}\omega_k(\tilde{A}\zeta)), \qquad |\zeta - \zeta_0| = \tilde{r}_1$$

and

$$|f(\zeta)| = |g(\zeta)/F(\zeta)| \leq |g|_n \exp(H(\zeta) - \omega_n(\zeta) + \tilde{A}\omega_k(\tilde{A}\zeta)), \qquad |\zeta - \zeta_0| = \tilde{r}_1.$$

Since $z \in B(\zeta_0, \tilde{r}_1) \subset B(z, \Delta^{-1}(z,\omega_k(z)))$, we get from 1.6(a)

$$\begin{aligned}\log|f(z)| &\leq \log|g|_n + h(H; z, \Delta^{-1}(z,\omega_k(z)))(z) - \omega_n(z/3) + \tilde{A}\omega_k(3\tilde{A}z)\\ &\leq \log|g|_n + H(z) + \omega_k(z) - \omega_n(z/3) + \tilde{A}\omega_k(3\tilde{A}z).\end{aligned}$$

If k is chosen sufficiently large, by 1.1, there is $C > 0$ such that

$$\log|f(z)| \leq C + \log|g|_n + H(z) - \omega_k(z), \quad z \in \Gamma_H.$$

Since it is well known that f is at most of order ρ and finite type, the Phragmén-Lindelöf property of $A_{H-\Omega}$ gives $f \in A_{H-\Omega}$.

2.9 THEOREM. *Let $\rho > 0$, let $H, L \in \mathcal{H}_\rho$, where H is plurisubharmonic, and let $\Omega = (\omega_k)_{k\in\mathbb{N}}$ be a weight system. We assume that $A^0_{H+\Omega}$ has the Phragmén-Lindelöf-property and that for each $n \in \mathbb{N}$ there are $k \in \mathbb{N}$ and $r_0 > 0$ with*

$$\omega_k(z)/|z|^\rho \leq \Delta(H; z/|z|, \omega_n(z)/|z|^\rho), \quad z \in \Gamma_H, |z| \geq r_0.$$

Then for each $F \in A^0_{L+\Omega}\backslash\{0\}$ the following are equivalent:

(i) Whenever $g \in A^0_{L+H+\Omega}$ *and* $f := g/F \in A(\mathbb{C}^N)$, *then* $f \in A^0_{H+\Omega}$.

(ii) $FA^0_{H+\Omega}$ *is a closed subspace of* $A^0_{L+H+\Omega}$.

(iii) For each $n \in \mathbb{N}$ *there are* $k \in \mathbb{N}$ *and* $C > 0$ *such that for all* $f \in A^0_{H+\Omega}$ *we have* $\|f\|_n \le C\|Ff\|_k$ *(the first norm is take in* $A^0_{H+\Omega}$*, the second one in* $A^0_{L+H+\Omega}$*).*

(iv) For each $k \in \mathbb{N}$ *there is* $R > 0$ *such that for each* $z \in \Gamma_H$ *with* $|z| \ge R$ *there exists* $w \in B(z, \Delta^{-1}(H; z, \omega_k(z)))$ *such that* $|F(w)| \ge \exp(L(w) - \omega_k(w))$.

(v) For each $k \in \mathbb{N}$ *there is* $R > 0$ *such that for each* $z \in \Gamma_H$ *with* $|z| \ge R$ *there exists* $w \in B(z, |z|^{1-\rho}\omega_k(z))$ *such that* $|F(w)| \ge \exp(L(w) - \omega_k(w))$.

The same assertion is true if "$A^0_{H+\Omega}$*" is replaced by "*$A_{H-\Omega}$*", "*$A^0_{L+H+\Omega}$*" is replaced by "*$A_{L+H-\Omega}$*", and (iii) is replaced by 2.4(iii), if we moreover assume that 1.7(*) holds.*

Proof. By the extra hypothesis on H and Ω, "(iv)⇒(v)" holds trivially. Thus the proof is the same as the one of Theorem 2.7.

The following is the result which has been announced in the introduction.

2.10 THEOREM. *Let* $\rho > 0$, *let* $H, L \in \mathcal{H}_\rho$, *where* H *is plurisubharmonic. Put* $\Omega := (\frac{1}{k}|z|^\rho)_{k\in\mathbb{N}}$. *If* $A^0_{H+\Omega}$ *has the Phragmén-Lindelöf-property, then the conditions (i)-(v) of Theorem 2.9 are equivalent for each* $F \in A^0_{L+\Omega}\setminus\{0\}$. *If* $A_{H-\Omega} \ne \{0\}$ *has the Phragmén-Lindelöf-property, then (i)-(v) of Theorem 2.9 are equivalent, with "*$A^0_{H+\Omega}$*" replaced by "*$A_{H-\Omega}$*", "*$A^0_{L+H+\Omega}$*" replaced by "*$A_{L+H-\Omega}$*", and (iii) replaced by 2.4(iii).*

Proof. By Theorem 2.9 and 1.7(a), we have only to note that by 1.6(b) and (c) for each $n \in \mathbb{N}$ there is $k \in \mathbb{N}$ with

$$\Delta(H; z/|z|, \omega_n(z)/|z|^\rho) = \Delta(H; z/|z|, \frac{1}{n}) \ge \frac{1}{k}, \quad z \in \Gamma_H\setminus\{0\}.$$

2.11 REMARK. Let $\rho > 0$, $L, H \in \mathcal{H}_\rho$, and let Ω be a weight system. If $F \in A^0_{L+\Omega}$ satisfies condition 2.9(iv), then we get for the regularized indicator of F that

$$\mathcal{L}_F(z) := \left(\limsup_{t>0, t\to\infty} t^{-\rho}\log|F(tz)|\right)^* = L(z), \quad z \in \Gamma_H.$$

Proof. Let condition 2.9(iv) hold. Since $F \in A^0_{L+\Omega}$, we have $\mathcal{L}_F \le L$ on $\mathbb{C}^N$. Assume now that there is $a_0 \in \Gamma_H \cap S$ with $\mathcal{L}_F(a_0) < L(a_0)$. Since $\mathcal{L}_F$ is upper semicontinuous and L is continuous, there is $\varepsilon > 0$ such that $\mathcal{L}_F(a) < L(a) - 2\varepsilon$ for all $a \in B(a_0, \varepsilon)$. By the Hartogs Lemma, there is $R > 0$ such that

$$t^{-\rho}\log|F(ta)| < L(a) - \varepsilon, \quad t \ge R, a \in B(a_0, \varepsilon),$$

and thus there is $\varepsilon' > 0$ with

$$\log|F(ta)| < L(ta) - \varepsilon'|ta|^\rho$$

for all $t \geq R$ and all $a \in \mathbb{C}^N$ with $|ta - ta_0| \leq \varepsilon |ta_0|$. Since by 1.1 and by 1.6(b) and (c), there is $k \in \mathbb{N}$ with

$$\Delta^{-1}(H; z, \omega_k(z)) = |z|\Delta^{-1}(H; z/|z|, |z|^{-\rho}\omega_k(z)) \leq |z|\varepsilon$$

for large $|z|$, this contradicts 2.9(iv).

3 APPLICATIONS

To apply the theorems 2.7 or 2.9, first one has to check the Phragmén-Lindelöf-property of $A^0_{H+\Omega}$ or $A_{H-\Omega}$, and secondly the lower bound for $\Delta(H; \cdot, \cdot)$ with respect to Ω.

To check the Phragmén-Lindelöf-property of certain weighted spaces $A^0_{H+\Omega}$ or $A_{H-\Omega}$, the following two results may be useful. The first one is contained in [12] (see also Krivosheev [9], Lemma 3.1). The second one is well known and applies to the case $N = 1$ or to functions $H(z) = \sum_{j=1}^N H_j(z_j)$, $z \in \mathbb{C}^N$, in the case $N \geq 1$.

3.1 PROPOSITION. *Let H be the support function of a bounded convex domain of $\mathbb{C}^N$. If u is plurisubharmonic on $\mathbb{C}^N$ with*

$$u(z) \leq C|z| + C, \quad z \in \mathbb{C}^N,$$

for some $C > 0$, then

$$u(z) \leq H(z) \text{ for all } z \in \Gamma_H,$$

implies

$$u(z) \leq H(z) \text{ for all } z \in \mathbb{C}^N.$$

3.2 LEMMA. *For $0 < \alpha < 2$ put $D := \{z \in \mathbb{C} \mid |\arg(z)| < \alpha\pi/2\}$. Let $\omega: \mathbb{R}_+ \to \mathbb{R}_+$ be a continuous nondecreasing function which vanishes in a neighborhood of zero, with $\omega(r) = o(r^{1/\alpha})$, and $\int_1^\infty \omega(r)/r^{1+1/\alpha} dr < \infty$. Put $\omega(z) := \omega(|z|)$, $z \in \mathbb{C}$. Then there is a continuous function $u: \overline{D} \to \mathbb{R}_+$, harmonic in D, with $u = \omega$ on ∂D, and*

$$u(z) \leq 9\sqrt{2}/\alpha \int_1^\infty \omega(4zt)/t^{1+1/\alpha} dt, \qquad z \in D.$$

(a) If there are $\varepsilon > 0$ and $s_0 > 0$ with $\omega(2s)/\omega(s) \leq 2^{1/\alpha - \varepsilon}$ for $s \geq s_0$, then there is $C = C(\varepsilon, \alpha) > 0$ with $u(z) \leq C\omega(z)$ for all $z \in D$ with $|z| \geq s_0$.

(b) If $x \mapsto \omega(e^x)$ is convex for $x \in \mathbb{R}$ then $u(z) \geq \omega(z)$ for all $z \in D$.

Proof. For sake of completeness we sketch a proof of this well known facts. We proceed as in Tsuji [19], Thm. III.70. For $r > 0$, let u_r denote the harmonic measure of $D \cap S(0, r)$ with respect to $D \cap U(0, r)$. As in [19] we get

$$u_r(z) \leq 9\sqrt{2}(4|z|/r)^{1/\alpha}, \qquad r > 4|z|.$$

Let v_r denote the bounded harmonic function on D with $v_r = 0$ on $\partial D \cap U(0,r)$ and $v_r = 1$ on $\partial D \cap (\mathbb{C}\backslash B(0,r))$. Then $0 \le v_r \le 1$. Since $v_r \le u_r$ on $D \cap U(0,r)$ and since

$$\int_0^\infty \frac{d\omega(r)}{r^{1/\alpha}} = \lim_{r\to\infty} \omega(r)/r^{1/\alpha} + \frac{1}{\alpha}\int_0^\infty \omega(r)/r^{1+1/\alpha}dr < \infty,$$

as in [19] we get a harmonic function u by

$$u(z) := \int_0^\infty v_r(z)d\omega(r), \qquad z \in D,$$

such that for all $z \in \partial D$

$$u(z) = \lim_{\zeta\to z}\int_0^\infty v_r(\zeta)d\omega(r) = \int_0^\infty \lim_{\zeta\to z} v_r(\zeta)d\omega(r) = \int_0^\infty v_r(z)d\omega(r) = \int_0^{|z|} d\omega = \omega(z).$$

Integration by parts gives for each $z \in D$

$$\begin{aligned} u(z) &\le \int_0^{4|z|} d\omega + 9\sqrt{2}(4|z|)^{1/\alpha}\int_{4|z|}^\infty \frac{d\omega(r)}{r^{1/\alpha}} \\ &\le 9\sqrt{2}(4|z|)^{1/\alpha}/\alpha\int_{4|z|}^\infty \frac{\omega(r)}{r^{1+1/\alpha}}dr \\ &= 9\sqrt{2}/\alpha\int_1^\infty \frac{\omega(4zt)}{t^{1+1/\alpha}}dt. \end{aligned}$$

(a) There is $K = K(\varepsilon,\alpha) > 0$ such that for all $s \ge s_0$

$$\int_1^\infty \frac{\omega(st)}{t^{1+1/\alpha}}dt \le \sum_{j=0}^\infty \int_{2^j}^{2^{j+1}} \frac{2^{(1/\alpha-\varepsilon)(j+1)}\omega(s)}{t^{1+1/\alpha}}dt \le K\omega(s).$$

(b) If $x \mapsto \omega(e^x)$ is convex then ω is subharmonic on $\mathbb{C}$. Thus because $\omega(r) = o(r^{1/\alpha})$, as in [19] we get $u \ge \omega$ on D.

The following results are useful to check the extra hypothesis on $\Delta(H;\cdot,\cdot)$ and Ω. The extra hypothesis of Theorem 2.7 is obviously fulfilled if H belongs to the following subclass of $\mathcal{H}_\rho$.

3.3 DEFINITION. Let $\rho > 0$. By $\mathcal{H}_\rho^* = \mathcal{H}_\rho^*(N)$ we denote the set of all plurisubharmonic functions $H \in \mathcal{H}_\rho$ such that there are $\varepsilon, R_0 > 0$ and $C \ge 1$ with

$$\Delta(H;a,R) \ge \varepsilon R^C, \qquad a \in S_H,\ 0 \le R \le R_0.$$

Since $\Delta(H;a,\cdot)$ is nondecreasing, (decreasing ε) we can assume R_0 to be arbitrary large.

3.4 LEMMA. *Let $H_l\colon \mathbb{C}^{N_l} \to \mathbb{R}_+$, $l = 1,\ldots,n$, be plurisubharmonic. For $H(z) := \sum_{l=1}^n H_l(z_l)$, $z \in \mathbb{C}^N = \prod_{l=1}^n \mathbb{C}^{N_l}$, $N = \sum_{l=1}^n N_l$, the following holds*

$$\min_{1\le l\le n} \Delta(H_l;a_l,R/\sqrt{n}) \le \Delta(H;a,R) \le \min_{1\le l\le n} \Delta(H_l;a_l,R), \qquad a \in \mathbb{C}^N,\ R \ge 0.$$

Let $\rho > 0$. If in addition for all $l = 1, \dots, n$ the functions H_l are positively homogeneous of order ρ and $\Gamma_{H_l} \neq \emptyset$, then $\Gamma_H = \prod_{l=1}^n \Gamma_{H_l}$. If in addition $H_l \in \mathcal{H}_\rho^*(N_l)$, $l = 1, \dots, n$, then $H \in \mathcal{H}_\rho^*(N) = \mathcal{H}_\rho^*$.

Proof. The proof of the first part is the same as the one of [13], Lemma 3.4. It remains to prove the last assertion. Let $H_l \in \mathcal{H}_\rho^*(N_l)$ for all $l = 1, \dots, n$. For $R \geq 2$ and $a_l \in S_{H_l}$, $l = 1, \dots, n$, we have $U(a_l, R) \supset U(0, R-1)$ and

$$\Delta(H_l; a_l, R) \geq \Delta(H_l; 0, R-1) = R^\rho(1 - \frac{1}{R})^\rho \Delta(H_l; 0, 1) \geq 2^{-\rho}\Delta(H_l; 0, 1)R^\rho.$$

Hence by Definition 3.2 we may assume that there are $\varepsilon > 0$ and $C \geq \max\{1, \rho\}$ with

$$\Delta(H_l; a_l, R) \geq \varepsilon \min\{R^\rho, R^C\}, \qquad a_l \in S_{H_l},\ R \in \mathbb{R}_+,\ l = 1, \dots, n.$$

Let $a \in S_H$ and $0 < R \leq 1$. Since $\Delta(H_l; \cdot, \cdot)$ is positively homogeneous of order ρ, if $a_l \neq 0$ we have

$$\begin{aligned} \Delta(H_l; a_l, R) &= |a_l|^\rho \Delta(H_l; a_l/|a_l|, R/|a_l|) \\ &\geq \varepsilon \min\{R^\rho, \frac{R^C}{|a_l|^{C-\rho}}\} \\ &\geq \varepsilon \min\{R^\rho, \frac{R^C}{|a|^{C-\rho}}\} \\ &= \varepsilon R^C. \end{aligned}$$

If $a_l = 0$, we have

$$\Delta(H_l; a_l, R) = R^\rho \Delta(H_l; 0, 1) > \varepsilon R^C.$$

Hence from the first part of the assertion, we get

$$\Delta(H; a, R) \geq \min_{1 \leq l \leq n} \Delta(H_l; a_l, R/\sqrt{n}) \geq \varepsilon n^{-C/2} R^C.$$

3.5 EXAMPLE. Let $\rho > 0$. If $H(z) = |z|^\rho$, $z \in \mathbb{C}^N$, and if $0 < q < \min\{1, \sqrt{\rho/2}\}$ then there are $R_0, \varepsilon > 0$ with

$$h(H; a, R)(z) - H(z) \geq \varepsilon R^2, \qquad a \in S,\ 0 \leq R \leq R_0,\ z \in B(a, qR).$$

In particular $S_H = S$ and $H \in \mathcal{H}_\rho^*$.

Proof. Fix $a \in S$ and $0 < R \leq R_0 < 1$, where $R_0 > 0$ will be chosen later. Fix $z \in B(a, R)$ and put $t := z - a$. By the Taylor series expansion, we have

$$\begin{aligned} H(z) &= H(a) + \sum_{j=1}^N \frac{\partial H(a)}{\partial z_j} t_j + \sum_{j=1}^N \frac{\partial H(a)}{\partial \bar{z}_j} \bar{t}_j \\ &\quad + \operatorname{Re}\Big(\sum_{j,k=1}^N \frac{\partial^2 H(a)}{\partial z_j \partial z_k} t_j t_k \Big) + \sum_{j,k=1}^N \frac{\partial^2 H(a)}{\partial z_j \partial \bar{z}_k} t_j \bar{t}_k + R_a(t) \end{aligned}$$

(see Hörmander [7], page 51), where

$$|R_a(t)| \leq \max_{|w|\leq|t|} \left\| \left(\frac{\partial^2 H(a+w)}{\partial x_i \partial x_j} - \frac{\partial^2 H(a)}{\partial x_i \partial x_j}\right)_{i,j=1}^{2N} \right\| |t|^2.$$

We abbreviate the four leading terms of this expansion by $P(z)$ and note that P is a pluriharmonic function of $z \in \mathbb{C}^N$. Hence

$$h(H;a,R) = P + h(H-P;a,R) = H + h(H-P;a,R) - (H-P).$$

We now estimate $H - P$. Note that

$$Q_a(t) := \sum_{j,k=1}^{N} \frac{\partial^2 H(a)}{\partial z_j \partial \bar{z}_k} t_j \bar{t}_k = \rho(\rho-2)/4 \Big| \sum_{j=1}^{N} a_j \bar{t}_j \Big|^2 + \rho/2|t|^2 \geq \min\{\frac{\rho^2}{2}, \frac{\rho}{2}\}|t|^2.$$

Choose $0 < c < (\min\{(\rho/2)^2, \rho/2\} - (q^2\rho/2))/(1+q^2)$. Since the second derivative of H is continuous, we can choose $0 < R_0 < 1$ only depending on H and c such that $|R_a(t)| \leq c|t|^2$ if $0 < R \leq R_0$. Thus we get

$$H(z) - P(z) = Q_a(t) + R_a(t) \leq (\rho/2 + c)|t|^2 = (\rho/2 + c)|t|^2$$

and

$$H(z) - P(z) \geq Q_a(t) - |R_a(t)| \geq (\min\{(\rho/2)^2, \rho/2\} - c)|t|^2.$$

This finally gives for each $z \in B(a, qR)$

$$h(H-P;a,R)(z) - (H-P)(z) \geq (\min\{(\rho/2)^2, \rho/2\} - c - (\rho/2 + c)q^2)R^2 = \varepsilon R^2.$$

3.6 LEMMA. *Let $\rho > 0$ and let $H \in \mathcal{H}_\rho$ be plurisubharmonic. Let $0 < q < \min\{1, \sqrt{\rho/2}\}$. Assume that there are $R_0, C, \varepsilon, \alpha > 0$ such that for each $a \in S$ and $0 < R \leq R_0$ there are $z_0 \in S$, $|z_0 - a| \leq qR$, a plurisubharmonic function v on $U(a,R)$ and some $\kappa = \kappa(a,R) \geq \varepsilon R^C$ such that*

$$\kappa|z|^\alpha + v^*(z) \leq H(z) \quad \text{on} \quad S(a,R), \qquad \text{and} \qquad \kappa|z_0|^\alpha + v(z_0) = H(z_0).$$

Then $S_H = S$ and $H \in \mathcal{H}_\rho^*$. In particular, this holds if $H \in \mathcal{H}_\rho$ is twice continuously differentiable on $\mathbb{C}^N \backslash \{0\}$ and strictly plurisubharmonic.

Proof. Choose $\varepsilon' > 0$ and $0 < R_0' \leq R_0$ according to 3.5. Let $0 < R \leq R_0'$. Since v is plurisubharmonic on $U(a,R)$, we have

$$h(H;a,R)(z) \geq \kappa h(|\zeta|^\alpha; a,R)(z) + v(z), \qquad z \in U(a,R).$$

Thus by 3.5, we get

$$\Delta(H;a,R) \geq h(H;a,R)(z_0) - H(z_0) \geq \kappa(h(|\zeta|^\alpha;a,R)(z_0) - |z_0|^\alpha) \geq \varepsilon\varepsilon' R^{C+2}.$$

If $H \in \mathcal{H}_\rho$ is twice continuously differentiable on $\mathbb{C}^N \backslash \{0\}$ and strictly plurisubharmonic, then there is $\kappa > 0$ such that $z \mapsto H(z) - \kappa|z|^\rho$ is plurisubharmonic.

The following two propositions have been proved in [13], Lemmas 3.3 and 3.4.

3.7 PROPOSITION. *Let $G = \prod_{l=1}^{n} G_l$, $G_l \subset \mathbb{C}^{N_l}$, $\sum_{l=1}^{n} N_l = N$, be a bounded convex domain in $\mathbb{C}^N$. Let H and H_l be the support function of G and G_l, respectively. Then*

$$\Gamma_H = \prod_{l=1}^{n} \Gamma_{H_l}.$$

If G_l, $l = 1, \ldots, n$, have a boundary of class C^1, then $S_H = S$. If moreover G_l, $l = 1, \ldots, n$, have a boundary of class $C^{1,\lambda}$ for some $0 < \lambda \leq 1$, then $H \in \mathcal{H}_\rho^*$ for $\rho = 1$.

3.8 PROPOSITION. *If H is the support function of a bounded open convex polyhedron in $\mathbb{C}^N$, then there exists $\varepsilon > 0$ with*

$$\Delta(H; a, R) \geq \varepsilon R, \quad R > 0,\, a \in \Gamma_H.$$

This section will be finished by giving two applications of Theorem 2.9. We leave it to the reader to formulate more theorems applying the results of this section to the theorems 2.7 and 2.9. See also [13] for further applications.

We obtain the following version of a result of Krivosheev [9]. In view of Remark 2.11, condition 3.9(iii) is equivalent to several notions of "regular growth" of the function F on the cone Γ_H (see Sigurdsson [16], Thm. 2.5).

3.9 THEOREM. *Let H be the support function of a bounded convex domain in $\mathbb{C}^N$, and let $L \in \mathcal{H}_\rho$ for $\rho = 1$. For $\Omega = (|z|/k)_{k \in \mathbb{N}}$, the following are equivalent for each $F \in A^0_{L+\Omega} \backslash \{0\}$.*

(i) Whenever $g \in A^0_{L+H+\Omega}$ and $f := g/F \in A(\mathbb{C}^N)$, then $f \in A^0_{H+\Omega}$.

(ii) $FA^0_{H+\Omega}$ is a closed subspace of $A^0_{L+H+\Omega}$.

(iii) For each $k \in \mathbb{N}$ there is $R > 0$ such that for each $z \in \Gamma_H$ with $|z| \geq R$ there exists $w \in B(z, |z|/k)$ such that $|F(w)| \geq \exp(L(w) - |w|/k)$.

The same assertion is true if "$A^0_{H+\Omega}$" is replaced by "$A_{H-\Omega}$" and "$A^0_{L+H+\Omega}$" is replaced by "$A_{L+H-\Omega}$".

Proof. We may assume that $H(z) > 0$ for all $z \in \mathbb{C}^N \backslash \{0\}$. Hence we may replace $(H - |z|/k)_{k \in \mathbb{N}}$ by $((1 - 1/k)H)_{k \in \mathbb{N}}$. Thus by Proposition 3.1 (or by Krivosheev [9], Lemma 3.1), the space $A^0_{H+\Omega}$ has the Phragmén-Lindelöf-property. The same arguments apply to $A_{H-\Omega}$. Obviously $A_{H-\Omega} \neq \{0\}$. Thus the theorem follows from Theorem 2.10.

The characterizing condition of the following theorem contains both the dependence on L (on the "support of the convolutor") as well as the dependence on the disturbance of H by a weight system Ω (on the "regularity" of the class of functions on which the convolution operator acts). In [13], Thm. 3.5, we prove a corresponding result for $L = 0$ under a much weaker assumption on H. We formulate both results only for the space $A_{H-\Omega}$, because an appropriate Phragmén-Lindelöf theorem for $A^0_{H+\Omega}$ is not available.

3.10 THEOREM. *Let H be the support function of an open bounded convex polyhedron in $\mathbb{C}^N$, let $L \in \mathcal{H}_\rho$, and let $\Omega = (\omega_k)_{k\in\mathbb{N}}$ be a weight system of concave functions $r \mapsto \omega_k(r)$, $r \geq 0$, and with convex functions $x \mapsto \omega_k(e^x)$, $x \in \mathbb{R}$, $k \in \mathbb{N}$. Then the following are equivalent for each $F \in A^0_{L+\Omega}\backslash\{0\}$.*

(i) Whenever $g \in A_{L+H-\Omega}$ and $f := g/F \in A(\mathbb{C}^N)$, then $f \in A_{H-\Omega}$.

(ii) $FA_{H-\Omega}$ is a closed subspace of $A_{L+H-\Omega}$.

(iii) For each $k \in \mathbb{N}$ there is $R > 0$ such that for each $z \in \Gamma_H$ with $|z| \geq R$ there exists $w \in B(z, \omega_k(z))$ such that $|F(w)| \geq \exp(L(w) - \omega_k(w))$.

Proof. By the concavity assumption, 1.7(∗) holds for $A_{H-\Omega}$. By the convexity assumption, ω_k is plurisubharmonic on $\mathbb{C}^N$ for each $k \in \mathbb{N}$. Hence by Proposition 3.1, $A_{H-\Omega}$ has the Phragmén-Lindelöf-property. The extra hypothesis of Theorem 2.9 holds by Proposition 3.8. Thus the theorem follows from 2.9.

4 WEIGHT SYSTEMS FOR WHICH DIVISION IS ALWAYS POSSIBLE

In this section we only deal with the case $L = 0$. Up to this place we did not check whether the "characterizing" condition 2.7(v) is always fulfilled. This happens (which is a classical result) in the limit case $\Omega = (|z|^\rho/k)_{k\in\mathbb{N}}$. To be more precise, in this case for all $H \in \mathcal{H}_\rho$ and all $F \in A^0_\Omega\backslash\{0\}$ the following holds: Whenever $f \in A^0_{H+\Omega}$ (or $f \in A_{H-\Omega}$) and $g := f/F$ is entire, then $g \in A^0_{H+\Omega}$ ($g \in A_{H-\Omega}$). In the present section we characterize all weight systems Ω with this property. It turns out that there are in fact some other (few) ones.

For most of the proofs of this section we refer to [13], where corresponding results are proved for $\rho = 1$. The proofs for arbitrary order $\rho > 0$ are quite the same.

4.1 REMARK. It is well known that for each nonzero polynomial P on $\mathbb{C}^N$ there is $\varepsilon > 0$ such that

$$\sup_{|w-z|\leq 1} |P(w)| \geq \varepsilon, \qquad z \in \mathbb{C}^N.$$

Hence by 2.7, in all spaces $A^0_{H+\Omega}$ and $A_{H-\Omega}$ division is possible by all nonzero polynomials.

4.2 PROPOSITION. *Let $\Omega = (\omega_k)_{k\in\mathbb{N}}$ be a weight system (possibly without 1.1(3)). Instead of 1.1(5) we even assume that there is $C > 1$ such that for each $k \in \mathbb{N}$ there is $R > 0$ with $\omega_k(2r) \leq C\omega_k(r)$ for all $r \geq R$. Let $(a_j)_{j\in\mathbb{N}}$ be a sequence in $\mathbb{C}\backslash\{0\}$ with $\lim_{j\to\infty} |a_j| = \infty$. Put $n(t) := \sum_{|a_j|\leq t} 1$ for $t \geq 0$. Assume that for each $k \in \mathbb{N}$ there is $r_0 > 0$ with*

$$N(r) := \int_0^r \frac{n(t)}{t}\, dt \leq \omega_k(r), \qquad r \geq r_0.$$

Then there exists a nonzero entire function $F \in A^0_\Omega$ of one variable which has at least the zeros $(a_j)_{j\in\mathbb{N}}$ (with respect to multiplicities).

4.3 LEMMA. *Let $\Omega = (\omega_k)_{k\in\mathbb{N}}$ be a weight system (possibly without 1.1 (3)). Let $(\rho_k)_{k\in\mathbb{N}}$ be a weight system with $\rho = 1$ in 1.1(3) (and possibly without 1.1(2) and (5)). Assume that for each $k \in \mathbb{N}$ there are $m \in \mathbb{N}$ and $R_0 > 0$ with*

$$\inf_{S>R} \frac{\omega_m(S)}{\log(S/R)} \le \frac{\omega_k(R)}{\log(R/\rho_k(R))}, \qquad R \ge R_0.$$

Then for each $F \in A^0_\Omega\backslash\{0\}$ the following condition holds: For each $k \in \mathbb{N}$ there is $R_0 > 0$ such that for each $z \in \mathbb{C}^N$ with $|z| \ge R_0$ there exists $w \in \mathbb{C}^N$ with $|w-z| \le \rho_k(z)$ and $|F(w)| \ge \exp(-\omega_k(w))$.

4.4 LEMMA. *Let Ω and $(\rho_k)_{k\in\mathbb{N}}$ be weight systems as in Lemma 4.3. Assume that in addition there is $C > 1$ such that for each $k \in \mathbb{N}$ there is $r_0 > 0$ with $\omega_k(2r) \le C\omega_k(r)$ for all $r \ge r_0$. Assume that each function $F \in A^0_\Omega\backslash\{0\}$ satisfies the condition given in Lemma 4.3. Then for each $k \in \mathbb{N}$ there are $m \in \mathbb{N}$ and $R_0 > 0$ with*

$$\inf_{S>R} \frac{\omega_m(S)}{\log(S/R)} \le \frac{\omega_k(R)}{\log(R/\rho_k(R))}, \qquad R \ge R_0.$$

4.5 THEOREM. *Let $\rho > 0$ and let Ω be a weight system with concave functions $t \mapsto \omega_k(t^{1/\rho})$, $t \ge 0$, $(k \in \mathbb{N})$. The following assertions are equivalent:*

(i) For each $H \in \mathcal{H}_\rho$ and each $F \in A^0_\Omega\backslash\{0\}$ the following holds: Whenever $f \in A^0_{H+\Omega}$ and $g := f/F$ is entire, then $g \in A^0_{H+\Omega}$.

(ii) For each $H \in \mathcal{H}_\rho$ and each $F \in A^0_\Omega\backslash\{0\}$ the following holds: Whenever $f \in A_{H-\Omega}$ and $g := f/F$ is entire, then $g \in A_{H-\Omega}$.

(iii) For each $k \in \mathbb{N}$ there are $m \in \mathbb{N}$ and $R_0 > 0$ with

$$\inf_{S>R} \frac{\omega_m(S)}{\log(S/R)} \le \frac{\omega_k(R)}{\log(R^\rho/\omega_k(R))}, \qquad R \ge R_0.$$

Proof. By the concavity assumption, we have $\limsup_{r\to\infty} \omega_k(2r)/\omega_k(r) \le 2^\rho$, $k \in \mathbb{N}$. Moreover by 1.7 and by the concavity assumption, there is $H \in \mathcal{H}^*_\rho$ with $S_H = S$ and 1.7(*), for example $H(z) = |z|^\rho$, $z \in \mathbb{C}^N$. If we apply 4.3 and 4.4 with $\rho_k(r) = r^{1-\rho}\omega_k(r)$ we get the assertion from 2.7.

4.6 EXAMPLES. Let $\rho > 0$.

(a) In the limit case $\omega_k(r) = \frac{1}{k}r^\rho$, $k \in \mathbb{N}$, the condition of Theorem 4.5(iii) holds which gives a well known result (see Lelong and Gruman [10], Theorem 9.11).

(b) If $\Omega = (\frac{1}{k}\omega)_{k\in\mathbb{N}}$ is a weight system with $\liminf_{r\to\infty} \omega(r)/r^\rho = 0$ and there are $A > 1$ and $r_0 > 0$ with $2\omega(r) \le \omega(Ar)$ for $r \ge r_0$, then 4.5(iii) does not hold. For in this case, there is $C > 0$ such that $\omega(R)/C \le \inf_{S>R} \omega(S)/\log(S/R) \le C\omega(R)$ for large $R > 0$.

(c) For $\Omega = (\omega_k)_{k\in\mathbb{N}}$ where $\omega_k(r) = \frac{1}{k}(\log(1+r))^d$ $(d > 1)$, the condition 4.5(iii) holds.

(d) Let $\Omega = (\omega_k)_{k\in\mathbb{N}}$ be a weight system such that for each $k \in \mathbb{N}$ there are $m \in \mathbb{N}$ and $r_0 > 0$ with $\omega_m(r)\log r \leq \omega_k(r)$ for all $r \geq r_0$ (this condition on Ω has been communicated by S. N. Melikhov). Then, choosing $S = 2R$ in 4.5(iii), one gets that Ω satisfies the condition 4.5(iii).

REFERENCES

[1] V. S. Azarin: On rays of completely regular growth of an entire function. *Math. USSR, Sb., 8*: 437–450 (1969).

[2] E. Bedford, B. A. Taylor: The Dirichlet problem for a complex Monge–Ampère equation. *Inv. Math., 37*: 1–44 (1976).

[3] L. Ehrenpreis: Solution of some problems of division. IV. *Am. J. Math., 82*: 522–588 (1960).

[4] O. V. Epifanov: On a convolution epimorphism in convex domains. *Sov. Math., Dokl., 15*: 994-996 (1974).

[5] O. V. Epifanov: The problem concerning the epimorphicity of a convolution operator in convex domains. *Math. Notes Acad. Sci. USSR, 16*: 837–841 (1974).

[6] L. Hörmander: On the range of convolution operators. *Ann. of Math., 76*: 148–170 (1962).

[7] L. Hörmander: *An Introduction to Complex Analysis in Several Variables.* North–Holland Publishing Company 1973.

[8] G. Köthe: *Topological vector spaces.* Springer Verlag 1969.

[9] A. S. Krivosheev: A criterion for the solvability of nonhomogeneous convolution equations in convex domains of $\mathbb{C}^N$. *Math. USSR, Izv., 36*: 497–517 (1991).

[10] P. Lelong and L. Gruman: *Entire functions of several complex variables.* Springer Verlag 1986.

[11] B. Ja. Levin: *Distribution of zeros of entire functions.* AMS Providence, Rhode Island 1980.

[12] S. Momm: A Phragmén- Lindelöf theorem for plurisubharmonic functions on cones in $\mathbb{C}^N$. *Indiana Univ. Math. J.*, to appear.

[13] S. Momm: A division problem in the space of entire functions of exponential type. Preprint.

[14] V. V. Morzhakov: On epimorphicity of a convolution operator in convex domains in $\mathbb{C}^N$. *Math. USSR, Sb., 60*: 347–364 (1988).

[15] V. V. Napalkov: Spaces of analytic functions of prescribed growth near the boundary. *Math. USSR, Izv., 30*: 263–281 (1988).

[16] R. Sigurdsson: Convolution equations in domains of $\mathbb{C}^N$. *Arkiv för Mat., 29*: 285–305 (1991).

[17] V. A. Tkačenko: On operators of convolution type in spaces of analytic functionals. *Sov. Math., Dokl., 15*: 1628–1632 (1974).

[18] V. A. Tkačenko: Equations of convolution type in spaces of analytic functionals. *Math. USSR, Izv., 11*: 361–374 (1977).

[19] M. Tsuji: *Potential theory in modern function theory.* Second edition. New York 1975.

On the Liouville Property of Section Spaces of Coherent Analytic Sheaves

V. P. PALAMODOV Mech. –Math. Faculty, University of Moscow, Moscow, Russia

Dedicated to the memory of Constantin Bănică

1 PROBLEMS AND STATEMENTS

Let X be a complex analytic space, whose topology admits a countable base, $\mathcal{F}$ be an analytic coherent sheaf on X. The space $\Gamma(X, \mathcal{F})$ of its sections is equipped with the canonical nuclear Fréchet topology [6]. We investigate here the following
PROBLEM: *When does the space* $\Gamma(X, \mathcal{F})$ *have the property* (DN)*?*

This property means that the topology of $\Gamma(X, \mathcal{F})$ can be given a logarithmically convex system of norms. This property is important in view of the theorem of Vogt–Wagner [13], [14]: $\mathrm{Ext}^1(F, E) = 0$ for Fréchet–Hilbert spaces F and E, if $E \in (\Omega)$ and $F \in (\mathrm{DN})$.
The space $\Gamma(X, \mathcal{F})$ always possesses the poperty (Ω), if X is a Stein space ([15]).

To illustrate this result suppose that $F = \mathcal{O}(X)/\mathcal{I}$, where $\mathcal{I}$ is the sheaf of germs of holomorphic functions, which vanish on a closed analytic subset Y of a Stein space X. There is an exact sequence of Fréchet spaces and linear continuous operators:

$$0 \to \Gamma(X, \mathcal{I}) \to \Gamma(X, \mathcal{O}(X)) \to \Gamma(X, \mathcal{F}) \to 0 \,. \tag{1}$$

The equation $\mathrm{Ext}^1(\Gamma(X, \mathcal{F}), \Gamma(X, \mathcal{I})) = 0$ implies a splitting of (1) i. e. existence of a linear continuous operator $\Gamma(X, \mathcal{F}) \to \Gamma(X, \mathcal{O}(X))$, which transforms any holomorphic function f on Y to a holomorphic function g on X such that $g \mid Y = f$. Thus the inclusion $\Gamma(X, \mathcal{F}) \in (\mathrm{DN})$ is sufficient for the splitting of (1) and is necessary as well, if $\Gamma(X, \mathcal{O}(X)) \in (\mathrm{DN})$ (cf. [8], Prop. 5. 3).

We show in this paper that this problem for an arbitrary coherent sheaf $\mathcal{F}$ can be brought to the same question for sheaves $\mathcal{O}(S)$ for $S \in \mathrm{Ass}\,(\mathcal{F})$. The latter denotes the set of closed irreducible analytic subsets of $\operatorname{supp}\mathcal{F}$, which are loci of primary components of the zero–subsheaf of $\mathcal{F}$ (section 3). In fact these are the components in the Lasker–Noether decomposition for this subsheaf ([17], [5], [10], [11]). Recall that a closed analytic set $Y \subset X$ is called irreducible, if there is no representation of $Y = Y' \cup Y''$, where Y' and Y'' are proper closed analytic subsets of Y. For any complex space X and any coherent sheaf $\mathcal{F}$ on X the set $\operatorname{supp}\mathcal{F}$ of points $x \in X$ such that the stalk $\mathcal{F}_x$ is equal to zero, is a closed analytic subset of X. Our main result is the following

THEOREM 1.1 *Let X be a complex space and $\mathcal{F} \in \operatorname{Coh}(X)$ such that either* supp *$\mathcal{F}$ is a Stein space or X is a projective space. Then for an arbitrary open subset $U \subset X$ the space $\Gamma(U, \mathcal{F})$ belongs to the class* (DN) *iff*

i) there is only a finite number of non–empty sets $U \cap S$, $S \in \operatorname{Ass}(\mathcal{F})$, and

ii) each space $\Gamma(U \cap S, \mathcal{O}(S))$ possesses the property (DN).

For a proof we use the fact that $\mathcal{F}$ possesses a global Noether operator (sect. 3). The condition ii) can be described in more analytic way.

DEFINITION 1.1 *We shall say that a complex space X is of* Liouville type, *if X is reduced (i. e. there are no nilpotent elements in the structure sheaf $\mathcal{O}(X)$), has only a finite number of irreducible components and any plurisubharmonic function on X, which is bounded from above, is locally constant.*

Recall that a function h on a reduced space X is *plurisubharmonic* if h is an upper semicontinuous mapping $X \to \mathbb{R} \cup \{-\infty\}$ and the inverse image of h under an arbitrary holomorphic mapping $U_1 \to X$ of the unit disc U_1 is a subharmonic function. For example, the complex plane is of Liouville type, but the circle is not.

THEOREM 1.2 *[16] The space $\Gamma(X, \mathcal{O}(X))$ for a Stein manifold X belongs to the class* (DN), *iff X is of Liouville type.*

We check one part of this statement for singular complex spaces, using the standard arguments.

PROPOSITION 1.3 *If X is a complex space of Liouville type, the space $\Gamma(X, \mathcal{O}(X))$ belongs to the class* (DN).

QUESTION 1 *Is the converse true for reduced singular complex spaces?*

COROLLARY 1.4 *Let $\mathcal{F}$ be a coherent sheaf on a complex space X such that* $\operatorname{supp}\mathcal{F}$ *is Stein, let U be an open part of X such that*

i) the number of non–empty sets $U \cap S$, $S \in \operatorname{Ass}(\mathcal{F})$, is finite,

ii) each set $U \cap S$ is of Liouville type.

Then the space $\Gamma(U, \mathcal{F})$ has the property (DN).

For a proof we combine Proposition 1.3 and Theorem 1.1.

COROLLARY 1.5 *Let X be a Stein space, $\mathcal{L}, \mathcal{F} \in \mathrm{Coh}(X)$, where $\mathcal{L}$ is a locally free sheaf and $\mathcal{F}$ is a subsheaf of $\mathcal{L}$. Then if for an open set $U \subset X$ the space $\Gamma(U, \mathcal{O}(S))$ has the property* (DN), *so does $\Gamma(U, \mathcal{F})$.*

This assertion is a generalization of the main argument of [1], Theorem III. 1.

PROOF of Corollary 1.5. It is sufficient to check that $\mathrm{Ass}(\mathcal{F}) \subset \mathrm{Ass}(\mathcal{O}(X))$. For any point $x \in X$ and set $S \in \mathrm{Ass}(\mathcal{F})$ any irreducible component of S_x is an element of $\mathrm{Ass}(\mathcal{F}_x)$ and vice versa. Hence we only have to prove the inclusion $\mathrm{Ass}(\mathcal{F}_x) \subset \mathrm{Ass}(\mathcal{O}_x(X))$ for any $x \in X$. According to the definition (cf. [18]), $\mathrm{Ass}(\mathcal{F}_x)$ is the set of all prime ideals $\mathfrak{p}$ of the algebra $\mathcal{O}_x(X)$ such that there exists an element $f \in \mathcal{F}_x$, whose annihilator ideal, denoted by $\mathrm{Ann}(f)$, coincides with $\mathfrak{p}$. It follows that $\mathfrak{p} \in \mathrm{Ass}(\mathcal{L}_x) = \mathrm{Ass}(\mathcal{O}_x^p(X))$ for some p, since $\mathcal{L} \cong \mathcal{O}^p(X)$ locally. Hence we may write $f = (f_1, \ldots, f_p)$, where $f_j \in \mathcal{O}_x(X)$ and $\mathfrak{p} = \bigcap\{\mathrm{Ann}(f_j), j = 1, \ldots, p\}$. It follows that $\mathfrak{p} = \mathrm{Ann}(f_j)$ for some j, since $\mathfrak{p}$ is a prime ideal. Therefore $\mathfrak{p} \in \mathrm{Ass}(\mathcal{O}_x(X))$ q. e. d. .

COROLLARY 1.6 *If X is an analytic subspace of $\mathbb{C}\mathrm{P}^n$, Y is a subspace of X, $\mathcal{F} \in \mathrm{Coh}(X)$ and $\mathcal{L}$ is a locally free coherent sheaf on X, then the space $\Gamma(X \setminus Y, \mathcal{F} \otimes \mathcal{L})$ belongs to* (DN), *where $\otimes$ means the tensor product over $\mathcal{O}(X \setminus Y)$.*

In particular, this statement implies that $\Gamma(\mathbb{C}\mathrm{P}^n, \mathcal{F})$ belongs to (DN) for any coherent analytic sheaf $\mathcal{F}$ on $\mathbb{C}\mathrm{P}^n$ ([5]). It implies as well [1], Corollary III. 3, if we set $Y = X \cap H$ and $\mathcal{F} = \mathcal{O}(\mathbb{C}\mathrm{P}^n)$.

PROOF of the Corollary 1.6. We have the following obvious inclusion:

$$\mathrm{Ass}(\mathcal{F} \otimes \mathcal{L}) \subset \mathrm{Ass}(\mathcal{F} \mid X \setminus Y) = \{S \setminus Y, S \in \mathrm{Ass}(\mathcal{F})\},$$

hence it is sufficient to check that $V = S \setminus Y$ is of Liouville type for any projective algebraic sets Y and S. We may take for Y a divisor on S, hence Y is the proimage of a hyperplane for a projective embedding of S. Therefore our assertion follows from

PROPOSITION 1.7 *Any affine algebraic set V is of Liouville type.*

PROOF. It is sufficient to consider the case, where V is an arbitrary affine algebraic curve V in $\mathbb{C}^n$. Choose a linear projection $\pi : \mathbb{C}^n \to \mathbb{C}$ such that the restriction $\pi \mid V$ is proper and finite (this is in fact the only property of V we need). Let h be a subharmonic function on V, which is bounded from above. We define a function on $\mathbb{C}$:

$$g_1(w) := \sum\{h(z), z \in V, \pi(z) = w\}.$$

It is easy to verify, that g_1 is subharmonic. It is evidently bounded from above, hence g_1 is constant. This implies that in fact h is continuous and harmonic outside of the set of critical points of π. This set is finite, hence h is harmonic everywhere on V. Then we consider the function

$$g_2(w) := \frac{1}{2}\sum\{h^2(z), z \in V, \pi(z) = w\}.$$

We may calculate its Laplacian at any point $w \in \mathbb{C}$, except for critical values of π as follows:

$$\frac{\partial^2}{\partial w \partial \bar{w}} g_2(w) = \sum \frac{\partial h}{\partial w}(z) \frac{\partial h}{\partial \bar{w}}(z) = \sum |\frac{\partial h}{\partial w}(z)|^2 \geq 0, \tag{2}$$

where the sum runs over the set of proimages z of w. This inequality implies that the function g_2 is subharmonic as well. It is bounded from above, hence is constant. It follows from (2) that $\partial h / \partial w = 0$, hence h is locally constant q. e. d. .

QUESTION 2 *Let X be a complex space, Y is a closed analytic subset of X and $\mathcal{F} \in \mathrm{Coh}(X)$. When is the local cohomology space $\mathrm{H}^i_Y(X,\mathcal{F})$ a Fréchet–space ?*

Note that this space possesses the canonical QFS–topology (quotient topology of a Fréchet–Schwartz space) such that all mappings in the standard exact sequence

$$\ldots \to \mathrm{H}^{i-1}(X,\mathcal{F}) \to \mathrm{H}^{i-1}(X\setminus Y,\mathcal{F}) \to \mathrm{H}^i_Y(X,\mathcal{F}) \to \mathrm{H}^i(X,\mathcal{F}) \to \mathrm{H}^i(X\setminus Y,\mathcal{F}) \to \ldots$$

are continuous. This topology can be given in the following way: Consider some Stein covering $\mathcal{U}$ of X and $\mathcal{V}$ of $X\setminus Y$, where $\mathcal{V}$ is a part of $\mathcal{U}$ and the canonical morphism of cochain complexes

$$\rho : \mathrm{C}^*(\mathcal{U},\mathcal{F}) \to \mathrm{C}^*(\mathcal{V},\mathcal{F}) .$$

Let K^* be the cone of this morphism:

$$K^i := \mathrm{C}^i(\mathcal{U},\mathcal{F}) \oplus \mathrm{C}^{i-1}(\mathcal{V},\mathcal{F}) , \; i = 0,1,\ldots$$

with the differential

$$\kappa(f,g) = (\delta f, \rho f - \delta g) .$$

Then (K^*,κ) is a complex of Fréchet–Schwartz spaces and

$$\mathrm{H}^*_Y(X,\mathcal{F}) \cong \mathrm{H}^*(K^*,\kappa) .$$

This isomorphism induces a QFS–topology on the $\mathrm{H}^*_Y(X,\mathcal{F})$.

Hence the answer to Question 2 is positive for the case $i = 0$, since $\mathrm{H}^0_Y(X,\mathcal{F})$ is a closed subspace in the Fréchet space $\Gamma(X,\mathcal{F})$. It is positive for any i, if Y is a holomorphically convex compact set in a Stein space X [2].

PROPOSITION 1.8 *If X is a Stein space, $\mathcal{F} \in \mathrm{Coh}(X)$, $Y = Z \cup K$, where Z is a closed analytic subset and K is a compact subset of X, then $\mathrm{H}^1_Y(X,\mathcal{F})$ is a Fréchet space.*

If Y is an analytic set, it is a special case of Siu–Trautmann's theorem [12] (see as well [2], II. 6. 8).

QUESTION 3 *If the answer to Question 2 is positive, when does the space $\mathrm{H}^i_Y(X,\mathcal{F})$ belong to the class* (DN)*?*

For the case $i = 0$, X is a Stein space and U is an open part of X, the space $\Gamma_Y(U,\mathcal{F})$ belongs to (DN), iff only finitely many sets $S \in \mathrm{Ass}(\mathcal{F})$ are contained in Y and for any such set S the space $\Gamma(S,\mathcal{O}(S))$ belongs to (DN).

For the case $i = 1$ Corollary 1.6 implies the positive answer to Question 3 if $X = \mathbb{C}\mathrm{P}^n$, where n is arbitrary. Really both spaces $\Gamma(X,\mathcal{F})$ and $\mathrm{H}^1(X,\mathcal{F})$ are of finite dimension and $\Gamma(X/Y,\mathcal{F})$ belongs to (DN).

Another example: X is a Stein neighbourhood of a point $Y \in \mathbb{C}^n$, $\mathcal{F} = \mathcal{O}(\mathbb{C}^n)$. Then $\mathrm{H}^i_Y(X,\mathcal{F}) = 0$ for $i \neq n$ and $\mathrm{H}^n_Y(X,\mathcal{F})$ is isomorphic to the space of Laurent series

$$\sum\{c_i z^i \, , \; i = (i_1,\ldots,i_n) \, , \; i_k < 0\} ,$$

which converge in the domain $z_1 \cdots z_n \neq 0$. This is evidently a space of the class (DN).

EXAMPLE. Let

$$\ldots \to \mathcal{L}_2 \to \mathcal{L}_1 \to \mathcal{L}_0 \to \mathcal{F} \to 0$$

be an exact sequence of coherent sheaves on a Stein space X, where all the sheaves $\mathcal{L}_i$ are locally free. This sequence gives rise to the following sequence of Fréchet spaces:

$$\ldots \to \Gamma(X,\mathcal{L}_2) \xrightarrow{\mathrm{d}} \Gamma(X,\mathcal{L}_1) \xrightarrow{\mathrm{d}} \Gamma(X,\mathcal{L}_0) \xrightarrow{\mathrm{d}} \Gamma(X,\mathcal{F}) \to 0 . \tag{3}$$

COROLLARY 1.9 *Suppose that any element $S \in \mathrm{Ass}\,(\mathcal{F}) \cup \mathrm{Ass}\,(\mathcal{O}(X))$ is of Liouville type. Then the sequence (3) is split, i. e. there exist linear continuous operators*

$$\mathrm{r}_i : \Gamma(X,\mathcal{L}_{i-1}) \to \Gamma(X,\mathcal{L}_i)\ ,\ i = 0,1,\ldots\ ,\ \mathcal{L}_{-1} := \mathcal{F}$$

such that

$$\mathrm{d}\mathrm{r}_1\mathrm{d} = \mathrm{d}\ .$$

PROOF. It follows from Corollary 1.4 that there exists an operator r_0, which is a right inverse to the surjection $\mathrm{d} : \Gamma(X,\mathcal{L}_0) \to \Gamma(X,\mathcal{F})$. Then the inclusion $\mathrm{Spec}\,\mathcal{L}_0 \subset \mathrm{Spec}\,\mathcal{O}(X)$ and again Corollary 1.4 imply that $\Gamma(X,\mathcal{L}_0)$ has the property (DN). We set $\mathcal{F}' := \ker p$. This is a coherent sheaf on X and the topological space $\Gamma(X,\mathcal{F}')$ coincides with the kernel of surjection d. Hence the subspace $\Gamma(X,\mathcal{F}') \subset \Gamma(X,\mathcal{L}_0)$ is closed and is of class (DN) as well. The operator $\mathrm{e} - \mathrm{r}_0\mathrm{d}$, where e is the identity mapping, is a projection onto this subspace. The property (DN) implies that the surjection $\Gamma(X,\mathcal{L}_1) \to \Gamma(X,\mathcal{F}')$ has a linear continuous right inverse s. Then the operator $\mathrm{r}_1 := \mathrm{s}(\mathrm{e} - \mathrm{r}_0\mathrm{d})$ satisfies the equation

$$\mathrm{d}\mathrm{r}_1\mathrm{d} = \mathrm{d}\mathrm{s}\mathrm{d} = \mathrm{d}\ , \text{and so on} \quad \text{q. e. d.}\ .$$

2 PROOFS

First we prove Proposition 1.3. Choose a sequence of non empty open subsets $V_0 \subset\subset V_1 \subset\subset \ldots V_j \subset\subset X$, whose union is equal to X and V_0 has non–empty intersection with any irreducible component of X. Define the following sequence of norms on $\Gamma(X,\mathcal{O}(X))$:

$$\|f\|_j := \sup\{|f(z)|\ ,\ z \in V_j\}\ ,\ j = 0,1,\ldots\ .$$

We show that for any j there exists a number k such that

$$\|f\|_j^2 \le \|f\|_0 \cdot \|f\|_k\ ,\ \forall f \in \Gamma(X,\mathcal{O}(X))\ . \tag{4}$$

This inequality will imply the inclusion $\Gamma(X,\mathcal{O}(X)) \in$ (DN). Consider the set P of plurisubharmonic functions h on X, which satisfy the conditions:

$$\sup\{h(z)\ ,\ z \in V_0\} = 0\ ,\ \sup\{h(z)\ ,\ z \in V_j\} = 1\ .$$

LEMMA 2.1 *There exists a number k such that*

$$\sup\{h(z)\ ,\ z \in V_k\} \geq 2\ ,\ \forall h \in P\ .$$

PROOF of Lemma. Suppose the converse. Then for any k there exists a function $h_k \in P$ such that

$$\sup\{h_k(z)\ ,\ z \in V_k\} \leq 2\ . \tag{5}$$

Take for each k the function

$$H_k(z) := \limsup\{\sup\{h_l(z')\ ,\ l \geq k\}\ ,\ z' \to z\}\ .$$

It belongs to P and satisfies the inequality $H_k \geq H_{k+1}$. Consequently $H_k \to H$ monotonically on X, where $H(z) > -\infty$. The function H is upper semi-continuous since so are the functions H_k. Hence H is plurisubharmonic and bounded by (5). It follows from the condition that H is locally constant and constant on any component of X. From the inclusion $H_k \in P$ then follows $H \leq 0$. Since $H_k \to H$ uniformly on $\bar{V}_j$ we have $H_k < 1$ on V_j for large k, which contradicts to the inclusion $H_k \in P$. This proves the Lemma.

Now check (4). It is evident for any constant function f. Let f be any non-constant holomorphic function on X. We may suppose that the supremum of $|f|$ on V_0 is equal to 1. Consider the plurisubharmonic function $h := \log|f|$ and set $s := \sup\{h(z),\ z \in V_j\}$. The number s is positive since otherwise the supremum of $|f|$ on V_j is equal to the supremum on $V_0 \subset\subset V_j$, which implies the equation $f \equiv \text{const}$ on V_0. It follows that f is a constant on X, since X is irreducible. The function $s^{-1} \cdot h$ belongs to P, hence

$$\sup\{s^{-1} \cdot h(z)\ ,\ z \in V_k\} \geq 2\ .$$

Therefore

$$\begin{aligned} \|f\|_j^2 &= \exp(2\sup\{h(z),\ z \in V_j\}) \\ &= \exp(2s) \\ &\leq \exp(\sup\{h(z),\ z \in V_k\}) \\ &= \|f\|_0 \cdot \|f\|_k\ , \qquad \text{q. e. d.}\ . \end{aligned}$$

PROOF of Theorem 1.1. According to Theorem 3.4 there exists an injective sheaf morphism (Noether operator)

$$\mathrm{n}\ :\ \mathcal{F} \to \prod\{[\mathcal{O}(S)]^{r(S)}\ ,\ S \in \mathrm{Ass}\,(\mathcal{F})\}\ ,$$

which generates for any open $U \subset X$ the mapping

$$\mathrm{N}\ :=\ \Gamma(U,\mathrm{n})\ :\ \Gamma(U,\mathcal{F}) \to \prod\{[\Gamma(V,\mathcal{O}(S))]^{r(S)}\ ,\ S \in \mathrm{Ass}\,(\mathcal{F})\}\ . \tag{6}$$

The latter is open for the canonical topologies, because of Theorem 3.9. The condition i) implies that the product on the right-hand side is finite. Hence the second space in (6) is a finite product of Fréchet spaces, which are of class (DN) by ii). Therefore the space $\Gamma(U,\mathcal{F})$ is isomorphic to a closed subspace of a space with (DN)-property and hence it has the property (DN). This proves the sufficiency of conditions i) and ii).

Now we prove the necessity of these conditions. For any $T \in \mathrm{Ass}\,(\mathcal{F})$ we consider the sheaf morphism

$$\mathrm{n}_T \,:\, \mathcal{F} \to [\mathcal{O}(T)]^{r(T)} ,$$

which is equal to the composition of n and of the projection of $\prod$ to the T–factor. This is a T–Noether operator (see Definition 3.5), hence $\ker \mathrm{n}_T$ is a coherent subsheaf of $\mathcal{F}$ and contains no elements f, whose annihilator ideal is equal to $I(Z)$ for a germ Z of the set T at a point $x \in T$. Therefore for any $S \in \mathrm{Ass}\,(\mathcal{F})$ the intersection

$$\mathcal{F}_S := \bigcap \{\ker \mathrm{n}_T \,,\, T \in \mathrm{Ass}\,(\mathcal{F})\,,\, T \neq S\}$$

is a coherent sheaf as well (since the intersection is locally finite on X) and $\mathrm{Ass}\,(\mathcal{F}_S) = \{S\}$. Therefore there exists an integer k such that

$$(I(S))^{k+1} \cdot \mathcal{F}_S = 0 \,, \tag{7}$$

and a restriction of the operator n_S defines an injective differential operator

$$\mathrm{n}'_S \,:\, \mathcal{F}_S \to [\mathcal{O}(S)]^{r(S)} .$$

Let k be the minimal number which satisfies (7). Then the sheaf $(I(S))^k \cdot \mathcal{F}_S$ does not vanish and is still coherent. Choose a non–zero section f_S of this sheaf on X. Then the formula

$$a \sim\!\!\to a \cdot f_S$$

defines an $\mathcal{O}(X)$–morphism of sheaves

$$\mathrm{e}_S \,:\, \mathcal{O}(S) \to \mathcal{F}_S \subset \mathcal{F} ,$$

since $I(S) \cdot f_s - 0$.

LEMMA 2.2 $\quad$ *e_S is an injection.*

PROOF. Suppose the contrary. We have $\ker \mathrm{e}_S \neq \mathcal{O}(S)$, since $\mathrm{e}_S(1) \neq 0$. Therefore $\ker \mathrm{e}_S$ is a proper ideal sheaf in $\mathcal{O}(S)$. Therefore any element of $\mathrm{Ass}\,(\mathrm{Coim}\, \mathrm{e}_S)$ is a proper analytic subset of S. On the other hand e_S induces a monomorphism of $\mathrm{Coim}\, \mathrm{e}_S$ to $\mathcal{F}_S$, hence $\mathrm{Ass}\,(\mathrm{Coim}\, \mathrm{e}_S) = \{S\}$. This contradiction proves Lemma 2.2.

Lemma 2.2 implies that the sequence

$$0 \to \Gamma(U \cap S, \mathcal{O}(S)) \xrightarrow{\mathrm{e}} \Gamma(U, \mathcal{F}) \xrightarrow{\mathrm{q}} \Gamma(U, \mathrm{Cok}\, \mathrm{e}_S)$$

is exact, since the functor $\Gamma(U, \cdot)$ is left exact. Both operators $\mathrm{e} = \mathrm{e}_S$ and q are continuous, hence $\mathrm{im}\, \mathrm{e} = \ker \mathrm{q}$ is a closed subspace of a Fréchet space. This implies that e is a topological isomorphism onto a closed subspace of $\Gamma(U, \mathcal{F})$, which is of class (DN) by the condition. It follows that the space $\Gamma(U \cap S, \mathcal{O}(S))$ has the property (DN) as well, which means that ii) is verified. To check i) we consider the product of injections e_S:

$$0 \to \prod \{\Gamma(U \cap S, \mathcal{O}(S))\} \to \Gamma(U, \mathcal{F}) .$$

Taking into account only constant sections of the sheaves $\mathcal{O}(S)$, $S \in \mathrm{Ass}\,(\mathcal{F})$, we construct an injection of the product of lines $\mathbb{C}$ in the space $\Gamma(U, \mathcal{F})$. This product is finite since otherwise the property (DN) of $\Gamma(U, \mathcal{F})$ is violated. This implies that the number of elements $S \in \mathrm{Ass}(\mathcal{F})$ with a non–void intersection $U \cap S$ is finite. Now to prove i) we need to check that for any S the number of irreducible components of $U \cap S$ is finite as well. Suppose that it is not true for an element $S \in \mathrm{Ass}(\mathcal{F})$ and $V_1, V_2, \ldots$ is a sequence of irreducible components of the set $U \cap S$. Note that each set V_i is the closure of the connected component $V_i \cap \mathrm{reg}\, S$ of the manifold $U \cap \mathrm{reg}\, S$, where $\mathrm{reg}\, S$ is the set of regular points of S (cf. [8]). The set $\mathrm{sing}\, S := S \setminus \mathrm{reg}\, S$ is a closed analytic subset of the Stein space S, hence there exists a non–zero holomorphic function s on S, which vanishes on $\mathrm{sing}\, S$. This function, when restricted to V_i, does not vanish identically, since S is irreducible, but vanishes on any intersection $V_i \cap V_j$, $j \neq i$, since this intersection is a part of $\mathrm{sing}\, S$. Therefore the function s_i, which is equal to s on V_i and $s_i = 0$ on V_j, $j \neq i$, is holomorphic on S. Consider the operator

$$\prod \mathbb{C} \to \Gamma(U \cap S, \mathcal{O}(S)) ,$$

acting by the formula

$$\{c_i\} \sim\!\!\rightarrow \sum c_i \cdot s_i \, .$$

This sum is locally finite and hence is well–defined and the operator is injective. This contradicts the property (DN) of the space $\Gamma(U \cap S, \mathcal{O}(S))$, and completes the proof.

3 ALGEBRAIC TOOLS

DEFINITION 3.1 *Let K be a field, A be a commutative K–Algebra, M a A–module. A prime ideal $\mathfrak{p}$ of A is called* associated *to M ([18], [4]), if there exists an element $m \in M$, whose annihilator ideal coincides with $\mathfrak{p}$, i. e.*

$$\mathfrak{p} = \{a \in A \, , \; a \cdot m = 0\} \, .$$

M is called $\mathfrak{p}$–coprimary, *if $\mathfrak{p}$ is the only prime ideal associated to M; we denote it by $\mathfrak{p}(M)$. A Lasker–Noether decomposition of M is a representation of the zero submodule of M in the form*

$$0 = M_1 \cap \ldots \cap M_J \, , \tag{8}$$

where all $M/M_1, \ldots, M/M_J$ are coprimary A–modules. This decomposition is called irreducible, *if no one of modules M_j in (8) can be omitted and all prime ideals $\mathfrak{p}_j = \mathfrak{p}(M/M_j)$, $j = 1, \ldots, J$ are different.*

If A is a Noetherian algebra, any A–module M of finite type admits an irreducible Lasker–Noether decomposition. The set of prime ideals $\{\mathfrak{p}_1, \ldots, \mathfrak{p}_J\}$, denoted $\mathrm{Ass}(M)$, is defined uniquely.

DEFINITION 3.2 *Let L, M be A–modules, $\mathrm{d} : M \to L$ be a K–linear mapping. This mapping is called an A–*differential operator of order $\leq k$ *if*

$$(\mathrm{ad} b)^{k+1} \mathrm{d} \equiv \sum (-1)^i \binom{k+1}{i} b^{k+1-i} \cdot \mathrm{d} \cdot b^i = 0 \tag{9}$$

for any $b \in A$, where $(\mathrm{ad} b)\mathrm{d} := \mathrm{d} b - b \mathrm{d}$.

DEFINITION 3.3 *Let* $\mathfrak{p}$ *be an ideal associated to an* A*-module* M*; we call*

$$\mathrm{n} : M \to [A/\mathfrak{p}]^r$$

a $\mathfrak{p}$–Noether operator, *if*

i) n *is an* A*-differential operator and* $r < \infty$*,*

ii) ker n *is a submodule of* M *and* $\mathfrak{p}$ *is not associated to* ker n.

A Noether operator *for a module* M *is a direct sum*

$$\mathrm{n} = \sum \mathrm{n}_{\mathfrak{p}} : M \to \prod\{[A/\mathfrak{p}]^{r(\mathfrak{p})} , \ \mathfrak{p} \in \mathrm{Ass}\,(M)\} ,$$

where $\mathrm{n}_{\mathfrak{p}}$ *is a* $\mathfrak{p}$*–Noether operator (cf. [10]).*

PROPOSITION 3.1 *If* A *is noetherian and* M *is of finite type, the kernel of any Noether operator is the zero submodule.*

PROOF. We have $\ker \mathrm{n} = \bigcap \ker \mathrm{n}_{\mathfrak{p}}$, hence this is a submodule of M. No one of ideals $\mathfrak{p} \in \mathrm{Ass}\,(M)$ is associated to ker n, because of Definition 3.3 ii). Hence the set Ass (ker n) is empty. This occurs only for the zero–module in virtue of the Lasker–Noether decomposition.

From now on we suppose that A is an analytic algebra, i. e. a quotient of the algebra $\mathbb{C}\{z_1, \dots, z_p\}$ for some p. It is a Noetherian algebra.

EXISTENCE THEOREM 3.2 *[10]. For any* A*-module* M *of finite type there exists a Noether operator* $\mathrm{n} = \prod \mathrm{n}_{\mathfrak{p}}$.

UNICITY THEOREM 3.3 *Let* M *be an* A*-module,* K, L *be* A*-modules of finite type,* $\mathrm{l} : M \to L$*,* $\mathrm{k} : M \to K$ *be differential operators such that* ker k *is a submodule,* L *is* $\mathfrak{p}$*-coprimary and* l *vanishes on* ker k*. Then there exist an element* $s \in A \setminus \mathfrak{p}$ *and a differential operator* $\mathrm{d} : K \to L$ *such that* $s\mathrm{l} = \mathrm{dk}$.

To check this statement we choose a Noether operator $\mathrm{n} : K \to N$ for K and apply the unicity theorem of [10] to the composition $\mathrm{nk} : M \to N$.

Let $(X, \mathcal{O})$ be a complex analytic space, $\mathcal{M}$, $\mathcal{L}$ are $\mathcal{O}$–sheaves; a sheaf morphism $\mathrm{d} : \mathcal{M} \to \mathcal{L}$ is called an *$\mathcal{O}$-differential operator*, if for any point $x \in X$ the fibre morphism $\mathrm{d}_x : \mathcal{M}_x \to \mathcal{L}_x$ is a differential operator over $\mathbb{C}$–algebra $\mathcal{O}_x$; $\mathrm{ord}(\mathrm{d}) := \sup \mathrm{ord}(\mathrm{d}_x)$.

For any coherent analytic sheaves $\mathcal{M}, \mathcal{L}$ on a complex space X and any point $x \in X$ any $\mathcal{O}_x$–differential operator $\mathrm{d}_x : \mathcal{M}_x \to \mathcal{L}_x$ admits a unique extension to an $\mathcal{O}$–differential operator $\mathrm{d} : \mathcal{M} \to \mathcal{L}$, defined on a neighbourhood of x. If d_x is a Noether operator, then d_y is a Noether operator for each y close to x.

Fix a point $x \in \mathbb{C}^p$ and consider the analytic algebra A of germs of holomorphic functions at x. Let G be an irreducible germ at x of an analytic set in $\mathbb{C}^p$. The ideal $I(G)$ in A, consisting of function germs, which vanish on G, is prime; vice versa, any prime ideal in A is equal to $I(G)$ for some irreducible germ G. We call such a germ G *associated to* an A–module M, if so is the ideal $I(G)$. For example, for an arbitrary analytic germ Y in $\mathbb{C}^l$ the germs $Z_1, \dots, Z_l$, associated to $\mathcal{O}(Y)$ are all the irreducible components of Y.

Now we pass to the global case and operate with closed irreducible analytic sets in X instead of germs. Notice that a germ of an irreducible analytic set may be reducible. Consider, for example, any connected algebraic curve with a point of self–intersection.

DEFINITION 3.4 *Let $(X, \mathcal{O})$ be a complex space, $\mathcal{F}$ be a coherent analytic sheaf on X. We call an analytic subset $Y \subset X$* associated to $\mathcal{F}$, *if*

i) Y is closed and irreducible,

ii) for any point $x \in Y$ its germ Y_x is the union of irreducible germs $Z_1, \ldots, Z_k$, $k > 0$ associated to the $\mathcal{O}_X$-module $\mathcal{F}_x$ (not necessarily all of them).

We denote by Ass$(\mathcal{F})$ *the collection of all analytic sets associated to $\mathcal{F}$.*

DEFINITION 3.5 *For any $Y \in$Ass$(\mathcal{F})$, we call Y*–Noether operator *for $\mathcal{F}$ any $\mathcal{O}$-differential operator $\mathrm{d} : \mathcal{F} \to [\mathcal{O}(Y)]^s$, $s < \infty$ such that for any point $x \in Y$ and any irreducible component Z of Y_x the composition $r_Z \cdot d$ is a Z-Noether operator, where $r_Z : [\mathcal{O}(Y)]^t \to [\mathcal{O}(Z)]^t$ for arbitrary t means the restriction morphism.*

THEOREM 3.4 *Let $\mathcal{F}$ be a coherent analytic sheaf on a complex space X such that* supp$\mathcal{F}$ *is a Stein space. Then for any set $S \in$ Ass$(\mathcal{F})$ there exists a S-Noether operator $\mathrm{n}_S : \mathcal{F} \to [\mathcal{O}(S)]^{r(S)}$ and*

$$\bigcap\{\ker \mathrm{n}_S\ ,\ S \in \mathrm{Ass}\,(\mathcal{F})\} = 0\ .$$

Note that it follows from the second assertion that for any point $x \in$supp$\mathcal{F}$ any set germ Z, associated to the stalk $\mathcal{F}_x$, is a component of the germ of a set S associated to $\mathcal{F}$. Hence Ass$\mathcal{F}$ coincides with the similar sets constructed in [5], [11].

LEMMA 3.5 *Let $\mathcal{F}$ be a coherent analytic sheaf on a Stein space $X =$ supp$\mathcal{F}$, Y be an irreducible component of X and $\mathrm{d} : \mathcal{F} \to [\mathcal{O}(Y)]^r$ for some $r < \infty$ be an $\mathcal{O}(X)$-diffential operator. Suppose that for a point $y \in Y$ and an irreducible component W of Y_y the composition $\partial := r_W \cdot \mathrm{d}_y$ is a W-Noether operator, satisfying the following condition: For any element $a \in \mathcal{O}_y(Y)$ there exists an $\mathcal{O}_y(Y)$-endomorphism b of $[\mathcal{O}_y(Y)]^r$ such that*

$$\partial \cdot a = b \cdot \partial\ . \tag{10}$$

Then d *is a Y-Noether operator.*

PROOF of Lemma 3.5. First we verify that the sheaf $\mathcal{K} = \ker \mathrm{d}$ is an $\mathcal{O}(X)$–subsheaf of $\mathcal{F}$. For this we consider the sheaf $\mathcal{D}(Y)$ of differential operators $\mathrm{e} : \mathcal{F} \to \mathcal{O}(Y)$ of order $\leq$ ord(d). It is a coherent sheaf (see [10], Prop. 11. 3). Let $\mathcal{I}$ be the subsheaf of $\mathcal{D}(Y)$, generated by the components $\mathrm{d}_1, \ldots, \mathrm{d}_r$ of the operator d. Take an arbitrary holomorphic function a on X and consider the subsheaf $\mathcal{A}$ in $\mathcal{D}(Y)$, generated by $\mathcal{I}$ and operators $\mathrm{d}_i \cdot a$, $i = 1, \ldots, r$. The sheaf $\mathcal{A}/\mathcal{I}$ is coherent and its support belongs to Y. The germ of supp$(\mathcal{A}/\mathcal{I})$ at y does not contain the germ W by (10). Hence supp$(\mathcal{A}/\mathcal{I})$ is a proper analytic subset of Y. Choose an arbitrary holomorphic function s, which belongs to the annihilator ideal of $\mathcal{A}/\mathcal{I}$ but does not vanish identically on Y. Then all operators $s \cdot \mathrm{d}_i \cdot a$ are sections of the sheaf $\mathcal{I}$, hence for any point $x \in Y$ we have

$$s \cdot \mathrm{d}_i \cdot a = \sum b_j \cdot \mathrm{d}_j$$

with some function germs b_j at the point x. Therefore the equation $\mathrm{d}(f) = 0$ for $f \in \mathcal{F}_x$, $x \in Y$ implies that $s \cdot \mathrm{d}(a \cdot f) = 0$ and hence $\mathrm{d}(a \cdot f) = 0$, q. e. d. .

The $\mathcal{O}(X)$–sheaf $\mathcal{K}$ is coherent in virtue of [10] (Th. 2). Hence $\operatorname{supp}\mathcal{K}$ is a closed analytic subset of Y. We claim that this set does not contain the germ W. Suppose the opposite: Then there exists an element $f \in \mathcal{K}_y$ such that $a \cdot f \neq 0$ for any holomorphic function germ $a \in \mathcal{O}_y(X) \setminus I(W)$. Choose $a \in \bigcap \mathfrak{q}$, where $\mathfrak{q}$ are all the elements of $\operatorname{Ass}(\mathcal{K}_y)$ except for $I(W)$. Then for large integer k and and any $\mathfrak{q}$ the function $\mathrm{n}_\mathfrak{q}(a^k \cdot f)$ vanishes $(\operatorname{mod}\mathfrak{q})$ by (9), where $\mathrm{n}_\mathfrak{q}$ is a $\mathfrak{q}$–Noether operator for $\mathcal{F}_y$. Then $\mathrm{d}(a^k \cdot f)$ implies $a^k \cdot f = 0$, because of Proposition 3.1. This contradicts the relation $a^k \in \mathcal{O}_y(X) \setminus I(W)$, which proves that $\operatorname{supp}\mathcal{K}$ does not contain W. Hence $\operatorname{supp}\mathcal{K}$ cannot contain any germ of Y. This completes the proof of Lemma 3.5.

PROOF of Theorem 3.4. We may assume that $X = \operatorname{supp}\mathcal{F}$. Otherwise we shrink X to $\operatorname{supp}\mathcal{F}$. Let X_j, $j \in J$ be the irreducible components of the space X (cf. [7],ch. V). Each of them is a Stein space and the covering $X = \bigcup X_j$ is locally finite. Therefore it is sufficient to prove the theorem for each sheaf $\mathcal{F} \otimes \mathcal{O}(X_j) \mid X_j$, $j \in J$. Hence we may suppose that $\operatorname{supp} X$ is an irreducible Stein space. Fix a point $x \in X$ and an irreducible component Y of the germ X_x. By Theorem 3.2 there exists a Y–Noether operator $\mathrm{m} : \mathcal{F}_x \to [\mathcal{O}(Y)]^r$. Consider the coherent sheaf $\mathcal{D}(X)$ of germs of differential operators $\mathrm{d} : \mathcal{F} \to \mathcal{O}(X)$ of order $\leq \operatorname{ord}(\mathrm{m})$. Let d_i, $i = 1, \ldots, r$ be its sections on X, which generate this sheaf at x. Define an operator

$$\mathrm{d} : \mathcal{F} \to [\mathcal{O}(X)]^r : \mathrm{d}(f) := (\mathrm{d}_1(f), \ldots, \mathrm{d}_r(f)) .$$

LEMMA 3.6 *The composition morphism* $\mathrm{d}_Y := r_Y \cdot \mathrm{d}$ *is a* Y*–Noether operator.*

PROOF. For this we check that $\ker \mathrm{d}_Y = \ker \mathrm{m}$. Once more we choose a function $a \in \mathcal{O}_x(X)$, which vanishes on all the components of the germ X_x, except for Y. For any component $\mathrm{m}_j : \mathcal{F}_x \to \mathcal{O}(Y)$ of m the composition $a \cdot \mathrm{m}_j$ belongs to the stalk $\mathcal{D}_x(X)$. Hence $a \cdot \mathrm{m}_j$ belongs to the $\mathcal{O}_X$–envelope of the set $\{\mathrm{d}_i, i = 1, \ldots, r\}$. This implies the inclusion $\ker \mathrm{m} \supset \ker \mathrm{d}_Y$. For the inverse inclusion we use Theorem 3.3. It follows, that $s \cdot \mathrm{d}_Y = \mathrm{b} \cdot \mathrm{m}$ for a germ $s \in \mathcal{O} \setminus I(Y)$ and for a $\mathcal{O}(Y)$–differential operator b in $[\mathcal{O}(Y)]^r$. This equation implies that $\ker \mathrm{m} \subset \ker \mathrm{d}_Y$, hence these modules coincide, q. c. d. .

Now Lemma 3.6 implies that d is a X–Noether operator for $\mathcal{F}$. Set $\mathcal{K} := \ker \mathrm{d}$; this is a coherent subsheaf of $\mathcal{F}$ and for any point x, the $\mathcal{O}_x(X)$–module $\mathcal{K}_x$ has no associated irreducible germs of dimension $\dim X$.

Now we argue using the induction on $\dim \operatorname{supp}\mathcal{F}$ and may suppose that Theorem 3.4 is proved for the sheaf $\mathcal{K}$.

LEMMA 3.7 *The equation* $\operatorname{Ass}(\mathcal{F}) = \{X\} \bigcup \operatorname{Ass}(\mathcal{K})$ *holds.*

PROOF. Evidently $\operatorname{Ass}(\mathcal{K}) \subset \operatorname{Ass}(\mathcal{F})$. For any point $x \in X$ any component of the germ X_x belongs to $\operatorname{Ass}(\mathcal{F}_x)$, since d is a X–Noether operator for $\mathcal{F}$. It remains to check that for any point $x \in X$ any germ $Y \in \operatorname{Ass}(\mathcal{F}_x)$, which is not a component of the germ X_x, belongs to $\operatorname{Ass}(\mathcal{K}_x)$. Choose an element $a \in I(Y)$, which does not vanish on any component of X_x, and take an element $f \in \mathcal{F}_x$, whose annihilator ideal is equal to $I(Y)$. We claim that $f \in \mathcal{K}_x$. In fact the equation $a \cdot f = 0$ implies in view of the Leibnitz formula (9) that $a^k \cdot \mathrm{d}(f) = 0$, for $k > \operatorname{ord}(\mathrm{d})$. It follows that $\mathrm{d}(f) = 0$, since a^k is not a zero–divisor in $\mathcal{O}_x(X)$, hence $f \in \mathcal{K}_x$. This implies that $Y \in \operatorname{Ass}(\mathcal{K}_x)$, q. e. d. .

To prove Theorem 3.4 we choose for any $Z \in \operatorname{Ass}(\mathcal{K})$ a regular point $z \in Z$ and a Z_z–Noether operator n_Z for $\mathcal{F}$. Consider the sheaf $\mathcal{D}(Z)$ of all differential operators

$\mathcal{F} \to \mathcal{O}(Z)$ of order $\leq k := \mathrm{ord}(\mathrm{n}_Z)$. It is a coherent sheaf on the Stein space X. Choose a finite set $\{\mathrm{e}_1, \ldots, \mathrm{e}_r\} \subset \Gamma(X, \mathcal{D}(Z))$, which generates the stalk $\mathcal{D}(Z)_z$. Consider the differential operator $\mathrm{e}(Z) : \mathcal{F} \to [\mathcal{O}(Z)]^r$ with the components $\mathrm{e}_1, \ldots, \mathrm{e}_r$, and the sheaf $\mathcal{G} = \ker \mathrm{e}(Z)$. The stalk $\mathcal{G}_z$ is a $\mathcal{O}_z(X)$–module, since for any number i and any function germ a at z the operator $\mathrm{e}_i \cdot a$ is of order $\leq k$ and hence belongs to the envelope of $\{\mathrm{e}_1, \ldots, \mathrm{e}_r\}$. The set $\mathrm{Ass}\,(\mathcal{G}_z)$ does not contain the germ Z_z, since this envelope contains all components of n_Z. Hence the germ of $\mathrm{e}(Z)$ at z is a Z_z–Noether operator, satisfying (10). Lemma 3.5 implies that $\mathrm{e}(Z)$ is a Z–Noether operator for $\mathcal{F}$, which completes the proof of Theorem 3.4.

THEOREM 3.8 *For any p and $\mathcal{F} \in \mathrm{Coh}\,(\mathbb{C}\mathrm{P}^p)$ there exist an integer m and a Noether operator*

$$\mathrm{n} : \mathcal{F} \to \prod \{[\mathcal{O}(S) \otimes \mathcal{O}(m)]^{r(S)}\}, \; S \in \mathrm{Ass}\,(\mathcal{F}),$$

where $\mathcal{O}(m)$ is the standard invertible sheaf on $\mathbb{C}\mathrm{P}^p$.

PROOF. The set $\mathrm{Ass}\,(\mathcal{F} \mid U)$ is well defined for any affine part U of $\mathbb{C}\mathrm{P}^p$. For any variety $S_U \in \mathrm{Ass}\,(\mathcal{F} \mid U)$ and another affine part U' of $\mathbb{C}\mathrm{P}^p$ there exists a variety $S_U \in \mathrm{Ass}\,(\mathcal{F} \mid U')$, which coincides with S_U in $U \cap U'$. Hence these varieties are parts of a projective variety S, which is associated to $\mathcal{F}$ and vice versa. Now for each $S \in \mathrm{Ass}\,(\mathcal{F})$ we choose an arbitrary point $s \in S$ and consider the coherent sheaf $\mathrm{Diff}^q(\mathcal{F}, \mathcal{O}(S))$, where the number q is equal to the degree of a local S–Noether operator $\nu : \mathcal{F}_s \to [\mathcal{O}_s(S)]^r$. According to Serre's theorem (see, for ex. [2]) for any sufficiently large integers m the space of sections of the sheaf

$$\mathrm{Diff}^q(\mathcal{F}, \mathcal{O}(S)) \otimes \mathcal{O}(m) \cong \mathrm{Diff}^q(\mathcal{F}, \mathcal{O}(S) \otimes \mathcal{O}(m))$$

generates this sheaf at the point s. Choose a basis $\sigma_1, \ldots, \sigma_\rho$ for the space $\Gamma(\mathrm{Diff}^q(\mathcal{F}, \mathcal{O}(S) \otimes \mathcal{O}(m)))$ and define the differential operator

$$\mathcal{F} \to [\mathcal{O}(S) \otimes \mathcal{O}(m)]^\rho$$

to be equal to the direct product of operators $\sigma_1, \ldots \sigma_\rho$. This is a S–Noether operator for $\mathcal{F}$, by the arguments of Theorem 3.4. Then we take the product n of these operators when S runs over the set $\mathrm{Ass}\,(\mathcal{F})$ and the number m is chosen to be sufficiently large for all S. This is a Noether operator for $\mathcal{F}$, which completes the proof of Theorem 3.8.

THEOREM 3.9 *Let $\mathrm{n} = \prod \mathrm{n}_S$ be a Noether operator for a sheaf $\mathcal{F}$ and for $S \in \mathrm{Ass}\,(\mathcal{F})$, let $Z(S)$ be a proper closed analytic subset of S such that* $\mathrm{sing}\, S \subset Z(S)$. *Then for any open set $U \subset X$ the linear operator*

$$\nu := \Gamma(U, \mathrm{n}) : \Gamma(U, \mathcal{F}) \to \prod \{[\Gamma(U \cap S \setminus Z(S), \mathcal{O}(S))]^{r(S)}, \; S \in \mathrm{Ass}\,(\mathcal{F})\}$$

is continuous and open, if the second space is endowed with the topology induced from the distribution space $\prod [\mathcal{D}'(U \cap S \setminus Z(S))]^{r(S)}$.

PROOF. It is injective by Theorem 3.4. Then we suppose that U is a closed subspace of the open unit polydisc Δ in a coordinate space $\mathbb{C}^n_z$ and there is a morphism $\alpha : \mathcal{K} \to \mathcal{L}$ of free coherent sheaves on Δ such that $\mathrm{Cok}\,\alpha \cong \mathcal{F}$. We have the following exact sequence

$$\Gamma(\Delta, \mathcal{K}) \xrightarrow{\alpha} \Gamma(\Delta, \mathcal{L}) \xrightarrow{\pi} \Gamma(\Delta, \mathcal{F}) \to 0,$$

where π is the canonical projection. The composition

$$\nu \cdot \pi \ : \ \Gamma(\Delta, \mathcal{L}) \to \prod \Gamma(\Delta \cap S, \mathcal{O}(S))$$

is a differential operator. Hence it can be written in the following explicit form:

$$\nu \cdot \pi(u) = \sum \nu_i \mathrm{D}_z^i u \ ,$$

where $\nu_i \in \Gamma(\Delta, \mathcal{L}')$, where $\mathcal{L}'$ is the dual to the sheaf $\mathcal{L}$ and the sum is finite (cf. [10]). Each D_z^i is a continuous operator by Cauchy's inequality, hence $\nu \cdot \pi$ is continuous. Then the same is true for ν, because π is open by the definition of the topology in $\Gamma(\Delta, \mathcal{F})$ (cf. [6]). In the general case the topology of $\Gamma(U, \mathcal{F})$ is the supremum of topologies, induced from all the spaces $\Gamma(\Delta, \phi_*(\mathcal{F} \mid Y))$, where Y runs over a set of relatively compact analytic polyhedrons, which covers U, $\phi : Y \to \Delta$ is a closed embedding. Hence the general case is reduced to the case $X = \Delta$, $\mathcal{F} \cong \operatorname{Cok} \alpha$. It is obvious that ν is still continuous in this case. To prove the openness of ν we use

LEMMA 3.10 *For any point $z \in \Delta$ and its neighbourhood $U \subset\subset \Delta$ there exist a neighbourhood $V \subset U$ of z, a neighbourhood $W(S)$ of $Z(S)$ and a constant C such that for any $f \in \Gamma(U, \mathcal{K})$ there exists a section $g \in \Gamma(V, \mathcal{L})$, which satisfies the inequality*

$$\sup\{|f + \alpha g| \, , \, z \in V\} \le C \max\{\sup(|\nu_S f(z)| \, , \, z \in U \cap S \setminus W(S)) \, , \, S \in \operatorname{Ass}(\mathcal{F})\} \, . \quad (11)$$

The maximum on the right–hand side is well–defined since $U \cap S$ is empty except for a finite subset of $\operatorname{Ass}(\mathcal{F})$. Lemma 3.10 implies Theorem 3.9, since we can choose a polyhedral covering for X, consisting of neighbourhoods V, which satisfy (11) and sup–norm in the right–hand side is majorized by the topology induced from the distribution space $\mathcal{D}'(U \cap S \setminus Z(S))$.

PROOF of Lemma 3.10. It is in fact proved in [9] (ch. IV) for a special Noether operator λ. By the unicity theorem 3.3 we have for any S:

$$s\lambda_S = \sum \sigma_{ST} \cdot \nu_T \ , \quad (12)$$

where λ_S is the S–component of λ and $\sigma_{ST} \, : \, [\mathcal{O}(T)] \to [\mathcal{O}(S)]$ is a differential operator in a neighbourhood of z, $s \in \mathcal{O}(X_z) \setminus I(S_z)$. Note that $\sigma_{ST} \neq 0$ only if $S \subset T$. Applying [9], we get the estimate

$$\sup(|f + \alpha g| \, , \, z \in V) \le C \max\{\sup(|\lambda_S f| \, , \, z \in U' \cap S \setminus W'(S)) \, , \, S \in \operatorname{Ass}(\mathcal{F})\}$$

for a section g of the sheaf $\mathcal{K}$, some neighbourhoods V, $U' \subset\subset U$ of z and an open set $W'(S) \supset\supset W(S) \cup s^{-1}(0)$. Combining this with the inequality

$$\sup(|\lambda_S f| \, , \, z \in U \cap S \setminus W'(S)) \le C' \max\{\sup(|\nu_T f| \, , \, z \in U \cap T \setminus W(T)) \, , \, T \supset S\} \, , \quad (13)$$

we get (11). To prove (13) we use (12) and the estimate

$$\sup(|\sigma_{ST} h| \, , \, z \in U' \cap S \setminus W(S)) \le C \cdot \sup(|h| \, , \, z \in U \cap T \setminus W(T))$$

for any holomorphic function h on $U \cap T$, which follows from Cauchy's inequality. The Lemma and hence Theorem 3.9 are proved.

PROOF of Proposition 1.8. Consider the commutative diagram

$$\begin{array}{ccc}
\Pi[\Gamma(S,\mathcal{O}(S))]^{r(S)} & \xrightarrow{\rho'} & \Pi[\Gamma(S\setminus Y,\mathcal{O}(S))]^{r(S)} \\
\uparrow{\scriptstyle \mathrm{n}} & & \uparrow{\scriptstyle \mathrm{n}'} \\
\Gamma(X,\mathcal{F}) & \xrightarrow{\rho} & \Gamma(X\setminus Y,\mathcal{F})
\end{array}$$

Figure 1

where the vertical arrows are generated by a Noether operator ν for the sheaf $\mathcal{F}$ and ρ, ρ' are the restriction mappings. We should prove that ρ is a strict morphism i. e. an open mapping to its image. The operators n and n$'$ are strict injections in virtue of Theorem 3.9, hence it is sufficient to check that the mapping ρ' is strict. This means that for any $S \in \mathrm{Ass}\,(\mathcal{F})$ the restriction mapping

$$\Gamma(S,\mathcal{O}(S)) \to \Gamma(S\setminus Y,\mathcal{O}(S)) \tag{14}$$

is strict. It is trivial in the case $S \subset Y$. Suppose that the set $S \setminus Y$ is non–empty. We claim that for any compact set $Q \subset S$ there exists a compact set $R \subset S \setminus K$, such that

$$\max\{|f(s)|\,,\ s\in Q\} \le \max\{|f(s)|\,,\ s \in R\} \tag{15}$$

for any holomorphic function f on S. Since S is a Stein space we may choose a compact analytic polyhedron P, which contains a neighbourhood of $Q \cup K$:

$$P := \{s \in S\ :\ |f_1(s)| \le 1,\dots,|f_p(s)| \le 1\}\ ,$$

where $f_1,\dots,f_p$ are holomorphic functions on S. For any holomorphic function f the function $|f(s)|$ attains its maximum on P at a point of its boundary ∂P. Hence (15) holds for $R := \partial P$. Now take an open neighbourhood $U \subset S \setminus K$ of R and apply Theorem 3.9 to U and $Z(S) := S \cap Z$. As the result we find that there exists a compact subset $T \subset U \setminus Z \subset S \setminus Y$ such that the right–hand side of (15) is majorized by $C{\cdot}\max\{|f(s)|\,,\ s\in T\}$. This implies that the mapping (14) is a strict injection, which completes the proof.

REFERENCES

[1] Aytuna A. : *The Fréchet space structure of global sections of certain coherent analytic sheaves*, Mathematica Balkanica **3**(1989), Fasc. 3–4, 311–324.

[2] Banica C. , Stanasila O. : *Algebraic method in the global theory of complex spaces*, John Wiley & Sons, 1976.

[3] Bourbaki N. : *L'algebre commutative*, Paris, Hermann, 1967.

[4] Djakov P. , Mitiagin B. S. : *The structure of polynomial ideals in the algebra of entire functions*, Studia Math. **68**(1980), 85–104.

[5] Forster O. : *Primärzerlegung in Steinschen Algebren*, Math. Ann. **154**(1964), 307–329.

[6] Grauert H. , Remmert R. : *Theorie der Steinschen Räume*, Springer, 1977.

[7] Gunning R. , Rossi H. : *Analytic functions of several complex variables*, ch. V, Englewood Cliffs, Prentice Hall, 1965.

[8] Mityagin B. S. , Henkin G. M. : *Linear problems of complex analysis*, Uspekhi Mat. Nauk **26**, N4(1971), 93–152; Russian Math. Surveys **26**(4)(1971), 99–164.

[9] Palamodov V. P. : *Linear differential operators with constant coefficients*, Moscow, Nauka, 1967; Springer–Verlag, 1970.

[10] Palamodov V. P. : *Differential operators in coherent analytic sheaves*, Mathematics of the USSR Sbornik, **77**, N3(1968), 390–422; **6**(1968), 365–391.

[11] Siu Y. T. : *Noether–Lasker decomposition of coherent analytic subsheaves*, Trans. of A. M. S. **135**(1969), 375–385.

[12] Siu Y. T. , Trautmann G. : *Gap–sheaves and extensions of coherent analytic subsheaves*, Lecture Notes 172, Springer Verlag, 1971.

[13] Vogt D. : *Charakterisierung der Unterräume von s*, Math. Z. **155**(1977), 109–117.

[14] Vogt D. , Wagner M. J. : *Charakterisierung der Unterräume von s und eine Vermutung von Martineau*, Studia Math. **67**(1980), 225–240.

[15] Vogt D. : *On the functors $Ext^1(E, F)$ for Fréchet spaces*, Studia Math. , **35**(1987), 163–197.

[16] Zakharyuta V. P. : *Isomorphism of spaces of analytic functions*, Dokl. Akad. Nauk **255**, N1(1980), 11–14; Soviet Math. Dokl. , **22**(1980), 631–634.

[17] Zariski O. , Samuel P. : *Commutative algebra*, Ch. IV, Van Nostrand, 1958.

Tame Splitting Theory for Fréchet-Hilbert Spaces

MARKUS POPPENBERG Fachbereich Mathematik, Universität Dortmund, Postfach 500500, D-4600 Dortmund, Germany

DIETMAR VOGT Fachbereich Mathematik, Bergische Universität-GH Wuppertal, Gaußstr. 20, D-5600 Wuppertal, Germany

0 INTRODUCTION

In the literature the splitting of short exact sequences of Fréchet spaces has been extensively studied. A central result is contained in [26] (see below for the notation): Let $0 \to F \to G \to E \to 0$ be an exact sequence of Fréchet-Hilbert spaces; if E has property (DN) and F has property (Ω), then the sequence splits.

Recently, in the theory of Fréchet spaces tameness conditions have become important, for instance when dealing with implicit function theorems in Fréchet spaces of so called Nash-Moser-type (see e.g. [4], [5], [10]), or in connection with basis problems in Fréchet spaces (e.g. [2], [3], [25]). Consequently, the tame splitting of exact sequences of Fréchet spaces was investigated.

In [18] a tame splitting theorem is proved under very general hypotheses. On the one hand, the continuity estimates of the above topological result are strongly improved, on the other hand, the tame result is applicable to power series spaces as well of infinite as of finite type which is in contrast to the topological case.

The present note is concerned with this tame splitting result (Theorem 4.1.). The purpose is to give a brief outline of the known related results and then to formulate the theorem, to clarify its hypotheses and to show how they can be met in applications, further to give an idea of the general proof (cf. [18]) by carrying it out in some particular case, and to state some consequences and corollaries.

In section 1 the class of graded Fréchet-Hilbert spaces is introduced including the power series spaces as most important examples. In section 2 the concept of a tame (resp. linearly tame) linear map between graded Fréchet spaces is defined.

In section 3 exact and tamely exact sequences of Fréchet spaces are considered. There is given a short summary on the known results on topological and tame splitting; in particular, the difference between the case of power series spaces of infinite and of finite type is explained. In the finite type case, a general topological splitting result fails (see Theorem 3.4.), and thus the tame splitting results become important also in the topological situation.

Section 4 contains the statement of the main result, the tame splitting Theorem 4.1. The hypotheses of this theorem are analyzed and evaluated for power series spaces.

In section 5 we try to give an idea of its fairly complicated proof given in [18]. One essential tool is a decomposition lemma for matrix operators on Hilbert scales formulated in Lemma 5.1. without proof (see [18], 1.3.). Applying this lemma we carry out the proof of Theorem 4.1. in the particular case of power series spaces deviating from the general proof in [18]. The proof of this special case given here is much easier than the general one; however, it is based on the same plan and can serve to clarify some of the main ideas of the proof in [18].

In section 6 we include a linear-tame splitting theorem based on the fundamental Proposition 5.4. (cf. [18], 5.2.). We also show how Proposition 5.4. can be used to obtain a new proof of the above topological splitting result [26]; this proof also gives some hints how to obtain further splitting results for spaces satisfying conditions like power series spaces of finite type.

Finally, in section 7 the fundamental theorem of structure theory of nuclear stable power series spaces is stated which in the infinite type case is based on splitting techniques and in the finite type case on different methods. The splitting Theorem 4.1. is an appropriate instrument to obtain a unified (and tame) structure theory for both cases which will be contained in a forthcoming paper.

Throughout this note we shall use common notation on Fréchet spaces, see [6], [7], [14] and concerning tameness [4], [24].

1 GRADED FRÉCHET-HILBERT SPACES

A Fréchet space E is a complete Hausdorff metrizable locally convex topological vector space. The topology can be given by a countable system of seminorms

$$\|\ \|_0 \leq \|\ \|_1 \leq \|\ \|_2 \leq \cdots$$

such that a sequence $x_j \to x$ if and only if $\|x_j - x\|_k \to 0$ for all k; a basis of neighborhoods of zero is then obtained by the sets

$$U_{k,\epsilon} = \{x \in E : \|x\|_k \leq \epsilon\} .$$

Such a fundamental system of seminorms is not at all uniquely determined by the topology; any choice of such a collection of seminorms is called grading. A graded Fréchet space is one equipped with a fixed choice of grading. Graded subspaces and graded quotient spaces are equipped with the induced gradings.

A seminorm $\| \ \|$ on E is called hilbertian if there is a semiscalar product $<,>$ on E such that $\|x\|^2 = <x,x>$ for all $x \in E$. A Fréchet-Hilbert space is a Fréchet space which admits a grading consisting of hilbertian seminorms, a graded Fréchet-Hilbert space is one equipped with such a grading.

Examples of Fréchet spaces which enjoy natural gradings are function spaces and sequence spaces. For instance, we mention the spaces of infinitely differentiable functions $\mathcal{C}^\infty(\Omega)$ for $\Omega \subset \mathbb{R}^N$ open, or $\mathcal{C}^\infty(K)$ for a compact $K \subset \mathbb{R}^N$ (e.g. $K = \overline{\overset{\circ}{K}}$ with smooth boundary) or for a compact $\mathcal{C}^\infty$-manifold K, or the spaces $\mathcal{H}(\Omega)$ of holomorphic functions for $\Omega \subset \mathbb{C}^N$ open; these spaces are all graded in a canonical way (of course, the gradings on $\mathcal{C}^\infty(\Omega)$ and $\mathcal{H}(\Omega)$ depend on a given compact exhaustion of the open set Ω).

A lot of function spaces are isomorphic to sequence spaces by means of series expansions (e.g. Fourier expansions), see for instance [30].

Let $A = (a_{j,k})$ be an infinite matrix such that $0 \leq a_{j,k} \leq a_{j,k+1}$ for all j, k and $\sup_k a_{j,k} > 0$ for all j. The Köthe sequence space $\lambda^2(A)$ is defined by

$$\lambda^2(A) = \{ x = (x_1, x_2, \ldots) : \|x\|_k^2 = \sum_j |x_j|^2 a_{j,k}^2 < \infty \text{ for all } k \} .$$

The space $\lambda^2(A)$ equipped with the seminorms $\| \ \|_k$ is a graded Fréchet-Hilbert space. A special and important class of Köthe sequence spaces is formed by the power series spaces:

For any sequence $0 \leq \alpha_1 \leq \alpha_2 \leq \ldots \nearrow +\infty$, for $R \in [0, \infty]$ and $r_0 < r_1 < \ldots \nearrow R$ we put $\Lambda_R^2(\alpha) = \lambda^2(A)$ where $a_{j,k} = e^{r_k \alpha_j}$. The spaces $\Lambda_R^2(\alpha)$ are called power series spaces of infinite type (if $R = \infty$) or of finite type (if $R < \infty$), respectively. For $R \in [0, \infty]$ we have the graded Fréchet-Hilbert spaces

$$\Lambda_R^2(\alpha) = \{ x = (x_1, x_2, \ldots) : \|x\|_k^2 = \sum_j |x_j|^2 e^{2 r_k \alpha_j} < \infty \text{ for all } k \} .$$

Usually, we consider the case $R \in \{0, \infty\}$ and the gradings defined by $r_k = k$ on $\Lambda_\infty^2(\alpha)$ and by $r_k = -\frac{1}{k}$ on $\Lambda_0^2(\alpha)$, respectively.

For a graded Fréchet space E let E_k denote the Banach space obtained by completion of $E/\ker\| \ \|_k$ with respect to the norm $\|\hat{x}\|_k = \|x\|_k$ for $\hat{x} = x + \ker\| \ \|_k \in E/\ker\| \ \|_k$. We then have the canonical connecting maps $i_k^m : E_k \to E_m$ for $m \leq k$ and $i^k : E \to E_k$. Of course, for a graded Fréchet-Hilbert space E the spaces E_k are Hilbert spaces.

For Fréchet spaces E, F let $L(E,F)$ denote the space of all continuous linear maps $E \to F$. The space $L(E,F)$ is a Banach space if E and F are Banach spaces. For a graded Fréchet space E equipped with a grading consisting of norms and a Banach space F we shall sometimes identify the spaces $L(E_k, F)$ and $\{ T \in L(E,F) : \sup \{ \|Tx\|^F : \|x\|_k^E \leq 1 \} < \infty \}$.

2 TAMENESS

A linear map $T : E \to F$ between two graded Fréchet spaces E and F is continuous if and only if there is a map $\sigma : \mathbb{N}_0 \to \mathbb{N}_0$ and a sequence of constants $c_k > 0$ such that

$$\|Tx\|_k \leq c_k \|x\|_{\sigma(k)}$$

for all $k = 0, 1, 2, \ldots$ and all $x \in E$; the linear map T is called tame if σ can be chosen so that $\sigma(k) \leq k + b$ for some $b \in \mathbb{N}_0$, it is called linear-tame if a choice $\sigma(k) \leq ak + b$ for some $a, b \in \mathbb{N}_0$ is possible.

A linear bijection $T : E \to F$ between two graded Fréchet spaces is called a tame (resp. linear-tame) isomorphism if both T and T^{-1} are tame (resp. linear-tame). Two gradings on E are called tamely (resp. linear-tamely) equivalent if identity is a tame (resp. linear-tame) isomorphism.

For instance, let $K \subset \mathbb{R}^N$ be compact with $\mathcal{C}^1$-boundary so that $K = \overline{\mathring{K}}$, and let $\mathcal{C}^\infty(K)$ be equipped with its natural grading

$$\|f\|_k = \sup_{|\alpha|\leq k} \sup_{x\in K} |\frac{\partial^\alpha}{\partial x^\alpha} f(x)| \,, \; f \in \mathcal{C}^\infty(K)$$

where $|\alpha| = \alpha_1 + \ldots + \alpha_N$ for a multiindex $\alpha = (\alpha_1, \ldots, \alpha_N)$. Then every linear partial differential operator $P : \mathcal{C}^\infty(K) \to \mathcal{C}^\infty(K)$ with coefficients $a_\alpha \in \mathcal{C}^\infty(K)$ defined by

$$P(f)(x) = \sum_{|\alpha|\leq m} a_\alpha(x) \frac{\partial^\alpha}{\partial x^\alpha} f(x)$$

is a tame linear map (we may choose $\sigma(k) = k + m$).

By means of the Sobolev Imbedding Theorem, a tamely equivalent grading on $\mathcal{C}^\infty(K)$ is defined by the hilbertian seminorms

$$|f|_k^2 = \sum_{|\alpha|\leq k} \int_K |\frac{\partial^\alpha}{\partial x^\alpha} f(x)|^2 \, dx$$

which make $(\mathcal{C}^\infty(K), |\;|_k)$ a graded Fréchet-Hilbert space.

We notice that every continuous linear map from $\Lambda_0^2(\alpha)$ into $\Lambda_0^2(\beta)$ is linear-tame (with respect to the gradings defined by $r_k = -\frac{1}{k}$), see e.g. [28], 2.3. or [1], while by a result of Zaharjuta [35] every continuous linear map from $\Lambda_0^2(\alpha)$ to $\Lambda_\infty^2(\beta)$ is even bounded (i.e. σ can be chosen bounded), cf. [22], 5.1. or [28], 2.2.

On the other hand, for continuous linear maps from $\Lambda_\infty^2(\alpha)$ to $\Lambda_\infty^2(\beta)$ such bounds to σ do in general not exist. For example, Dubinsky constructs in [1] for the case $\alpha_j = j$ for an arbitrarily given map $\sigma : \mathbb{N}_0 \to \mathbb{N}_0$ an endomorphism T of $\Lambda = \Lambda_\infty^2(j)$ such that $\sup_{x\in\Lambda} \frac{\|Tx\|_k}{\|x\|_{\sigma(k)}} = \infty$ for all k.

Dubinsky and Vogt characterize in [2], [3] the class of power series spaces of infinite type such that every continuous endomorphism is linear-tame; for instance, this holds if α is strictly unstable which means $\lim_j \frac{\alpha_{j+1}}{\alpha_j} = +\infty$, while power series spaces occurring in analysis (e.g. in sequence space representations of function spaces) do in general not own this property.

3 SHORT EXACT SEQUENCES OF FRÉCHET SPACES

Let E, F, G be Fréchet spaces and $j : F \to G$, $q : G \to E$ be continuous linear maps. Then the sequence

$$(*) \qquad 0 \longrightarrow F \xrightarrow{\;j\;} G \xrightarrow{\;q\;} E \longrightarrow 0$$

is called exact if j is injective, q is surjective and $\operatorname{im} j = \ker q$. In this case $\operatorname{im} j$ is closed, and by means of the open mapping theorem j is a topological isomorphism onto $\ker q$, and q is open, hence a topological quotient map. Therefore, the maps j and q induce the canonical isomorphisms

$$\overline{j} : F \xrightarrow{\;\cong\;} \operatorname{im} j \ , \ \overline{q} : G/\ker q \xrightarrow{\;\cong\;} E \,.$$

Now let E, F, G be graded Fréchet spaces and let $(*)$ be exact. Then the subspace $\operatorname{im} j$ of G is a graded Fréchet space equipped with the restricted grading of G, and the quotient space $G/\ker\, q$ is a graded Fréchet space endowed with the induced quotient grading

$$\|\hat{g}\|_k = \inf\{\, \|\xi\|_k : q\xi = qg \,\} \ , \ \hat{g} = g + \ker q \in G/\ker q \,.$$

The sequence $(*)$ is called tamely exact (resp. linear-tamely exact), if $\overline{j} : F \to \operatorname{im} j$ and $\overline{q} : G/\ker q \to E$ are tame (resp. linear-tame) isomorphisms. More explicitely, the exact sequence $(*)$ is tamely exact, if and only if there exist $b \in \mathbb{N}_0$ and constants $c_k > 0$ such that

$$\|jf\|_k^G \le c_k \|f\|_{k+b}^F \ , \ \|f\|_k^F \le c_k \|jf\|_{k+b}^G$$

$$\|qg\|_k^E < c_k \|q\|_{k+b}^G \ , \ \inf\{\, \|\xi\|_k^G : q\xi = e \,\} \le c_k \|e\|_{k+b}^E$$

for all k and all $e \in E$, $f \in F$, $g \in G$.

We say that the exact sequence $(*)$ splits if q admits a continuous linear right inverse, or, equivalently, if j admits a continuous linear left inverse. Analogously, we say that the tamely (resp. linear-tamely) exact sequence of graded Fréchet spaces $(*)$ splits tamely (resp. linear-tamely) if q admits a tame (resp. linear-tame) right inverse, or, equivalently, if j admits a tame (resp. linear-tame) left inverse.

Now let $(*)$ be an exact sequence, let $\| \ \|_k$ denote a grading on G, and let $| \ |_k$ denote the gradings on F and E induced by j and q from $(G, \| \ \|_k)$ as graded subspace and quotient space, respectively. Let $F_{(k)}$ and $E_{(k)}$ denote the Banach spaces belonging to $(F, | \ |_k)$ and $(E, | \ |_k)$. Then we obtain exact sequences of Banach spaces

$$(**) \qquad 0 \longrightarrow F_{(k)} \xrightarrow{\;j_k\;} G_k \xrightarrow{\;q_k\;} E_{(k)} \longrightarrow 0$$

where j_k and q_k denote the corresponding continuous extensions of j and q.

In the literature, there are given a lot of examples of exact sequences $(*)$ which do not split even though all the corresponding 'local exact sequences' $(**)$ do split (see e.g. 3.4., [22], 5.4., [21], 3.1., [28], 3.). In fact, there exist exact sequences $(*)$ of Fréchet-Hilbert spaces which do not split; in this case the exact sequences $(**)$ of Hilbert spaces obviously split.

These observations caused several authors to investigate the splitting problem, i.e. the question under which conditions sequences of the form $(*)$ split; more recently, the corresponding tame splitting problem has been studied as well, i.e. the problem of the existence of a tame resp. linear-tame splitting.

The following notation is useful (cf. [27]):

A pair (E, F) of Fréchet spaces is called a splitting pair if all exact sequences of the form $(*)$ split; a pair (E, F) of graded Fréchet spaces is called a tame (linear-tame) splitting pair, if all tamely (linear-tamely) exact sequences of the form $(*)$ split tamely (linear-tamely).

In the same way, we call a pair (E, F) of (graded) Fréchet-Hilbert spaces a (tame / linear-tame) Hilbert splitting pair, if for all (graded) Fréchet-Hilbert spaces G all (tamely / linear-tamely) exact sequences of the form $(*)$ split (tamely / linear-tamely).

In 1977 and 1980 Vogt and Wagner used in [20], [33], [21] the topological invariant conditions (DN) and (Ω) for the characterization of subspaces, quotient spaces and complemented subspaces of power series spaces and proved in this context the following fundamental result:

3.1. THEOREM. ([20], [33], [21]) *Let E, F be Fréchet spaces one of which is nuclear. If E has property* (DN) *and F has property* (Ω), *then the pair (E, F) is a splitting pair.*

Properties (DN) and (Ω) can be defined as follows (cf. [26], [31]), for different equivalent descriptions we refer to [20], [33], [21], [29].

DEFINITION. Let E be a Fréchet space and $\| \ \|_0 \leq \| \ \|_1 \leq \dots$ a fundamental system of seminorms. Then E has property (DN) resp. (Ω) if

$$(\mathrm{DN}) \ \exists\, p \,\forall\, k \,\forall\, 0 < \tau < 1 \,\exists\, K \,\exists\, c > 0 \ : \ \| \ \|_k \leq c \| \ \|_p^{1-\tau} \| \ \|_K^{\tau} \,,$$

resp.

$$(\Omega) \ \forall\, p \,\exists\, q \,\forall\, Q \,\exists\, 0 < \theta < 1 \,\exists\, c > 0 \ : \ \| \ \|_q^* \leq c \| \ \|_p^{*\,1-\theta} \| \ \|_Q^{*\,\theta} \,.$$

Here $\| \ \|_k^*$ denotes the dual (extended real valued) norm

$$\|\phi\|_k^* = \sup \{ |\phi(x)| : \|x\|_k \leq 1 \} \in \mathbb{R} \cup \{\infty\} \,, \ \phi \in E' \,.$$

REMARK. ([29]). Property (DN) is equivalent to the existence of a grading $\| \ \|_k$ on E such that

$$\| \ \|_k^2 \leq \| \ \|_{k-1} \| \ \|_{k+1} \,, \quad k = 1, 2, \dots$$

while property (Ω) is equivalent to the existence of a grading $\| \ \|_k$ on E such that

$$\| \ \|_k^{*\,2} \leq \| \ \|_{k-1}^* \| \ \|_{k+1}^* \,, \quad k = 1, 2, \dots$$

The latter conditions are the so called standard forms of properties (DN) and (Ω).

In 1985 Krone and Vogt characterized in [9] the Köthe sequence spaces such that the pair $(\lambda(A), \lambda(B))$ is a splitting pair. In 1986 Vogt [26] proved a splitting theorem for Fréchet-Hilbert spaces without nuclearity assumptions and without supposing the existence of a basis:

3.2. THEOREM. ([26]). *Let* E, F *be Fréchet-Hilbert spaces. If* E *has property* (DN) *and* F *has property* (Ω) *then the pair* (E, F) *is a Hilbert splitting pair.*

We notice that conditions (DN) and (Ω) are also connected in a necessary way to the above splitting theorems (cf. e.g. [23]).

The topological splitting theorems 3.1., 3.2. are far away from producing tame or linear-tame right inverses. Recently, some particular results in this direction have been proved.

For $R \in \{0, \infty\}$ we define

$$\Lambda_R^1(\alpha) = \{ x = (x_1, x_2, \ldots) : \|x\|_k = \sum_j |x_j| \, e^{r_k \alpha_j} < \infty \text{ for all } k \}$$

where $r_k = k$ if $R = \infty$ and $r_k = -\frac{1}{k}$ if $R = 0$.

In 1989 Nyberg [12], [13] and Vogt [27] proved:

3.3. THEOREM. *The pair* $(\Lambda_R^1(\alpha), \Lambda_R^1(\beta))$ *is a tame and a linear-tame splitting pair for* $R = 0$ *and for* $R = \infty$.

However, the pair $(\Lambda_0^1(\alpha), \Lambda_0^1(\beta))$ is in general not a splitting pair:

3.4. THEOREM. ([22], 5.4.). *Suppose that* $\lim_n \frac{\log n}{\alpha_n} = \lim_n \frac{\log n}{\beta_n} = 0$ *(nuclearity) and* $\sup_n \frac{\alpha_{2n}}{\alpha_n} < \infty$, $\sup_n \frac{\beta_{2n}}{\beta_n} < \infty$ *(stability). Then there exists an exact sequence*

$$0 \longrightarrow \Lambda_0^1(\beta) \longrightarrow G \longrightarrow \Lambda_0^1(\alpha) \longrightarrow 0$$

which does not split.

Therefore, a topological counterpart to the tame / linear-tame splitting result Theorem 3.3. does not hold for power series spaces of finite type. Thus, in this case tameness conditions are also essential if one is solely interested in the existence of a continuous linear right inverse.

Vogt proves in [24] tame splitting theorems for spaces admitting smoothing operators in the sense of Moser [11] or Sergeraert [19], see also [25], [15], [16] for related results. A further particular result is contained in

3.5. THEOREM. ([17]). *Suppose that* $\sup_n \frac{\log n}{\alpha_n} < \infty$ *(nuclearity). If the graded Fréchet-Hilbert space* F *is tamely nuclear (in the sense of [17], 2.) and satisfies property* (Ω) *in standard form (as above), then the pair* $(\Lambda_\infty^1(\alpha), F)$ *is a tame splitting pair.*

All the above tame results need the assumption either of the existence of a basis or of nuclearity; in addition, the results for power series spaces of infinite type and of finite type, respectively, are verified by different proofs.

In the following section we state a very general tame splitting theorem for graded Fréchet-Hilbert spaces which has been proved in [18]. This result does not need any nuclearity assumption nor the existence of a basis. It may be applied to power series spaces of infinite type and of finite type as well, based on a common proof for both cases. From the proof, we also get back the topological result 3.2. (see 6.3.) and obtain a linear-tame splitting theorem (see sec. 6). It should be emphasized that this tame splitting Theorem 4.1. may also serve as a 'topological splitting theorem' for spaces satisfying conditions like power series spaces of finite type (cf. 3.4.).

4 A TAME SPLITTING THEOREM FOR FRÉCHET-HILBERT SPACES

Let E, F, G be graded Fréchet-Hilbert spaces, which means Fréchet spaces equipped with fixed fundamental systems of hilbertian seminorms

$$\| \ \|_0 \leq \| \ \|_1 \leq \| \ \|_2 \leq \dots$$

We assume that E and F satisfy the following conditions of (DN)-type and (Ω)-type, respectively.

(1) There are numbers $0 < \tau_k < 1$ and constants $c_k > 0$ such that

$$(\mathrm{DN}_\tau) \qquad \| \ \|_k \leq c_k \| \ \|_{k-1}^{1-\tau_k} \| \ \|_{k+1}^{\tau_k}$$

on E for all k.

(2) There are numbers $0 < \theta_k < 1$ and constants $c_k > 0$ such that

$$(\Omega_\theta) \qquad \| \ \|_k^* \leq c_k \| \ \|_{k-1}^{*\,1-\theta_k} \| \ \|_{k+1}^{*\,\theta_k}$$

on $(F, \| \ \|_{k-1})'$ for all k.

Moreover, we suppose the numbers τ_k and θ_k to be linked as follows:

(3) There is p_0 such that $\theta_k \geq \tau_{k+p}$ for all $p \geq p_0$ and all k.

Under these assumptions the following tame splitting theorem has been proved in [18]:

4.1. THEOREM. ([18], 6.1.). *Let E, F, G be graded Fréchet-Hilbert spaces, let E satisfy (1) for suitable τ_k and let F satisfy (2) for suitable θ_k, let τ_k, θ_k satisfy (3). Then every tamely exact sequence*

$$0 \longrightarrow F \longrightarrow G \longrightarrow E \longrightarrow 0$$

splits tamely.

In other words, the pair (E, F) is a tame Hilbert splitting pair.

REMARKS.

(i) ([33]). By means of the bipolar theorem condition (Ω_θ) in (2) is equivalent to the condition

$$U_k \subset c'_k \{ r^{n_k} U_{k+1} + \tfrac{1}{r} U_{k-1} \} \quad \text{for all } r > 0$$

where $U_k = \{ x \in F : \|x\|_k \leq 1 \}$ and $n_k = \frac{1-\theta_k}{\theta_k}$.

(ii) For instance, condition (3) holds if $\theta_k \geq \tau_k$ for all k and one of the sequences (θ_k), (τ_k) is monotonically decreasing.

We analyse the hypotheses of the theorem in the case of power series spaces. Let $0 \leq \alpha_1 \leq \alpha_2 \leq \ldots \nearrow +\infty$, let $R \in [0,\infty]$ and $r_0 < r_1 < \ldots \nearrow R$. We define the graded Fréchet-Hilbert space $\Lambda_R^2(\alpha)$ as in section 1 and put

$$|x|_t^2 = \sum_j |x_j|^2 e^{2t\alpha_j}$$

for $x \in \Lambda_R^2(\alpha)$ and $t < R$. We remember that the grading in $\Lambda_R^2(\alpha)$ is defined by

$$\|x\|_k = |x|_{r_k} .$$

It is easy to verify that for $t_0 < t_1 < t_2 < R$ and $0 < t < 1$ with $t_1 = (1-t)t_0 + tt_2$ the following holds

$$|\ |_{t_1} \leq |\ |_{t_0}^{1-t} |\ |_{t_2}^{t} \quad \text{on } \Lambda_R^2(\alpha)$$

and

$$|\ |_{t_1}^* \leq |\ |_{t_0}^{*\,1-t} |\ |_{t_2}^{*\,t} \quad \text{on } \Lambda_R^2(\alpha)' .$$

Therefore, the space $\Lambda_R^2(\alpha)$ satisfies conditions (1) and (2) for $c_k = 1$ and

$$\tau_k = \theta_k = \frac{r_k - r_{k-1}}{r_{k+1} - r_{k-1}} ,$$

hence

$$\frac{1}{\tau_k} - 1 = \frac{1}{\theta_k} - 1 = \frac{r_{k+1} - r_k}{r_k - r_{k-1}} .$$

Now let $F = \Lambda_R^2(\alpha)$ and $E = \Lambda_S^2(\beta)$ where $\Lambda_S^2(\beta)$ is defined by a sequence $0 \leq \beta_1 \leq \beta_2 \leq \ldots \nearrow +\infty$, $S \in [0,\infty]$ and $s_0 < s_1 < \ldots \nearrow S$.

4.2. COROLLARY. *If there is p_0 such that*

$$(*) \qquad \frac{r_{k+1} - r_k}{r_k - r_{k-1}} \leq \frac{s_{k+p+1} - s_{k+p}}{s_{k+p} - s_{k+p-1}}$$

for all $p \geq p_0$ and all k, then the pair $(\Lambda_S^2(\beta), \Lambda_R^2(\alpha))$ is a tame Hilbert splitting pair.

Condition $(*)$ is e.g. fulfilled if one of the sequences $(\frac{r_{k+1}-r_k}{r_k-r_{k-1}})_k$, $(\frac{s_{k+1}-s_k}{s_k-s_{k-1}})_k$ is monotonically increasing and $\frac{r_{k+1} - r_k}{r_k - r_{k-1}} \leq \frac{s_{k+1} - s_k}{s_k - s_{k-1}}$ for all k (cf. [28]).

In the case $r_k = s_k$ we conclude:

4.3. COROLLARY. *If the sequence* $\Delta_k = r_k - r_{k-1}$ *is logarithmically convex, i.e.* $\Delta_k^2 \le \Delta_{k-1}\Delta_{k+1}$ *for all* k *and* $r_k \nearrow R \in [0,\infty]$, *then the pair* $(\Lambda_R^2(\beta), \Lambda_R^2(\alpha))$ *is a tame Hilbert splitting pair.*

For the standard gradings defined by $r_k = k$ if $R = \infty$ and $r_k = -\frac{1}{k}$ if $R = 0$ we obtain

4.4. COROLLARY. *Let* G *be a graded Fréchet-Hilbert space. Then all tamely exact sequences of the form*

$$0 \longrightarrow \Lambda_\infty^2(\alpha) \longrightarrow G \longrightarrow \Lambda_\infty^2(\beta) \longrightarrow 0$$

$$0 \longrightarrow \Lambda_0^2(\alpha) \longrightarrow G \longrightarrow \Lambda_0^2(\beta) \longrightarrow 0$$

split tamely.

5 THE PROOF OF THE TAME SPLITTING THEOREM

The proof of the general tame splitting Theorem 4.1. is given in [18]; it is rather complicated and involved. In this section, we shall sketch the main idea of the proof; besides, we carry out the proof for the particular situation of Theorem 5.3. Although this proof is much easier than the general one, it explains the strategy and clarifies some of the ideas used in [18].

Roughly speaking, the proof of Theorem 4.1. given in [18] is divided into two main parts. The first part is a decomposition lemma for matrix operators acting on Hilbert scales. The second - and most puzzling - part is concerned with the following geometrical problem: Let $\|\ \|_0 \le \|\ \|_1 \le \ldots \le \|\ \|_K$ be a finite number of hilbertian seminorms on a linear space E satisfying inequalities of type (DN_τ) resp. (Ω_θ) (cf. sec. 4). Then the purpose is to represent E equipped with these seminorms as good as possible as a subspace resp. quotient space of a suitable Hilbert scale.

We begin with discussing the first part of the proof:

Let I, J be two index sets and $(\alpha_i)_{i\in I}$, $(\beta_j)_{j\in J}$ be families of nonnegative integers. For any real number t we set

$$G_t = \{ x = (x_i)_{i\in I} \in \mathbb{C}^I : |x|_t^2 = \sum_{i\in I} |x_i|^2 e^{2t\alpha_i} < +\infty \},$$

$$H_t = \{ x = (x_j)_{j\in J} \in \mathbb{C}^J : |x|_t^2 = \sum_{j\in J} |x_j|^2 e^{2t\beta_j} < +\infty \}.$$

The spaces G_t, H_t are Hilbert spaces equipped with their natural scalar products. We have $G_s \subset G_t$ for $s \ge t$, and $G_t' = G_{-t}$ in a canonical way.

The following decomposition Lemma is proved in [18] (cf. [26]).

5.1. LEMMA. ([18], 1.3.). *Let* $T \in L(G_{s_0}, H_{t_0})$, *let* $\gamma > 0$ *be a rational number and let* $t_1 > 0$, $\epsilon > 0$. *Then there exist linear operators* T^+ *and* T^- *with* $T^+, T^- \in L(G_s, H_{t_0})$ *for all* $s > s_0$ *such that*

$$T = T^+ + T^- \text{ on } G_s \text{ when } s > s_0$$

and

$$T^+ : G_{s_0+\gamma t} \longrightarrow H_{t_0+t} \ , \ T^- : G_{s_0-\gamma t} \longrightarrow H_{t_0-t}$$

are continuous for every $t \geq t_1$ *with*

$$\|T^-\|_{G_{s_0-\gamma t} \to H_{t_0-t}} \leq \epsilon^t .$$

We now deviate from our proceeding in [18]. From Lemma 5.1. we immediately obtain the following fundamental proposition (cf. [28], 4.1.).

5.2. PROPOSITION. *Let* (r_k) *and* (s_k) *be sequences of rational numbers that satisfy*

$$r_{k-1} < r_k \text{ and } \frac{s_k - s_{k-1}}{r_k - r_{k-1}} \leq \frac{s_{k+1} - s_k}{r_{k+1} - r_k}$$

for all $k = 1, 2, \dots$, *let* $\sigma_k > s_k$. *Let* $A_k \in L(G_{s_k}, H_{r_k})$ *for* $k = 0, 1, \dots$ *Then there exist* $B_k \in L(G_{\sigma_k}, H_{r_k})$ *such that* $A_k = B_{k+1} - B_k$ *(on* G_{σ_k}*) for all* k.

Proof: We put $\gamma_k = \frac{s_k - s_{k-1}}{r_k - r_{k-1}}$ and $\epsilon_k = 2^{-\frac{k}{r_k - r_{k-1}}}$. By means of the previous Lemma we can choose linear operators A_k^+ and A_k^- satisfying $A_k = A_k^+ + A_k^-$ on G_s when $s > s_k$ such that

$$A_k^+ : G_{s_k+\gamma_k t} \longrightarrow H_{r_k+t} \ , \ A_k^- : G_{s_k-\gamma_k t} \longrightarrow H_{r_k-t}$$

are continuous for $t \geq r_k^1 = \min(r_{k+1} - r_k, r_k - r_{k-1})$ and

$$\|A_k^-\|_{G_{s_k-\gamma_k t} \to H_{r_k-t}} \leq \epsilon_k^t .$$

For $n > k$ we have $s_k + \gamma_k(r_n - r_k) \leq s_n$, hence

$$A_k^+ : G_{s_n} \longrightarrow H_{r_n} \ , \ n > k$$

is continuous, and for $n < k$ we have $s_k - \gamma_k(r_k - r_n) \leq s_n$, hence

$$\|A_k^-\|_{G_{s_n} \to H_{r_n}} \leq \epsilon_k^{r_k - r_n} \leq 2^{-k} \ , \ n < k .$$

For $k = 0, 1, 2, \dots$ we define

$$B_k = \sum_{n=0}^{k-1} A_n^+ - A_k^- - \sum_{n=k+1}^{\infty} A_n^- .$$

A formal calculation shows that $A_k = B_{k+1} - B_k$ for all k. Further we see that $\sum_{n=0}^{k-1} A_n^+ \in L(G_{s_k}, H_{r_k})$, $A_k^- \in L(G_{\sigma_k}, H_{r_k})$, and

$$\sum_{n=k+1}^{\infty} \|A_n^-\|_{G_{s_k} \to H_{r_k}} \leq \sum_{n=k+1}^{\infty} 2^{-n} < \infty ,$$

thus $B_k \in L(G_{\sigma_k}, H_{r_k})$ which proves the assertion.

The previous proposition enables us to prove the following particular case of Theorem 4.1. The Theorem below includes Corollary 4.2.; its proof uses standard techniques (see e.g. [18], 6.1.) and will be included only for sake of completeness.

5.3. THEOREM. *Let E, F, G be graded Fréchet-Hilbert spaces and let*

$$0 \longrightarrow F \longrightarrow G \xrightarrow{q} E \longrightarrow 0$$

be tamely exact. Let E be tamely isomorphic to a graded subspace of $\Lambda_S^2(\beta)$ and let F be tamely isomorphic to a graded quotient space of $\Lambda_R^2(\alpha)$ where R, $S \in [0,\infty]$. Assume that there is p_0 such that for all $p \geq p_0$ and all k we have

$$\frac{r_{k+1} - r_k}{r_k - r_{k-1}} \leq \frac{s_{k+p+1} - s_{k+p}}{s_{k+p} - s_{k+p-1}} .$$

Then the sequence splits tamely.

Proof: We may assume that E is a graded subspace of $\Lambda_S^2(\beta)$ and that F is a graded quotient space of $\Lambda_R^2(\alpha)$ where (s_k) and (r_k) are rational numbers. Let E_k and F_k denote the Hilbert spaces associated to the corresponding gradings. We then have canonical maps $j_k : E_k \to H_{s_k}$ and $l_k : G_{r_k} \to F_k$ (with H_s and G_r defined as above). On the other hand, using the notation of section 3 let $E_{(k)}$ and $F_{(k)}$ denote the Hilbert spaces attached to the gradings induced on E resp. F by $\| \ \|_k^G$ from the exact sequence. Since the sequence is tamely exact, we can choose numbers b, c, $d \geq 0$ such that the canonical maps

$$i_{k+b}^{(k)} : F_{k+b} \to F_{(k)} , \ i_{(k+c)}^{k} : F_{(k+c)} \to F_k , \ i_{k+d}^{(k)} : E_{k+d} \to E_{(k)}$$

are continuous. For every k we have an exact sequence of Hilbert spaces

$$0 \longrightarrow F_{(k)} \longrightarrow G_k \xrightarrow{q_k} E_{(k)} \longrightarrow 0 .$$

We choose $r_k \in L(E_{(k)}, G_k)$ such that $q_k \circ r_k = id$.

We set $\tilde{A}_k = i_{k+1}^k \circ r_{k+1} - r_k$. Then we have $q_k \circ \tilde{A}_k = 0$, hence $\tilde{A}_k \in L(E_{(k+1)}, F_{(k)})$, and $\bar{A}_k = i_{(k)}^{k-c} \circ \tilde{A}_k \in L(E_{k+d+1}, F_{k-c})$, $k \geq c$. We choose $A_k \in L(H_{s_{k+d+1}}, G_{r_{k-c}})$ such that $l_{k-c} \circ A_k \circ j_{k+d+1} = \bar{A}_k$.

Putting $a = \max(p_0 + b, b + c + d + 1)$ we then can apply Proposition 5.2. with $A_{k+b+c} \in L(H_{s_{k+a}}, G_{r_{k+b}})$ to find $B_k \in L(H_{s_{k+a+1}}, G_{r_{k+b}})$ satisfying $A_{k+b+c} = B_k - B_{k+1}$.

We set $\tilde{B}_k = i^{(k)}_{k+b} \circ l_{k+b} \circ B_k \circ j_{k+a+1} \in L(E_{k+a+1}, F_{(k)})$ and $R_k = i^k_{k+b+c} \circ r_{k+b+c} + \tilde{B}_k \in L(E_{k+a+1}, G_k)$.

It is easy to verify that $i^k_{k+1} \circ R_{k+1} - R_k = 0$. Hence, there is $R \in L(E,G)$ such that $R_k = i^k \circ R$, and R is a tame right inverse for q.

In Theorem 5.3. we have proved Theorem 4.1. for the particular case that E is a subspace of $\Lambda^2_S(\beta)$ and F is a quotient space of $\Lambda^2_R(\alpha)$. The assumption on s_k and r_k in 5.3. corresponds to condition (3) in 4.1. The above proof illustrates the role played by that condition:

Picking out suitable numbers s_k and r_k of the Hilbert scales (H_s) and (G_r), we may apply Proposition 5.2. in order to get the desired decomposition $A_k = B_k - B_{k+1}$.

The general proof of Theorem 4.1. originates in the same idea, we include here a short sketch of it:

Let E and F be graded Fréchet-Hilbert spaces satisfying conditions (DN_τ) and (Ω_θ) of section 4, respectively. It would be enough to represent E as a subspace of a suitable $\Lambda^2_S(\beta)$ and F as a quotient space of some $\Lambda^2_R(\alpha)$ where the numbers s_k and r_k are calculated from the numbers τ_k and θ_k as in section 4. However, this problem of structure theory is as hard to handle as the splitting theorem itself (cf. [20], [33], or [15], [16]). Thus, the following weaker problem is studied:

Let E and F be linear spaces both equipped with a finite number of hilbertian seminorms $\| \;\|_0 \leq \| \;\|_1 \leq \ldots \| \;\|_K$. Assume that for suitable $0 < \tau_k < 1$ and $0 < \theta_k < 1$ we have

$$(1) \qquad \| \;\|_k \leq \| \;\|_{k-1}^{1-\tau_k} \| \;\|_{k+1}^{\tau_k} \quad \text{on } E$$

$$(2) \qquad \| \;\|_k^* \leq \| \;\|_{k-1}^{*\,1-\theta_k} \| \;\|_{k+1}^{*\,\theta_k} \quad \text{on } (F, \| \;\|_{k-1})'$$

for $k = 1, \ldots, K-1$. Then the purpose is the following:

Construct Hilbert scales (H_s) and (G_r) and find numbers $s_0 < \ldots < s_K$ and $r_0 < \ldots < r_K$ such that E (resp. F) can be considered as a subspace of H_{s_K} (resp. as a quotient space of G_{r_K}) such that the by H_{s_k} (resp. G_{r_k}) induced seminorms $| \;|^\wedge_{s_k}$ on E (resp. $| \;|^\wedge_{r_k}$ on F) are good approximations for the given seminorms $\| \;\|^E_k$ (resp. $\| \;\|^F_k$), $k = 0, 1, \ldots, K$.

In sections 2, 3, 4 of [18] this problem is investigated and solved in a fashion sufficiently good for our purposes. The construction given there produces small error terms $\epsilon > 0$ occurring in the indices of the scales; very roughly speaking, an appropriate representation is possible such that the seminorms on E (analogously: on F) satisfy estimates like

$$c_k^{-1} | \;|^\wedge_{s_k-\epsilon} \leq \| \;\|^E_k \leq c_k | \;|^\wedge_{s_k+\epsilon}, \quad k = 0, \ldots, K.$$

The numbers s_k and r_k are calculated from the numbers τ_k and θ_k in the same way as for power series spaces (see section 4). For the details of the construction we refer to [18].

Summarizing, the following fundamental Proposition is proved in [18].

5.4. PROPOSITION. ([18], 5.2.). *Let E and F be graded Fréchet-Hilbert spaces satisfying (1) and (2) for suitable $0 < \tau_k < 1$ and $0 < \theta_k < 1$, respectively, $k = 1, 2, \ldots$ Suppose that $\theta_k \geq \tau_k$ for all k.*

Then for every sequence $A_k \in L(E_k, F_k)$, $k = 0, 1, \ldots$, *there is a sequence* $B_k \in L(E_{k+3}, F_k)$, $k = 0, 1, \ldots$, *such that*

$$A_k \circ i_{k+4}^k = B_k \circ i_{k+4}^{k+3} - i_{k+1}^k \circ B_{k+1}, \quad k = 0, 1, \ldots$$

Based on this proposition, Theorem 4.1. is proved by the same arguments as 5.3.

6 LINEAR-TAME AND TOPOLOGICAL SPLITTING

The following linear-tame splitting theorem is easily deduced from Proposition 5.4. by slightly modifying the proof of Theorem 5.3. (applying 5.4. instead of 5.2. there).

6.1. THEOREM. *Let* E, F, G *be graded Fréchet-Hilbert spaces, let* E *satisfy condition (1) of section 4 for suitable* τ_k *and let* F *satisfy condition (2) of section 4 for suitable* θ_k. *Assume that*
$(3)'$ *There are* p_0 *and* $q \geq 1$ *such that* $\theta_k \geq \tau_{qk+p}$ *for all* $p \geq p_0$ *and all* k.
Then every linear-tamely exact sequence

$$0 \longrightarrow F \longrightarrow G \longrightarrow E \longrightarrow 0$$

splits linear-tamely.

Of course, Theorem 6.1. implies linear-tame splitting results for power series spaces by analogy with section 4. We here only state:

6.2. COROLLARY. *If the sequence* $(\frac{r_{k+1}-r_k}{r_k - r_{k-1}})_k$ *is monotonically increasing, then the pair* $(\Lambda_R^2(\beta), \Lambda_R^2(\alpha))$ *is a tame and a linear-tame Hilbert splitting pair,* $R \in [0, \infty]$. *In particular, this holds for the standard cases* $(R = \infty, r_k = k)$ *and* $(R = 0, r_k = -\frac{1}{k})$.

We now consider the topological case. From Proposition 5.4. we also can easily obtain the topological splitting result Theorem 3.2. which first has been proved in [26].

6.3. THEOREM. *Let* E *and* F *be graded Fréchet-Hilbert spaces. If* E *has property* (DN) *and* F *has property* (Ω), *then the pair* (E, F) *is a Hilbert splitting pair.*

Proof: Let $0 \to F \to G \xrightarrow{q} E \to 0$ be an exact sequence of Fréchet spaces equipped with hilbertian gradings $\| \; \|_k$. We first consider the grading on F. Inductively applying the (Ω)-condition (cf. sec. 3), choosing an appropriate subsequence and then relabelling the new grading, we may assume that

$$\| \; \|_k^* \leq c_k \| \; \|_{k-1}^{*\,1-\theta_k} \| \; \|_{k+1}^{*\,\theta_k} \text{ on } (F, \| \; \|_{k-1})'$$

for certain $0 < \theta_k < 1$ and suitable constants $c_k > 0$, $k = 1, 2, \ldots$

Let F_k denote the Hilbert space attached to $(F, \| \; \|_k)$, and let $F_{(k)}$ denote the Hilbert space belonging to the by $\| \; \|_k^G$ on F induced grading (as subspace). Omitting a lot of seminorms of the grading on G and then changing the numbering we may further assume that the canonical maps $F_{(k)} \to F_k \to F_{(\phi(k))}$ are continuous for some

$\phi(k) \leq \phi(k+1) \to \infty$, $\phi(k) \leq k$. Let $E_{(k)}$ denote the Hilbert space belonging to the by $\| \ \|_k^G$ and q induced grading on E (as quotient space).

We now change the given grading on E by omitting a lot of seminorms: We choose $0 < \tau_k < 1$ so that $\tau_{k+1} \leq \theta_k$. Inductively applying the (DN)-condition and then relabelling we may assume that

$$\| \ \|_k \leq c_k \| \ \|_{k-1}^{1-\tau_k} \| \ \|_{k+1}^{\tau_k} \quad \text{on } E$$

for suitable constants $c_k > 0$, $k = 1, 2, \ldots$, and, moreover, that the canonical maps $E_k \to E_{(k)}$ are continuous; here E_k denotes the Hilbert space associated to $(E, \| \ \|_k)$.

With the above choices we consider the exact sequences of Hilbert spaces

$$0 \longrightarrow F_{(k)} \longrightarrow G_k \xrightarrow{q_k} E_{(k)} \longrightarrow 0$$

and proceed as in the proof of 5.3.: We choose $r_k \in L(E_{(k)}, G_k)$ such that $q_k \circ r_k = id$, and put $\tilde{A}_k = i_{k+1}^k \circ r_{k+1} - r_k \in L(E_{(k+1)}, F_{(k)})$. Then $A_k = i_{(k)}^k \circ \tilde{A}_k \in L(E_{k+1}, F_k)$. Since $\theta_k \geq \tau_{k+1}$ we may apply Proposition 5.4. to find $B_k \in L(E_{k+4}, F_k)$ such that $A_k = B_k - i_{k+1}^k \circ B_{k+1}$, and define $R_k = i_k^{\phi(k)} \circ r_k + i_k^{(\phi(k))} \circ B_k$. We observe that $i_{\phi(k+1)}^{\phi(k)} \circ R_{k+1} - R_k = 0$. Hence there is $R \in L(E, G)$ such that $R_k = i^{\phi(k)} \circ R$. The map R defines a continuous linear right inverse for q.

7 STRUCTURE THEORY OF POWER SERIES SPACES

In structure theory of power series spaces, splitting theory is applied to characterize the subspaces and quotient spaces of power series spaces of infinite type by properties (DN) and (Ω), respectively, while the subspaces and quotient spaces of power series spaces of finite type are characterized by properties ($\underline{\text{DN}}$) and ($\overline{\Omega}$) defined as follows (see [22], [25]).

DEFINITION. Let E be a Fréchet space and $\| \ \|_0 \leq \| \ \|_1 \leq \ldots$ a fundamental system of seminorms. E has property ($\underline{\text{DN}}$) resp. ($\overline{\Omega}$) if

$$(\underline{\text{DN}}) \ \exists\, p \,\forall\, k \,\exists\, 0<\tau<1 \,\exists\, K \,\exists\, c>0 \,:\, \| \ \|_k \leq c \| \ \|_p^{1-\tau} \| \ \|_K^{\tau},$$

resp.

$$(\overline{\Omega}) \ \forall\, p \,\forall\, 0<\theta<1 \,\exists\, q \,\forall\, Q \,\exists\, c>0 \,:\, \| \ \|_q^* \leq c \| \ \|_p^{*\,1-\theta} \| \ \|_Q^{*\,\theta}.$$

We state the fundamental theorem of structure theory of nuclear stable power series spaces. Let $\Lambda_R^1(\alpha)$ be stable, i.e. $\sup\limits_n \frac{\alpha_{2n}}{\alpha_n} < \infty$, and nuclear, i.e. $\sup\limits_n \frac{\log n}{\alpha_n} < \infty$ if $R = \infty$ resp. $\lim\limits_n \frac{\log n}{\alpha_n} = 0$ if $R < \infty$. In generalization of the Komura imbedding theorem [8] a Fréchet space E is then called $\Lambda_{\mathbb{N}}(\alpha)$-nuclear if E can be imbedded into $\Lambda_\infty^1(\alpha)^{\mathbb{N}}$, and E is called $\Lambda_1(\alpha)$-nuclear if E can be imbedded into $\Lambda_0^1(\alpha)^{\mathbb{N}}$ (see [34], [32] also for different descriptions based on Kolmogorov diameters).

7.1. THEOREM. *Let* E *be a Fréchet space and let* $\Lambda_R^1(\alpha)$ *be nuclear and stable,* $R \in \{0, \infty\}$.

(i) ([34]). *E is isomorphic to a subspace of* $\Lambda_\infty^1(\alpha)$ *iff* E *is* $\Lambda_{\mathbb{N}}(\alpha)$*-nuclear and has property* (DN).

(ii) ([34]). *E is isomorphic to a quotient space of* $\Lambda_\infty^1(\alpha)$ *iff* E *is* $\Lambda_{\mathbb{N}}(\alpha)$*-nuclear and has property* (Ω).

(iii) ([22],[32]). *E is isomorphic to a subspace of* $\Lambda_0^1(\alpha)$ *iff* E *is* $\Lambda_1(\alpha)$*-nuclear and has property* $(\underline{\mathrm{DN}})$.

(iv) ([22]). *E is isomorphic to a quotient space of* $\Lambda_0^1(\alpha)$ *iff* E *is* $\Lambda_1(\alpha)$*-nuclear and has property* $(\overline{\Omega})$.

While the proofs of 7.1. (i), (ii) are based on splitting theorems, these methods do not fit in cases (iii), (iv) since a general splitting theorem for properties $(\underline{\mathrm{DN}})$ and $(\overline{\Omega})$ fails (see Example 3.4.); in the literature, 7.1. (iii), (iv) are thus proved by different techniques (e.g. so called three-spaces-property).

When trying to adapt the proof of Theorem 6.3. in the situation that E has property $(\underline{\mathrm{DN}})$ and F has property $(\overline{\Omega})$, we observe that this is in general not possible; however, the proof works in particular situations, e.g. if $E = \Lambda_0^2(\alpha)$, $F = \Lambda_0^2(\beta)$ and the sequence is tamely exact. More general, properties $(\underline{\mathrm{DN}})$ and $(\overline{\Omega})$ have to be given in form of compatible conditions (DN_τ) and (Ω_θ), and the exactness of the sequence has to be controlled. Therefore, we see that tameness conditions are essential for topological splitting.

The splitting Theorem 4.1. is the suitable tool to set up a unified structure theory for power series spaces based on splitting theory with identical proofs both for power series spaces of infinite and of finite type. Moreover, no nuclearity assumptions are needed, and a delicate control of continuity estimates is possible in order to obtain tame or linear-tame counterparts to Theorem 7.1. This will be done in a forthcoming paper which contains for instance the following Theorem 7.2.

For any linear space E and absolutely convex sets $A \subset B \subset E$ we define

$$\delta_n(A, B) = \inf\{\delta > 0 : A \subset \delta B + F,\ F \subset E \text{ subspace with } \dim F \le n\}.$$

For properties of the Kolmogorov numbers $\delta_n(A, B)$ we refer to [14].

Let $R \in \{0, \infty\}$ and $\Lambda_R^2(\alpha)$ be equipped with the grading defined by $r_k = k$ if $R = \infty$ resp. $r_k = -\frac{1}{k}$ if $R = 0$, let $\tau_k = \frac{1}{2}$ if $R = \infty$ resp. $\tau_k = \frac{1}{2}(1 + \frac{1}{k})$ if $R = 0$.

7.2. THEOREM. *Let E be a graded Fréchet-Hilbert space and* $U_k = \{x \in E : \|x\|_k \le 1\}$. *Let* $(\alpha_n) \nearrow \infty$ *satisfy* $\lim_n \frac{\alpha_{2n}}{\alpha_n} = 1$. *Assume that for* $0 < \tau_k < 1$ *and constants* $c_k > 0$, $c > 0$ *we have*

(1) $\|\ \|_k \le c_k \|\ \|_{k-1}^{1-\tau_k} \|\ \|_{k+1}^{\tau_k}$, $k = 1, 2, \dots$

(2) $\delta_n(U_1, U_0) \le c e^{-\alpha_n}$, $n = 0, 1, 2, \dots$

Then E is linear-tamely isomorphic to a graded subspace of $\Lambda_\infty^2(\alpha)$ *if* $\tau_k = \frac{1}{2}$ *resp. of* $\Lambda_0^2(\alpha)$ *if* $\tau_k = \frac{1}{2}(1 + \frac{1}{k})$.

References

[1] E. Dubinsky, Non-linear analysis in different kinds of Fréchet spaces, preprint 1982.

[2] E. Dubinsky and D. Vogt, Bases in complemented subspaces of power series spaces, *Bull. Acad. Polon. Sci.* , *34:* 65-67 (1986).

[3] E. Dubinsky and D. Vogt, Complemented subspaces in tame power series spaces, *Studia Math.*, *93:* 71-85 (1989).

[4] R. S. Hamilton, The inverse function theorem of Nash and Moser, *Bull. Amer. Math. Soc.*, *7:* 65-222 (1982).

[5] L. Hörmander, The boundary problems of physical geodesy, *Arch. Rational Mech. Anal.* , *62:* 1-52 (1976).

[6] G. Köthe, Topological vector spaces I, Berlin Heidelberg New York: Springer 1983.

[7] G. Köthe, Topological vector spaces II, Berlin Heidelberg New York: Springer 1979.

[8] T. Komura and Y. Komura, Über die Einbettung der nuklearen Räume in $(s)^A$, *Math. Ann.*, *162:* 284-288 (1966).

[9] J. Krone and D. Vogt, The splitting relation for Köthe spaces, *Math. Z.*, *190:* 387-400 (1985).

[10] S. Lojasiewicz Jr. and E. Zehnder, An Inverse Function Theorem in Fréchet-Spaces, *Journ. of Funct. Anal.*, *33:* 165-174 (1979).

[11] J. Moser, A new technique for the construction of solutions of nonlinear differential equations, *Proc. Nat. Acad. Sci. USA*, *47:* 1824-1831 (1961).

[12] K. Nyberg, Tameness of pairs of nuclear power series spaces and related topics, *Trans. Amer. Math. Soc.*, *283:* 645-660 (1984).

[13] K. Nyberg, A tame splitting theorem for Köthe spaces, *Arch. Math.* , *52:* 471-481 (1989).

[14] A. Pietsch, Nuclear locally convex spaces, Ergebnisse der Math. 66, Berlin Heidelberg New York: Springer 1972.

[15] M. Poppenberg, Characterization of the subspaces of (s) in the tame category, *Arch. Math.*, *54:* 274-283 (1990).

[16] M. Poppenberg, Characterization of the quotient spaces of (s) in the tame category, *Math. Nachr.* , *150:* 127-141 (1991).

[17] M. Poppenberg, A sufficient condition of type (Ω) for tame splitting of short exact sequences of Fréchet spaces, *Manuscr. Math.* , *72:* 257-274 (1991).

[18] M. Poppenberg and D. Vogt, A Tame Splitting Theorem for Exact Sequences of Fréchet Spaces, preprint.

[19] F. Sergeraert, Un théorème de fonctions implicites sur certains espaces de Fréchet et quelques applications, *Ann. scient. Éc. Norm. Sup., IV. Sér.*, *5:* 599-660 (1972).

[20] D. Vogt, Charakterisierung der Unterräume von s, *Math. Z.* , *155:* 109-117 (1977).

[21] D. Vogt, Subspaces and quotient spaces of s, in Bierstedt K.D., Fuchssteiner B. (eds), Functional Analysis: Surveys and Recent Results, North–Holland Math. Studies, 27: 167-187 (1977).

[22] D. Vogt, Eine Charakterisierung der Potenzreihenräume von endlichem Typ und ihre Folgerungen, *Manuscr. Math.* , *37:* 269-301 (1982).

[23] D. Vogt, Some results on continuous linear maps between Fréchet spaces, in Bierstedt K.D., Fuchssteiner B. (eds.), Functional Analysis: Surveys and Recent Results II, North–Holland Math. Studies, 90: 349-381 (1984).

[24] D. Vogt, Tame spaces and power series spaces, *Math. Z.* , *196:* 523-536 (1987).

[25] D. Vogt, Power series space representations of nuclear Fréchet spaces, *Trans. Amer. Math. Soc., 319:* 191-208 (1990).

[26] D. Vogt, Interpolation of nuclear operators and a splitting Theorem for exact sequences of Fréchet spaces, preprint.

[27] D. Vogt, Tame splitting pairs, preprint.

[28] D. Vogt, Operators between Fréchet spaces, preprint.

[29] D. Vogt, On two classes of (F)-spaces, *Arch. Math.* , *45:* 255-266 (1985).

[30] D. Vogt, Sequence space representations of spaces of test functions and distributions, Lecture Notes in Pure and Applied Math. , 83, Dekker New York: 1983.

[31] D. Vogt, Structure Theory of Nuclear Stable Power Series Spaces, Proc. of the confer. on vector measures and integral representations of operators and on functional analysis, Essen : 1982. .

[32] D. Vogt, Charakterisierung der Unterräume eines nuklearen stabilen Potenzreihenraumes von endlichem Typ, *Studia Math., 71:* 251-270 (1982).

[33] D. Vogt and M.J. Wagner, Charakterisierung der Quotientenräume von s und eine Vermutung von Martineau, *Studia Math.* , *67:* 225-240 (1980).

[34] D. Vogt and M. J. Wagner, Charakterisierung der Unterräume und Quotientenräume der nuklearen stabilen Potenzreihenräume von unendlichem Typ, *Studia Math.* , *70:* 63-80 (1981).

[35] V.D. Zaharjuta, Isomorphism of spaces of analytic functions (Russian), *Docl. Akad. Nauk. SSSR, 255* (1980), English transl. *Soviet Math. Dokl.* , *22:* 631-634 (1980).

Fréchet Spaces Between which All Continuous Linear Maps Are Compact

M.S. RAMANUJAN[1] Department of Mathematics, University of Michigan, Ann Arbor, MI 48109

DIETMAR VOGT Fachbereich Mathematik, Bergische Universität–GH Wuppertal D–5600 Wuppertal 1, Germany

The conditions under which pairs of specific Fréchet spaces F, E have the property that all linear continuous maps from F to E are compact have been obtained by several authors (see [4]). In all these cases both spaces are Köthe spaces. Frequently E is also Montel, so that the conditions obtained were really sufficient to guarantee that linear continuous maps were indeed bounded. The problem of when $L(E,F) = LB(E,F)$, i.e., all continuous linear maps from E to F are bounded, has been completely resolved for pairs of arbitrary Fréchet spaces in [4]. Here we consider the question as to when the set $LB(F,E)$ of all bounded linear maps and the set $K(F,E)$ of all compact linear maps coincide and we consider the cases where E and F are both Fréchet–Hilbert spaces or Köthe spaces.

PRELIMINARIES: We consider Fréchet spaces F equipped with a fundamental sequence of seminorms $\| \ \|_1 \leq \| \ \|_2 \leq \ldots$. F_n denotes the local Banach space, which is the completion in the obvious quotient norm, of $F/ker\| \ \|_n$; i_n denotes the quotient map $F \longrightarrow F_n$ and $i_l^k : F_l \longrightarrow F_k$ is the continuous extension (to the completions) of the canonical injection. $LB(E,F)$ (resp. $K(E,F)$) denotes the linear space of all bounded (resp. compact) operators from E to F, i. e, those linear maps which send some neighbourhood of zero into a bounded (resp. compact) set. We refer the reader to [2] and [3] for notations and terminology regarding Köthe spaces $\lambda^1(A)$ and $\lambda^p(A)$

[1]Support from the Alexander von Humboldt Stiftung is gratefully acknowledged.

and regarding nuclearity, approximation numbers etc.

We start with some preliminary Lemmas.

LEMMA 1 *For F a Banach space and E a Fréchet space $K(F,E)$, with the topology of uniform convergence on the unit ball of F, is a Fréchet space.*

PROOF: It is easy to see that a map $A \in L(F,E)$ is compact if and only if $i^k \circ A : F \longrightarrow E_k$ is compact for all k; this is a consequence of Tychonoff's theorem. Then the proof follows since $K(F, E_k)$ is a Banach space. □

LEMMA 2 *$L(F,E) = K(F,E)$ if and only if $L(F,E) = LB(F,E)$ and $LB(F,E) = K(F,E)$.*

LEMMA 3 *$LB(F,E) = K(F,E)$ if and only if for every n there is N such that $A \circ i_N^n \in K(F_N, E)$ for every $A \in L(F_n, E)$.*

PROOF: By the mapping $A \longrightarrow A \circ i^n$ the space $L(F_n, E)$ is imbedded in $LB(F,E)$ and in this sense $LB(F,E) = \bigcup_n L(F_n, E)$ and the same identification also gives $K(F,E) = \bigcup_n K(F_n, E)$. Thus both $LB(F,E)$ and $K(F,E)$ are unions of Fréchet spaces and then by Grothendieck's factorisation theorem ([1], I, p.16) it follows that $LB(F,E) \subset K(F,E)$ if and only if for each n there exists N such that $LB(F_n, E) \subset K(F_N, E)$. The Lemma now follows if we trace back this identification. □

LEMMA 4 *Let H_1, H_2 be two infinite dimensional Hilbert spaces and $T : H_1 \longrightarrow H_2$ be linear, continuous but not compact. Then there exist an infinite dimensional closed subspace G of H_1, and an $\varepsilon > 0$ such that for all $x \in G$, $\|Tx\| \geq \varepsilon \|x\|$.*

The proof easily follows from the spectral representation of the linear, continuous, self–adjoint and non–compact operator $A = T^*T$ and the fact that since A is not compact it cannot be the uniform limit of finite rank operators.

THEOREM 5 *In each of the following cases the relation $LB(F,E) = K(F,E)$ holds if and only if either F is a Schwartz space or E is a Montel space:*

1. *F is a Köthe space $\lambda^1(A)$,*
2. *both F and E are Fréchet–Hilbert spaces,*
3. *both F and E are Köthe spaces $\lambda^p(A)$ and $\lambda^p(B)$ respectively.*

PROOF: In all cases the proof of the sufficiency is clear. For the proof of necessity in each case we assume that F is not a Schwartz space and prove that E has to be Montel.

1. Since $F = \lambda^1(A)$ is assumed to be not a Schwartz space, there exists n such that for each N, (a_j^n / a_j^N) is not a null sequence and therefore there exist an $\varepsilon = \varepsilon(N) > 0$ and an infinite subset $J = J(N)$ of $\mathbb{N}$ such that for each $j \in J$, $a_j^n / a_j^N \geq \varepsilon$. We choose the N according to Lemma 3. Consider

$$G_n = \left\{ x = (x_j)_{j \in J} : \|x\| = \sum |x_j| a_j^n < \infty \right\}$$

and denote by U_n, V_n the unit balls of F_n, G_n respectively. Consider an arbitrary separable bounded set B in E and let (b_n) be dense in B. The map $T\colon (a_j^n)^{-1}e_j \to b_j$, $j \in J$, can be linearly and continuously extended to $T\colon G_n \to E$ and $\overline{T(V_n)} \supset B$; by setting $\widehat{T}x = T(x\mid_J)$ we have $\widehat{T} \in L(F_n, E)$, $\overline{\widehat{T}(U_n)} \supset B$. Now

$$b_j = \widehat{T}\left((a_j^n)^{-1}e_j\right) \in \varepsilon^{-1}\widehat{T} \circ i_N^n(U_N)$$

and then $B \subset \overline{\varepsilon^{-1}\widehat{T} \circ i_N^n(U_N)}$, with $\widehat{T} \circ i_N^n$ compact. Thus B is relatively compact. Since the space E in question is Fréchet there is no loss of generality in having considered B to be separable, and the proof is complete.

2. Since by assumption the Fréchet–Hilbert space F is not a Schwartz space, there exists n such that for each N, the map $i_N^n : F_N \longrightarrow F_n$ is not compact; we choose N according to Lemma 3, so that $L(F_n, E) \subset K(F_N, E)$ in the sense that for $A \in L(F_n, E)$ we have $A \circ i_N^n \in K(F_N, E)$. Since i_N^n is not compact, by Lemma 4, there exists a countable infinite dimensional subspace G of F_N such that on G, $\varepsilon\|x\|_N \leq \|i_N^n x\|_n$. But on F_N, $N \geq n$, $\|i_N^n x\|_n \leq \|x\|_N$ and thus on G, the two norms $\|x\|_N$ and $\|i_N^n x\|_n$ are equivalent. Let P denote the orthogonal projection onto G from F_n. If we consider now a separable bounded set $B \subset E$, which without loss of generality is a Hilbert ball, so that $E_B = \bigcup_{k\in\mathbb{N}} kB$ is complete with B as its unit ball we have an isometry $\widetilde{A} : G \longrightarrow E_B$. Define $A = \widetilde{A} \circ P$, then $B = \widetilde{A} \circ P(U_n)$ and $B \subset \varepsilon^{-1}\widetilde{A} \circ P \circ i_N^n(U_N)$ with $\widetilde{A} \circ P \in L(F_n, E)$ and $\widetilde{A} \circ P \circ i_N^n \in K(F_N, E)$. Thus B is precompact and E is Montel.

3. The proof is similar and omitted. □

One might next consider the question of $LB(F, E) = s(F, E)$ where $s(F, E)$ is the space of s–nuclear maps. Recall briefly,

$$s = \left\{(x_n) : n^k|x_n| \to 0 \quad \text{for each } k = 1, 2, \ldots\right\}$$

and is a Fréchet space, for example, with the seminorms $p_k(x) = \sum_n n^k|x_n|$ A subset M of E is said to be a s–nuclear set if it is contained in the weakly closed absolute convex hull of a sequence $(x_n) \subset E$ with $(q_j(x_n))_n \in s$ for each q_j of the fundamental system of seminorms on E. A linear (continuous) map $T : F \longrightarrow E$ (both Fréchet) is said to be s–nuclear if T carries a neighbourhood of $0 \in F$ into an s–nuclear set. For $T : X \longrightarrow Y$ (both Banach) T is s–nuclear if and only if the sequence of its approximation numbers $(a_n(T)) \in s$. The space $s(X, Y)$ of s–nuclear maps can be given a complete metric topology (not necessarily locally convex) and Lemma 1 and Lemma 3 will have their obvious analogues. Then one can prove, exactly as before, that if $LB(F, E) = s(F, E)$ with F and E as in the different cases of Theorem 5, and if F is not a Schwartz space then E has to be such that each bounded set in E is s–nuclear or equivalently (see [3]) E is a nuclear Fréchet space.

The natural analogue suggested by Theorem 5 would of course be, in all cases in question, $LB(F, E) = s(F, E)$ if and only if F is s–nuclear or E is nuclear. However the following example in which both F and E are Schwartz and neither is nuclear proves that the "only if" part above is false.

EXAMPLE: Let (n_k) be a subsequence of the positive integers:

$$1 = n_0 < n_1 < n_2 < \ldots$$

Define exponent sequences (α_n) and (β_n) as follows:

$$\left.\begin{array}{llll} \text{for} & n_{2k} \le n < n_{2k+1}, & \alpha_n = n, & \beta_n = n_{2k} \\ \text{for} & n_{2k+1} \le n < n_{2k+2}, & \alpha_n = n_{2k+1}, & \beta_n = n \end{array}\right\}, k = 0, 1, \ldots$$

Let $F = \Lambda_0(\alpha)$, $E = \Lambda_\infty(\beta)$. Suppose a map $A: F \longrightarrow E$ maps some neighbourhood of zero, say, $U_{s_0} = \{x : |x|_{s_0} \le 1\}$, $s_0 < 0$, into a bounded set in E, so that for each t, there exists a constant C_t and $|Ax|_t \le C_t |x|_{s_0}$ holds for all x. Then we choose $0 > s > s_0$ and for arbitrary t we estimate the approximation numbers of the induced map $A_{s,t} : F_s \longrightarrow E_t$:

$$\begin{aligned} a_{2n}(A_{s,t}) &\le \|A_{s_0,t+\varepsilon}\| a_n(i_s^{s_0}) a_n(i_{t+\varepsilon}^t) \\ &\le \|A_{s_0,t+\varepsilon}\| e^{-(s-s_0)\alpha_n - \varepsilon\beta_n} \end{aligned}$$

so, for $s - s_0 \ge \varepsilon$ we obtain $a_{2n}(A_{s,t}) \le Ce^{-\varepsilon n}$. We then have that $A_{s,t}$ is s–nuclear. Now by choosing (n_k) so that $\limsup_n \alpha_n^{-1} \log n = \limsup_n \beta_n^{-1} \log n = \infty$, we see that each of F and E is non–nuclear. Notice that by Zaharjuta [5] in our example $LB(F, E) = L(F, E)$, so that in fact we have $L(F, E) = s(F, E)$.

REFERENCES

[1] A. Grothendieck: *Produits tensorieles topologiques et espaces nucléaires,* Mem. Amer. Math. Soc. 16 (1955).

[2] G. Köthe: *Topologische lineare Räume I,* Berlin–Göttingen–Heidelberg 1960.

[3] A. Pietsch: *Nukleare lokalkonvexe Räume,* Berlin 1965.

[4] D. Vogt: *Fréchetträume zwischen denen jede stetige lineare Abbildung beschränkt ist,* J. reine, angew. Math. 345 (1983), 182–200.

[5] V. P. Zaharjuta: *On the isomorphism of cartesian products of locally convex spaces,* Studia Math. 46 (1975), 201–221.

Spaces of Harmonic Functions

V. ZAHARIUTA Mechanics and Mathematics Department, Rostov State University, Rostov-on-Don, Russia

0 INTRODUCTION

For the F-space of all harmonic functions in an open set in $\hat{\mathbb{R}}^n$ there are considered the following interconnected questions: interpolation properties, Hilbert scales, existence and construction of bases, isomorphic classification. Following the view developed in papers [20], [21], [23], [25] for spaces of analytic functions and in [22] for spaces of harmonic functions, we essentially improve the results from [22]. We restrict ourself to the case of harmonic functions, but it is not difficult to transfer most of the results to the general case of elliptic partial differential equations with constant coefficients.

The results of [13], [1] on the characterizations of power series spaces of finite and infinite types are substantially used in the present paper. Mityagin's work [3] renders a great influence on the investigation of concrete functional spaces and contains a lot of important initial ideas.

Other results concerning analogous questions for spaces of solutions of differential equations see, for example, in [14], [2], [19] and the references there.

The present paper is a first part of our publications on spaces of harmonic functions; later we shall consider further developments and applications as e.g. construction of special continuable bases, best approximation by harmonic polynomials, harmonic extension.

Acknowledgement: This paper has been written during the author's stay at the University of Wuppertal. The author wishes to thank DAAD and the University of Wuppertal for their support and to Prof. Vogt for fruitful discussions and his essential assistance in the preparation of the final form of this paper, and to Mrs. Z. Tuğlu for the printing of the Latex-version.

1 NOTATION

$\hat{\mathbb{R}}^n$ is the real n-dimensional Euclidean space compactified by the unique point ∞; $F^* := \hat{\mathbb{R}}^n \setminus F$ for every set $F \subset \hat{\mathbb{R}}^n$; $\Omega_s \uparrow \Omega$ means that Ω_s, Ω are open sets in $\hat{\mathbb{R}}^n$, $\Omega = \cup_{s=1}^{\infty} \Omega_s$ and $\Omega_s \subset\subset \Omega_{s+1}$, $s \in \mathbb{N}$; $K_s \downarrow K$ means that K_s, K are compact sets in $\hat{\mathbb{R}}^n$, $K = \cap_{s=1}^{\infty} K_s$ and $K_{s+1} \subset \overset{\circ}{K}_s$, $s \in \mathbb{N}$; for every set F and function $f : F \to \mathbb{R}$ we denote:

$$|x|_F = \sup\{|x(t)| : t \in F\}.$$

For a pair of locally convex spaces X, Y the notation $X \subset Y$ usually means a linear continuous imbedding. For every function $x \in C^1(\Omega)$ we denote:

$$\Lambda_x(t) := \max\left\{|x(t)|, \left|\frac{\partial x(t)}{\partial t_1}\right|, \ldots, \left|\frac{\partial x(t)}{\partial t_n}\right|\right\}.$$

By $\Gamma(X)$ we denote the diametral dimension of the F-space X [3].

2 PRELIMINARIES

2.1 Spaces of Harmonic Functions

For an open set $\Omega \subset \hat{\mathbb{R}}^n$ we denote by $h(\Omega)$ the space of all harmonic functions (with real or complex values) in Ω equipped with the usual F-topology of uniform convergence on compact subsets of Ω. If $\infty \in \Omega$ we assume in the case of $n \geq 3$ that for all $x \in h(\Omega)$ the following holds:

$$x(t) = O(|t|^{2-n}), \quad t \to \infty.$$

For $n = 2$ we admit logarithmic singularities in the point ∞ assuming the following asymptotics:

$$x(t) = \alpha \, \ln \, |t| + o(1), \quad t \to \infty, \quad \alpha = \alpha(x).$$

Topologically the space $h(\Omega)$ is defined in this case as a subspace of the space $h(\Omega \setminus \{0\})$.

For an arbitrary set $F \subset \mathbb{R}$ we consider the space $h(F)$ of all harmonic germs on F with the usual inductive topology:

$$h(F) = \lim_{\Omega \in \Gamma(F)} \text{ind } h(\Omega),$$

where $\Gamma(F)$ is the collection of all open neighbourhoods of F, such that every connected component intersects F.

For an arbitrary pair of functions $x \in h(F)$, $x' \in h(F^*)$ we consider the following differential form defined in the common range of definition of the both functions x, x':

$$\omega(x,x') = \sum_{k=1}^{n}(-1)^k \left(x(t)\frac{\partial x'(t)}{\partial t_k} - x'(t)\frac{\partial x(t)}{\partial t_k} \right) [dt]_k,$$

where $[dt]_k = dt_k \wedge \cdots \wedge dt_{k-1} \wedge dt_{k+1} \wedge \cdots \wedge dt_n$, $k = 1, \ldots, n$. We consider the following bilinear integral form, defined on $h(F) \times h(F^*)$:

$$\langle x, x' \rangle = \int_\Gamma \omega(x,x'),$$

here Γ is any surface consisting of a finite set of smooth closed $(n-1)$-dimensional surfaces separating the singularities of the functions x and x'. The integral does not depend on the choice of Γ.

The following duality is very important for the investigation of harmonic function spaces.

PROPOSITION 2.1 (H.G. Tillmann [10]) The formula $x^*(x) = \langle x, x' \rangle$ defines a natural isomorphism $J : h(F)^* \to h(F^*)$, $x' = Jx^* \in h(F^*)$, $x \in h(F)$ if F satisfies some additional restrictions (for details, see [10]) which are satisfied if F is a compact or open set in $\hat{\mathbb{R}}^n$.

For a compact set K we denote by $hE^{(1)}(K)$ the closure of $h(K)$ in the Whitney space $E^{(1)}(K)$. It is a Banach space.

2.2 Spaces of Analytic Functions and Related Facts

We shall denote by $A(\Omega)$ the F-space of all analytic functions in an open set $\Omega \subset \mathbb{C}^n$ with the topology of uniform convergence on compact subsets of Ω; for an arbitrary set F in $\mathbb{C}^n$ we denote by $A(F)$ the space of all germs of analytic functions on F, defined as the following inductive limit:

$$A(F) := \lim \operatorname{ind}_{G \in \Gamma(F)} A(G).$$

The study of such spaces is connected with the consideration of the following maximal plurisubharmonic functions (see, for example, [23], [25]; compare with below mentioned considerations for spaces of harmonic functions).

For any open set Ω in $\mathbb{C}^n$ and a subset K we define:

$$\begin{aligned} \omega(\Omega, K; t) &:= \overline{\lim}_{\tau \to t} \sup\{u(\tau) : u \in P(K,\Omega)\}, \\ \gamma(\Omega, K; t) &:= \overline{\lim}_{\tau \to t} \sup\{u(\tau) : u \in P_a(K,\Omega)\}, \end{aligned} \tag{2.1}$$

where $P(K,\Omega)$ is the set of all plurisubharmonic functions $u(\tau)$ in Ω such that $u|K \leq 0$ and $u(\tau) < 1$ in Ω; $P_a(K,\Omega)$ is the subset of $P(K,\Omega)$ which contains only functions of the form

$$u(\tau) = \alpha \ln |f(t)|, \; \alpha > 0, \quad f \in A(\Omega). \tag{2.2}$$

We remark that to obtain the class $P_a(K,\Omega)$ it is enough to consider only functions (2.2) which satisfy the additional requirement: $0 < \alpha < \epsilon$ with any fixed $\epsilon > 0$. It is true because each function (2.2) can be represented in the form

$$u(\tau) = \frac{\alpha}{m} \ln |g_m(\tau)|, \quad g_m(\tau) = f(\tau)^m$$

for every $m \in \mathbb{N}$, thus each function from the class $P_a(K,\Omega)$ can be expressed in the form (2.2) with an arbitrary small α.

LEMMA 2.2 *For any pseudoconvex open set $\Omega \subset \mathbb{C}^n$ and its compact subset K the following identity holds*

$$\omega(\Omega, K; t) = \overline{\lim}_{\tau \to t} \lim_{r \to \infty} \lim_{s \to \infty} \gamma(\Omega_s, K_r; \tau). \tag{2.3}$$

This fact is a consequence of [23], Lemma 3.1.

2.3 Power Spaces

A power space (or power series space) is a Köthe sequence space of the following form:

$$\Lambda_\alpha(a) := \left\{ \xi = (\xi_k) : |\xi|_p := \sum_{k=1}^{\infty} |\xi_k| \exp \alpha_p a_k < \infty, \quad p \in \mathbb{N} \right\},$$

where $a = (a_k), \alpha_p \uparrow \alpha \leq \infty$, with the topology defined by the system of norms $|\xi|_p, p \in \mathbb{N}$. We say that the power space has a finite type, if $\alpha < \infty$, and infinite type, if $\alpha = \infty$.

2.4 Kolmogorov Diameters

Let $X_1 \subset X_0$ be normed spaces with linear continuous imbeddings. The Kolmogorov diameters of X_1 (or the unit ball B_{X_1}) with respect to X_0 (or to the unit ball B_{X_0}) are defined by the formula (see, for example, [3]):

$$d_k(X_1, X_0) = d_k(B_{X_1}; B_{X_0}) := \sup_L \inf\{\alpha > 0 : B_{X_1} \subset \alpha B_{X_0} + L\},$$

where L runs the set of all subspaces of X_0 such that $\dim L = k$.

Let $H_1 \subset H_0$ be Hilbert spaces with a dense completely continuous imbedding. Then there exists the common orthogonal basis $\{e_k\}$ which can be normalized so that

$$\|e_k\|_{H_0} = 1, \quad \|e_k\|_{H_1} = \mu_k \uparrow \infty. \tag{2.4}$$

The sequence $\mu_k = \mu_k(H_0, H_1)$ is defined uniquely and has the following expression by means of the Kolmogorov diameters:

$$\mu_k(H_0, H_1) = d_{k-1}(H_1, H_0)^{-1}, \quad k \in \mathbb{N}.$$

From simple properties of the Kolmogorov diameters follows

LEMMA 2.3 *Let H_0, H_1, G_0, G_1 be Hilbert spaces with completely continuous imbeddings: $H_1 \subset G_1 \subset G_0 \subset H_0$. Then there exists a constant c such that: $\mu_k(G_0, G_1) \leq c\mu_k(H_0, H_1)$, $k \in \mathbb{N}$.*

By $H^\alpha = (H_0)^{1-\alpha}(H_1)^\alpha$, $\alpha \in \mathbb{R}$, we shall denote the Hilbert scale spanned by a given pair $H_1 \subset H_0$ of Hilbert spaces with continuous imbeddings (see, for example, [3]). In the above mentioned case of completely continuous imbeddings the space H^α is generated by the weighted ℓ^2-norm:

$$\|x\|_{H^\alpha} = \left(\sum |(x, e_k)_{H_0}|^2 \mu_k^{2\alpha}\right)^{1/2}, \quad \alpha \in \mathbb{R}.$$

2.5 Dini's Theorem

We need the following simple variation of the classical theorem.

LEMMA 2.4 *Let u_k be a non-increasing sequence of upper-semicontinous functions on a compact set K such that $\lim_{k\to\infty} u_k(t) \leq u(t)$, where u is a continous function on K. Then for any $\epsilon > 0$ it exists a number m such that $u_k(t) < u(t) + \epsilon$ for all $t \in K$, $k \geq m$.*

3 MAXIMAL Lh-FUNCTIONS

Let Ω be an open set in $\hat{\mathbb{R}}^n$. We define the class $\tilde{L}h(\Omega)$ (resp. $\tilde{L}h_\epsilon(\Omega)$) as the minimal class of functions which contains all functions $\alpha \ \ln|x(t)|$, $x \in h(\Omega)$, $\alpha > 0$ (resp. $0 < \alpha < \epsilon$) and is closed with respect to the operation of taking the regularized upper envelopes of arbitrary families of functions:

$$\overline{\lim}_{\tau\to t} \sup\{u_i(\tau) : i \in I\}. \tag{3.1}$$

By $Lh(\Omega)$ we denote the minimal class of functions, defined on Ω, which (a) contains all functions $u(t) = \lim_{s\to\infty} u_s(t)$, where $u_s \in \tilde{L}h_{\epsilon(s)}(\Omega_s)$ is any sequence of functions such that: $\Omega_s \uparrow \Omega$; $u_{s+1}(t) \leq u_s(t), t \in \Omega_s, s \in \mathbb{N}; \epsilon(s) \downarrow 0$ if $s \uparrow \infty$; (b) is closed with respect to the operation (3.1).

Now we introduce special functions from the class $Lh(\Omega)$ which are in some sense maximal. It is not difficult to recognize an analogy with some notions of so called Complex Potential Theory (Lelong, Bremermann, Siciak, Zahariuta, Bedford-Taylor, Sadullayev, etc.) which prove to be effective tools in investigations of analytic function spaces (see, for example, [23], [25], [6], and 2.1 above).

Because of the close connections between analytic and harmonic functions in $\mathbb{C} = \mathbb{R}^2$ it is possible to use the same methods of potential theory for studying both classes (see, for example, Walsh [17], [18], Siciak [7], Nguyen [4], Skiba [9], Nguyen and Djebbar [5], Zeriahi [28]).

For $n \geq 3$ the traditional classical method of complexification allows to realize a space $h(\Omega)$ (in this context it is more natural to consider complex-valued harmonic functions) as subspace $A_h(\Omega_c)$ in the space $A(\Omega_c)$, where Ω_c is such a maximal open set in $\mathbb{C}^n = \mathbb{R}^n + i\mathbb{R}^n$ so that every function $x \in h(\Omega)$ can be extended in a unique way to a function $x_c \in A(\Omega_c)$, which satisfies the complexified Laplace equation in Ω_c. After this we are ready to apply methods of complex potential theory, but unfortunately their direct utilization can get us only some rough estimations for harmonic functions, because they constitute a very small part of $A(\Omega_c)$.

It is shown in the present paper (see also [22], [27]) that, if we consider harmonic functions in $\mathbb{R}^n$ instead of analytic functions in $\mathbb{C}^n$, then we have to consider Lh-functions instead of plurisubharmonic functions and maximal Lh-functions (see below) instead of maximal plurisubharmonic functions.

First we define the functions (compare [22]):

$$\begin{aligned}
\chi_\epsilon^{(1)}(\Omega,K;t) &= \overline{\lim}_{\tau\to t}\sup\{v(\tau)=\alpha\ln|x(\tau)|: \\
&\qquad 0<\alpha<\epsilon, |x|_K\leq 1, |x(\tau)|\leq \exp\frac{1}{\alpha} \text{ in } \Omega\} \qquad (3.2)\\
\chi_\epsilon^{(2)}(\Omega,K;t) &= \overline{\lim}_{\tau\to t}\sup\left\{\frac{\ln|x(\tau)|}{\ln|x|_\Omega}: x\in h(\Omega),\ |x|_K\leq 1,\ |x|_\Omega\geq \exp\frac{1}{\epsilon}\right\} \qquad (3.3)\\
\psi_\epsilon^{(1)}(\Omega,K;t) &= \overline{\lim}_{\tau\to t}\sup\{v(\tau)=\alpha\ln\Lambda_x(\tau): \\
&\qquad x\in h(\Omega),\ 0<\alpha\leq\epsilon, v|K\leq 0; v<1 \text{ in } \Omega\}\\
\psi_\epsilon^{(2)}(\Omega,K;t) &= \overline{\lim}_{\tau\to t}\sup\left\{\frac{\ln\Lambda_x(\tau)}{\ln|\Lambda_x|_\Omega}: x\in h(\Omega), \Lambda_x|K\leq 1,\ |\Lambda_x|_\Omega\geq\exp\frac{1}{\epsilon}\right\}.
\end{aligned}$$

We consider also two Lh-functions:

$$\begin{aligned}
\chi_0^{(i)}(\Omega,K;t) &= \lim_{\epsilon\to 0}\chi_\epsilon^{(i)}(\Omega,K;t),\\
\psi_0^{(i)}(\Omega,K;t) &= \lim_{\epsilon\to 0}\psi_\epsilon^{(i)}(\Omega,K;t).
\end{aligned}$$

LEMMA 3.1 *The following relations are valid:*

(i) $\chi_0^{(1)}(\Omega,K;t)\equiv\chi_0^{(2)}(\Omega,K;t)=:\chi_0(\Omega,K;t)$

(ii) $\psi_0^{(1)}(\Omega,K;t)\equiv\psi_0^{(2)}(\Omega,K;t)=:\psi_0(\Omega,K;t)$

(iii) $\psi_0(\Omega,K;t)\leq\chi_0(\Omega,K;t)$.

(iv)

$$\chi_\epsilon^{(1)}(\Omega,K;t)=\overline{\lim}_{\tau\to t}\sup\{v(\tau): v\in\tilde{L}_\epsilon h(\Omega), v|K\leq 0, v(\tau)<1 \text{ in } \Omega\} \qquad (3.4)$$

Proof: (i) Let us fix $\alpha: 0<\alpha\leq\epsilon$. The set $M=M(\alpha)$ of all x, satisfying the conditions in (3.2) with this α, we divide into two subsets:

$$M_1=\left\{x:|x|_\Omega\leq\exp\frac{1}{\delta(\alpha)}\right\},\quad M_2=\left\{x:|x|_\Omega>\exp\frac{1}{\delta(\alpha)}\right\},$$

where $\delta(\alpha) = \alpha/\sigma(\alpha)$ such that $\sigma(\alpha) \downarrow 0$ and $\delta(\alpha) \downarrow 0$ when $\alpha \downarrow 0$. For all $x \in M_1$ the inequality

$$\alpha \ln |x(\tau)| \leq \sigma(\epsilon), \quad \tau \in \Omega, \tag{3.5}$$

holds and for $x \in M_2$ we have

$$\alpha \ln |x(\tau)| \leq \frac{\ln |x(\tau)|}{\ln |x|_\Omega} \text{ and } \ln |x|_\Omega > \frac{1}{\delta(\alpha)}.$$

Therefore for $x \in M_2$ the inequality

$$\alpha \ln |x(\tau)| \leq \chi^{(2)}_{\delta(\alpha)}(\Omega, K; t) \tag{3.6}$$

holds. Thus from (3.5), (3.6) it follows that

$$\chi^{(1)}_\epsilon(\Omega, K; t) \leq \max\left\{\sigma(\epsilon), \chi^{(2)}_{\delta(\epsilon)}(\Omega, K; t)\right\}.$$

Letting $\epsilon \to 0$ in this inequality we get the inequality:

$$\chi^{(1)}_0(\Omega, K; t) \leq \chi^{(2)}_0(\Omega, K; t). \tag{3.7}$$

On the other side let x satisfy the conditions in (3.3). Then if we take $\alpha := 1/\ln |x|_\Omega$ we obtain that the pair (α, x) satisfies the conditions in (3.2). Therefore

$$\chi^{(2)}_\epsilon(\Omega, K; t) \leq \chi^{(1)}_\epsilon(\Omega, K; t), \quad t \in \Omega.$$

From here, due to (3.7), we have (i).

The relation (ii) can be proved in the same way; (iii) is clear because all derivatives $\frac{\partial x}{\partial x_\nu}$ are also harmonic functions.

The relation (3.4) follows from the obvious fact that when we use the operation (3.1) of regularized upper envelope more than one time it is the same if we first take the supremum of all considered functions and only then take the regularization $\overline{\lim}_{\tau \to t}$.

Let $\Omega_s \uparrow \Omega$, $K_s \downarrow K$. We define also the following Lh functions which should be the main tools in our further considerations:

$$\begin{aligned}
\chi(\Omega, K; t) &= \lim_{s\to\infty} \chi_0(\Omega_s, K; t) \\
h(\Omega, K; t) &= \overline{\lim}_{\tau\to t} \lim_{s\to\infty} \chi(\Omega, K_s; \tau) \\
\psi(\Omega, K; t) &= \lim_{s\to\infty} \psi_0(\Omega_s, K; t) \\
g(\Omega, K; t) &= \overline{\lim}_{\tau\to t} \lim_{s\to\infty} \psi(\Omega, K_s; \tau)
\end{aligned}$$

It is clear that these functions do not depend upon the choice of the sequences $\{\Omega_s\}, \{K_s\}$.

LEMMA 3.2 *$h(\Omega, K; t) \equiv g(\Omega, K; t)$ in Ω.*

Proof: Let us consider sequences $\Omega_s \uparrow \Omega$, $K_s \downarrow K$ such that $K_1 \subset \Omega_1$. Then for every $s \in \mathbb{N}$ there exists a constant M_s such that

$$|\Lambda_x|_{\Omega_s} \leq M_s |x|_{\Omega_{s+1}}, \; |\Lambda_x|_{K_{s+1}} \leq M_s \, |x|_{K_s}$$

for any $x \in h(\Omega_{s+1})$; we can assume that M_s is non-decreasing. Therefore for any α and x from (3.2) (with Ω_{s+1} and $K_r, r \leq s$, instead of Ω and K respectively) and $y = x/M_s$ we obtain the estimations:

$$\begin{aligned} \alpha \ln|x(\tau)| &\leq \alpha(\ln|y(\tau)| + \ln M_s) \leq \alpha \ln|y(\tau)| + \epsilon \ln M_s \\ |\Lambda_y|_{\Omega_s} &\leq \exp\frac{1}{\alpha}, \ |\Lambda_y|_{K_{r+1}} \leq \frac{M_r}{M_s} \leq 1. \end{aligned}$$

From here it follows that

$$\chi_\epsilon^{(1)}(\Omega_{s+1}, K_r; t) \leq \psi_\epsilon^{(1)}(\Omega_s, K_{r+1}; t) + \epsilon \ln M_s, \quad t \in \Omega_s.$$

Letting first $\epsilon \to 0$, then $s \to \infty$, then $r \to \infty$, we show that $h(\Omega, K; t) \leq g(\Omega, K; t), t \in \Omega$. Because of Lemma 3.1 (iii) we have the other inequality, which finishes the proof.

REMARK From this lemma it follows that $h(\Omega, K; t) \leq \psi(\Omega, K; t)$ in Ω. We do not know if the identity $h(\Omega, K; t) \equiv \psi(\Omega, K; t)$ is true, although such a supposition should be reliable.

The following fact justifies in some sense the term "maximal Lh-function".

PROPOSITION 3.3 *The following equation holds:*

$$h(\Omega, K, t) = \overline{\lim}_{\tau \to t} \sup\{u(\tau) : u \in Lh(\Omega),\ u|K \leq 0; u(\tau) \leq 1 \text{ in } \Omega\}, \quad t \in \Omega. \tag{3.8}$$

Proof: Let $\Omega_s \uparrow \Omega$, $K_s \downarrow K$, $K_1 \subset \Omega_1$. Let us denote the right part of (3.8) by $b(t)$.
a) First we prove that $h(\Omega, K; t) \leq b(t)$. For every sequence $\epsilon(s) \downarrow 0$ it is clear that the sequence of functions $\chi_{\epsilon(s)}(\Omega_s, K_m; t)$ is non-increasing with respect to $s \uparrow \infty$ and the inequalities

$$\chi_0(\Omega_s, K_m; t) \leq \chi_{\epsilon(s)}(\Omega_s, K_m; t) \leq \chi_{\epsilon(s)}(\Omega_r, K_m; t)$$

are valid, for $s \geq r$ and every $m \in \mathbb{N}$. Therefore $\chi(\Omega, K_m; t) \equiv \lim_{s\to\infty} \chi_{\epsilon(s)}(\Omega_s, K_m; t)$. From there it follows that the function $\chi(\Omega, K_m; t)$ belongs to the class, defined in the brackets of (3.8), hence $\chi(\Omega, K_m; t) \leq b(t)$ for every $m \in \mathbb{N}$. Thus $h(\Omega, K; t) \leq b(t)$.

b) Let now u be an arbitrary function which satisfies the conditions of (3.8). Then because of upper semi-continuity of u we get that for every $\epsilon > 0$ there exists $m \in \mathbb{N}$ such that

$$u(\tau) < \frac{\epsilon}{2}, \quad \tau \in K_m. \tag{3.9}$$

On the other hand, assuming the consideration from the proof of (iv) in Lemma 3.1., the function u can be represented by the expression (3.1) with some family of functions $\{u_i, i \in I\}$ such that every function u_i can be expressed by the following formula:

$$u_i(\tau) = \lim_{s\to\infty} v_{i,s}(\tau), \quad \tau \in \Omega, \tag{3.10}$$

where $v_{i,s} \in Lh_{\epsilon(s)}(\Omega_s)$ is a non-increasing sequence of functions and $\epsilon(s) \downarrow 0$ if $s \uparrow \infty$. By Lemma 2.4 we conclude from (3.9), (3.10) that for every $i \in I, r \in \mathbb{N}, \epsilon > 0$ there exists a number s_0 such that $v_{i,s}(\tau) \leq \epsilon,\ \tau \in K_m;\ v_{i,s}(\tau) \leq 1 + \epsilon,\ \tau \in \Omega_r; \epsilon(s) < \epsilon$ for $s \geq s_0$. Therefore in view of (3.4)

$$v_{i,s}(\tau) - \epsilon \leq \chi_\epsilon^{(1)}(\Omega_r, K_m; \tau), \quad \tau \in \Omega_r, r \in \mathbb{N}.$$

Letting first $s \to \infty$, then $\epsilon \to 0$ we obtain the inequality: $u_i(\tau) \le \chi_0(\Omega_r, K_m; \tau), \tau \in \Omega_r$. From here by letting $r \to \infty, m \to \infty$ and taking (3.1) we get that $u(t) \le h(\Omega, K; t), t \in \Omega$. Because of the arbitrariness of u we have $b(t) \le h(\Omega, K; t), t \in \Omega$, which finishes the proof.

The following assertion can be useful for the calculation of h in some special cases.

PROPOSITION 3.4 *Let K be a compact set whose complement $D = K^*$ consists of a finite number of connected components: $D_j, j = 1, \ldots, m$. Let Ω be an open neighbourhood of K such that its complement $F = \Omega^*$ has the same number of connected components F_j and $F_j \subset D_j$, $j = 1, \ldots, m$. Then*

$$h(\Omega, K; t) = \max\{h(\Omega_j, K_j; t) : j = 1, \ldots, m\} \tag{3.11}$$

where $\Omega_j = F_j^*, K_j = D_j^*, j = 1, \ldots, m$.

Proof: Let $K_s \downarrow K, \Omega_s \uparrow \Omega, K_i \subset \Omega_i$. We can assume that each of the boundaries $\partial K_s, \partial\Omega_s$ consists of m closed smooth surfaces. Let us denote by $b(t)$ the right side of (3.11). Because the inequality $b(t) \le h(\Omega, K; t)$ is obvious we need only to prove the inequality:

$$h(\Omega, K; t) \le b(t). \tag{3.12}$$

Let $D_{s,j} := K_s^* \cap D_j$, $F_{s,j} := \Omega_s^* \cap D_j$. Let $u(\tau) = \alpha \ln |x(\tau)|$ be a function which satisfies the conditions in (3.1) with Ω_{s+1} and K_{r-1} instead of Ω and K respectively. Then using Green's formula we can represent $x \in h(\Omega)$ in the form

$$x(t) = \sum_{j=1}^{m} x_j(t),$$

where $x_j \in h(\Omega_j)$ satisfies the estimates

$$|x_j|_{\Omega_{j,s}} \le C_s |x|_{\Omega_{s+1}}, \quad |x_j|_{K_{j,s}} \le C_s |x|_{K_{s+1}}, \quad s \in \mathbb{N},$$

where the constants C_s do not depend on x. Therefore

$$\begin{aligned} \alpha \ln |x(\tau)| &\le \alpha \ln \sum |x_j(\tau)| \le \alpha \ln(m \max\{|x_j(\tau)| : j = 1, \ldots, m\}) \le \\ &\le \alpha \max\{\ln |x_j(\tau)| : j = 1, \ldots, m\} + \epsilon \ln m \end{aligned}$$

and the functions x_j satisfy the following estimates

$$\begin{aligned} \alpha \ln |x_j(\tau)| - \alpha \ln C_s &\le 1, \quad \tau \in \Omega_{j,s}, \\ \alpha \ln |x_j(\tau)| - \alpha \ln C_r &\le 0, \quad \tau \in K_{j,r}. \end{aligned}$$

From here we have that

$$\chi_\epsilon^{(1)}(\Omega_{s+1}, K_{r-1}; t) \le \max\left\{\chi_\epsilon^{(1)}(\Omega_{j,s}, K_{j,r}; t) : j = 1, \ldots, m\right\} + \epsilon \ln M,$$

where $M = \max\{mC_s, mC_r\}$. Letting in the last inequality first $\epsilon \to 0$, then $s \to \infty$, $r \to \infty$ and taking $\overline{\lim}_{\tau \to t}$ we obtain the inequality (3.12) and, consequently, the identity (3.11).

REMARK Under the conditions of the proposition we can rewrite the formula (3.11) in the following equivalent form:

$$h(\Omega, K; t) = h(\Omega_j, K_j; t), \quad t \in \Omega_j \setminus \tilde{K}_j,\ j = 1, \ldots, m,$$

where by $\tilde{K}_j$ we denote the harmonic envelope of K_j with respect to Ω_j. To explane this we use the notations of the proof of the proposition. Let $\tilde{K}_{j,s}$ be the harmonic envelope of $K_{j,s}$. Then $\chi(\Omega_{j,s}, K_{j,r}; t) \equiv \chi(\Omega_{j,s}; \tilde{K}_{j,r}; t)$ in $\Omega_{j,s}$ for every $r, s \in \mathbb{N}$. Because $\chi(\Omega_{j,s}, K_{j,s}; t) \equiv 0$ in $K_{j,s}$ and $\Omega_{i,s} \subset \tilde{K}_{j,s}$ for $i \neq j$, we obtain that the right side of the equality (3.11) coincides with $h(\Omega_j, K_j; t)$ on the set $\Omega_j \setminus \tilde{K}_j$.

From the well known result on "three balls" Hadamard-type estimates for harmonic functions follows the next expression for the maximal Lh-functions in the case when $\Omega = B(a, R)$, $K = \overline{B}(a, r)$:

$$h(\Omega, K; t) \equiv \chi(\Omega, K; t) \equiv \chi_0(\Omega, K; t) \equiv \begin{cases} \frac{\ln|t| - \ln r}{\ln R - \ln r} & \text{for} \quad |t - a| > r \\ 0 & \text{for} \quad |t - a| \leq r \end{cases} \tag{3.13}$$

In the case $n = 2$ it follows from the results of Nguyen Than Van [4] that the function $h(\Omega, K; t)$ coincides with the corresponding usual harmonic measure.

PROPOSITION 3.5 *For an open set $\Omega \subset \hat{\mathbb{R}}^2$ and its compact subset K the following identity holds: $h(\Omega, K; t) \equiv \omega(\Omega, K; t)$ in Ω.*

Proof: We consider $\Omega_s \uparrow \Omega, K_s \downarrow K$ such that $K_1 \subset \Omega_1$ and every set $\partial\Omega_s, \partial K_s$ consist of finite number of closed Jordan curves. Then by [4] it exists a constant $C = C(\alpha, \epsilon, s)$ such that

$$|x(t)| \leq C(|x|_{K_s})^{1-\alpha-\epsilon}(|x|_{\Omega_s})^{\alpha+\epsilon}$$

for every $x \in h(\Omega_s)$ and

$$t \in \Omega_\alpha^{(s)} = \{t \in \Omega_s : \omega(\Omega_s, K_s, t) < \alpha\}, \quad s \in \mathbb{N}.$$

This means that $\chi_0(\Omega_s, K_s, t) \leq \omega(\Omega_s, K_s, t)$, $t \in \Omega_s$. Therefore by the definition of the function h and simple properties of the function ω we get the inequality:

$$h(\Omega, K; t) \leq \omega(\Omega, K; t), \quad t \in \Omega. \tag{3.14}$$

To prove the other-side inequality

$$\omega(\Omega, K; t) \leq h(\Omega, K; t), \quad t \in \Omega, \tag{3.15}$$

we use Lemma 2.2 and the remark before it. Indeed for every $s, r \in \mathbb{N}$ we have

$$\begin{aligned} \gamma(\Omega_s, K_r; t) &= \overline{\lim}_{\tau \to t} \sup\{v(\tau) = \alpha \ln|f(\tau)| : v \in P_a(K_r, \Omega_s),\ 0 < \alpha < \epsilon\} \\ &\leq \chi_\epsilon^{(1)}(\Omega_s, K_r; t) + \epsilon \ln \sqrt{2}, \end{aligned}$$

from which follows (3.15). The inequalities (3.14), (3.15) get us the desired identity.

Now we introduce some notions in terms of the above considered maximal functions.

DEFINITION 1 An open set $\Omega \subset \hat{\mathbb{R}}^n$ is said to be *Lh-regular*, if for every compact set $K \subset \Omega$ the identity $h(K^*, \Omega^*; t) \equiv 0$ holds on Ω^*. A compact set $K \subset \hat{\mathbb{R}}^n$ is said to be *Lh-regular*, if its complement K^* is a Lh-regular open set, i.e. for every open neighbourhood D of K the identity $h(D, K; t) \equiv 0$ on K holds.

DEFINITION 2 A compact set $K \subset \hat{\mathbb{R}}^n$ is said to be *strongly Lh-polar*, if there exists an open neighbourhood Ω such that $h(K^*, \Omega^*; t) \equiv 0$ in K^*. An open set $\Omega \subset \hat{\mathbb{R}}^n$ is said to be *strongly Lh-copolar (Lh-parabolic)*, if its complement Ω^* is a strongly Lh-polar compact set, i.e. there exists a compact set $L \subset \Omega$ such that $h(\Omega, L; t) \equiv 0$ in Ω.

In the particular case $n = 2$ we get from Proposition 3.5.

PROPOSITION 3.6 *An open set $\Omega \subset \hat{\mathbb{R}}^2$ is Lh-regular if and only if it is regular in the sense of potential theory.*

It is not difficult to see that the Lh-regularity for a compact set K coincides with the weak logarithmic regularity [22]. We compare now the notion of Lh-regularity with the notion of logarithmic regularity [22] which means that for every open neighbourhood Ω of K the identity $\chi_0(\Omega, K; t) \equiv 0$ on K holds.

PROPOSITION 3.7 *From the logarithmic regularity of a compact set $K \subset \hat{\mathbb{R}}^n$ in the sense of* [22] *follows Lh-regularity of K, but the inverse conclusion is not true even for the case of two variables.*

Proof: The first conclusion follows from the evident inequality $h(\Omega, K; t) \leq \chi_0(\Omega_1, K; t)$ for every open neighbourhoods Ω, Ω_1 of K such that Ω_1 is relatively compact in Ω. On the other hand the existence of Lh-regular but not logarithmic regular compact sets follows from the next example.

EXAMPLE 3.8 Let $K = \{(t_1, 0) \in \mathbb{R}^2 : -1 \leq t_1 \leq 1\}$, Ω is any bounded open neighbourhood of K in $\mathbb{R}^2$. Then $\chi(\Omega, K; t) \equiv \chi_0(\Omega, K; t) \equiv 1$ in Ω, but $h(\Omega, K; t) \equiv 0$ on K and $0 < h(\Omega, K; t) < 1$ in $\Omega \setminus K$.

Indeed, the function $x(t) \equiv t_2$ is harmonic in Ω and $x(t) \equiv 0$ on K. Let $M = \max\{|t_2| : t = (t_1, t_2) \in \Omega\}$. Then the function

$$v_k(t) := \frac{\ln |kx(t)|}{\ln kM} = \frac{\ln |t_2| + \ln k}{\ln M + \ln k}$$

satisfies the condition in (3.2) with $\epsilon = \epsilon(k) = \frac{1}{\ln M + \ln k}$. Therefore

$$\chi^{(1)}_{\epsilon(k)}(\Omega, K; t) \geq \frac{\ln |t_2| + \ln k}{\ln M + \ln k}.$$

Letting here $k \to \infty$ we obtain that $\chi_0(\Omega, K; t) \geq 1, t \in \Omega$, for every open neighbourhood Ω. Because the inequality $\chi_0(\Omega_1, K; t) \leq \chi(\Omega, K; t) \leq 1, t \in \Omega$ is fullfilled for any open set $\Omega_1 : K \subset \Omega_1 \subset \Omega$, we have $\chi(\Omega, K; t) \equiv \chi_0(\Omega, K; t)$ in Ω_1. On the other hand the compact K is regular in $\mathbb{R}^2$, hence due to Proposition 3.6 it is Lh-regular, which means that $h(\Omega, K; t) \equiv 0$ on K for every Ω. The inequality $0 < h(\Omega, K; t) < 1, t \in \Omega \setminus K$, follows from Proposition 3.5.

PROPOSITION 3.9. *For an open set $\Omega \subset \hat{\mathbb{R}}^n$ to be strongly Lh-copolar it is necessary and sufficient that Ω satisfies the Liouville principle for Lh-functions, i.e. if $v \in Lh(\Omega)$ is bounded in Ω, then v has to be a constant in every connected component of Ω.*

We consider now some general sufficient conditions for Lh-regularity of open (compact) sets. The following geometrical characteristic of a bounded open set Ω (and also of the corresponding compact set $K = \Omega^*$) is a suitable notion to describe such conditions [22]:

$$\delta(\tau) = \delta(\tau, \Omega) := \inf\{|\tau - t| : t \in \Omega\}, \quad \tau \in \Omega^*.$$

THEOREM 3.10 *Let $\Omega \subset \hat{\mathbb{R}}^n$ be a bounded open set such that for every point $t \in \partial\Omega$ the following relation be fullfilled:*

$$\underline{\lim}_{\epsilon \to 0} \inf \left\{ \frac{|\ln \delta(\tau)|}{|\ln \epsilon|} : |\tau - t| = \epsilon, \ \tau \in \Omega^* \right\} =: d(t) = 1. \tag{3.16}$$

Then for every compact set $K \subset \Omega$ we have $\chi(K^*, \Omega^*; t) \equiv 0$ on Ω^*.

Proof: By condition (3.16) for any point $t_0 \in \partial\Omega$ there exist sequences $\delta_s \downarrow 0, \epsilon_s \downarrow 0, a_s \in \Omega^*, s \in \mathbb{N}$, such that

$$\lim \frac{|\ln \delta_s|}{|\ln \epsilon_s|} = d(t_0) = 1, \quad \epsilon_s = |t_0 - a_s| \tag{3.17}$$

and

$$L_s := \{t : |t - a_s| \leq \delta_s\} \subset \Omega^*. \tag{3.18}$$

Let us choose $r > 0$ so that $\rho(t_0, K) > r/2$. Then there exists $s_0 \in \mathbb{N}$ such that

$$t_0 \in U_s := \{t : |t - a_s| < r\} \subset K^*, \quad s \geq s_0. \tag{3.19}$$

From (3.18), (3.19) we get the inequality

$$\chi(K^*, \Omega^*; t) \leq \chi(U_s, L_s; t), \quad t \in U_s, \ s \geq s_0.$$

Using the formula (3.13) and the relations (3.17), (3.19) we obtain from here that

$$0 \leq \chi(K^*, \Omega^*; t_0) \leq \lim_{s \to \infty} \frac{\ln \epsilon_s - \ln \delta_s}{\ln r - \ln \delta_s} = 0.$$

Therefore $\chi(K^*, \Omega^*; t) \equiv 0$ on $\partial\Omega$. But for $t_0 \in \Omega^* \setminus \partial\Omega = \hat{\mathbb{R}}^n \setminus \overline{\Omega} = \operatorname{int} \Omega^*$ we have that $\chi(K^*, \Omega^*; t_0) = 0$. Thus the identity $\chi(K^*, \Omega^*; t) = 0$ on Ω^* has been proved.

REMARK Theorem 3 in [22] is true only under the restriction $d = 1$, when it coincides with Theorem 3.10.

4 INTERPOLATIVE CLASSES OF F-SPACES

Let X be an F-space, $\{|x|_p, p \in \mathbb{N}\}$ a system of seminorms defining the topology of X. Let

$$|x'|_p^* := \sup\{|x'(x)| : x \in X, |x|_p \leq 1\}, \quad x' \in X^*, p \in \mathbb{N},$$

be a polar system of (usually non-bounded) norms (conorms) in the strong dual space X^*. Let us consider four classes of such spaces by means of the following relations, which describe the different kinds of multiplicative estimates of an intermediate norm (conorm) by marginal "small" and "big" norms (conorms):

$$X \in D_1 \quad := \quad \exists p \, \forall \mu > 0 \; \forall q \exists r \exists c : |x|_q \leq c(|x|_p)^{1-\mu}(|x|_r)^{\mu}, \; x \in X, \tag{4.1}$$
$$X \in D_2 \quad := \quad \forall p \, \forall \mu > 0 \; \exists q \forall r \exists c : |x'|_q^* \leq c(|x'|_p^*)^{\mu}(|x'|_r^*)^{1-\mu}, \; x' \in X^*, \tag{4.2}$$
$$X \in \Omega_1 \quad := \quad \forall p \exists q \forall r \exists \mu < 1 \exists c : |x'|_q^* \leq c(|x'|_p^*)^{\mu}(|x'|_r^*)^{1-\mu}, \; x' \in X^* \tag{4.3}$$
$$X \in \Omega_2 \quad := \quad \exists p \forall q \exists r \exists \mu < 1 \exists c : |x|_q \leq c(|x|_p)^{1-\mu}(|x|_r)^{\mu}, \; x \in X. \tag{4.4}$$

These classes where concidered in [11], [15], [24], [26] (in the first two references there were used other notations for these classes, respectively: $DN, \overline{\Omega}, \Omega, \underline{DN}$). We shall use the following general results, describing in terms of classes (4.1)-(4.4) spaces, which are isomorphic respectively to finite and infinite power spaces.

PROPOSITION 4.1 *Let X be a nuclear F-space. For X to be isomorphic to some finite type power space $\Lambda_0(a)$ it is necessary and sufficient that $X \in D_2 \cap \Omega_2$.*

This statement was proved in [13]; with the additional requirement of the existence of a basis in X it was proved also in [23].

PROPOSITION 4.2 *Let X be a nuclear F-space such that $\Gamma(X) = \Gamma(X \times X)$. For X to be isomorphic to some infinite type power space $\Lambda_\infty(a)$ it is necessary and sufficient that $X \in D_1 \cap \Omega_1$.*

This result is proved in [1], earlier in [15] it was proved that condition $X \in D_1 \cap \Omega_1$ is necessary and sufficient for X to be isomorphic to some topologically complemented subspace of some infinite type power space.

It is very useful the following equivalent characterization of the class D_2 due to Vogt [13] (for spaces with ℓ^2-basis see also [23]).

PROPOSITION 4.3 *For a Schwartz F-space X belong to class D_2 it is necessary and sufficient that there exists a Banach space $E \subset X$ with continuous imbedding such that*

$$\forall p \, \forall \mu > 0 \; \exists q \; \exists c : |x'|_q^* \leq c(|x'|_p^*)^{\mu}(\|x'\|_{E^*})^{1-\mu}, \quad x' \in X^*.$$

DEFINITION We shall write $E \in V(X)$ for every Banach space what satisfies the condition of Proposition 4.3.

5 INTERPOLATIVE PROPERTIES OF SPACES OF HARMONIC FUNCTIONS

Now we study the question: for which open sets $\Omega \subset \hat{\mathbb{R}}^n$ the space $h(\Omega)$ belongs to one of four classes considered in the previous section.

THEOREM 5.1 *For every open set $\Omega \subset \hat{\mathbb{R}}^n$ with a finite set of connected components the space $h(\Omega)$ belongs to both of the classes Ω_1 and Ω_2.*

Proof: a) First we prove that $h(\Omega) \in \Omega_2$. Without loss of generality we can assume that $\infty \overline{\in} \Omega$ and consider the case of complex valued harmonic functions. For every point $t \in \Omega$ we take some ball $U_t = B(t, \rho(t))$ which is contained in Ω. By [8] every harmonic function $x \in h(U_t)$ can be continued in a unique way to a function $\tilde{x} \in A(G_t)$, where G_t is a domain in $\mathbb{C}^n$ which contains U_t and is contained in the complex ball $\{z \in \mathbb{C}^n : |z - t| < \rho(t)\}$. For every $x \in h(\Omega)$ we define the function

$$\tilde{x}(z) = \tilde{x}_t(z), \quad z \in G_t, \ t \in \Omega. \tag{5.1}$$

The construction of the continuation in [8] is such that formula (5.1) defines an univalent analytic function $\tilde{x}(z)$ in the domain $G = \cup_{t \in \Omega} G_t$ and the map $x \to \tilde{x}$ is an isomorphic imbedding of $h(\Omega)$ into $A(G)$. Therefore the space $h(\Omega)$ is realized as a subspace of the space $A(G)$, which belongs to class Ω_2 (see, for example, [12]). Since the property $X \in \Omega_2$ is inherited by subspaces we get that $h(\Omega)$ belongs to Ω_2.

b)Without loss of generality we can assume that $K := \Omega^*$ does not contain the point ∞ and that Ω is connected. Then by using once more the result from [8] we can show that it exists the isomorphic linear imbedding of the space $h(K)$ into the space $A(L)$ for some compact set $L : K \subset L \subset \mathbb{C}^n$. Because the relation $X \in \Omega_2$ is inherited by quotient spaces and the space $A(L)^*$ belongs to Ω_1 for every compact set L in $\mathbb{C}^n$ we get that $h(\Omega) \in \Omega_1$.

REMARK The fact that $h(\Omega) \in \Omega_2$ can be obtained as a particular case from the more general result of Vogt [13].

THEOREM 5.2 *For $h(\Omega) \in D_1$ it is necessary and sufficient that a compact set K exists such that*

$$h(\Omega, K; t) \equiv 0 \quad \text{in} \quad \Omega. \tag{5.2}$$

Proof: Let $\Omega_s \uparrow \Omega$.

a) In view of the definition the relation $h(\Omega) \in D_1$ means that $\exists p \forall \mu > 0 \forall q \exists r \exists c = c(p, q, \mu)$:

$$|x|_{\Omega_q} \leq c(|x|_{\Omega_p})^{1-\mu}(|x|_{\Omega_r})^{\mu}, \quad x \in h(\Omega), \tag{5.3}$$

or in an equivalent form (with the same quantifiers):

$$\frac{\ln|x(t)| - \ln|x|_{\Omega_p}}{\ln|x|_{\Omega_r} - \ln|x|_{\Omega_p}} \leq \mu + \frac{\ln c}{\ln|x|_{\Omega_r} - \ln|x|_{\Omega_p}}, \quad t \in \Omega_q.$$

Without loss of generality we can require that $\ln |x|_{\Omega_p} = 0$, hence the last condition is equivalent to the inequality: $\chi_\epsilon^{(2)}(\Omega, K; t) \leq \mu + \epsilon \ln c, t \in \Omega_q$. The next inequality follows from here: $\chi_0(\Omega_r, \overline{\Omega}_p; t) \leq \mu, t \in \Omega_q$, hence we get for every $q \in \mathbb{N}, \mu > 0$, that $\chi(\Omega, \overline{\Omega}_p; t) \leq \mu,\ t \in \Omega_q$. After taking some compact set K in Ω such that $\Omega_p \subset \operatorname{int} K$ we obtain the identity (5.2).

b) Let now identity (5.2) hold. Taking $p \in \mathbb{N}$ such that $K \subset \Omega_p$ we obtain $\chi(\Omega, \overline{\Omega}_p; t) \equiv 0$ in Ω, because of the obvious inequality $\chi(\Omega, \overline{\Omega}_p; t) \leq h(\Omega, K; t), t \in \Omega$. Then by Lemma 2.4 for every $q \in \mathbb{N}, \mu > 0$ there exists $r \in \mathbb{N}$ such that $\chi_0(\Omega_r, \overline{\Omega}_p; t) < \mu/2$ for every $t \in \overline{\Omega}_q$. Using the same lemma once more we get that for every $t \in \overline{\Omega}_q$ and $\epsilon \leq \epsilon_0 = \epsilon_0(p, q, \mu)$ the inequality $\chi_\epsilon^{(2)}(\Omega_r, \overline{\Omega}_p; t) < \mu$ holds. This means that

$$\frac{\ln |x(t)| - \ln |x|_{\Omega_p}}{\ln |x|_{\Omega_r} - \ln |x|_{\Omega_p}} < \mu, \quad t \in \overline{\Omega}_q,$$

if

$$\ln |x|_{\Omega_r} - \ln |x|_{\Omega_p} \geq M := (\epsilon_0)^{-1}. \tag{5.4}$$

Hence the relation (5.3) is true with $c = 1$ if (5.4) holds. But if the requirement (5.4) is not valid we have (assuming $p \leq q \leq r$)

$$\frac{|x|_{\Omega_q}}{|x|_{\Omega_p}} \leq \frac{|x|_{\Omega_r}}{|x|_{\Omega_p}} < \exp M \leq \exp M \left(\frac{|x|_{\Omega_r}}{|x|_{\Omega_q}}\right)^\mu .$$

Thus for every $x \in h(\Omega)$ we have the relation (5.3) with $c = \exp M$.

THEOREM 5.3 *For $h(\Omega) \in D_2$ it is necessary and sufficient that for every compact set K in Ω the next relation holds:*

$$h(K^*, \Omega^*; t) \equiv 0 \quad \text{on} \quad \Omega^*. \tag{5.5}$$

Proof: Let $\Omega_s \uparrow \Omega$, $\Delta_s = (\overline{\Omega}_s)^*$. By definition because of Proposition 2.1 the condition $h(\Omega) \in D_2$ means that $\forall p\, \forall \mu > 0\ \exists q\, \forall r\, \exists c = c(p, r, \mu)$:

$$|x|_{\Delta_q} \leq c(|x|_{\Delta_r})^{1-\mu}(|x|_{\Delta_p})^\mu, \quad x \in h(\Omega^*), \tag{5.6}$$

or in the equivalent form (with the same quantifiers):

$$\frac{\ln |x(t)| - \ln |x|_{\Delta_r}}{\ln |x|_{\Delta_p} - \ln |x|_{\Delta_r}} \leq \mu + \frac{\ln c}{\ln |x|_{\Delta_p} - \ln |x|_{\Delta_r}}, \quad t \in \overline{\Delta}_q$$

for all $x \in h(\Omega^*)$ (if the left side of this inequality is not defined we assume it to be equal to 0). From here the inequality follows: $\chi_\epsilon^{(2)}(\Delta_p, \overline{\Delta}_r; t) \leq \mu + c\epsilon,\ t \in \overline{\Delta}_q$, for all $x \in h(\Delta_p)$ such that

$$\ln |x(t)|_{\Delta_p} - \ln |x|_{\Delta_r} > \frac{1}{\epsilon}.$$

Letting $\epsilon \to 0$ we obtain $\chi_0(\Delta_p, \overline{\Delta}_r; t) \leq \mu,\ t \in \overline{\Delta}_q$.

Now for any compact set K in Ω we choose $p \in \mathbb{N}$ such that $\overline{\Delta}_p \subset K^*$, hence the inequality

$$\chi(K^*, \overline{\Delta}_r; t) \leq \chi(\Delta_p, \overline{\Delta}_r; t) \leq \mu,\ t \in \Delta_q,$$

from which we get (by letting $r \to \infty$) $h(K^*, \Omega^*; t) \leq \mu$, $t \in \Omega^*$. Since $\mu > 0$ is arbitrary we get finally (5.5).

To prove the inverse direction we first take for every $p \in \mathbb{N}$ a compact set K in Ω such that $\overline{(K^*)} \subset \Delta_p$. From (5.5), taking into account the upper semi-continuity of the function h, we get that for every $\mu > 0$ there exists $q > p$ such that the inequality $h(K^*, \Omega^*; t) \leq \mu/2$, $t \in \Delta_q$ holds. From this it follows that for every $r > q$ we have $\chi_0(\Delta_p, \Delta_r; t) \leq \chi(K^*, \Delta_r; t) \leq h(K^*, \Omega^*; t) \leq \mu/2$ for each $t \in \Delta_q$. By using Lemma 2.2 like in the proof of the previous theorem we obtain that there exists $\epsilon_0 = \epsilon_0(p, r, \mu)$ such that $\chi_\epsilon^{(2)}(\Delta_p, \Delta_r; t) \leq \mu$, $t \in \Delta_q$, if $\epsilon \leq \epsilon_0$. Hence the inequality

$$\frac{\ln |x(t)| - \ln |x|_{\Delta_r}}{\ln |x|_{\Delta_p} - \ln |x|_{\Delta_r}} \leq \mu, \quad t \in \Delta_q,$$

holds, if the condition

$$\ln |x|_{\Delta_p} - \ln |x|_{\Delta_r} \geq \frac{1}{\epsilon_0}$$

is fulfilled. Further, like in the proof of Theorem 5.2 we obtain that the condition (5.6) is fullfilled with the constant

$$c = \exp \frac{1}{\epsilon_0(p, r, \mu)}.$$

6 HILBERT SCALES AND COMMON BASES

Let K be an Lh-regular compact set in an Lh-regular open set $\Omega \subset \hat{\mathbb{R}}^n$. We suppose that K is harmonically convex with respect to Ω and that each connected component of Ω has non-empty intersection with K.

By this requirement we shall construct a common basis for the spaces $h(K), h(\Omega)$ that should be realized as a common orthogonal basis for a suitable pair of Hilbert spaces H_0, H_1 which have to be very closely connected whith the spaces $h(K), h(\Omega)$. Such a construction was considered in [20], [23], [25] for spaces of analytic functions and in [22] for spaces of ' armonic functions. It is necessary to note that an abstract scheme for the application of a common orthogonal basis for the construction of a basis in so called centers of Riesz scales was first suggested in [3], §5.

The question on a common basis is connected closely with the studying of the behaviour of the Hilbert scale $H^\alpha = (H_0)^{1-\alpha}(H_1)^\alpha, \alpha \in \mathbb{R}$, spanned on a given pair of Hilbert spaces $H_1 \subset H_0$ which are connected with the spaces $h(\Omega), h(K)$ respectively.

THEOREM 6.1 *Let Ω be an open set in $\hat{\mathbb{R}}^n$, K a compact subset in Ω, which has a non-empty intersection with every connected component of Ω. Let the compact sets K, Ω^* are such that the imbeddings $h(K) \to hE^{(1)}(K)$, $h(\Omega^*) \to hE^{(1)}(\Omega^*)$ are injective. Let H_0, H_1 be Hilbert spaces having dense continuous imbeddings:*

$$h(K) \subset H_0 \subset hE^{(1)}(K), \tag{6.1}$$

$$h(\Omega^*) \subset H_1' \subset hE^{(1)}(\Omega^*) \tag{6.2}$$

where H_1' is the realisation of the Hilbert space $H_1 \subset h(\Omega)$ according to Tillmann duality. Then for the Hilbert scale $H^\alpha = (H_0)^{1-\alpha}(H_1)^\alpha$ the continuous imbeddings

$$h(F_\alpha) \subset H^\alpha \subset h(D_\alpha), \quad 0 < \alpha < 1, \tag{6.3}$$

are valid, where

$$D_\alpha := \{t \in \Omega : \psi(\Omega, K; t) < \alpha\},\ F_\alpha := \hat{\mathbb{R}}^n \setminus \{t \in K^* : \psi(K^*, \Omega^*; t) < 1 - \alpha\}, 0 < \alpha < 1.$$

Proof: We consider a common orthogonal basis $\{e_k\}_{k\in\mathbb{N}}$ for the spaces $\hat{H}_0, H_1$, where $\hat{H}_0$ is the closure of H_1 in H_0. We can normalize and arrange this basis in such a way that the following conditions hold:

$$\|e_k\|_{H_0} = 1,\ \|e_k\|_{H_1} = \mu_k \uparrow \infty. \tag{6.4}$$

Let $\Omega_s \uparrow \Omega$. Then taking into account the conditions (6.1), (6.2), (6.4) we obtain

$$\begin{aligned} |\Lambda_{e_k}|_K &\leq C\|e_k\|_{H_0} = C \\ |\Lambda_{e_k}|_{\Omega_s} &\leq C_s\|e_k\|_{H_1} = C_s\mu_k \end{aligned}$$

for some constants $C, C_s, s \in \mathbb{N}$. By definition of the function $\psi_\epsilon^{(1)}$ we have for $\mu_k \geq \frac{C}{C_s}\exp\frac{1}{\epsilon}$ the inequality:

$$\frac{\ln|\Lambda_{e_k}(\tau)| - \ln C}{\ln C_s\mu_k - \ln C} \leq \psi_\epsilon(\Omega_s, K; \tau), \quad \tau \in \Omega_s. \tag{6.5}$$

Further in the proof we need only the following information on the behaviour of the sequence $\{\mu_k\}$:

$$\forall\sigma > 0 : \sum_{k=1}^{\infty} \mu_k^{-\sigma} < \infty \tag{6.6}$$

It follows from the imbeddings (6.1), (6.2) and from the fact of the nuclearity of the imbedding $h(D) \subset h(L)$ for any open set D and any of its compact subsets L (the question on more precise asymptotic behaviours of $\{\mu_k\}$ we discuss below).

By the definition of the function h we have the equality:

$$D_\alpha = \bigcup_{\epsilon>0}\bigcup_{s} D_\alpha(\epsilon, s) \tag{6.7}$$

where

$$D_\alpha(\epsilon, s) := \{t \in \Omega_s : \psi_\epsilon^{(1)}(\Omega_s, K; t) < \alpha\},\ 0 < \alpha < 1,\ \epsilon > 0,\ s \in \mathbb{N}.$$

For any compact set $M \subset D_\alpha$ we have from (6.5), taking into account that $|x(t)| \leq \Lambda_x(t)$:

$$\overline{\lim}_{k\to\infty} \frac{\ln|e_k|_M}{\ln\mu_k} \leq \alpha.$$

From here it follows that there exists a constant $A = A(M, \mu, \delta)$ such that

$$|e_k|_M \leq A\mu_k^{\alpha+\delta}. \tag{6.8}$$

Let $x \in H^{\alpha+\epsilon}$, $\epsilon > \delta$, $\xi_k = (x, e_k)_{H_0}$. Then from (6.6), (6.8) we have

$$\begin{aligned} |x|_M &\le \sum |\xi_k|\, |e_k|_M \le A \sum |\xi_k| \mu_k^{\alpha+\delta} \\ &\le A \left(\sum |\xi_k|^2 \mu_k^{2(\alpha+\epsilon)}\right)^{1/2} \left(\sum \mu_k^{-2(\epsilon-\delta)}\right)^{1/2} = A_1 \|x\|_{H^{\alpha+\epsilon}}. \end{aligned}$$

Because of the arbitrariness of the compact set $M \subset D_\alpha$ and of the constant $\epsilon > 0$, the last inequality means that we have continuous imbeddings at the right hand side of (6.3). Due to Tillmann duality (Proposition 2.1) the continuity of the imbeddings on the left hand side of (6.3) can be proved in the same way.

With the help of Proposition 4.3 we describe now the general conditions on Hilbert spaces H_0, H_1 which provide imbeddings like (6.3) but in terms of the function $h(\Omega, K; t)$.

THEOREM 6.2 *Let K, Ω satisfy the conditions which were considered in the beginning of this section. Let $H_1 \in V(h(\Omega))$ and $H_0' \in V(h(K^*))$, where $H_0' \subset h(K^*)$ is the realization of the ¿dual space H_0^* according to Proposition 2.1. Then the following continuous imbeddings hold*

$$h(K_\alpha) \subset H^\alpha \subset h(\Omega_\alpha), \quad 0 < \alpha < 1. \tag{6.9}$$

Here $H^\alpha = (H_0)^{1-\alpha}(H_1)^\alpha$ is the Hilbert scale spanned on H_0, H_1, and

$$\Omega_\alpha := \{t \in \Omega : h(\Omega, K; t) < \alpha\}, \ K_\alpha := \mathbb{R}^n \setminus \{t \in \hat{K}^* : h(K^*, \Omega^*; t) < 1 - \alpha\}. \tag{6.10}$$

Proof: It is necessary first to remark that the existence of spaces H_1, H_0 satisfying the conditions of the theorem follows from Theorem 5.3 and Proposition 4.3 taking into account the nuclearity of the spaces $h(\Omega), h(K)$. Let $\{e_k\}$ be the common orthogonal basis for the spaces H_0, H_1 satisfying the conditions (6.4), where the non-decreasing sequence $\mu_k = \mu_k(H_0, H_1)$ is defined uniquely and coincides with λ_k^{-1}, where λ_k are the s-numbers of the operator of imbedding $J : H_1 \to H_0$. Let us consider two sequences $K^{(r)} \downarrow K, \Omega^{(s)} \uparrow \Omega$. From the condition $H_0' \in V(h(K^*))$ it follows that for every $\epsilon > 0$ there exist $r = r(\epsilon)$ and $c > 0$ such that the following estimate holds

$$|x|_{K^{(r)}} \le c(\|x\|_{H_0})^{1-\epsilon} (\|x\|_{H_1})^\epsilon, \quad x \in h(K).$$

Due the conditions (6.4) we have from here that $|e_k|_{K^{(r)}} \le c\mu_k^\epsilon$. Because of the condition $H_1 \in V(h(\Omega))$ the imbedding $H_1 \subset h(\Omega)$ is continuous, hence for every $s \in \mathbb{N}$ there exists a constant C_s such that taking into account (6.4) we have the estimate $|e_k|_{\Omega^{(s)}} \le C_s \mu_k$. From the last two estimates for e_k we obtain that

$$\frac{\ln |e_k(t)| - \ln c - \epsilon \ln \mu_k}{(1-\epsilon)\ln \mu_k + \ln C_s - \ln c} \le \chi_\delta^{(2)}(\Omega^{(s)}, K^{(r)}; t), \quad t \in \Omega^{(s)}, \tag{6.11}$$

if the denominator of the left side in (6.11) is more then $\frac{1}{\delta}$. Because of $\mu_k \uparrow \infty$ we have from here the inequality

$$\overline{\lim}_{k\to\infty} \frac{\ln |e_k|_M}{\ln \mu_k} \le \epsilon + (1-\epsilon)\alpha \tag{6.12}$$

for every compact set

$$M \subset \Omega_\alpha^{(r)} := \{t \in \Omega : \chi(\Omega, K^{(r)}; t) < \alpha\}.$$

Letting $\epsilon \downarrow 0$ we get that (6.12) is valid with $\epsilon = 0$ for every compact set $M \subset \Omega_\alpha$. Going on like in the proof of Theorem 6.1 we obtain the right imbeddings in (6.9). Because of the Tillmann duality the left imbeddings can be proved in the same way.

We discuss now the asymptotic behaviour of the sequence $\mu_k = \mu_k(H_0, H_1)$ which was defined in the proofs of the previous theorems. It is connected with the s-numbers $s_k(J)$ of the linear imbedding $J : H_1 \to H_0$ or with the Kolmogorov diameters $d_k(B_{H_0}, B_{H_1})$ (see, for example, [3]) by the following relations:

$$\mu_k(H_0, H_1) = 1/s_k(J) = 1/d_{k-1}(B_{H_1}, B_{H_0}), \quad k \in \mathbb{N}. \tag{6.13}$$

First we consider the following

EXAMPLE 6.3 *Let $B(a,r) := \{t \in \hat{\mathbb{R}}^n : |t-a| < r\}$, and $G(r)$ be the closure of $h(\overline{B(a,r)})$ in the Hilbert space $L^2(\overline{B(a,r)})$. Then for any $r_0 < r_1$ we have*

$$(G(r_0))^{1-\alpha}(G(r_0))^\alpha = G(r(\alpha)),$$

where $r(\alpha) = (r_0)^{1-\alpha}(r_1)^\alpha, \alpha \in \mathbb{R}$, and

$$\ln \mu_k(G(r_0), G(r_1)) \sim \sigma_n \ln \frac{r_1}{r_0} \cdot k^{1/n-1}, \quad k \to \infty, \tag{6.14}$$

where σ_n depends only on the dimension n.

This facts follow from well known properties of the basis of orthogonal harmonic polynomials.

PROPOSITION 6.4 *Let the conditions of Theorem 6.2 are valid. Then for the sequence (6.13) the following weak asymptotic formula holds:*

$$\ln \mu_k(H_0, H_1) \approx k^{1/n-1}, \quad k \to \infty, \tag{6.15}$$

which means that there exists a constant $c > 0$ such that

$$\frac{1}{c} k^{1/n-1} \le \ln \mu_k(H_0, H_1) \le c k^{1/n-1}, \quad k \in \mathbb{N}.$$

If we take another pair of Hilbert spaces G_0, G_1 satisfying the condition of Theorem 6.2 (G_i instead of $H_i, i = 0, 1$) then the corresponding sequence (6.13) has the same exact asymptotic behaviour:

$$\ln \mu_k(G_0, G_1) \sim \ln \mu_k(H_0, H_1), \quad k \to \infty. \tag{6.16}$$

Proof: a) The weak asymptotics (6.15) can be concluded from the well known asymptotics of Kolmogorov diameters for spaces of solutions of elliptic partial differential equations. We give below a simple proof of this relation (compare with the case of analytic functions [23]) with the help of Theorem 6.2 for our concrete situation. We use only the simplest fact considered in Example 6.3. Let us take in Example 6.3 such a point $a \in \Omega$ and such a radius r_1 that $\overline{B(a, r_1)} \subset \Omega$. Let us compare now the following Hilbert scales:

$$H^\alpha = (H_0)^{1-\alpha}(H_1)^\alpha, \; G^\alpha = (G(r_0))^{1-\alpha}(G(r_1))^\alpha, \; F^\alpha = (G(r_0))^{1-\alpha}(H_1)^\alpha.$$

Using three times Theorem 6.2 we can choose numbers α_i, β_i in the interval $(0,1)$ in such a way that the following continuous imbeddings hold:

$$F^{\alpha_1} \subset H^{\alpha_2} \subset F^{\alpha_3} \subset F^{\alpha_4} \subset H^{\alpha_5} \subset F^{\alpha_6}, \tag{6.17}$$

$$G^{\beta_1} \subset F^{\beta_2} \subset G^{\beta_3} \subset G^{\beta_4} \subset F^{\beta_5} \subset G^{\beta_6}. \tag{6.18}$$

Taking into account that for every Hilbert scale the following equality holds:

$$\ln \mu_k(H^\delta, H^\epsilon) = (\epsilon - \delta) \ln \mu_k(H_0, H_1), \quad 0 < \delta < \epsilon < 1,$$

and by using Lemma 2.3 we obtain from (6.14), (6.17), (6.18) that

$$\ln \mu_k(H_0, H_1) \approx \ln \mu_k(F_0, F_1) \approx \ln \mu_k(G_0, G_1) \approx k^{1/n-1}, \quad k \to \infty.$$

b) We prove now the second statement (6.16). From Theorem 6.2 it follows that for every $\delta > 0$ there exists $\epsilon > 0$ such that the following continuous imbeddings hold: $H_1 \subset G^{1-\epsilon} \subset H^{1-\delta} \subset H^\delta \subset G^\epsilon \subset H_0$. From here using again Lemma 2.3 we get the estimates

$$c\mu_k(H_0, H_1)^{1-2\delta} \le \mu_k(G_0, G_1)^{1-2\epsilon} \le C\mu_k(H_0, H_1)$$

with some constants c, C. Because $\delta > 0$ was arbitrary we obtain (6.16).

COROLLARY 6.5 *Let Ω be an open set in $\hat{\mathbb{R}}^n, \Omega^* \neq \emptyset$, and K be a compact subset of Ω. For every Banach spaces X_i having the continuous imbeddings: $X_1 \subset h(\Omega) \subset X_2 \subset X_3 \subset h(K) \subset X_4$ the following inequalities hold:*

$$\overline{\lim}_{k\to\infty} \frac{-\ln d_k(X_4, X_1)}{k^{1/n-1}} > 0,$$

$$\overline{\lim}_{k\to\infty} \frac{-\ln d_k(X_3, X_2)}{k^{1/n-1}} < \infty.$$

THEOREM 6.6 *Let the conditions of Theorem 6.2 are valid. Then the common basis for the spaces H_0, H_1 is also a common basis for spaces $h(K), h(\Omega)$. For this basis and its biorthogonal system $\{e_k'\} \subset h(K^*)$ the following estimates hold: for every compact sets $L \subset \Omega_\alpha, L' \subset K_\alpha^*$, $0 < \alpha < 1$, there exist constants $c = c(L, \alpha), c' = c(L', \alpha)$ such that*

$$|e_k|_L \le c\mu_k^\alpha \cdot |e_k'|_{L'} \le c'\mu_k^{-\alpha}, \quad k \in \mathbb{N}, \tag{6.19}$$

where the sequence $\{\mu_k\}$ satisfies the condition (6.15).

Proof: Because of Lh-regularity of Ω and K we have the inclusions $K \subset \Omega_\alpha, \Omega^* \subset K_\alpha^*, 0 < \alpha < 1$. On the other hand from the maximum principle for harmonic functions we have that

$$\Omega = \bigcup_{\alpha<1} \Omega_\alpha, \; K = \bigcap_{\alpha>0} K_\alpha.$$

Therefore it follows from the imbeddings (6.9) that

$$h(\Omega) = \lim \operatorname{proj}_{\alpha\to 1} H^\alpha, \tag{6.20}$$

$$h(K) = \lim \operatorname{ind}_{\alpha \to 1} H^{\alpha}. \tag{6.21}$$

Hence $\{e_k\}$ is a basis in the both of spaces $h(\Omega)$ and $h(K)$. The estimates (6.19) follow from imbeddings (6.9).

REMARK In contrast to the case of spaces of analytic functions [23], [25] we do not obtain from the results on Hilbert scales that the system $\{e_k\}$ is a basis in some intermediate domain $G : K \subset G \subset \Omega$. Because of the possible existence of a gap between the sets K_α and Ω_α it is possible to deduce only the following

COROLLARY 6.7 *Every function $x \in h(K_\alpha)$ can be represented by the series*

$$x(t) = \sum_{k=1} \xi_k e_k(t)$$

which convergs in the topology of the space $h(\Omega_\alpha)$.

7 ISOMORPHY OF SPACES OF HARMONIC FUNCTIONS

We consider now criteria for the isomorphy of spaces $h(\Omega)$ to canonical spaces $h(B^n), h(\mathbb{R}^n)$. They will be got as simple consequences from the previous considerations.

THEOREM 7.1 *Let Ω be an open set in $\hat{\mathbb{R}}^n$ with a finite set of connected components. Then the following conditions are equivalent:*

(i) $h(\Omega) \simeq h(B^n)$,

(ii) $h(\Omega) \simeq \Lambda_0(k^{1/n-1})$,

(iii) *$h(\Omega)$ is isomorphic to some finite type power space $\Lambda_0(a)$,*

(iv) *Ω is Lh-regular,*

(v) *For every Lh-regular compact set $K \subset \Omega$ which intersects with each connected component of Ω and which is a Runge set in Ω (i.e. $h(\Omega)$ is dense in $h(K)$) there exists a common basis $\{e_k\}$ for the spaces $h(\Omega)$ and $h(K)$ which satisfies the estimates (6.19).*

Proof: From Example 6.3 it follows that $h(B^n) \simeq \Lambda_0(k^{1/n-1})$, therefore (i) is equivalent to (ii); (ii) $\Rightarrow$ (iii) is obvious; (iii) $\Rightarrow$ (iv) follows from Theorem 5.3; (iv) $\Rightarrow$ (v) follows from Theorem 6.6. The implication (v) $\Rightarrow$ (ii) is true because of the existence of a basis with the properties described in (v) provides that the operator $T : h(\Omega) \to \Lambda_0(k^{1/n-1})$, defined by the following formula:

$$T(x) := (\mu_k e_k'(x))_{k \in \mathbb{N}}$$

is an isomorphism.

REMARK The requirement of finiteness of the set of connected components of Ω is necessary for the isomorphy $h(\Omega) \cong h(B^n)$ because otherwise the space $h(\Omega)$ has no continuous norm.

THEOREM 7.2 *Let Ω be an open set in $\hat{\mathbb{R}}^n$. Then the following assertions are equivalent:*

(i) $h(\Omega) \simeq h(\hat{\mathbb{R}}^n)$,

(ii) $h(\Omega) \simeq \Lambda_\infty(k^{1/n-1})$,

(iii) $h(\Omega)$ *is isomorphic to some infinite type power space* $\Lambda_\infty(a)$,

(iv) Ω^* *is strongly Lh-polar,*

(v) Ω *satisfies the Liouville principle for the class* $Lh(\Omega)$.

Proof: Because of previous considerations we need only to prove the conclusion (iii) $\Rightarrow$ (ii). Let $T : \Lambda_\infty(a) \to h(\Omega)$ be an isomorphism. We denote by Y_p the completion of $h(\Omega)$ with respect to the norm $\|x\|_p := |T^{-1}x|_p$ (definition of the norm $|\xi|_p$ above in 2.3), $p \in \mathbb{N}$. Because $d_{k-1}(Y_s, Y_r) = exp(\alpha_p - \alpha_r)a_k, k \in \mathbb{N}, p < r$, [3], we have from Corollary 6.5 that $a_k \approx k^{1/n-1}, k \to \infty$. This means that $h(\Omega) \simeq \Lambda_\infty(a) \simeq \Lambda_\infty(k^{1/n-1})$.

8 FINAL OBSERVATIONS AND UNSOLVED PROBLEMS

Here we give a short description of further results without proof and discuss questions, which should be considered in the future.

8.1 Lh-Green Functions (LhG-Functions)

For a compact set $K \subset \mathbb{R}^n$ we define two natural versions of LhG-function with the positive or negative normalized singularity in the infinite point ∞:

$$h^+(K,t) := \overline{\lim}_{\tau\to t} \sup\{u(\tau) : u \in Lh(\mathbb{R}^n), u|K \le 0, u(\tau) - \ln(1+|\tau|) \le C \text{ in } \mathbb{R}^n\}, \tag{8.1}$$

$$h^-(K,t) := \overline{\lim}_{\tau\to t} \sup\{u(\tau) : u \in Lh(K^*), u(\tau) \le \min\{0, -\ln|\tau| + C\} \text{ in } K^*\}, \tag{8.2}$$

where $C = C(u)$ are constants which depend only on u.

Somewhat different LhG-functions were considered in [27]:

$$\lambda_K^\pm(t) = \lim_{\epsilon\to 0} \overline{\lim}_{\tau\to t} \sup\{v(\tau) : v \in \mathcal{H}_\epsilon^\pm(K)\}, \tag{8.3}$$

where the class $\mathcal{H}^{\pm}_{\epsilon}(K)$ consists of the functions $v(\tau) = \alpha \ln|x(\tau)|, 0 < \alpha < \epsilon$, such that a) $x \in h(\mathbb{R}^n), v|K \leq 0, v(\tau) \leq \ln(1+|\tau|) + C$ in $\mathbb{R}^n$, if we take " + ", and b) $x \in h(K^*), v(\tau) \leq \min\{0, -\ln|\tau| + C\}$ in K^*, if we consider " − "; here $C = C(v) =$ const.

For $K \subset \mathbb{R}^2$, which is considered in Example 3.8 we have,

$$\begin{aligned} \lambda^{+}_{K}(t) &\equiv +\infty; \\ h^{+}(K,t) &\equiv -h^{-}(K,t) \equiv -\lambda^{-}_{K}(t) = \ln|z + \sqrt{z^2-1}|, z = t_1 + it_2, t = (t_1, t_2) \in K^*. \end{aligned}$$

8.2 Bernstein-Walsh Theorem (BWT)

LhG-functions (8.1)-(8.3) are useful for considering of BWT about relations between the rate of the best approximation by harmonic polynomials and the harmonic extension. The functions (8.3) were used in [27] to get a version of BWT for the classical case of uniform approximation on K. Here we consider another version of BWT in terms of the LhG-function $h^{\pm}(K,t)$.

Let E be a Banach space, such that $h(K)$ is continuously and densely imbedded in $E : h(K) \subset E$. We assume that the natural realization of the conjugate space $E^* \subset h(K^*)$ is a Vogt space for $h(K^*)$, i.e. $E^* \in V(h(K^*))$ (see definition in Section 4). We use the notation:

$$\delta_s(x, E) = \inf\{\|x - p\|_E : p \in \mathcal{H}_s\},$$

where $\mathcal{H}_s$ is the set of all harmonic polynomials with a degree $\leq s$.

THEOREM 8.1 *Let K be a Lh-regular compact set in $\mathbb{R}^n$ and E satisfy the above mentioned conditions. If $x \in E$ satisfies the condition* $(0 < \sigma < \infty)$:

$$\delta_s(x, K) = 0(\exp(-\sigma + \epsilon), \quad \forall \epsilon > 0 \tag{8.4}$$

then x has a harmonic extension to the open set

$$\Omega^{+}_{\sigma} = ht \in \mathbb{R}^n : h^{+}(K,t) < \sigma\}.$$

On the other hand for (8.4) to be valid it is sufficient that x has a harmonic extension to the open set:

$$\Omega^{-}_{\sigma} = (\{t \in \mathbb{R}^n : h^{-}(K,t) \leq -\sigma\})^*.$$

For $n = 2$ always $h^{+}(K,t) \equiv -h^{-}(K,t)$ and we have no gap between sufficient and necessary conditions. But for $n = 3$ such a gap can be realized for very smooth compact sets. Thus, in Example 2 [27] we have $-\lambda^{-}(K,t) \equiv -h^{-}(K,t) < \lambda^{+}(K,t) \equiv h^{+}(K,t)$.

REMARK In [27] both theorems need the following corrections: it is necessary to consider the above defined function $h^{-}(K,t)$ instead of the function $\lambda^{-}_{K}(t)$.

8.3 Orthogonal Harmonic Polynomial Basis for Compact Sets

Let $\{e_i\}_{i\in\mathbb{N}}$ be the canonical system of all spherical harmonic polynomials, which are enumerated in such a way that the sequence of degrees $s(i)$ is non-decreasing. Let K satisfy the conditions of Theorem 8.1 and $H = E$ be a Hilbert space, satisfying the conditions in the beginning of the previous section. Let us consider the sequence of orthonormal polynomials

$$p_i(t) = \sum_{j=1}^{i} a_{ji} e_j(t), \tag{8.5}$$

which are obtained by Gram-Schmidt orthogonalization. By a reasoning, similar to the case of analytic functions [28], the following fact can be proved (compare [27]).

THEOREM 8.2 *In the above mentioned conditions the system (8.5) is the common basis in spaces $h(K)$ and $h(\mathbb{R}^n)$.*

For any $\sigma > 0$ we can conclude only that the system (8.5) is a basis for the space $h(\Omega_\sigma^-)$ in topology of the space $h(\Omega_\sigma^+)$, which means that every function $x \in h(\Omega_\sigma^-)$ can be expanded in a unique way in a series:

$$x(t) = \sum \xi_i e_i(t),$$

which converges in the topology of the space $h(\Omega_\sigma^+)$.

Problem 1 Is the system (8.5) a basis in the spaces $h(\Omega_\sigma^+), h(\Omega_\sigma^-)$, $0 < \sigma < \infty$?

8.4 Other Questions

Finally we bring yet two problems arising naturally in the context of the present paper.

Problem 2 How to describe the class $Lh(\Omega)$ in terms of Partial Differential Equations Theory?

Problem 3 Maximal plurisubharmonic functions are described by the complex Monge-Ampere equation (Bedford-Taylor). What do we have to consider instead of the Monge-Ampere operator to describe the maximality of Lh-functions?

REFERENCES

[1] Aytuna, A., Krone, J., Terzioğlu, T. Complemented infinite type power series subspaces of nuclear Fréchet spaces. *Math. Ann.*, 283 (1989) 193-202.

[2] Langenbruch, M. Bases in solution sheaves of systems of partial differential equations. *J. Reine Angew. Math.*, 373 (1987), 1-36.

[3] Mityagin, B.S. Approximative dimension and bases in nuclear spaces. *Russian Math. Surveys.* 26 (1961), 59-127.

[4] Nguyen Thanh Van. Bases communes pour certains espaces de fonctions harmoniques. *Bull Sc. math.*, 2^e^ serie, 97 (1973), 33-49.

[5] Nguyen Thanh Van and Djebbar. Proprietes asymptotiques d'une suite orthonormale de polynomes harmoniques. *Bull Sc. math.* (5) 113 (1989), 239-251.

[6] Siciak, J. On some extremal functions and applications in the theory of analytic functions of several complex variables. *Trans. Amer. Math. Soc.*, 105 (1962), 322-357.

[7] Siciak, J. Asymptotic behavior of harmonic polynomials bounded on a compact set. *Ann. Polon. Math.*, 29 (1968), 267-278.

[8] Siciak, J. Holomorphic continuation of harmonic functions. *Ann. Polon. Math.*, 29 (1974), 67-73.

[9] Skiba, N.I. On Hilbert scales of spaces of harmonic functions of two real variables. *Actual Problems of Mathematical Analysis.* Rostov-on-Don University, 1978, 161-166 (Russian).

[10] Tillmann, H.-G. Dualität in der Potentialtheorie. *Portugal Math.*, 13 (1954), 55-86.

[11] Vogt, D. Charakterisierung der Unterräume von s. *Math. Z.* 155 (1977), 109-117.

[12] Vogt, D. Charakterisierung der Unterräume eines nuklearen stabilen Potenzreihenräumes von endlichen Typ. *Studia Math.* 71 (1982), 251-270.

[13] Vogt, D. Eine Charakterisierung der Potenzreihenräume von endlichem Typ und ihre Folgerungen. *Manuscripta Math.*, 37 (1982), 269-301.

[14] Vogt, D. Sequence space representations of spaces of testfunctions and distributions. Functional analysis, holomorphy and approximation. *Proc. Sem. Rio de Jeneiro 1979, LN Pure Appl. Math.*, 83 (1983), 405-443.

[15] Vogt, D., Wagner, M.J. Charakterisierung der Quotientenräume von s und eine Vermutung von Martineau. *Studia Math.* 67 (1980), 225-240.

[16] Vogt, D., Wagner, M.J. Charakterisierung der Unterräume von endlichen Typ. *Studia Math.* 70 (1981), 63-80.

[17] Walsh, J.L. Maximal convergence of sequences of harmonic polynomials. *Ann. of Math.*, 38 (1937), 321-364.

[18] Walsh, J.L. Interpolation and approximation by rational functions in the complex domain. *Amer. Math. Soc. Coll. Publ.*, 20, Third Edition, 1960.

[19] Wiechert G. Dualität- und Strukturtheorie der Kerne von linearen Differentialoperatoren. Dissertation, Wuppertal 1982.

[20] Zahariuta, V.P., Continuable bases in spaces of analytic functions of one and of several variables. *Siberian Math. J.* 8 (1967), 204-216 (Russian).

[21] Zahariuta, V.P., Spaces of functions of one variable, analytic in open sets and on compacta. *Math. USSR Sbornik* 11(1970), 75-88.

[22] Zahariuta, V.P. Isomorphism of spaces of harmonic functions. *Math. Analysis and its applications, Rostov University, Rostov-on-Don.* 3 (1971), 152-158 (Russian).

[23] Zahariuta, V.P., Extremal plurisubharmonic functions, Hilbert scales and the isomorphism of spaces of analytic functions of several variables, I, II. *Teor. Funk. Funk. Anal. i prylozen*, 19 (1974), 133-157, 21 (1974), 65-83 (Russian).

[24] Zahariuta, V.P. Some linear topological invariants and isomorphism of tensor products of centers of scales. *Notes of North Caucausian Sc. center.* 4 (1974), 62-64 (Russian).

[25] Zahariuta, V.P. Extremal plurisubharmonic functions, orthogonal polynomials and Bernstdein-Walsh Theorem for analytic functions of several complex variables. *Ann. Pol. Math.* 33 (1976), 137-148 (Russian).

[26] Zahariuta, V.P. Isomorphisms of spaces of analytic functions. *Sov. Math. Dokl.*, 22 (1980), 631-634.

[27] Zahariuta, V.P., Skiba, N.I. Orthogonal harmonic polynomials and Bernstein-Walsh Theorem in $\mathbb{R}^n$. *Doğa - Tr. J. of Mathematics*, 16 (1992), 46-49.

[28] Zeriahi, A. Bases communes dans certains espaces de fonctions harmoniques et fonctions separement harmoniques sur certains ensembles de $\mathbb{C}^n$. *Ann. Fac. Sc. Toulouse*, 6 (1982), 75-102.

Index